QH      Grant, Philip
491
.G7     Biology of
          developing systems

| DATE | | | |
|------|---|---|---|
| AUG 19 '81 | | | |
| OCT 14 '81 | | | |
| NOV 16 '81 | | | |
| DEC 17 '81 | | | |
| FEB 3 '82 | | | |
| DEC 07 1993 | | | |
| APR 2 3 1996 | | | |
| | | | |
| | | | |
| | | | |
| | | | |

# BIOLOGY OF
# DEVELOPING SYSTEMS

# BIOLOGY OF DEVELOPING SYSTEMS

## PHILIP GRANT
### UNIVERSITY OF OREGON

HOLT, RINEHART AND WINSTON

New York   San Francisco   Chicago   Atlanta   Dallas   Montreal   Toronto   London   Sydney

**Library of Congress Cataloging in Publication Data**

Grant, Philip.
  Biology of developing systems.

  Bibliography:  p.
  Includes index.
  1.  Biology of developing systems.  I.  Title.
[DNLM:  1.  Biology.  2.  Growth.  QH308.2 G762d]
QH491.G7        574.3        76-39965

ISBN 0-03-077515-9

Printed in the United States of America
8  9  0  1   032  9  8  7  6  5  4  3  2

*To Rose —*
*who knows why*

# PREFACE

If there is any paradigm of developmental biology today, it is that developing systems are essentially programs of gene expression. Although these systems are as diverse as species, many biologists believe that simple rules of gene expression are shared by all living organisms. Any insight into gene regulation in prokaryotes, for example, also tells us something about gene action in growth and morphogenesis in eukaryotes. In fact, if E.B. Wilson were to write *The Cell in Development and Heredity* today, he might entitle the book *DNA in Development and Heredity* instead.

In spite of recent advances in molecular biology, we still do not understand how genes act in development. Mechanisms of prokaryote gene regulation cannot explain all modes of gene expression in the rich developmental repertoire of eukaryotes. Before a "modern synthesis" of development can be written, mechanisms of cell and tissue interaction must also be found to explain the development of whole body patterns.

The successes of molecular biology have also stimulated major changes in the teaching of biology. Core programs have been introduced, with developmental biology occupying an equal place with cell biology and genetics. Developmental biology courses have been substituted for traditional embryology courses in recognition of the fact that embryos are but one of many developing systems. Biochemistry has also become basic to any program in biology. But textbooks have not kept up with these changes; no text, for example, blends classical embryology with biochemistry and molecular genetics.

In this text I have tried to record the fusion of cell biology, genetics, and development by focusing on genetic information and its role in programming development. In the first few chapters I describe the nature and organization of developmental information in cells and the mechanisms for its retrieval. The bulk of the book explores how this information is utilized in growth, morphogenesis, and differentiation in a variety of developmental systems. The final chapters describe how information is misused in such phenomena as congenital abnormalities and neoplasia.

One of the most frustrating aspects of writing a textbook is keeping pace with information that is turning over virtually every two years. Only a few paradigms have emerged; principles of development are not as easy to identify as are principles of genetics. I often felt while I was writing that no developmental biology text should be written at this time—the field did not stand still long enough for concepts to be verified and widely accepted. What was "true" at the time of writing was not necessarily true one year later. Moreover, new areas of interest developed too rapidly for any one person to keep up with them. For example, the cell membrane and extracellular matrices only became important within the past five years, and new findings are taking place in this field at an exponential rate.

To keep up with the flood of new information I have sacrificed some traditionally important topics. Discussion of descriptive morphogenesis, for instance, is very much reduced compared to most embryology texts, but current trends seem to be directed away from descriptive embryology, and we are leaning more heavily on cellular, genetic, and biochemical mechanisms. The somewhat superficial treatment of the early material on genome organization is also a matter of deliberate choice. To present a balanced view of a field that overlaps both cell biology and genetics, compromises had to be made to limit the size of the book. Only the information from cell and molecular biology that seems relevant to our understanding of development is included; I have tried to give greater emphasis to the more important areas of morphogenesis and differentiation.

Mysteries have always been an exciting ingredient of embryology; they have accounted for much of the field's attraction in the past. But as we learn more about developing systems we can no longer wallow in the sheer mystery of it all. We study developing systems at all levels of organization with the underlying faith that simple, unifying explanations will be found. Even the complexities of pattern development will be resolved in terms of gene action systems. It is the exquisite simplicity of developmental mechanisms that evoke the most awe. Unraveling these mechanisms should be our principal goal.

An informational approach to development gives the teacher a framework on which developmental phenomena can be hung. Many concepts in genetics and cell biology extend logically into developmental phenomena and facilitate the organization of cohesive curricula. This book is designed for upper-division undergraduates and first-year graduate students. Although a one-year course in biology will suffice as background, the book is also suitable for a core program in which development follows segments on cell biology and genetics. It is also useful for one-semester or one-year courses in developmental biology or embryology. In the latter case, supplemental descriptive morphogenesis would have to come from additional readings or from laboratory exercises.

Only now, after having written the text, do I see flaws and lacunae. I am sure many more will be brought to my attention by students and teachers. One anticipated criticism is the absence of literature citations in the text. But I have eliminated them in an effort to simplify reading because I find that frequent citations tend to disrupt the flow of ideas. All relevant citations are listed in the bibliography that follows each chapter.

The mistakes in this book are mine. Many of the positive aspects of the book are due to my wife, Rose, who served as friend, advisor, editor, and sounding board while I was writing. She read every chapter and helped improve my academic style considerably. Each chapter had to make sense to her even though she has no training in biology. We assumed that if she could follow the arguments, so could a student, and much material was adjusted to suit this requirement.

An acknowledgment of my children's patience and understanding is also called for. They grew up with "the book," and I'm sure the image of their father rooted to his desk each morning is an indelible part of their memory. I apologize for the many family weekends that were sacrificed so that work on the book could continue.

I would like to acknowledge the assistance provided by Jean Shindler, who went beyond her responsibilities as editor. She introduced me to the world of textbook publishing, and I have relied on her advice and suggestions during the editorial and production phases of this book.

The many readers who identified substantive errors early enough for me to make corrections deserve special thanks. I particularly welcomed the insights provided by James Ebert, David Epel, Charles Kimmel, Howard Schneiderman, James Weston, and Charles Wyttenbach.

I also wish to thank the following persons, who made many valuable suggestions: John R. Baker, Iowa State University; William Ball, University of Washington; John W. Brookbank, University of Florida; Mary Ann Clark, University of Texas, Arlington; Richard Davenport, University of Illinois; Craig Davis, San Diego State University; Gary Freeman, University of Texas, Austin; Robert D. Grey, University of California, Davis; Jack Lilien, University of Wisconsin; R. W. McGaughey, Arizona State University; Florence Moog, Washington University; John D. O'Connor, University of California, Los Angeles; Ralph Quatrano, Oregon State University; Dorothea Rudnick, Albertus Magnus College; Sidney B. Simpson, Jr., Northwestern University; Louis Stearns, California State University, Los Angeles; J. Ross Stevenson, Kent State University; Leah W. Williams, West Virginia University.

Finally, I must thank James Ebert, who, in addition to my wife, was influential years ago in persuading me to write this book, much against my inclinations at the time. Although I complained loudly during the more frustrating periods, I am convinced that it was all worthwhile. I have felt very much a part of the excitement of a rapidly moving field, and I have managed to keep up with diverse areas that normally would have been neglected. It has been a true learning experience for me. I hope it will prove equally valuable for the student.

**Eugene, Oregon**
**June 1977**

**Philip Grant**

# CONTENTS

xi

# Part IV
# DIFFERENTIATION

539

# Part V
# DEVELOPMENTAL DEFECTS    657

# Developmental Systems

Throughout the history of biology philosophical arguments have created an impression that developmental phenomena are just too complicated to be understood. In the past we could attribute this somewhat romantic view to our ignorance about living systems. But in the midst of the current biological revolution, as we barely keep pace with the accumulation of new information, this attitude is no longer appropriate. Though many basic problems remain unsolved, we do have greater insight into the nature of the cell and its behavior in development.

Most concepts of developmental systems focus on the egg as it becomes a fully formed organism at hatching or birth. But embryogenesis is only an arbitrary time slice of a developing system. A more realistic developmental period is the life cycle, from the moment of fertilization to a reproductively mature adult, when new life cycles are initiated. After reproduction, organisms mature, age, and die. Aging results in part from lesions in developmental processes such as growth and regeneration; as lesions accumulate, function declines, and entropy increases.

The embryo is only one kind of developing system. More primitive systems, such as replicating viruses or sporulating bacteria and yeast, pose similar problems. Asexual buds in invertebrate organisms consist of cells that develop directly into the adult body plan and do it more efficiently than an egg. The emergence of a fungal hyphae into a fruiting body, the growth of a root, or the formation of a flower are all developmental systems that have acquired unique adaptations through evolution. There are as many different developmental "programs" as there are species. Nevertheless, amidst this diversity we seek unifying principles—rules of development that explain all systems.

A close examination of these systems reveals some common features. All seem to exploit the replicative potential of living material to generate new levels of organizational complexity. Each is a process of hierarchical change, beginning from a state of relatively low order (a DNA double helix or a single cell) to more complex arrangements as macromolecules are synthesized or cells accumulate. David Bohm, a physicist, offered a musical analogy to describe this process: "There may be a short set of notes in a given order. This order changes, then changes again and again. But all the changes of order form a yet higher order, which constitutes a part of the development of the overall theme. Each order of development itself changes in an ordered way to form a still higher order of development."[1]

Development encompasses all life processes proceeding along an extended time line. In contrast to physiological processes, which produce reversible changes in seconds or minutes, most developmental changes are irreversible and are measured in hours, days, and years. In development metabolic reactions transform the chemical nature, spatial relationships, and physical state of living molecules. Though development is unidirectional, it is also cyclical since events are repeated from one life cycle to the next.

## DEVELOPMENT: A HIERARCHICAL PROCESS

As a historical process, each step in the developmental sequence emerges from previous events. Though long-term changes in the physical world follow the second law of thermodynamics and dissipate towards more disordered states, developmental processes are constructive. New levels of biological order are created, new molecules synthesized, new structures assembled, and new properties acquired in a hierarchy of increasingly ordered arrangements. What starts out as simple and homogenous emerges as complex, heterogenous, and functionally adapted.

As a hierarchical process, developmental events at one level are translated into new properties at a next higher level. Large macromolecules, for example, exhibit properties of viscosity, elasticity, information storage, and replication far greater than a summation of subunit properties. Transitions from one organizational level to another always increase the informational content of a system. Energy is transformed into new information as macromolecules organize into cell organelles, cells into tissues,

[1] C. H. Waddington, ed., *Towards a Theoretical Biology*, vol. 2. Edinburgh: Edinburgh University Press, 1969.

and tissues into organs and organ systems. Development is epigenetic; it is a coming into being, a progressive materialization of new levels of organization.

The interface between organizational levels is indistinct. No sharp stepwise transitions occur; rather, one organizational level merges into another. To understand these relationships we isolate and analyze "time slices," that is, small segments of connected life cycles. However, in so doing we often lose the essence of change since only static images of a process can be studied. The role of developmental biologists is to reconstruct the sense of flow. It is like charting an immense river from its many sources in the mountains to its mouth by navigating only small segments; mistakes are inevitable and misconceptions arise.

## PREFORMATION VERSUS EPIGENESIS

In the past embryologists, overwhelmed by the "wonder, beauty, and mystery of it all," found themselves at philosophical dead ends. From the very moment we became aware of eggs and embryos, the history of embryology has been torn by controversies centering on such philosophical issues as *preformation* and *epigenesis*. Preformationists believed that embryogenesis was an unfolding or enlargement of a fully formed but minute organism, whereas epigeneticists claimed that the embryo arose from a progressive acquisition of new properties. Depending on the fashion of the times, this polemic persisted, and it continues even today in a more sophisticated vocabulary of cybernetics and genetics. Embryologists also became involved in questions of vitalism and mechanism, two opposing philosophical views of the organism. The former explained embryogenesis in terms of vital "principles," found only in living systems, that accounted for purposeful behavior. Mechanists described life in terms of known laws of physics and chemistry, implying that no new "life forces" need be invoked to explain vital phenomena. Hans Driesch, an eminent embryologist at the turn of this century, formulated a metaphysical concept called *entelechy* to explain the harmonious nature of embryonic development. Like most philosophical approaches to development, it was a dead end since no meaningful experiments could be designed to define entelechy nor its mode of action. As embryologists probed developmental systems, a cult of the complex evolved, explaining phenomena in terms of ambiguous generalities such as "morphogenetic fields," "organizers," "inducers," and "gradients," among others. Though these concepts are still used today, we admit that they do not explain phenomena; they simply reflect our inability to define them in more simple, physical-chemical terms.

Since the revolution in genetics in recent decades, developmental biologists have been looking for biochemical and genetic explanations of complex morphogenetic phenomena. What molecular events are causally related to cell and tissue behavior in developmental systems? How is gene expression translated into the diversity of form and pattern?

Though we know more about cells, chromosomes, and genes, most problems that challenged the classical embryologists such as Roux, Weismann, Boveri, Driesch, and Spemann at the turn of the century still confront us today. For example, we still do not know how a single egg can produce populations of different specialized cells, all carrying identical genetic information. Our methods are more sophisticated and our conceptual framework more secure, but our progress is still slow.

## COMPONENTS OF DEVELOPMENT

Developmental systems at all levels of organization, from viruses to man, share certain fundamental features that include (1) storage and transfer of developmental information, (2) molecular, cell, or organismic growth, (3) morphogenesis or changes in form and shape, and (4) differentiation, the emergence of a functionally specialized state. Since these processes mutually interact, it is difficult to perturb one without interfering with another. In a sense we are confronted with a biological "indeterminancy principle" whereby we cannot measure two properties simultaneously. To perform our analysis we must isolate each process and study it independently. In so doing we diminish the informational content of the entire system and lose some critical relationship between components. Unless we understand the behavior and interactions of parts, we cannot expect to understand the whole. Although we analyze isolated components of a system, we assume, almost as an article of faith, that the relationship of one part to another and to the whole can be reconstructed as each process is understood. This reductionist approach has proven successful in genetics and biochemistry. A similar analysis may succeed with developing systems, although some biologists claim that reductionism can furnish only partial answers. The view expressed by the latter biologists is a holistic concept that explains the principles of development in the rules of interaction and communication among parts. These cannot be studied by isolation and homogenization since they are properties of the intact organism. But we find that unless we isolate, the system is far too complex to unravel. Clearly we must do both; we must never lose sight of the fact that the whole is indeed greater than the sum of its parts.

The knowledge explosion in genetics and biochemistry has taught us that nature is basically simple; that is, simple mechanisms carry out complicated processes. Now that we understand the nature of informational macromolecules, we can work out some of the basic concepts of biological information storage and retrieval. Models of gene action derived from primitive microbial systems are used because we know that information flow in all biological systems is basically the same. But we also know that the developmental repertoire of multicellular systems is several orders of magnitude greater than that in simple microorganisms. The genome of higher organisms is larger and more complex and may be governed by different rules. Let us not assume, however, that such rules are beyond analysis. As we study growth, morphogenesis, and differentiation, we want to know how each is explained by the information encoded in macromolecular tapes. Using the new techniques of modern biology, we assume that what now seems impenetrable will turn out to be simple and comprehensible.

# DEVELOPMENTAL INFORMATION

A developmental system is analogous to a program, a sequence of prescribed events following a temporal order toward a goal. A set of coded instructions responsible for its execution resides in the totality of genetic information in developing cells. This information is stored as informational tapes, nucleic acid macromolecules assembled in chromosomes and cytoplasmic organelles. Since these molecules code for all proteins in an organism, a developmental program may be defined as a timed sequence of synthesis and assembly of new protein populations in cell clones.

Developmental information, however, is not restricted to genetic macromolecules; it also resides in the spatial arrangement of cellular metabolic systems and their interaction network. The asymmetrical packing of enzyme arrays in membranes contributes to a directed flow of developmental information. Moreover, metabolic oscillations and mitotic rhythms are used as clocks ticking off developmental events.

As cells proliferate into multicellular sheets, the informational content of the system increases. Information processed by one group of cells is communicated to adjacent and distant cells by diffusible molecules. The information of a multicellular system is now defined in terms of supracellular properties, the positional relationships of cells interacting within tissue fabrics.

Most questions about developmental information relate to its organization, storage, and use as macromolecular tapes. How is genetic information organized in chromosomes? What is the nature of information flow between different cellular compartments, and how are feedback circuits stabilized into patterns of cell behavior? Which developmental phenomena are preprogrammed and which are epigenetic?

5

# 2

# Organization of Developmental Information in Cells

The cell is the basic unit for carrying out developmental programs, a concept recognized by E. B. Wilson in his first edition of *The Cell in Development and Inheritance* (1896). At that time the nature of genetic information was unknown, but now we know that developmental processes depend on the organization and use of information encoded in giant nucleic acid macromolecules. These "instructions" are carried out within cells whose organization is itself a reservoir of information; cellular constituents are spatially arranged into functional units, and molecular traffic is directional, following prescribed channels between compartments. Such ordering of cellular elements is also part of a cell's information store, its so-called "structural information." Both genetic and structural information are utilized as a developmental program unfolds. We will see that virtually all the genetic information encoded in macromolecular tapes is preformistic while the structural information undergoes epigenetic changes.

## DEVELOPMENTAL INFORMATION: PREFORMISTIC OR EPIGENETIC?

All developmental systems start out with intrinsic genetic and structural information. Higher levels of biological order emerge during development; the system seems to gain information as it develops. Development is both preformed, in terms of its initial set of genetic instructions, and epigenetic, because new rules, or boundary conditions, evolve as the original instructions are carried out. This concept resolves an age-old controversy that has plagued embryologists since Aristotle, that is, preformation versus epigenesis. Preformationists believed that all parts of the organism are preformed in the egg from the very beginning. Development merely converts latent properties into manifest differences without increasing intrinsic complexity or creating new properties. The viewpoint reached its logical conclusion in the view of the spermists, a sect of preformationists (circa 1600) that stressed the importance of the sperm

7

**Figure 2-1** *Spermists' view of development. (After Hartsoeker's drawing of a human spermatozoon.)*

to development; examining the head of a sperm through their crude microscopes, they were able to see a complete human fetus curled up which they called the *homunculus* (Figure 2-1). On the other hand, most epigeneticists, beginning with Aristotle, claimed that the egg is of limited complexity and that new properties and complexities must arise during development. The egg, they postulated, is a formless mass that gradually acquires order and form (Figure 2-2). These same issues are still controversial, though today they are couched in the language of information theory.

Some biologists have gone so far as to compare the amount of information contained in an egg with that in an adult. Such calculations invariably show that the adult contains much more information. For example, if information is defined as the total number of molecules in a system, then an adult organism will necessarily have more information than an egg; the adult is larger and has many more molecules. According to one estimate, an organism contains $5 \times 10^{25}$ informational bits as molecules (or $2 \times 10^{28}$ bits when calculated as atoms) while the egg has only $10^{10}$ significant molecular bits of information. Other researchers have come up with similar numbers using different criteria such as the information stored in DNA molecules.

Acceptance of these numbers inevitably leads to a philosophical cul-de-sac of the preformation versus epigenesis controversy since we must then explain how the adult contains more information than the egg. The argument requires that certain rules be followed. For example, information, like energy, obeys the second law of thermodynamics and cannot be created de novo. On one hand, the epigeneticist claims that information is introduced exogenously since it increases during embryogenesis. A preformationist, on the other hand, assumes that the egg is encoded with all the information necessary for developing into an adult. Development is simply a decoding process; the increased information in the adult is not new but only a reiteration or amplification of preexisting information. For example, genetic information in the form of DNA molecules is replicated accurately in each cell through development. This means, however, that an encoding process is simultaneously stamping out informational

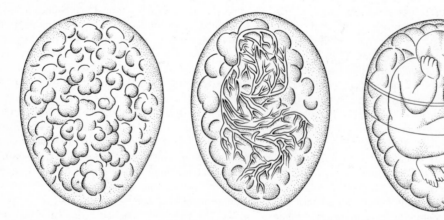

**Figure 2-2** *Epigeneticists' view of development. Aristotelian coagulum of blood and seed in the uterus after Jacob Rueff's drawing* De Conceptu et Generatione Hominis *(1554).*

copies for each cell as the system is decoding. Although organisms can do this, some basic assumptions of information theory are contradicted, since a decoder cannot simultaneously behave as an encoder.

It has been pointed out that assumptions about the nature of information are entirely arbitrary, have no relevance to biological systems, and often lead to some trivial results. Since there is nothing absolute about informational content as measured in this way, we cannot build arguments on meaningless numbers. Most assumptions are unrealistic; all calculations neglect "structural and temporal information" contained in spatial ordering of functional components of living systems. In other words, we need more than an estimate of informational bits; we need, as Apter and Wolpert suggest, an organizational plan to show how parts are related in space and time.

To date information theory has not helped us to understand developing systems. Cells do contain redundant information as identical DNA molecules, and new structural information does emerge as cells organize into tissues and organs. Information is acquired as new hierarchies of biological order emerge from interacting cell populations. Yet we have no rules governing these ordering processes, nor have we successfully translated spatial and temporal qualities into "bits" of information. Even if we could, it is doubtful whether the analysis would bring us closer to an understanding of developmental systems.

We will use the term "information" primarily as an organizational blueprint for developmental programs. This blueprint includes instructions encoded in linear macromolecules for making all proteins necessary for survival, growth, and morphogenesis. Also included is information residing in the spatial-temporal organization of cells; that is, cell structure, in itself, has informational content. The primary informational source is genetic since all proteins are made according to gene instructions. But once a protein is synthesized, its deployment into cellular architecture depends upon interactions between its intrinsic steric properties and the intracellular microenvironment. Thus intracellular conditions of ion flux, pH, charge distribution, and so on, will determine the final functional configuration of a protein in a cell. These conditions or metabolic states are *epigenetic*, undefined by specific gene instructions, yet are essential informational components of the cell. Finally, as cells interact in populations, new rules emerge to govern and integrate their behavior. These too are epigenetic events. We will see that development of hierarchical organization results from interactions between genetic and epigenetic components of cellular information systems.

## PREFORMED INFORMATION AS GENETIC MACROMOLECULES

Eons ago, in the primordial soup, life evolved from the nonliving world. At some point in chemical evolution, large macromolecules (polynucleotides) acquired information for their own replication. Once this occurred, organic evolution was possible. Since the replication process is not error-free, mistakes or mutations arise which are also replicated. Accordingly, heterogeneous populations of informational molecules arose and through natural selection evolved to produce macromolecular and cellular diversity.

Secondarily (or perhaps simultaneously) these polynucleotides became messages, encoded with information for protein structure. Because of their size and spatial configuration proteins catalyze specific chemical reactions. How polynucleotides became messages is not clear, but their ability to store and transmit structural information accounts for the infinite variety of protein catalysts that make life possible. Such molecules possess all properties of a gene: they replicate, mutate, recombine, and code for protein — and have evolved as the preformed information in developing systems.

Life's dynamism and diversity reside in the catalytic and structural properties of proteins. They carry out all vital activities and are building blocks for all levels of biological organization. They act as molecular switches and turn biochemical reactions on or off. An organism is but the functional expression of its own constellation of proteins. Protein diversity resides in linear sequences of amino acids, and these in turn are encoded in nucleic acid messages, or genes. Development therefore may be viewed (at the molecular level) as a spatiotemporal ordered appearance of new proteins while temporal regulation of the expression of different sets of genes may be defined as a developmental program.

## GENES AS DEVELOPMENTAL INSTRUCTIONS

A fabric of interacting genes underlies all developmental processes, from a simple change in protein structure to the tissue movements involved in organ growth. Bristle formation in *Drosophila* is a case in point, where the phenotype of only two progenitor cells requires coordinated expression of many genes (Figure 2-3). Two bristle cells first appear in the hypodermis, or skin, of the pupa after division of an enlarged stem cell. Thereafter these paired cells grow rapidly, attaining a volume about a thousand times greater than that of neighboring epidermal cells. One cell, the *tormogen,* or socket cell, is displaced above and to the side of the *trichogen,* or bristle cell, as a bristle is formed.

Every step is under genetic control from determination of the location of a stem cell to bristle growth. Mutation of any gene in the sequence results in abnormal bristles. For example, the first division of hypodermal cells, to establish the primary precursor, or stem cell, is modified by mutant genes. Both the *scute* and *hairy* genes control these proliferative events and thereby determine the number of stem cells. The *scute* mutant inhibits cell division so that fewer bristles are formed, whereas the *hairy* gene stimulates division to produce many stem cells and, of course, many more bristles than normal.

The subsequent division of the stem cell to set up the paired bristle cells is critical. If more than one division occurs or if the division plane shifts during this step to change the relative positions of the cells, either more bristles appear or abnormal bristles emerge. The mutant gene *split* induces two divisions rather than one to yield four irregularly arranged cells which differentiate into a duplicate bristle. Other mutant genes, like *stubble,* give rise to short,

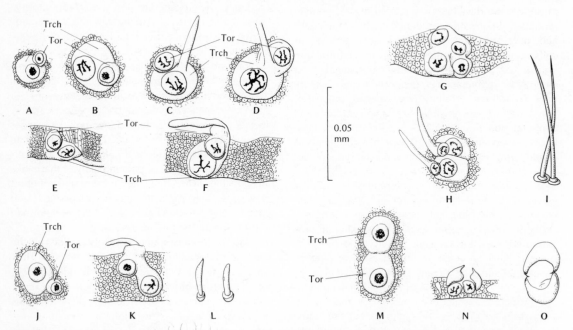

**Figure 2-3**   *Some mutants of bristle development in* Drosophila. *(A–F) Normal sequence of development of wild type bristles. (A–D) Surface view of development from 18 to 40 hours. (E, F) Cross-section of epidermis shows arrangement of tormagen, a collar cell to trichogen, a bristle cell. (G–I) Development of split mutant. (G, H) Section shows extra division with two tormagens and two trichogens. (I) Split bristle. (J–L) Development of stubble; tormagen is displaced from trichogen after division. Compare to B and F in normal sequence. (M–O) Development of hairless. Division is equal, and each cell secretes an irregular mass to produce dual sockets (O). (From A. D. Lees and C. H. Waddington,* Proc. Roy. Soc. London Ser. B *131:87–110, 1942.)*

thick bristles by shifting spindle axes so that the tor-mogen cell occupies the wrong side of the trichogen. Finally, synthesis of chitinlike proteins and their secretion from the trichogen cell as a bristle are also under gene regulation. Genes like *spineless,* for example, reduce the amount of secretion from the trichogen, and short, thin bristles develop.

The program for bristle development from stem cell determination to bristle differentiation is en-coded in a group of related genes linked on one chromosome. Normal bristle formation coordinates mitosis with cell position and protein secretion. These processes are imprinted in sets of regulatory and structural genes. Whereas structural genes code for specific proteins involved in bristle formation and assembly, regulatory genes usually determine when a structural gene is "on" or "off." Some regulatory genes determine when division begins and ends. Others influence the plane of division or the time of synthesis of specific proteins. Together these gene sets program a harmonious morphogenetic pattern. But even if we knew how all these genes functioned, we still would not understand why bristles are ar-ranged in definite patterns on the body surface. Bristles are not set in the epidermis at random but occupy a precise site as if they "know" exactly where they are in space. Do "patterning genes" assign the position of cells into an overall, sym-metrical body plan (see Chapter 16)? Evidently what appears to be a simple process involving two cells and a single product is in fact a complex interacting system of structural and regulatory genes controlling cell position, timing of cell division, and protein synthesis.

Some biologists believe that complete under-standing of embryogenesis will come only after all genes are inventoried, their time and place of ac-tion mapped, and their products followed into cell organization and behavior. This goal is clearly un-realistic for organisms with many thousands of genes. A simpler task is to discuss the nature of genes and how they are organized in cells to store, replicate, and retrieve developmental information.

## GENES CODING FOR PROTEINS

Intuitively we believe that most developmental information is encoded in genetic macromolecules because we have learned that sequences of subunits in polynucleotides code for protein structure. We will briefly review some aspects of gene-protein relationships that are relevant to the flow of informa-tion in developing systems. Early in the history of genetics, it was noted that Mendelian genes in hu-mans seem to act at the biochemical level to affect cell metabolism. Later, metabolic cycles in bacteria, yeast, and molds were shown to be controlled by enzymes whose syntheses were governed by specific genes segregating and recombining in typical Mendelian fashion. Biochemical mutants were ob-tained which either did not possess a functional en-zyme or produced inactive proteins structurally related to the enzyme. Evidently mutant genes were responsible for the production of abnormal pro-teins.

The analysis of the genetics of human hemo-globins showed how genes code for proteins. Nor-mal hemoglobin, or hemoglobin A, the oxygen-combining protein in red cells, is a complex protein consisting of four polypeptide chains—two alpha and two beta chains. Individuals suffering the heredi-tary disease known as sickle cell anemia are homozy-gous for the sickle cell gene Hb *S* and have abnormal "sickled" erythrocytes which contain a different hemoglobin, hemoglobin S. The latter has different physical-chemical properties (for example, electro-phoretic mobility) and is less efficient in combining with oxygen, which accounts for the abnormal syn-drome. Individuals heterozygous for the sickle cell gene (Hb *A*/Hb *S*) carry both hemoglobins A and S and exhibit a much milder form of the disease. The amino acid sequences of both chains have been analyzed; the alpha chain has 141 amino acids and the beta chain has 146 amino acids. A comparison of the amino acid sequences of hemoglobin A with hemoglobin S shows that the sequence in alpha chains is identical but that one amino acid, valine, has been substituted for a glutamine in a section of the hemoglobin S beta chain. This single amino acid substitution out of 146 amino acids is enough to change all properties of sickle cell hemoglobin. The substitution alters the three-dimensional shape of the molecule and thereby affects its oxygen-transporting capability. A mutant gene changing only one amino acid in a polypeptide sequence means that a gene contains information for the primary amino acid se-quence of a protein.

## DNA AND INFORMATION STORAGE

Before the relationship between genes and proteins was worked out, the chemical nature of the gene was described in 1953 in the now classic Watson-Crick model for DNA which shows that genetic information of all living systems, except for some RNA viruses, is encoded in DNA. This nucleic acid is comprised of long-chain polymers, sequences of four different subunits—adenine, guanine, thymine (or uracil), and cytosine nucleotides. Some DNAs contain methylated bases as well, such as 5-methyl-cytosine.

With molecular weights of the order of 10 to 20 million, DNA must contain thousands of nucleotide subunits in linear array. These are arranged as two interwoven helical chains held together by hydrogen bonds between nucleotides stacked like poker chips. Viral RNA genomes, on the other hand, are single-stranded chains of ribonucleotides sometimes loosely organized as helices.

Because of their molecular configuration, nucleotides form complementary pairs; adenine (A) nucleotides on one chain of the double helix form hydrogen bonds only with thymine (T) nucleotides on the other chain, and guanine (G) pairs only with cytosine (C) nucleotides. This strict stereochemical pairing is the principal "grammar rule" of "communication" or information transfer between each strand of the double helix. Nucleotide sequences on one helix automatically specify the sequence on the other. If one strand has the sequence ATCGAAG . . . ,

the opposite strand must have the sequence TAGCTTC . . . . DNA molecules are polarized; they have a beginning and an end. Whatever information is contained in the polynucleotide sequence is "read" in one direction.

Complementarity of nucleotide sequences in double-stranded DNA is a universal property of all DNAs. Linearity of strands, however, is not. Some viral and bacterial DNAs are circular and may be isolated as intact circles from infected cells. Other viral DNAs are single-stranded and become double-stranded only during their replication in infected cells.

These large macromolecules undergo configurational changes over the entire molecule or only in localized regions. The primary linear sequence of mononucleotides held together by covalent bonds may be folded or coiled as helices, but this is only an idealized structure of a dynamic molecule that may assume shapes from a fully extended state to a highly compacted, almost crystalline arrangement, when packaged in sperm nuclei (Figure 2-4). The supercoiled, compacted, or extended state of the molecule is a function of intramolecular, inter-molecular, and local forces impinging on the molecule. Covalent interactions with other ions and molecules, primarily proteins, determine the shape and relative state of packing of DNA within chromosomes. Packing DNA into small volumes such as the head of a viral particle is a problem in view of DNA length. Lengths range from a viral DNA of 1 $\mu$ (single-stranded DNA of $\phi$X174) to DNA from ani-

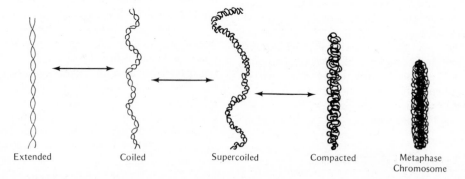

Extended            Coiled            Supercoiled            Compacted            Metaphase Chromosome

**Figure 2-4**  *Various transitions in the physical state of DNA. Parts of chromosomes or whole chromosomes may contain DNA in any one of these states during interphase. At metaphase most DNA is in the compacted state.*

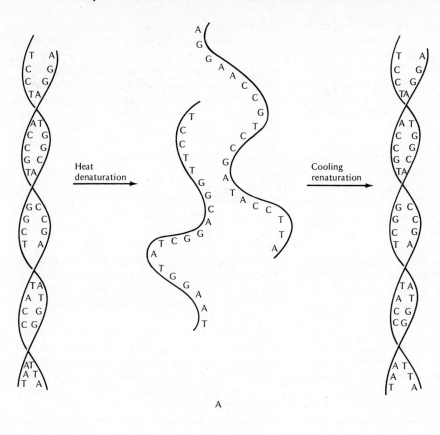

Heat
denaturation

Cooling
renaturation

A

**Figure 2-5** *Melting and reannealing of double helical DNA. (A) Heat denaturation of a DNA double helix results in strand separation. Strands reassociate spontaneously when the mixture is cooled according to base-pairing ru'es. (B) UV absorption curves for DNAs extracted from different bacteria and an artificial DNA containing no guanine or cytosine (AT). As temperature increases, the absorption remains constant until melting occurs. A rapid rise in UV absorption occurs when strands separate. Notice that melting temperature (in paren heses for each curve) varies directly with the percentage of GC bases (also in parentheses). Melting temperature is computed when the steeply rising part of the curve reaches half the maximum absorbance. (After J. Marmur and P. Doty, Nature 183:1427, 1959. Reproduced by permission of Nature.)*

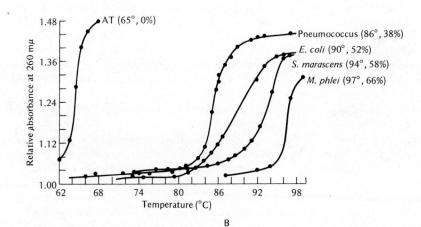

B

mal and plant cells that is several centimeters long. In all instances, the size of the DNA when extended is considerably greater than the structure in which it is contained, which means that the molecule undergoes tight packing (supercoiling?) to fit into small volumes. In animal and plant cells changes in configuration take place within chromosomes and correlate with cyclic changes in cell activity such as mitosis. Such structural changes are bound to have profound effects on the accessibility of information encoded in these sequences.

Pairing of complementary nucleotide sequences occurs spontaneously by hydrogen bonding. When these weak bonds are broken by gentle heating, a denaturation process known as *melting,* the strands of a double helix separate and transform into two random coils (Figure 2-5). The melting of DNA can be followed experimentally because separated strands absorb more readily in the ultraviolet than double-stranded DNA; this is the so-called hypochromicity effect. The transition in UV absorbance occurs sharply at certain temperatures, depending on nucleotide composition. DNA with abundant guanine-cytosine (GC) pairs usually melts at higher temperatures than DNA rich in adenine and thymine (AT). Even within the molecule, regions rich in GC melt at higher temperatures than regions rich in AT, thereby influencing the shape of the melting curve. Melting is reversible if the temperature is lowered slowly. The separated strands spontaneously reassociate (anneal) into stable, biologically active double helices, with nucleotide sequences in correct register, associating according to the "grammar rule" previously mentioned. Reversible melting-reannealing is useful in comparing DNA nucleotide sequences obtained from different species of organisms.

Since genes are composed of DNA, each species contains DNAs with unique nucleotide compositions. Some DNAs are rich in adenine and thymine; others, such as the DNAs of viruses and bacteria, have a high guanine-cytosine content. The degree of relatedness between two species is determined by melting and reannealing mixtures of their DNAs. Single strands sharing the same or similar nucleotide sequences tend to reanneal quickly into double-stranded hybrids, and these are easily separated from single strands. Since the DNAs of closely related species share many nucleotide sequences, they form hybrid helices more readily than distantly related species. This technique of DNA-DNA hybridization

has high resolving power and can show close phylogenetic relationships. Thus human DNA has more nucleotide sequences in common with DNAs of monkeys and apes than with DNAs of other mammals, while mouse DNA is more closely related to rat and hamster DNA. Close morphological and physiological resemblances among primates derive from the fact that they draw on a common gene pool. Sharing similar genetic information, they exhibit similar developmental programs.

## THE GENETIC CODE

The preceding discussion points to the species-specificity of nucleotide sequences in DNA. The analysis of human hemoglobins also suggests that a gene codes for the primary amino acid sequence of a protein. A nucleotide sequence in DNA encodes the amino acid sequences in polypeptide chains. Since there are 4 different nucleotides in DNA and 20 amino acids in proteins, how can an alphabet based on 4 letters be translated into an alphabet of 20 letters? Obviously combinations of more than 1 nucleotide must specify a single amino acid.

More than a decade of genetic and biochemical studies have successfully unraveled the genetic code. A code word in DNA is a triplet; an arrangement of three nucleotides defines a single amino acid in a polypeptide chain. This furnishes 64 different code-word combinations, or *codons,* more than enough to specify 20 amino acids. It also implies code degeneracy; that is, several trinucleotide combinations code interchangeably for one amino acid. Thus valine is coded by the codons GUU, GUC, GUA, GUG. The genetic tape, therefore, is a linear sequence of triplet codons, the instructions for ordering amino acid sequences in proteins. Codons are nonoverlapping; that is, each is contiguous with its neighbor. The code is read in one direction; it has no commas or spaces between each unit.

The code allows us to define a mutation as a change of one nucleotide in a codon that produces an amino acid substitution in a protein. Thus changing the codon GAA to GCA converts glutamic acid to alanine. If changed to CAA, it becomes glutamine. Chemical mutagens act by changing codons, and substances like 2-amino-purine or nitrous acid induce *transition mutations,* or nucleotide substitutions. Nitrous acid, for example, results in transitions of cytosine to uridine ($C \rightarrow U$) and adenine to guanine ($A \rightarrow G$). Such transitions occur spontaneously during DNA replication at a very low frequency and may account for spontaneous mutations.

With our codon catalog genetic messages can be read. Substitution, deletion, or addition of only one nucleotide suffices to change the meaning of a primary amino acid sequence. Because the codon is a triplet, a message reader or decoder "jumps" from one triplet to another as it scans the message tape to translate a genetic message. If the mode of reading the message changes by shifting only one base, a completely different protein is produced.

It has been shown in studies of mutations of the polypeptide subunits of the enzyme tryptophane synthetase in *Escherichia coli* that there is a point-to-point correspondence between the position of mutations on the genetic map and the position of amino acid substitutions in the mutant proteins. This finding established the principle of colinearity between the gene and its protein product, and it showed that the direction of reading instructions in the gene corresponds to the ordering of amino acids in the protein; that is, linear sequences of nucleotides in DNA code for the amino acid sequences in a protein.

Summarizing, a gene is a small segment of a DNA helical molecule consisting of a sequence of nucleotide pairs. Approximately 600 to 1200 nucleotides code for average-sized proteins of 20,000 molecular weight. A gene occupies a specific site in a chromosome, or *locus*, and may associate with neighboring genes to form a gene complex functioning as an integrated unit of regulation, or *operon*. An operon may consist of several structural genes, or *cistrons*, coding for specific proteins plus a few regulatory genes that control the expression of the operon.

## EUKARYOTE AND PROKARYOTE CELLS

Biological systems are organized on a cellular plan; that is, the basic functional unit of protoplasm is the cell. Two different cellular organizations are found: a primitive prokaryote plan in bacteria and blue-green algae and a more complicated eukaryote plan in all other organisms (Figure 2-6). The size and complexity of genomes differ within these two cell types, and for this reason they differ with respect to developmental potential. Eukaryote genomes are more complicated, contain more information, and account for a rich repertoire of developmental programs.

Prokaryotes ("before a nucleus") lack a well-defined nucleus bounded by a membrane. A clear nuclear area is seen in such cells merging directly with the surrounding cytoplasm. Eukaryote genomes, on the other hand, are segregated from the cytoplasm by a nuclear envelope creating two separate but interacting compartments, each with a unique ionic environment.

Prokaryote cells are small, ranging in length from 0.2 to 3 $\mu$, a size that favors diffusion-mediated exchanges of metabolites and gene products. Cell organization is simple and relatively homogeneous with little organelle differentiation. Energy-transducing enzyme systems are either dispersed in the cytoplasm, confined to the plasma membrane, or associated with extensions of the outer membrane thrown into the cytoplasm. They contain no chloroplasts or mitochondria; bacterial chlorophyll is dispersed in the cytoplasm rather than assigned to membranes. Cytoplasmic ribosomes are smaller than eukaryote ribosomes and are distributed uniformly within the cytoplasm and nuclear areas. The life cycle of these cells is too short to support complicated developmental programs. Accordingly, the developmental repertoire of most prokaryotes is modest. Some sporulate and pass through a cycle of dormancy followed by germination. Others assemble flagella, the principal motile organelles.

The genetic systems of viruses and prokaryotes are relatively simple. In general they are haploid, consisting of a single chromosome or linkage group. The chromosome is a "naked" DNA double helix relatively free of protein. A viral chromosome from lambda bacteriophage, a virus which infects the bacterium *E. coli*, is a circular DNA helix about 17$\mu$ in length, consisting of approximately 50,000 nucleotide pairs. A molecule of this size is estimated to have only about 50 to 65 genes, enough for a few proteins for viral replication and development. Chromosomes of other bacterial viruses, the so-called T-even phages, are much longer double helices, approximately 56 $\mu$ long, and they contain over 100 genes. Since all viruses are minute, these long genetic molecules are accommodated in their small volumes by tight coiling and compaction.

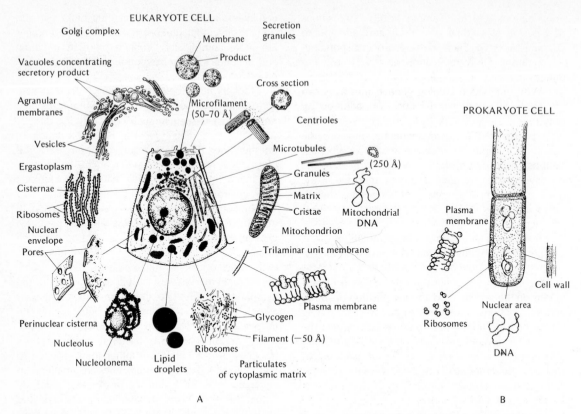

**Figure 2-6** *Comparison of eukaryote and prokaryote cell organization. The ultrastructure of cell organelles is shown in the periphery around each cell type. (Based on W. Bloom and D. W. Fawcett,* A Textbook of Histology, *9th ed. Philadelphia: Saunders, 1968.)*

DNA, partially extracted from viral particles and examined in the electron microscope, spills out as a network of tangled threads.

Bacterial cells are more complex and contain much more DNA. Like viruses, they also have a single chromosome, a DNA double helix, either linear or circular. The *E. coli* chromosome, 1000 $\mu$ long, is isolated as a circular DNA double helix, estimated to carry up to 1000 genes.

Cell division is also primitive in these organisms; the naked chromosome unwinds for replication. There is no mitosis, and there are no centrioles, spindles, or spindle fibers. Chromosomal attachment to the cell membrane coordinates cell division with chromosomal segregation (see Chapter 5).

Sexual recombination, defined as exchanges of DNA between cells, occurs in prokaryotes probably not too differently from its origins in primitive organisms. DNA exchange in prokaryotes is generally incomplete since only random portions of the genome are exchanged between conjugating cells. Sexual recombination, therefore, is not a major feature of prokaryote evolution. Mutations are the principal source of genetic variability, and these are immediately expressed and tested in their haploid state. The short cell cycle and prolific reproductive capabilities of these organisms usually preserve adaptive gene combinations, yet enough genetic variations arise through mutation to permit evolutionary modification.

By contrast, eukaryotes display an almost infinite variety of developmental patterns as free-living single cells or as multicellular organisms. Cell size is larger and more diverse than in prokaryotes, with lengths ranging from 0.2 $\mu$ for small protists through 10 to 20 $\mu$ for most somatic cells of multicellular organisms to 1000 to 1500 $\mu$ for large free-living protists. Larger size means more genetic information, a longer growth period, and an extended life cycle. The element of time becomes integrated into the life of eukaryote cells more so than in prokaryotes. In the latter time seems unimportant; growth and division are measured in minutes and hours. In the longer-lived eukaryote cells, on the other hand, lives are measured in days and years. These two parameters, size and time, contribute to the greater developmental potential of eukaryotes.

The eukaryote genome is sequestered in its own nuclear environment, communicating with surrounding cytoplasm through a semipermeable nuclear envelope. The envelope contains pores through which large macromolecules pass to facilitate nucleo-cytoplasmic interactions. Set apart by a boundary layer, the genome is screened from more volatile changes in the cytoplasm. Eukaryote cells contain many chromosomes to accommodate an increased gene load and higher levels of DNA. Chromosomes are large, each bearing thousands of genes. Chromosome replication and segregation are accomplished by mitosis with its spindle apparatus, asters, and centrioles. The chromosome is adapted as a packaging device to faithfully transmit portions of a large complicated genome to daughter cells.

Large cell size in eukaryotes makes diffusion-mediated exchanges of gene products less efficient. Metabolic needs for gene products are satisfied by cellular compartmentalization, the evolution of membrane-bound organelles such as chloroplasts and mitochondria. As in prokaryotes, the metabolic needs of these small organelles are met by diffusion-mediated flow of substrate. Moreover, their membranes behave as metabolic pumps that facilitate metabolite and ion flow. These organelles contain DNA genomes that resemble in size and shape those found in prokaryotes. Some claim these structures have evolved from primitive prokaryote symbionts that invaded evolving eukaryote cells. Their DNA templates code for organelle-specific proteins; hence the total genetic information of a eukaryote cell is partitioned into several metabolic compartments. This places additional demands for regulatory machinery to control the flow of information within the cell.

The cytoplasm of eukaryote cells is highly ordered, consisting of membrane-bound compartments, tubules, vesicles, filaments, and so on. Structural information is a major part of the total informational content of these cells. Communication between compartments is membrane-regulated, and metabolite flow is polarized.

The nuclear genome of eukaryotes is usually diploid, a condition with high selective advantage; it allows for the evolution of meiotic mechanisms of sexual recombination. Sets of homologous chromosomes in the same nuclear compartment become available for chromosome pairing and crossing over, a source of new gene combinations in meiosis.

Most primitive eukaryotes (algae and fungi) are haploid, a karyotypic state usually correlated with morphological simplicity and limited developmental potentialities. Diploidy arises at fertilization after gamete fusion, and the diploid zygote is usually adapted for dormancy and long-term survival under unfavorable conditions. Haploidy is usually restored shortly after fertilization, and before development of a new organism, when the zygote undergoes meiotic recombination. Most eukaryote life cycles alternate between haploid and diploid phases, with the latter evolving as the more dominant condition. But this is only possible if meiosis is suppressed in the zygote. When suppressed, the diploid zygote is able to grow, express its more versatile genome, and develop into an adult phenotype. It is during the state of suppressed meiosis in the diploid phase that the tremendous diversity of developmental programs has evolved. In higher plants and animals meiosis is usually suppressed until gametes differentiate in mature adults. Hence only the gametes display the haploid phase.

Most mutations are concealed in diploids and are carried as recessive genes as part of the genetic load of the population. Accordingly, a great source of potential variability accumulates, leading to a high degree of heterozygosity. Diploidy favors gene interactions and dosage effects in phenotypic expression, in the sense that more than one copy of a gene affects the expression of a trait. Such complex interactions also account for enriched developmental expression.

## ORGANIZATION OF GENETIC INFORMATION IN EUKARYOTES

Two distinct genetic compartments exist in eukaryotes, nuclear and cytoplasmic. Though the genetic information is organized differently in each, both are mutually dependent and exchange information con-

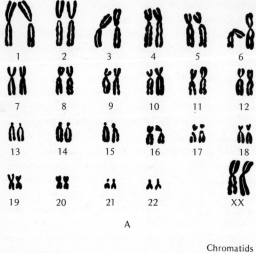

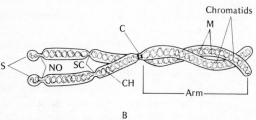

**Figure 2-7**   *Normal human karyotype (female). (A) Appearance of chromosomes in metaphase. Karyotype obtained after matching chromosome pairs. (B) Diagram of chromosome showing two chromatids: C = centromere (kinetochore) or primary constriction; CH = chromonema or coiled DNA helix; M = matrix; SC = secondary constrictions, often absent; NO = nucleolus organizer region, sometimes known as a secondary constriction and found on only one pair of chromosomes; S = satellites, or short segments.*

## The metaphase chromosome

Each species has a characteristic karyotype appearing at metaphase; the number of chromosomes and their morphology are generally the same in all cells. In humans the karyotype of 46 chromosomes includes 22 pairs of autosomes of distinctive shape and 1 pair of sex chromosomes (Figure 2-7). The basic unit of the chromosome is the *chromatid,* a fibrous DNA element combined with various proteins. Most metaphase chromosomes are made up of two chromatids, but some, like the giant polytene chromosomes in *Drosophila,* have hundreds of chromatids in parallel array. Until a chromosome replicates, we may assume it is composed of one chromatid. Since the number of chromatids per chromosome is constant and since each somatic cell has an identical karyotype and a constant amount of DNA, each chromatid probably has a constant amount of DNA. This so-called DNA constancy rule states that the amount and kind of DNA per somatic nucleus is constant and a function of the number of chromosomes (chromatids). Each cell of an organism contains the same amount and the same kind of genetic information. Although there are exceptions to this rule, it is true in general.

The simplest model of a chromatid is the *unineme,* or single-strand model; that is, a single DNA double helix makes up the principal structural axis. The *polyneme* model proposes a multistranded chromatid made up of many double helices held together at a common site. Both models are compatible with the accepted mode of DNA replication, but the polynemic model runs into some difficulties with respect to gene function. For instance, if all strands in a polynemic chromatid are functional, recessive mutations would never be expressed since it is unlikely that the same gene would mutate simultaneously in all strands. Complicated control mechanisms would have to be invoked to explain phenotypic expression of recessive mutations, inasmuch as genes in all but one or two DNA helices would have to be inactivated. The cytological evidence does not help us in resolving this problem, since there are no satisfactory preparations of chromosome organization. Some studies of *Drosophila* and yeast chromosomal DNA support the single-strand model of chromatid structure. Measurements of the amount of DNA in chromosomes correlate with the contour length of extracted DNA, as seen in the electron microscope,

tinuously throughout the life of the cell (see Chapter 3). Most of the cell's genetic information is organized as chromosomes in the nuclear compartment whereas cytoplasmic genes are distributed as small circles of DNA within mitochondria and chloroplasts.

Chromosomes are not visible in most interphase cells and are difficult to study. Mitotic chromosomes, on the other hand, are identifiable entities, particularly at metaphase, and we know more about their structure. Genetic material in metaphase chromosomes is organized for transport and segregation to daughter cells and is inactive.

suggesting that only one DNA duplex is found in each chromatid.

The *centromere,* or *kinetochore,* is a constricted region on the metaphase chromosome separating it into two distinct arms. It serves to identify each chromosome pair since the relative lengths of chromosome arms are determined by the location of this primary constriction. The centromere is the site of chromosomal fiber attachment, without which engagement on the mitotic spindle and segregation into daughter cells is impossible. Fragments of chromosomes without kinetochores wander aimlessly in the cytoplasm during mitosis and do not move to the daughter cells at anaphase. A larger constriction appears on some pairs of metaphase chromosomes: the secondary constriction or *nucleolar organizer region.* Generally the number of such constrictions corresponds to the number of nucleoli in a somatic cell. These SAT chromosomes (the region distal to the secondary constriction is referred to as a *satellite*) generate new nucleoli at the end of mitosis.

Chromosome morphology is species-specific. Each chromosome pair is characterized by overall length, secondary constrictions, and position of kinetochores. Each chromosome carries a different constellation of genes; it is a linkage group. Ideally we would like to know the total informational content of each chromosome, but this is almost impossible to obtain.

DNA in metaphase chromosomes is exceedingly condensed into tight supercoils. When metaphase chromosomes are partially unraveled, they appear as "coiled coils," a major helically coiled axial filament which is twisted into minor coils (Figure 2-8). Coiling is detected at the fine structural level after metaphase chromosomes are exploded into a tangle of 200- to 300-Å fibers. Each fiber is itself coiled into a helical configuration. Irrespective of the method of preparation, a 200- to 300-Å fiber seems to be a constant feature of metaphase chromosome organization, which suggests this may be the intact chromatid.

Metaphase chromosomes have been isolated in relatively pure form from a variety of cells, including human cells (Table 2-1). Here we see one major difference between the eukaryote and prokaryote

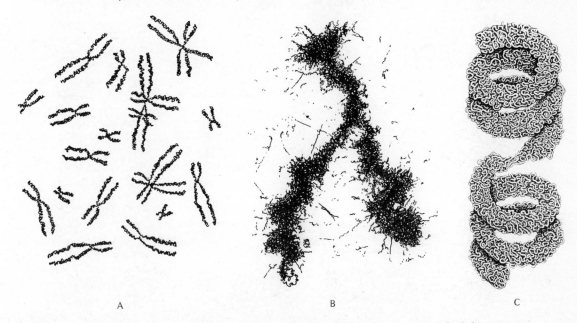

A                              B                              C

**Figure 2-8**  *"Coiled-coil" structure of human metaphase chromosomes. (A) coiling as seen in the light microscope. (B) Coiling as seen in the electron microscope showing numerous 200- to 300-Å fibers packed in a coiled-coil configuration. (C) Interpretation of fine structure showing how quaternary coiling is superimposed on a folded-fiber chromatid. (From E. J. Dupraw,* Cell and Molecular Biology. *New York: Academic Press, 1968; based on E. J. Dupraw,* Nature *206:338, 1965.)*

**TABLE 2-1   Average Composition of Isolated Mammalian Metaphase Chromosomes**

| Substance | Percent Dry Weight |
|---|---|
| DNA | 16.1 |
| RNA | 12.5 |
| Protein (total) | 70.6 |
|    Histone (acid soluble) | 30.8 |
|    Nonhistone (acid insoluble) | 40.0 |

Source: Modified from N. Salzman, D. Moore, and J. Mendelsohn, *Proc. Nat. Acad. Sci. U.S.* 56:1449, 1966.

chromosome; eukaryote DNA is covalently bound to a number of different proteins. In fact more than 70 percent of the dry weight of the chromosome is protein, which means these macromolecules must play an important role in chromosome structure or function, or both. In addition to histones and non-histone proteins, large amounts of RNA are also present. We will have more to say about these components later.

There is no general agreement about chromosome organization. Many models of metaphase chromosome structure have been proposed; one of the simpler ones is the folded-fiber model (Figure 2-9). It is a unineme chromatid model; an interphase chromosome is made up of two chromatids, each a 230-Å extended DNA double helix wrapped in proteins, held in a random secondary coil by additional proteins. A chromatid replicates, mutates, segregates, and recombines as though it were a single DNA double helix.

A cell in metaphase is committed to cell division. Its genetic information is silent, wrapped in compressed DNA coils, which tells us nothing about its functional state. To examine this more closely we will have to look at interphase chromosomes.

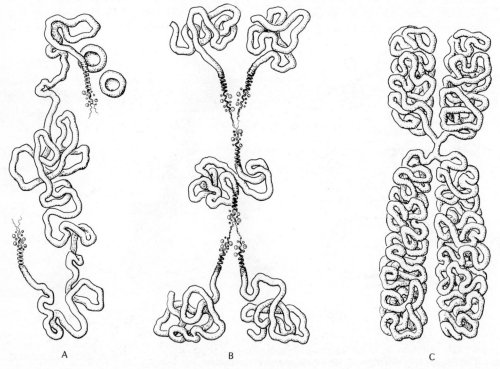

A                 B                 C

**Figure 2-9** *The folded-fiber model of chromosome structure as proposed by Dupraw (1965). (A) An interphase chromosome consists of a single DNA helix wrapped with protein into a 230-Å fiber. (B) Simple model of replication. Two replicating forks proceed from ends toward centromere. (C) At metaphase, sister chromatids compacted into coiled coils, possibly held together by unreplicated segments in the centromere. (From E. J. Dupraw,* Cell and Molecular Biology. *New York: Academic Press, 1968; based on E. J. Dupraw, Nature 209:577, 1966.)*

## Giant chromosomes in Diptera: active interphase chromosomes

What is the appearance of a chromosome when it is active? Most chromosomes in living interphase nuclei are difficult to see because they are extended into fine threads and are diffusely arranged. As frequently happens, we learn about a general phenomenon by studying the unusual. In this case the giant interphase chromosomes found in some insect larval tissues (salivary glands of most dipterans) supply us with clues about functional chromosome organization. These can be seen in living cells with the phase microscope (Figure 2-10). When isolated and stained, these chromosomes display a regular pattern of dark and light bands along their length. Each chromosome pair is aligned side by side in these preparations with its unique banding pattern. Chromosome banding patterns are identical in all tissues examined.

The large size of these chromosomes stems from the peculiarities of larval growth. Tissues grow by cell enlargement rather than by cell division. To accommodate larger cytoplasmic mass, DNA replicates without chromosome duplication, forming multistranded *polytene* chromosomes. Polytene chromosomes therefore consist of bundles of DNA helices, up to 1024 strands per mature chromosome, the result of $2^{10}$ doublings of an originally two-stranded structure. They are banded throughout their 2000-$\mu$ length. The dark band delineates coiled *chromomere* regions lined up in parallel and in register with interbands (interchromomeric regions) made up of extended parallel strands (Figure 2-11). Most chromosome DNA is localized in tightly packed chromomeres while the amount in interband regions suggests that each strand is one or two DNA double helices. In various regions the chromosome is expanded into lateral enlargements known as *puffs* and *Balbiani rings*. In these sites condensed chromomeric strands seem to open out into large lateral loops. Structure in these regions is dynamic; puffs appear and regress at different times as DNA strands undergo reversible condensation. A simple model for this chromosome envisions lateral packing of hundreds of DNA double helices with coiled and uncoiled regions in alignment. This exaggerated condition probably resembles most smaller, interphase chromosomes in other tissues, that is, alternating sequences of coiled and extended DNA strands.

Genetically, a chromomere behaves like a single gene locus. Genetic recombination experiments in *Drosophila* show that the linear order of genes corresponds in a one-to-one relationship to the visible band patterns on the giant chromosomes. Accordingly, a specific band deletion behaves as a gene mutation whereas visible addition of bands, as in gene duplications, also has predictable dosage

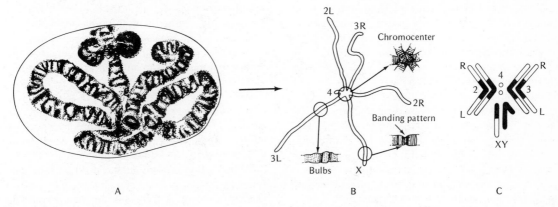

A                                    B                                    C

**Figure 2-10**   *Giant interphase chromosomes of* Drosophila. *(A) Appearance of in erphase nuclei in salivary gland (phase contrast). (B) Chromosomes released from nuclei in squash preparation. Homologous chromosomes are intimately paired along their lengths while centromeres and centric heterochromatin fuse to form a chromocenter, a densely staining heterochromatic blob. The banding pattern is shown in the X chromosome while the structure of a bulb is illustrated in the left arm of chromosome 3. (C) A diagram of mitotic chromosomes in a dividing cell of the male. The heterochromatic regions are shaded and the centromeres are shown as clear zones set off by heavy lines. Mitotic chromosomes are 100 times smaller than giant chromosomes. R = right arm; L = left arm. Numbers identify chromosomes.*

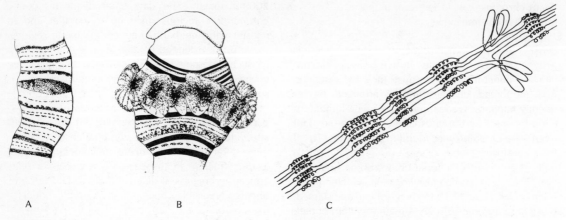

A                              B                              C

**Figure 2-11**   *Model of fine structure of giant chromosome in* Chironomus. *(A) A section of one chromosome from* Chironomus *larval salivary glands enlarged to show the banding pattern. (B) An enlarged view of a puff region on one chromosome (Balbiani ring). (C) A diagram to interpret banding patterns and puffs. Only six strands of a multistranded (over 1000 strands) chromosome are shown. Each strand is assumed to be a single chromatid, a single DNA double helix associated with protein. The band region is made up of coiled sections of each strand (chromomere) aligned in register; an interband region is a section of extended strands. Band thickness is correlated with the length of strand in the coiled or compacted configuration. A puff is a region where chromomeres have extended into large lateral loops.*

effects. The band pattern, then, is a visible expression of the linear arrangement of genes in the chromosome with each chromomeric region a locus of a Mendelian gene.

The haploid genome of *Drosophila melanogaster* contains about 0.14 picograms of DNA or about $1.4 \times 10^8$ nucleotide pairs. Assuming 1000 nucleotide pairs code for a typical protein, there is enough DNA to code for 100,000 genes. Both cytological and genetic analyses show, however, 5000 bands and probably no more than 5000 Mendelian functions. Since interbands have few genes, how do we explain this apparent paradox?

There is another paradox. Chromomere DNA is a complex genetic unit consisting of about 25,000 nucleotide pairs, enough to code for 25 proteins. How can this be reconciled with genetic and cytological evidence showing each chromomere as a site of only one Mendelian gene presumably coding for only one function? Is the chromomere a giant cistron responsible for a single giant protein, or is the chromomere a site where many copies of the same gene are clustered together? Neither one of these alternatives is consistent with the data on the frequency and nature of mutations appearing in specific bands. For instance, the frequency of any

gene mutation in *Drosophila* in response to mutagenic agents is far too high if many copies are present. To produce mutation would require simultaneous changes in all copies of the gene, an event that is highly improbable.

One model of the chromosome suggests that more than one gene is included in the chromomere region and concerned with the same function. This means that many regulatory genes along with structural genes control a single essential function and implies that mutations in any one structural or regulatory gene can produce a loss in the same function. There is some evidence in support of this model. In an analysis of the *zeste-white* region of the X chromosome of *D. melanogaster* (a region controlling eye color), 130 different lethal mutations were found, all falling into 16 distinct complementary groups. This observation corresponds to the 15 bands seen in this region. Likewise in chromosome 4 (the smallest chromosome), 200 different mutations could be distributed among 37 loci. This plus 6 other mutations makes a total of 43 fourth-chromosome functions, which is a fairly close match to the total of 50 bands in the chromosome. It suggests that there are many more regulatory than structural genes in a eukaryote genome; many mutations obtained at each of these loci were at nonstructural gene sites. It is probably this feature of eukaryote genomes—the high ratio of regulatory to structural genes—that accounts for their greater developmental versatility.

In contrast to metaphase chromosomes, these interphase chromosomes are metabolically active and carry out the primary gene functions of replication (DNA synthesis) and message production, or the synthesis of RNA (see Chapter 3). When mitotic cells are exposed to radioactive precursors for nucleic acid synthesis (thymidine $^3$H as the precursor for DNA and uridine $^3$H as the precursor for RNA), no or little precursor is incorporated into metaphase chromosomes. On the other hand, salivary glands incubated in these precursors will incorporate large amounts into chromomeric bands on chromosomes. Thymidine incorporation implies DNA replication since thymidine is a specific precursor of DNA synthesis. Radioactive uridine is also incorporated into chromosomes, entering DNA as well as RNA. Cells possess enzymes that convert some uridine into thymidine and cytosine, each acting as a precursor for DNA synthesis. RNA is synthesized at various regions of the chromosome, indicating that the genome is active in message production.

The most active region of a chromosome is a puff, particularly during the period of its formation. Radioactive precursors for RNA and protein synthesis are rapidly incorporated into these regions. Ribonucleoproteins are synthesized and accumulated in large amounts in a puff, which accounts in part for their large size. Puffing occurs in different tissues at different times in development and seems to correlate with cell function. It is generally believed that puffs are visible manifestations of genes in action (see Chapter 17). Since most interphase nuclei exhibit similar patterns of precursor incorporation when chromosomes are in a highly diffuse state, we assume that the functional state of DNA is the extended, open configuration exemplified by a puff in a giant chromosome.

### Heterochromatin and euchromatin

The band pattern indicates that portions of DNA assume alternative tertiary configurations, condensed or extended. Such gross physical changes in chromosome coiling or compaction often appear as differential staining patterns in both metaphase and interphase chromosomes. As early as 1928 it was observed by the German cytologist E. Heitz that extended chromosome strands in interphase nuclei stain diffusely except for a few regions which consistently appear as deeply staining clumps or "blobs." These regions of condensed chromatin did not undergo the usual coiling-uncoiling changes but remained noticeably clumped throughout mitosis. Heitz called these compacted, heavily staining regions *heterochromatin* to distinguish them from the diffuse, lightly staining *euchromatin* strands. Originally heterochromatin was conceived as a qualitatively different material from euchromatin, but we now recognize that both contain DNA and protein in different states of compaction.

Two types of heterochromatin exist, constitutive and facultative. *Constitutive heterochromatin* seems to be a fixed "structural" component of chromosomes, localized in or near centromeres (sometimes called *centromeric heterochromatin*) and nucleolar organizers, with small amounts intercalated in other chromosomal regions. Homologous chromosomes display identical patterns of constitutive heterochromatin in all physiological states of the cell. *Facultative heterochromatin,* on the other hand, is a more labile state of the DNA, found in certain chromosomes (usually sex chromosomes). Its state of compaction depends on the physiological state of the cell. It differs from constitutive heterochromatin inasmuch as homologous chromosomes are not equally heterochromatic. Usually one member of the pair, as in the X chromosome, is facultatively heterochromatic.

Some newer techniques of staining have led to a burst of interest in the organization of heterochromatin in metaphase chromosomes. Mild treatment of metaphase chromosomes with heat, or low concentrations of acid or alkali, followed by staining with a Giemsa stain has yielded consistent patterns of staining of centromeric regions (Figure 2-12). Each homologous pair of chromosomes in a karyotype has a distinctive C-banding pattern of centromeric heterochromatin.

Another technique, such as short treatment with trypsin or urea, followed by Giemsa staining, produces a different but reproducible chromosome pattern. In this case a series of light and dark bands (G bands) are seen in each chromosome arm. These bands are called *intercalary heterochromatin* to distinguish them from centromeric heterochromatin. Similar patterns are obtained if chromosomes are

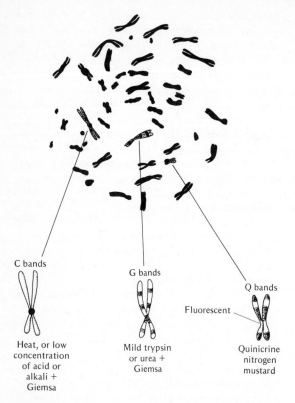

C bands

G bands

Q bands

Fluorescent

Heat, or low
concentration
of acid or
alkali +
Giemsa

Mild trypsin
or urea +
Giemsa

Quinicrine
nitrogen
mustard

**Figure 2-12** *Various "heterochromatic" banding patterns
in human metaphase chromosomes. The banding patterns
are diagnostic for each pair of homologous chromosomes.
A metaphase plate is usually stained for only one of the
three patterns shown here.*

stained with a fluorescent dye, quinacrine mustard.
The band appears a bright yellow green as seen by
ultraviolet microscopy. Here too each homolog has
an identical pattern which identifies it as constitutive heterochromatin. Facultative heterochromatin
is also stained with quinacrine and serves as a means
of distinguishing the heterochromatic X chromosome in human cells. Each species has a distinctive
G- or Q-banding pattern in its chromosomes, and its
reproducibility makes it useful for identifying individual chromosomes and for studying any changes
associated with genetic abnormalities. The staining
reactions are not understood. It is uncertain whether
specific DNA sequences are staining (GC-rich) or
whether DNA structure as modified by proteins is
responsible.

In interphase constitutive heterochromatin is
usually detected as a deeply staining blob or mass
that is variable in different tissues. For example, the
vole, *Microtus agrestis,* possesses giant X and Y sex
chromosomes bearing large regions of constitutive
heterochromatin (centromeric type). These show up
in interphase nuclei as spherical blobs, crescent
shapes, or irregular threads in different tissues (Figure 2-13). The intercalary heterochromatin appears
associated with the nuclear membrane. The reason
for these differences is not known but may reflect
different nuclear ionic environments or differences
in a cell's metabolic state. In the salivary gland
nuclei of *Drosophila* heterochromatic ends of the
autosomes aggregate with the heterochromatic arms
of the Y and X chromosomes and combine with

**Figure 2-13** *Giant heterochromatic sex
chromosomes in various interphase and
dividing nuclei of Microtus agrestis (shrew).
(A) Interphase nucleus from fibroblast. (B)
Hepatocyte interphase nucleus. (C)
Endothelial nucleus. (D) Neuron. (E) Oogonial
prophase. (F) Oogonial metaphase. (G)
Spermatogonial anaphase. (H) Spermatogonial telophase. Note that heterochromatic
sex chromosomes retain their distinctive
staining properties throughout mitosis.
(From J. J. Yunis and W. G. Yasminehi,
Science 174:1200–1209, 1971. Copyright ©
1971 by The American Association for the
Advancement of Science.)*

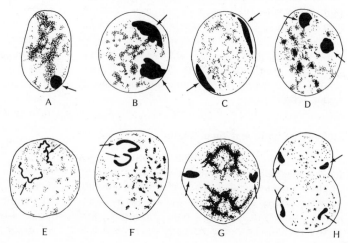

A        B        C        D

E        F        G        H

nucleoli to form a large blob called the *chromo-center* (see Figure 2-10).

Centromeric heterochromatin does not code for proteins; it is genetically deficient. Its localization in centromere regions suggests a structural role, perhaps in the organization of the kinetochore, the region for spindle fiber attachment. Constitutive heterochromatin has also been implicated in pairing of homologous chromosomes in meiosis. Intercalary heterochromatin (G or Q band) is also genetically inactive, although its presence in cells may have a regulatory role in the expression of neighboring genes.

Facultative heterochromatin is restricted to specific chromosomal regions or to whole chromosomes, such as the Y chromosome in male *Drosophila* or one X chromosome in humans. Even whole chromosome sets may be facultatively heterochromatic. Facultative heterochromatin is also genetically silent, but in contrast to centromeric heterochromatin it contains genes coding for proteins. Genes in heterochromatic X or Y chromosomes are rarely, if ever, expressed in most somatic cells but become active when gametes differentiate. When active, these regions revert to a euchromatic condition. In human females both X chromosomes are euchromatic in the developing oocyte, and in *Drosophila* the Y chromosome becomes visibly active during spermatogenesis (see Chapter 17).

Compaction or heterochromatization of an X chromosome seems to be a general feature of mammalian development. In human females one of the X chromosomes is wholly heterochromatic in all tissues, forming a condensed, easily recognizable body (the *Barr body*) in interphase nuclei. The genes associated with an X chromosome are inactivated by random compaction during the first few divisions of the zygote. Although one X chromosome is inactivated, this does not mean that its genes are never expressed. X-chromosome heterochromatization is random in a tissue; that is, one X chromosome is nonfunctional in some cells while the homologous X is inactive in others. But because one X is functional in each cell, the total phenotypic effect for the whole tissue is the same as if both X chromosomes were active in all cells. The heterozygosity of the diploid state is fully expressed. In males the X chromosome is completely euchromatic and functional while the Y chromosome is heterochromatic. Both sexes, therefore, possess only one functional X

chromosome. This acts as a dosage compensation mechanism for all X-chromosome genes; males and females have the same levels of X-chromosome gene products.

The best example of whole sets of chromosomes becoming facultatively heterochromatic is found among the *Coccidae,* or wingless mealy bugs (Figure 2-14). In males the entire paternal set of chromosomes becomes heterochromatic in the first divisions of the zygote whereas in females they remain euchromatic. These condensed chromosomes are

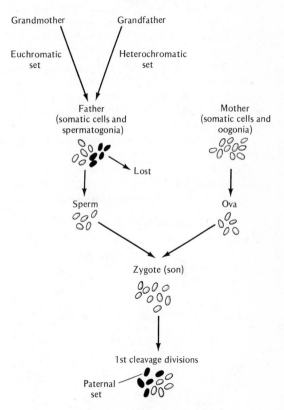

**Figure 2-14** *Heterochromatic chromosomes in the mealybug system. In males the maternal chromosome set is euchromatic and functional (light chromosomes) whereas the paternal set is heterochromatic and genetically silent (dark chromosomes). During sperm formation the heterochromatic set is lost; all sperm carry only the maternal set of chromosomes. In females all chromosomes are euchromatic and functional. In male zygotes the euchromatic paternal set becomes heterochromatic in the first few divisions. Thus males only transmit their maternal chromosome set to their offspring.*

genetically silent since none of their genes is ever expressed in somatic cells. Irradiation of fathers with increasing doses of X-rays before mating produces zygotes that carry a heavily irradiated paternal set of chromosomes, including many dominant lethal mutations. Yet male survival is unaffected even when high doses are used, since this set becomes heterochromatic in the first cleavage division. In females, where the paternal set remains euchromatic, even low doses reduce survival since many dominant lethal mutations are expressed during early development. In contrast to females, which are diploid, males are physiologically haploid since heterochromatic paternal chromosomes are genetically "shut down," leaving only a functional maternal chromosome set. All genetic mutations in these species follow a similar pattern of inheritance; that is, males express and transmit only the genes received from their mothers since the paternal set of chromosomes is consistently heterochromatic and genetically inactive.

On the other hand, heterochromatic chromosomes become active during spermatogenesis, reverting to the euchromatic state. Germ-cell differentiation is causally related to activation of genes carried by a set of heterochromatic chromosomes.

**Figure 2-15**  *Late-replicating regions of chromosomes in dorsal mesoderm cells of a Rana pipiens neurula embryo. Note that silver grains are clustered over the ends of most chromosomes, which indicates that most late-replicating heterochromatin is localized in these regions. (Based on photograph in P. J. Stambrook and R. A. Flickinger, J. Exp. Zool. 174:101–114, 1970.)*

The behavior of these chromosomes indicates that expression of large blocks of genetic information is controlled by changes in the physical state of DNA helices within the chromosome.

Not only is heterochromatin genetically inactive, it is also metabolically inert, just like the metaphase chromosome. For example, the large nuclei found in thymocytes of the calf thymus gland contain regions of condensed chromatin distinguishable from the more diffuse euchromatic regions. Exposure of these cells in situ to a radioactive RNA precursor such as $^3$H-uridine results in a characteristic incorporation pattern. The diffuse, euchromatic regions incorporate more precursor into RNA than do the condensed clumps. In vitro diffuse euchromatin is far more effective in stimulating RNA synthesis than is condensed heterochromatin, suggesting that open, extended DNA molecules are a necessary condition for maximum template activity.

The replicative behavior of DNA also depends upon its physical state in the chromosome. We know that DNA and chromosome replication take place asynchronously during interphase in most dividing cells; some chromosome regions seem to replicate first before any other region (see Chapter 5). Using thymidine radioautography, we can distinguish *late* from *early* replicating chromosomal regions within a metaphase nucleus. A population of dividing cells is exposed to radioactive thymidine and then treated with colchicine to trap them in metaphase for chromosome analysis. Radioautographs are made, and the distribution of silver grains over the chromosomes is determined (Figure 2-15). The first labeled metaphase cells to appear must have been in the final phases of DNA synthesis when the cells were exposed to radioactive thymidine. The chromosome regions labeled at this time are the late replicating regions. In cultured human cells one of the two X chromosomes is late replicating along its entire length whereas the other replicates early. The late replicating X chromosome corresponds to the heterochromatic X. In addition, each chromosome pair exhibits a pattern of late replicating regions that is diagnostic. In fact the pattern is reproducible. A karyotype may be identified according to the distribution of late replicating regions. Centromeric heterochromatin is also late replicating, as are most G bands.

In certain human clinical conditions involving multiple X chromosomes such as XXX, XXXX, XXXY, and XXXXX, the number of Barr bodies in cell nuclei is one less than the total X chromosomes present, leaving only one functional X. Here too the number of late replicating X chromosomes is equivalent to the number of Barr bodies. These and other findings have led to the hypothesis that late replication of DNA correlates with heterochromatization or compaction of DNA. In other words, gene inactivation, heterochromatization, and delayed replication are all related phenomena. In general compacted chromatin regions are relatively inactive with respect to RNA synthesis and DNA replication; these regions must first uncoil to replicate or to synthesize primary gene products.

Heterochromatin seems to be a chromosomal mechanism of wholesale silencing of large blocks of genes. It is not found in prokaryote genomes. The information in whole chromosome sets or parts of chromosomes is not required by all eukaryote cells at all times. When certain cells, such as gametes, undergo maturation, such heterochromatic regions may become actively euchromatic. Though heterochromatin lacks the precision of timing and specificity required for the control of single genes, physical contiguity of a gene with a heterochromatic region exercises a strong effect on its expression, shutting it off at one time or even turning it on.

## Position effect

For certain genes, their precise location within a chromosome is indispensable to function, as if contact with neighboring genes is a necessary condition for activity. Changing the position of such genes by transfer to another chromosome or by inversion within the chromosome is enough to alter their phenotypic effects. This phenomenon, known as the *position effect* for gene expression, may arise when chromosomes break and rejoin in aberrant combinations, usually during meiosis. Inversions of whole segments upset the linear order of genes within the chromosome. Other changes involve *translocations*, where chromosome segments, or whole chromosome arms, are exchanged between nonhomologous chromosomes, rearranging the original order of genes within the genome. Chromosome breakage and refusion are not usually accompanied by gene loss; the total amount of genetic material remains the same without any effect on expression of most genes. On occasion a chromosomal aberration does lead to a mutant expression of a gene or to a developmental abnormality.

It is not clear why changing the position of some genes and not others alters their expression. A possible explanation is that some genes are organized into functional operons of many interacting genes. Breaking the physical integrity of the complex interferes with the expression of component genes, as if gene action depends upon spatial continuity.

Sometimes the action of a normal wild type gene will change after it is transposed to a heterochromatic region within the chromosome to produce a *variegated position effect* expression of a mutant gene. A variegated position effect involves proximity of wild type genes to heterochromatin and results in a type of somatic mosaicism that may spread to more than one gene. An example of somatic mosaicism is the "white-mottled four" eye-color pattern in *Drosophila*. Normally eye color is red, but in this mutant, heterozygous flies containing the dominant wild type allele for red eye color develop a mottled eye with irregular red and white patches. This is explained by the fact that in these flies the dominant eye-color gene ($w+$) on the X chromosome has been translocated from a euchromatic region to a heterochromatic region to form an abnormal chromosome designated as $R(w+)$. If this chromosome is in heterozygous combination with a chromosome containing the white recessive gene ($w$) in its normal position, the designation is $R(w+)/w$, and the variegated condition is expressed. When the wild type gene is close to heterochromatin, it does not function normally. Its proximity to heterochromatin determines the magnitude of the variegated change. Other X-chromosome genes transposed to the vicinity of heterochromatin also exhibit varying degrees of inactivation so that complex mosaic patterns are expressed phenotypically when these are heterozygous. These changes in gene expression are controlled by heterochromatin, possibly through a contagious pattern of chromosome condensation which shuts down gene-product synthesis. More distant stretches of euchromatin are less affected than those proximate to heterochromatin.

Not all genes are affected when translocated to heterochromatic regions, nor are all genes organized into functional groupings. In fact most genes seem to act autonomously, functioning equally well in any portion of the chromosome, unaffected by the structure of neighboring regions. Though the randomness

**TABLE 2-2**  Chromosome Number and Genome Size in Amphibians as Expressed as a Percentage of the Human Genome

| Species | 2N Number | Genome Size, % |
|---|---|---|
| *Urodeles* | | |
| Axolotl mexicanum | 26 or 28 | 939 |
| Plethodon cinereus | 26 or 28 | 705 |
| Triturus cristatus | 24 | 830 |
| Triturus viridescens | 22 | 1300 |
| Amphiuma means | 28 | 2700 |
| Necturus maculosus | 38 | 2780 |
| *Anura* | | |
| Pipa pipa | 22 | ±40 |
| Xenopus laevis | 36 | ±85 |
| Bufo americanus | 22 | 140 |
| Rana catesbiana | 26 | 180 |
| Rana pipiens | 26 | 220 |

Source: Adapted from S. Ohno, *Evolution by Gene Duplication.* New York: Springer-Verlag, 1970.

of variegated position effect minimizes its importance as a general "programming" mechanism for gene action during development, it does suggest that the physical state of DNA, its relative position in the genome, and its sequence organization affect its informational role during development.

**Sequence organization of DNA in the eukaryote genome**

Most eukaryote cells are diploid, containing large amounts of DNA, thousands of times that found in most prokaryotes. *Drosophila,* we have seen, has enough DNA to code for approximately 100,000 genes, while man, with 20 times the amount of DNA, has enough for 2 million genes. Nevertheless, bacterial cells manage to carry on all vital activities with much less genetic information. Large amounts of DNA correlate with the larger, more complex eukaryote cells, differentiated into hundreds of specialized phenotypes with unique proteins and organelles. More genes are needed to code for and regulate this cellular diversity. Besides, prokaryotes, with a limited developmental repertoire, probably do not need as many regulatory genes to monitor complex developmental programs. All this may explain in part the higher DNA content of animal and plant cells. But this does not account for some of the bizarre differences in genome content found, for

example, in amphibians where some species possess almost 50 times the DNA content of others, in spite of possessing similar or even lower chromosome numbers (Table 2-2). In other words, it is not clear why salamanders should need 50 to 100 times more genetic information than frogs.

A comparison of the annealing behavior of prokaryote and eukaryote DNA provides a partial answer to the problem of an apparent excess of DNA in many eukaryote cells. By fragmenting long DNA molecules into short 500- to 1000-nucleotide pieces, increasing the concentration of DNA fragments and changing the time of incubation, we can vary the annealing pattern (Figure 2-16). When we do this with prokaryote DNA, we find that increasing incubation time and DNA concentration gives a single smooth curve of the type shown in the figure. Most, if not all, DNA anneals at high DNA concentrations and long incubation times. DNA from eukaryotes, however, shows a different pattern. The annealing curves are more complex with several distinct transitions. At low DNA concentrations and short incubation times, a large proportion of single DNA strands reanneal into double-stranded helices, and as the concentration-time values (CoT values) increase, additional double-stranded molecules are formed to produce a biphasic curve. Evidently DNA strands in early forming hybrids share many identical sequences since they anneal rapidly, even under conditions most unfavorable for double-strand formation. It has been shown that rapidly renaturing fragments are actually identical to one another, and large numbers of them are present, even at low DNA concentration. Slowly annealing duplexes obtained at higher CoT values are present only in few or single copies. In other words, the greater the number of identical DNA fragments the more rapidly they reanneal at low DNA concentration and short incubation times, since every strand collision leads immediately to double helix formation. Hence low CoT values for DNA annealing curves are interpreted as a sign of sequence redundancy in DNA fragments. On the other hand, high CoT values (long incubation times and high concentrations of DNA) mean that most DNA strands are dissimilar. Not enough similar sequences are present to reanneal unless the concentration of DNA is very high.

With this technique we can determine the relative proportion of redundant to nonredundant re-

**Figure 2-16**  *Melting and annealing procedure to demonstrate repeated copies of DNA in a genome. (A) An extract of DNA is fragmented into many pieces of varying lengths. Three copies of a repeated sequence AA' are shown along with single-copy species BB', CC', and DD'. (B) Heating the DNA above melting temperature separates double helices into separate strands. (C) Upon cooling, repeated DNA copies will reassociate first because there are many more copies and they will collide more frequently than single-copy sequences. The rate of reassociation is, therefore, proportional to the degree of repetition. (D) The time course of reassociation is illustrated in the graph for prokaryote (E. coli) DNA and for eukaryote (calf) DNA. The percentage reassociated is plotted against CoT values, a notation that refers to the product of DNA concentration and time. E. coli DNA plots as an S-shaped curve, characteristic of single-copy DNA. Calf DNA plots into two components, an early (with low CoT values) reassociating family of repeated sequences followed later by a single-copy component (with high CoT values). (Based on R. J. Britten and D. E. Kohne, Sci. Am. 222:24–31, 1970.)*

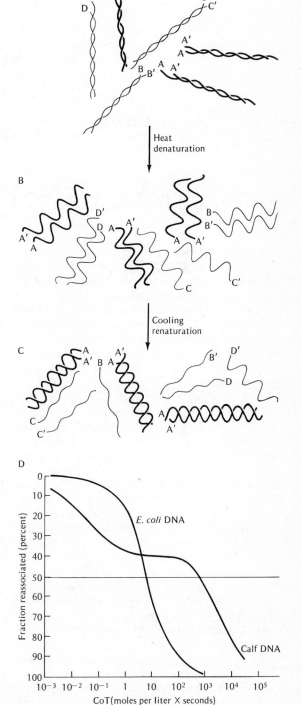

gions in a genome. These annealing curves tell us that eukaryote genomes are more complex than prokaryote genomes; they contain two or more families of DNA sequences with different degrees of sequence redundancy. One family (highly repetitive) is made up of rapidly annealing DNA sequences which are present in the genome as 100,000 to $10^6$ copies; another is an intermediate, repetitive class of DNAs of 1000 to 100,000 copies; there is also a moderately repetitive class (20 to 50 copies) and a so-called single-copy region of the DNA where only a few copies—1 or 2—are represented (nonrepetitive). We see that eukaryote genomes do not necessarily contain that much more information than prokaryotes, since a large proportion of the information is redundant.

Variation in the amounts of DNA in a species are now attributed to varying amounts of repetitive DNA (Table 2-3). The genomes of most eukaryote species may be characterized in terms of the percentage of classes of repeated sequences within their genomes. For example, 25 percent of the sea urchin genome is moderately repetitive while 27 percent of the genome is repetitive. *Xenopus* shows a higher repetitive class (31 percent). In *Drosophila*, on the other hand, 75 percent of the genome is made up of unique sequences, 10 percent is highly repetitive, and 15 percent moderately repetitive. Each species seems to

**TABLE 2-3   Complexity of Some Eukaryote Genomes**

| Species | Percent in Each Frequency Class | | | |
| --- | --- | --- | --- | --- |
| | Nonrepetitive | Moderately Repetitive[a] | Repetitive[b] | Highly Repetitive[c] |
| *Nassaria* | | | | |
| *obsoleta* (snail) | 38 | 12 | 15 | 18 |
| Calf | 55 | — | 38 | 5 |
| *Xenopus laevis* | 54 | 6 | 31 | 9 |
| *Strongylocentrotus* | | | | |
| *purpuratus* (sea urchin) | 38 | 25 | 27 | 10 |
| *Drosophila* | 75 | — | 15 | 10 |

[a] 20–50 copies.
[b] 250–60,000 copies.
[c] Up to 1 million copies, including satellite DNAs.
Source: Based in part on table in E. H. Davidson and R. J. Britten, *Quart. Rev. Biol.* 48:565–613, 1973.

have its own unique class of repeated sequences while it may share some repetitive sequences with other classes of organisms. During the course of eukaryote evolution, the genome has experienced frequent episodes of gene duplication, some genes undergoing this process more frequently than others. After duplication, mutations may accumulate in these reiterated sequences, resulting in some divergence between species but not enough to conceal many shared sequences.

In some instances, we can actually localize some highly reiterated regions within chromosomes. For example, the most highly repetitive DNAs are the so-called satellite DNAs that are obtained by density-gradient centrifugation of total DNA. Because they are present in such large amounts relative to other regions of the genome, satellite DNAs can be isolated from bulk DNA. Mouse satellite DNA is unusually rich in adenine and thymine bases (its AT content is almost twice its GC content), which means it is less dense than the bulk DNA of the cell with a higher proportion of guanine and cytosine. Consequently, when total mouse DNA is centrifuged to equilibrium, the smaller satellite DNA separates from the bulk DNA as a distinct density band. It represents 10 percent of the total genome and turns out to be highly redundant, a class of small nucleotide sequences (the repeat units as short as 7 to 12 nucleotides), with more than 1 million identical copies within the genome.

The satellite DNA can be purified and used as a template in vitro to make highly radioactive complementary RNA. Since this RNA is an exact complementary copy of satellite DNA sequences, it is used to determine where satellite DNA is located within the chromosome set. Is it distributed at random over all chromosomes, or is it confined to a specific chromosome or a specific region within a chromosome? To answer this question a DNA-RNA hybridization annealing experiment in situ is carried out (Figure 2-17). A smear of mouse metaphase chromosomes is prepared on a slide, and DNA in chromosomes is denatured into single strands by treatment with alkali. To this chromosome preparation we add radioactively labeled RNA, and we incubate it under conditions of time, temperature, and salt concentration which favor binding of radioactive RNA strands with complementary DNA sequences in intact chromosomes. Regions of the chromosome where labeled double-stranded DNA-RNA hybrids have formed are visualized by radioautography. We find that only the centromere regions of each chromosome have become labeled; no other region is radioactive. From this we conclude that satellite DNA is localized to the centromere regions of the chromosomes. The most highly reiterated sequences of DNA are clustered in tandem array within the centromere regions, the points of spindle fiber attachment during mitosis. Centromeric localization suggests that centromeric heterochromatin and satellite DNA are one and the same material.

The function of satellite DNA within the eukaryote chromosome is still not understood since it does not code for any proteins within the cell. Such DNA may have a structural role in the chromosome or may be important in chromosome pairing during meiosis. In any event it is evident that 10 to 15 percent of the total genome of some eukaryote species does not code for any specific developmental information. It has a noninformational role, perhaps related to chromosome behavior and structure during the complex processes of mitosis and meiosis in eukaryote cells.

This accounts for the highly repetitive DNA in eukaryote genomes. We now ask: are the other classes of redundant genes also clustered or are they dispersed throughout the genome?

### Repetitive ribosomal RNA genes

A clustered arrangement is also found in the genes coding for ribosomal RNA in virtually all cells. Ribosomal RNA (rRNA), comprising approximately 85 to 90 percent of a cell's total RNA, is found in cytoplasmic ribosomes, the organelles involved in protein synthesis. The eukaryote ribosome consists of many proteins (30 to 50) combined with three different size classes of rRNA—28S, 18S, and 5S.[1] The ribosome consists of two subunits of different size, 60S and 40S. The 28S rRNA is found in the larger 60S subunit whereas the 18S rRNA is associated with proteins to make the smaller 40S unit. The 5S RNA is found only in the large subunit. Although differing in size, rRNAs from different species have typically high guanine + cytosine base compositions (for example, 65 percent for 28S, 59 percent for 18S, and 60 percent for 5S in *Xenopus* ribosomes) as contrasted with bulk DNA, which is richer in adenine

---

[1] S is an abbreviation for sedimentation constant (measured as Svedbergs), which is an indication of the size of a molecule (molecular weight) as determined by the ultracentrifugation formula:

$$M = \frac{RTs}{D(1-V\rho)}$$

M = molecular weight
R = gas constant (83.2 × 10⁵ ergs per degree)
T = absolute temperature
s = sedimentation constant
D = diffusion constant
V = partial specific volume of particle
ρ = density of solvent

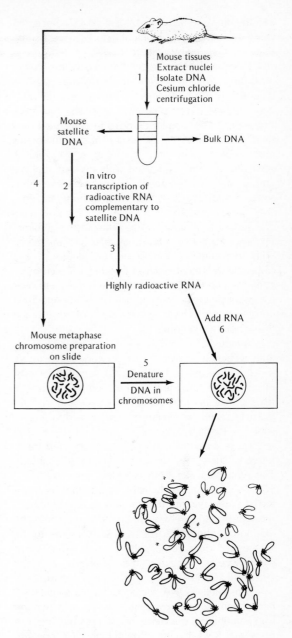

**Figure 2-17** *Procedure for in situ hybridization of mouse satellite DNA. Sequence of steps given by the numbers 1–6. In radioautograph below most silver grains are concentrated over centromeres on each chromosome, indicating that satellite DNA is localized in the centromere region.*

and thymine and has only 42 percent GC. Since 28S rRNA has a higher GC content than 18S, we assume that each is coded by a different gene.

Ribosomal RNAs are synthesized within nuclei according to a complex assembly-line process (Figure 2-18). In most cells the nucleolus is the principal organelle for ribosome synthesis. Nucleoli, however, do not appear de novo in cells; they owe their origins to a specific region of a homologous chromosome pair, the nucleolar organizer. A nucleolus appears associated with this region at the end of mitosis. Apparently genes coding for ribosomal RNA are included in newly forming nucleoli.

The direct methods of in situ hybridization have allowed us to pinpoint these genes in the genome. Highly radioactive ribosomal RNA is incubated with denatured chromosome preparations. The radioactivity is localized over the nucleoli or chromocenter region in *Drosophila* chromosomes and over a few specific regions in human nucleolar organizer chromosomes. These genes are not distributed throughout the genome.

Although 28S and 18S RNA are coded by separate genes, it appears that both genes are closely linked, since the first step in ribosomal RNA synthesis is the formation in nucleoli of large 45S RNA precursors that contain the 18S and 28S sequences. The large precursor is processed to smaller subunits through a series of steps. First, the 18S portion is removed (along with protein) and transferred to the cytoplasm as the smaller 40S subunit. The remaining 32S RNA is further processed to a 28S component which, with its proteins, is transferred to the cytoplasm as the larger 60S subunit. In the cytoplasm the two subunits combine into the functional ribosome. The 5S RNA seems to be synthesized on a completely separate gene, not at the nucleolar organizer, and combines with the 60S particle before it is transferred to the cytoplasm.

Like satellite DNA, the genes for ribosomal RNA have been isolated by centrifugation and purified from amphibian oocytes. This is possible because ribosomal RNA, having a much higher GC content, means that the DNA coding for these molecules is equally rich in these bases. In addition, ribosomal RNA genes are present in many copies (several thousand in the amphibian oocyte, as we will see), making it possible to separate these "heavier" gene sequences from bulk DNA by density gradient centrifugation. Having a collection of relatively pure

genes has made it possible to analyze their organization and arrangement in the intact genome.

When isolated, the 18S and 28S genes are found to be arranged in a sequence separated by a small amount of DNA that is included as part of the large 45S precursor. In addition, there is a long piece of DNA, a "spacer," that is attached to the genes but is never synthesized as an RNA copy. Four hundred fifty copies of these genes are arranged in tandem in the *Xenopus* genome within the nucleolar organizer region, and many additional copies are included in nucleoli.

These studies demonstrate that certain reiterated genes are clustered in specific regions of the genome. Some, like the 5S RNA genes in *Xenopus*, are distributed in different sites in the genome. This process of ribosomal RNA gene amplification has selective advantages since it furnishes many templates for ribosome synthesis and diminishes the deleterious effect of mutations accumulated within genes coding for this vital cell organelle. RNA genes are highly conservative and show little change in evolution.

How are structural genes organized within the genome? Are genes coding for specific proteins such as hemoglobin or myosin also amplified? Only a few genes, principally those for hemoglobin, silk fibroin, and the genes coding for histone proteins, have been isolated in sufficient amount to answer this question. The structural genes for hemoglobin and fibroin are present as single copies, at the most two to five copies within a haploid genome. In both cases there is no evidence that these genes are amplified even in those cells exhibiting the highest levels of specific protein synthesis. Thus in the silkworm the number of fibroin genes in silk glands is the same as the number in all other tissues. These preliminary studies suggest that extensive amplification of structural genes is not the case.

Only in the case of histones do we find evidence of amplification. In sea urchin embryos, histone genes may be present in 400 to 1200 copies, and these genes seem to be clustered in tandem. Histone proteins are relatively homogeneous. Like ribosomal RNA, their sequences have been conserved during the course of evolution and show considerable homology within species and between species. The essential information for ribosome and chromosome structure is stored in reiterated sequences and is organized in a similar manner within the genome.

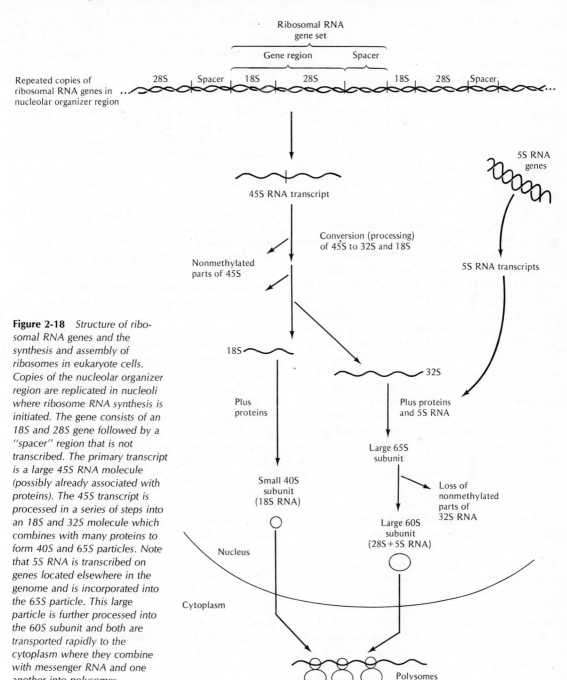

**Figure 2-18** *Structure of ribosomal RNA genes and the synthesis and assembly of ribosomes in eukaryote cells. Copies of the nucleolar organizer region are replicated in nucleoli where ribosome RNA synthesis is initiated. The gene consists of an 18S and 28S gene followed by a "spacer" region that is not transcribed. The primary transcript is a large 45S RNA molecule (possibly already associated with proteins). The 45S transcript is processed in a series of steps into an 18S and 32S molecule which combines with many proteins to form 40S and 65S particles. Note that 5S RNA is transcribed on genes located elsewhere in the genome and is incorporated into the 65S particle. This large particle is further processed into the 60S subunit and both are transported rapidly to the cytoplasm where they combine with messenger RNA and one another into polysomes.*

## Other repetitive DNA sequences

We have seen that genes coding for chromosome and ribosome structure are clustered whereas single-copy genes are distributed throughout the genome. The arrangement of other moderately redundant sequences in the genome may tell us something about their function. For instance, if the same sequence is found in many different places, perhaps interspersed among single-copy sequences, it may be serving as a regulatory gene simultaneously controlling many structural genes in different regions of the genome.

Once again the technique of DNA-DNA hybridization provides some answers. In these experiments we determine the annealing patterns between large DNA fragments (about 4000 nucleotides long) and smaller fragments (about 300 to 450 nucleotides long) at conditions when only highly reiterated sequences should anneal into double-stranded duplexes. The annealing pattern tells us whether highly clustered sequences make up the larger segments of DNA or whether the reiterated sequences are interspersed among single-copy components. Studies of *Xenopus* and sea urchin DNA indicate that most of these repetitive sequences are in fact interspersed among nonrepetitive DNA. Moreover, the data show that more than 90 percent of these reiterated sequences are not clustered within the larger strands, since only a few short reiterated sequences (about 300 nucleotides in length) could be found annealed to each large fragment. Only a small percentage of repetitive sequences, about 6 percent, seems to be organized as tandem clusters.

It has been postulated that the moderately redundant fractions of the genome are regulatory genes interspersed among the single-copy structural genes. According to this model of the eukaryote genome, different structural genes distributed throughout the genome are regulated by identical regulatory genes. This explains why many different proteins are synthesized in a coordinated manner when a tissue is functionally specialized. In muscle, for example, besides the muscle proteins actin and myosin that make up the contractile apparatus, there are many enzymes such as creatine phosphatase which are synthesized at high levels when muscle cells differentiate (see Chapter 17). Presumably each gene coding for one of many structural proteins is activated by the same regulatory gene, which explains the need for many identical regulatory genes. The interspersion of small repetitive sequences within larger single-copy segments is consistent with the model.

Summarizing, we may attribute the rich developmental potential of the eukaryote genome to its complex organization. Several classes of genes exist, from single-copy structural genes to highly reiterated small sequences of centromeric DNA clustered in specific regions. The excess of DNA in eukaryote genomes is in part explained by the tremendous amplification of the genome that seems to be concerned with either structural features of the chromosome and its behavior in mitosis and meiosis or with the fundamental aspects of protein synthesis and cell growth. Other redundant sequences of DNA are dispersed within the single-copy component of the genome and may play a regulatory role in gene expression during development. Finally, the major single-gene fraction is present in one or a few copies as structural genes.

The sequence organization of DNA is its primary structure. Some regions, such as centromeric heterochromatin, are rich in AT base pairs, or, as in ribosome RNA genes, GC base pairs. This clustering may account in part for the differential staining properties of G, C, and Q bands of heterochromatin. It does appear that G bands are rich in AT. In addition, chromosome structural heterogeneity is due to secondary and tertiary configurations of the DNA that reflect coiling and uncoiling cycles within small chromosome regions, parts of chromosomes, or whole chromosomes. These visible expressions of the kinetic properties of DNA in eukaryote chromosomes probably depend upon proteins that reversibly associate with DNA during a cell's life cycle.

## Chromosomal proteins

Evidently the physical organization of DNA in the chromosome correlates with its functional activity. Gene action seems to depend on the state of compaction of DNA helices, and this in turn is governed by protein-DNA interaction.

Prokaryote chromosomes are generally free of protein, although some proteins are definitely bound to DNA for gene regulation. These are few compared to the large mass of proteins associated with eukaryote chromosomes. All along the grooves in the DNA double helix, certain sites on purine and pyrimidine bases are accessible for covalent and hydrogen bonding interactions. Phosphate groups along the helical backbones give DNA a strong negative charge, making it reactive to basic, positively charged molecules to form salt linkages or direct covalent bonds. Since many proteins are positively charged molecules, they do react readily with DNA.

Several classes of proteins have been obtained from isolated interphase nuclei. Histones, a class of basic proteins, are released by dilute acids or concentrated salt solutions. The remaining nonhistone proteins are mostly acidic; they are extracted by treatment with alkali. The proteins insoluble in acid, alkali, or salt solutions are lumped into an ambiguous class of "residual" chromosomal proteins. Of the three classes, only the histones have been well characterized.

Histones are relatively small proteins (about 10,000 to 20,000 mol wt), rich in basic amino acids like arginine and lysine and generally lacking in tryptophan. The complete amino acid sequence structure has already been obtained for two different histones from calf thymus and pea seed nuclei. These molecules are polycationic; that is, they contain many positively charged groups such as $HN_2^+$ which combine readily with acidic phosphate groups of the DNA chain. Fewer than 10 different molecular species of histones have been identified, and these seem to be structurally similar in most organisms. Histones from different cells are also similar. The amino acid sequences of histone in pea seed and calf thymus differ by only 2 residues out of a total of 102, showing a remarkable evolutionary conservation; the molecule has undergone very little mutational change. Presumably the entire amino acid sequence is essential to histone function. If histones regulate gene expression, we would expect a greater degree of heterogeneity; each gene would be regulated by a specific histone. But histones exhibit no great selectivity in binding to DNA, which suggests they are not involved in specific gene regulation.

The various histones are grouped into the following classes: *arginine-rich histones, moderately lysine-rich histones, very lysine-rich histones,* and *protamines,* a special class of basic proteins obtained from fish sperm. The first three can be obtained from all plant and animal tissues. The arginine-rich histones consist of only two major classes, an 11,000 mol wt class and a 20,000 mol wt class of structur-

ally similar molecules. Arginine-rich and moderately lysine-rich histones are found in all tissues and are easily modified by secondary addition of methyl, acetyl, and phosphate groups. These side-chain modifications may play an important role in histone function. For example, more histones are acetylated in rapidly dividing tissues than in slowly dividing cells, and acetylation usually precedes or accompanies increases in RNA synthesis.

The very lysine-rich class of histones differs significantly from the other two because it is not found in all tissues. It is obtained from most mature tissues but is lacking in some embryos. As tissues mature, amounts of lysine-rich histones increase. These histones are resolvable into four different components in terms of amino acid composition and other physical properties. They may have sufficient heterogeneity to function as a regulator of classes of genes rather than of individual genes. Approximately half the molecule is strongly polycationic, like most histones, but the other half, about 100 amino acids, resembles most other nonhistone proteins. The latter region may have binding sites conferring specificity on histone molecules. Though 85 percent of the molecule is unchanging, there is enough variation in the remaining 15 percent to give each a unique specificity. Very lysine-rich histones may have two active regions, the "invariant" polycationic portion of the molecule, which may account for strong, nonselective binding to DNA, and the variable portion, which may confer specificity to the molecule in its interactions with other macromolecules (RNA or protein?).

The amount of histone in chromatin is correlated with the amount of DNA; the ratio of histone to DNA in different chromatins is relatively constant (Table 2-4) whereas nonhistone proteins show variations. This one-to-one relationship suggests a structural role for histones. Like DNA, histones are also metabolically stable; compared with most other proteins they do not turn over rapidly. The observation that histone synthesis coincides with the synthesis of DNA during the cell cycle is also consistent with a structural role. We would expect structural chromosomal proteins to be synthesized simultaneously with DNA, whereas regulatory proteins might be synthesized independently at other times in the cell cycle.

We find that all histones are synthesized in the cytoplasm at similar rates, even in cells undergoing

**TABLE 2-4  Chemical Composition of Interphase Chromatins**

| Tissue Source | DNA | Ratio Relative to DNA | |
| | | Histone | Nonhistone Protein |
| --- | --- | --- | --- |
| Pea embryo | 1.0 | 1.03 | 0.29 |
| Pea bud | 1.0 | 1.30 | 0.10 |
| Pea cotyledon | 1.0 | 0.76 | 0.36 |
| Rat liver | 1.0 | 1.0 | 0.67 |
| Rat ascites tumor | 1.0 | 1.16 | 1.00 |
| HeLa cells | 1.0 | 1.02 | 0.71 |
| Cow thymus | 1.0 | 1.14 | 0.33 |
| Sea urchin blastula | 1.0 | 1.04 | 0.48 |
| Sea urchin pluteus | 1.0 | 0.86 | 1.04 |

Source: Modified from J. Bonner et al., *Science* 159:47–56, 1968. Copyright © 1968 by the American Association for the Advancement of Science.

physiological changes. Here too there seems to be no selectivity. If histones do regulate gene function, we would expect synthesis of certain histones to correlate with cell metabolism. After a histone polymer is synthesized, it may be methylated by reactions with methione derivatives, acetylated by interacting with acetyl CoA, the direct acetyl donor, or phosphorylated by ATP. Such reactions undoubtedly alter histone structure and its association with DNA. In each case the degree of methylation, acetylation, or phosphorylation does correlate with histone activity. For example, the hormone cortisone inhibits histone acetylation in thymus and stimulates it in liver, and these reactions are accompanied by corresponding effects on RNA synthesis in each of these tissues. It has also been shown that histones in active euchromatin are more highly acetylated than histones extracted from heterochromatin.

Heterochromatin and histone levels seem to be correlated. The paternal chromosomes of the mealy bug are heterochromatic, as we have seen, and do not synthesize RNA. They contain high levels of very lysine-rich histones. Moreover, many parts of these chromosomes are inaccessible to specific reagents that combine with DNA. For example, heterochromatic chromosomes do not bind actinomycin D, an antibiotic that combines specifically with DNA in guanine-cytosine rich regions. The drug fits in between the grooves, prevents attachment of the RNA polymerase to DNA, and blocks RNA synthesis.

This reagent has proven most useful in studying DNA template functions. Failure to bind actinomycin means that GC regions are inaccessible to the drug in heterochromatic chromosomes. If, however, histones are extracted from the heterochromatic chromosomes, RNA synthesis is activated and the DNA becomes accessible to actinomycin D. Histone removal transforms compacted DNA into a more active structure, suggesting that histones are involved in cross-linking DNA strands.

Because histones and DNA are charged molecules, they inevitably form strong electrostatic associations. Histones combine readily with DNA in vitro. Such nonspecific combinations grossly alter DNA structure and behavior. In vitro very lysine-rich histones combine preferentially with high AT sequences in DNA whereas arginine-rich histones prefer GC regions. Such specificity is still of a general nature, however, lacking the high resolving power needed to single out specific genes for regulation.

We still do not understand how histone is associated with DNA, whether it lies within the groove, lies across the strands, or encases the DNA as a sheath. Interactions of histones among themselves with nonhistone proteins and with RNA in the chromosome probably occur and make all sorts of complicated associations possible. These may acquire the specificity required for regulation of portions of the genetic information. More significantly, periodic coiling and uncoiling of chromosomes during mitosis may be linked to cyclic interactions of various histones with DNA strands. Compaction of DNA into small volumes, as in sperm heads, is also associated with metabolic changes in histones. In these instances, histones transform into protamines which seem to be more effective in DNA condensation. Histones are capable of helix formation in solution and become more helical in the presence of DNA.

In summary, histones are important components of the eukaryote genome that combine nonspecifically with DNA and seem to regulate its secondary and tertiary structure. Such interactions undoubtedly affect DNA coiling behavior and may even alter its activity. Most evidence indicates that histones lack the recognition specificity essential to fine-tuning regulation required for gene expression. Instead, these reactive proteins may regulate genetic information in whole blocks of genes, parts of chromosomes, or even whole chromosome sets by controlling the extent of DNA coiling. More precise recognition of specific DNA sequences and fine-tuning regulation may be a property of nonhistone proteins.

### Nonhistone acidic proteins

The nonhistone proteins associated with DNA form a heterogeneous class of varying sized molecules. Molecular weights range from 7000 to 15,000, and hundreds of different proteins can be separated from chromatin by gel electrophoresis. Most of these proteins are phosphorylated; in fact, 95 percent of all nuclear-bound phosphate is found in the acidic proteins. The amount of phosphorylation of these proteins changes during the cell cycle and also during development. Phosphorylation is reversible, probably resulting in conformational changes by the addition of negatively charged phosphate groups. Such conformational changes could alter binding affinities for DNA, histone, or both.

Because of their diversity and metabolic activity these proteins may play a major role in eukaryote gene regulation. In contrast to histones, acidic proteins from different somatic tissues are variable and tissue-specific. Moreover, their nature and amount change during early embryogenesis and also at later stages when somatic cells differentiate. Specific acidic proteins are augmented in chromatin by hormones known to activate genes, such as estradiol or cortisol.

These proteins also display tissue-specific binding to DNA. Nonhistone proteins from rat liver chromatin bind more effectively with rat liver DNA than with kidney DNA. Sequence specificity of binding is another property of these molecules. If calf thymus DNA is fractionated into different repetitive classes based on CoT values, the pattern of nonhistone proteins binding to each of these is also different, indicating that these heterogeneous molecules have a high degree of sequence preference.

The extent of phosphorylation of these proteins also correlates with gene activity. Active euchromatin, for instance, contains more protein-bound phosphorous per milligram of DNA than heterochromatin. Phosphorylation is brought about by enzymes known as protein kinases, with many different kinds found in nuclei, suggesting that each phosphorylates a different class of proteins. The phosphate from ATP (or other trinucleotides) is transferred to the serine hydroxyl group on the protein. The molecule becomes more acidic, increasing its binding affinity for such basic molecules as histones.

Included among these proteins are the various DNA and RNA polymerases which recognize and copy DNA sequences. These also show preferences for certain sequences or DNA configurations before becoming active. Other specific acidic proteins isolated from prokaryote cells exhibit gene regulatory functions (see Chapter 3), but what is most significant are recent observations that eukaryotic chromatin may contain contractile proteins actin, myosin, tropomyosin, and troponin. In some cases these may be cytoplasmic contaminants, but most evidence indicates they are major constituents of chromatin and may regulate transitions from extended euchromatin to condensed heterochromatin. Most vary in amount as cells shift from proliferative to nonproliferative states. For example, as proliferating cells with large nuclei and diffuse chromatin leave the mitotic cycle, tropomyosin disappears from nuclei, actin accumulates, and chromosomes condense. At the same time, many high-molecular-weight acidic proteins leave the nucleus. The converse happens as cells become mitotically active after a period of inactivity. Such reversible transitions in contractile proteins could control coiling-condensation cycles of chromosomes during the cell cycle.

Summarizing, a variety of enzymatic, structural, and regulatory proteins are associated with chromatin. These change DNA conformation and alter gene expression. They also modify histone structure and probably affect its interactions with DNA. A system of protein kinases phosphorylates both histones and nonhistone proteins and alters their conformation and interaction with DNA. The nature of these interactions is not known, but the diversity of DNA-protein interactions predicts that genome modulation in eukaryotes is more complicated than in prokaryotes.

## CYTOPLASMIC GENOMIC ORGANIZATION

The cytoplasm is also a repository of genetic systems equipped with informational DNA tapes. Such extranuclear genetic systems contribute to the total developmental information of a cell. Many cell properties are not inherited according to Mendelian segregation but are transmitted by cytoplasmic genes which, in contrast to nuclear genes, do not obey strict rules of segregation and transmission.

As early as 1909 examples of non-Mendelian heredity were correctly interpreted as having an extrachromosomal basis. Extrachromosomal genes, or plasmagenes, are now known to reside primarily in mitochondria and chloroplasts as DNA helices. Moreover, exogenous genomes (that is, RNA and DNA viruses) may become more or less permanently integrated into cytoplasmic or nuclear genomes as part of the total information available to a cell.

Cytoplasmic gene systems are less well understood than nuclear genomes. They are semiautonomous and dependent upon some nuclear information for their own formation. However, once mature they are autonomous and exhibit hereditary transmission, even in the absence of nuclei.

An extrachromosomal genetic system must meet certain criteria for cytoplasmic inheritance. First, it must consistently determine a measurable phenotypic trait. Variegated patterns of yellow and green patches in the leaves of many higher plants are good examples. These are attributed to normal and mutated chloroplasts that segregate randomly during division. Second, the transmitted trait must exhibit genetic continuity from cell to cell over many generations; it is stable and resists dilution by cell multiplication. An extrachromosomal entity therefore must self-replicate to keep pace with cell division. If lost, such entities cannot be replaced from chromosomal material since they originate from a preexisting extrachromosomal gene system. Third, if a plasmagene replicates, it must also mutate and transmit this stable change to progeny. Plasmagenes, like chromosomal genes, are identified only by mutations.

Finally, extrachromosomal systems segregate independently of chromosomal genes. The most crucial test for extrachromosomal inheritance is a pattern of non-Mendelian segregation; extrachromosomal systems are not transmitted according to Mendelian laws. Cytoplasmic inheritance is assumed when persistent differences between reciprocal crosses are observed in genetic experiments. For instance, we know that the manner in which nuclear (Mendelian) genes are combined in a cross has no effect on the phenotype; identical progeny result whether genes come from the father or mother. On the other hand, the source of genetic material is crucial in extrachromosomal systems; the progeny differ depending upon parental origin. Maternal inheritance is the most common feature of reciprocal crosses. The egg, being larger than the sperm, contributes virtually all cytoplasmic constituents of the zygote, and the relevant phenotype always resembles the maternal parent, irrespective of its nuclear genes.

An example of maternal inheritance is the poky mutation in *Neurospora*. Wild type strains complete growth in the so-called growth tube assay in 3 to 4 days while poky mutants require 10 to 12 days under similar conditions. Some poky mutants exhibit Mendelian inheritance and are nuclear, while others are transmitted through cytoplasmic genomes, probably mitochondria, and exhibit maternal inheritance (Figure 2-19).

Mendelian segregation of nuclear genes is easy to follow in *Neurospora* crosses. The life cycle begins when asexual haploid spores give rise to mold mycelia, each bearing female fruiting bodies, or *protoperitheca*, and *conidia*, the male gametangia. Mating occurs only between two different mating types, controlled by the nuclear genes *A* and *a*. If a female of mating type *A* is crossed with conidia from an opposite mating type *a*, diploid zygotes develop within the mature perithecium. A single zygote nucleus occupies one of many sacs, or *asci*, within the perithecium, and this nucleus undergoes meiosis, producing four haploid nuclei. Each divides mitotically to yield eight haploid ascospores arranged in the order in which they segregate in meiotic division. All nuclear genes, such as mating type, segregate according to Mendelian laws, and the genetic outcome of each cross is easily scored by isolating and germinating the ascospores. Fifty percent of the progeny will be mating type *A* and fifty percent mating type *a*. Poky mutants, however, do not segregate in this manner; instead, they persistently follow the maternal phenotype through many generations and through any combinations of crosses with wild type. The male conidial parent contributes virtually no cytoplasm to this cross, which suggests that the extrachromosomal genetic systems responsible for transmission of the poky trait are the mitochondria of the perithecial cells.

Usually the behavior of visible cytoplasmic elements such as mitochondria correlates with transmission and segregation of the extrachromosomal genetic system. On the other hand, if no visi-

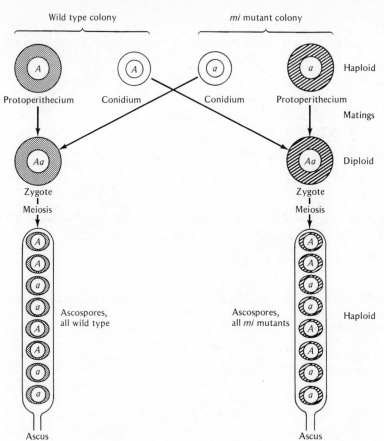

**Figure 2-19** *Cytoplasmic inheritance of poky mutation* mi *in* Neurospora. *The female parent in each case is the protoperithecium while the male gamete is derived from a conidium. Note that inheritance of the mutant is entirely maternal. If the maternal parent is wild type, all ascospores will be wild type, irrespective of the nuclear genotype. Likewise, if the maternal parent is mutant, all ascospores are mutant. (From J. L. Jinks,* Extrachromosomal Inheritance. © *1964 by Prentice-Hall, Inc.; based on M. B. Mitchell and H. K. Mitchell,* Proc. Nat. Acad. Sci. U.S. *38:442–449, 1952.)*

ble entity correlates with the behavior of the genetic system, then some other entity, a virus or *episome,* is postulated. An episome is defined as an informational unit that can either be incorporated into the chromosomal genome or exist as an independent self-replicating entity in the cytoplasm. It has a dual existence, behaving like the host genome when physically integrated and exhibiting an autonomy when in the cytoplasm. Examples of episomic behavior are the symbiotic viruses in certain bacteria such as lambda phage in *E. coli* (see Chapter 17).

### Mitochondria as extrachromosomal genetic systems

Mitochondria meet the first test of an extrachromosomal system; they determine a measurable cell-phenotypic trait. All aspects of oxidative metabolism in the cell come under mitochondrial control, in-

asmuch as they contain the respiratory enzymes and regulate cellular energy transformations. Most hereditary changes in a cell's respiratory behavior may be traced to modifications in the structure or physiological activity of mitochondria.

Mitochondria show genetic continuity. They are transmitted from cell to cell and do not arise de novo but always come from preexisting mitochondria. Mitochondrial segregation occurs during mitosis; that is, bundles of mitochondria surrounding the spindle segregate at random into two groups at anaphase and pass into the newly forming daughter cells.

Mitochondrial continuity was demonstrated by experiments on the bread mold *Neurospora. Neurospora* normally grows on a medium containing ammonium sulfate as a nitrogen source. By shifting the mold from a medium containing ammonium sulfate with heavy nitrogen ($^{15}N$) to one with $^{14}N$, it was possible to follow mitochondrial synthesis

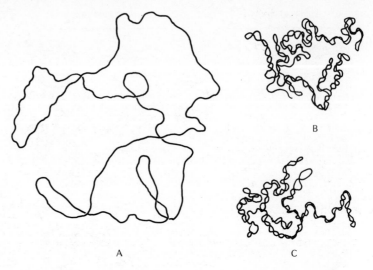

**Figure 2-20** *Molecules of mitochondrial DNA extracted from Xenopus oocytes. (A) Open circles. (B, C) Twisted circles. (From I. B. Dawid and D. R. Wolstenholme. J. Mol. Biol. 28:233–245, 1967.)*

and distinguish old from newly formed mitochondria. If mitochondria arose de novo, then all new mitochondria made in the $^{14}$N medium should contain $^{14}$N exclusively in their proteins. Under these conditions two discrete classes—new, light $^{14}$N mitochondria and old, heavy $^{15}$N mitochondria—should be separable by differential centrifugation. But at no time were discrete density classes of mitochondria obtained. As growth proceeded, mitochondria of intermediate density were isolated, containing $^{15}$N and $^{14}$N components, and these became progres-

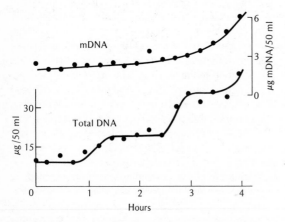

**Figure 2-21** *Comparison of mitochondrial and nuclear DNA synthesis in a synchronously dividing culture of Saccharomyces cerevisiae. (From D. H. Williamson and E. Moustachi,* Biochem. Biophys. Res. Commun. *42:195–201, 1971.)*

sively lighter as more $^{14}$N was incorporated. It was concluded that old mitochondria grow by the insertion of proteins synthesized in $^{14}$N medium into their structure. Mitochondria in progeny cells do not arise de novo but come from preexisting mitochondria.

The molecular basis for these genetic properties probably resides in organelle DNA. Uniform circles of DNA, 5 $\mu$ in contour length, have been isolated from mitochondria of many plant and animal tissues (Figure 2-20). Within the intact organelle DNA appears as fibrils embedded in the matrix. Organelle DNAs can be separated from nuclear DNAs as distinct molecules because of their different buoyant densities when centrifuged. They also differ with respect to average nucleotide composition, which means few sequences are shared in common. Organelle DNAs are uniquely different from nuclear genomes in the same cell.

The uniformity of mitochondrial DNA in all organisms suggests a common informational molecule coding for similar mitochondrial proteins. Like nuclear DNA, organelle DNA behaves as a template molecule in its own replication. Replication is independent of nuclear DNA replication. Nuclear DNA synthesis is discontinuous, occurring only in interphase, while mitochondrial DNA is synthesized continuously through the cell cycle (Figure 2-21).

Plastid circular DNA resembles the DNA circles found in viruses and bacteria. In fact there is almost as much DNA in chloroplasts and mitochondria as there is in a bacterial genome, which has led to the hypothesis that these organelles were originally bacterial symbiotes that infected some primitive cell early in evolution.

Mitochondria contain plastid-specific DNA and RNA polymerase which recognize mitochondrial DNA. They also contain small ribosomes, similar to those in prokaryotes, and all the enzymes necessary for protein synthesis. Accordingly, mitochondrial protein synthesis also exhibits differences in sensitivity to inhibitors. Like bacterial ribosomes, they are sensitive to chloramphenicol and insensitive to cyclohexamide. The reverse is true for cytoplasmic ribosomes.

If a stretch of three nucleotides in a DNA chain codes for one amino acid, then a $5\text{-}\mu$ circle of mitochondrial DNA could code for 20 different proteins with an average molecular weight of 30,000. Since there are far more than 20 different proteins in mitochondria, their DNA is insufficient to account for all their proteins. Most mitochondrial respiratory enzyme systems are probably coded by the nuclear genome; only their ribosomes, transfer RNAs, and polymerases, and perhaps some membrane proteins, may be directly coded by organelle DNA.

Mitochondria carry energy-transducing systems organized into rows of membrane lamellae (*cristae*) embedded in a matrix. The membranes contain crystalline arrays of enzyme complexes involved in oxidative phosphorylation and electron transport. Because of this respiratory function, mitochondrial morphology may vary under different physiological conditions. For example, a rat liver cell contains 1000 mitochondria as compared with the 300 or so mitochondria in a kidney epithelial cell. Cristae are short and few in number in liver mitochondria, projecting inside from the inner membrane (Figure 2-22). In kidney they are more numerous and regularly arranged with only thin attachments to the inner membrane. Cristae morphology seems to be related to the respiratory activity of a cell, with many more fenestrated cristae found in a vigorously respiring heart muscle than in an epithelial cell. Mitochondria tend to localize in energy-requiring regions of the cell, along the sarcomeres of muscle myofibrils,

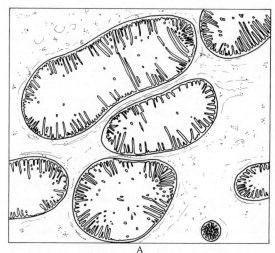

A

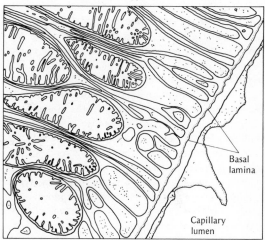

Basal lamina

Capillary lumen

B

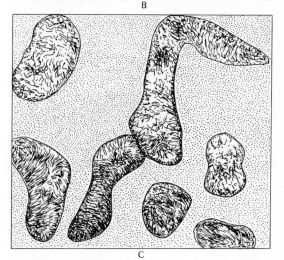

C

**Figure 2-22** *Tissue-specific variation in mitochondria structure. (A) Liver mitochóndria from a bat. (B) Distal convoluted tubule of guinea pig kidney. (C) Ventricular heart muscle of a cat. Note differences in shape, numbers of cristae, and branching.*

wound around the midpiece of sperm tails or aligned in parallel in kidney tubule cells with their long axes in the direction of active transport.

Changes in organelle morphology and position are often diagnostic of cell function. Mitochondria exhibit osmotic changes; they swell and shrink during respiratory metabolism as precursors and product molecules exchange across their membranes. Changes in mitochondrial structure may also arise as a result of mutations in mitochondrial DNA.

Mutant mitochondria exhibit structural defects; they appear abnormal in the electron microscope, with fewer cristae. Biochemically, poky mutants are deficient in cytochrome enzymes; they lack cytochrome a, and some strains contain an excess of cytochrome c with no detectable cytochrome b. The absence of cytochromes does not mean that poky genes control cytochrome synthesis; in fact, we know that cytochrome enzymes are coded by the nuclear genome. Their absence in poky mitochondria, even in the presence of a normal nuclear genome, suggests that the mutation may involve a regulatory rather than a structural gene.

Mutations in mitochondrial genes also correlate with changes in mitochondrial DNA. This observation was made in an analysis of the "petite" mutants in yeast. These, like the poky mutants in *Neurospora,* are respiratory-deficient, slow-growing mutations that result in very small colonies when grown on agar. Most are mitochondrial mutants exhibiting typical extranuclear segregation and transmission genetics. DNA extracted from strains of petite mitochondria exhibits different buoyant densities from wild type mitochondrial DNA, which means that the mutant DNA has undergone changes in nucleotide composition.

## Chloroplasts as extrachromosomal genetic systems

Genetically, plant plastids behave like mitochondria. Chloroplasts are semiautonomous membranous organelles involved in photosynthesis supplying cells with carbohydrates and ATP. Photosynthetic enzyme systems are organized in crystalline array on the lamellar membranes, closed, flattened vesicles known as discs, or *thylakoids.* These are embedded in a matrix, or *stroma,* surrounded by a membrane. Chloroplasts contain self-replicating DNA templates along with complete self-contained enzyme systems for replication and protein synthesis. They possess ribosomes and presumably manufacture their own precursor, but, like mitochondria, they are not completely autonomous and must depend on nuclear genes for information.

Many abnormal phenotypes of photosynthesis owe their origin to mutant chloroplasts that are transmitted faithfully from cell to cell as extrachromosomal determinants. The pattern of segregation and the persistent differences displayed in reciprocal crosses confirm a plasmagene basis for heredity. The variegation phenomenon in higher plants with leaves showing yellow and green patches is an example of extrachromosomal inheritance. Variegated inheritance is explained by assuming two kinds of plastids, normal and mutant. Starting with an ovule with both types of plastids after fertilization, it is easy to see how multiplication and segregation of plastids during cell division will yield some cells with only mutant plastids, others with normal plastids, and a third group with mixed plastids. This gives rise to the variegated phenotype: the first class of cells produces white or yellow sectors; the second class, green sectors; and the third class, a region of variegated sectors capable of further sectoring. In almost all cases the phenotypic trait at the cell or tissue level corresponds to a defect in chloroplast structure and behavior. Some of these phenotypic effects result from mutations in chloroplast DNA.

We are only beginning to unscramble the network of feedback relationships operating between nuclear and cytoplasmic genomes. But this circuitry is part of a larger story of intracellular communication. The information in linear macromolecules must be retrieved and translated into cell structure and behavior as cells carry out developmental programs.

## SUMMARY

Information for development is both preformed and epigenetic, with the former arranged as systems of genetic macromolecules in cells whose organization is itself a store of preformed structural information. New information is acquired epigenetically as organizational hierarchies emerge in multicellular systems.

Two cellular plans have evolved, prokaryote and eukaryote. Prokaryote cells are simple with genetic information organized as single, naked DNA helices. These cells display limited developmental repertoires. Eukaryote cells are more complex, containing large amounts of DNA in nuclei and in cytoplasmic organelles such as mitochondria and chloroplasts, which explains their diverse developmental expression.

Eukaryote DNA is packaged into chromosomes with histone and nonhistone proteins. DNA-protein interactions induce dynamic changes in chromosome structure, from condensed coils (heterochromatin) to extended threads (euchromatin). These changes affect the functional expression of genetic information in development; only when extended does the information become available. Such transitions are cyclic during mitotic episodes and also occur during interphase when genetic information is utilized.

The sequence organization of eukaryote DNA is also more complex. It is composed of nucleotide sequences with varying degrees of redundancy, including highly repetitive, moderately repetitive, and single-copy classes. These occupy different regions in the genome, with the most highly repetitive classes arranged in tandem in centromeres, where they have a structural rather than an informational role. Moderately repetitive sequences are interspersed among single-copy sequences, suggesting that the former act as regulatory genes controlling single-copy structural genes.

Extrachromosomal genetic systems in mitochondria and chloroplasts contribute to the developmental versatility of eukaryotes. These semiautonomous organelles contain complete informational systems, and interactions between nuclear and cytoplasmic compartments are essential to information retrieval during development.

**REFERENCES**      Apter, M. J., and Wolpert, L. 1965. Cybernetics and development. I. Information theory: A critique of cybernetic models as applied to developmental biology. *J. Theoret. Biol.* 8:244.

Back, F. 1976. The variable condition of euchromatin and heterochromatin. *Int. Rev. Cytol.* 45:25–64.

Britten, R. J., and Kohne, D. E. 1968. Repeated sequences in DNA. *Science* 161:529–540.

Brown, D. D. 1973. The isolation of genes. *Sci. Am.* 229:20–29.

Brown, S. W. 1966. Heterochromatin. *Science* 151:417.

Chromosome structure and function. *Cold Spring Harbor Symp. Quant. Biol.* Vol. XXXVIII. New York: Cold Spring Harbor Laboratory of Quantitative Biology, 1974.

Davidson, E. H., and Britten, R. J. 1973. Organization, transcription, and regulation in the animal genome. *Quart. Rev. Biol.* 48:565–613.

Elgin, S. C. R., and H. Weintraub. 1975. Chromosomal proteins and chromatin structure. *Ann. Rev. Biochem.* 44:725–774.

Elsasser, W. M. 1958. *The physical foundation of biology.* London: Pergamon Press.

Flamm, W. G. 1972. Highly repetitive sequences of DNA in chromosomes. *Int. Rev. Cytol.* 32:2–52.

Hennig, W. 1973. Molecular hybridization of DNA and RNA in situ. *Int. Rev. Cytol.* 36:1–44.

Hsu, T. C. 1973. Longitudinal differentiation of chromosomes. *Ann. Rev. Genet.* 7:153–176.

Ingram, V. 1957. Gene mutation in human hemoglobin: The chemical difference between normal and sickle cell hemoglobin. *Nature* 180:326–328.

Inman, R. B., and Baldwin, R. L. 1964. Helix-random coil transitions in DNA homopolymer pairs. *J. Mol. Biol.* 8:452–469.

Kornberg, R. D., and Thomas, J. O. 1974. Chromatin structure. *Science* 184:865–871.

Lees, A. D., and Waddington, C. H. 1942. The development of the bristles in normal and some mutant types of *Drosophila melanogaster. Proc. Roy. Soc. London Ser. B* 131:87–110.

Lestourgeon, W. M., Totten, R., and Forer, A. 1974. The nuclear acidic proteins in cell proliferation and differentiation. *Acidic proteins of the nucleus,* ed. I. L. Cameron and J. R. Jeter, Jr. New York: Academic Press, pp. 159–190.

Marmur, J., and Doty, P. 1962. Determination of the base composition of deoxyribonucleic acid from its thermal denaturation temperatures. *J. Mol. Biol.* 5:109–118.

Pardue, M. L., and Gall, J. G. 1970. Chromosomal localization of mouse satellite DNA. *Science* 168:1356–1358.

Phillips, D. M. P., ed. 1971. *Histones and nucleohistones.* New York: Plenum.

Raven, C. P. 1958. Information versus preformation in embryonic development. *Arch. neérl. Zool.* 13(Suppl. 1):185.

Sager, R. 1972. *Cytoplasmic genes and organelles.* New York: Academic Press.

Tartof, K. D. 1975. Redundant genes. *Am. Rev. Genet.* 9:355–385.

Watson, J. D. 1965. *Molecular biology of the gene.* Menlo Park, Calif.: Benjamin.

Wilson, E. B. 1924. *Cell in development and heredity.* New York: Macmillan.

# 3

# Retrieval and Exchange
# of Developmental Information

Development may be thought of as a decoding process wherein information in one-dimensional DNA tapes is translated into the functional behavior of three-dimensional protein molecules. The process is sequential and orderly, and the use of information is carefully regulated. In developmental programs only a small proportion of the total genetic information is expressed at one time. Information is selectively retrieved and utilized, a problem complicated in eukaryotes by the fact that two separate genetic compartments, nuclear and cytoplasmic, exist, and they must interact and cooperate to produce a functional cell phenotype. Hence we find informational circuits operating at different organizational levels with information flowing between macromolecules within compartments, between organelles, and between whole nuclear and cytoplasmic compartments. As the hierarchy becomes more complex, information is exchanged between cells and tissues. Some rules governing information flow are common to all levels, but certain rules of

interaction, or "languages," are specific to a given level. All rules determine what and how much information is retrieved and when and where it is utilized in the cell. In this chapter we examine information flow at three levels: the molecular level where macromolecules exchange information, the organelle level of nuclear and cytoplasmic genome interaction, and the level of interaction between whole nuclear and cytoplasmic compartments.

## MOLECULAR INFORMATION FLOW

Information retrieval at the molecular level involves protein synthesis, the translation of information encoded in DNA nucleotide sequences into linear amino acid sequences. We know that the information in DNA tapes is not directly involved in protein synthesis since removal of nuclei from a cell or inhibition of DNA synthesis has no effect on protein synthesis. On the other hand, inhibition of RNA synthesis does block protein synthesis, suggesting that RNA and not DNA is directly involved. The genetic information in DNA is first copied on linear RNA molecules which function as the informational templates in protein synthesis. The metabolic role of the genome is assigned to RNA copies or messages, while the more conservative information storage role is retained by DNA. DNA is more stable than RNA; its double helix gives it a highly regular three-dimensional structure held together by hydrogen bonds. RNA is a single-stranded polynucleotide with a ribose-phosphate "backbone" instead of a deoxyribose phosphate chain. The single-stranded structure probably makes RNA less stable. The subunits of RNA are ribonucleotides of adenine, guanine, cytosine, and uracil in place of thymine. Nevertheless, the same "grammar rules" are followed when a copy of RNA is made on a DNA template; guanine pairs with cytosine, and adenine pairs with uracil instead of thymine. An RNA copy of a DNA nucleotide sequence reading ATCCGGTTAG would be UAGGCCAAUG. . . . Such RNA copies will combine with single-stranded DNA wherever the sequences are complementary and form DNA-RNA hybrids, just as complementary sequences of DNA strands reanneal. The facility with which DNA-RNA hybrid molecules are formed is also a measure of the sequence homologies between the two macromolecules.

The information stored in DNA molecules is retrieved by two processes, replication and transcription (Figure 3-1). In each case the DNA molecule acts as a template on which polynucleotide complementary copies are made, DNA in the former and RNA in the latter. Replication is a means of conserving and transmitting whole copies of the genome to cell progeny; transcription generates working RNA copies for translation into a cell's protein machinery. The replicating mode of the template is determined by the specific copying enzyme, DNA

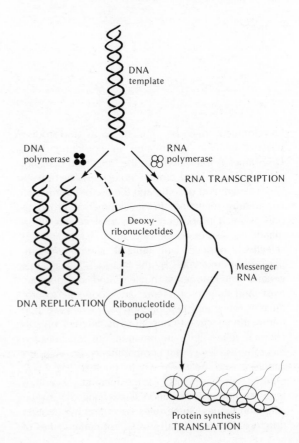

**Figure 3-1**  *General scheme of information flow in eukaryote cells. The same DNA template is used for replication and transcription, depending on the polymerase enzyme involved. Note that a common nucleotide pool is utilized.*

polymerase, drawing on a pool of deoxyribonucleotide precursors, while the transcriptive mode is defined by its copying enzyme, RNA polymerase, drawing on the pool of ribonucleotides. In fact a common ribonucleotide pool is available since ribonucleotides are first reduced to deoxyribonucleotides before incorporation into DNA. The primary direction of information flow is DNA → RNA → protein, the so-called central dogma of molecular biology. This rule is true for prokaryotes as well as eukaryotes, except for some animal RNA viruses in which information flow is RNA → DNA → RNA and then to viral protein (see Chapter 21).

The specific cellular organization of prokaryotes and eukaryotes accounts for some basic differences in information flow. Transcription and translation occur simultaneously in prokaryotes, whereas they are compartmentalized and separated spatially and temporally in eukaryotes (Figure 3-2). Transcription occurs within the nucleus, and the RNA message is first processed before it is released to the cytoplasm, where it combines with ribosomes to organize into polyribosomes for translation. Because of RNA processing, the primary nuclear RNA product differs significantly from the cytoplasmic RNA message ultimately utilized for translation into polypeptide sequences.

## Transcription: RNA synthesis and the making of message tapes

Information about transcription has come from studies of in vitro systems for RNA synthesis. The enzyme, RNA polymerase, catalyzes the polymerization of ribonucleotide triphosphates into RNA chains on DNA templates. The base-pairing rule of the nucleic acid language is followed, as G pairs with C and A pairs with U or uracil nucleotides. RNA strands appear with base sequences complementary to only one strand of the DNA double helix. We know this because newly synthesized RNA forms hybrid complexes with only one strand of the double-stranded DNA template. This means the strands of the DNA double helix separate to allow the polymerase to read a nucleotide sequence on one strand and insert the proper purine and pyrimidine nucleotides in a growing RNA chain (see Figure 3-2). If this were not so, different copies would be made at the same gene locus and would yield two different proteins. Transcription resembles replication only in

the requirement for strand separation to accommodate the polymerizing enzyme. During replication, however, both strands are copied simultaneously as template strands separate. Within a short segment of a transcribing template, at the point of enzyme attachment, the nascent RNA exists temporarily as a DNA-RNA hybrid. This unwinds, a single-stranded RNA copy is released, and the DNA strands reassociate into a double helix.

DNA strands are read in only one direction, from the 5' end of the chain to the 3' end. This means that messenger RNA sequences are translated into amino acid sequences in the same direction. If the direction of transcription was not polarized, then a stretch of a DNA double helix would code for two different messenger RNAs, which is obviously not the case. A genetic message, like a sentence, has a beginning and an end, marked off by appropriate punctuation marks.

This model imposes several informational constraints on RNA polymerases; they must recognize nucleotide sequences marking the beginning and end of a specific gene. Indeed, certain nucleotide sequences identify initiation points (known as *promotors*) where polymerases attach to start transcription, while other sequences at the terminal point say "stop, get off here." Sequence recognition resides in RNA polymerase structure. These enzymes are different in each species and may even be different within the same cell. Some polymerases exist in nucleoli and others are associated with chromosomes in the nucleoplasm; organelle-specific polymerases have been found that recognize only DNA templates of mitochondria and chloroplasts.

RNA polymerases, purified from microorganisms, turn out to be large, complex proteins made up of several polypeptide subunits, called alpha, beta, and sigma chains. The alpha and beta chains make up the core enzyme and are responsible for its catalytic action, whereas sigma chains retain the recognition function to initiate transcription on intact DNA templates. Several sigma factors have been identified in prokaryotes, each conferring a different specificity of action on alpha-beta subunits. For example, bacteriophage infecting *E. coli* induce synthesis of their own sigma factors, which combine with host RNA polymerase to initiate transcription on viral rather than host DNA templates.

Recognition specificity and faithful transcription require the double-stranded helix. If the DNA helix

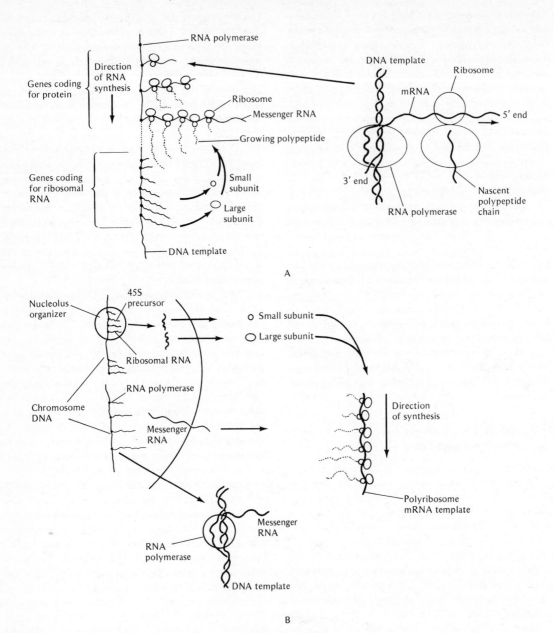

A

B

**Figure 3-2**  *Comparison of information flow in eukaryote and prokaryote cells. (A) Simultaneous transcription and translation in prokaryote cells. As RNA polymerase "reads" the DNA template, it assembles messenger RNA which immediately associates with ribosomes for translation. (B) In eukaryotes, transcription in the nucleus is segregated from translation in the cytoplasm by a nuclear membrane. Messenger RNAs appearing in the nucleus are processed before transport to the cytoplasm where they combine with ribosomal subunits for translation on polysomes. (Based on O. Miller, Sci. Am. 228:34–42, 1973.)*

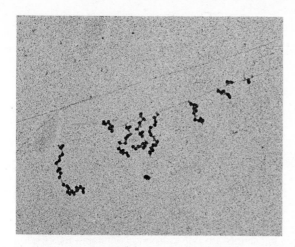
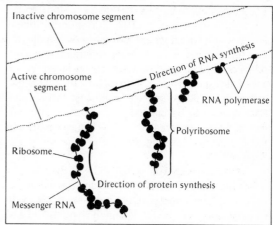

**Figure 3-3** *Visualization of transcription and translation of bacterial gene in the electron microscope (left). The interpretation is shown on the right. The lower chromosomal segment is active and the pattern of transcription and translation is identical to proposed model shown in Figure 3-2. (Photo from O. L. Miller, Jr., et al.,* Science 169:392– 395, 1970. Copyright © 1970 by the American Association for the Advancement of Science.)

is melted into two strands, the enzyme attaches at many points and copies both strands at random rather than limiting itself to the "correct" informational strand. A double helix probably has a fixed number of initiating points (promotors) defined by a recognizable cluster of nucleotides. Such clusters have been identified in viral genomes as sectors in which one strand has a sequence of pyrimidines (the informational strand) while the opposite inactive strand has a sequence of purines in the same region. Presumably sigma factors recognize such clusters and bind the polymerase. Transcription may be blocked completely by the addition of certain drugs, particularly actinomycin D. This molecule has a high binding affinity for GC base pairs and intercalates into the double helix at these points. This either prevents the RNA polymerase from attaching to the template or interferes with enzyme movement along the strand and thus stops RNA synthesis.

Transcription of prokaryote and eukaryote genes has been studied by O. Miller, who isolated segments of the interphase genome, spread them by osmotic shock, and photographed them in the act of transcription. His electron micrographs of DNA molecules transcribing RNA are virtually identical to textbook diagrams of the process. Masses of extruded *E. coli* DNA fibers, when photographed at high magnification, showed fine threads of DNA to

which were attached small granules in ever increasing numbers (Figure 3-3). The granules are ribosomes, attaching to the RNA message at the same time that it is being transcribed along a DNA template. The RNA polymerase can be identified at the point of junction with the axial thread of the DNA template. Presumably growing protein chains are associated with each ribosome as it attaches to the growing message and begins to translate. Eukaryote chromosomes, isolated by osmotic shock, also show DNA strands being transcribed by individual polymerases (see Chapter 11). In this case no ribosomes are seen along the RNA strand; instead, as each polymerase moves along the DNA chain, it transcribes RNA molecules which increase progressively in length, just as predicted in the diagram in Figure 3-2. In both cases the direction of RNA synthesis is polarized as polymerases read a strand of the double helix in one direction.

## Messenger RNA: the informational tape

The translatable, or *messenger RNA* (mRNA), though present as a small fraction of total cellular RNA (roughly 1 to 2 percent), is produced by most of the total genome; accordingly, its base ratios are similar to those of DNA. Heterogeneity is the main feature of this RNA fraction, since it codes for as many as

1000 different proteins in a cell like *E. coli.* Different nucleotide sequences in DNA correspond to different proteins, which means that messenger-molecule size distribution extends from sedimentation constants of 8S to 30 to 40S and sometimes larger. When a cell is exposed for a short period to radioactive precursors of RNA, mRNA is the first cytoplasmic RNA species to be labeled; it is more rapidly synthesized than ribosomal RNA. Size heterogeneity and rapid precursor labeling are two criteria for identifying mRNA.

Bacterial mRNA is quite labile; it decays with a half-life of less than 20 minutes and sometimes as little as 2 to 5 minutes. During the short life cycle of a bacterial cell, mRNA is rapidly synthesized and degraded. In longer-lived eukaryotic cells, mRNA is more stable, its half-life extending to hours and even months. For example, the mammalian reticulocyte, the red-cell precursor, lacks a nucleus and is without functional DNA templates, yet it continues to synthesize hemoglobin for days. Hemoglobin mRNAs transcribed prior to nuclear degeneration must be long-lived to sustain hemoglobin synthesis. This difference in message stability between prokaryotes and eukaryotes may account for fundamental differences in genome regulation between the two cell types (see Chapter 17).

Ribosomal RNAs are stable molecules that survive for long periods, whereas messenger RNAs, even in eukaryote cells, are rapidly degraded by nucleases after use. Ribosomal RNAs are folded into stable secondary structures, made possible by the inclusion of methylated nucleotides in their sequences. Since mRNAs contain no methylated bases, they are not folded and can be degraded by nucleases.

Most mRNAs extracted from eukaryote cells come off polyribosomes, which means the message is in a functional state with ribosomes attached, in various stages of protein synthesis. When mRNA is extracted, it contains protein contaminants which are sites of attachment of message to the small ribosome subunit. Further purification of mRNA reveals that the attachment site is a unique sequence of about 200 adenine nucleotides, a poly A sequence. All messages seem to be equipped with a poly A sequence at their 3' end. Since these poly A stretches readily associate with proteins, they are assumed to play a role in attaching the message to the ribosome in preparation for translation. Messenger RNAs,

bearing such sequences, are not readily destroyed by ribonucleases.

## Heterogeneous nuclear RNA

Eukaryote messages are first transcribed in the nucleus, and before they become functionally associated with ribosomes in the cytoplasm, they must be processed. The message transcripts appearing in the nucleus are giant molecules, much larger than any messages found on cytoplasmic polysomes. They vary in size from several hundred to 50,000 nucleotides in length and have been called *heterogeneous nuclear RNA,* or hnRNA. Some of the larger molecules must contain several mRNA sequences; such transcripts are *polycistronic,* containing information for more than one gene. For most cells we can assume that hnRNA is a mixture of large polycistronic messages and smaller single-gene messages. In addition, about one quarter of the hnRNA is composed of very small low-molecular-weight RNAs, 200 to 300 nucleotides in length. These are not degradation products, since they survive as long as most mRNA, but their function and relation to the larger hnRNAs are not known.

Heterogeneous nuclear RNA resembles bulk DNA in base composition and contains repetitive as well as nonrepetitive sequences. It is extremely labile, turning over within the nucleus in a matter of minutes, although in tissue culture cells the half-life is of the order of 20 to 60 minutes. Most of the hnRNA is degraded within the nucleus with fewer than 5 percent of its sequences appearing as part of the cytoplasmic message population attached to polysomes.

It has been suggested that hnRNA is a precursor for cytoplasmic messages. Just as ribosomal RNA is carved out of a large precursor in a stepwise manner, large hnRNA molecules are assumed to pass through an assembly-line process of message production. One step that has been proposed is a posttranscriptional addition of polyadenylic acid (poly A) sequences to the 3' ends of some of the messages. These poly A sequences are not coded in the genome but are synthesized by a nontranscriptional process and added after the primary transcripts are made.

The model further proposes that hnRNA is subsequently degraded by intranuclear nucleases into smaller mRNA molecules and small nucleotides. The message molecules with poly A are protected from degradation, perhaps by combining with protein for transport from the nucleus to the cytoplasm. Some of these messages associate with ribosomes and become functional messages while others are perhaps degraded in the cytoplasm.

The model described seems to explain most of the data available. Poly A message molecules are found in the cytoplasm. Complex ribonucleoprotein units have been obtained from nuclei, but exceptions have been noted, and the issue is still cloudy. As more evidence is collected, the model will certainly change. We still cannot decide whether all hnRNA molecules contain one or a few structural genes that are preserved while a larger sequence is rapidly degraded, or whether some hnRNA molecules are direct precursors of cytoplasmic messages while a larger proportion of hnRNA has some other function. It seems that some, if not all, translatable cytoplasmic messages may pass through several discrete processing steps, each of which serves as a site for regulating information flow.

Which genes are transcribed into primary transcripts? Which of these transcripts receive poly A sequences? Which hnRNA sequences are degraded? Which processed RNA messages are transported to the cytoplasm? Each question marks a switch point that determines which mRNA molecules are released into the cytoplasm for translation into proteins. Another switch in the cytoplasm determines which messages associate with ribosomes and which are degraded. Various cells and tissues may differ as to which posttranscriptional controls are most important at different stages of growth and development. Each "decision" will determine the kinds and amounts of specific proteins made at any time.


## Translation or tape reading

In general RNA message tapes are read by ribosomes, the tape readers moving along the tape ratchet-fashion, exposing a codon at a time, while activated transfer RNAs read the codon and insert an amino acid into a growing polypeptide chain (Figure 3-4). The ribosome is a complex organelle made up of two basic subunits, which assemble into a functional ribosome after mRNA attaches to the smaller subunit. The ribosomes are used over and over again as tape readers as subunits undergo cyclic changes of association and dissociation with mRNAs and polypeptide chains. Ribosomal proteins act as enzymes

in different steps in the complicated machinery of polypeptide synthesis.

As we have seen, the time of translation differs in prokaryotes and eukaryotes. In the former, message transcription and translation occur simultaneously. In eukaryotes, transcription of a message always precedes its translation, sometimes by hours, days, or months. This lag between transcription and translation plays an important role in the regulation of information flow in eukaryote developing systems, as we will see.

During translation protein and nucleic acid macromolecules match specific nucleotide and amino acid sequences. The key recognition molecules are the transfer RNAs, the small adaptor molecules that carry specific amino acids for insertion into a growing polypeptide chain.

In 1958 Crick postulated such an adaptor molecule for the purpose of assembling amino acids into polypeptide sequences specified by the mRNA. He conceived of the adaptor as having two recognition sites, one for a specific amino acid and the other, a nucleotide triplet or anticodon, complementary to a codon sequence on a mRNA template specifying the position of the amino acid in a polypeptide chain.

Crick's prediction has been proven since populations of small transfer RNA molecules approximately 80 nucleotides in length have been obtained from many cells. They are coded on DNA templates in different parts of the genome and contain several methylated nucleotides which do not form complementary base pairs with other nucleotides. These nucleotides must arise by secondary methylation after the original nucleotide sequence is transcribed.

Transfer RNAs are cloverleaf molecules, twisted into a three-dimensional configuration by intramolecular complementary base pairing (AU and GC) (Figure 3-5). Each transfer RNA is specific for a single amino acid with an anticodon triplet in one of the cloverleaf loops. For each codon there is a transfer RNA, which means that there are at least 64 different transfer RNAs.

A first recognition step involves "charging" a transfer RNA with its specific amino acid by an enzyme, aminoacyl-tRNA synthetase. There is a different enzyme for each amino acid. These enzymes have two recognition sites, one combining with a specific activated amino acid and the other with the acid's corresponding tRNA. The enzyme catalyzes

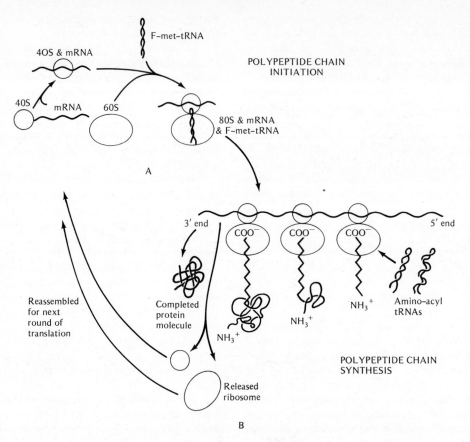

**Figure 3-4** *General scheme of translation or protein synthesis. (A) Initiation requires the association of messenger RNA with the smaller 40S ribosomal subunit. This, together with the larger 60S subunit and the initiating transfer RNA, F-methionine–tRNA, assembles into the complete functional ribosome. (B) Polypeptide chain synthesis on polyribosomes followed by termination and release of the completed polypeptide. The chain grows by stepwise additions of activated amino acids brought in by transfer RNAs.*

the association of amino acid with transfer RNA to form charged transfer RNA. The complex then diffuses to the ribosome, the tape reader, where it recognizes a codon triplet in the message and inserts the amino acid into the growing chain.

Three critical steps occur during tape reading: chain initiation, polypeptide elongation, and chain termination. Each process involves protein-RNA interactions and many enzyme-mediated steps within the ribosome. Initiation and termination are single events that call for recognition specificities of the "start here, get off here" type. "Start here" is correlated with a specific codon AUG, which codes for the amino acid methionine that is tacked on to the

beginning of each message. This sequence is the first to attach to the 40S ribosomal subunit which then couples with the larger 60S subunit to form the ribosome. Initiation requires at least three different ribosomal proteins organized as an initiation complex.

Elongation of polypeptides includes peptide-bond synthesis, movement of ribosomes along the tape, and simultaneous insertion and release of the transfer RNAs. These events also require elongation factors acting catalytically with such co-factors as GTP and magnesium ions. Finally, chain termination and release is dependent on several chain termination proteins which facilitate release of the completed polypeptide chain. The ribosomal subunit dissociates, and the mRNA is released for reuse or degradation.

Once proteins are formed, the flow of information from the gene is completed. Polypeptide chains proceed to fold up into more stable three-dimensional configurations; they may do this independently, in combination with other polypeptides forming complex proteins, or with lipids or polysaccharides to form lipo- or glycoprotein complexes. In any event, it is these final structural changes that confer functional activity and specificity on a protein. The specific catalytic or structural role assigned to the protein is expressed only after it acquires its three-dimensional configuration, a transformation that depends first on its amino acid sequence and second on its immediate environment (ions, pH, temperature, surfaces, membrane, and so on). The gene specifies only the amino acid sequence of a protein; additional information for its three-dimensional structure comes from its place in cell architecture.

## Regulation of information flow in prokaryotes

Earlier we mentioned that only a small proportion of the total information of a eukaryote cell is utilized at any time. This conclusion is based on a variety of observations (see Chapter 17), one of which is that proteins in different tissues are tissue-specific. Insulin is found only in pancreas cells and hemoglobin only in red blood cells, in spite of the fact that both types of cell contain identical genes for each of these proteins. In addition, the amount of different proteins varies in tissues. Some proteins are present in low amounts, as little as 10 molecules per cell, while others, such as ribosomal proteins, may accumulate up to 100,000 molecules per cell. Proteins may be absent at some developmental stages and appear only at certain times. Economical use of energy and metabolites dictates that only those proteins that are needed at any one stage should be synthesized. Information flow must be regulated to account for sequential changes in type and amount of protein populations in developing cells. This implies that at any given time only some genes are expressed while the majority are silenced. How is this accomplished?

Most information about gene regulatory networks comes from viral and microbial systems. In

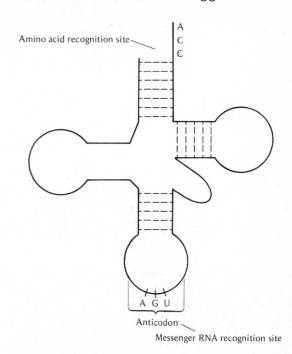

**Figure 3-5** *The cloverleaf model of transfer RNA molecules. Most transfer RNAs are composed of 70 to 80 nucleotides held into the cloverleaf configuration by hydrogen bonds, shown as dotted lines. These regions are believed to be double-stranded helical RNA. The anticodon site contains a series of three nucleotides which match the complementary codon on the messenger RNA. The amino acid attachment site occupies the opposite end of the molecule.*

viral-infected bacterial cells, for example, a sequence of synthesis and assembly of more than 100 different viral proteins is completed within 20 minutes (Chapter 7). These developmental programs are easily manipulated by gene mutations and are accessible to biochemical analysis. On the basis of such studies, Jacob and Monod postulated in 1961 the now classic operon model to explain how *E. coli* structural genes are regulated. The model is also the best for information retrieval in eukaryote gene action systems.

Most prokaryote genomes are small, containing 100 to 1000 (at most) genes, usually organized into functional clusters of genes controlling a metabolic pathway (operons). An operon may contain several structural genes (cistrons), in tandem, coding for related enzymes in a common metabolic pathway.

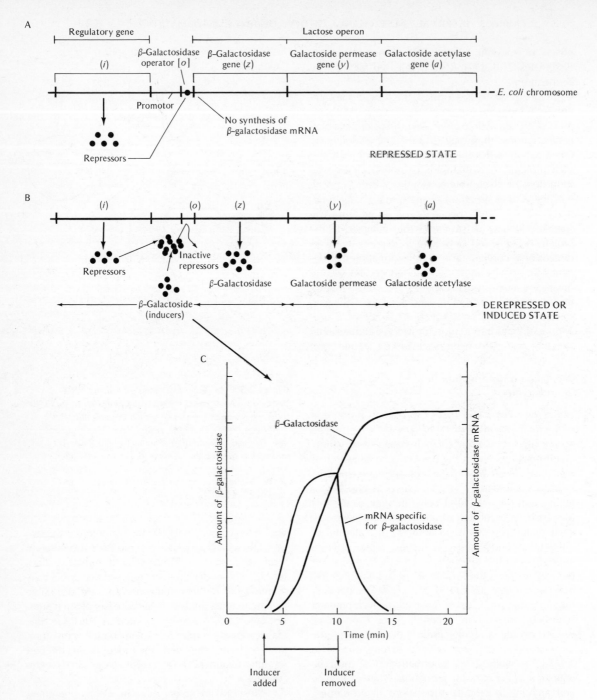

**Figure 3-6** *The lactose operon in E. coli and its regulation. (A) Genes of the operon are drawn to scale and relative position in the repressed state. The exact distance between regulatory gene and lactose operon is uncertain. An active regulatory gene codes for synthesis of repressor proteins which combine with operator site and prevent transcription of the operon. (B) The depressed state. The addition of inducer inactivates the repressor protein by allosteric interaction, releases repressor from the operator site, and transcription and translation of all protein of the operon follow. (C) The time course of appearance of mRNA and β-galactosidase after induction.*

These genes are adjacent to controlling genes, operators and promotors, which govern the expression of all genes in the complex. The operon behaves as a coordinated physiological unit; all genes are transcribed at the same time, producing messages that code for complete metabolic reaction sequences.

The classic example is the lactose operon consisting of four separate cistrons in tandem which direct lactose metabolism (Figure 3-6). The $z$ gene, the structural gene for the enzyme $\beta$-galactosidase, which metabolizes galactose; the $y$ gene is a structural gene for a $\beta$-galactoside permease, an enzyme which actively transports substrate molecules into the cell and regulates its intracellular concentration; finally, there is the $AC$ gene for a $\beta$-galactoside transacetylase enzyme which transfers an acetyl group from an acetyl donor to $\beta$-galactoside sugar. At the beginning of the cluster is the $o$, or operator, gene which directs the expression of the multigene complex, and the promotor, $p$, the point of attachment of the transcribing RNA polymerase. Nearby and separate from the operon is the $i$, or regulatory, gene, another controlling element, which, as we will see, regulates expression of the entire gene cluster.

The expression of this and other operons is regulated by small-molecule inducers entering the cell from the environment. These molecules are usually nutrient substrates involved in both catabolic and anabolic metabolism. A sugar — $\beta$-galactose or even one of its structural analogs (molecules that resemble it structurally but need not be metabolites) —when added to *E. coli* previously growing in glucose, will switch on the entire set of genes in the lac operon, and the battery of galactose enzymes is synthesized de novo. These are synthesized as long as inducer remains in the medium; however, as soon as it disappears, the genes are shut off, enzyme synthesis stops, and the level of enzyme activity declines. Prior to enzyme synthesis we can detect the synthesis of $\beta$-galactosidase mRNA, and this declines abruptly when inducer is removed. The inducer molecule activates a silent set of genes which transcribe RNA messages only so long as inducer is present. An entire metabolic pathway for handling a new substrate is turned on as an integrated unit only when it is needed. Since most bacteria are saprophytic, living in a fluctuating nutrient environment, they must have a mechanism for choosing from their enzymatic repertoire only

those enzyme systems essential for survival at the moment while conserving all others. This metabolic economy is achieved by control systems which select and coordinately switch on only those groups of enzymes needed to metabolize the nutrients available.

The principal regulatory gene in this circuitry is the $i$ gene. Mutations of the $i$ gene affect appearance of the $\beta$-galactosidase enzymes without affecting their structure or activity. The gene is regulatory and controls inducibility of all genes in the operon, exerting its effect without physical association with the structural gene. Instead, it codes for a protein product, a repressor protein, which diffuses through the cytoplasm and combines specifically with the operon gene site to regulate its transcription.

Thus inducible enzymes usually are not synthesized unless inducer is present; their operons are always in a state of repression by a repressor coded by a regulatory gene. The addition of inducer relieves this repressed state to allow the cistrons to act; that is, the genes are *derepressed*.

The key controlling element is the repressor protein encoded in the $i$ gene. The repressor recognizes and interacts directly with the nucleotide sequences that make up the operator locus, inhibiting transcription of all cistrons. An RNA polymerase may attach to the promotor site, but reading stops at the operator locus if repressor protein is attached.

Repressor protein exists in two functional states in equilibrium, active and inactive. The active configuration combines with the operator and represses it. As inducer enters the cell, it combines with repressor, alters its structural configuration, and transforms it to the inactive state. Repressor is thereby released from the operator gene, and the operon is available for transcription. Repressors therefore are examples of *allosteric* proteins which exhibit different functional states depending on their three-dimensional configuration. Combination of an allosteric protein with small effector molecules such as inducers shifts the protein from an active to an inactive state. Each allosteric repressor has two recognition sites, an operator-gene combining site and a site for attachment of the small effector molecule. Ability of repressor to recognize the nucleotide sequence of an operator gene depends on its three-dimensional structure, and this in turn is determined by the nature of the molecule at the effector combining site.

The repressor protein regulates by combining with nucleotide sequences in DNA, changing template structure enough to interfere with transcription. The isolated protein has high specific binding affinities for lac operator DNA. Mutations in the operator locus alter the binding of repressor to the operator gene.

The repressor is an acidic protein, a tetramer composed of four identical subunits with a molecular weight of approximately 150,000 to 200,000. Its two binding sites can be altered independently by different mutations in the *i* gene. Since the amino acid sequence of parts of the molecule has been determined, we know that the amino-terminal portion of the molecule binds to lac operator DNA, a region of known sequence rich in AT base pairs. If parts of this amino acid sequence are lacking as a result of deletion mutations, the repressor cannot bind to the operator DNA although it retains its tetrameric structure and continues to bind to inducer.

The molecule shows a high binding affinity for the operator region but also binds to artificial polynucleotides with high AT base pairs. If BUDR (5-bromodeoxyuridine) is substituted for thymidine in the lac operator DNA, the binding of repressor to operator increases tenfold. This suggests that BUDR inhibits DNA expression in cell differentiation by increasing the binding of its regulatory molecules (see Chapter 17). Repressor proteins are not alone in recognizing stretches of DNA; the enzymes DNA and RNA polymerase also possess precise recognition affinity with DNA templates. They identify specific promotor or initiator sites to start transcription or replication. The recognition attachment of RNA polymerase to the lac operon promotor site requires another small molecule, cyclic AMP. This ubiquitous molecule is involved in a wide range of eukaryotic regulatory phenomena, from cell motility to hormone activation, and it also plays an important role at the site of gene transcription in *E. coli*. Since it may also function at the template site in eukaryotes, let us examine this molecule and its behavior in *E. coli* in more detail.

Cyclic AMP is a nucleotide in which the phosphate group of the molecule forms a ring with the carbon atoms to which it is attached (Figure 3-7). It was isolated in 1958 by E. Sutherland and is found in virtually all cells. It is synthesized from ATP by an enzyme, adenylcyclase, bound to the plasma

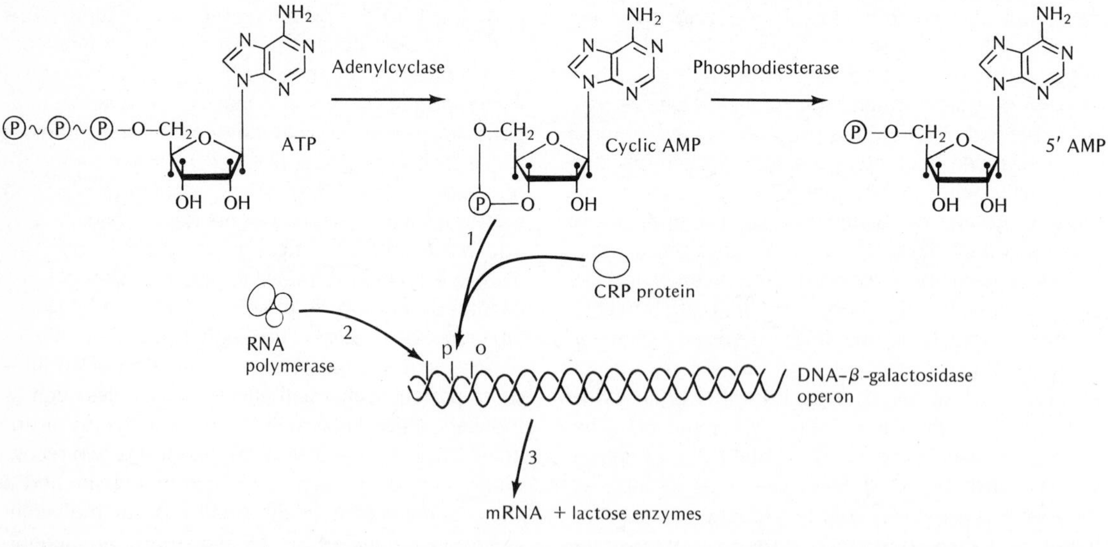

**Figure 3-7** *Cyclic AMP and its role in the regulation of the lactose operon. A membrane-bound adenylcyclase converts ATP into cyclic AMP. The levels of cyclic AMP are, in turn, regulated by phosphodiesterase, which converts cyclic AMP into 5' AMP in the cytoplasm. Cyclic AMP facilitates expression of the lactose operon by combining with the CRP protein. This complex then changes the configuration of the promotor (p) so that RNA polymerase may attach and begin transcribing the structural genes of the operon. The label ''o'' refers to the operator site.*

membrane. Localization of the enzyme at the cell surface accounts for this system's responsiveness to hormone signals. Many hormone effects on eukaryote cells are mediated by the adenylcyclase–cyclic AMP system. Cyclic AMP is an effective regulatory substance, and its concentration in the cell is monitored by another enzyme, phosphodiesterase, which degrades cyclic AMP to AMP. Thus the relative activities of two enzymes control the equilibrium levels of cyclic AMP in a cell.

Cyclic AMP participates in numerous metabolic reactions in cells, largely because of its ability to activate various enzymes. Here we will discuss only its role in the regulation of gene transcription in *E. coli*. *E. coli* cells grown in glucose do not have the enzymatic machinery for metabolizing other sugars such as lactose. It was found that addition of cyclic AMP activated the expression of the β-galactosidase gene. Lactose could now be metabolized under these conditions. Furthermore, in a complex cell-free preparation of *E. coli* in which both transcription and translation take place, the addition of cyclic AMP to DNA templates enriched in β-galactosidase genes stimulated the transcription of specific RNA messages for β-galactosidase which in turn were translated into the enzyme.

The precise mode of operation of cyclic AMP in control of transcription of the lac gene was determined from a study of mutants which produced large amounts of cyclic AMP but still could not synthesize the lactose enzyme. Such mutants lacked a specific binding protein that combined with cyclic AMP (CRP, or cyclic receptor protein). Accordingly, the following model emerged. Cyclic AMP first combines with CRP. This complex then combines specifically with the promotor site of the operon, and only then can the copying enzyme, RNA polymerase, attach and begin to read the template. Presumably configurational changes induced by cyclic AMP at the promotor site are essential to the attachment of RNA polymerase and subsequent transcription.

Other acidic proteins terminate transcription at specific sites. The rho factor in *E. coli* is a polymeric protein of six subunits organized in a single plane with an empty core. This protein binds by slipping around DNA strands so that the helix occupies the central core and thereby stops transcription of nascent RNA chains. At high rho concentrations, for example, only the first structural gene of the galactose operon is transcribed. Still other proteins act to destabilize the DNA double helix to stimulate replication. These DNA unwinding proteins also must recognize specific sequences to act. We see that several acidic proteins do regulate the prokaryote genome at the template level, repressing, activating, and terminating transcription.

The repressor model proposed by Jacob and Monod emphasizes regulation at the level of transcription. The selection of genes to be transcribed is made directly on the DNA template. The model predicts that only during induction of β-galactosidase should the specific messenger RNA molecules be detected. A DNA-RNA hybridization study demonstrated that induced cells contain a greater quantity of RNA molecules that hybridize with *E. coli* DNA than do noninduced cells, confirming the Jacob-Monod prediction. Repressors regulate genome expression by controlling transcription.

Not all prokaryote genomes are organized as operons; many function as single genes, which suggests that other kinds of regulation may be operating in bacteria besides coordinate operon control. For example, genes for arginine synthesis in *E. coli* are not closely linked in a single operon but are separated into five distinct groups, each regulated by the substrate arginine. This means simultaneous control of five different operator genes by the same controlling molecule, a situation similar to eukaryote cells. Still other cases have been found in which more than one regulatory gene controls a single enzymatic behavior. These more complicated gene-action systems begin to resemble gene expression in eukaryotes. Although the operon model can account for simple changes in enzyme synthesis in response to environmental signals in bacteria, can it also explain complicated multigene networks that code for the diversity of developmental programs in eukaryotes?

## Operon model in eukaryotes

Patterns of gene expression in eukaryotes suggest that some features of the model have changed during evolution. Operons are less common in eukaryotes. Even in yeasts and molds, which are closely related to bacteria, the operon-repressor model for coordinate gene control is found in only a few cases. In *Drosophila* and maize, whose genetics are best understood, few gene clusters have been detected. Generally eukaryote genes concerned with the same phenotypic trait are scattered through many chromosomes. If operons are prevalent in eukaryotes, they would be more difficult to detect, since most eukaryotes are diploid and gene expression is often masked by dosage and dominance effects.

Some eukaryote genes, however, do behave as if they are part of an operon; transposing them in the genome alters their expression. The position-effect phenomenon (Chapter 2) is an example of a phenotypic change in response to relocation of a gene in a chromosome set. For example, the *Bar* mutant, which decreases the number of ommatidia in the eye, exhibits a position effect. When two doses of *Bar* are present in separate chromosomes, as in *B/B*, the mean number of ommatidia is 68, but if these are on the same chromosome as in *+/BB*, the mean number is 45; the effect is more severe. Without any change in dosage, the phenotype is altered simply by shifting a gene from one chromosome to another, as if the rearrangement separates the gene from a controlling element (operator?).

### Transcriptive regulation in eukaryotes

Differences in the life style of eukaryote and prokaryote cells imply differences in gene regulation. Eukaryote cells have much longer generation times, 24 to 48 hours compared with the 20 minutes of *E. coli*. Some eukaryote cells do not divide and persist as postmitotic cells for years. Accordingly, their proteins have time to degrade and turn over in equilibrium with amino acids. Protein turnover is far more important in regulating protein content in eukaryotes than it is in prokaryotes.

Eukaryote genomes exhibit patterns of information flow not found in simple prokaryotes. Eukaryote genomes are large and heterogeneous with different families of reiterated sequences interspersed among single-copy genes. The role of reiterated sequences is still not understood, but specific gene amplification, as in ribosomal genes, may require new modes of regulation. Some repetitive sequences may be regulatory genes that control many gene sets simultaneously throughout the genome. The appearance of heterochromatin in eukaryote chromosomes suggests that large blocks of genes are shut down and made inaccessible to regulation. Such gross chromosomal mechanisms of control are not found in prokaryote genomes. Delayed utilization of message is the rule in eukaryotes; genes transcribed at one time are translated at later developmental stages.

As we have seen, spatial separation of nucleus from the cytoplasm and long-lived messages introduces additional sites for posttranscriptional regulation such as message processing, transport, storage, and utilization. Hence developmental versatility of eukaryote systems may stem from a diversity of gross (chromosomal) and fine (genetic) mechanisms of regulation that have evolved in large, complex genomes.

In spite of these differences in genome organization and life styles, we assume, as a first approximation, that Jacob-Monod types of gene regulation are also operative within eukaryote genomes. Gross chromosomal mechanisms may shut down large blocks of genes in cells, leaving only a small proportion of the total genetic information available for expression. The small amount of genetic information that is read is probably modulated by fine-tuning regulation of the Jacob-Monod type.

Since transcription is an important site of regulation in eukaryotes, we would expect RNA message transcripts to be qualitatively different in various cell types of the same organism. The populations of RNA molecules extracted from cells or from nuclei should show tissue specificities; the RNA transcripts in liver should differ from those in kidney. To test this prediction, we carry out DNA-RNA hybridization studies of RNAs extracted from different tissues and compare their sequence homologies (Figure 3-8). This can be done under conditions that detect repetitive as well as nonrepetitive DNA sequences. In these experiments labeled RNA is extracted from a tissue and hybridized with DNA obtained from sperm. An example of a hybridization analysis for repetitive sequences in mouse tissues is shown. Although tissues share some sequences, they do exhibit distinct differences; tissues from the same organism contain qualitatively different repetitive and nonrepetitive message transcripts.

Related hybridization experiments designed to assess the percentage of the single-copy component of eukaryote genomes that are transcribed reveal another important feature of eukaryote genome regulation. Of the total single-copy sequences in various cells approximately 1 to 28 percent are transcribed into RNA messages (Table 3-1). If only one DNA strand is actually copied (as is generally assumed), then this means 2 to 46 percent of the working genome is transcribed. For *Dictyostelium*, a slime mold with a relatively small genome, this represents a sizeable proportion transcribed through its life cycle (see Chapter 9). But for most tissues only a relatively small proportion of the genome is in fact available for transcription; the remainder is inaccessible. But even though a small proportion is transcribed, this may represent thousands of different genes. Some tissues (for example, mouse brain) display an enormous complexity of informational molecules, which means that an equally large number of discrete regulatory events at the template level must have taken place. Whatever circuits operate to regulate eukaryote information flow, we

**TABLE 3-1  The Extent of Transcription in Eukaryote Cells**

| Species and Tissue | RNA Fraction | Percent Transcribed[a] |
|---|---|---|
| *Dictyostelium discoideum* | | |
|   Amoebae | Total | 15 |
|   Whole | Total | 28 |
| *Drosophila melanogaster* | | |
|   Cultured cells | Total | 15 |
|   Embryos | Total | 15 |
|   Adults | Total | 10 |
| *Xenopus laevis* | | |
|   Oocytes | Total | 0.6 |
| Chicken | | |
|   Red cells | Total | 2 |
|   Liver | Total | 15 |
| Mouse | | |
|   Brain | Nuclear | 9–13 |
| | Total | 8–11 |
|   Liver | Nuclear | 3–5 |
| | Total | 2–4 |
|   Kidney | Nuclear | 3–5 |
|   Spleen | Total | 2 |
|   Embryo | Total | 8–12 |
|   Hepatoma | Total | 2 |
| Rat | | |
|   Ascites tumor | Nuclear | 4 |
| Cow | | |
|   Liver | Total | 5 |
|   Brain | Total | 2 |

[a] Based on direct single-strand measured values. Actual transcription level obtained by multiplying by 2, assuming transcription is asymmetric.

Source: Combined data from E. H. Davidson and R. J. Britten, *Quart. Rev. Biol.* 48:565–613, 1973; and B. J. McCarthy and M. Janowski, *Macromolecules Regulating Growth and Development,* ed. E. D. Hay, T. J. King, and J. Papaconstantinou. New York: Academic Press, 1974, pp. 201–216.

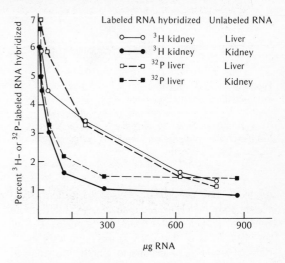

| Labeled RNA hybridized | Unlabeled RNA |
|---|---|
| $^3$H kidney | Liver |
| $^3$H kidney | Kidney |
| $^{32}$P liver | Liver |
| $^{32}$P liver | Kidney |

**Figure 3-8** *DNA-RNA competition assay to compare RNA transcripts (repetitive DNA) from different mouse tissues. A mixture of $^3$H-labeled kidney RNA and $^{32}$P-labeled liver DNA was incubated with mouse embryo DNA in the presence of varying amounts of unlabeled RNA from kidney or liver. The percentage of $^3$H- and $^{32}$P-labeled RNA hybridized is plotted against the amount of unlabeled RNA added. Both kidney and liver share some RNA since there is evidence of competition, but each also produces transcripts that are not found in the other tissue. (From B. H. McCarthy and B. H. Hoyer, Proc. Nat. Acad. Sci. U.S. 52:915–922, 1964.)*

conclude that control at the level of transcription occurs for both repetitive and nonrepetitive DNA sequences in most eukaryote cells.

Since large portions of the eukaryote genome are not transcribed, can we assume that inactivation is associated with the large amount of chromosomal proteins combined with DNA? Does it follow that eukaryote genomes consist of masked and unmasked regions?

One attempt to answer this question is to isolate nuclei from tissues and then extract chromatin, a complex mixture of DNA, histones, acid proteins, and possibly some RNA. The chromatin preparation is then treated with an enzyme, a nuclease that digests naked DNA. Results show that only 50 percent of the total DNA in the chromatin preparation can be digested while the remainder seems to be protected by protein. It can be shown that virtually all the protein in chromatin is bound to only 50 percent of the DNA; the rest of the DNA seems to be

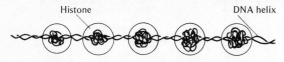

**Figure 3-9** *Model of chromatin strand based on differential enzyme susceptibility. Approximately 50 percent of the DNA is accessible, shown as strands between the beads on the necklace. The beads contain histone-DNA complexes with the DNA tightly compacted within a histone coat.*

relatively free of protein. The DNA-protein complexes may be organized into a series of discrete spherical particles, like beads on a necklace (Figure 3-9). Most of the DNA in the particle is small, only 110 to 120 base pairs in length, too small to code for a protein. The particles seem to be interspersed with short spacer naked DNA of only 50 base pairs in length. Roughly 85 to 95 percent of total cell DNA seems to be organized in this manner, leaving 10 to 15 percent of the DNA unaccounted for. It is this nonprotein-bound DNA that may be active in gene expression, and the proportion is in agreement with the proportion of DNA that is actively transcribed in most eukaryote cells (see Table 3-1).

It appears that various histones, bound to each other, make up the protein in the particles. Is packaging of DNA into nucleohistone particles a mechanism for silencing most of the eukaryote genome? A study of genome masking is based on the in vitro transcribing ability of chromatin preparations as compared with that of naked DNA. In such a cell-free system containing DNA, RNA polymerase, and radioactive ribonucleotide precursors, complementary radioactive RNA copies are synthesized on any accessible DNA template. All masked sites (that is, those combined with proteins) are inaccessible to the polymerizing enzyme. The sequence complementarity of the RNA synthesized in vitro is determined by hybridization with denatured naked DNA used as a template. This tells us what proportion of the redundant genome has been transcribed and what kinds of RNA transcripts are made.

For many tissue chromatins tested in this system, only a small percentage of the total redundant DNA is transcribed. RNA synthesized on mouse-liver chromatin hybridizes with a little more than 3 percent of the total DNA, while RNA synthesized on mouse-kidney chromatin hybridizes with about 7 percent of the total mouse genome. More than 90 percent of the DNA in chromatin is inaccessible for transcription, probably because it is complexed with protein. More to the point, the RNAs made in each case are tissue-specific; different portions of the genome are transcribed by each chromatin preparation as seen in DNA-RNA competition hybridization experiments. In addition, the RNAs made in vitro on isolated chromatins share many sequences with RNAs made by these same tissues in vivo. This suggests that isolated chromatin is organized in vitro not very differently from its state in intact cell nuclei; that is, similar informational templates are available for transcription.

Masking is accomplished by chromatin-associated proteins. If chromatin is divested of all histones and acid proteins, the remaining naked DNA exhibits an enhanced in vitro transcriptive ability; virtually the entire genome is transcribed in this case (Table 3-2). When chromatin is reconstituted from naked DNA by the addition of histones and/or acid proteins, the strands are compacted, and transcription is once again inhibited. Thus we can determine which protein component is more effective in masking DNA transcription. If both are added to naked DNA, chromatin with the original transcribing capacity is reconstituted; the same species of RNA are synthesized. When DNA is recombined only with

**TABLE 3-2  Role of Proteins in Masking Transcription of Calf Thymus Chromatin**

| Template | Percent of Genome Transcribed in Vitro |
|---|---|
| DNA | 35–40 |
| Deproteinized chromatin | 35 |
| Whole chromatin | 5 |
| Chromatin minus histones | 10–16 |
| DNA plus histones | 0 |
| Dehistonized chromatin plus histones | 4 |
| DNA plus total chromosomal proteins | 5 |
| DNA plus histones plus nonhistone proteins | 5 |
| DNA plus nonhistone protein | 14 |

Source: Modified from J. Paul, *Current Topics in Developmental Biology*, ed. A. A. Moscona and A. Monroy, vol. 5. New York: Academic Press, 1970, pp. 317–352.

histones, an abnormal chromatin is made. DNA is virtually precipitated, and its transcriptive properties are reduced or abolished. The small amount of RNA made in vitro is not related to natural RNA. When nonhistone acid proteins are added, we obtain an intermediate type of chromatin, one that is between naked DNA and native chromatin. While certain sequences of DNA are not transcribed, more of the DNA seems to be accessible to RNA polymerase than in native chromatin. In some cases chromatin denuded of histones but with some acid proteins exhibits template activity that is essentially identical to that of DNA alone.

These experiments suggest that histones have a nonspecific masking effect on DNA transcription while acidic proteins retain some masking specificity. This hypothesis was tested by partially reconstituting rabbit thymus DNA with its own histone fraction. To one portion of this partially reconstituted chromatin nonhistone proteins from thymus were added and to another portion, nonhistone proteins from bone marrow; then both reconstituted chromatins were tested for in vitro transcription. Chromatin with thymus proteins transcribed RNAs resembling those directly extractable from thymus tissues, while chromatin with bone marrow proteins synthesized RNAs similar to those found in bone marrow. Transcription specificity of reconstituted chromatin seems to reside in the nonhistone protein fraction and not in the histones. Some speculate that histones combine with DNA and cause supercoiling or compaction, thereby making RNA-polymerase binding sites inaccessible. Acidic proteins could protect DNA double helices from histones, thereby making the template more available to the enzyme for transcription.

Barrett et al. have demonstrated specific regulation of globin genes by nonhistone proteins in chick erythrocytes which synthesize hemoglobin. They isolated chromatin from chick erythrocytes, removed the proteins from DNA, and made a reconstituted chromatin by recombining DNA only with histones. In addition, pure RNA messages for globin were isolated from these cells and used in vitro to make DNA copies (cDNA) complementary to globin messenger RNA. This was carried out with viral RNA transcriptases or polymerases that make DNA on RNA templates, forming double-stranded DNA copies of RNA messages (see Chapter 21). They showed that cDNA specifically coded for globin since it hybridized only with globin RNA transcripts. These DNA copies were, in a sense, pure globin genes.

With this system the effect of nonhistone proteins on in vitro transcription by reconstituted chromatin was tested. When nonhistone proteins from chick reticulocytes were added to reconstituted chromatin, RNA transcripts specific for globin genes were transcribed; the newly synthesized RNA hybridized readily with cDNA. The total and fractional yields of globin RNA from reconstituted chromatin were similar to yields from native chromatin. But when liver nonhistone protein was used instead, no globin RNA messages were transcribed. Only erythrocyte nonhistone proteins confer transcription specificity with respect to globin synthesis. These acidic proteins may form tight bonds with associated histones and remove them to expose naked DNA. They are negatively charged proteins (probably phosphorylated) and may have greater affinity for positively charged histones than histones have for DNA.

We still do not know how DNA, histone, and acid proteins interact in intact nuclei, nor do we know much about their function. But whatever the mechanism of genome masking, these studies suggest that most informational templates are shut down by proteins, leaving only small segments available for transcription. The regulation of these segments seems to be under transcriptive control very much like prokaryote operons.

### A model of eukaryote genome regulation

Many models of gene regulation in eukaryotes have been proposed. All try to explain the role of repetitive sequences interspersed in the genome as well as the nature of hnRNA and its high turnover in the nucleus. The models are essentially more complicated versions of the Jacob-Monod model for operon regulation in prokaryotes.

Georgiev proposed a relatively simple model of the eukaryote operon consisting of two regions, a proximal regulatory gene segment and a distal structural gene segment (Figure 3-10). The regulatory end contains a promotor site for RNA polymerase attachment followed by sequences of many different acceptor sites, some copied in tandem. Acceptor sites make up the redundant portions of the genome since many different operons share comparable acceptor-

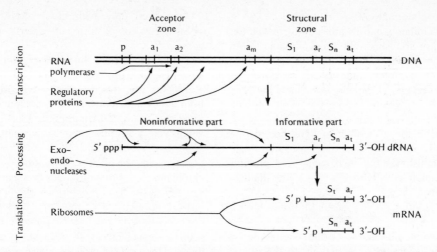

**Figure 3-10**  *Georgiev model of eukaryote genome transcription and regulation. The label p is the promotor site; $a_1$–$a_m$ are various acceptor sites responsive to a variety of different signals; $a_r$ and $a_t$ are other regulatory components; $S_1$ . . . $S_n$ represent a series of structural genes. The $a_1$–$a_m$ regions are part of the redundant or repetitive component of the genome, presumably interspersed among many different structural genes. Large polycistronic messenger transcripts are synthesized, but during processing, the regulatory, noninformative part of the molecule is degraded by endonucleases, leaving the small informative part for transfer to the cytoplasm where it is translated on polyribosomes. (From G. P. Georgiev,* Current Topics in Developmental Biology, *ed. A. A. Moscona and A. Monroy, vol. 7. New York: Academic Press, 1972, pp. 1–60.)*

site sequences. Structural genes in the operon make up the single-copy component. The regulatory portion of the genome is assigned to redundant genes, the intermediate class present at 100 to 100,000 copies. For any operon there will be sets of different acceptor sites with each acceptor responsive to a different regulatory signal: a histone repressor protein, a nonhistone activator protein, a hormone, an activator RNA, or an inducing agent. Hence many different operons distributed throughout the genome are accessible to and regulated by identical regulatory genes.

Transcription of an operon begins at the acceptor end; the RNA polymerase transcribes the long regulatory portion of the genome before structural genes are transcribed. A large polycistronic nuclear RNA molecule results, composed of nontranslatable sequences in the acceptor region and a translatable portion at the distal end. Transcription of the entire operon depends on the combination of repressor, activator, and protein associations at acceptor sites. If a histone, or repressor protein, is attached at any site, then RNA polymerase stops transcribing at that point. If, however, the sites are derepressed, then the entire message is read and released.

Only the structural gene portion of this polycistronic message will be translated in the cytoplasm; the large regulatory piece must be removed before it leaves the nucleus. This processing is completed in the nucleus by enzymes that degrade the regulatory segment. The message is simultaneously prepared for transport by addition of poly A sequence and is transferred to the cytoplasm through the nuclear membrane.

This accounts for the high turnover of large nuclear RNA within the nucleus and the absence of most reiterated sequences in the cytoplasm inasmuch as they are degraded during processing.

The informative part of the transcribed message combines with ribosomes in the cytoplasm and forms a functional polysome. The process is now subject to a barrage of translational controls.

Since acceptor loci are multiple and control a variety of structural genes, we see how many different genes are controlled by the same signal. Herein may lie the explanation for many different tissues responding to the same hormone signal, as is true for thyroxin control of amphibian metamorphosis (see Chapter 19).

Gene expression in eukaryotes seems to be regulated according to a Jacob-Monod scheme; the only difference is that the reiterated regulatory segment of an operon is transcribed along with the structural component. The model predicts accurately that redundant regulatory portions should be interspersed among single-copy components. We have already seen that some moderately redundant DNA is not organized in tandem clusters but seems to be distributed throughout the eukaryote genome among larger single-copy segments. We still have to learn more about nuclear RNAs (hnRNA) within eukaryote cells and, more specifically, about the nature of redundant gene sequences before any model is accepted.

## INFORMATION FLOW BETWEEN NUCLEAR AND CYTOPLASMIC GENOMES

The second level of information flow involves exchanges between nuclear and extrachromosomal genomes. We have shown that genetic systems in mitochondria and chloroplasts are virtually autonomous; they contain DNA templates, ribosomes, and polymerases and probably obey the same rules for macromolecular information flow as described in the previous section. We would expect information in DNA templates to be copied into RNA messages and these translated into proteins, all within the organelle. These genomes probably code for organelle-specific ribosomal and transfer RNAs. In addition, they contain structural genes coding for polymerases, some membrane proteins of the organelle, and organelle-specific (ribosomal) proteins.

Information is exchanged between nuclear and cytoplasmic gene systems. Nuclear genes may code for proteins in mitochondria and chloroplasts or control the stability, expression, and replicative behavior of plasmagenes. Plasmagenes, on the other hand, may alter the expression of nuclear genes and in several cases will cooperate with nuclear genes in the development of a specific cell phenotype.

We know that nuclear genes control proteins and enzyme systems normally found in cytoplasmic organelles because several nuclear gene mutations will mimic the behavior of cytoplasmic mutations. In *Neurospora,* for instance, nuclear mutations have effects identical to the slow-growing poky mutants of extrachromosomal origin. These chromosomal mutants are also deficient in cytochromes b and a, which suggests that both mutant types affect similar

organelle components. Though the end result is the same, they do not share identical routes of action.

Nuclear mutants may be combined with extrachromosomal mutants to study their interactions. An example is the combination of the cytoplasmic poky mutant of *Neurospora* with a nuclear mutant which also results in deficient mitochondria. Poky mitochondria lack cytochromes b and a but have an excess of cytochrome c. Mitochondria of the nuclear mutant lack cytochrome a but have a normal amount of cytochrome b and a slight excess of cytochrome c. In combinations, the result is not additive; that is, the cytochrome contribution of the nuclear mutant is not added to the products of the mitochondrial mutant. Rather, the interaction inhibits the expression of either or both when both genomes occupy the same cell. Cytochrome b is absent in the joint mutant, a result consistent with the hypothesis that without proper mitochondrial proteins controlled by the cytoplasmic "poky" determinant, a nuclear product (cytochrome b) is lost.

According to one hypothesis, mitochondrial genes code for mitochondrial membrane proteins that act as a scaffolding for the insertion and support of enzymes coded by the nuclear genome. The hypothesis predicts that membrane proteins isolated from mutant mitochondria would differ from wild type proteins, particularly in their ability to bind a nuclear-controlled enzyme such as malate dehydrogenase. Such experiments are difficult to carry out because uncontaminated membrane preparations are not easily isolated, nor is it simple to remove membrane proteins. Conformational and functional properties of an enzyme may be greatly altered by attachment to mutant mitochondria. In this way changes in a mitochondrial membrane, controlled by the mitochondrial genome, would cause malfunction of a normal nuclear product. Alternatively a mutant enzyme coded by the nuclear genome might itself be so altered as to bind irregularly to normal mitochondrial membranes and thereby function abnormally in vivo. Thus interaction between a nuclear gene product and a plasmagene-mediated structural protein could alter the functional expression of an enzyme system. The end result of mutation in either of the genetic systems may be total absence of nuclear products in the organelle.

The importance of cooperation is highlighted in certain plant systems where nuclear and cytoplasmic genomes may be mixed to study their interactions. Races of the plant *Oenothera* are characterized by possession of different types of chloroplasts, which by genetic crosses can be combined with different nuclear genomes. Each combination produces a different effect on chloroplast development. Certain plastids are green and functionally normal in most chromosomal backgrounds. In some crosses, however, these same plastids fail to develop properly. For example, plastids from the *Hookeri* strain which are normal when interacting with *Hookeri* chromosomes are deleteriously affected and lose chlorophyll when in a hybrid cell containing the *Hookeri* and the *Parviflora* genomes. Other hybrid chromosome combinations are equally deleterious to these plastids, but plastids from other strains are unaffected. In the presence of its normal genome a plastid can function at maximum level. Combined with a noncooperative genome, chloroplasts do not synthesize chlorophyll nor do they replicate. No mutation in the genome or plasmagene is responsible for the abnormal phenotype, but abnormalities arise because of incompatibilities between an organelle genome and a foreign nuclear genome; the two cannot cooperate in supporting plastid development.

In some instances the nucleus dominates the expression of extrachromosomal systems. A nuclear suppressor mutation may transform a mutant phenotype controlled by cytoplasmic plasmagenes to the wild or normal type. Only mutant plasmagene expression is modified by suppressor genes. For example, a nuclear suppressor gene restores a slow-growing *Neurospora* mutant to normal growth; its mitochondria acquire a complete cytochrome system. During its association with the nuclear suppressor gene the mutant plasmagene does not function, though it is stable and is transmitted to progeny. In subsequent segregations the plasmagene may escape from nuclear control to resume its original mutant behavior. In some way a nuclear gene compensates for the action of a mutant plasmagene by contributing essential components for normal mitochondrial function.

Some nuclear genes control the stability of extrachromosomal systems rather than their expression. An example is the *lojap* gene in maize, a recessive gene producing white-striped plants when homozygous (Figure 3-11). When present, this nu-

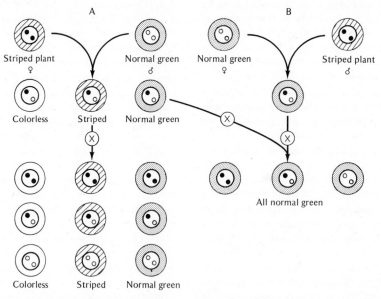

**Figure 3-11** *Cytoplasmic inheritance of the lojap striped condition in maize. Maize plants homozygous for the lojap allele (black circles) exhibit the striped or variegated plastid condition. Once the plastids have been altered by the gene, however, their subsequent inheritance is maternal. Thus a cross of a striped female plant with a normal male will give rise to three different phenotypes, depending on the sector of the plant used for pollination: white sectors will give rise to white plants, variegated sectors will yield striped, while normal sectors produce normal plants, irrespective of the genotype. An F₂ mating of striped plants is shown giving all phenotypes. The variegated condition persists in the absence of the lojap allele. (From R. Sager,* Cytoplasmic Genes and Organelles. *New York: Academic Press, 1972.)*

**Figure 3-12** *Nuclear control over replication of plasmagenes in the red variant of the mold* Aspergillus nidulans. *(A) The red variant with wild type nuclear genes exhibits persistent segregation of normal and red variants because cytoplasmic determinants randomly segregate within uninucleate asexual spores. The rare normal variant continues to produce a majority of normal phenotypes and few red variants. The red variant segregates normal and red variants indefinitely in the proportion shown. (B) Two nuclear mutations result in different segregation patterns. One mutation ($m_1$) favors the normal plasmagene and a stable normal phenotype arises via asexual reproduction. The other ($m_2$) favors replication of the red cytoplasmic determinant at the expense of the normal and a stable red nonsegregating phenotype emerges.*

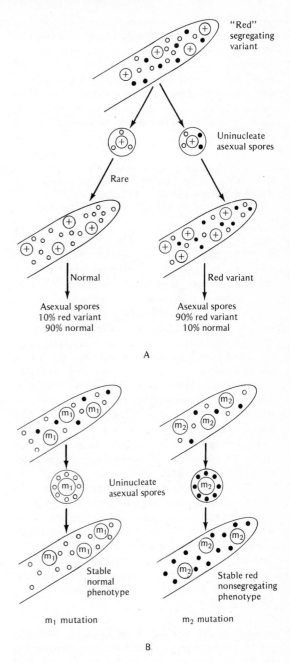

clear gene induces irreversible changes in plastids that are retained even in the absence of *Iojap* genes. Chloroplasts mutate in the presence of the homozygous *Iojap* allele and produce the white, chlorophyll-free phenotype. These mutant chloroplasts are then transmitted to subsequent generations in a typical nonchromosomal pattern of segregation and retain their aberrant phenotype irrespective of the nuclear genome. The phenotype is maternally inherited; that is, striping is now controlled by chloroplast genes. Again, we know little about the mechanisms of interaction, but this example demonstrates a nuclear-induced mutation of a cytoplasmic system. Normal chloroplasts are somehow irreversibly modified by products of the nuclear genome or possibly by failure of certain nuclear products to be synthesized.

Finally, the replicative behavior of an extrachromosomal system may also be controlled by nuclear genes. Colonies of the red variants of the fungus *Aspergillus nidulans* (a nuclear mutation) display a pattern of persistent segregation of red and white morphological types (Figure 3-12). During growth red and white variants seem to appear at random. Segregation occurs in spite of the fact that all colonies are clonally derived from single asexual spores. Haploid spores germinate into normal white or mutant red colonies, but as growth proceeds, each spontaneously segregates and sectors into red and white phenotypes at each generation. Since all nuclei are identical and there is no evidence of any high spontaneous mutation rate or other chromosomal abnormalities, the only explanation is that normal and red mutant cytoplasmic entities are seg-

regating. Some mutant nuclear genes favor replication of one or the other cytoplasmic entity. Figure 3-12 illustrates how two different plasmagenes in the same cytoplasm in fungi could result in persistent segregation and even lead to two distinct clones with different properties.

Other nuclear mutations prevent the red variant from showing persistent segregation. In their presence red phenotypes breed true from generation to generation, as if the nuclear mutations favor the replication of only the red cytoplasmic variant. In this case replication of normal cytoplasmic determinants is suppressed whereas in other cases the white cytoplasmic determinant is promoted at the expense of the red.

## INFORMATION FLOW BETWEEN NUCLEAR AND CYTOPLASMIC COMPARTMENTS: NUCLEOCYTOPLASMIC INTERACTIONS

Several examples of nuclear genome regulation of organelle properties have been described, but no instances of control by organelle genomes over nuclear function have been mentioned. We cannot assume a priori that this does not occur, but so far it has been difficult to demonstrate. At a level of interaction between nuclear and cytoplasmic compartments, however, the cytoplasmic state has been shown to affect nuclear behavior in cell function. In a sense, at this level it is the cell's structural information that is involved since different cytoplasmic or nuclear "physiological states" are implicated. As cells adjust to an environment, they develop stable metabolic states that are transmissible to their progeny. Such metabolic states reflect networks of interactions between nuclear and cytoplasmic compartments.

As we acquire more information about such vaguely defined physiological states, we will probably learn that specific gene products or regulatory molecules are involved. At the present state of our knowledge, however, we may assume that the "informational" state of the cytoplasm or nucleus at any instant is an expression of spatial and temporal ordering of metabolic systems, that is, the cell's structural information.

In most instances we define these states operationally, usually by one of the following methods. (1) We may remove a nucleus or some cytoplasm to see what effect its absence has on cell survival, motility, growth, or morphogenesis. Enucleation experiments tell us something about the role of the nucleus in cell behavior and its effect on the cytoplasm. (2) We may transfer nuclei from one cytoplasmic state to another to see how nuclear behavior changes in response to new sets of cytoplasmic signals. (3) We may fuse cells in different states to form a hybrid in which two different genomes interact in a common but mixed cytoplasm. The results of these experiments show whether nucleus and cytoplasm are mutually dependent and if certain cytoplasmic states do, in fact, regulate nuclear behavior.

### Enucleation experiments

We know that cell survival and reproduction are dependent on the nucleus. Removing a nucleus from a cell disrupts cytoplasmic activities and ultimately leads to cell death. But the nucleus exercises more subtle control over cellular activities as well.

Most studies of nucleocytoplasmic interactions are carried out on large single-celled organisms such as *Amoeba*, the large ciliate *Stentor*, and the unicellular algae *Acetabularia*. These cells can be cut into nucleated and enucleated fragments, and because of their large size, nuclei are easily transplanted between cells.

Since most genetic instructions come from the nucleus, enucleation of a cell obviously eliminates a primary source of informational molecules and gene products. And yet removal of the nucleus does not always result in an abrupt cessation of cell function. Many cell activities persist over long periods of time in the complete absence of a nucleus. Photosynthesis, protein synthesis, and even cell division continue in enucleated cells, suggesting that nuclear control over cell function is not immediate but delayed.

One of the immediate effects of removing a nucleus from an amoeba is cessation of motility. Pseudopodia do not appear, and the cell rounds up into a quiescent sphere. Since it does not form pseudopodia, it cannot feed and thus survives for less than two weeks. By contrast, an amoeba with its nucleus intact survives for three weeks, even though it is starved. Enucleated halves are unable to metabolize glycogen and fat reserves and utilize a declining protein reserve instead. The most immediate change after enucleation, however, is a dramatic loss in total cell RNA. Though protein synthesis continues, its rate declines as RNA is exhausted and is not replenished by the nucleus.

Transfer of a nucleus from a normal amoeba into a quiescent or nonmotile, enucleated cell restores normal function. The recipient cell begins movement and resumes feeding. The nuclear signal for reactivation is not species-specific since a nucleus from an unrelated species is equally capable of restoring function.

Certain cells survive for much longer periods without a nucleus. *Acetabularia* is a large unicellular algae with a prominent nucleus in its rhizoid, many chloroplasts in its stalk, and a unique "cap" at the tip (see Figure 3-13). After enucleation, the enucleated stalk, if kept in the light, continues to photosynthesize and survives for periods up to seven months. It synthesizes cell wall, grows several millimeters in length, and in many cases will even regenerate a new cap. Survival time depends upon photosynthetic capacity; enucleated fragments in the dark survive for shorter periods.

Survival of an enucleated cell is in part a function of feeding behavior. An enucleated amoeba, which must move to feed, exhibits a shorter life than an enucleated *Acetabularia* stalk, which is still able to synthesize its own food. Chloroplast turnover in *Acetabularia* is quite low; these organelles function without nuclear information for long periods of time, relying entirely on instructions and products from their own genomes.

In the absence of a nucleus cytoplasmic systems begin to degenerate, some more rapidly than others. These degenerative changes are reversible since early reintroduction of a nucleus into enucleated cytoplasm restores vitality. A point of no return is reached when degenerative changes become irreversible and nuclear transplants do not resurrect the cell. Nuclei probably synthesize informational molecules to balance their utilization by cytoplasmic organelles and metabolic machinery.

## Nuclear control of cap regeneration in Acetabularia

Nuclei are also essential for the development of cell organelles. The best case of nuclear control of cell morphogenesis is Haemmerling's classic study of *Acetabularia* (Figure 3-13). In the normal life cycle of this unicellular algae a diploid zygote grows into a cylindrical cell with a rhizoid. The stem grows in length and when mature forms a cap or umbrella-like structure at its apical end for the production of sexual spores. The shape of the cap varies in different species of *Acetabularia*; that of *A. crenulata* is lobate while *A. mediterranea* is disclike. When the cap is mature, the large primary nucleus in the rhizoid becomes smaller and undergoes a series of mitotic divisions, forming smaller secondary nuclei. These migrate up the stalk and collect in the cap region where they are eventually gathered in cysts. Cysts are protective, and when they germinate, uninucleate gametes are discharged into sea water. They fuse to form zygotes, and the cycle is therefore complete.

Cap regeneration is under nuclear control, but control need not be exercised immediately. If a cap is cut from a nucleated cell, it regenerates a new cap at the apical end. But an enucleated fragment also regenerates a new, albeit, smaller cap as long as photosynthesis continues (Figure 3-14). An enucleated stem, however, can regenerate only one cap whereas a nucleated stem regenerates caps each time one is surgically removed. A continuous supply of nuclear substances is not essential for formation of the first cap, suggesting that some morphogenetic substances are in the cytoplasm but seem to be exhausted after the first cap regenerates.

Are morphogenetic substances distributed evenly throughout the stem? If enucleated stems are cut into three equal segments, each fragment exhibits different regenerative behavior. Apical segments regenerate more rapidly and completely than mid-segments, and basal segments do not regenerate at all. There is an apico-basal gradient of cap regenerative capabilities residing within the cytoplasm, with the apex exhibiting the highest regenerative potential. Presumably the highest concentration of "cap morphogens" is at the apex. If, however, a basal nucleated fragment is kept in the dark for five days or more and transferred to light and enucleated, a cap regenerates just as if the fragment was apical. In the dark the nucleus synthesizes morphogenetic substances that diffuse in gradient fashion in the cytoplasm of intact stems with the highest concentration at the apex. A basal nucleated fragment in the dark accumulates enough morphogenetic materials (they cannot diffuse to the apex) to regenerate as rapidly as an apical segment.

Cap regeneration itself is a light-dependent phenomenon; only actively photosynthesizing stems

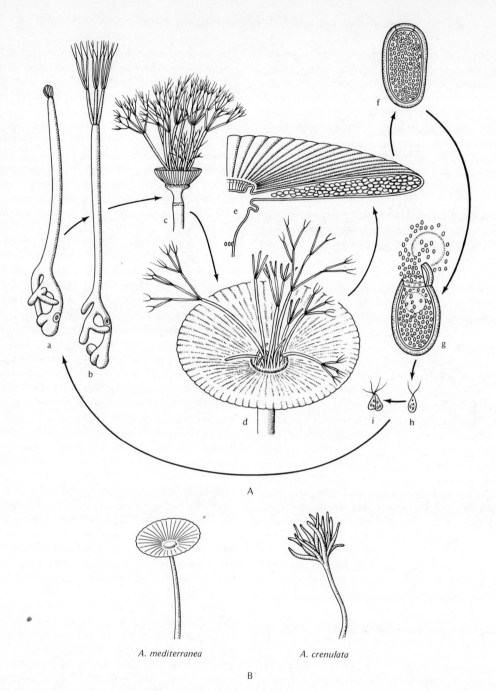

**Figure 3-13** *(A) Life cycle of* Acetabularia mediterranea. *(a) Young stem showing rhizoid with large nucleus. (b) Growth of whorls at cap region. (c) Beginning of reproductive cap as a flattened disc beneath the whorls. (d) Maturing cap. (e) Cross section of mature cap showing large numbers of cysts containing many uninucleate spores. (f) Cysts burst, releasing spores which transform into motile, haploid gametes. (h) These fuse into a diploid zygote which germinates into a stalk, thus completing the cycle. (B) A comparison of mature cap structure of* A. mediterranea *and* A. crenulata. *(From A. Kühn,* Lectures in Developmental Physiology. *New York: Springer-Verlag, 1971.)*

regenerate. Regeneration does not occur in weak light or in light of wavelengths incapable of supporting photosynthesis; evidently some products of chloroplast photosynthesis are also required for cap regeneration. In this case cap formation depends on a cooperative interaction between nuclear substances generated in the dark and materials synthesized by cytoplasmic genetic systems in the light.

Haemmerling's experiments on interspecific nuclear transplantations showed that nuclear instructions are species-specific in *Acetabularia*. An enucleated stem of one species may be grafted onto a nucleated fragment of another species to determine the respective contributions of nucleus and cytoplasm to cap formation. The cap of *Acetabularia crenulata* consists of many fingerlike projections distributed in a circle around the central axis while that of *Acetabularia mediterranea* is smooth and circular. Cap morphology is stable; all newly regenerated caps always resemble the typical species pattern.

An example of a uninucleate graft, a combination of an *A. mediterranea* nucleated stem with an enucleated *A. crenulata* stalk, is shown in Figure 3-15. The first cap to regenerate is an intermediate cap; morphological features of both parents are expressed in part. If the newly regenerated cap is removed, the next cap to appear resembles *A. mediterranea* in morphology. The reciprocal experiment of an *A. crenulata* nucleus grafted onto an *A. mediterranea* stem also yields an intermediate cap with a tendency toward *A. crenulata* characteristics, while subsequent regenerates always exhibit the *A. crenulata* pattern. Since the first regenerated cap is invariably of an intermediate morphology in each case, we conclude that cytoplasm contributes information to regeneration of the first cap, a result consistent with the behavior of enucleate stems already described. Species-specific information stored in cytoplasm is exhausted in the first regeneration, and all subsequent regenerations are characteristically determined by new information flowing from the resident nucleus. Cap morphology is determined by nuclear information.

## Nuclear control of RNA synthesis

What is the nature of the nuclear information stored in the cytoplasm? In the absence of a nucleus, total protein synthesis is unaffected for many weeks,

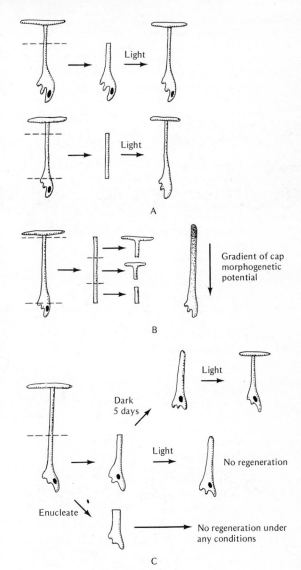

**Figure 3-14** *Experiments illustrating cap regeneration in* A. mediterranea. *(A) Removal of a cap results in regeneration from enucleated and nucleated fragments in the light. (B) Gradient of morphogenetic potential. Enucleated fragments from different regions exhibit varying degrees of cap morphogenesis. The apical region is most capable of forming caps, while regions near rhizoids exhibit the least regenerative potential. (C) Cap morphogenesis depends upon products of nuclear metabolism in the dark. A proximal nucleated fragment will regenerate if maintained in the dark for 5 days before exposure to light.*

**Figure 3-15**  *The nucleus is a source of morphogenetic information. (A) An interspecific graft between a nonnucleated stem of A. crenulata and onto a nucleated proximal fragment of A. mediterranea. The first cap to regenerate is intermediate between the two species and, if it is removed, a second and all subsequent caps are of the A. mediterranea type. (B) Irradiation of nucleated stems with UV light at a wavelength of 2650 Å inhibits cap regeneration. An irradiated enucleated stem will regenerate if a nonirradiated nucleated fragment is grafted. An irradiated nucleated fragment cannot support regeneration. (C) Actinomycin D treatment of nucleated and enucleated fragments. The apical enucleated fragment regenerates slightly because its rich supply of morphogenetically active substances (RNA?) is unaffected by actinomycin D. On the other hand, the nucleated proximal fragment, with little stored morphogenetic materials, cannot regenerate in the presence of actinomycin but does so as soon as the inhibitor is washed out. This suggests that RNA is the morphogenetic material.*

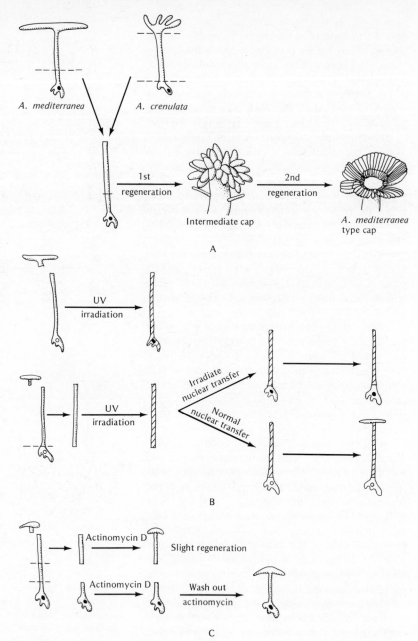

which explains why enucleated stems continue growth and morphogenesis. Part of this synthesis may be attributed to mitochondria and chloroplasts which, as we have seen, contain all the machinery for protein synthesis. One process is immediately affected by enucleation, however: the synthesis of RNA. In spite of continued growth, total cell RNA in enucleated stems either does not increase or accumulates at much slower rates than in nucleated stems. Removing a nucleus alters RNA metabolism, suggesting that nuclear instructions are in the form of RNA molecules.

Any comparison of RNA metabolism in nucleated and enucleated stems is complicated by the fact that mitochondria and chloroplasts also synthesize RNA, and we cannot easily distinguish between nuclear and cytoplasmic RNA. The importance of RNA synthesis in cap regeneration is shown by inhibiting RNA synthesis.

An *Acetabularia* nucleus is visibly large with a swollen nucleolus, rich in RNA. If nucleated stems are exposed to radioactive uridine, the label first appears in nuclear and nucleolar RNA and then seems to be transferred to the cytoplasm, accumulating at the apex of the stalk. Any interference with RNA synthesis and accumulation prevents regeneration (see Figure 3-15B). If, for example, stems are irradiated with UV light, nuclear RNA synthesis is inhibited, none accumulates at the apex, and such stems do not regenerate caps. If, however, an unirradiated nucleus is transplanted into an irradiated enucleate fragment, then both RNA synthesis and morphogenetic capacity are restored. A transplanted irradiated nucleus has no effect. The UV wavelength primarily responsible for the effect is 254 m$\mu$, the wavelength preferentially absorbed by nucleic acid. These experiments indicate that cap regeneration depends on nuclear RNA synthesis.

Which species of nuclear RNA is involved, ribosomal RNA, mRNA, transfer RNA, or all three? It is unlikely that ribosomal or transfer RNA possesses the specificity for cap morphology which leaves mRNA. The fact that enucleated stems regenerate a cap over a period of weeks suggests that nuclear information for cap formation is relatively stable. Does this mean that we are dealing with a stable species of mRNA? One way of determining mRNA stability is to use actinomycin D, the inhibitor of transcription (Figure 3-15C). If nucleated basal stems and enucleated apical stems are immersed in actinomycin, both stop synthesizing RNA. Nucleate basal stems, as we have seen, do not regenerate, but enucleate apical stems, containing a higher amount of stored morphogenetic substance, do regenerate small abnormal caps. On the other hand, regeneration of basal nucleate fragments is prevented (even if they are kept in the dark) as long as actinomycin remains in the medium. Presumably large amounts of a stable, preexisting RNA, stored in apical cytoplasm, can support regeneration. Regeneration of basal

fragments occurs only after actinomycin is washed out, and nuclei resume RNA synthesis. We see that nuclear synthesis of long-lived mRNAs is necessary for normal morphogenesis. As the nucleus manufactures RNA, it diffuses up the stem, accumulating at the apex. This concentration gradient correlates with the gradient of morphogenetic potential in the stem.

Actinomycin D experiments are difficult to interpret because chloroplast and mitochondria synthesize RNA and are equally sensitive to actinomycin. Under conditions of optimal illumination, enucleate stems synthesize small amounts of RNA; a twofold net increase in total RNA is frequently observed within five to seven days after removal of the nucleus. This level of synthesis must be associated with chloroplast DNA since synthesis declines when stems are placed in the dark. RNAs synthesized in chloroplasts include transfer RNAs, ribosomal RNAs, and small amounts of mRNA. Thus it is possible that some stable messages from the chloroplast genome cooperate with nuclear information during cap morphogenesis.

Cap removal stimulates regeneration. Stored informational molecules in the cytoplasm are activated and translated into proteins required for cap morphogenesis. This implies that the process is under translational control. Little is known about such translational controls in *Acetabularia*. Protein synthesis continues in enucleate stems (as we have seen) at rates only slightly different from those in nucleated stems, utilizing stable messages in the cytoplasm. Inhibitors of protein synthesis such as puromycin and cyclohexamide prevent cap regeneration in both nucleated and enucleated stems. Enzymes of carbohydrate metabolism are synthesized at slightly lower levels in enucleate stems. Each enzyme accumulates at its own rate, which may be different from the overall pattern of protein synthesis. This means that the synthesis of each enzyme is under a different mode of translational control. This is to be expected if some of the enzymes involved are coded by the chloroplast genome. The relative contribution of nuclear and cytoplasmic genomes to cap morphogenesis is still uncertain since substances specific for cap morphogenesis have not been isolated, stable message populations have not been characterized, nor have the complex interactions between nuclear and cytoplasmic genomes been worked out.

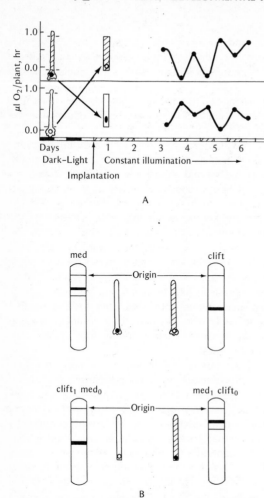

A

B

**Figure 3-16**  *Nuclear control of cytoplasmic processes.*
*(A) Nuclear control of photosynthetic rhythms. Two sets*
*of stems are entrained on completely inversed light-dark*
*cycles before nuclei are reciprocally transplanted. The*
*entrained photosynthetic rhythms (as measured by*
*oxygen evolution) characteristic of the nuclear donor*
*persist even in the presence of constant illumination.*
*(Based on E. Schweiger et al., Science 146:658–659,*
*1964. Copyright © 1964 by The American Association*
*for the Advancement of Science.) (B) Influence of nucleus*
*on synthesis of malic dehydrogenase, a mitochondrial*
*enzyme. Two species of Acetabularia differ with respect*
*to MDH isozyme patterns. After reciprocal nuclear*
*exchange, the isozyme pattern of the nuclear donor is*
*expressed. (From H. G. Schweiger, R. W. P. Master, and*
*G. Werz, Nature 216:554–557, 1967. Reproduced by*
*permission of Nature.)*

For example, the nucleus exercises control over several aspects of organelle function, including photosynthesis. By manipulation of light-dark cycles the photosynthetic rhythm of *A. mediterranea* can be modified and entrained so that in constant light a functional rhythm characteristic of the entrainment regime is sustained over several days (Figure 3-16A). If nuclei are reciprocally exchanged between cells with rhythms differing by 12 hours, the acceptor cell soon acquires the rhythm typical of the donor nucleus. It is noteworthy that the synthesis of mRNA is also subjected to a diurnal rhythm. A nuclear product, possibly mRNA, probably codes for enzymes involved in photosynthesis. This enzyme might induce quantitative changes that would shift the photosynthetic rhythm.

Some organelle enzymes controlled by the nuclear genome include malic and lactic dehydrogenase, which are localized in mitochondria and may also be found in chloroplasts (Figure 3-16B). Each species of *Acetabularia* has a characteristic pattern of these enzymes (isozymes) that can be characterized by gel electrophoresis (see Chapter 17). When nuclei are exchanged between cells of different species, the isozyme pattern of the acceptor cell disappears and is replaced by a pattern characteristic of the nuclear donor. The enzymes coded by the nucleus seem to dominate organelle structure. These same cytoplasmic enzymes are retained and even increase in enucleated cells, which suggests they may also be coded by plasmagenes or may be synthesized on stable messages. But in the presence of a foreign nucleus the stable mRNAs coding for these enzymes disappear and are replaced by newly synthesized messages from the nucleus. Some form of protection is afforded organelle mRNAs in enucleated cells that is lost when a foreign nucleus is introduced.

These studies demonstrate that stable informational mRNAs are produced by nuclear and plastid genomes, stored in the cytoplasm in a gradient manner, and summoned to action at critical periods in the life cycle by some translational triggering mechanism. The informational molecules are destroyed as they are used in cytoplasmic syntheses and are replenished by continued transcription on nuclear and organelle DNA templates. Once in the cytoplasm, however, the activity and fate of these molecules are independent of the nucleus and dependent only upon cytoplasmic control systems.

Later we will see that this pattern of long-term template storage and message regulation is found in many other eukaryote developmental systems, particularly eggs, seeds, and spores.

## Cytoplasmic signaling circuits: somatic cell heterokaryons

The nucleus is not autonomous; its behavior and properties rely on a continuous supply of energy and precursors from the cytoplasm. Along with its general "nutritional" or "housekeeping" role, the cytoplasm also regulates the nuclear genome. Cytoplasmic signals affect gene replication and transcription, chromosome kinetics during mitosis, and morphogenetic expression of nuclei.

A cell divides after completing a sequence of nuclear and cytoplasmic syntheses, among which is DNA replication (see Chapter 5). Replication depends on nucleotide precursors supplied by the cytoplasm. By regulating precursor pool size and the amount of energy funneled into the nuclear machinery, we can see how a cytoplasmic state would affect replication. But cytoplasmic regulation of DNA synthesis is far more subtle than simple mass-action effects. Before a cell nucleus is triggered into a replication mode, one of several alternative steady states—a "division readiness" state—must be achieved by the cytoplasm. Once this has happened, then virtually any nucleus introduced into the cytoplasm, irrespective of its own physiological state, is switched to DNA replication.

Nondividing nuclei such as those of the red blood cell are stimulated to synthesize DNA when placed in the cytoplasm of a proliferating cell. This has been shown by the technique of somatic cell fusion (Figure 3-17). Certain infecting animal viruses (Sendai virus) alter the outer surfaces of cultured cells and vastly increase the frequency with which cells fuse. The resulting fusion product is a cell containing two or more nuclei derived from different cells. If the cells are from the same population, then the fused cell is a *homokaryon,* or two nuclei of similar genotype in the same cytoplasm. If two different genomes share a common cytoplasm, they are called *heterokaryons.* In most experiments the virus is inactivated by ultraviolet light (to prevent its replication) and then incubated with a mixture of cells. Viruses attach to cell surfaces, and the viral

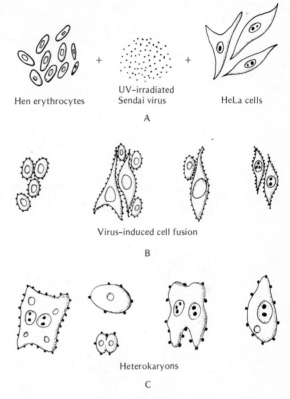

Hen erythrocytes    UV–irradiated Sendai virus    HeLa cells

A

Virus–induced cell fusion

B

Heterokaryons

C

**Figure 3-17**  *Somatic cell fusion with inactivated Sendai virus. (A) Human HeLa cells and chicken erythrocytes are treated with UV-inactivated Sendai virus. Without injuring the cells, the virus changes the surface, making them sticky. (B) Various cell fusions take place to yield heterokaryons. (C) Heterokaryons, consisting of genetically different nuclei sharing a common cytoplasm.*

membrane bridges the gap between membranes of two adjacent cells, which fuse. In this way nucleated chicken erythrocytes have been fused with human cancer cells (a strain of HeLa cells) to form heterokaryons, each cell containing chicken and HeLa cell nuclei. Most cytoplasm in these heterokaryons comes from the HeLa cell since much less erythrocyte cytoplasm is contributed during fusion. Fusion is a random process; heterokaryons with one or more nuclei from each parental cell may be found.

The erythrocyte nucleus is normally inactive, destined to disintegrate when the cell dies. Within hours after fusion, however, the compact red blood

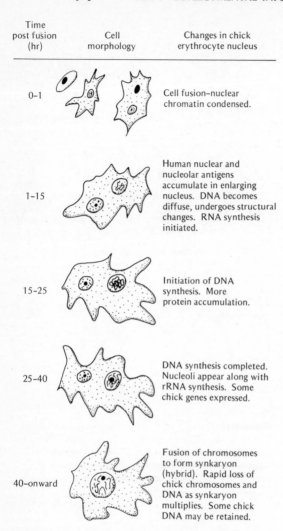

| Time post fusion (hr) | Cell morphology | Changes in chick erythrocyte nucleus |
|---|---|---|
| 0–1 | | Cell fusion–nuclear chromatin condensed. |
| 1–15 | | Human nuclear and nucleolar antigens accumulate in enlarging nucleus. DNA becomes diffuse, undergoes structural changes. RNA synthesis initiated. |
| 15–25 | | Initiation of DNA synthesis. More protein accumulation. |
| 25–40 | | DNA synthesis completed. Nucleoli appear along with rRNA synthesis. Some chick genes expressed. |
| 40–onward | | Fusion of chromosomes to form synkaryon (hybrid). Rapid loss of chick chromosomes and DNA as synkaryon multiplies. Some chick DNA may be retained. |

**Figure 3-18**  *Activation of chick erythrocyte nuclei after introduction into mammalian cell cytoplasm. (Based on R. Appels and N. R. Ringertz,* Current Topics in Developmental Biology, *ed. A. A. Moscona and A. Monroy, vol. 9. New York: Academic Press, 1975, pp. 137–166.)*

cell nucleus increases rapidly in volume (twenty- to thirtyfold), taking up proteins from the surrounding cytoplasm while its chromatin uncoils and becomes more disperse (Figure 3-18). Such heterokaryons actively incorporate radioactive thymidine into DNA of both types of nuclei. This experiment shows that an active cytoplasmic "state" can trigger DNA replication in nuclei which normally would never syn-

thesize DNA again. Furthermore, we see that cytoplasmic signals for DNA synthesis are not species-specific; a foreign erythrocyte genome can respond to a HeLa cytoplasmic signal.

Once initiated, DNA synthesis in heterokaryon nuclei is synchronized. Within two days most erythrocyte nuclei have completely doubled their DNA, and nucleoli have appeared. In some cells a fusion nucleus arises but only HeLa chromosomes complete mitosis; most chicken chromosomes are unable to enter mitosis and break up into fragments.

Before replication begins, the physical state of DNA changes; it exhibits a marked increase in affinity for dyes that intercalate into DNA helices. Reactivation of chromatin involves "loosening" or uncoiling of compact templates to render them more accessible to all binding molecules. Nuclear enlargement seems to be the primary event as HeLa proteins flow into nuclear compartments.

Protein exchanges between nucleus and cytoplasm have been studied by immune fluorescence. In this procedure fluorescent antibodies specific for nucleolar or cytoplasmic proteins are used as markers to localize HeLa and chicken cell proteins in heterokaryons. The fluorescent antibodies enable us to observe that human nucleolar antigens appear in newly forming chick nucleoli in a matter of hours. In addition, human nucleoplasm antigens can be detected within chick nuclei as they enlarge, but no human cytoplasmic antigens are detectable in chick nuclei in these early stages. The proteins that enter chick nuclei are highly specific nuclear proteins. Evidently nuclear swelling does not result from a random uptake of cytoplasmic proteins but rather from uptake of selected nuclear proteins.

Uptake of human proteins by chick nuclei may be related to the activation of the chick genome. For instance, gene expression is also stimulated since RNA synthesis begins in chick nuclei. Adult erythrocyte nuclei have stopped transcribing RNA along with shutting down DNA replication. HeLa cell nuclei synthesize both RNA and DNA. Shortly after fusion, as the erythrocyte nucleus enlarges and DNA becomes diffuse, RNA synthesis begins. Apparently total RNA synthesis is stimulated, including ribosomal RNA, which explains the early appearance of chick nucleoli in two to four days. Transcription of the erythrocyte genome is completely reactivated along with replication.

Shortly after the heterokaryon is made, chick-specific antigens are detectable on the heterokaryon cell surface. The heterokaryon plasma membrane is a mosaic of human and chick plasma membranes. Within hours, however, as RNA synthesis begins, the chick-specific antigens gradually disappear from the surface, only to reappear again shortly after nucleoli become visible in chick nuclei. It is only after RNA synthesis has continued for a day or so and nucleoli have organized that chick antigen messages begin to be translated. This suggests that elements of the nucleolus, perhaps a ribosomal RNA or protein subunit, are needed to transport chick messages to the cytoplasm.

It is possible that human nucleolar proteins, incorporated into a chick genome, act as regulatory molecules to turn on transcription of chick genes. On the other hand, activation of RNA synthesis could result from the flow of HeLa cell RNA polymerases into chick nuclei to initiate transcription.

This system of shuttling nuclear proteins tells us that nuclei adjust to their cytoplasmic environment by feedback exchanges of specific proteins. These proteins may be regulatory molecules that adapt the nuclear genome to the physiological state of the cytoplasm. In Chapter 5 we will see that such proteins may regulate DNA synthesis and the sequence of changes leading to cell division.

### Shuttling proteins in Amoeba

Shuttling proteins may be the vehicles for cytoplasmic control of nuclear activities. Such proteins may include enzymes (DNA and RNA polymerases), histones, and other regulatory molecules which interact with DNA by recognizing specific nucleotide sequences. Regulatory proteins probably exchange continually between nucleus and cytoplasm, even in nondividing cells.

Goldstein and Prescott have shown that in Amoeba, nuclei may be transferred between cells to make nucleocytoplasmic combinations used to study the exchange of proteins between compartments. In one experiment all Amoeba proteins are first labeled by incorporation of a radioactive amino acid (Figure 3-19). If such labeled amoebae are grown for many cell generations in the absence of a labeled amino acid, the labeled cytoplasmic proteins are rapidly turned over and replaced by unlabeled proteins, but the proteins in the nuclei retain their high level of radioactivity. This suggests that the

nuclear proteins are quite stable and are degraded more slowly than cytoplasmic proteins. These nuclear proteins seem to remain in the nucleus throughout most of the cell cycle but are released into the cytoplasm at mitosis, after the nuclear membrane breaks down. At the completion of mitosis these same nuclear proteins migrate back into telophase nuclei of daughter cells.

These proteins exhibit high recognition specificity; they return rapidly to nuclei at the end of division. Moreover, if a labeled nucleus is transferred into an intact unlabeled cell, 90 percent of the labeled nuclear proteins are distributed between the two nuclei with very little accumulating in the cytoplasm. A constant ratio of labeled proteins (2.6:1) is always found between the original grafted nucleus and the host nucleus. Approximately 40 percent of the labeled nuclear proteins do not migrate but remain bound to the original nucleus, while 60 percent freely exchange between nuclei without localizing in the cytoplasm.

The proteins are first synthesized in the cytoplasm before being transferred to the nucleus. Enucleated amoebae incubated with radioactive amino acid synthesize labeled cytoplasmic proteins. An unlabeled nucleus introduced into such a cell incorporates this cytoplasmic labeled protein, and if it is transferred into an intact unlabeled cell, some labeled nuclear proteins migrate into the host nucleus according to the usual 2.6:1 ratio. Both classes, migrating and nonmigrating proteins, are synthesized in the cytoplasm and associate with the nuclei, only to shuttle between nucleus and cytoplasm at each mitosis. Because of their heterogeneity, these proteins have not been characterized, nor is their function understood. They may include polymerases to initiate replication and transcription; others may control the physical structure of the chromosome.

### Cytoplasmic control of nuclear and chromosome morphology

The cytoplasm also exercises control over gross aspects of nuclear and chromosome behavior before and during mitosis. Nuclear morphology, chromosome replication, and coiling are all controlled by the "state of the cytoplasm." An example of such cytoplasmic control of nuclear morphology is seen in Stentor, a large ciliated protozoon.

**Figure 3-19** *Experiments on "shuttling proteins" in* Amoeba. *(A) Amino acids incorporated into* Amoeba *proteins are progressively diluted out of the cytoplasm during multiplication. Persistently labeled proteins are confined to the nucleus. (B) When the nuclear membrane breaks down during mitosis the nuclear proteins are released to the cytoplasm and return to the daughter nuclei at the completion of division. (C) The nuclear proteins are localized exclusively in nuclei. Transplantation of a labeled nucleus into a nonlabeled cell is followed, in 20 hours, by accumulation of labeled nuclear proteins in the host nucleus at a ratio of 2.6:1 (donor to host). (D) Nuclear proteins are synthesized in the cytoplasm before segregating into nuclei. An enucleated cell is labeled with amino acids and a nonlabeled nucleus is transplanted into the labeled cytoplasm. Radioactive proteins in the nucleus will diffuse into an unlabeled host nucleus at the 2.6:1 ratio.*

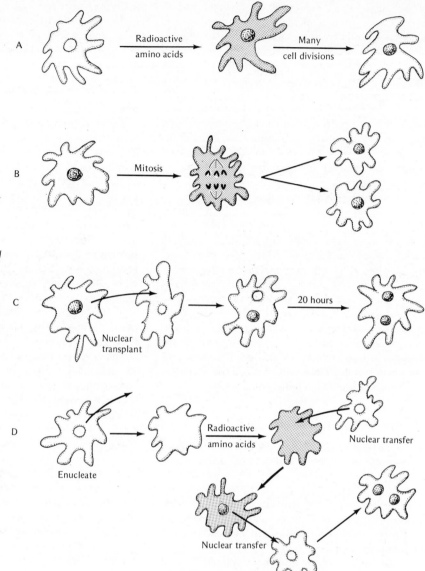

In *Stentor* rows of cilia run the length of the body separated by alternate bands of pigment (Figure 3-20). The bands are narrow on one side and increase in width around the cell. The oral region at the cell's anterior is surrounded by a band of large cilia—the oral apparatus—which beat vigorously during feeding. The nuclear genome comprises a large, beaded macronucleus that governs growth and morphogenesis and two smaller diploid micronuclei involved in sexual reproduction. The macronucleus is polyploid, each nuclear node containing many copies of the diploid genome.

Cell division in *Stentor* is amitotic although the macronucleus passes through a series of morphologic stages. Division readiness is first indicated by the appearance of a cleft in the cortex at the oral primordium site, at the border between thick and thin stripes. This future oral apparatus of the posterior daughter cell enlarges and acquires ciliary rows that expand into a membranellar band. Meanwhile the beaded macronucleus, having replicated its DNA, coalesces into an irregular mass beneath the cortex. Shortly thereafter the macronucleus elongates into a rod and renodulates; as the new oral apparatus takes up an anterior position in the midline, the cleavage furrow constricts cytoplasm and macronucleus into two daughter cells.

Each macronuclear change is controlled by a specific cytoplasmic state. For example, grafting a dividing cell (before macronuclear coalescence) to a nondividing interphase cell causes both macronuclei to coalesce at later stages. Transfer of nuclear nodes from a nondividing cell to one in division induces nodes to coalesce in synchrony with the host nucleus. In both cases a cytoplasmic state induces coalescence in nuclei normally not prepared for it.

Nuclear elongation at a later stage is also mediated by the cytoplasm. A compacted nucleus from a dividing cell will not elongate if transferred into a nondividing interphase cell. It does so, however, after transfer into another dividing cell in the elonga-

tion phase. Evidently a cytoplasmic factor in the elongation phase exerts a positive effect on the grafted nucleus. On the other hand, elongation is inhibited if a partially elongated nucleus is placed in an interphase cell. Negative control is displayed here. The cytoplasm seems to pass through transient states from interphase to division, each one of which conveys a different signal to control nuclear morphology.

The nature of these nuclear morphological changes is not well understood, but electron micrographic studies of nuclei in other ciliates reveal many intranuclear microtubules, cylindrical organelles that resemble spindle fibers. In response to cytoplasmic signals, these may change nuclear morphology and operate, perhaps, as an intranuclear spindle.

### Nucleocytoplasmic incompatibility: species hybrids

For normal chromosome replication the cytoplasm must provide a "compatible environment," whatever

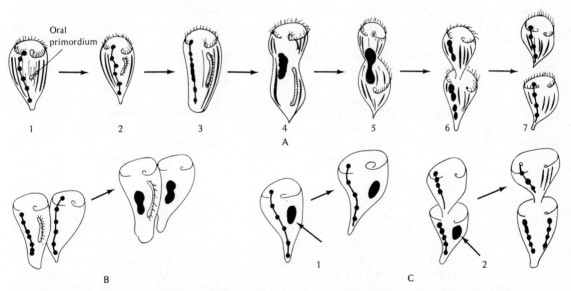

**Figure 3-20** *Cytoplasmic control of nuclear morphology in* Stentor. *(A) Normal amitotic division pattern in* Stentor: *1. First sign of division is appearance of oral primordium in region of stripe contrast. 2. Oral primordium enlarges. 3. Macronuclear nodes begin to coalesce as cell lengthens. 4. Complete coalescence of macronucleus. Constriction beginning. Oral primordium complete. 5. Constriction of cell and macronucleus into two cells. Oral primordium assumes position in posterior cell. 6. Elongation of macronucleus into discrete nodes in each daughter cell. 7. Separation of daughter cells. (B) Cytoplasmic control of nuclear coalescence. Fusion of a dividing cell (with oral primordium) with nondividing cell. Both macronuclei coalesce at the same time. (C) Cytoplasmic control of macronuclear elongation: 1. A coalesced macronucleus grafted to a nondividing cell has no effect on the physical state of the grafted nucleus. 2. Insertion of a similar macronucleus into a cell with an elongating macronucleus induces the graft nucleus to elongate as well.*

that means in molecular terms. Compatibility, for example, may mean the presence of a species-specific DNA polymerase which recognizes certain initiation points on DNA templates. Rates of DNA replication are carefully adapted to the physiological state of the cell. When nucleus and cytoplasm are made incompatible (for example, when chromosomes are placed in a foreign cytoplasm), chromosome replication and segregation are completely upset. Such nucleocytoplasmic incompatibility arises when foreign sperm fertilize eggs of another species to form species hybrids.

Normally cross-fertilization does not occur in nature. If, by chance, it does, the results are usually lethal, since foreign chromosomes in an egg evoke nucleocytoplasmic incompatibilities, and the egg develops abnormally. In the laboratory we can overcome the natural barriers to interspecific fertilization and artificially fertilize eggs of one species with sperm of closely or even distantly related species from other genera. The species hybrids that arise carry two unrelated haploid sets of chromosomes in a common nucleus, interacting with one another and with egg cytoplasm from only one species.

Development of species hybrids, if it occurs at all, is usually abnormal. In many instances the events of fertilization are disturbed, and eggs do not enter first cleavage. In those instances where cleavage occurs, the paternal chromosomes cannot survive in foreign egg cytoplasm and are unable to replicate. Chromosomes break, and many are eliminated during the first few divisions of the fertilized egg. Nonreplicated chromosomes fail to attach to a spindle and wander into the cytoplasm where they are lost (digested by nucleases?). Sometimes a few paternal chromosomes remain, creating a lethal unbalanced genome.

The abnormal behavior of foreign chromosomes is seen in sea urchin hybrids where cross-fertilization is easily carried out in the laboratory. Development varies, depending on which species is the egg donor. For example, when eggs of *Paracentrotus* are fertilized by the sperm of *Sphaerechinus*, development stops at an early stage. The reciprocal cross is more successful; *Sphaerechinus* eggs fertilized by *Paracentrotus* sperm develop into an intermediate type of larva bearing some resemblance to both parents.

As early as 1910 F. Baltzer showed that 16 of 38 chromosomes are eliminated at the first division from the first hybrid cross. Most eliminated chromosomes are paternal since an enucleated *Paracentro-*

*tus* egg fertilized by *Sphaerechinus* sperm starts out with 20 paternal chromosomes but only 4 of these survive first cleavage. This means that the original cross is not a true diploid hybrid since most paternal chromosomes are lost. The reciprocal cross is a true hybrid; no chromosomes are eliminated, as if *Sphaerechinus* cytoplasm is compatible with *Paracentrotus* chromosomes and permits normal replication. We do not know why the cytoplasm of one species is more tolerant of foreign chromosomes than that of the other, but the reason may relate to cytoplasmic factors controlling replication.

A comparative analysis of amphibian species hybrids indicates that closely related species produce more viable hybrids, whereas hybrids between distantly related species or even distant geographic varieties of the same species usually develop abnormally. Depending upon the cross, some fertilized eggs fail to divide while others develop to an early embryonic stage and stop.

Incompatibilities may arise in diploid hybrids because (1) two different haploid chromosome sets share the same nucleus or (2) interactions between egg cytoplasm and foreign chromosomes are abnormal. This can be resolved by transplanting a diploid nucleus from a *Rana pipiens* embryo into the enucleated egg cytoplasm of *Rana sylvatica* (Figure 3-21). (See Chapter 11 for technique of nuclear transplantation.) In contrast to a species hybrid with two different haploid genomes in the same cytoplasm, in this case a diploid set of *R. pipiens* chromosomes are in a foreign cytoplasm. Development is also abnormal, with embryos stopping at an early embryonic stage. The *R. sylvatica* cytoplasm has a deleterious influence on replicating *R. pipiens* chromosomes. When a *R. pipiens* nucleus is returned to its own compatible egg cytoplasm after replicating in *R. sylvatica* cytoplasm, it cannot support normal development because its karyotype is abnormal with many fragmented chromosomes. Some factor in *R. sylvatica* egg cytoplasm (or its absence) prevents a balanced replication of *R. pipiens* chromosomes early in development, leading to chromosomal breaks and aberrations. These lesions are irreversible: they cannot be repaired, even when replicating in their own cytoplasm. We do not understand the basis for these incompatibilities; they may involve specific regulatory molecules or the enzymes involved in DNA replication and chromosome segregation.

Some species hybrids develop abnormally without any apparent alteration of chromosome number

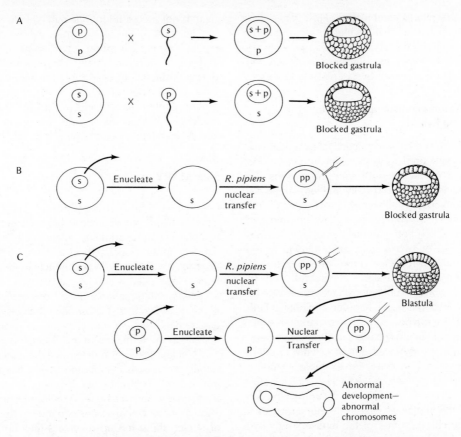

**Figure 3-21** *Nucleocytoplasmic incompatibilities in amphibian interspecific hybrids between* Rana pipiens *and* Rana sylvatica. *(A) Fertilization of a* R. pipiens *egg with* R. sylvatica *sperm leads to a blocked gastrula. The reciprocal cross of* R. pipiens *sperm in* R. sylvatica *egg is also unable to develop beyond an early gastrula stage. (B) Nuclear transfer of a diploid* R. pipiens *nucleus into enucleated* R. sylvatica *egg cytoplasm also leads to a blocked early gastrula. A diploid set of* R. pipiens *chromosomes is unable to overcome the block. (C) Replication of diploid* R. pipiens *nucleus in* R. sylvatica *cytoplasm after nuclear transfer produces abnormal chromosomes. If such nuclei are returned to* R. pipiens *egg cytoplasm (donated from a blastula stage), the resulting development is abnormal and chromosomal aberrations are seen.*

or structure. In these instances more subtle incompatibilities exist between the genome of one species and the cytoplasm of another. The species hybrid *Rana pipiens* male × *Rana palustris* female gives rise to normal tadpoles that metamorphose into frogs. These two species are closely related. No nucleocytoplasmic incompatibilities arise when a *R. pipiens* haploid paternal set of chromosomes together with a *R. palustris* maternal set share *R. palustris* cytoplasm. Incompatibilities develop, however, if a diploid *R. pipiens* nucleus is transplanted into an enucleated *R. palustris* egg. In the absence of *R. palustris* chromosomes, a nucleocytoplasmic incompatibility re-

sults in abnormal development, even though the karyotype of these embryos is normal and chromosomes also replicate normally. What accounts for abnormal development in this case?

The nucleocytoplasmic incompatibility disappears in the presence of a host haploid set of chromosomes. This is shown by experiments in which triploid embryos are made by transplanting diploid *R. pipiens* nuclei into normal *R. palustris* eggs, retaining their maternal haploid nucleus. Most embryos develop normally, indicating that the haploid set of *R. palustris* chromosomes, even in the presence of a double dose of foreign *R. pipiens* chromosomes, salvages this hybrid combination. Host genes seem to be essential to normal development and expression of foreign chromosomes. The experiments also suggest that the expression of genes in a foreign cytoplasm may be influenced by cytoplasmic factors.

## Nuclear cytoplasm incompatibility in somatic cell hybrids

Fusion of somatic cells with inactivated Sendai virus has also been used to create somatic cell hybrids for analysis of nucleocytoplasmic interactions. The first product of fusion is usually a heterokaryon but, if heterokaryons are allowed to grow, a small fraction, about 1 in 10,000, may develop into *synkaryons* in which both parental nuclei merge. Chromosome sets of two species are included in one nucleus and will replicate and function together for many cell generations as a true cell hybrid. Many interspecific hybrid combinations have been made between cultured mouse, rat, hamster, and human cells. Since the karyotypes of any two parental strains differ morphologically, changes in chromosome constitution may be followed during in vitro proliferation of these hybrid cells. For example, a mouse-human hybrid between a HeLa cell and a mouse fibroblast contains the large metacentric chromosomes of the human as well as the numerous small telocentric chromosomes of the mouse. The karyotype usually contains both intact chromosome sets, at least in the early stages of hybrid cell growth, but as cells proliferate, certain chromosomes are lost. A karyotypic analysis reveals that human chromosomes are the ones to disappear until only a few of the original 46 chromosomes remain. In these hybrids one parental set of chromosomes is unable to keep pace with cell proliferation, and chromosomes are lost in a manner similar to paternal chromosome loss in a species egg-sperm hybrid described previously.

Such somatic cell hybrids are useful in a genetic analysis of linkage of human genes, since the loss of specific chromosomes may be correlated with the loss of certain human phenotypic traits. In this instance, however, we need only point out that chromosome stability, replication, and segregation seem to be under the influence of the cytoplasm. When genomes from different species fuse and share a common cytoplasm, they adjust their respective rates of DNA replication and cell division to an intermediate hybrid level. It appears, however, that this adjust-ment is not successful for one parental chromosomal set; its constituent chromosomes exhibit mitotic irregularities in replication, attachment to spindles, and segregation, and many chromosomes are lost during proliferation. Because of the ease with which large populations of somatic cell hybrids can be produced, they have been invaluable in analyzing nucleocytoplasmic interactions during cell division and differentiation (see Chapters 5 and 17).

## SUMMARY

Information flow in eukaryotic cells is a complicated network of interactions between nuclear and cytoplasmic compartments. Many cell traits, particularly those associated with energy transformations, are controlled by extranuclear genetic systems. What a cell is and how it behaves depend on a specific "informational state" of the cell as determined by nucleocytoplasmic exchanges of regulatory signals. Nuclear and cytoplasmic genomes cooperate in the formation of a cell phenotype.

Nuclear genes also determine the character and stability of various cytoplasmic states. Nuclear control is not immediate but involves long-term replenishment and repair of cytoplasmic machinery. Though many cell functions continue in the absence of nuclei, these run down unless they are restored by nuclear products.

Some nuclear products include stable, species-specific informational molecules that are stored in cytoplasm. They may contribute to the stability of the cytoplasmic state and may play significant roles in cell morphogenesis.

Cytoplasmic factors also regulate the flow of cell information. Once a cytoplasmic state has been stabilized, it may exercise control over nuclear behavior. Both replication and transcription respond to signaling circuits in the cytoplasm. These circuits may involve small regulatory molecules that impinge on the cell surface and are transduced in the cytoplasm, or they may involve large macromolecules that shuttle back and forth between compartments. Structural elements in the cytoplasm may also affect nuclear morphology and position within the cell. Cytoplasmic cues exert positive control of nuclear behavior; they may activate nuclear systems that are usually inactive. Later we will see that cytoplasmic systems exercise negative control as well, shutting down the expression of nuclear genes.

**REFERENCES**   Allfrey, V. G. 1974. DNA-binding proteins and transcriptional control in prokaryotic and eukaryotic systems. *Acidic proteins of the nucleus,* ed. I. L. Cameron and J. R. Jeter, Jr. New York: Academic Press, pp. 1–27.

Appels, R., and Ringertz, N. R. 1975. Chemical and structural changes within chick erythrocyte nuclei introduced into mammalian cells by cell fusion. *Current topics in developmental biology,* ed. A. A. Moscona and A. Monroy. Vol. 9. New York: Academic Press, pp. 137–166.

Axel, R., Cedar, H., and Felsenfeld, G. 1974. Chromatin template activity and chromatin structure. *Cold Spring Harbor Symp. Quant. Biol.* 38:773–783.

Barrett, T., Maryanka, D., Hamlyn, P. H., and Gould, H. J. 1974. Non-histone proteins control gene expression in reconstituted chromatin. *Proc. Nat. Acad. Sci. U.S.* 71:5057–5061.

Brachet, J. 1968. Synthesis of macromolecules and morphogenesis in *Acetabularia. Current topics in developmental biology,* ed. A. A. Moscona and A. Monroy. Vol. 3. New York: Academic Press, pp. 1–36.

Britten, R. J., and Davidson, E. H. 1969. Gene regulation for higher cells. *Science* 165:349–357.

Carpenter, G., and Sells, B. H. 1975. Regulation of the lactose operon in *E. coli* by cAMP. *Int. Rev. Cytol.* 41:29–58.

Darnell, J. E. 1976. mRNA structure and function. *Prog. Nuc. Acid. Res. and Mol. Biol.* 19:493–511.

Darnell, J. E., Philipson, E. L., Wall, R., and Adesnik, M. 1971. Polyadenylic acid sequences: role in conversion of nuclear RNA into messenger RNA. *Science* 174:507–510.

De Terra, N. 1969. Cytoplasmic control over the nuclear events of cell reproduction. *Int. Rev. Cytol.* 25:1–29.

Ephrussi, B. 1953. *Nucleocytoplasmic relations in microorganisms.* New York: Oxford University Press.

Georgeiv, G. P. 1972. The structure of transcriptional units in eukaryotic cells. *Current topics in developmental biology,* ed. A. A. Moscona and A. Monroy. Vol. 7. New York: Academic Press, pp. 1–60.

Goldstein, L., and Prescott, D. M. 1967. Proteins in nucleocytoplasmic interactions. I. The fundamental characteristics of the rapidly migrating proteins and the slow turnover proteins of the *Amoeba proteus* nucleus. *J. Cell Biol.* 33:637–644.

Hämmerling, J. 1963. Nucleocytoplasmic interactions in *Acetabularia* and other cells. *Ann. Rev. Plant Physiol.* 14:65.

Harris, H. 1970. *Cell fusion.* Oxford: Clarendon Press.

———. 1974. *Nucleus and cytoplasm.* 3d ed. Oxford: Clarendon Press.

Jacob, F., and Monod, J. 1961. On the regulation of gene activity. *Cold Spring Harbor Symp. Quant. Biol.* 26:193–211.

Jinks, J. L. 1964. *Extrachromosomal inheritance.* Englewood Cliffs, N. J.: Prentice-Hall.

Lohr, D., et al. 1977. Comparative subunit structure of Hela, yeast, and chicken erythrocyte chromatin. *Proc. Nat. Acad. Sci.* 74:79–83.

Miller, O. L., and Hamkalo, B. A. 1972. Visualization of RNA synthesis on chromosomes. *Int. Rev. Cytol.* 33:1–25.

Paul, J. 1970. DNA masking in mammalian chromatin: A molecular mechanism for determination of cell type. *Current topics in developmental biology,* ed. A. A. Moscona and A. Monroy. Vol. 5. New York: Academic Press, pp. 317–352.

Penman, S., Vesco, C., and Penman, M. 1968. Localization and kinetics of formation of nuclear heterodisperse RNA, cytoplasmic heterodisperse RNA and polyribosome-associated messenger RNA in HeLa cells. *J. Mol. Biol.* 34:49–69.

Rao, P. N., and Johnson, R. T. 1970. Mammalian cell fusion: Studies on the regulation of DNA synthesis and mitosis. *Nature* 225:159–164.

Schweiger, H. G. 1969. Cell biology of *Acetabularia. Current Topics Microbiol. Immunol.* 50:1.

Stein, G. S., Spelsberg, T. C., and Kleinsmith, L. J. 1974. Nonhistone chromosomal proteins and gene regulation. *Science* 183:817–824.

Subtelny, S. 1974. Nucleocytoplasmic interactions in development of amphibian hybrids. *Int. Rev. Cytol.* 39:35–88.

Wiseman, A., Gillham, N. W., and Boynton, J. E. 1977. Nuclear mutations affecting mitochondrial structure and function in Chlamydomonas. *J. Cell Biol.* 73:56–77.

# GROWTH

Growth—an increase in the size or mass of a developing system—is an irreversible process that occurs at all organizational levels. It is difficult to define because it is multifactorial, that is, a composite of interacting growth patterns including increases in cell mass (auxetic growth), increases in cell number by cell division (multiplicative growth), and accumulation of extracellular products (accretionary growth). Organismic size increase results from a balanced expression of each growth process during the life cycle.

As cells enlarge they usually divide. Such cells are in growth-duplication cycles, and it is the regulation of these cycles in developing systems that accounts for the growth of parts in relation to the whole, as well as growth of the entire organism. Most organisms grow because their cells proliferate, not because cells enlarge.

Auxetic and multiplicative growth are processes underlying all other developmental phenomena. The acquisition of form (morphogenesis) and the appearance of specialized cell types (differentiation) derive from the growth behavior of cells. Problems of growth center on the regulation of genes concerned with cell enlargement and cell division. We must ask how cells achieve balanced growth and what determines when a cell divides and stops dividing. For ordered development to occur, systems of growth control must operate at all stages of an organism's life cycle. The growth of parts is coordinated by neural and hormonal modes of integration coupled with seasonal environmental cues that govern reproduction and development. When growth controls break down, cells may proliferate abnormally as a tumor and destroy the organism.

# 4

# Growth of Single Cells

An increase in active cell mass is the result of synthetic and degradative processes acting simultaneously. Factors controlling auxetic growth probably play an important role in cell division and accretionary growth. Because protoplasmic syntheses are regulated, cell populations may switch from auxetic growth to division or differentiation. From the point of view of development, it is the regulation of switch mechanisms that is most important. We want to know how cell enlargement and cell replication are related, what terminates growth when a cell reaches maturity, what factors initiate division, and why some cell populations stop dividing, begin to grow, and synthesize functional organelles.

## THE GROWTH-DUPLICATION CYCLE

Cell growth has been studied in cell populations in a growth-duplication cycle, a life cycle circumscribed by division periods (Figure 4-1). Single-celled organisms and somatic cells in culture are

examples. A cell, born after a division, proceeds to grow by macromolecular synthesis, reaches a species-determined division size, and replicates to repeat the cycle. This cycle acts as a unit of biological time and defines the life history of a cell. Most somatic cells of an adult organism, such as neurons, have left the growth-duplication cycle and continue to grow without any further mitoses.

Mitosis, or nuclear division, marks the end of the cycle in eukaryote cells and is usually preceded by DNA replication. Mitosis is absent in prokaryote cells, and in these cells DNA replication is continuous throughout the cycle, stopping before cell division.

The cell cycle in eukaryotes includes a long interphase when most growth occurs, followed by mitosis, a short period of chromosome condensation, segregation, and cytoplasmic division. In contrast to prokaryotes, DNA synthesis in eukaryote cells is discontinuous; DNA doubling usually occurs in a restricted period of interphase. DNA synthesis may be studied by measuring the DNA content of cell

**85**

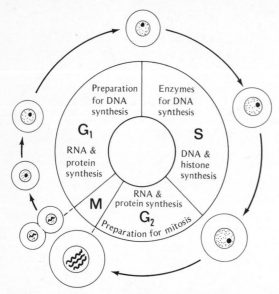

**Figure 4-1** *The growth-duplication cycle. After completion of mitosis (M), daughter cells enter the G₁ phase. Toward the end of G₁, syntheses preparing for DNA and chromosome replication take place. At the beginning of S, DNA and histone synthesis is initiated and chromosomes are replicated. After DNA is replicated, the cell with double its DNA content enters the G₂ phase, a short period in preparation for chromosome condensation. At mitosis, nucleoli disappear and prophase begins as chromosomes condense.*

nuclei or by following the incorporation of radioactive thymidine into interphase nuclei. This unique precursor for DNA synthesis is incorporated into DNA (and no other macromolecule) only when chromosomes replicate. For practically all eukaryote cells, thymidine incorporation into nuclear DNA occurs only during a small portion of interphase. Accordingly, Howard and Pelc proposed in 1953 that the cell cycle should be based on this discrete period of chromosome replication, designated as "S" for DNA synthetic period. For most cells, the S phase is preceded by a period of no DNA synthesis or "G₁" (G refers to a "gap" in which unknown processes are taking place). It was noted that cells do not enter mitosis immediately after completing DNA synthesis in the S phase. Another interphase gap or "G₂" precedes mitosis, during which the nucleus contains double the diploid amount of DNA while the cell prepares for mitosis.

Most dividing cell populations are asynchronous; that is, individual cells are in different phases of the cell cycle at any time. The duration of each phase of the cell cycle can be measured in such populations by following the kinetics of incorporation of radioactive thymidine or by measuring the amount of DNA in cell nuclei. An example of the latter is shown in Figure 4-2 for mouse fibroblasts. These dividing cells have a generation time of 18 to 20 hours, a G₁ of 8 hours, an S of 6 hours, and a G₂ of 5 hours. Mitosis is short and is completed in one hour. These times for phases of the cell cycle are representative of most mammalian cells in vitro.

The thymidine incorporation procedure, as applied to most animal and plant cells, shows that phases of the cell cycle vary considerably (Table 4-1). An extreme example is mouse ear epidermis with a G₁ of 22 days and an S of 30 hours. Plant cells have a more variable S phase and a longer mitotic time (3 to 4 hours) than animal cells to accommodate cell plate formation after telophase. In higher plants, the duration of the cell cycle and the length of the S phase vary directly with the amount of DNA in the species; the greater the DNA content, the longer the

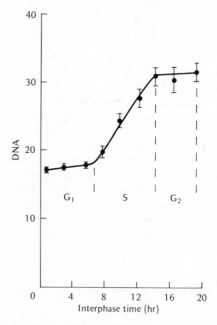

**Figure 4-2** *Pattern of DNA synthesis during interphase of mouse fibroblasts in a growth-duplication cycle. DNA is measured as relative Feulgen dye staining of nuclei in a microspectrophotometer. (From D. Killander and A. Zetterberg, Exp. Cell Res. 38:272–284, 1965.)*

**TABLE 4-1  Duration of Phases of Cell Cycle (in hours)**

| Cell Type | $G_1$ | S | $G_2$ | M | T |
|---|---|---|---|---|---|
| **In vitro** | | | | | |
| Human embryonic fibroblast | 2.5 | 11.5 | 4.5 | ? | 18.5 |
| Human kidney | 13.6 | 8.0 | 4.6 | 0.8 | 27.0 |
| HeLa | 12–16 | 8.5 | 3–7.5 | ? | 28.0 |
| Human skin | 11.2 | 5.4 | 3.9 | 1.2 | 21.7 |
| Fetal lung | 6.0 | 6.0 | 4.0 | 0.8 | 16.8 |
| Mouse L cell | 4.0 | 13.5 | 4.8 | 0.9 | 23.2 |
| Mouse fibroblast | 8.0 | 6.0 | 5.0 | ? | 19.0 |
| Rabbit kidney | 32–60 | 10.0 | 5.0 | ? | 47–57 |
| **In vivo** | | | | | |
| Mouse intestinal epithelium | 9.0 | 7.5 | 1.5 | 1.0 | 19.0 |
| Mouse uterine epithelium | 31.5 | 8.5 | 1.0 | 1.0 | 42.0 |
| Mouse ear epidermis | 22 days | 30.0 | 6.5 | 3.8 | 24 days |
| Mouse embryonic neural tube | 2.3 | 4.0 | 0.8 | 1.3 | 8.4 |
| Mouse ascites tumor | 2.5 | 8.5 | 7.0 | ? | 18.0 |
| Rat liver (1 day old) | 5.0 | 7.0 | 1.5 | 0.3 | 13.8 |
| Rat liver (8 weeks old) | 28.0 | 16.0 | 1.8 | 1.7 | 47.5 |
| Chick embryo retina | 4.0 | 4.0 | 2.0 | ? | 10.0 |
| Grasshopper neuroblast | 0.0 | 1.5 | 0.0 | 2.0 | 3.5 |
| Bullfrog lens epithelium | 30 days | 48.0 | 7.0 | ? | 33 days |
| *Tradescantia* root tip | 1.0 | 10.5 | 2.5 | 3.0 | 17.0 |
| Onion root tip | 10.0 | 7.0 | 3.0 | 5.0 | 25.0 |

Source: Based on J. E. Cleaver, *Thymidine Metabolism and Cell Kinetics*. Amsterdam: North-Holland, 1967.

S period, suggesting that rates of DNA strand replication are the same in all species. Some trends are apparent. The S and $G_2$ phases are relatively constant in all species, but the $G_1$ phase shows the greatest variability in response to physiological conditions. We will see later that it is probably in the $G_1$ phase that the principal events of division and growth regulation occur.

Cells in early embryos are almost entirely in S since their generation times are quite short, often equivalent to the 15- to 20-minute cycles of prokaryotes. And like prokaryotes, they may have no $G_1$ or $G_2$ phases. In the early frog embryo (*Xenopus*) the total cell cycle is 20 to 25 minutes with no $G_1$, a 17-minute S, a 4-minute $G_2$, and a mitotic time of 4 minutes. But this changes during development; a substantial $G_1$ and $G_2$ become apparent as the cell cycle gradually lengthens. Since cell division rates decline during development with increases in the length of S to several hours, this means that total genome DNA is replicated 100 times faster in early embryos than in late embryos or adult tissues. (See Chapter 5.)

Most somatic cells of the developing embryo leave the growth-duplication cycle as they acquire their specialized functions. Some adult cell types such as neurons, having left the growth-duplication cycle, also abdicate division potential. It is during this period that cells undergo maximum growth, usually by intense synthesis and accumulation of cell-specific proteins and organelles.

## KINETICS OF AUXETIC GROWTH

For many years, the only approach to studying single cell growth was to measure fixed cells at different times during growth. Linear dimensions were used to calculate cell volumes, but such studies were unreliable because of shrinkage, loss of low-molecular-weight compounds, and other fixation artifacts. More recently growth measurements have been made on living cells, but these too are subject to error since cell volume and mass fluctuate with water or nutritional uptake. Modes of nutrition differ for each cell and must be carefully controlled.

One problem is to determine the kinetics of cell enlargement between divisions. Is there a basic pattern common to all cells? The answer to this question is not simple because the criteria for measuring growth (for example, dry mass, volume, and linear dimensions) do not behave consistently even within the same cell. Early studies on fixed cells revealed two patterns of cell volume increase: linear and exponential. These have been confirmed in more recent studies on living cells (Figure 4-3). Each of these patterns for mass increase tells us something different about the nature of "true" growth. An exponential pattern means that growth rate is a function of total mass; as mass increases, growth rate increases accordingly. The curve is sigmoidal, behaving as an autocatalytic process with growth rate proportional to the amount of active protoplasm or replicating entities.

Linear kinetics, on the other hand, means that growth rate is constant throughout the cell cycle and does not increase. Growth rate is independent of cell mass but is related to a constant number of elements (synthetic sites?), the activities of which remain unchanged throughout the growth cycle. The distinction is basic and implies fundamental differences in the mechanisms underlying growth.

Sometimes it is difficult to distinguish linear from exponential growth kinetics. For example, some experiments on *E. coli* cells suggest that growth, as measured by cell volume, is linear with growth rate doubling after each division. Other measurements of bacterial growth suggest exponential kinetics. The amounts of DNA, RNA, and protein in *E. coli* increase exponentially during synchronous growth. These studies are complicated by short cell cycles which limit the number of measurements needed to resolve linear from exponential kinetics.

Eukaryotic cell growth is easier to measure; life cycles are longer, cells are larger, and the mass of a single cell may be followed continuously throughout the growth cycle. Moreover, cells with different modes of nutrition can be compared.

Mitchison measured the dry mass and volume of a budding yeast cell (Figure 4-4) and demonstrated that mass increases linearly after division without a lag and continues at a constant rate until it doubles before the next division; that is, the combined growth rate of two daughter cells is double the original rate of the mother cell. Cell volume, on the other hand, follows a roughly exponential curve for the first three quarters of the cell cycle, reaching a plateau prior to division. The kinetics for volume and dry

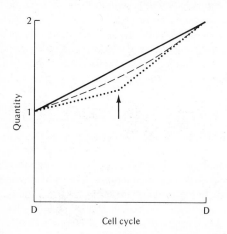

**Figure 4-3** *Patterns of macromolecular synthesis during a division cycle (D-division). Continuous line shows a linear increase; dashed line shows an exponential increase; dotted line shows a linear increase with doubling of rate in the middle of the cycle.*

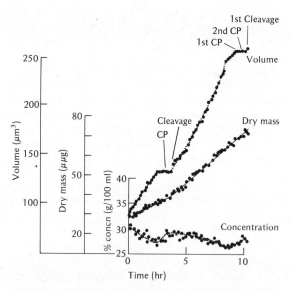

**Figure 4-4** *Growth curves of yeast cells (Schizosaccharomyces pombe) as measured with the interference microscope. Volume, dry mass, and concentration are compared at 23°C as one cell is followed through division into two daughters and subsequently to division of one of the daughters. [From J.M. Mitchison, J. Cell Comp. Physiol. 62 (Suppl. 1):1–31, 1963.]*

mass differ, suggesting that factors regulating one do not operate for the other.

Yeast growth kinetics points up several important features common to the growth of cells that absorb nutrients through their surface: (1) growth rate is constant between divisions and doubles immediately after; (2) cell mass doubles between divisions; (3) mass increase is correlated with nuclear changes rather than cytoplasmic, a strong argument in favor of the nuclear control of growth; (4) there is no lag period—linear growth begins immediately after division.

Exceptions to this pattern are found in the growth of some protozoa, such as *Amoeba*. Though mass in *Amoeba* doubles between divisions, its growth shows a diminishing rate of increase, reaching a plateau several hours before mitosis (Figure 4-5).

Though these results suggest that genetic factors determine growth patterns, some differences are nutritional. Yeast cells absorb exogenous nutrients directly across the cell membrane while *Amoeba* engulfs solid food. Contractile vacuole changes in *Amoeba* introduce extraneous perturbations into cell mass measurements. Furthermore, a diminishing growth rate and a period of constancy at the end of *Amoeba* growth reflect a decline in feeding before division. Such nutritional differences must be controlled when comparing growth kinetics in different cell systems.

Studies of single cells can only provide us with

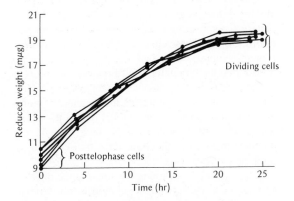

**Figure 4-5** *Growth curve of several individual amoebas as measured by reduced weight. Note the decline in growth and plateau before division. (From D.M. Prescott,* Exp. Cell Res. *9:328–337, 1955.)*

information about gross changes in cell mass, since it is difficult to study the biochemical changes in single cells. To examine macromolecular changes in growth duplication cycles, we need large synchronized cell populations.

## SYNCHRONOUS GROWTH

Regulation of the cell cycle is best studied in cell populations, all members of which are in synchrony with respect to growth and division. Most naturally growing populations are too heterogeneous because at any one moment cells are in different phases of growth and cell division. Biochemical analyses of such populations yield only an average metabolic state and tell us little about the physiological state of cells as they pass through each phase.

Synchronous growth is seen in some natural populations such as cleaving embryos. Though total cell mass is actually decreasing in these systems, cells divide synchronously and become smaller at each division.

Synchronously growing cell populations are produced artificially in the laboratory by techniques that yield mass populations exhibiting balanced growth over one or two generations. Most cells (90 to 100 percent) grow and divide at the same rate during the synchronous period; that is, they are in phase with respect to their morphological state, synthetic activity, and growth potential. Though perfect synchrony is seldom obtained because of inherent variability of cell generation times, changes in synthesis and turnover of macromolecules can be correlated directly with specific phases in the growth cycle.

An example of synchronous growth of yeast cells is shown in Figure 4-6. Synchrony is perfect over the first two cycles but is soon lost as generation time heterogeneity takes over. Though cell mass increases linearly between divisions, total protein per cell increases exponentially throughout the cell cycle, doubling just before division. Total cell RNA also increases exponentially throughout the cell cycle, running parallel with protein synthesis and the linear increase in mass. DNA, however, shows a different pattern entirely; it is synthesized discontinuously, usually in the middle of interphase, long before mitosis and cell division begin.

Many eukaryotic cells exhibit similar patterns

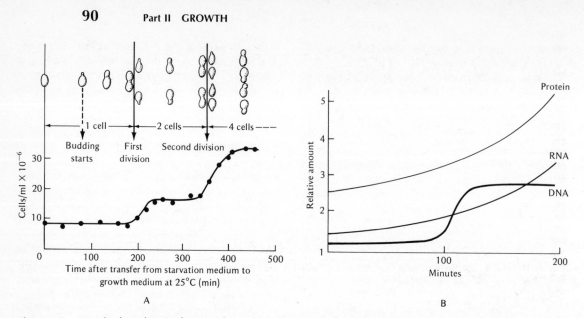

**Figure 4-6**    *Growth of synchronized yeast culture of* Saccharomyces cerevisiae *after inoculation into a synthetic medium. (A) Changes in cell number and appearance of cells during two division cycles. (Based on D.H. Williamson and A.W. Scopes, Symp. Soc. Gen. Microbiol., vol. 11. New York: Cambridge University Press, pp. 217–242, 1961.) (B) Changes in macromolecular synthesis during the first division cycle of a synchronized culture. Note that RNA and protein syntheses follow similar exponential patterns while DNA shows a stepwise increase midway in the cycle.*

of macromolecular synthesis. In most cases, whenever there is a doubling of growth rate, it occurs midway in the cycle. There seems to be no one pattern of cell growth common to all cells. Exponential and linear kinetics may be distinct for each cell type, or we may not be able to resolve the two patterns, given the fact that the shape of growth curves varies from cell to cell with changes in medium and temperature.

## REGULATION OF CELL GROWTH

Some linear growth patterns, particularly in cultured mammalian cells, display a doubling in growth rate, usually midway in the cycle during the S phase. Doubling of growth rate during the period of DNA replication suggests that control of cell growth depends on factors regulating DNA synthesis. Though this may be true for mammalian cells, it is not the case in yeast and protozoa where doubling of growth rate occurs at the end of the cycle, long after DNA synthesis has been completed. If DNA replication is not responsible for growth regulation, what is?

Mammalian cell growth seems to involve a continuous feedback interaction between the two major compartments. The dry mass of nucleus and cytoplasm of synchronized mammalian cells exhibit different growth kinetics (Figure 4-7). Most nuclear growth occurs in the latter half of the cycle, whereas cytoplasmic mass increases more rapidly at the beginning of the cycle and falls off at the end. A similar pattern is seen in the volume of these compartments during growth. These curves suggest that cytoplasmic proteins synthesized in the latter half of the cycle are transported to the nucleus. Furthermore, amino acid incorporation into nuclear and cytoplasmic proteins during these phases indicates that nuclear proteins flow into the cytoplasm in the first half of the cycle, during $G_1$. These proteins may have a regulatory function in DNA replication, RNA transcription, or protein synthesis.

One factor minimizing the importance of DNA replication in cell growth regulation is that replication is less sensitive to nutritional changes than is growth rate. Moreover, growth may be dissociated from replication by inhibition of cell division; growth will continue even if DNA replication is inhibited.

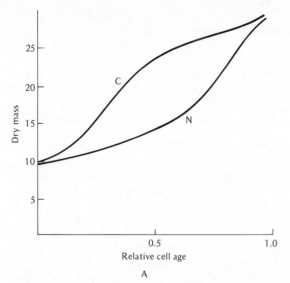

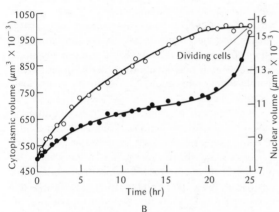

**Figure 4-7**  *Different patterns of growth in nucleus (N) and cytoplasm (C). (A) Comparative increase in dry mass of cytoplasm and nucleus in mouse fibroblasts. The S-shaped curve for the nucleus correlates with the cytoplasmic curve; proteins shift from cytoplasm to nucleus toward the end of the cycle. (From A. Zetterberg, Exp. Cell Res. 42:500–511, 1966.) (B) Volume changes during the amoeba cell cycle. Open circles, cytoplasm; closed circles, nucleus. Here too nuclear volume increases markedly before division. (From D.M. Prescott, Exp. Cell Res. 9:328–337, 1955.)*

In such cells, RNA and protein synthesis are unaffected, which implies that control of growth rate is probably more directly concerned with these processes than with DNA replication.

## RNA and protein synthesis as sites of growth regulation

RNA and protein make up the bulk of cell mass, and their synthesis is more responsive to changes in nutritional conditions than is DNA replication. Accordingly we would expect synthesis of these macromolecules to be more directly linked to cell growth regulation. Both follow similar kinetics in all cells and are similar to the kinetics of cell mass increase in linear and exponential growth.

Generally, ribosomal and transfer RNAs comprise more than 90 percent of total extractable cell RNA. Both RNAs are synthesized continuously throughout the eukaryote cell cycle. The rate of synthesis may increase during the cycle, but there is no uniform pattern except in mammalian cells when the rate of ribosomal RNA synthesis doubles, presumably after ribosomal RNA genes replicate during the S phase along with all other genes. The pattern for mRNA is not known since different species of mRNA are synthesized at different periods of the cycle, perhaps at different rates. It seems likely that total mRNA synthesis resembles the pattern for stable RNA.

The correspondence between overall cell growth patterns and ribosomal RNA synthesis suggests that the production of ribosomes might be an important site for growth regulation. Apparently, the total number of ribosomes in a bacterial cell controls the rate of synthesis of all proteins during growth; that is, "the number of ribosomes per DNA genome is proportional to the rate of growth and protein synthesis" (O. Maaløe). If this is true for eukaryote cells, then growth rate should correlate with the total number of cytoplasmic ribosomes per cell, and these are controlled by the nucleolus, the seat of ribosome synthesis.

## The nucleolus: a cell growth regulator?

Nucleoli are semiautonomous; that is, they contain DNA templates for ribosomal RNA synthesis and are rich in protein. The number of nucleoli corresponds to the number of nucleolar organizers in the genome which in turn determines the number of ribosomal RNA genes per cell. The nucleolar organizer as a site for ribosomal RNA genes is seen in *Drosophila* where strains with different numbers of nucleolar organizer regions per cell genome can be obtained. The nucleolar organizer is found in the heterochromatic regions of both X and Y chromosomes. Using flies with X chromosome inversions, stocks whose cells contain one, two, three, or four nucleolar organizer regions can be produced. Radioactively labeled ribosomal RNA is prepared and combined with DNA extracted from each of the stocks in DNA-RNA hybridization studies to determine the relative levels of DNA saturation. The results show a direct correlation between the number of nucleolar organizers present and the levels of RNA saturation of the DNA (Figure 4-8). The saturation level of DNA from flies with four nucleolar organizers is twice that of flies with two and four times those with only one. This means flies with four nucleolar organizers have twice the number of DNA sequences coding for ribosomal RNA than flies with two nucleolar organizers. The analysis shows that each nucleolar organizer region contains approximately 130 copies of the ribosomal RNA genes.

Cyclic changes in nucleolar behavior correlate with cell growth. During interphase, when cells are

actively growing, nucleoli are prominent and synthesize ribosomal RNA at a high rate. In prophase, when growth stops, nucleoli disappear, emptying their contents into the nucleoplasm. Nucleoli are absent in metaphase and anaphase but reappear early in telophase at twice their original number, as new nucleoli organize at nucleolar organizer sites in each daughter nucleus. The combined growth rate of the daughter cells increases to twice the rate of the original mother cell though total protoplasmic mass has not changed. The transition from a state of physiological "oneness" to that of physiological "twoness" is most closely coordinated with duplication of nucleolar organizer regions and formation of nucleoli. Conceivably, growth rate doubles after nucleolar organizer regions are replicated during the S phase when twice the number of ribosomal RNA cistrons begin to transcribe ribosomal RNA.

Even a small portion of the organizer region can generate a functional nucleolus. Breaks in the nucleolar organizer region have been induced by X-rays, and the fragments have been translocated to other chromosome regions. Each of the many small fragments in its new location is capable of producing a fully functioning nucleolus, testifying to its polycistronic nature.

Cell growth may be related to the number of nucleolar organizer regions. Longwell and Svihla in 1960 studying dry-mass changes of pollen precursor cells in wheat showed that an increase in the number of nucleolar chromosomes from four to six resulted in a 50 percent increase in cytoplasmic dry mass and a corresponding increase in cytoplasmic RNA (ribosomal RNA) content. No such increase was seen when the dosage of nonnucleolar organizing chromosomes was increased. We would expect an elevated synthesis of ribosomal RNA if the number of ribosomal RNA genes were to increase but evidence for this is still incomplete.

The nucleolus is a dynamic organelle, attuned to metabolic demands, responding rapidly to changing needs for new patterns and rates of growth. It becomes large and metabolically active in most growing and proliferating cells, such as tumors, and disappears in cells not active in protein synthesis. When there is a heavy demand for ribosomes, as in maturing oocytes, nucleolar function is amplified by making many additional nucleoli (up to 1000 in some species), each equipped with a segment of

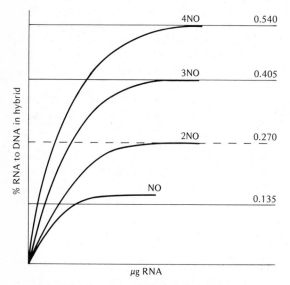

**Figure 4-8**  *Saturation levels of Drosophila DNA containing different dosages of nucleolar organizer (NO) regions with RNA. NO dosage is indicated for each curve. Dotted horizontal line is a calculated estimate for a dosage of 2, and solid lines indicate predicted levels for dosages of 1, 3, and 4, respectively. Numerical values of saturation are given at right. (From F.M. Ritossa and S. Spiegelman,* Proc. Nat. Acad. Sci. U.S. *53:737–745, 1965.)*

DNA containing many copies of the ribosomal RNA cistrons (see Chapter 11).

Growth regulation seems to hinge upon the nucleolus and its control over ribosome synthesis. Ribosome production and turnover may determine cell growth, with the nucleolus as the principal "flow-through" center or "valve" regulating the entire process. In eukaryotic cells growing at a linear rate, we assume that the rate of production of ribosomes is constant. When the genome replicates, the number of copies doubles but the controlling assembly center, the nucleolus, does not duplicate at the completion of the DNA synthetic phase, and rates of ribosome synthesis may not necessarily change after DNA replication. Not until new nucleoli (or new centers of ribosome assembly) are reconstituted at the end of mitosis does growth rate double.

### Protein synthesis during a cell cycle

Though useful as a working hypothesis, the above version of growth control oversimplifies a complicated network of macromolecular synthesis during cell growth. Most eukaryote cell growth results from total protein accumulation—the net balance between total protein synthesis and protein degradation. Both processes are subject to different modes of regulation, the former largely dependent upon the availability of ribosomes. Total protein synthesis is not regulated en masse; it is a net average of different patterns of regulation for each cell protein. For example, though total cell protein may seem to increase continuously through the cell cycle, some individual proteins may be constant, others may be decreasing, and still others may be increasing in a stepwise fashion. If we assume that levels of enzyme activity correlate with actual amount of enzyme present, then four different patterns of enzyme synthesis can be detected during a cell cycle (Figure 4-9). (1) Synthesis may be periodic like DNA synthesis and increase rapidly during one phase of the cycle. Enzymes involved in DNA synthesis, such as thymidine kinase, do show a stepwise pattern. (2) Continuous synthesis of enzymes, either linear or exponential, is typical of many respiratory enzymes in mouse fibroblasts. (3) Some enzymes show a peak pattern. They increase rapidly during the cycle, then disappear, presumably as a consequence of degradation and turnover. Within the same cell, each

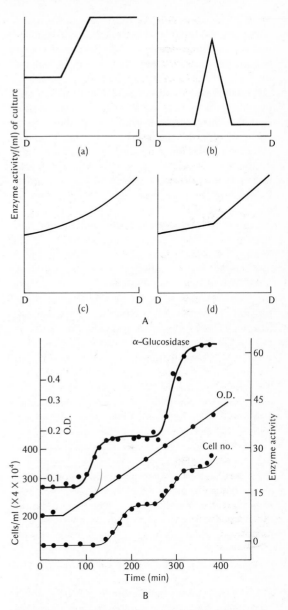

**Figure 4-9**  *Patterns of enzyme synthesis during the cell cycle. (A) Various patterns seen in different cells: (a) a step synthesis, (b) peak synthesis followed by decline, (c) continuous exponential, and (d) continuous linear. (B) Example of step synthesis of α-glucosidase during synchronous growth of the yeast Saccharomyces cerevisiae. O.D. = optical density, a measure of cell concentration. (B from P. Tauro and H.O. Halvorson, J. Bacteriol. 92:652–661, 1966.)*

protein may be regulated independently of others. Some sets of proteins may be coordinately regulated, particularly those proteins that characterize a cell phenotype. Moreover, patterns seem to be different for each cell type, often depending on the number of DNA templates and their availability for transcription. In prokaryotes, where many enzymes show stepwise increases, the synthesis of the enzyme coincides with the replication of its specific coding gene. When the latter doubles, the rate of enzyme synthesis also doubles. This is particularly true for inducible enzymes in *E. coli* where rates of enzyme induction double after the gene for $\beta$-galactosidase replicates.

Most prokaryote genes are available for transcription and translation at all stages of the cell cycle. During the growth of *E. coli* almost 100 percent of the genome is transcribed as measured by DNA-RNA hybridization studies. Even in sporulating *Bacillus subtilus,* where a sequential developmental program is expressed as spores are formed (see Chapter 17), all genes are transcribed. This is in contrast to the eukaryote genome, where only 10 to 20 percent may be active (see Chapter 3). A large proportion of the genome is silent, but even those genes that are active are not transcribing throughout the cell cycle. For example, the enzyme tyrosine amino transferase of rat hepatoma cells cannot be induced during $G_2$, M, and the first three hours of $G_1$, but may be induced during most of $G_1$, and S, representing 65 percent of the cell cycle.

Regulation of protein synthesis is complicated since changes in many "step" and "peak" enzymes in prokaryotes as well as eukaryotes seem to be unrelated to DNA replication. DNA synthesis may be inhibited without affecting the pattern of enzyme synthesis. This implies that gene dosage does not influence the observed pattern and other types of regulation must be involved.

Two hypotheses have been proposed to explain such patterns of protein synthesis. One model for control of periodic enzyme synthesis, called "oscillatory repression," invokes end product (the enzyme) repression of its own synthesis by negative feedback. When the enzyme pool is high, synthesis is repressed; when low, synthesis is increased. This pattern will lead to stable oscillations which need not correspond in frequency to other events in the cell cycle, such as DNA synthesis. Oscillations are entrained by an event independent of the cycle, such

as a pulse of mRNA synthesis. The second model states that a chromosome is itself programmed for sequential expression of genes at different stages in the cell cycle. Here an ordered reading of genes is emphasized. The RNA polymerase moves along the genome transcribing genes in sequence. Hence genes are available for transcription only at specific periods of the cell cycle. For example, linear reading of genes has been reported for synchronized growing yeast. The time of synthesis of 12 different enzymes in yeast corresponds to the position of their respective genes in the chromosome. Each enzyme is synthesized in a stepwise manner with the position of the gene seemingly dictating the order of its expression during the cell cycle. The enzymes encoded in genes linked on the same chromosome are synthesized in a definite order depending on the distance of respective genes from the centromere. The gene closest to the centromere is the earliest to be expressed. In this sense, the chromosome is truly a programmed tape, read in a temporal sequence which could explain the sequential appearance of enzymes in several developing systems. (See Chapters 9 and 17.)

By no means do these two models account for the many aspects of protein regulation during the cell cycle. Though oscillatory repression may be more applicable to prokaryotes, both mechanisms are probably found in all cells. The models do not explain enzymes that increase continuously throughout the cycle nor those that respond to changes in gene dosage. Besides, many proteins in eukaryote cells are subject to translational or posttranslational controls, and some show high turnover rates. Furthermore, enzyme activity is not always a measure of enzyme content; many factors such as activators, inhibitors, and degree of compartmentalization also affect enzyme activity. We see, then, that linear or exponential increases in cell mass resulting from synthesis of total protein in the cell cycle conceal a host of independently controlled patterns of enzyme synthesis.

Most mature somatic cells are functionally specialized and are not in a growth-duplication cycle. For these cells, protein synthesis is not coupled to cell division but to maintenance, replacement, and repair of functional cytoplasmic systems—for example, the contractile systems of muscle or the secretory mechanisms of pancreas cells. Patterns of regulation in macromolecular syntheses need not resemble regulation of cell growth in proliferating cells. Protein synthesis in specialized cells, directed to synthesis and assembly of cell organelles, is balanced by protein turnover and degradation. Such cells normally do not grow; or if they do, they do so in response to functional demands—for example, muscle cells hypertrophy by synthesizing more myofibrils and contractile proteins.

## THE SIZE OF CELLS

There is a striking size difference among free-living single cells, as contrasted with most somatic cells in multicellular organisms. The range of size from the smallest to the largest single cell is several orders of magnitude greater than the size range for somatic cells (Table 4-2). The largest protozoan cell is approximately $10^{11}$ times the smallest bacterium, while among somatic cells volumes vary between $10^3$ and $10^6$ from the smallest to the largest. The majority of somatic cells in multicellular organisms is uniformly small in spite of a considerable variation in function and morphology.

What accounts for these size differences? Though genetic factors are implicated, we know that large cells require more gene products than small cells because a greater volume of cytoplasmic machinery must be maintained. To achieve large size, the amount of gene material is usually increased. This suggests that cell size regulation is tied to the nucleus or, more directly, to the total amount of DNA.

Hertwig in 1903 proposed a nuclear control of cell size in his hypothesis of the karyoplasmic or nucleocytoplasmic ratio; that is, the ratio of nuclear volume to cytoplasmic volume is specific for each cell type and is constant during cell growth. Nuclear volume was thought to be responsible for determining final cell size. Hertwig believed that a given nucleus had a limited sphere of influence over the cytoplasm and a cell normally stopped growing after

**TABLE 4-2   Cell Volumes in Different Species**

| Species | Cell Type | Cell Volume, $\mu^3$ |
| --- | --- | --- |
| Prokaryotes | | |
| *Pseudomonas indigofera* | | $3.0 \times 10^{-4}$ |
| *Bacillus megatherium* | | $0.3 \times 10^2$ |
| *E. coli* | | $0.25$ |
| Eukaryotes | | |
| *Schizosaccharomyces pombe* | | $0.8 \times 10^2$ |
| Single-Celled Protists | | |
| *Frontonia* | | $2.9 \times 10^6$ |
| *Actinosphaerium* | | $6.2 \times 10^6$ |
| *Spirostonum* | | $13.1 \times 10^6$ |
| *Opalina* | | $38.8 \times 10^6$ |
| Multicellular Organisms | | |
| *Homo sapiens* | Amnion | $3.0 \times 10^3$ |
| | Epithelium | $6.0 \times 10^3$ |
| | Cerebral cortex | $2.3 \times 10^3$ |
| *Mus* | Tumors | $2.0 \times 10^3$ |
| *Rattus* | Liver | $1.5 \times 10^3$ |
| *Rana* | Liver | $8.7 \times 10^3$ |
| *Triturus* | Epidermis | $3.2 \times 10^3$ |
| *Crepidula (mollusk)* | Epithelium | $1.7 \times 10^3$ |
| *Fucus* | Hypha | $7.0–95.0 \times 10^3$ |
| *Acer* | Epithelium | $7.3 \times 10^3$ |
| *Allium* | Root tip meristem | $1.2–10.0 \times 10^3$ |
| | Mature cell | $9.0–950.0 \times 10^3$ |

reaching this limit. If cytoplasmic growth continued beyond the limit, the ratio was disturbed; the cell became unstable and divided. Constancy of the karyoplasmic ratio insured balanced cell growth. Constancy, however, did not hold in many systems, particularly cleaving eggs where cells decrease in size without concomitant adjustment of nuclear volume. But we now know that cell size correlates with chromosome number and DNA content rather than with nuclear volume. The latter, sensitive to physiological change, is not always an accurate indicator of nuclear control.

As far back as 1902, Boveri demonstrated that nuclear size is directly proportional to chromosome number in sea urchin embryos. Nuclei were largest in tetraploid larvae and smallest in haploids. Since Boveri's study the relation between ploidy and cell size has been well established. In the amphibian embryo, for example, where ploidy may be experimentally manipulated, a progressive increase in cell size in comparable tissues of haploid, diploid, and polyploid tadpoles can be demonstrated. In virtually all cells, any increase in the number of chromosome sets is accompanied by an increase in cell size.

It is not chromosome number but nuclear DNA content that correlates with cell size. For example, larval growth in most *Diptera* occurs by cell enlargement (cell hypertrophy) between each molting period and not by cell division. Both cell and nuclear volume are proportional to different stages of DNA replication by polytenization. DNA strands replicate eight to ten times without any chromosome replication. Growth of larval epidermis in *Calliphora* (the blowfly) is accompanied by continuous replication of chromosomal strands, reaching a state of polyteny equivalent to 1024 times the haploid set. We find that cell size corresponds directly to the degree of polyteny as measured by the amount of DNA per nucleus (Figure 4-10). Nuclear volume, cell volume, and the amount of DNA per nucleus fall into distinct classes as multiples of the haploid condition. Each duplication of the genome is accompanied by a balanced adjustment of cytoplasmic volume; that is, as DNA content increases, cell size increases proportionally.

Amplification of the genome is accomplished by a balanced increase in the number of chromosomes (polyploidy), a balanced increase in the amount of DNA per chromosome (polyteny), or a selected replication of different portions of the genome (gene redundancy). If the number of gene copies is greatly increased, rates of gene product synthesis increase concordantly.

Certain genes rather than total DNA content seem to be most important in controlling cell growth rate and size. Von Wettstein demonstrated that cell size in haploid and diploid phases of many moss species does not always show a 1:2 ratio. Diploid cells are often 1.5 to 8 times larger than haploid cells of the same strain, and this ratio persists from generation to generation. Since the number of haploid sets does not correlate with cell size it suggests that specific genes are concerned with size regulation.

The presence of specific chromosomes also affects cell size. In the plant *Crepis tectorum* additions of certain single chromosomes increase or decrease total cell mass. In addition, when some accessory chromosome fragments are included in some plants, cell size increases without showing any other phenotypic change. These observations imply that certain genes, but not all, are directly concerned with cell size regulation.

It is tempting to speculate that the genes controlling cell size and growth rate may be the reiterated sequences of ribosomal RNA cistrons in nucleolar organizer regions. These would double in each polyploid or polytene series and stimulate protein synthesis by increasing the population of ribosomes.

## SUMMARY

As we have seen, a most important feature of organismic growth is the manner in which cell mass increases. For cells in a growth-duplication cycle, a period of mass increase precedes cell division. For somatic cells, mass increase is usually associated with specialization of cell function. Control of cell enlargement is probably coupled to factors that regulate cell division and differentiation in developing systems.

Cells in a growth-duplication cycle pass through different phases in relation to the period of DNA replication, which is discontinuous in most cells. For such cells, growth may be linear or exponential, depending on cell type. In general, cells grow to double their mass before they divide. In each case,

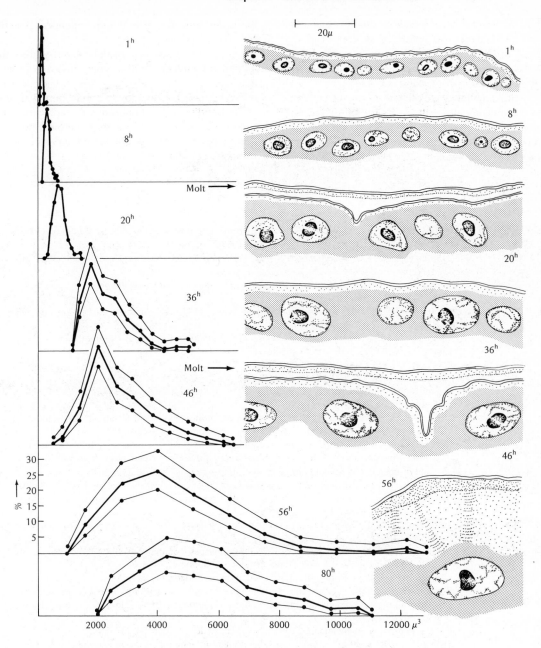

**Figure 4-10**  *Growth of epidermal nuclei of* Calliphora erythrocephala *during larval development. Nuclear volume changes in the graphs are separated into discrete, evenly spaced classes. Chromosomes replicate without mitosis (polyteny), and the amount of DNA per nucleus is doubling within each nuclear size class. Light lines indicate 95.5 percent confidence limits. Histological appearance of epidermis is shown on the right. h = hours after hatching from egg. (Based on G. Wagner, Zeit. Naturforsch. 6b:86–90, 1951.)*

the kinetics of mass increase is usually matched by parallel syntheses of proteins and RNA. The synthesis of DNA, being discontinuous, is not directly related to the kinetics of cell growth. It is, however, more involved in controlling cell size.

The doubling of growth rate during the cell cycle seems to correlate with the behavior of nucleoli, sites of ribosome synthesis. Cell growth depends, in part, on the total population of ribosomes, suggesting that growth regulation is linked to ribosome synthesis and turnover. This, in turn, influences patterns of protein synthesis during the cell cycle. Growth regulation is, however, complicated because the patterns of synthesis and turnover for each protein may vary. This is particularly true for specialized cells that have left the growth-duplication cycle and are responding to functional cues instead. Sometimes such cells are induced to enter the growth-duplication cycle, and we now ask, what happens in a cell to trigger cell division?

**REFERENCES**

Adams, J. 1977. The interrelationship of cell growth and division in haploid and diploid cells of *Saccharomyces cerevisiae. Exp. Cell Res.* 106:267–275.

Adolph, E. F. 1931. *The regulation of size as illustrated in unicellular organisms.* Baltimore, Md: Thomas.

Balls, N., and Billett, F. S., eds. 1973. *The cell cycle in development and differentiation.* New York: Cambridge University Press.

Cameron, I., Padilla, G. M., and Zimmerman, A. M. 1971. *Developmental aspects of the cell cycle.* New York: Academic Press.

Cleaver, J. E. 1971. *Thymidine metabolism and cell kinetics.* Amsterdam: North-Holland.

Conklin, E. G. 1912. Body size and cell size. *J. Morphol.* 23:159–188.

Enger, M. D., and Tobey, R. A. 1969. RNA synthesis in Chinese hamster cells. II. Increase in rate of RNA synthesis during $G_1$. *J. Cell Biol.* 42:308–315.

Halvorson, H. O., Carter, B. L. A., and Tauro, P. 1971. Synthesis of enzymes during the cell cycle. *Adv. Microb. Physiol.* 6:47–106.

Howard, A., and Pelc, S. R. 1953. Synthesis of deoxyribonucleic acid in normal and irradiated cells and its relation to chromosome breakage. *Heredity* Suppl. 6:261–273.

Killander, D., and Zetterberg, A. 1965. Quantitative cytochemical studies on interphase growth. I. Determination of DNA, RNA and mass content of age-determined mouse fibroblasts in vitro and of intercellular variation in generation time. *Exp. Cell Res.* 38:272–284.

Kubitschek, H. E. 1970. Evidence for the generality of linear cell growth. *J. Theoret. Biol.* 28:15–29.

Maaløe, O. 1969. An analysis of bacterial growth. *Communication in development,* ed. A. Lang. New York: Academic Press.

Mitchison, J. M. 1957. The growth of single cells. I. *Schizosaccharomyces pombe. Exp. Cell Res.* 13:244–262.

————. 1963. Pattern of synthesis of RNA and other cell components during the cell cycle of *Schizosaccharomyces. J. Cell Comp. Physiol.* 62 (Suppl. 1):1–13.

————. 1971. *Biology of the cell cycle.* New York: Cambridge University Press.

Prescott, D. M. 1955. Relation between cell growth and cell division. I. Reduced weight, cell volume, protein content and nuclear volume of *Amoeba proteus* from division to division. *Exp. Cell Res.* 9:328–337.

Ritossa, F. M., and Spiegelman, S. 1965. *Proc. Nat. Acad. Sci. U.S.* 53:737–745.

Szarski, H. 1976. Cell size and nuclear DNA content in vertebrates. *Int. Rev. Cytol.* 44:93–113.

Williamson, D. H. 1966. Nuclear events in synchronously dividing yeast cultures. *Cell synchrony,* ed. I. L. Cameron and G. M. Padilla. New York: Academic Press.

Zetterberg, A., and Killander, D. 1965. Derivation of synthesis curves from the distribution of DNA, RNA, and mass values of individual mouse fibroblasts in vitro. *Exp. Cell Res.* 39:22–32.

# 5

# Cell Division: Multiplicative Growth

Small size and short life cycles may account for the limited developmental repertoires of single-celled organisms. Full developmental potential is realized only with the evolution of multicellularity where cell division has become a primary growth force. Growth patterns depend upon precise control of the timing and location of mitotic activity. In higher plants, for instance, proliferating cell populations are confined to discrete growth zones, or meristems, where cells divide, elongate, and generate the longitudinal growth patterns of shoots and roots.

Though "growth centers" are uncommon in animals, temporal ordering of mitotic potential is at the basis of the growth of parts in relation to the whole. An egg proliferates into a multicellular aggregate before morphogenesis can begin while spatially ordered patterns of cell division affect an organism's size and symmetry. The regulation of proliferative potential becomes a major problem in our efforts to understand development.

Early in development all cells divide, but later, as the embryo grows, replication slows down and cells specialize. Some cells irreversibly lose the capacity to divide, as if division genes are permanently shut off. Other cells, though functionally differentiated and nondividing, never lose mitotic potential. They can be induced to proliferate when isolated in a nutrient culture medium or exposed to strong stimuli such as wounding. Another class of adult somatic cells remains mitotically active throughout the life of the organism, responding to functional demand for new cells in such tissues as bone marrow and intestine. Since mitotic mechanisms are similar in most eukaryote cells, we assume, as a first approximation, that a common control system is shared by all. Cells escaping these controls, either by mutation or by accident, proliferate into a neoplastic growth (Chapter 21). In each case we try to determine how a cell becomes committed to divide and what turns off the division mechanism.

**99**

## CELL DIVISION AND CELL MASS

In a growth-duplication cycle mitosis is the culmination of auxetic growth. Proliferating cells possess a "division clock" that is coupled to cell growth in most cases. Growth and division are independent, dissociable processes, however, inasmuch as cells may grow without dividing or divide without growing. Nevertheless, the two are so frequently interrelated as to suggest that achievement of a critical cell mass triggers a cell to divide.

The relationship between cell size, growth rate, and division is illustrated in *Amoeba* (Figure 5-1). Before an amoeba divides, it first grows to a maximum size. Small and large cells each grow to the same mature division size before they enter mitosis and do so in approximately the same length of time by adjusting their respective growth rates. Division may be inhibited by surgically removing cytoplasm of fully grown cells. After the loss of cytoplasm, the cell resumes growth until full size is restored. But if the operation is repeated before the cell enters mitosis, it does not divide and begins growing again. This procedure may be continued for more than 50 growth cycles without cell division and without any

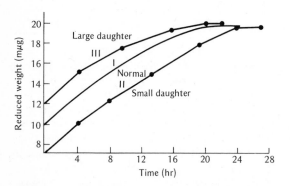

**Figure 5-1**   *Cells of different size reach the same division mass in* Amoeba. *Curve I illustrates the normal pattern for a typical cell. Cell mass doubles before division, reaching a plateau several hours before mitosis. Curves II and III are patterns for small and large cells obtained after an unequal division induced by strong illumination. Though initially at different masses, each reaches the division size characteristic of the species rather than doubling in mass. (From D.M. Prescott, Exp. Cell Res. 1:86–98, 1956.)*

reduction in growth potential. In this case, surgical removal of cytoplasm substitutes for cell division and sustains auxetic growth. It suggests that a critical cell mass must be attained before a cell can divide.

A similar relation between the division clock and critical cell mass is found in mouse fibroblasts. Small cells with a low initial mass spend a longer time in $G_1$ than large cells before reaching a critical mass and pass through a short S and $G_2$ before division. Initial size of a cell determines the length of its $G_1$ phase, as if cell mass acts as a mitotic trigger.

A cell grows by synthesizing macromolecules, and before it divides it completes a "division readiness program" that prepares it for mitosis. Synthetic activities during $G_1$ and S are part of this program. Chromosomes are replicated, which means that DNA and chromosomal proteins are synthesized. Spindle proteins are synthesized and assembled into a mitotic apparatus to segregate chromosomes. Regulatory proteins must also be on hand to couple cytoplasmic to nuclear events. Respiratory energy is funneled into the division cycle to support macromolecular synthesis, chromosome coiling, and assembly of the division apparatus. All preparations for division are coordinated to facilitate an orderly progression through mitosis. Sometime during this period, a cell is committed to mitosis; having reached a point of no return, a cell will complete division even if protein and RNA synthesis are inhibited or the cell's supply of energy is cut off.

Though the amputation experiment suggests that division readiness is achieved when a cell grows to a critical mass, other interpretations are possible. Surgical removal of cytoplasm does more than reduce cell mass; it also decreases the number of regulatory molecules which may have to be at threshold levels before division can occur. Besides, division readiness can be dissociated from cell mass. For example, binucleate amoebae can be produced by growing cells to full size and allowing mitosis to occur while experimentally preventing cell division. These fully grown amoebae possess two nuclei in a mature cytoplasm. If one nucleus is subsequently removed, cells with a mature cytoplasmic mass containing a newborn nucleus result. Nevertheless, these fully grown cells will begin to divide only after 16 hours. During this time they do not grow, which suggests that approximately 16 hours of the normal

24-hour cell cycle are necessary to prepare a cell for mitosis. Additional events in both nucleus and cytoplasm besides mass increase must be completed before a cell is ready to enter mitosis. Part of that time is devoted to DNA replication during a 6-hour S phase and a long $G_2$ when other regulatory events are completed. Growth to mature size is a necessary but not a sufficient condition for mitosis.

Is DNA replication the trigger to mitosis? Here, too, we find that DNA replication is necessary but not sufficient. Cells complete DNA synthesis in the S phase long before they enter mitosis. Though DNA synthesis always precedes mitosis, it happens too early in the cycle to act as a trigger. The entire $G_2$ period seems to be devoted to setting the mitotic clock into motion. Moreover, DNA replication is dissociable from mitosis in polyploid or polytene cells where DNA replication occurs without any mitosis or cell division.

Though the two events are not causally related, it is generally assumed that when most cells divide, DNA replication has occurred. It is probably true that once a cell begins DNA synthesis, it completes cell division without further interruption. Completion of DNA synthesis may be the point of no return in the cell cycle.

## REPLICATION OF PROKARYOTE DNA

Division of a cell is a reproductive act that expresses at the cell level the self-replication properties of nucleic acid macromolecules. These molecules serve as templates for their own replication by base sequence copying (Figure 5-2). But the DNA molecule cannot replicate without the polymerase enzyme that recognizes the template and "selects" the nucleotide precursors for synthesis of new copies. The point at which the helix unwinds, the replicating fork, is the site at which DNA polymerase attaches and makes complementary copies of each strand, according to base-pairing rules. The replicating form of the molecule is double-stranded; single-stranded molecules are not replicated. Even if the original parental template molecule is a single strand, as in some DNA viruses, a double-stranded molecule is first made into a replicating template. Attachment of the polymerizing enzyme and stability of replication seem to require the double-stranded condition.

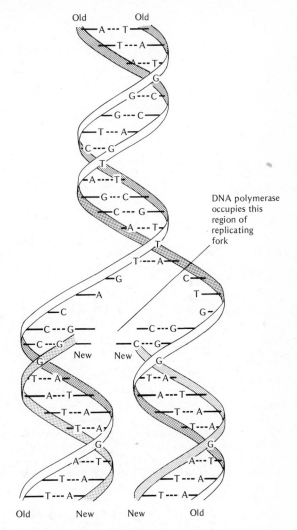

**Figure 5-2**   *Replication of DNA according to the Watson-Crick semiconservative model.*

To study prokaryote DNA replication, J. Cairns developed a procedure for extracting intact replicating DNA molecules from *E. coli*. Cells are first grown in a medium containing $^3$H-thymine for slightly less than two generations to label their DNA. The DNA is isolated and examined by autoradiography (Figure 5-3). We see the *E. coli* DNA as a circular structure defined by silver grains. The density of silver grains along the length of the chromosome is proportional to the number of $^3$H-thymine molecules

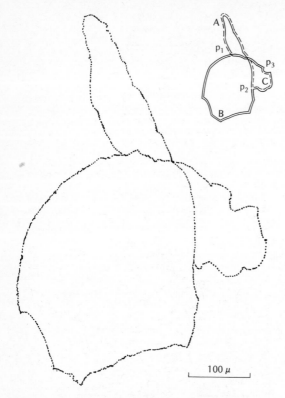

**Figure 5-3** *Radioautographs of replicating DNA in prokaryotes. Radioautograph of E.coli chromosome that has incorporated ³H-thymine for almost two generation periods. The grain pattern is obtained after 2 months exposure of the radioautograph and shows different grain densities over double- and single-stranded portions of the molecule. The diagram in the inset illustrates single-stranded portions A ($p_1$–$p_2$) and C ($p_2$–$p_3$) and a double-stranded sector B ($p_1$–$p_2$). (From J. J. Cairns,* Cold Spring Harbor Symp. Quant. Biol. *28:43–46, 1963.)*

incorporated into a region, which, in turn, is related to the number of rounds of replication completed by the molecule during the period of exposure to the precursor. In other words, grain counts show which segments have replicated. Sector B plus part of sector C from $p_1$ to $p_3$ has twice as many grains as that part of sector C from $p_3$ to $p_2$ and sector A. At the moment of isolation this circle of DNA had replicated about two-thirds of its length, that is, region B from $p_1$ to $p_2$ and region A from $p_1$ to $p_2$. Region C is still in the process of replication from the replicating fork at $p_2$ to $p_1$, the starting point for replication (see Figure 5-4 for an analysis of the replication history of the molecule).

The analysis is consistent with a model of semi-conservative replication that is polarized with the polymerase starting at a fixed initiating point and progressing sequentially along the circular chromosome until two daughter circles are produced. It also shows that one round of DNA replication is completed before another one begins.

The initiation of replication depends upon protein synthesis. If protein synthesis is inhibited replication already in progress is completed but no new rounds of replication are initiated. This suggests that a protein (initiator) synthesized during the cell cycle is required to initiate replication. Presumably it is used up during replication since the next round of replication depends upon another cycle of protein synthesis. The timing of initiation and DNA replication seems to depend upon accumulation of threshold amounts of initiator proteins. This may explain why cell division is linked to cell mass; a critical ratio of initiator protein to DNA may trigger replication. If this is so, then blocking DNA synthesis while protein synthesis continues should accumulate initiator protein, and release of the block should be followed by a resumption of DNA synthesis at a faster rate.

This prediction has been verified in several prokaryote systems. We find, for example, that cell mass at the time of DNA replication is the same for cells growing at different rates, which is consistent with the concept of a threshold level of initiator protein to trigger DNA replication. Some bacterial mutants fail to initiate DNA synthesis because they lack initiator proteins. These proteins have been isolated with solubility properties similar to membrane proteins which suggests that replication of DNA involves the cell membrane. According to one hypothesis, the signal for replication in *E. coli* comes from a specific attachment site on the cell surface known as a mesosome. At least one strand of the double helix attaches to the polymerase complex which, in turn, is affixed to the mesosome in the cell membrane. Electron micrographs of replicating bacteria show DNA attached to the outer membrane and newly synthesized DNA is usually found attached to isolated membrane fractions. Presumably, a DNA strand attaches to the initiator protein in the membrane during replication and when replication is complete, each double helix is subsequently segregated by membrane growth and synthesis of a cross-wall between the two cells (Figure 5-5). This serves as a primitive mitotic mechanism since the chromosomes condense and extend during membrane growth and nuclear formation.

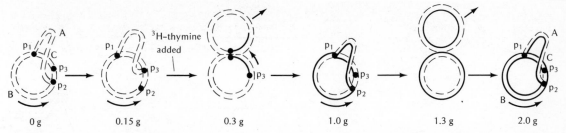

**Figure 5-4**   *Diagram of the history of the chromosome labeling pattern shown in Figure 5-3. A double line represents two strands of the double helix, a solid line is a labeled strand, and a dashed line is unlabeled. Two generations before the extraction of the chromosome (at 0 g), the DNA of the grandparent bacterium was at exactly the same state of replication as its granddaughter DNA when extracted (2. 0 g) as in Figure 5-3. $p_1$ is the initiation point and $p_2$ is the replicating fork with the small arrow indicating the direction of replication. At 0 g, sectors B and A are replicated while sector C is still unreplicated. At 0.15 g, the replicating fork has moved from $p_2$ to $p_3$ and ³H-thymine is added and all newly synthesized DNA strands are labeled. By 0.3 g, replication of the original chromosome is completed and two daughter circles of DNA are present, only a small part of each being labeled (from $p_3$ to $p_1$). In the next round of replication we follow only one circle unit until the replicating fork has reached $p_2$, approximately two-thirds of a generation later (1.0 g). At this point sector B, sector A, and part of sector C have half-labeled double helices while the sector from $p_2$ to $p_3$ is entirely unlabeled. At 1.3 g, the replication is complete with the fork at $p_1$. Two granddaughter circles are produced with fully labeled double helix from $p_3$ to $p_1$ and the remainder only half-labeled. We now follow one granddaughter chromosome two-thirds of a generation later (2.0 g) to give rise to the labeled chromosome as seen in Figure 5-3. (From Molecular Genetics: An Introductory Narrative by Gunther S. Stent. W.H. Freeman Company. Copyright © 1971.)*

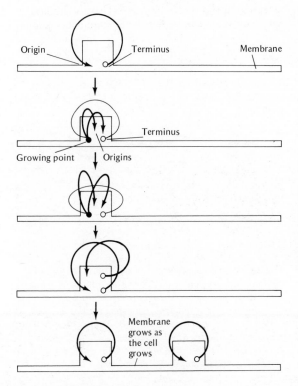

## DNA REPLICATION IN EUKARYOTES

Eukaryote chromosomes are larger than those in prokaryotes and contain a thousand times more DNA. In spite of their greater complexity, they also replicate semiconservatively according to the Watson-Crick model. This was shown by Taylor in studies of chromosome replication in root tip meristems of *Vicia fabia* (Figure 5-6). He exposed rapidly dividing cells in root tips to radioactive thymidine for several generations. In autoradiographs of dividing cells chromatids in each metaphase chromosome were uniformly labeled. The root tips were then placed in a nonradioactive medium for a new cycle of division. Both parental chromatids in this case were labeled but all newly replicated chromatids

**Figure 5-5**   *A model of DNA replication in E. coli linking division of the cell with the DNA through a replicating apparatus or mesosome on the membrane. The origin, terminus, and growing point are all linked to the mesosome which segregates at the end to yield separate chromosomes. This may act as the signal for cell wall formation and separation into two cells. (Based on N. Sueoka and W.G. Quinn, Cold Spring Harbor Symp. Quant. Biol. 33:695–705, 1968.)*

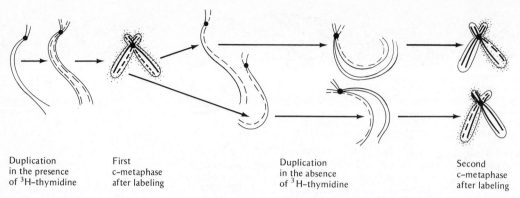

| Duplication in the presence of ³H–thymidine | First c–metaphase after labeling | | Duplication in the absence of ³H–thymidine | Second c–metaphase after labeling |

**Figure 5-6** *Replication of eukaryote chromosomes follows a semiconservative model. The diagram illustrates the steps in chromosome replication and segregation in an experiment on* Vicia fabia. *The dashed line represents labeled chromatids while the continuous line indicates unlabeled units. The dots are silver grains in the radioautograph. (Based on J.H. Taylor, P.S. Woods, and W.L. Hughes,* Proc. Nat. Acad. Sci. U.S. *53:122–127, 1957.)*

were synthesized from nonradioactive precursors and were unlabeled. Autoradiographs showed each chromosome as a hybrid consisting of one labeled (parent) and one unlabeled (daughter) chromatid. At each successive division the number of silver grains over a chromatid was always reduced by one-half, suggesting that each whole chromatid is replicating semiconservatively, just as a DNA double helix.

Calculations show that it takes 30 to 40 minutes for a replicating fork to complete a round of replication in *E. coli*. Assuming the unineme model for a eukaryote chromosome, the average chromatid of the broad bean contains one meter length of DNA, approximately 800 times the contour length of *E. coli* DNA. Yet total DNA synthesis in the bean chromosome is completed in 8 hours at 25°C, which is 100 times faster than the replicating rate in *E. coli*. If DNA polymerase moves along the eukaryote DNA at the same rate as it does in *E. coli*, then we must conclude that a eukaryote chromosome has many initiation points and many replicating forks. This model has been confirmed in whole-cell radioautographs of replicating chromosomes and in gentle extractions of replicating DNA from cells as viewed by electron microscopic radioautography (Figure 5-7). DNA synthesis does not begin at one end of the chromosome and proceed sequentially along its length to the other end, nor do all chromosomes replicate synchronously. Rather, DNA synthesis occurs simultaneously at scattered points in the chromosome. For example, up to 50 regions in the giant salivary chromosomes of *Drosophila* replicate at any one time. At the beginning of exposure to thymidine,

only a few sites show any incorporation, but these lengthen and new replicating regions appear. Chromosome labeling patterns are asynchronous; each chromosome has a characteristic replication pattern, reproducible from cycle to cycle. Some regions replicate early in the S phase and complete synthesis well before the end of the S phase. Other regions begin replication about the same time or later but complete DNA synthesis at the end of S. Usually the last regions to replicate are the heterochromatic regions.

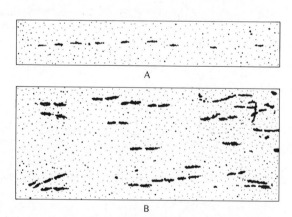

**Figure 5-7** *Radioautographs of replicating DNA strands from dissociated embryonic cells from* Triturus vulgaris. *(A) Labeled for 1 hour at 18°C. Note the discontinuous labeling pattern as different replicating regions initiate DNA synthesis at the same time. (B) A similar preparation labeled for 2 hours. The labeled regions have increased in number and size. (From H.G. Callan,* Cold Spring Harbor Symp. Quant. Biol. *38:195–203, 1974.)*

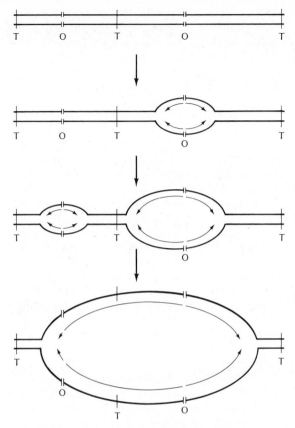

**Figure 5-8**  *Replication of mammalian DNA. Two adjacent replicons are shown, each sharing an origin (O) with one neighbor and a terminus (T) with another. Replication proceeds in both directions from adjacent origins until the terminus is reached. In this case, replication starts at different times in each replicon. (Based on J.A. Huberman and A.D. Riggs, J. Mol. Biol. 32:327–341, 1968.)*

A eukaryote chromosome seems to consist of many independently replicating segments or replicons, each with its own origin and terminus. Such multiple replicating patterns have led to a bidirectional model for DNA replication, with strands in adjacent replicons replicating in opposite directions (Figure 5-8). In such chromosomes initiating intervals seem to range between 15 and 120 $\mu$ with a mean of 50 $\mu$. Though replicons initiate replication at different times, the entire chromosome completes replication by the end of S.

One problem of eukaryote replication is the variation in the duration of S phases in different cells

of the same organism and at different times during development. For example, in the newt blastula embryo, the S phase is 1 hour but it is 8 hours in some somatic cells of the adult and even 200 hours in the premeiotic S phase of spermatocytes, yet the rate of strand replication is the same in all cells. The variation in S phase duration is explained by differences in the number of initiating points active at any stage. In early newt embryonic cells initiation points are only 10 $\mu$ apart, whereas they are 40 $\mu$ apart in somatic cells in vitro. There are many more initiating points per unit length of DNA from early embryonic cells. So far, we do not understand how this change in distribution of initiator sites comes about during development.

Chromosome proteins must also be synthesized to complete chromosome replication. Since histones and acidic proteins are not self-replicating, how and when are they synthesized and how do they join with DNA in the new chromosome?

For most cells, histone synthesis is synchronized with DNA replication during the S phase and ceases when DNA synthesis stops. Histone synthesis, like DNA synthesis, is absent during $G_1$, an expected result since histones are an integral part of chromosome organization. In some mammalian cells, however, histones are synthesized continuously through the cycle, suggesting that exceptions to the rule may be found. Though histone synthesis follows the DNA pattern, it occurs in a dispersive rather than semiconservative fashion. Most histones are probably synthesized on polysomes in the cytoplasm and transferred to the nucleus. We do not know how this transfer takes place.

Apparently newly made histones are essential to continued replication of DNA. If replicating cells are treated with an inhibitor of protein synthesis such as cyclohexamide, DNA chain elongation is inhibited to about half its original rate. This is maintained for about 30 minutes (in chick erythroblasts) and is followed by an exponential decline in synthesis. New histones seem to occupy the replicating fork region immediately after DNA is replicated which may stabilize the replicating complex. It is still not clear how histones interact with newly synthesized DNA strands, how old histones come off the original strands, and what happens to old histone molecules. There is some indication that "old histone" is recycled to daughter strands and attaches to one of two daughter helices synthesized in the absence of protein synthesis.

How is replication of many separate chromosomes synchronized? In what manner is the initiating signal transmitted throughout the nucleus so that all chromosomes complete replication in time for mitosis? Somatic cell fusion experiments point to the nuclear membrane as a controlling element. As we have seen, fused mouse, rat, and hamster cells form somatic cell hybrids in which the complete genomes of two parental cell types replicate together over several cell generations. (See Chapter 3.) A "cross" between hamster cells with a generation time of 20 hours and mouse cells with generation times of 60 hours leads to a hybrid line of cells with an intermediate generation time of 35 hours. Chromosomes from two different species, occupying one nucleus, replicate synchronously for several generations, although each genome comes from parents with totally different replicative histories. Replicating signals are not species-specific and a new hybrid cell cycle is established. Coordinate replication often fails after a while since chromosomes of one parental species fail to keep up with the new cycle and are progressively lost. According to Ephrussi and Weiss, coordinate chromosome replication is made possible by the attachment of each chromosome to the nuclear membrane. Each chromosome receives its replicating signal from the membrane, just as the prokaryote chromosome initiates replication at the mesosome.

Recent studies show that newly synthesized DNA is, in fact, associated with the nuclear membrane. DNA-nuclear membrane complexes have been isolated from sea urchin embryo nuclei at various stages of the cell cycle. Only nuclei obtained from cells in S yield membrane DNA complexes with newly synthesized DNA. Nuclei in $G_2$ or nuclei from unfertilized eggs which are not synthesizing DNA do not have such complexes, an indication that DNA replication in eukaryotes is linked to the nuclear membrane. To initiate DNA replication, an initiator complex seems to associate periodically with the nuclear membrane during synthesis and detaches at its completion.

## CHROMOSOME SEGREGATION: MITOSIS

Mitosis marks a key point in the cell cycle; growth has reached its maximum and the cell is prepared to divide. It entails chromosome coiling-condensation cycles which package large amounts of DNA into discrete, transportable units. These are then mobilized for error-free segregation to daughter cells by a spindle mechanism. Mitosis accomplishes for the eukaryote cell what the simple mesosome and expanding cell membrane achieve in prokaryotes. The spindle apparatus with its microtubules, centrioles, and asters is the principal mechanism for chromosome segregation and cell division. In plant cells, this spindle mechanism is intrinsically the same as it is in animals but modified to accommodate to rigid cell walls. Accordingly, no centrioles appear and daughter cells separate by synthesis of a new cell wall, or cell plate.

Most of the preparative activities for mitosis are completed during interphase, in the periods $G_1$, S, and $G_2$. During the long $G_2$ phase, the initiating events for chromosome condensation begin. Nothing is known about chromosome coiling, but the appearance of condensing chromosomes marks the beginning of mitosis. This is usually accompanied by appearance of the mitotic apparatus, consisting of the central spindle and centrioles at the poles with a radial arrangement of astral rays. This apparatus furnishes the motive force for chromosome segregation (Figure 5-9).

Two kinds of fibers are embedded in the spindle —spindle fibers which extend continuously from pole to pole, and chromosomal fibers which reach from the kinetochore region of the chromosome to the poles. Spindle and chromosomal fibers are composed of bundles of microtubules, arranged with their long axes parallel to one another. The microtubules are rodlike structures or hollow cylinders of uniform diameter (140 to 200 Å) and indefinite length, composed of globular subunits of a protein tubulin. (See Chapter 7.) Assembly of microtubules from tubulin is dependent on low $Ca^{++}$ ion concentration; in high calcium, they dissociate into subunits. In most cells, including nondividing cells, soluble tubulin and rodlike microtubules are in a dynamic equilibrium. Because of this reversible dissociation and reassociation of microtubules, the spindle is a dynamic structure, always poised between formation and dissolution from its microtubule subunits.

Presumably the motive force for chromosome segregation is provided by the microtubules, but to date there is no agreement as to the exact mechanism involved. Some believe that localized assembly and disassembly of the microtubules at the poles and

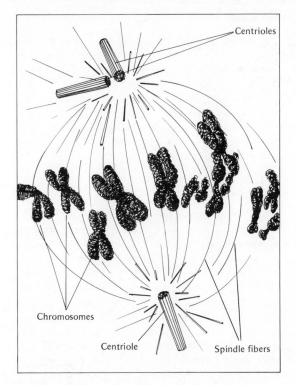

Centrioles

Chromosomes

Centriole

Spindle fibers

**Figure 5-9**   *The mitotic apparatus in dividing sperma-togonial cells in a cock testis as seen in the electron microscope. The centrioles are clearly seen at the poles of the spindle, one of which has already sprouted a daughter centriole at right angles to the original. Chromosomes and spindle fibers (microtubules) are also visible.*

kinetochore regions might be enough to generate forces to move chromosomes. Others contend that the microtubules behave as sliding muscle filaments in muscle contraction. Spindles can now be isolated from dividing cells and reversibly assembled and disassembled in vitro. This system is useful in studying mechanisms of spindle dynamics in chromosome segregation.

The centrioles at the poles orient the spindle, and act as a nucleating site for spindle microtubule assembly. The centriole in resting cells is a double structure. One part, the "mother" centriole, is a completely formed organelle. The other part, a *procentriole,* is a smaller cylindrical element projecting at right angles from one end of the mother centriole. This is a "young" centriole in the process of being formed. Two such bipartite centrioles are usually

found in interphase cells in the vicinity of the nucleus.

Centrioles reproduce by budding; that is, the new centriole grows by the assembly of microtubules from the basal region. The cartwheel structure at the basal end probably acts as an organizer or "seed" for the growth of procentriole microtubules.

At the beginning of mitosis, the centrioles separate and each one migrates to opposite poles of the cell. As they move apart, systems of microtubules form around them and assemble into astral rays. Centriole separation may derive its motive force from the lengthening of the microtubules in the astral region. Bundles of microtubules are also generated between diverging centrioles, extending from one to the other to form the spindle fibers in the center of the mitotic apparatus. As the centrioles separate, procentriole growth is stimulated. By the time the poles of the spindle are established, the centrioles have completely duplicated so that there are now two at each pole. Orientation of the main spindle axis in relation to the cell axis determines the constriction site dividing a cell into two.

Though the mitotic cycle is a continuous process, it is arbitrarily divided into phases (Figure 5-10).

**Prophase**

At some point at the end of the $G_2$ period, the long threadlike chromosome strands contract in a helical fashion, forming what is referred to as "coiled coiled coils." Although the nature of the coiling mechanism is obscure, it probably derives from the helical structure of DNA. Changes in the bonds between histones and DNA could alter the three-dimensional conformation of DNA and generate tight coils. The condensation process concentrates the DNA into a smaller area and chromosomes gradually become visible. The contraction is not completed at prophase but continues through metaphase and even well into anaphase.

**Metaphase**

Engagement of the chromosomes on the spindle takes place during metaphase. After the nuclear membrane breaks down, the chromosomes move into the equatorial plate in position for segregation. Migration is promoted entirely by the kinetochore,

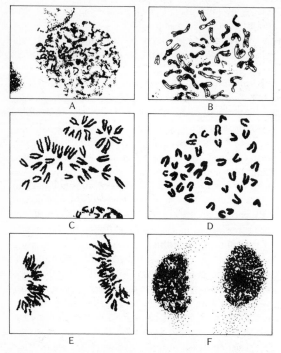

**Figure 5-10** *Mitosis in mouse fibroblasts in vitro. (A) Early prophase, beginning of chromosome condensation. (B) Late prophase–early metaphase. Individual chromosomes begin to take form as condensation nears completion. (C) Prometaphase. Chromosomes (two chromatids) still coiled as they become compacted into final metaphase configuration. (D) Metaphase. Chromosomes fully condensed and aligned on metaphase plate. (E) Anaphase. Segregation of chromosomes. (F) Formation of telophase nuclei.*

### Anaphase

The transition from metaphase to anaphase takes place abruptly as the sister chromatids separate at the kinetochore. Chromosome arms are also involved in the sudden separation, as if repelled by one another. The chromosomes move almost synchronously to opposite poles at speeds varying from about 1 to 8 $\mu$ per minute, depending on the cell type. This movement is slow compared with usual rates of protoplasmic movement.

### Telophase

When the chromosomes reach the poles, they are still tightly coiled. In telophase compact chromosomes unwind into active strands in daughter nuclei. A nuclear membrane forms from the fusion of vesicles which appear around each chromosome. Uncoiling is probably a reverse of the coiling process. The chromosomes swell, becoming diffuse and difficult to stain. The signal for nuclear reconstitution is not restricted to the polar region since single, detached chromosomes without a kinetochore will often form microvesicles and unwind as separate micronuclei whether or not they are at the pole.

Telophase marks the end of mitosis. A complete genome has been replicated and segregated into daughter nuclei, but the division process is not completed until the cytoplasm also divides.

### CYTOKINESIS: CONSTRICTION OF A CELL INTO TWO

Both DNA synthesis and mitosis are coupled to cytoplasmic division, or cytokinesis, the constriction of cytoplasm into two separate cells. Cytokinesis is an autonomous process, independent of DNA synthesis and mitosis since it may occur without DNA replication. Nevertheless, it is important to know how the cytokinetic clock is synchronized with the mitotic clock during the normal division cycle. In prokaryotes, as we have seen, coupling is accomplished by physically linking DNA replication to the cell membrane. As the membrane grows, replicated DNA circles are segregated into daughter cells. The mechanism in eukaryote cells, though more complex, is basically the same since the mitotic apparatus which segregates chromosomes also controls the events of cytokinesis.

since destruction of kinetochores by UV microbeam irradiation prevents chromosome engagement on the spindle. It is not clear how chromosome fibers connect to the poles, but at the end of prophase the chromosomes are already split, each sister chromatid equipped with its own kinetochore. The paired kinetochores face in opposite directions and insure that its chromosome fiber attaches each chromatid of a chromosome to an opposite pole.

The system seems momentarily poised, under extreme tensions, spatially positioned for the dynamic events to follow in anaphase. The spindle, focused from pole to pole, is in an active turnover state as new tubules form and others dissociate. Metaphase is the "point of no return" since division may no longer be blocked after this stage.

When an egg or cell divides, a constriction appears in the cell surface, usually perpendicular to the spindle axis, deepens into a furrow, and slowly cuts the cell into two daughter cells. The entire process involves changes of cell shape and contraction of cytoplasm, as if the motile machinery of the cell has been mobilized. Two features of this process have been studied: (1) the factors that determine the site of the constriction furrow and (2) the nature of the forces that constrict the cell into two. A third factor, perhaps more difficult to unravel, concerns the regulatory machinery coordinating temporal events of mitosis with those of cytokinesis and will be examined later.

## Determination of the site of the constriction furrow

Most studies have been carried out on large echinoderm eggs although the few studies on fibroblasts in vitro seem to be consistent with the mechanisms operating in eggs. The simplest working hypothesis is that what is true for cleaving eggs is also applicable to most dividing animal cells.

One of the first problems was to determine the location of the force-generating system in the egg. The experiments have been simple; organelles and cytoplasm within cleaving eggs have either been removed, shifted about with a needle, or inhibited chemically. For example, if the entire spindle apparatus is sucked out of the cell at late metaphase the egg continues to divide into two cells (Figure 5-11). Disruption of the mitotic apparatus after anaphase, by injection of a salt solution, also fails to inhibit constriction. And, finally, stirring the cytoplasm within the equatorial plane directly beneath a constriction furrow does not prevent division. These experiments, among others, indicate that the constriction mechanism is not localized in the spindle, nor the internal cytoplasm, but seems to be in or attached to the cell surface.

The mitotic apparatus, or more specifically the asters, are essential in determining the site of the cleavage furrow at some stage in the mitotic process. This has been shown in another series of experiments by Rappaport. He placed a small glass ball in the center of a sand dollar egg at first cleavage so as to make a torus-shaped cell (Figure 5-12). The spindle is displaced to the edge and after cleavage a binucleate, horseshoe-shaped cell is formed. At the

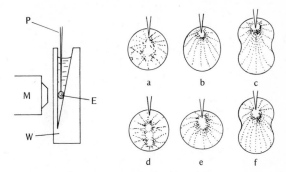

**Figure 5-11** *Effect of the mitotic apparatus on furrowing in echinoderm eggs. On the left is the micromanipulation chamber with egg (E), microscope objective (M), micropipette (P), and wall made of coverslip (W). To the right is shown the relation between position of mitotic figure and cleavage plane. Diagrams a through c show stages after spindle is removed: (a) before removal; (b) after removal; (c) furrow still appears at predetermined position. Diagrams d through f show sequence after displacement of spindle by removing some egg cytoplasm: (d) before displacement; (e) after displacement — astral rays are elongated; (f) furrow appears at a predetermined position. In both cases, cleavage planes are independent of shifted position of asters. (From Y. Hiramoto, Exp. Cell Res. 11:630–636, 1956.)*

next division two spindles appear, one in each arm of the horseshoe, and constrictions develop at the near surfaces. At the same time, a third constriction appears between two asters in the bend of the horseshoe, where no spindle existed. A constriction furrow is determined in a region of the surface where two astral rays have intersected, even in the absence of a spindle. Both asters must be present to fix the position of the future furrow; removal of one aster or increasing the distance between asters prevents furrow formation. A joint stimulus from interacting asters seems to transform the egg surface into a new physiological state and organize a constriction mechanism.

The surface of the echinoderm egg as used in the present discussion consists of a plasma membrane and a subjacent gel-like cortical cytoplasm 3 to 4 $\mu$ thick. The cortex is usually thickest just beneath the cleavage furrow where the constriction mechanism is probably localized. The interaction between the mitotic apparatus and the cortex must occur before late metaphase when the egg is still

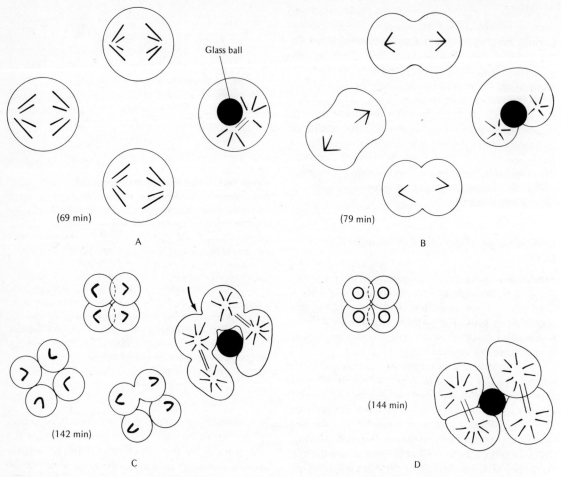

**Figure 5-12**   *Cleavage of a torus-shaped cell. State of mitotic apparatus shown with spindle defined by two straight lines. Surrounding eggs act as control of synchronous cleavage pattern. Time begins at fertilization. (A) Glass ball is placed in center of one cell immediately before furrowing. Spindle displaced to edge. (B) First cleavage produces a binucleate cell. (C) Second cleavage: two cells have divided from free ends of horseshoe with proper spindles between them. At the same time, the cell divides between the asters (at the base of the horseshoe arrow) where no spindle is present. (D) Division complete, each cell with one nucleus. (From R. Rappaport, J. Exp. Zool. 148:81–89, 1961.)*

spherical. The proximity of the mitotic apparatus to the cortex seems to determine furrow formation in normal cleavage of different eggs. In those eggs in which the spindle occupies the cell center, the furrow appears simultaneously at opposite poles. If, however, the normal position of the spindle is eccentric, closer to one pole, then a constriction furrow appears only in the surface close to the mitotic apparatus.

The interaction establishing the constriction mechanism is brief. It is enough to displace a mitotic apparatus (in late prophase or early metaphase) and hold it near an egg surface for only 1 minute for a furrow to appear 2½ minutes later. If held less than a minute, then no furrow develops. The signal from the asters requires a minute to act and another 2½ minutes to recruit the cortical mechinery for constriction.

## Mechanism of constriction

Once the furrow site is fixed, constriction into two cells depends upon the cortical layer. The cell surface is a kinetic structure that generates tension during cell constriction. Measurements of surface tension give values varying between 0.2 and 0.8 dyne/cm² reaching a maximum at the time of furrowing. This is enough to constrict a cell into two. When this surface force is counteracted by experimentally imposed hydrostatic pressure, the surface gel dissolves, tension is reduced, and division is inhibited.

Tension around the surface is not uniform during cytokinesis but is maximal at the equatorial region, beneath the constriction, and minimal at the poles. This led Marsland in 1960 to propose that the motive force for constriction is produced by a contractile ring located at the base of the cleavage furrow. Closure of the contractile ring is assumed to constrict the cell into two in those eggs exhibiting equilateral cleavage. The structure is transitory, present for only ten minutes during cleavage.

Filamentous organelles have been seen beneath the constriction furrow, in the cortical gel layer (Figure 5-13). These microfilaments are 30 to 70 Å in diameter and are arranged in bundles, like purse strings around the furrow. In the uncleaved egg similar microfilaments are seen within the microvilli on the egg surface, but as the furrow appears, the microfilaments collect as circumferential bundles in the furrow, perpendicular to the spindle axis. As the furrow gets smaller the microfilaments seem to disassemble until they disappear at the end of cleavage. The importance of microfilaments to constriction has also been shown by treating eggs with

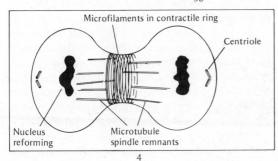

**Figure 5-13** *Formation of contractile ring in a cleaving Arbacia egg. (1a) Meridonal section of egg at late anaphase about 30 sec before furrowing begins. A is yolk-free zone of mitotic aster. (1b) Surface as seen in the electron microscope. No evidence of contractile ring is seen. Microfilaments are present only in microvilli. Y = yolk; M = mitochondria. (2a) Meridonal section 4 min after cleavage has begun. Chromosome vesicles (CV) indicate reforming nuclei. (2b) Cleavage furrow enlarged. Note accumulation of cross-sectional profiles of microfilaments in contractile ring beneath membrane in region of furrow. (3a) Meridonal section of late cleavage. Daughter nuclei reformed. (3b) Enlargement showing remnants of contractile ring in constricting cytoplasm. Complex protrusions appear. (4) Three-dimensional view of dividing cell showing position of microfilaments in contractile ring and spindle remnants. (Based on T. Schroeder, J. Cell Biol. 53:419–434, 1972, and T. Schroeder, Am. Zool. 13:949–960, 1973.)*

cytochalasin B, a drug known to interfere with cytoplasmic division of cultured mammalian cells. In the presence of the drug, furrowing stops but resumes as soon as the drug is washed out. Simultaneously, microfilaments in the cortex disappear and reappear. Disassembly of microfilaments in the drug is accompanied by cleavage inhibition.

Microfilaments have properties similar to the contractile muscle protein actin. Like actin, filaments in the cortex of eggs combine with another muscle protein, heavy meromyosin (HMM), to form the typical "arrowhead" patterns seen in the electron microscope. In addition, the tension generated beneath the constriction in a cleaving egg is of the same order of magnitude exerted by myofilaments during muscle contraction. Both actin and myosin have been extracted from the cortex of sea urchin eggs and most other cells and seem to be involved in cell mobility. (See Chapter 8.) Their presence in eggs suggests that cleavage furrowing may depend on contractile mechanisms similar to the sliding filament models for muscle contraction. Although other models have been proposed to explain cytokinesis, the behavior of microfilaments within the furrow supports the constriction-ring hypothesis suggested by Marsland.

There are some difficulties, however. The drug cytochalasin is not specific to microfilaments; it affects other metabolic activities including respiration, glucose uptake, and protein synthesis. Hence the effect on microfilaments may be only secondary. In addition, cytochalasin does not inhibit all contractile systems. It does not, for example, prevent the assembly of actin thin filaments in developing muscle, nor does it inhibit constriction in all eggs. Though amphibian eggs stop cleaving in its presence, microfilaments are unaffected. In this case, the drug may be inhibiting some other metabolic system essential to furrowing.

We will see later (Chapter 8) that microfilaments are ubiquitous organelles that seem to be involved in many types of protoplasmic movement, from simple changes in cell shape to amoeboid movement and cytoplasmic streaming. The constriction of an egg or any cell at division is only one expression of a cell's contractile machinery. We may assume that the interaction between the mitotic apparatus and the cortical gel in siting a furrow is only one regulatory step in the assembly of microfilaments. The system seems to be in an equilibrium state between soluble subunits and polymerized microfilaments. The nature of the contractile molecules, their localization in the cell, and their mode of action are problems that go beyond cell division. (See Chapter 8.) They include virtually all types of protoplasmic movement.

## ENDOGENOUS CONTROL OF THE CELL CYCLE

The sequential ordering of events in a growth-duplication cycle reflects regulatory mechanisms that (1) coordinate DNA synthesis with mitosis, (2) link cytokinesis to mitosis, and, most importantly, (3) integrate cell growth with cell division. It is during growth that a cell is triggered to divide.

DNA synthesis is one of many metabolic activities that prepare a cell for division, and it may be the key point of no return in the cell cycle. A cell seems to be committed to mitosis and cell division once DNA is replicated, although, as we have pointed out, replication and division can be dissociated experimentally. In most growth-duplication cycles, however, a biochemical event in $G_1$ seems to trigger the replicative machinery.

Synthesis of one or several regulatory proteins during $G_1$ may initiate DNA synthesis. One such protein could be an initiator that facilitates attachment of DNA polymerase to the template. But at this time we know nothing about the events in $G_1$ that initiate replication. One attempt to unravel the trigger mechanism has been the somatic cell fusion technique which makes it possible to determine whether controls of DNA synthesis depend upon positive or negative feedback signals.

### Cell fusion homokaryons

In Chapter 3 we saw that the cytoplasm of vigorously proliferating HeLa cells can stimulate terminal, inactive erythrocyte nuclei to synthesize DNA. The same technique of somatic cell fusion has also been used in studying the regulatory events that precede mitosis in proliferating cells. For example, two HeLa cells, each in a different phase of the cell cycle, may be fused. Cells in $G_1$ can be fused with cells in S, $G_2$, or even in mitosis (Figure 5-14) to form what are called heterophasic homokaryons, or fusion of cells with identical genotypes but in different physiological states. Such nucleocytoplasmic combinations

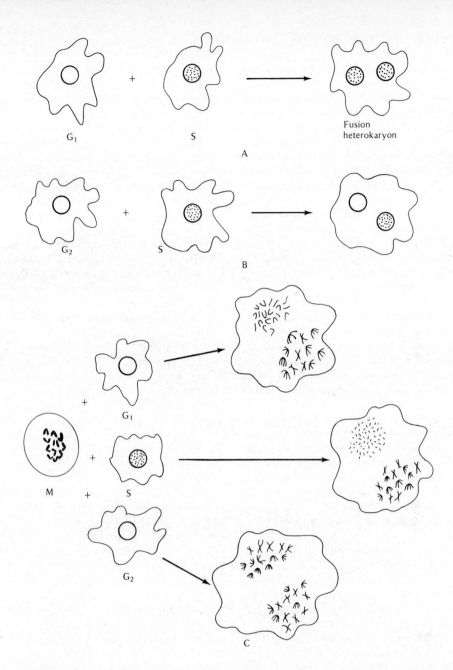

**Figure 5-14** *Fusion homokaryons made between synchronized HeLa cells at different phases of the cell cycle. (A) A cell in $G_1$ is fused with a cell in S. $G_1$ nucleus is induced to synthesize DNA prematurely by S cell cytoplasm. (B) A $G_2$ cell fused with an S cell. The $G_2$ nucleus is not stimulated to replicate DNA again, having already completed its own DNA synthesis. The S nucleus continues to synthesize DNA, indicating that no factor in $G_2$ cytoplasm shuts off replication. (C) A cell in mitosis (M) is fused with $G_1$, S, and $G_2$ cells, respectively. In all cases the chromosomes of the latter cells are induced to condense according to their state. In $G_1$ cells, chromosomes appear as condensing single strands because replication has not occurred as yet. Replicating S cell chromosomes, induced to condense prematurely, fragment and are pulverized. Chromosomes of a $G_2$ cell appear double-stranded when they condense, having already completed replication.*

show how cytoplasms in different states of the cell cycle can affect nuclear behavior.

The results for human HeLa cells may be summarized as follows: a $G_1$ nucleus in S cytoplasm is stimulated to enter S much sooner than it would have normally. If more than one S cell is fused with a $G_1$ cell, the stimulation of DNA synthesis is accelerated; the greater the dose of S cell cytoplasm the earlier a $G_1$ nucleus enters S. This suggests that an inducer of DNA synthesis, present in S cells, precociously stimulates DNA synthesis in $G_1$ nuclei. The result also implies that a $G_1$ cell does not inhibit DNA synthesis after it has begun in an S cell. As long as inducer is present, DNA synthesis continues.

Fusion of a $G_2$ cell with an S cell also has no effect on DNA synthesis. No matter what ratio of $G_2$ to S cells is used, synthesis in S nuclei continues but $G_2$ nuclei are not induced to enter DNA synthesis. We may conclude that absence of DNA synthesis in $G_1$ and $G_2$ nuclei is not due to a cytoplasmic inhibitor. Control of replication is positive; a cytoplasmic inducer is necessary for DNA synthesis to begin. Presumably this is synthesized late in $G_1$ or early in S.

The experiments suggest that DNA changes after replication; it becomes unresponsive to further replication cues during $G_2$. Furthermore the inducer is diffusible, passing through cytoplasm from one nucleus to another. Finally, dosage effects suggest that levels of inducer during replication are exhausted (or inactivated?) during $G_2$. Fusion of early $G_1$ and $G_2$ cells results in no DNA synthesis since no inducer is present.

Somatic cell fusion demonstrates that DNA replication is coupled to mitosis. For example, a synchronized population of HeLa cells enters mitosis at a specific time; within 5 to 6 hours after beginning $G_2$, 50 percent of the cells are in mitosis. A synchronized population of cells in S, further behind in the cell cycle, requires 13 to 14 hours to reach this same 50 percent mitotic level. Fusion of a $G_2$ cell to an S cell shifts the mitotic time of the S cell only slightly, but if the ratio of $G_2$ to S cells is increased, the time for 50 percent mitosis of the S nuclei is reduced by almost half. $G_2$ cytoplasm speeds up the rate at which S nuclei enter mitosis, and here too a diffusible factor is implicated which shortens the $G_2$ phase in nuclei further behind in the cycle.

Chromosome coiling and condensation may also be regulated by diffusible cytoplasmic factors. A cell in $G_2$ fused with a cell already in mitosis (for example, metaphase, M cell), is precociously stimulated to leave $G_2$ and its chromosomes condense into a typical metaphase state. Likewise, cells in $G_1$, fused with M cells, are stimulated to chromosome condensation, but these are single rather than double as in $G_2$. This seems to be another case of positive regulation; a factor (protein?) accumulates in the cytoplasm and triggers the coiling changes preparatory to mitosis. What is surprising is that an S cell fused with an M cell is also induced to chromosome condensation, but its chromosomes appear as a number of fragments. Coiling is another point of no return in the cell cycle, and this too seems to be mediated by a diffusible factor.

It seems that phosphorylation of very lysine-rich histones may be the trigger for chromosome coiling prior to mitosis. This histone class is more highly phosphorylated during $G_2$ than in S in the cellular slime mold *Physarum*, a multinucleated mass of protoplasm in which all nuclei go through the cell cycle in synchrony. Histone phosphokinase is most active in $G_2$ nuclei, shortly before chromosomes condense, which suggests that chromosome coiling is initiated after a phosphokinase phosphorylates histones synthesized in S and already bound to the DNA.

Cell fusion experiments help us define those cytoplasmic states (see Chapter 3) which regulate nuclear morphology and function. It would seem that two, or perhaps three, types of inducer substances are synthesized in the cytoplasm of proliferating cells, one initiating DNA replication, another inducing mitosis presumably by moving a cell out of $G_2$, and a third involved in chromosome coiling. Is the inducer for DNA synthesis similar to the initiator-membrane protein postulated for prokaryote DNA synthesis? Though we have no answer to this question, we do know that eukaryote DNA synthesis is also inhibited after protein synthesis is blocked by such inhibitors as puromycin or cyclohexamide. In synchronized HeLa cells, inhibition of protein synthesis is more effective at the beginning of S than later on, which suggests that synthesis of an initiator protein during $G_1$ or early in S is essential for DNA replication. DNA synthesis is also inhibited if cells are exposed to actinomycin D during $G_1$. Evidently expression of specific regulatory genes (initiator) is required during $G_1$ for DNA synthesis to begin.

## Division genes in yeast

Hartwell's study of mutants of the budding yeast, *Saccharomyces cerevisiae*, is an example of the efficacy of the genetic analysis in resolving many steps involved in regulation of the cell cycle. Synchronized cultures of yeast were examined by time-lapse microphotography as they progressed through the cell cycle. Cells in different stages of a normal wild type cycle exhibit a characteristic morphology when synthesizing DNA, initiating a bud, or undergoing nuclear division and cell separation (Figure 5-15). Each of these events occurs only once in the

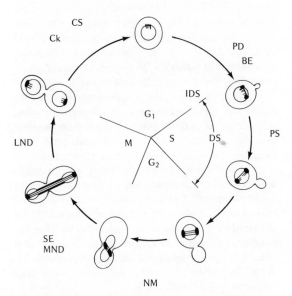

**Figure 5-15** *The cell cycle in yeast* (Saccharomyces cerevisiae). *Various landmarks in the cell cycle serve to define the site of mutant gene action: PD = plaque duplication; BE = bud emergence; IDS = initiation of DNA synthesis; DS = DNA synthesis; PS = plaque separation; NM = nuclear migration; MND = medial nuclear division; SE = spindle elongation; LND = late nuclear division; Ck = cytokinesis; CS = cell separation. Classes of mutants may be defective at any one of these points in the cycle and will accumulate in the medium. (From L.H. Hartwell,* Bacteriol. Rev. *38:164–198, 1974.)*

cycle and is called a landmark by which a cell's position in the cycle is assessed.

Many conditional lethal mutants have been obtained; these are known as "cdc" mutants or cell division cycle mutants. At a low temperature (23°C), the *permissive* temperature, these mutants grow normally. If, however, these cells are transferred to a higher temperature (36°C), the *restrictive* temperature, their mutant nature is revealed and they develop abnormally. They may continue through only part of the cycle and stop at one or another landmark. Evidently mutant genes specify a temperature-sensitive protein that functions normally at permissive temperatures but is destroyed or cannot function at the restrictive temperature. The system's value lies in the fact that when transferred to the restrictive temperature, a synchronously growing population of mutant cells expresses its mutant phenotype. All mutant cells stop at the same landmark, indicating they are unable to complete a metabolic reaction required for the next stage in the cycle and all exhibit the same abnormal morphology. In this way we obtain mutant classes whose genes are expressed at different stages of the cell cycle. With this technique the action of structural and regulatory genes at each step of the cell cycle may be studied.

Two classes of conditional mutants have been described depending on the number of cell cycles they complete after transfer to restrictive temperatures. Both control DNA synthesis, but each leads to a different abnormal morphological pattern; that is, the genes are expressed at different times in the cycle. Mutants of the first class act before the midpoint of the cycle. When shifted from permissive to restrictive temperatures, these mutants stop development at the single-cell stage with an undivided and constricted nucleus in the isthmus between parent cell and bud. On the other hand, mutants of the second class continue to divide into two cells if transferred in the early half of the cycle, but buds grow abnormally at the restrictive temperature without nuclear division and remain attached to a mother cell with an undivided nucleus.

Each mutant class interferes with a different step in DNA replication. The gene product of the first mutant class seems to be required for the initiation of DNA synthesis (synthesis of initiator protein?) while that of the second mutant class is necessary for the continuous propagation of DNA replication.

The block to DNA synthesis in each mutant prevents nuclear division. If DNA replication is completed in these mutants at permissive temperatures, then nuclear division is normal, even at restrictive temperatures, indicating that nuclear division depends on successful DNA replication. Likewise, completion of DNA synthesis is a prerequisite for cell separation but not for bud initiation or growth. Bud initiation is periodic, suggesting that it is controlled by a cellular clock independent of DNA replication. Nonetheless, this cytoplasmic clock is synchronized with nuclear events in the normal cell cycle, linked somehow to another clock governing DNA synthesis.

It may now be possible to isolate specific gene products regulating key switch points that move cells smoothly through the cell cycle.

In growth duplication, we may imagine two coupled cycles—one for DNA replication and the other for cytoplasmic division—operating semiindependently but dissociable from each other and from growth. Each cycle is attuned to its own set of regulatory signals, yet overall control is coordinated to insure orderly progression of the cell through its life history. If a set of conditions has reached a threshold value to initiate and complete mitosis, then the cell is committed to nuclear and cytoplasmic division.

## EXOGENOUS CONTROL OF THE CELL CYCLE

Proliferating cells, though responsive to intrinsic division clocks, are also sensitive to many exogenous cues. The duration of different phases of the cycle may be increased, decreased, or entirely inhibited by nutritional or hormonal factors. All responses to exogenous signals are screened first by the cell surface and second by the cytoplasm before reaching the nuclear genome. Membranes, by modulating all environmental signals, may act as the first "control gate" in cell function. Accordingly, modification of membrane structure may prevent cell proliferation in some instances or induce cells to grow as malignant neoplasms in others. Most information about membrane regulation of DNA synthesis has come from studies of normal and neoplastic cells in vitro (see Chapter 21).

A note of caution about cell culture in vitro should be added. Cells in vitro are challenged by a hostile environment. They are torn from their normal tissue relationships of intimacy with their neighbors and dissociated by lethal doses of proteolytic enzymes. Afterwards, they are immersed in a mixture of salts, amino acids, and vitamins, including foreign sera with a novel assortment of proteins, hormones, and inhibitors. Initially, many cells die, while damaged survivors attach to a glass or plastic substratum and spread into flattened, almost two-dimensional sheets, barely contacting other cells. Under these artificial conditions, we study cell growth and its regulation. Since most cells seem to be mobilizing metabolic machinery to adapt to an alien environment, much of their behavior may be artificial. Even permanent cell lines, grown for many years, exhibit behavior that may be true only for specific conditions in vitro.

At the outset, in vitro conditions lead to proliferation behavior characteristic of cells with damaged membranes. For example, it is easier to start a culture by inoculating with a dense population of cells. Cell density, not total cell number, is important, suggesting that growth is dependent on close cellular proximity. This behavior is interpreted as a cell surface artifact since membranes are damaged as cells make and break adhesive contacts with the glass substratum. Repeated "tears" in the membrane tend to make cells leaky and they lose metabolites, some probably essential to proliferation. The greater the open space surrounding a cell, the more vulnerable it is to surface lesions; and in large volumes of media, materials lost from cells are diluted below critical levels. Cell proliferation improves if cells are grown in small volumes or inoculated at high density conditions. Dense conditions favor survival and growth because membrane damage is probably reduced by cell-cell contacts and by microexudates surrounding each cell.

Cell populations below the critical density, which do not grow, are stimulated to do so if supplied with a "conditioning factor" released into media from densely grown cells. A membrane lipoprotein or glycoprotein from actively dividing cells may accumulate in conditioned media and stimulate growth by facilitating membrane repair.

But many factors influence growth of cells in vitro. Slight changes in pH of the medium, for instance, affect growth rate and the maximum size of the population. Ions, particularly divalent ions, are also important. Cells treated with low concentrations of a cation-binding agent, EDTA, do not grow, but the addition of zinc ions restores growth. Zinc ions seem to be more effective than $Ca^{++}$ or $Mg^{++}$ in this case. Effects of pH and ions are but a few of many complicated in vitro interactions between cells and serum factors in most growth media. Their significance with regard to growth regulation is still not understood.

## CONTACT INHIBITION OF CELL DIVISION

Many cells in vitro stop multiplying when population density is so high that all are in contact with one another, spread out as a monolayer. Growth ceases though the medium still has enough nutrients to support division. Replenishment with fresh medium usually does not relieve growth inhibition. Monolayer cultures may be stimulated to proliferate when a free edge is made in the cell layer by wounding or by surgical removal. DNA synthesis (as measured by radioactive thymidine incorporation) and cell division are initiated along the free surface until it is repaired and a confluent sheet is again restored. Since cells in the monolayer and those at the free edge occupy the same medium, it seems likely that stimulation of growth in the latter is due to release from intimate contact with neighbors. This phenomenon of density-dependent control of DNA synthesis and cell division has been called contact inhibition of growth, or *topoinhibition*. Surface contact may, in fact, be an important growth regulatory mechanism in vivo, since most tissue cells, in intimate contact with their neighbors, are also mitotically inactive.

Different cells exhibit different degrees of topoinhibition. Though most normal cells are contact-inhibited, neoplastic cells, spontaneous or viral-induced, exhibit uncontrolled proliferation and do not display density-dependent growth; that is, they form no monolayers. Surfaces of neoplastic cells differ from those of normal cells; they are less adhesive, much more motile, and usually are thrown up into numerous folds and microvilli (see Chapter 21). They do not form a mutual surface on contact but continue to migrate over one another, or over normal cells, and pile up. Again, this suggests that some membrane property may be important in regulating DNA synthesis and cell division.

Topoinhibition of DNA synthesis and cell division are still poorly understood. Though it seems that close contact of cell surfaces prevents cells from entering the S period it may not be the primary factor in growth regulation. Many conditions not necessarily related to cell contact relieve growth inhibition of cells in monolayers. For example, contact-inhibited cell lines may be stimulated to proliferate by slight changes in pH. More significantly, the amount and kind of serum in the medium also affects topoinhibition. For example, 3T3 mouse fibroblasts reach the same population size irrespective of density as long as they are grown in high serum concentrations. On the other hand, growth of a strain of epithelial cells which has a low serum requirement is highly dependent on cell density. These cells exhibit high topoinhibition, suggesting that cell multiplication is limited by different mechanisms in fibroblasts and epithelial cells. The former are medium-dependent; the latter seem to be cell-density-dependent, but this too may be related to the availability of medium rather than surface contact.

Growth-inhibited monolayers of fibroblasts, for example, can be stimulated to grow if fresh medium flows over the cell sheet. Under these conditions, the highest number of mitoses are found in the region of the moving stream. Topoinhibition of growth may result when a diffusion barrier develops at the cell surface at high density and limits the uptake of nutrients. If the barrier imposed by the diffusion boundary layer is released by high-velocity flow of medium, the cells can take up the nutrients required for growth. Even the wounding experiment mentioned earlier can be explained by the release of the diffusion barrier; cells at the wound edge have a new surface exposed for nutrient uptake and are released from the diffusion barrier imposed by close cell contact in the confluent region of a monolayer. Nutrient availability rather than surface contact seems to be more important in growth regulation in vitro.

## CELL SURFACE IN GROWTH REGULATION

Though topoinhibition of growth may be an artifact of in vitro culture conditions, it still appears that the cell surface is involved in controlling DNA replication and cell growth. Surface regulation of growth has distinct advantages since cell behavior can be dictated by environmental conditions. The cell surface is the first to detect nutrient or hormonal changes in the environment and is able to monitor the proximity of neighboring cells as well as any regulatory signals they transmit.

The membrane as a regulator of growth and cell division is important in other ways, particularly if signaling circuits between the membrane and the nucleus are present. According to a hypothesis proposed by Burger, DNA content and number of nuclei per cell are regulated by the membrane's role in cell division. After mitotic segregation of the nuclear material into daughter nuclei, the membrane-mediated cytokinetic events divide the cytoplasm into two cells and only after this occurs is the nucleus signaled to enter a new round of DNA replication. This prevents formation of polynucleated cells and insures that the ratio of DNA to cytoplasmic mass is kept within strict limits. Furthermore, growth increases cell volume and surface disproportionately; the latter grows as the square of the radius and does not keep pace with the volume increasing as the third power of the radius. Such changes in the surface volume ratios may also act as a feedback control of DNA replication, perhaps as a consequence of some structural modification of the cell surface.

Cells in vitro display dynamic surface changes during the cell cycle. Upon entering mitosis cells detach from the glass or plastic substratum and become spherical. When examined by time-lapse cinemicrography, their surfaces seem to undergo a violent bubbling activity with blebs and filamentous projections extending and contracting. In the scanning electron microscope, for example, the surface of a hamster ovary cell at this stage is seen to be covered with numerous long, thin microvilli.

As these cells complete mitosis and enter $G_1$, they begin to flatten out and attach to the substratum. The surface is covered with bubblelike projections and fewer and shorter microvilli. Cells entering the S or DNA replication phase are considerably more flattened, have fewer microvilli, and have smoother surfaces. Finally, in $G_2$, cells become thicker, microvilli reappear, and occasionally blebs appear as cells approach mitosis.

These gross changes indicate that the cell surface is changing continuously, and we now ask whether changes in the architecture of the surface, or in its constituents, or in its function, act to trigger some of the main events in the cell cycle such as DNA replication and chromosome condensation.

Experimental modification of the cell surface affects a cell's growth behavior in that topoinhibited cell monolayers are stimulated to replicate DNA if exposed to proteolytic enzymes such as trypsin. DNA synthesis in such cultures is stimulated over the entire monolayer. (Figure 5-16.) Enzymatic alteration of the membrane is enough to stimulate DNA synthesis in topoinhibited cell populations.

A clue to the nature of the membrane change has come from a most unlikely source, plants. Some substances extracted from seeds have cell-agglutinating properties. For instance, some, known as phytohemagglutinins, will agglutinate red blood cells, and many other such substances will agglutinate a variety of cultured animal cells with a high degree of specificity. These agglutinins are known as plant lectins, and one in particular, concanavalin A (Con A) extracted from jack beans, has been useful in studying membrane structure of cells under different conditions of multiplicative growth.

Agglutination comes about because Con A is a multivalent molecule with four reactive sites which

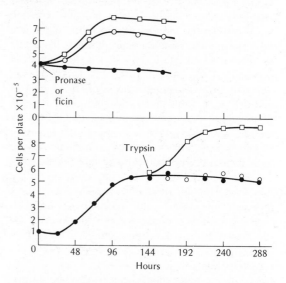

**Figure 5-16**  *Growth stimulation by proteolytic enzymes. (Upper curve) Mouse 3T3 fibroblasts were grown to saturation density before pronase or ficin was added. Control = solid circles; pronase = open circles; ficin = squares. (Lower curve) Cells were inoculated heavily at zero time and allowed to grow to saturation density before trypsin was added at the arrow. Trypsin = squares; inactivated trypsin = open circles; control without protease = closed circles. (Based on M.M. Burger, Nature 227:170–171, 1970.)*

combine with receptor sites of the surfaces of adjacent cells and act as cell bridges, causing them to clump. The cell-surface site must carry mannosides or glucosides for Con A to function inasmuch as addition of these substances to the medium competitively inhibit Con A agglutination. This suggests that the receptor is one of the glycoprotein constituents of the plasma membrane.

Normal cells which are topoinhibited are not agglutinated by Con A, but nontopoinhibited tumor cells do agglutinate. It seems that high mitotic activity correlates with sensitivity to Con A. In fact, normal cells in mitosis or normal cells transformed into a tumor by a virus are agglutinated by Con A, and their growth is no longer topoinhibited. In other words, two agents, proteolytic enzymes and a transforming virus, modify a normal cell so that it interacts with an agglutinating agent, loses growth topoinhibition, and is stimulated to replicate its DNA. It would seem that Con A receptor sites on the cell surface regulate cell division. But in spite of a strong correlation, there is no good evidence that the Con A receptor site is, itself, the regulator of cell replication nor that it is implicated in contact inhibition of growth. Its behavior in these experiments is more of a probe that reveals changes in cell surface architecture.

The cell surface is a complicated organelle that seems to regulate cell replication as well as many other cell functions that are important in development such as motility, cell recognition, adhesive behavior, and cell-cell interactions. It is possible that the same surface-signaling mechanism may be implicated in all these processes and Con A has been useful in its analysis. Since Chapter 10 includes a detailed review of the surface in morphogenesis, the discussion here will be limited to some phenomenological features of surface modulation of cell division as modified experimentally by Con A.

The surface behavior of lymphocytes has been one of the most promising systems for this analysis. B lymphocytes are a class of immune competent cells that synthesize antibody in response to an antigenic stimulus (see Chapter 18). The lymphocytes detect antigen because they possess antibody molecules on their surfaces which recognize and combine with antigen. This interaction stimulates the lymphocyte to enter DNA replication and produce many cells, each capable of synthesizing specific antibody against the antigen. This mitogenic or mitotic stimulating action by antigen is a key feature of the antibody response.

Attachment of the antigen also affects the fluidity of the lymphocyte surface. Many studies have shown that all cell surfaces are fluid and undergo translational displacement. The process is visualized with fluorescent antibodies specific for lymphocyte surface receptors. When added to lymphocytes, they combine with the receptors, staining them with a fluorescent label. Within a few minutes the diffuse fluorescence over the surface collects into large bright patches that move rapidly to one end of the cell, accumulating as a fluorescent "cap." Patching and cap formation result from a lateral displacement of surface receptors within the plane of the membrane. If surface-bound antigen undergoes a similar displacement then membrane movement and growth stimulation may be causally related.

Con A treatment suggests that some type of motile mechanism at the surface is involved. Lymphocytes treated with Con A before exposure to surface receptor antibodies exhibit no patching or capping. Capping occurs only after Con A is washed out. This means that surface binding of Con A inhibits membrane mobility. If a small portion of the lymphocyte surface is exposed to Con A, the mobility of membrane receptors is still blocked, which means that a local response to Con A is propagated over the entire surface.

Blocking of mobility is probably caused by cross-linking of membrane receptor glycoproteins by the tetravalent Con A. If Con A is converted into a bivalent or monovalent derivative, it cannot cross link adjacent receptors in the surface and does not inhibit displacement. Furthermore it can be shown that mobility is modulated by subsurface components and not by receptors embedded in the membrane. Lymphocytes blocked by Con A are released from this inhibition by treatment with colchicine; patching and capping occur. Since colchicine is specific in its action (it combines with the tubulin subunits and prevents microtubule assembly), its ability to reverse Con A cross-linking suggests that microtubules localized beneath the membrane are part of a motile system concerned with the translational movements of receptors in the cell surface. Disassembly of microtubules seems to free Con A-bound receptors.

Recent experiments suggest that receptor mobility may be a primary step in mitotic stimulation.

For example, low concentrations of Con A which do not block surface movement stimulate lymphocytes to synthesize DNA. But slightly higher concentrations, particularly those which prevent receptor displacement, rapidly eliminate the mitogenic action of

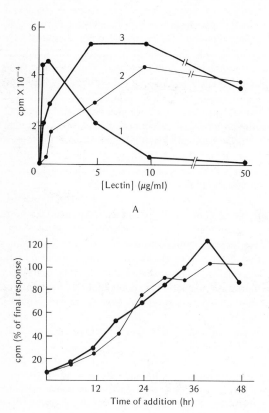

A

B

Con A. (Figure 5-17.) Moreover, the bivalent and monovalent Con A derivatives that do not block movement have a sustained mitogenic action on lymphocytes even at high concentrations. Thus surface movement of receptors and mitogenic action go hand in hand. Colchicine treatment is consistent with this interpretation. When added to cells stimulated to divide by low doses of Con A, colchicine blocks further division of cells. Since colchicine does not combine with Con A, nor does it alter its attachment to cell surface glycoproteins, the simplest hypothesis is that colchicine has affected microtubules and somehow inhibited mitosis.

Edelman's interpretation of these experiments is that microtubules are involved in the regulation of early biochemical events that commit a cell to DNA replication and mitosis. In support of this hypothesis are observations of abnormal microtubule assembly in many neoplastic cell lines which exhibit uncontrolled proliferation. Where normal cells have well-formed microtubules extending from the center to the periphery, neoplastic cell lines, though capable of synthesizing the microtubular protein, tubulin, are unable to assemble normal arrays of microtubules.

What kind of receptors are localized in the surface? One receptor may be the enzyme adenylcyclase which is found within the membranes of most cells. As we have seen (Chapter 3), this enzyme catalyzes the conversion of ATP to cyclic AMP, a regulatory molecule participating in gene transcription in *E. coli*. But this action on transcription is only one of many functions in eukaryote cells. Perhaps its most ubiquitous role is that of a second messenger, a regulatory molecule that is produced in response to many hormonal signals impinging on cell surfaces. The action of such hormones as glucagon seems to be mediated by the secondary production of cyclic AMP which, in turn, activates phosphorylating enzymes that metabolize glycogen. Likewise the action of neural transmitter substances such as noradrenalin is also secondarily mediated by cyclic AMP. In all cases the signaling molecules bind to the cell surface and activate adenylcyclase, which catalyzes a rapid rise of intracellular cyclic AMP.

It now appears that cyclic nucleotides play an important role in modulating cell division. For example, neoplastic cells which proliferate vigorously and are not contact-inhibited, will, in the presence of derivatives of cyclic AMP, exhibit a more normal

**Figure 5-17**  *(A) Mitogenic dose-response curves for Con A and its derivatives of different valence. Curve 1 = native Con A (tetravalent); curve 2 = monovalent Con A; curve 3 = divalent Con A. Note stimulation by low dose of Con A followed by rapid decline as concentration increases. Other derivatives sustain mitotic stimulation even at high concentrations. (B) Inhibition of Con A mitogenic action after removing Con A by competing mannosides (light line) or after adding $10^{-6}M$ colchicine. Cells initially stimulated at optimal dose of Con A, and each inhibitor was added at different times after Con A addition (heavy line). cpm = counts per minute of $^3$H-thymidine incorporated into DNA. (From G.M. Edelman, Science 192:218–226, 1976. Copyright © 1976 by the American Association for the Advancement of Science.)*

proliferative pattern and display contact inhibition of growth. (See Chapter 21.) It seems that high levels of cyclic AMP inhibit cell division, although it is not clear why. One possibility is that the assembly of the spindle apparatus is inhibited since it has been shown that microtubules disassemble in the presence of high levels of cyclic AMP. The amount of cyclic AMP in proliferating tumor cells is also considerably lower than in normal cells which is consistent with the hypothesis of growth inhibition by cyclic AMP. Moreover, measurement of cyclic AMP during the cell cycle of normal cells shows it fluctuating in a periodic manner, with its lowest levels during mitosis.

There are other possibilities for its action. Cyclic AMP also increases the flow of calcium ions that are essential cofactors in many metabolic systems including nerve conduction and muscle contraction. Alterations in the distribution of calcium could affect microtubules and contractile microfilaments, both seemingly important in surface mobility (Chapter 8).

Since many reactions in the cell respond to slight fluctuations in small molecule regulators, it is difficult to work out the metabolic sequence involved in triggering a cell to replicate. We can see that signals impinging on the cell surface from the outside could set off a train of reactions that lead to cell division. Each cell type is equipped with a different set of reaction systems poised in different states of readiness. The nature of the response to a hormone signal, for example, will vary depending on the cell state; some cells may stop mitotic activity while a different cell population will respond to the same signal by increasing mitotic activity.

## SUMMARY

The division of cells, an expression of replicating DNA, is the most important aspect of protoplasmic growth. Before a cell can divide it achieves a state of division readiness during auxetic growth when DNA replicates and RNA and proteins are synthesized. Although DNA replication can occur without mitosis or cell division, most cells that have initiated DNA replication are committed to a mitotic cycle. Actually no single metabolic sequence studied so far can be characterized as the trigger to cell division, although synthesis of an initiator protein seems essential. Many separate but interrelated events must be completed before a cell can divide. Nuclear and cytoplasmic events seem to follow independent clocks that are synchronized to create a harmonious growth-duplication cycle.

Control of replication is multifaceted; many sites for regulation exist, but the initiation of DNA replication seems to be the key element. DNA replication and mitosis respond to cytoplasmic signals. The exact signaling circuits have not been worked out, but the evidence is accumulating that these cytoplasmic states may be determined by the nature of the outer cell membrane, mediated perhaps by the action of intracellular levels of cyclic nucleotides on a motile mechanism beneath the surface.

Control involves inhibition and stimulation. Growth of organs and tissues is regulated by exogenous signals such as tissue-specific mitotic inhibitors and hormones. The action of these substances on a systemic level integrates growth of parts in relation to the whole.

**REFERENCES**   Bombik, B., and Burger, M. M. 1973. cAMP and the cell cycle. Inhibition of growth stimulation. *Exp. Cell Res.* 80:88–94.

Bradbury, E. M., Ingles, P. J., and Mathews, H. P. 1974. Control of cell division by very lysine rich histone ($F_1$) phosphorylation. *Nature* 247:257–261.

Burger, M. M. 1974. Role of the cell surface in growth and transformation. *Macromolecules regulating growth and development*, ed. E. D. Hay, T. J. King, and J. Papaconstantinou. New York: Academic Press, pp. 3–24.

Cairns, J. 1963. The bacterial chromosome and its manner of replication as seen by autoradiography. *J. Mol. Biol.* 6:208–213.

Cande, W. Z., Lazarides, E., and McIntosh, J. R. 1977. A comparison of the distribution of actin and tubulin in the mammalian mitotic spindle as seen by indirect immunofluorescence. *J. Cell Biol.* 72:552–567.

Dulbecco, R., and Elkington, J. 1973. Conditions limiting multiplication of fibroblastic and epithelial cells in dense cultures. *Nature* 246:197–199.

Edelman, G. M. 1976. Surface modulation in cell recognition and cell growth. *Science* 192:218–226.

Ephrussi, B. 1972. *Hybridization of somatic cells.* Princeton, N.J.: Princeton University Press.

Ephrussi, B., and Weiss, M. C. 1967. Regulation of the cell cycle in mammalian cells: Inferences and speculations based on observations of interspecific somatic hybrids. *Control mechanisms in developmental processes,* ed. M. Locke. New York: Academic Press, pp. 136–169.

Gefter, M. L. 1975. DNA replication. *Ann. Rev. Biochem.* 44:45–78.

Hartwell, L. H. 1974. *Saccharomyces cerevisiae* cell cycle. *Bacteriol. Rev.* 38:164–198.

Holley, R. W. 1975. Control of growth of mammalian cells in cell culture. *Nature* 258:487–490.

Huberman, J. A., and Riggs, A. D. 1968. On the mechanism of DNA replication in mammalian chromosomes. *J. Mol. Biol.* 32:327–341.

Infante, A. A., Navata, R., Gilbert, S., Hobart, P., and Firshein, W. 1973. DNA synthesis in developing sea urchins: Role of a DNA-nuclear membrane complex. *Nature, New Biology* 242:5–8.

Mazia, D. 1961. Mitosis and the physiology of cell division. *The cell,* ed. J. Brachet and A. E. Mirsky. Vol. 3. New York: Academic Press, pp. 77–412.

Pasten, I. H., Johnson, G. S., and Anderson, W. B. 1975. Role of cyclic nucleotides in growth control. *Ann. Rev. Biochem.* 44:491–522.

Porter, K., Prescott, D., and Frye, J. 1973. Changes in surface morphology of Chinese hamster ovary cells during the cell cycle. *J. Cell Biol.* 57:815–836.

Prescott, D. M. 1976. The cell cycle and the control of cellular reproduction. *Adv. Genet.* 18:100–177.

————. 1966. The synthesis of total macronuclear protein, histone and DNA during the cell cycle in *Euplotes eurystomus. J. Cell Biol.* 31:1–9.

Rao, P. N., and Johnson, R. T. 1974. Induction of chromosome condensation in interphase cells. *Adv. Cell Mol. Biol.* 3:138–189.

Rappaport, R. 1975. Establishment and organization of the cleavage mechanism. *Molecules and cell movement,* ed. S. Inoué and R. E. Stephens. New York: Raven Press, pp. 287–303.

Rebhun, L. 1977. Cyclic nucleotides, calcium and cell division. *Int. Rev. Cytol.* 49:1–54.

Rubin, H. 1971. pH and population density in the regulation of animal cell multiplication. *J. Cell Biol.* 51:686–702.

Schroeder, T. E. 1975. Dynamics of the contractile ring. *Molecules and cell movement,* ed. S. Inoué and R. E. Stephens. New York: Raven Press, p. 304.

Sueoka, N., and Quinn, W. C. 1968. Membrane attachment of the chromosome replication origin in *Bacillus subtilis. Cold Spring Harbor Symp. Quant. Biol.* 33:695–705.

Taylor, J. H. 1974. Units of DNA replication in chromosomes of eukaryotes. *Int. Rev. Cytol.* 37:1–20.

# 6

# Organismic Growth

Organisms grow because cells replicate during growth episodes in the life cycle. Accretionary growth, which also contributes to size increase, depends on the number of cells and the products they synthesize and accumulate. As organisms become larger, greater demands are made on their homeostatic mechanisms. Most cells in multicelluar organisms are not in direct contact with the environment and rely on other systems for nutrients, energy, and information about short- and long-term environmental perturbations. Both the nervous and endocrine systems have evolved as the principal signaling networks controlling short-term physiological responses to the environment. These systems also adapt the organism's response to long-term diurnal and seasonal rhythms of light, temperature, and humidity that trigger growth and reproduction.

Growth by cell proliferation is itself a homeostatic mechanism. It is one way an organism produces large numbers of functional units. A functional unit may be defined as a cell organelle or macro-

molecule performing a specialized function. Myofibrils in muscle cells and chloroplasts in leaf parenchyma are examples of visible functional units. An organism responds to functional demands either by making more functional units through cell proliferation or by increasing the number of units in cells by protein synthesis and cell growth. Both contribute to total growth of the organism.

Local patterns of cell replication are controlled so that demands for new functional units are satisfied without creating disharmonies in the growth of parts in relation to the whole. Most growth regulation depends on local tissue interactions via diffusible molecules or on systemic distribution of hormone signals from endocrine tissues. In addition, nerves penetrate most tissues and perform a dual function of growth regulation and nervous integration. Neurosecretory cells combine endocrine with nervous functions and act as communication links between the two systems. Such cells display two modes of chemical communication—neuronal, or cell to cell, via small molecule transmitters that diffuse across synaptic membranes, and systemic, via large macromolecules that enter the circulation and interact with distant target tissues. The neurosecretory circuit is functionally coupled to the endocrine system, chemically through diffusible molecules, and anatomically by direct connections. The pituitary gland, for example, the principal endocrine organ of the vertebrate body, is an embryonic derivative of the developing nervous system and retains physical continuity with the brain in the adult. This intimate connection between brain and endocrine system is also found in more primitive animals where intricate feedback networks have evolved to regulate changing seasonal demands for growth and reproduction.

Chemical regulators may stimulate or inhibit growth depending on the physiological state of the target tissue. Specificity in growth regulation arises because responding tissues possess receptor molecules that bind readily to hormone or neurosecretory molecules. The specific growth response of cells to these regulatory signals is an integral part of their behavioral repertoire.

Organismic growth is governed by positive and negative feedback control of the proliferative activity of cell populations. Before we look into some control mechanisms, we should examine the kinetics of population growth of cells committed indefinitely to a growth-duplication cycle, such as bacteria. These cells grow exponentially and are capable of unlimited proliferation. We will then consider the determinate growth of multicellular organisms, where exponentially growing embryonic cells are subjected to regulatory systems that switch them from growth-duplication cycles into nonproliferating modes of specialized function.

## EXPONENTIALLY GROWING CELL POPULATIONS

All cells in exponentially growing population retain reproductive potential. Each division product is itself capable of division; a daughter cell resumes growth, reaches a terminal size, and in turn is triggered to divide. No cells are lost, and the population increases exponentially. Examples of such populations are the free-living unicells *E. coli*, yeast, somatic cells in culture, and cleaving cells in an early embryo. Theoretically these systems are immortal; given proper conditions they could exhibit

unlimited growth. It has been calculated, for example, that if no factors limit growth, then the mass of *E. coli* arising from one cell dividing every 30 minutes would, in a few days, come to equal that of the solar system. Fortunately, the natural conditions of growth prevent this from happening.

During unlimited growth each cell divides at regular intervals, though not necessarily in synchrony with all other cells. Though each cell may have a different generation time, an average cell cycle can be obtained for the population. If all cells remain in the duplication phase, then at each interval equivalent to the average cell cycle time, the cell population doubles its size. Since growth rate is proportional to cell number, exponential growth kinetics is seen. The number of cells in the population $N$ at any time $t$ is proportional to the original population size before growth, $N_0$, and the exponential factor $e^{kt}$ where $e$ is the base of natural logarithms and $k$ is the instantaneous relative rate of growth in a given unit of time. The formula $N = N_0 e^{kt}$, when plotted, gives the exponential curve shown in Figure 6-1. The formula is similar to the compound interest law which shows a similar increase with time, only in this case the "interest" is added to the "capital" continuously from moment to moment rather than annually. When the formula is converted to its logarithmic form, it becomes $N = \ln N_0 + kt$, which plots on semilog paper as a straight line. The constant $k$ is the slope of the line and has a perfectly defined meaning; it is the instantaneous growth rate which relates to the mean generation time of the cell population, or doubling time.

Most populations of cells growing under natural or laboratory conditions begin to grow with exponential kinetics. As more cells accumulate, growth rate increases, but soon the rate begins to decline as growth-limiting conditions arise. Cells begin to leave the division cycle and the shape of the growth curve changes into the sigmoidal pattern so characteristic of biological population growth. For example, a small inoculum of cells introduced into a growth medium of constant volume begins to multiply at the instantaneous growth rate after a brief lag period. During the lag cells adjust their metabolism to the new medium, possibly inducing new enzymes to utilize the nutrients. The lag period may be eliminated if cells are transferred into a medium identical to the one in which they had been growing. The lag, in a sense, is an artifact of culture conditions. After the lag, cells multiply at an exponential rate as described by the equations above. But growth rate declines; in nature, as in the laboratory, nutrients are exhausted and waste products accumulate. These limit growth; a period of declining growth follows the exponential phase until growth stops at the stationary phase. The sigmoidal nature of population growth reflects these changing exogenous conditions.

A population of bacteria can be maintained indefinitely in an exponential or unlimited proliferative phase. This requires continuous replenishment of nutrients at a defined rate, removal of cells to maintain a constant population size, and elimination of toxic products. Such a system, called a *chemostat*, is used to study bacterial growth and metabolism. New medium is continuously added to dilute the cell population at a rate which exactly balances the rate of increase in cell number. Thus the total number and mass of cells per unit volume remains constant. By varying the nutrient flow rate or the nature of one nutrient, we can study changes in the instantaneous growth rates, $k$, and relate this information to the metabolism of growth. Given these optimal conditions, such populations remain in the exponential phase indefinitely; no intrinsic mechanisms seem to inhibit proliferation. The sigmoidal nature of cell population growth is due entirely to exogenous factors such as depletion of nutrients, waste product accumulation, and crowding, and not to intracellular signals to stop dividing.

The growth of a multicellular organism is also sigmoidal. All criteria for assaying organismic growth, such as mass, cell number, volume, or even linear dimensions, follow an S-shaped curve or some modification thereof. Early growth in the development of an organism is exponential, but this too reaches a plateau as several emergent conditions limit proliferation.

For multicellular organisms an important condition is the development of contact interactions between neighboring cells. For instance, growth may be limited by topoinhibition as described in Chapter 5, or by the establishment of communication across extracellular spaces. Cells acquire a social behavior; by exchanging information a population of independent cells is transformed into a tissue governed by new sets of growth rules. Determining what these rules are is one of the major problems of understanding organismic growth.

The early exponential phase in embryos is usually characterized by synchronously dividing populations with short cell cycles lacking a $G_1$ phase (Chapter 11). After a programmed number of divisions, synchrony ceases and generation time increases. The population becomes heterogeneous, S

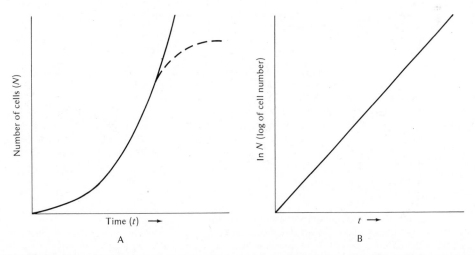

**Figure 6-1** *Exponential growth curve for increase in population size. (A) Arithmetic plot of increase in numbers of cells as a function of time. Dotted portion represents growth decline phase usually seen when growth is limited. (B) On a logarithmic scale the curve plots as a straight line. The slope of the line is the instantaneous growth rate for the population.*

and $G_2$ increase in duration, and a $G_1$ appears. Cell cycle time may increase from 30 minutes in early cleavage to 24 hours in the late embryo or larva. Exponential growth is soon replaced by limited or determinate growth.

As the average division cycle lengthens, cells shift from proliferation and DNA synthesis to morphogenesis. Such transitions from exponential to limited growth often mark critical growth episodes. For example, mammalian exponential growth slows down considerably after birth. Birth represents a crucial break in the growth rates of rats where the embryonic $k$ value of 0.53 several days before birth drops dramatically to a $k$ value of 0.13 immediately after birth.

Two factors account for the decline in growth rate; one is a progressive lengthening of the cell cycle, as we have seen, and the second is caused by cells leaving the division cycle as they functionally mature. Most cells stop dividing after they differentiate, but others are removed from growth by random cell injury or death. All these cells no longer contribute to the overall growth of the organism, and growth rate declines. In some cell populations if the

cell removal rate exceeds 50 percent of the cells per cycle, the total population actually declines.

Populations of cells leaving the division cycle in the embryo become, in the adult, tissues exhibiting different proliferative potential. These include (1) expanding cell populations, (2) static or postmitotic populations, and (3) renewal populations (Figure 6-2).

## EXPANDING CELL POPULATIONS

In expanding cell populations only some cells retain reproductive potential and continue to divide. Total cell number increases even after the initial exponential phase has passed. An embryo consists of expanding populations containing mixtures of mitotically active and inactive cells. In time most cells suppress division activity, leaving a progressively smaller number of cells to replicate. Most tissues in immature animals are made up of expanding cell populations.

Such populations are also found in the adult kidney, liver, pancreas, thyroid, and adrenal, among

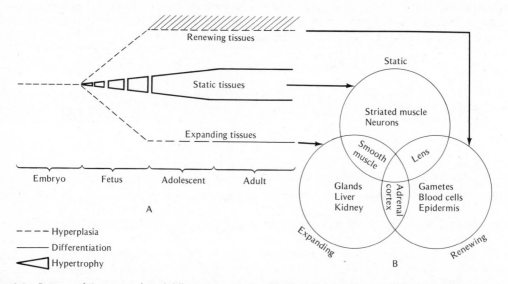

**Figure 6-2** *Patterns of tissue growth and differentiation during development of an organism. (A) In the embryo and fetus most tissues are capable of mitotic activity and proliferate (hyperplasia). A small population differentiates early and loses the capacity to divide. These static tissues grow by hypertrophy. Expanding tissues continue to grow until adult size is attained. These, though differentiated, retain mitotic potential. A small group of renewal tissues are mitotic throughout adult life. (B) The different classes of adult differentiated tissues are shown, including some tissues that overlap two classes and exhibit dual potentialities. (From R. Goss,* Control of Cellular Growth in Adult Organisms, *ed. H. Teir and T. Rytömaa. New York: Academic Press, 1967, pp. 3–27.)*

others. In general the mitotic index in such tissues is low, usually less than 1 percent (Table 6-1), which is not significantly lower than several renewal cell populations. A small proportion of the cells in these organs will incorporate ³H-thymidine after its injection into the animal, indicating that some cells are synthesizing DNA. The labeled cells are often retained for long periods, which suggests that cell turnover is low.

Most cells in these tissues have left the growth-duplication cycle; that is, they are in a $G_0$ functional phase. Their mitotic potential, generally suppressed by systemic signals, can often be stimulated to evoke new growth. In the case of liver, for example, a wounding stimulus or removal of part of the liver stimulates mitotic activity and regenerates new tissue by rapid multiplication. In response to physiological needs for additional functional units, expanding cell populations proliferate actively, exhibiting *hyperplasia*, the formation of new cells. They may also proliferate after isolation in vitro and become established as a primary cell culture.

Some tissues, such as uterine and mammary gland epithelium, respond periodically to hormonal signals by episodes of proliferation. A few mitotically active cells begin vigorous multiplication after hormone stimulation but revert to their original slow proliferating state as soon as hormone concentration is reduced.

## STATIC OR NONGROWING CELL POPULATIONS

During development cells are removed from the division cycle by death, injury, or differentiation. Cells in static tissues irreversibly lose the capacity to replicate once they differentiate. These mature somatic cells do not incorporate radioactive thymidine into their DNA and cannot be induced to proliferate under most conditions. Neurons, muscle tissues, as well as erythrocytes, fall into this category. (Erythrocytes, of course, are terminal cells destined to die within a short time.) Although cell fusion experiments indicate that erythrocytes may be stimulated to synthesize DNA again (see Chapter 3), these postmitotic cells never divide under normal conditions.

Most static tissues mature early in development and have, except for the red cell, an unusually long life, persisting until death of the organism. Static

**TABLE 6-1   Mitotic Activity in Adult Tissues**

| Organism | Tissue | Percent Mitotic Cells |
|---|---|---|
| Man | Bone marrow (R)[a] | 0.5–2.2 |
| | Duodenum (R) | 2.9–3.7 |
| | Epidermis (R) | 0.1–0.6 |
| Cat | Lymph node (R) | 0.5–3.0 |
| | Spleen (R) | 0.1–1.0 |
| Mouse | Intestine (R) | 0.9–1.0 |
| | Adrenal (E)[a] | 0.2 |
| Rat | Thymus (R) | 1.3–13.0 |
| | Liver (E) | 0.1–1.9 |
| | Submaxillary gland (E) | 0.02–0.1 |
| | Ileum (R) | 9.0 |
| | Epidermis (R) | 0.3–1.7 |
| | Cornea (E) | 0.6 |

[a] R = renewal; E = expanding.

tissues such as nerve and muscle grow only by cell enlargement or *hypertrophy*. Physiological demands for new functional units are met by synthesis of more units within the cell (for example, myofibril synthesis in muscle) rather than by cell division. In other words, growth of static tissues is principally by cell enlargement.

Genetic restriction of mitotic potential occurs when cells differentiate, but static cells go beyond the point of mitotic competence. In only a few instances are exceptions to this irreversible state found in static tissues. Restoration of mitotic activity is observed in muscle during limb regeneration in amphibians and is preceded by disassembly of functional organelles, the myofibrils. In other words, these cells dedifferentiate before they divide (Chapter 15).

Differences in organelle specialization may explain mitotic potential differences between static and expanding tissues. Differentiation commits a cell to synthesize specific proteins and organelles. According to Goss, cells such as muscle that contain complex structural proteins usually forfeit their capacity to divide. Function depends upon the integrity of cell organization. For mitosis to occur, these complex organelles must break down, leading to a temporary loss of function. Since function of a nervous system depends upon uninterrupted lines of communication, it is not surprising that neurons do not undergo mitosis. If they did, important nerve functions would be obliterated during division, and new synaptic contacts would have to be reestablished. On the other hand, expanding tissues such as pancreas, liver, or endocrine glands produce secretory proteins or enzymes. Proteins leave the cell rather than assemble into structural units within the cytoplasm. These cells are specialized for a chemical rather than for a structural role; their cytoplasmic machinery makes exportable products. Although the cytoarchitecture of such cells can be quite complex, the cell does not lose its specialized organelles when it divides. Expanding cells divide without disturbing specialized functions; secretory granules, chloroplasts, and pigment granules remain intact. A few structures, such as the Golgi apparatus, are more labile and may disappear, which means that some functional activity is partially or completely suppressed during mitosis.

## RENEWAL CELL POPULATIONS

Some cell populations retain a high rate of proliferative activity in the adult animal in order to replace terminal cell populations continuously shed by the organism, such as blood cells, skin, intestinal mucosa, and gametes. These renewal cell populations are in a steady state; cell gain equals cell loss and no increase in tissue size occurs. Their mitotic index fluctuates in response to functional demands (see thymus in Table 6-1). When such tissues are exposed to radioactive thymidine, a very large proportion of the cells is labeled, but in contrast to adult expanding tissues these labeled cells turn over rapidly and are lost either by dilution of label through successive divisions until undetectable or by migration to other parts of the organism.

Mature, functional products of renewal systems have only a limited lifetime (blood cells, spermatozoa, intestinal villi cells, and so on) and must be replenished by proliferation and maturation from undifferentiated *stem cells*. These cells are mitotically active progenitors whose daughters undergo stepwise maturation before becoming functionally specialized. At any one time a renewal population consists of subpopulations in different stages of maturation from undifferentiated stem cells to functionally mature cells. A steady state is achieved when all subpopulations or "compartments" are of constant size. The compartments may be separate as in intestinal epithelia, or they may be intermingled as in bone marrow or spleen. In the former, cells in different stages of maturation migrate from one compartment to the next and are spatially separated. Among the latter, the different maturation stages distinguish the compartments within a tissue; only mature blood cells occupy a separate compartment. As blood cells mature through various compartments, they gradually lose proliferative potential and eventually die.

The rate of renewal tissue loss is high. LeBlond and Walker estimate that approximately a half-pound of epithelial cells of the gastric mucosa is eliminated daily. About $15 \times 10^{10}$ granulocytes are produced daily, equivalent to about 75 ml of cells; these are also eliminated. In all renewal tissues the rate of production is balanced by an equivalent rate of cell destruction. These systems are sensitive to an organism's functional demands and respond to several systemic controls that regulate proliferation. When bleeding occurs, for example, the loss of blood stimulates mitotic activity in bone marrow and thereby replenishes lost erythrocytes.

We cannot easily study exponentially growing populations in embryonic or immature phases of growth. Too many processes interact during embryogenesis, and regulatory mechanisms are difficult to resolve. On the other hand, proliferating cell populations in adult organisms respond to growth regulation, perhaps by mechanisms similar to those operating during exponential growth. Although net growth has stopped in the adult, a study of renewal or expanding cell populations may indicate how organismic growth is regulated since these populations respond to local and systemic signals that are easily manipulated. From such studies we may learn how exponentially growing embryonic populations are shifted out of the division cycle.

## THE STEADY STATE

The adult animal is in a steady state with regard to growth, a period of size constancy resulting from a balance between incremental and decremental processes. Although total mass does not increase, growth continues. The two major regions of growing cells are expanding and renewal tissues, and the equilibrium between rates of cell gain and loss in these tissues is controlled by feedback control mechanisms. These mechanisms are models of organismic growth control.

In a model of growth regulation Bullough identifies four stages in differentiation of expanding and renewal tissues. These are progenitor cells (P), immature cells (I), mature cells (Ma), and dying or terminal cells (T) (Figure 6-3). According to Bullough, progenitor cells proliferate and give rise to maturing cells. Progenitor cells may be in any phase of the cell cycle ($G_1$, S, $G_2$, or M), whereas immature cells, though mitotically competent (they may even divide one or two times), are in the process of leaving the mitotic cycle to enter the initial steps of differentiation. They differentiate into the mature mass of the adult tissue. Some mature cells, particularly in expanding tissues, are still capable of reverting to a mitotically active state when stimulated. A proportion of mature cells "differentiates" further in the irreversible direction toward death, the tissue cell loss that balances the dividing population. Terminal cells such as erythrocytes, though committed to the death pathway, are functional for a time but are incapable of reverting to mitosis under most conditions.

Figure 6-3 illustrates a flow-through pattern of cells. The turnover behavior of any tissue is determined principally by the size of the progenitor cell population, its rates of mitosis and maturation, and the life span of cells in the mature state. Some cells in the P compartment may pass into a $G_0$ state to serve as a reserve of cells capable of a quick response to physiological demand. These cells are shunted out of the cycle (either from $G_1$ or even from $G_2$) but are sensitive to tissue loss, injury, or systemic factors signaling the need for cell replacement. These $G_0$ cells represent the reserve of cells needed for compensatory replenishment of red cells after bleeding.

Tissues like liver have an extremely small progenitor population, a large, long-lived mature population, and a relatively small dying population. A $G_0$ reserve population may exist, or regenerative cells may come from the functionally mature population instead, most of which are in an arrested $G_1$. It is difficult to distinguish one from the other in this case. Bone marrow tissue, on the other hand, has a large, rapidly dividing progenitor population of undifferentiated stem cells that are mitotically competent throughout the life of the organism. The life span of mature erythrocytes, however, is very short (about 60 days in the rat), which means that the mature population must be constantly renewed. The flow-through rate of cells in this tissue is far more rapid than that found in liver. Any factors (local or sys-

temic) affecting the flow-through rate of cells in a tissue will determine its population size.

Localized tissue growth is regulated by humoral agents acting as mitotic inhibitors or stimulators. Levels of humoral agents in the tissues are adapted to the physiological demands of the organism. The size of a circulating erythrocyte population, for example, is sensitive to slight changes in blood oxygen tension. Decreased oxygen tension leads to hormone stimulation of red cell production in blood-forming tissues. Likewise, periodic bursts of mitotic activity in vertebrate uterine epithelium are synchronized with ovulation and preparations for implantation of an embryo. The larval growth of insects is monitored

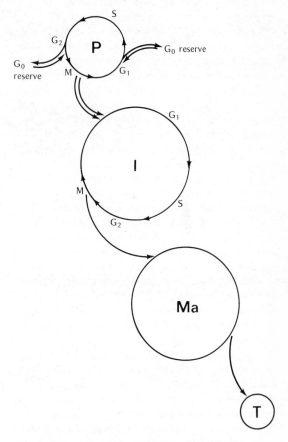

**Figure 6-3**  *Flow-through pattern of cells in different tissues. Each tissue consists of four "compartments" or cell populations in different phases of growth and differentiation. See text for details.*

by an intricate network of neurohumoral factors that regulate the specific growth patterns of many tissues. Most cases of local and organismic growth are governed by humoral agents affecting proliferation of expanding and renewal populations.

## HORMONE FEEDBACK LOOPS IN GROWTH REGULATION

The simplest mode of communication between tissues is via chemical signals. Hormones stimulate proliferative and hypertrophic growth of cells in most multicellular organisms. In this way the behavior of the *target tissue* is controlled by the physiological activity of an endocrine tissue. For growth to be adaptive to the needs of the organism, the two tissues are linked into a feedback regulatory loop so that the target tissue does not grow in an uncontrolled manner. In other words, the growth response of the target tissue must somehow be communicated to the endocrine tissue to modulate the latter's secretion of growth-promoting substances. Such feedback loops are as fundamental to the regulation of man-made machines and computers as they are to biological systems.

A simple negative feedback loop is illustrated in Figure 6-4. It consists of an input signal, two circles representing the regulator and effector systems, and an output behavior. The familiar "black box" of this system includes all unknown reaction mechanisms, the "hardware" that detects the feedback signal, compares it with a predetermined reference (the regulator), and signals the result to the output circuit (the effector). In this way the system is capable of modulating its own activity. When activity becomes excessive (more output), a feedback signal is generated as part of the output information which is fed back to and shuts off the input signal and thereby prevents further activity.

An example of a feedback loop is the interaction between the vertebrate pituitary and thyroid glands. Growth of the thyroid (cell proliferation) is stimulated by a thyrotrophic hormone, TSH, produced by the pituitary and secreted into the circulation. This hormone also stimulates maturation of the thyroid stem cell population. Mature thyroid tissue secretes thyroxine, which acts as a feedback regulator of thyroid growth by inhibiting the formation and secretion of TSH by the pituitary gland. Thus as thyroid cells

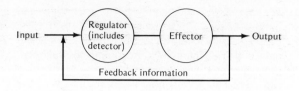

A

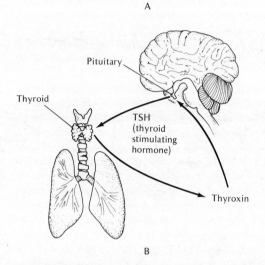

B

**Figure 6-4**  *Examples of feedback control circuits. (A) A model of a feedback control circuit as viewed by an engineer. It consists of a regulator device (with detector) and an effector device which produces some measurable output and is coupled to the regulator as shown. The input to the regulator plus any feedback information from the output are integrated by the regulator and transmitted to the effector to modulate its output. (B) A simple feedback circuit in an organism. The interaction between the pituitary and thyroid glands is shown as a negative feedback circuit. The pituitary gland (the regulator) secretes TSH, which controls the synthesis and secretion of thyroxin by the thyroid gland (the effector). Thyroxin percolating through the pituitary circulation, combined with other stimuli, is able to modify the levels of pituitary secretion of TSH.*

proliferate and mature under the action of TSH, more thyroxine accumulates in the circulation until its concentration is high enough to inhibit further TSH production. A delicate balance between TSH and thyroxine levels in the blood maintains the thyroid at an optimal functional size and state. If three-fourths of a rat thyroid is surgically removed, the amount of circulating thyroxine is reduced, and the pituitary responds to this deficit by raising its output of TSH. The remaining thyroid tissue is thereby stimulated to proliferate and grow. A mitotic stimulant, TSH, is produced indirectly by a deficit of the overall activity of the responding cell population. Growth regulation resides in the hormone thyroxine acting as the negative signal in this feedback loop.

The control system turns out to be far more complicated than is indicated by the simple diagram linking pituitary and thyroid. The pituitary is stimulated to secrete TSH in response to a polypeptide "releasing factor" secreted by neurosecretory cells of the hypothalamic region in the brain. Releasing factors enter the portal circulation that connects the hypothalamus with the pituitary. The polypeptide reacts with the surface of pituitary cells and activates adenylcyclase, which stimulates synthesis of intracellular cyclic AMP. It is this second signal, an elevated level of cyclic AMP, that evokes the pituitary response. Thyroid stimulating hormone diffuses to the thyroid, interacting with receptors on the surface where it also activates adenylcyclase to increase intracellular content of cyclic AMP, causing thyroxine release. Endocrine tissues act as both receptors and effectors; they are intermediates in the flow of metabolic information from primary receptor cells in the nervous system to the tissues in which the final chemical reaction takes place.

### Hormone control of an expanding cell population

For many hormone feedback loops the response of the effector system often includes cell proliferation. The hormone may act on a small proliferating subpopulation and increase its rate of progression through a cell cycle. Or it may act directly on nondividing functional cells blocked in either $G_1$ or $G_2$ by inducing them to resume the cycle. These alternatives have been examined in studies of the effect of estrogens on chick oviduct.

The oviduct of a 7-day-old chick is immature, possessing no glandular epithelium or ciliated cells, nor does it secrete ovalbumin, avidin, or lysozyme. Maturation is normally slow and is completed much later in the adult. Daily injection of estrogen into such chicks precociously stimulates a program of development. Within 2 to 4 days tubular glands appear which secrete ovalbumin and lysozyme, ciliated cells appear at 6 days, and at 7 to 9 days after a single supplementary injection of progesterone, goblet cells secreting avidin appear. The developmental program,

including cell division, changes in cell shape, and differentiation, depends on hormonal regulation of information flow from gene to protein. Here we examine only its effect on cell proliferation.

Mitotic activity in the immature oviduct is normally low with a mitotic index of 0.43, which means that less than .05 percent of all cells are in mitosis. The mitotic index is constant over a 24-hour period, indicating there is no diurnal rhythm. The duration of the $G_2$ and S phases of the cell cycle is 1.5 and 7.3 hours, respectively.

The effect of a single dose of estrogen on cell proliferation was determined by injecting chicks with estrogen, progesterone, or a combination of both (Figure 6-5). Estrogen stimulates mitosis of the oviduct cells, increasing the mitotic index to a peak of almost 3.0 by 18 hours after injection. An early lag of approximately 9 hours is followed by a rapid rise in mitotic index in the next 3 hours. Although the mitotic index declines by 48 hours, it is still more than twice that found in unstimulated oviducts. Progesterone alone exhibits low levels of stimulation; in combination with estrogen it does not enhance the effect—it may even inhibit.

A second dose of estrogen at 24 hours stimulates activity again, and a second rise in mitotic activity is seen. This suggests that the second dose keeps a single cell population progressing through the cycle at maximal rates; otherwise the rate declines.

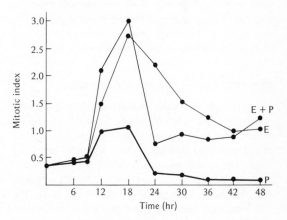

**Figure 6-5** *Mitotic index in oviduct surface epithelium following a single injection of either 17B-estradiol (E), progesterone (P), or 17B-estradiol and progesterone (E + P). (From S.H. Socher and B.W. O'Malley, Develop. Biol. 30:411–417, 1973.)*

How is the hormone acting in this system? Is it stimulating $G_2$ cells to resume the cell cycle and undergo mitosis, or is it inducing $G_1$ cells to enter S? To answer the question an inhibitor of DNA synthesis, FUDR (5-fluorodeoxyuridine), was used. If hormone induces a $G_2$ population to enter mitosis, then the addition of FUDR should have no effect on the hormone-stimulated mitotic index since only cells in S are inhibited. If, however, it stimulates $G_1$ cells, then FUDR should prevent these from completing S, and the mitotic index should be reduced. The evidence indicates that mitotic index is not increased in the presence of the inhibitor, which means that the hormone acts by stimulating $G_1$ cells to move into the S phase.

Similar patterns of mitotic stimulation are found in many other systems and suggest that embryonic cells leave the cycle in $G_1$ to differentiate and carry out specialized functions. $G_1$ is indeed the period of maximum protein and RNA synthesis and is most responsive to physiological modulation. The oviduct cells in $G_1$ seem incapable of initiating DNA synthesis; the hormone overcomes the block, possibly by derepressing the regulatory genes controlling DNA replication.

These experiments also demonstrate that the immature oviduct is already primed to respond to estrogen. Maturation is normally delayed because production of the triggering hormones depends upon an activated pituitary-ovarian feedback loop which matures in the adult. It should also be noted that the first stage in this complicated developmental program is preceded by cell proliferation, a common feature in most differentiating systems (see Chapter 17).

## Hormonal control of a renewal population

A renewal or stem cell population is self-maintaining; removal of cells from the population is balanced by proliferation. Population size is controlled by humoral and physiological factors acting in feedback loops that are still not understood. The best examples of stem cell populations are the intestinal crypt cells and the bone marrow blood precursor cells. We will briefly consider the latter system.

The differentiated products of the bone marrow stem cell may be one of three classes of blood cells, erythroid (red blood cells), myeloid (lymphocytes), and thrombocytic (platelets). Undifferentiated stem cells are committed to one of these pathways in response to exogenous signals and are removed from

the cell cycle. Removal of cells is brought about by differentiation, which often leads to cell death, as is the case for the erythrocyte. Cell removal may be continuous or discontinuous, its rate depending upon functional needs of the organism. A receptor system in the population is sensitive to signals calling for increases or decreases in new cells of each class. Changes in oxygen availability, for example, caused by excessive blood loss, signals renewal of erythrocytes.

One of many models to account for the kinetics of removal assumes that bone marrow stem cells make up two subpopulations, one in a $G_0$, or resting, state, and another in a cell cycle state C (Figure 6-6). $G_0$ cells are a reserve from which cells may either be triggered to return to the cell cycle or proceed to differentiate and are removed from the population. Conceivably, such a $G_0$ population is composed of cells that are in an arrested $G_1$, very much resembling the $G_1$ cells in immature oviduct. Each daughter in the cell cycle population either remains in the cycle or may enter the $G_0$ reserve.

$G_0$ cells are removed in response to a hormone signal and enter one of the three differentiation pathways. Most evidence suggests a common stem cell precursor for each pathway, which implies that information for specific programs is carried by the signaling molecules. For example, a glycoprotein hormone, erythropoietin, triggers red cell differentiation and stimulates red cell production in response to oxygen deprivation. Animals in a steady state can be induced to produce large amounts of red cells within 3 days after injection of the hormone. The nature of the

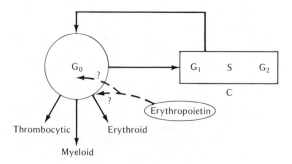

**Figure 6-6**   *Model of cell population size regulation in a bone marrow stem cell population. See text for details. (Based on L.G. Lajtha,* Control of Cellular Growth in Adult Organisms, *ed. H. Teir and T. Rytömaa. New York: Academic Press, 1967.)*

target cell and the mechanism of hormone action in programming a cell for differentiation will be discussed in Chapter 17. Similar hormones may exist for myeloid and thrombocytic programs.

As new populations of erythrocytes are produced, the proportion of cells in $G_0$ decreases while that in the cell cycle increases. This is because many cells differentiate while others return to the cell cycle at increased rates. It is not clear how $G_0$ cells are switched into either of these programs, however. For feedback control of population size we see that the cell cycle population is irrelevant and the population size that has any meaning in regulation is that of the $G_0$ compartment; it alone is responding to input signals. A further observation relating to population size control is that the rate of triggering into the cell cycle is related to the $G_0$ population size by exponential kinetics.

Some aspects of this model have been confirmed by experimental data and computor simulations; many features are still to be explained. It does not seem to apply to the intestinal crypt cells which seem to be in a continuous maximal steady state of differentiation. Approximately 50 percent of the stem cell population leave and become goblet cells in the intestinal villi. This stem cell population is, in a sense, in continuous growth that would exhibit exponential kinetics were it not balanced by cell removal. Most of the differentiated cells in the intestinal lining are terminal and after a period of function are sloughed off.

This model is a useful working hypothesis, but we have to resolve the mechanisms that regulate these stem cell populations. What happens at the critical switch points? How does the hormone control the path taken by $G_0$ cells? Answers to these questions may point to control loops that regulate embryonic cells in exponential cycles as they are committed to specific pathways of differentiation. What properties must these cells display in order to "choose" one of several pathways at switch points? We will see that one critical property is the ability to respond to humoral signals—to behave as a target cell.

### Target tissues and their receptors

Organismic control of growth by humoral factors depends on the ability of target tissues to respond to specific chemical signals. A target tissue is so designated because it, in contrast to most other tissues, detects low titers of a circulating hormone. Its sensitivity appears to reside in its possession of receptors with high binding affinities for specific hormones. These receptors "recognize" the hormone, combine specifically with it, and transport it to sites of action. Hence growth regulation turns out to be a two-component process: production of hormonal signals and their detection by receptors in a target tissue.

In adult mammals, estrogen titer in the circulation increases periodically, each time eliciting a growth response in uterine epithelium. This periodicity is brought about by a gonadal stimulating hormone from the pituitary that induces ovarian tissue to secrete estrogens. One target tissue for estrogen action is the uterine epithelium which exhibits mitotic stimulation as the titers of hormone increase. This response is part of a uterine program preparatory to receiving an implanting embryo.

The initial step in the rat uterus is activation of nuclear RNA synthesis within 2 minutes after estrogen injection. If this RNA synthesis is inhibited by the addition of actinomycin D, then mitotic stimulation is not observed. As in the chick oviduct, the hormone seems to act by derepressing sets of genes which are essential to DNA replication and mitosis. Does this mean the hormone is acting directly on gene templates to initiate transcription?

As a target tissue, the uterus is only one of a few tissues that respond to the hormone. The specificity of the response is probably due to binding of estrogen by uterine epithelium. The incorporation of radioactively labeled estradiol into various rat tissues shows uterus and vagina (another target tissue) exhibiting different patterns from most other tissues (Figure 6-7). Most tissues either incorporate very low amounts (muscle) or bind substantial amounts of hormone but release it in less than an hour (liver and kidney). Uterus and vagina bind the hormone more slowly and retain high levels for many hours. The incorporation pattern in target tissues seems to be a two-step process: an initial rapid uptake of estrogen that continues irrespective of dose (it is nonsaturable) and a retention stage which becomes saturated as the dose of the hormone exceeds physiological levels. Such binding kinetics are diagnostic of receptor substances with high-affinity binding constants for estrogen.

An in vitro system is preferred for studying the intracellular binding of estrogen inasmuch as isolated uterine tissues exhibit similar estrogen-binding curves as those in vivo. If uterine tissue is incubated with radioactive estradiol at 37°C, 20 to 25 percent of the hormone is bound to cytoplasmic macromolecules, and the rest is bound to the nucleus (Figure 6-8). On

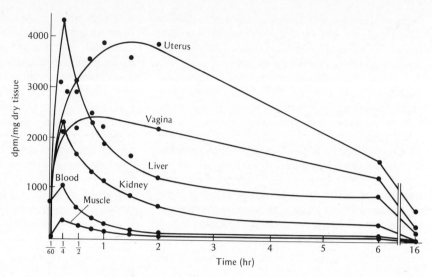

**Figure 6-7**  *Binding of tritium-labeled estradiol by different tissues after a single subcutaneous injection into immature rats. Although several tissues show an initial rapid incorporation of the hormone, only uterus and vagina retain relatively high levels of bound hormone for longer periods of time. (Based on E.V. Jensen et al., Gynecol. Invest. 3:108, 1972.)*

the other hand, if incubation is carried out in the cold in vitro, most estradiol binds to the cytoplasm and shifts to the nucleus only after the temperature is raised to 37°C. The cytoplasmically bound estradiol activity is associated with a cytoplasmic protein while estradiol in the nucleus seems to be bound to a different smaller protein. Specific binding of estradiol by this target tissue is probably due to its high content of a cytoplasmic receptor protein; no tissue has as high a content of this cytoplasmic protein as the uterus.

After binding to the cytoplasmic receptor, the hormone is metabolically transferred to the nucleus. Isolated uterine nuclei incubated with estradiol in the presence of a cytoplasmic fraction will yield a nuclear protein-hormone complex, but in the latter's absence no estradiol-bound nuclear protein is found. Furthermore, transfer of hormone from the cytoplasmic to the nuclear receptor occurs at 37°C and not at 2°C, which agrees with the in vitro observations on whole uteri mentioned above.

As a working hypothesis (see Figure 6-8B), the circulating hormone is presumed to have a strong affinity for a cytoplasmic receptor, either a 4S monomer or an 8S dimer, depending on ionic strength. The hormone-bound 4S complex is converted into a 5S complex in the cytoplasm before entering the nucleus. It is not clear whether the 5S complex combines directly with a DNA template or whether it combines with a repressor to activate RNA transcription.

In the last few years other hormones have been

shown to act on target tissues by combining first with receptor molecules. Insulin and most peptide hormones of the pituitary, such as growth hormone, follicle-stimulating hormone, vasopressin, prolactin, and thyrotropin combine with receptors in the target cell membrane rather than the cytoplasm, as in estrogen binding by the uterus. The mode of action of these hormones differs from steroids since the primary site of action is the membrane. Each hormone interacts with a specific receptor, and since some target tissues respond to more than one hormone (fat cells, for example, respond to eight different hormones), a target cell may possess many different membrane receptors. Sometimes two hormones may bind to a common receptor, or formation of a hormone-receptor complex may alter the distribution of other receptors in the membrane and change their binding properties. In this way two hormones may interact to produce a single response.

The action of nonpeptide hormones such as catecholamines and acetylcholine, the neurotransmitters of the nervous system, also involves specific binding to membrane receptors within target cells which include postsynaptic membranes in neuron-neuron synaptic junctions and muscle cell membranes at neuromuscular junctions. We find that both short-term and long-term response patterns in the neuroendocrine systems are mediated by an initial combination of the signaling molecule with a specific membrane receptor. The response of the target cell, whether it be transmission of a nerve impulse, synthesis and secre-

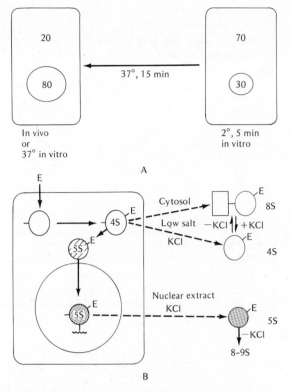

demonstrate (1) saturability, indicating a finite number of binding sites; (2) strict structural and steric specificity—that is, only substances structurally similar to the hormone can compete for binding sites; (3) a tissue specificity similar to biological specificity of the hormone; (4) high affinity, consonant with a biological response to low physiological concentrations of hormone; and (5) reversibility, with kinetics similar to those observed physiologically. The term "receptor" is used loosely in most cases since it usually implies ignorance about the chemical nature or locus of action of these molecules. When these are known, the term is usually replaced by the chemical name.

A most important unifying concept about hormone action has emerged from these studies. The biological effects of many hormones are modulated by the second messenger system, membrane adenylcyclase regulation of intracellular cyclic AMP (see Chapter 3). Most target cells will mimic specific hormone responses when exposed to physiological concentrations of cyclic AMP or its derivatives. Moreover, one of the first detectable metabolic events of hormone action is a change in intracellular levels of cyclic AMP. These results suggest that most hormone action is mediated through cyclic AMP. Furthermore, this small molecule can regulate many different metabolic reactions such as activation of protein kinases (enzymes that phosphorylate and activate proteins), $Ca^{++}$ flux and compartmentalization, histone phosphorylation, and gene transcription (Figure 6-9). It has been suggested that cyclic AMP acts on each of these many processes by first activating a specific protein kinase for each. A membrane signal is transduced into a specific set of metabolic reactions depending on the target cell and its physiological state at the moment of hormone interaction.

The receptor proteins probably do not include adenylcyclase itself. Nor is it likely that the hormone receptor and the enzyme exist as a complex in the membrane. A fat cell, for example, is responsive to eight different hormones, all modulated secondarily by adenylcyclase. Each hormone combines with a different membrane receptor and it is difficult to imagine a giant multireceptor–cyclase complex in the membrane to accommodate these hormones. A mobile receptor hypothesis (based on fluid mosaic membrane models), proposed by Cuatrecasas, attempts to explain the mechanism of modulation of cyclase activity by hormone-receptor complexes (Figure 6-10). A two-step reaction system for activation or inhibition is postulated. Initially, in the inactive state receptor

**Figure 6-8** *Mechanism of estrogen action. (A) Distribution of bound radioactive estradiol after isolated rat uteri are exposed to the labeled hormone under conditions as shown. Most activity is found in the nucleus under physiological conditions. The low-temperature in vitro experiment suggests a two-step process of hormone binding, an initial cytoplasmic binding followed by transfer to the nucleus when temperature is raised. (B) Suggested model of interaction pathways of estradiol in uterine tissues. E is estrogen. Behavior of hormone receptor complex in cytoplasmic and nuclear extracts is shown on right. See text for explanation. (Based on E.V. Jensen and E.R. DeSombre, Science 182:126–134, 1973. Copyright © 1973 by the American Association for the Advancement of Science.)*

tion of a protein, or cell division, occurs only after the hormone-receptor complex is formed.

Receptor interactions are studied by following the kinetics of binding of a radioactively labeled hormone to intact target cells or to isolated membrane preparations. A receptor is defined operationally as those molecules that recognize and bind hormone and evoke a specific biological response. To qualify as a receptor system, the interactions must

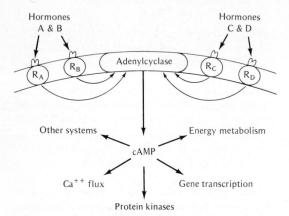

Figure 6-9   *Generalized scheme for second messenger role of cyclic AMP in hormone action. A cell membrane contains adenylcyclase and many different hormone receptors, each specific for a hormone. Hormone binding to receptor forms a hormone-receptor complex in the membrane which activates the enzyme in some manner to increase or decrease intracellular cAMP. The latter, as a regulatory molecule, controls many different metabolic systems. $R_A$, $R_B$, $R_C$, and $R_D$ are hormone receptors.*

and enzyme are segregated in the target cell membrane. In the first step a stimulatory or inhibitory hormone combines with the receptor, and this complex, having acquired structural specificity, secondarily combines with the cyclase, a process facilitated by the mobile membrane (see Chapters 5 and 10). In this combined state the cyclase is activated or inhibited. It should be recalled that membrane fluidity seems to be modulated by a microtubule-microfilament system which transduces membrane signals into the cytoplasm.

Inhibitory and stimulatory hormone-receptor complexes may compete for the same binding site on the enzyme, or each may combine with a different site to affect allosteric configurational changes in the enzyme. The model may also explain mechanisms of cyclase activation in cells exhibiting morphological polarity. For example, epithelial cells of the liver, kidney, or intestine usually border a lumen and function in a polarized manner. The hormone receptors are assumed to be localized on the basal surface (near the capillary) while the effector molecules (cyclase) are assigned to opposite functional surfaces facing the lumen. A hormone-receptor complex, forming at

the basal surface, would migrate rapidly by lateral displacement in the membrane toward the site of action. An equilibrium would be stabilized in the membrane between active forms of the receptor at the apical end of the cell, with its inactive free form still able to combine with hormone at the basal end.

This model tends to bring together many cell phenomena into a unified concept of regulation of growth, proliferation, cell interactions, and morphogenesis by linking hormone action to receptors in fluid membranes which activate an enzyme system controlling intracellular concentrations of cyclic AMP. One leg of the feedback loop in growth regulation is secretion of signaling molecules by neuroendocrine tissues; the other involves a growth response in specific target tissues by virtue of their content of receptor molecules. Any cells unable to synthesize receptors do not respond to these hormones and may behave as postmitotics, never to divide again. Other cells may lose their receptors by some metabolic lesion, exhibit uncontrolled proliferation, and develop into a malignant neoplasm (Chapter 21).

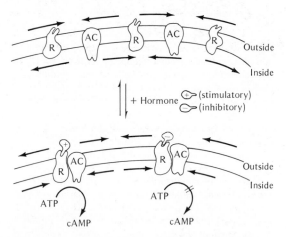

Figure 6-10   *Mobile receptor hypothesis to explain hormone modulation of adenylcyclase activity. In upper diagram, membrane contains receptor (R) and adenylcyclase (AC) in a resting, or low-activity, state. Arrows indicate membrane mobility. After addition of hormone, receptor-hormone complex changes receptor configuration so that it can combine with enzyme either to stimulate it or to block its action completely. Hormone binding sites on receptors exposed to outside, catalytic sites of enzyme exposed to cytoplasm. (Based on P. Cuatrecasas et al., Recent Prog. Hormone Res. 31:37–84, 1975.)*

## LOCAL TISSUE REGULATION OF GROWTH

Organ growth is also regulated by local production or release of mitotic stimulators or inhibitors from neighboring tissues. The concept of growth stimulation by factors derived from dead or dying cells is an old one dating back to 4000 B.C. The Egyptian Imhotep believed that injury produces its effects by "not something from the outside" but by "the intrusion of something which the flesh engenders." Virchow in 1858 proposed that formative substances from damaged tissues stimulated adjacent cells to proliferate. And in recent years organ-specific stimulation of growth by substances derived from necrotic cells has been demonstrated. For example, if necrosis and inflammation are induced in one outer orbital gland of the rat, the contralateral gland exhibits a burst of mitotic activity. The effect is tissue-specific; only orbital gland tissue responds, and only orbital gland tissue extracts will stimulate mitosis after injection. These observations suggest that a circulating mitogenic factor is derived from dying orbital gland cells.

Though factors from dead cells stimulate mitosis, often no specificity is shown. Chick embryo extracts are essential components of most tissue culture media and promote growth of a variety of different vertebrate tissues. Tissue products may have a general nutritional role. Moreover, it is difficult to understand how physiological cell death can evolve as a factor in growth regulation, particularly when experiments show that products of injured cells often inhibit local growth and often without specificity. Specific negative feedback loops of growth regulation have been studied, but these do not derive from necrotic cells.

### Compensatory growth

When up to 70 percent of a rat liver is removed or damaged, the remaining tissue responds by an abrupt rise in mitotic activity to make up the deficit. In a 4- to 8-month-old rat, more than 90 percent of the original liver mass is restored in 7 days, a response called *compensatory growth*. A wave of mitotic activity is noted 24 hours after partial hepatectomy and then declines, only to be followed by additional smaller waves of proliferation in step with the diurnal mitotic rhythms typical of normal liver. Many hours before DNA synthesis begins, widespread changes in cell architecture are noted. The endoplasmic reticulum becomes disrupted, membranes are disorga-

nized, glycogen disappears, and many free ribosomes accumulate. These organizational changes precede the initiation of RNA and protein synthesis and the first wave of mitotic activity. Similar structural changes are noted in normal liver during fasting or when large concentrations of toxic substances are in the circulation. Hence cytoplasmic events accompanying compensatory growth responses are comparable to changes associated with *functional overloading;* new liver cells seem to be produced to satisfy a need for additional functional units.

Two explanations for growth stimulation have been suggested: a sudden local release of mitotic stimulators from dead and dying cells or disappearance of normal circulating growth inhibitors. Support for the latter comes from several observations:

1. Serum from partially hepatectomized rats is less inhibitory of growth of liver tissue in vitro than is serum from normal animals.
2. Dilution of blood plasma in vivo by circulating saline through the portal system induces mitosis in intact rat liver tissue.
3. Concentration of plasma in vivo by fluid restriction inhibits mitosis in regenerating liver.

These data point to a humoral factor passing through the liver circulation that normally inhibits cell proliferation. When this factor is diluted, mitosis is stimulated; when its concentration increases, mitosis is inhibited. According to the specific inhibitor hypothesis, a compensatory growth response is signaled by a tissue deficit. Since the tissue normally adds some tissue-specific mitotic inhibitor to the circulation, the inhibitor concentration is reduced after tissue is surgically removed. Liver cells respond by increased mitotic activity, and growth ensues until most liver tissue is restored. Presumably growth stops after new liver tissue becomes functional and threshold levels of inhibitor are again attained.

The finding that liver homogenates can both stimulate and inhibit mitotic activity in regenerating liver (depending on the time of injection), and the evidence of changes comparable to functional overloading have complicated what was at one time thought to be a simple model of autoregulation of growth. According to this model, all tissues produce diffusible substances called *chalones* that self-regulate growth. These metabolic products of differentiated cells act as tissue-specific mitotic inhibitors.

Negative feedback growth inhibitors have been extracted from various organs. For example, an epidermal chalone extracted from skin, with properties of a glycoprotein, specifically inhibits mitosis of epidermal cells in mouse ear. No substance from other tissues has this effect. Chalone inhibition of proliferation is complicated because hormones are involved; the epidermal chalone interacts with adrenalin for maximum inhibition. Presumably the free chalone is unstable and disappears, a situation which stimulates mitosis. Combination of chalones with hormones may stabilize the inhibitor and its action. The hypothesis of autoregulation of growth predicts that tissue extracts should inhibit growth of specific organs, but, as we have seen, growth stimulation is also observed.

Several instances of apparent growth stimulation turn out to be immunological phenomena instead. This is because immune competent cells are distributed in many tissues besides spleen, thymus, and bone marrow. Since antigens always stimulate mitotic activity of these cells before they secrete antibody (see Chapter 18), such proliferative responses to a foreign tissue extract could be mistaken for a specific growth response. Moreover, caution is necessary when immune-competent organs (spleen or lymph nodes) are used in growth studies. For example, implanting adult spleens on the egg membranes of chick embryos stimulates enlargement of the host embryonic spleen. Kidney and liver explants have no effect. We find, however, that specific growth enhancement in not involved. Host spleen cells are not growing; instead, the adult spleen graft has released cells which colonize the immunological immature host spleen and respond to their "foreign" neighbors by producing antibodies, a process accompanied by massive multiplication of grafted cells. This is not a case of specific growth stimulation of host tissue; rather, it is a "graft against host" immunological reaction by donor spleen cells.

### The nerve growth factor

One of the best cases of specific tissue substances other than hormones that have growth-promoting effects is the *nerve growth factor* (NGF) first detected by Bueker in 1948. He found that a fragment of a mouse tumor (sarcoma 180) implanted on the body wall of a 3-day chick embryo stimulated massive nerve outgrowth from adjacent sensory and sympathetic ganglia. Levi-Montalcini and Hamburger repeated the experiments with sarcomas derived from

mouse salivary glands and noted a two- to fivefold increase in size of the medio-dorsal region of sensory ganglia. The reaction seemed to be highly specific since no other nerve tissues responded. Furthermore, contact between tumor and ganglion cells was unnecessary since ganglia explanted in vitro grew vigorously if incubated with tumor extracts. Obviously a diffusible factor was responsible.

The active tumor factor was characterized initially as a nucleoprotein. To distinguish the active component a snake venom was used to digest nucleic acids, but this produced a surprising result. The snake venom proved to be a far more potent NGF than the original tumor extract! Other venoms and even mouse salivary glands turned out to be rich sources of NGF, with submaxillary glands the richest source of all.

A protein purified from these glands is a dimer of 27,000 mol wt with 10,000 times more NGF activity than the original sarcoma. The primary amino acid sequence of the subunit shows a resemblance to the structure of proinsulin. Some believe that this structural relationship indicates that both regulatory molecules have evolved from a common precursor sometime in the past.

Although NGF has been purified, we still do not understand how it functions in the nervous system, nor do we understand why such large amounts accumulate in mouse salivary glands. Though it can be extracted from embryonic and adult sympathetic ganglia, its role in neural development has been demonstrated only indirectly. For instance, antibodies prepared against purified NGF selectively destroy sympathetic nerve cells in newborn mice. The antibodies are highly specific since neurons normally insensitive to NGF are unaffected, which suggests that NGF (or molecules antigenically similar to NGF) are indeed present in specific regions of developing nervous systems. Presumably, NGF is most active during maturation of the sympathetic and sensory ganglia, but which cell properties are affected?

Cell proliferation within developing ganglia may be the target of NGF action; that is, NGF-treated ganglia contain approximately twice the number of cells as do untreated ganglia. Or NGF may protect neurons normally destined to die, since many neurons that normally fail to make end organ connections degenerate, and this may not occur if NGF is present. NGF may promote axon regeneration in vivo since most axons of isolated ganglia have been cut. All these are equally valid sites for NGF action, and so far the evidence does not favor any one.

One of the most impressive effects of NGF is its stimulation of large amounts of neurotubules and neurofilaments in nerve cells. The cell hypertrophies as fibrous elements are assembled, and they extend into the growing axons. These organelles, a diagnostic feature of nerve cell differentiation, probably participate in axoplasmic flow, the movement of materials up and down the axon from cell body to terminal. It is still not clear whether the increase of fibrous elements is dependent on de novo synthesis of tubule precursors or whether NGF promotes the assembly and stability of neurotubules from a pre-existing pool of precursors.

We know that RNA and protein synthesis are stimulated in ganglia treated with NGF, and microtubule precursor proteins are synthesized. NGF is not acting at the level of transcription, however, since inhibition of RNA synthesis by actinomycin D does not inhibit axon growth. Besides, NGF has other effects on cell metabolism; glucose oxidation and lipid synthesis are also heightened in its presence. Since its metabolic effects are nonspecific, NGF may be acting on membrane permeability, thereby facilitating entrance of small molecule precursors. In fact, many metabolic effects of NGF on nerve cells resemble the effect of insulin on its target cells.

Like insulin, there is some preliminary evidence that NGF combines with a specific surface receptor in the nerve cell membrane. Specific binding of $^{125}$I-labeled NGF was observed on the surfaces of neuroblastoma cells in vitro with kinetics that resemble, in part, the kinetics of insulin binding to its receptor. Thus it is possible, although by no means proven, that NGF, like many hormones, combines with a surface receptor on nerve cell membranes to initiate the first step in a sequence of membrane-mediated reactions. Is cyclic AMP involved here too?

The effect of NGF may be mimicked slightly by exogenously applied cAMP or one of its derivatives (Figure 6-11). But NGF does not change the intracellular content of cyclic AMP, nor does it have any effect on the enzyme adenylcyclase. This evidence suggests that NGF does not employ cyclic AMP as a second messenger and must have some other effect on the cell. This is somewhat different from insulin which is mediated by cAMP. These differences do not exclude the possibility that insulin and NGF may have evolved from the same precursor regulatory molecule but diverged into different modes of action.

Another specific growth factor has been isolated from mouse salivary glands that stimulates mitotic

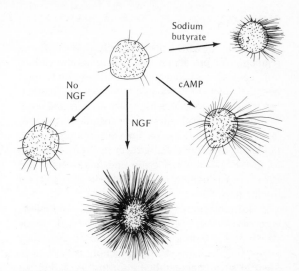

**Figure 6-11**   *Diagram showing behavior of sympathetic ganglion explants in the presence of NGF or a cAMP derivative, dibutyryl adenosine 3'-5' monophosphate and sodium butyrate. In the presence of NGF, a halo of outgrowing axons surrounds the explant after 24 hr at 37°C. Cyclic AMP derivative has only a slight effect on axon outgrowth, not significantly greater than that produced by sodium butyrate. (Based on photograph from W.A. Frazier et al., Proc. Nat. Acad. Sci. U.S. 70:2448–2452, 1973.)*

activity of epidermis and is known as *epidermal growth factor* (EGF). This small protein of molecular weight approximately 6500 has several effects: it enhances epidermal growth and keratin formation, stimulates precocious eyelid opening in newborn rats, and promotes epithelial cell proliferation in vitro. Like NGF, it also stimulates RNA and protein synthesis of epithelial cells.

These two factors are the best examples of growth-promoting substances that act locally. In some ways they resemble hormone action inasmuch as they seem to bind to target cell receptors and exhibit a high degree of specificity. There are many aspects that are difficult to understand, particularly the fact that these substances are produced in such large amounts in salivary glands. Can we assume that in these cases the salivary gland is behaving as an endocrine gland secreting growth-promoting hormones? Some studies of NGF effects on microtubule assembly suggest that the substance may, in fact, be signaling the target cell via systems that mediate membrane mobility and the transduction of signals from the surface into the interior.

# SUMMARY

Organismic growth depends largely upon systemic regulation of proliferating cell populations. Embryonic cells grow exponentially, but their growth soon becomes limited by emergent conditions in interacting cell populations such as surface contact and communication. The kinetics of organismic growth reflects this underlying pattern of limiting growth potential. As exponential growth declines, cells are removed from the cell cycle and differentiate into classes of functionally specialized cells with different mitotic potentials.

The proliferative behavior of these populations is controlled by neuroendocrine feedback loops which coordinate growth of parts into a functioning whole and adapt growth, reproduction, and development to environmental rhythms. Most hormones act on target cells by combining with specific receptors in the membrane or cytoplasm. Steroid hormone-receptor complexes seem to act on gene templates. Others, particularly peptide hormones and neurotransmitters, act through a second messenger system involving cyclic AMP.

Growth may also be regulated locally by tissue factors with stimulatory or inhibitory action. The mode of action of most of these substances is not understood, but in a few cases, such as NGF, it appears that specificity of action also requires combination with target cell receptors on the surface. The action of activated surface receptors may be transduced into the cell cytoplasm by microtubule systems beneath a mobile membrane. Thus assembly and disassembly of such membrane-associated structures becomes an important level of intracellular regulation. In the following chapter we will see how organelles are automatically assembled from soluble precursors in response to intrinsic structural properties and local conditions of ionic strength and pH.

**REFERENCES**

Abercrombie, M. 1957. Localized formation of new tissue in an adult animal. *Symp. Soc. Exp. Biol.* 11:235–254.

Bradshaw, R. A., and Young, M. 1976. Nerve growth factor—recent developments and perspectives. *Biochem. Pharmacol.* 25:1445–1449.

Brody, S. 1945. *Bioenergetics and growth.* New York: Reinhold Publishing.

Bullough, W. S., and Laurence, E. B. 1966. Tissue homeostasis in adult mammals. *Advances in biology of skin,* ed. W. Montagna and R. L. Dobson. *Carcinogenesis.* Vol. 7. Elmsford, N.Y.: Pergamon.

Cautrecasas, P., Hollenberg, M. D., Chang, K. J., and Bennett, V. 1975. Hormone receptor complexes and their modulation of membrane function. *Recent Prog. Hormone Res.* 31:37–84.

Goss, R. J. 1964. *Adaptive growth.* New York: Academic Press.

Houck, J. C., ed. 1976. *Chalones.* Amsterdam: North-Holland.

Jensen, E. V., Numata, M., Smith, S., Suzuki, T., Brecher, P. I., and DeSombre, E. R. 1969. Estrogen-receptor interaction in target tissues. *Communication in development,* ed. A. Lang. New York: Academic Press, pp. 151–171.

LeBlond, C. P., and Walker, B. E. 1956. Renewal of cell populations. *Physiol. Rev.* 36:255–276.

Levi-Montalcini, R., Revoltella, R., and Calissano, P. 1974. Microtubule proteins in the nerve growth mediated response. *Recent Prog. Hormone Res.* 30:635–668.

Liao, S. 1975. Cellular receptors and mechanism of action of steroid hormones. *Int. Rev. Cytol.* 41:87–172.

McCulloch, E. A. 1971. Control of hematopoiesis at the cellular level. *Regulation of hematopoiesis.* Vol. 1. Ed. A. S. Gordon. New York: Appleton-Century-Crofts, pp. 133–159.

Medawar, P. 1945. Size, shape and age. *Essays on growth and form,* ed. W. E. Le Gros Clark and P. B. Medawar. London: Oxford University Press, pp. 157–187.

Socher, S. H., and O'Malley, B. W. 1973. Estrogen mediated cell proliferation during chick oviduct development and its modulation by progesterone. *Develop. Biol.* 30:411–417.

Teir, H., Lahtiharju, A., Alho, A., and Forsell, K. J. 1967. Autoregulation of growth by tissue breakdown products. *Control of cellular growth in adult organisms,* ed. H. Teir and T. Rytömaa. New York: Academic Press, pp. 67–82.

Thompson, D'A. W. 1942. *Growth and form.* New York: Cambridge University Press.

Weiss, P. 1949. Differential growth. *Chemistry and physiology of growth,* ed. A. K. Parpart. Princeton: Princeton University Press, pp. 135–186.

# MORPHOGENESIS

Biological organization is hierarchical, following the "box within box" principle. Each component occupies a definable spatial relation to and behaves as a subunit of a component at the next higher level. Molecules and macromolecules, cells and tissues, organs and organ systems are such related entities. Transitions from one level to the next higher level are accompanied by the emergence of new properties. Only macromolecules and not their constituent subunits are capable of contraction, replication, and catalysis. Each quantum jump between macromolecules and cells and cells and organisms involves a thousandfold increase in informational content.

Although living organization changes continuously, changes are imperceptible over short time periods. Change in organization becomes apparent only over long time periods characterizing growth and development. This process is called *morphogenesis,* a term meaning the "origin or coming into being of form." Morphogenesis occurs at all levels in the hierarchy and follows ordering rules that are specific for each level. Molecular configuration is an ordering principle for self-directed assembly of macromolecules and organelles from identical subunits. Cellular space-filling principles assign organelles to specific sites in cell architecture, polarize metabolic function, and determine cell shape. Cell contact, motility, and communication govern tissue and organ patterning. Any attempt to understand this multileveled aspect of morphogenesis calls for a reductionist approach. We isolate and examine the properties at each level, hoping to discover how the ordering forces at one level are translated into new principles at the next higher level.

# 7

# Macromolecular Assembly:
# The Ontogeny of Cell Organelles

An important architectural principle of living organization is the concept of the subunit as a basic building block in morphogenesis. Structures at each organizational level are built of identical subunits, be they molecules, macromolecules, or even cells. Each subunit is equipped with specific binding sites, like the knobs and grooves on the pieces of a puzzle. Subunits interlock easily, and only a few basic subunits are required for assembly of many types of complicated structures. This conserves genetic information since a small amount of information for making identical subunits may be amplified into a diversity of structures. Moreover, abnormal subunits may be rejected during the construction process; because of their unrecognizable configuration, they cannot assemble with the others. Even if a few faulty subunits are incorporated into a larger structure, it still functions, because most units are active, and slight imperfections do not alter the overall functional configuration.

## PROTEINS AS SUBUNITS

A basic subunit in biological organization is the protein molecule, a linear macromolecule made up of various combinations of amino acids joined by peptide bonds. The diversity of biological organization stems from an almost infinite variety of proteins, ranging in size from about 10,000 to approximately 1,000,000 molecular weight. An average-sized protein with a molecular weight of 20,000 is composed of 150 to 200 individual amino acid residues. These are arranged in polypeptide chains, each characterized by a different primary sequence. Most small proteins are organized as single polypeptide chains (monomers), but large proteins (greater than 50,000 mol wt) are composed of two or more polypeptide chains (dimers, tetramers, and so on) joined into a complex molecule. The function of such molecules depends on a composite structure; activity disappears when chains are separated.

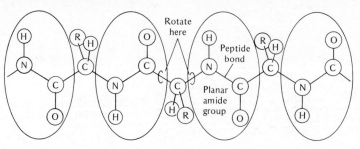

A

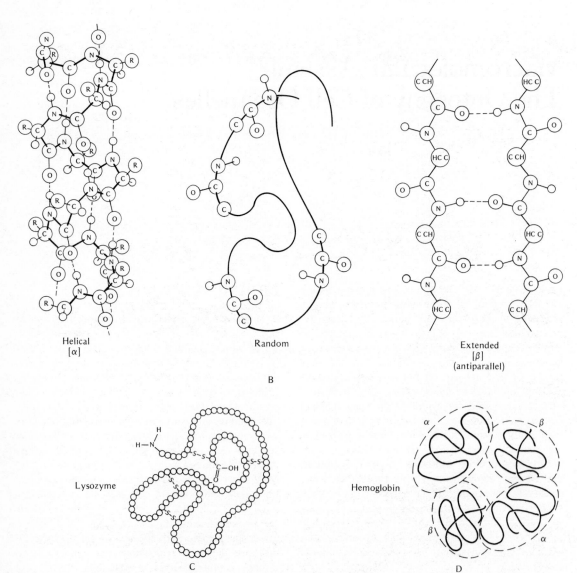

Helical
[α]

Random

Extended
[β]
(antiparallel)

B

Lysozyme

C

Hemoglobin

D

Most proteins do not remain in a linear configuration since covalent bonds are flexible (Figure 7-1). Groups within the molecule tend to associate through weak interactions and twist the molecule into a secondary structure, often into a helical shape. Weak forces within the molecule, such as hydrogen bonding and hydrophobic interactions, fold the linear chain spontaneously into tertiary, three-dimensional configurations. In biological systems all native proteins (as they exist in the cell) are stabilized into such folded structures, and it is in this state that they are functionally active, either as enzyme catalysts or as subunits for assembly into more complex structures. Larger protein molecules consist of several polypeptide chains organized as a quaternary structure and are functional only in this state.

The ordering principles in these macromolecules reside in the strength and distribution of intermolecular forces within the chain. Most forces are weak, but since they act together, they bring about major conformational changes of the molecule. Hydrogen bonding is the most important of these weak forces. These bonds can be represented as interactions between a basic atom B: with an unshared electron pair in association with a weak acid HA or B:...HA. Strength of the bond becomes greater with increasing acidity of HA and increasing proton-attracting power of B. When this proceeds too far, a proton is transferred to B, resulting in B—H$^+$. . . :A$^-$, the hydrogen bond between the BH$^+$ ion and the :A$^-$ ion. Only one hydrogen bond forms between two groups, and its energies are less than a covalent bond.

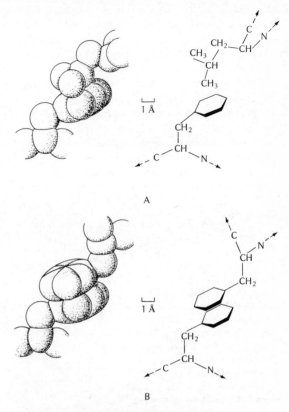

**Figure 7-2** *One type of hydrophobic bond (Van der Waals forces) between nonpolar groups of amino acids. The hydrogens are not indicated individually. The space-filling drawings indicate the relative three-dimensional shape of the residues in the molecules. The structural formulae show the arrangement of the different atoms. (A) Phenylalanine-leucine bond. (B) Phenylalanine-phenylalanine bond. (Based on H.A. Scheraga,* The Proteins, *ed. H. Neurath, vol. 1, 2d ed. New York: Academic Press, 1963, p. 527.)*

**Figure 7-1** *(Opposite) Conformational states of polypeptide chains. (A) The primary sequence of amino acids joined by peptide bonds. R represents a different amino acid side chain, and each is free to rotate as shown. (B) Various periodic repeating structures assumed by single polypeptide chains as a result of intramolecular hydrogen bonds (shown as dotted lines) and hydrophobic interactions. These include α helices and β helices. The random coil is the state assumed by denatured proteins. (C) Tertiary configuration of a protein, lysozyme. The folded globular shape is held together by four disulfide bonds between portions of the chain. (D) Quaternary configuration of a hemoglobin molecule: two alpha chains and two beta chains, folded into their respective globular forms held together by hydrogen bonds. (Based in part on E.R. Blout,* The Neurosciences, *ed. G.C. Quarton et al. New York: Rockefeller University Press, 1967, pp. 57–66.)*

A hydrophobic interaction refers to the tendency of nonpolar side chains of molecules (Figure 7-2) to associate with one another in an aqueous environment rather than interact with polar water molecules. In any aqueous medium nonpolar groups tend strongly to interact with one another; the strong association of water molecules "squeezes" nonpolar groups into an oriented arrangement.

A linear polypeptide possesses many exposed charged groups as well as hydrophobic and hydrophilic sites. This is not a stable configuration in a medium surrounded by polar water molecules and ions. Accordingly, the molecule tends to "seek out" a thermodynamically more stable configuration, twisting and folding into a globular shape. Stability is achieved when all charged molecules have interacted and when nonpolar groups have freed themselves from water molecules by hydrophobic interactions. For example, in myoglobin most hydrophobic amino acid residues in these highly folded molecules face away from the aqueous environment to interact with one another within the molecule (Figure 7-3), while all polar side chains are exposed. The structure is compact with few, if any, water molecules inside. Outside polar groups interacting with water molecules may form a shell of structured water or "ice" around the molecule.

The most stable configuration of a protein is usually its native functional state. An enzyme, such as lysozyme, possesses catalytic surfaces and reactive groups that are spatially oriented, an arrangement that confers high substrate specificity on enzyme catalysis. Only certain substrates have the correct size and order of side groups to make a stereospecific fit with the enzyme. The enzyme molecule is not rigid; its shape is dynamic and actually adjusts to the shape of the substrate molecule as they interact. Any molecule that shares the size and reactive sites of a

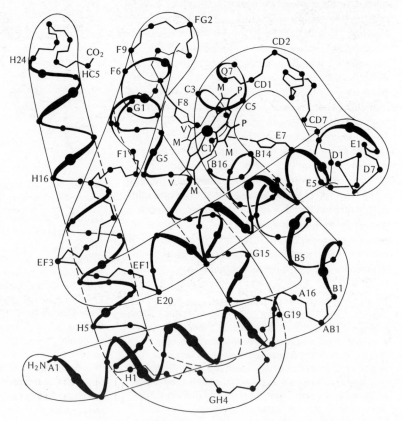

**Figure 7-3**  *Three-dimensional structure of myoglobin molecule based on X-ray analysis. Large dots are carbon positions. Helical regions within the molecule are given a letter-number designation (for example, H1–H24 helix, or F1–F9). The heme group is noted by the cluster labeled M, V, P, top center. The amino acid residues in the a-helix segment (E1 to E20) are listed below with their inside-outside relationships. The hydrophobic amino acids leucine and valine are internal whereas the hydrophilic alanine and lysine are outside, exposed to the medium. (Based on E. Blout,* The Neurosciences, *ed. G.C. Quarton et al. New York: Rockefeller University Press, 1967, pp. 57–66.)*

| | Amino acid | Environment | | Amino acid | Environment |
|---|---|---|---|---|---|
| E7 | His | Heme | E14 | Ala | out |
| E8 | Gly | in | E15 | Leu | in |
| E9 | Val | in | E16 | Gly | (?) |
| E10 | Thr | out | E17 | Ala | out |
| E11 | Val | in | E18 | Ileu | in |
| E12 | Leu | in | E19 | Leu | in |
| E13 | Thr | out | E20 | Lys | out |

substrate molecule can fit the enzyme surface in a functional configuration. Such molecules may act as competitive inhibitors of enzyme action, since they compete with the natural substrate for the reactive site.

The globular form of a protein molecule such as ribonuclease is the native configuration, but it can be unfolded into a linear polypeptide chain by denaturation, a process of breaking hydrogen bonds or other stabilizing bonds such as disulfide bonds (S—S) (Figure 7-4). As hydrogen or disulfide bonds are broken by such agents as urea, the atoms are exposed to and interact with water molecules. All hydrophobic interactions are eliminated, and the chain unfolds, which suggests the main driving forces responsible for stability of native proteins are hydrophobic interactions. Urea acts by increasing the solubility of those amino acids with nonpolar side chains such as phenylalanine and leucine, while decreasing the solubility of those with polar side chains such as threonine and histidine. The process is reversible; after urea is removed, the linear molecule folds up and assumes its original native configuration and function. Thus, by a process of controlled denaturation and renaturation, the activity of a protein molecule can be "turned off and on." No additional information is required to reestablish the molecule; the extended linear polypeptide chain automatically folds into its tertiary configuration because of weak interactions within the chain.

The amino acid sequence determines a protein's three-dimensional configuration since the position of each residue defines hydrophobic and hydrophilic sites. Substitution of a polar residue by a nonpolar residue may change the shape of a protein and modify or destroy its activity. We can now explain a mutation at the molecular level; it is a replacement, or deletion, of at least one amino acid from the linear polypeptide which alters intramolecular interactions and folding tendencies. If the polypeptide is functioning as an enzyme, it can no longer organize into a catalytic surface and accept substrate molecules. If the polypeptide is a subunit of a biological structure, it will not have the correct shape, and either it will not interact with other subunits to assemble a larger structure, or, more likely, it will organize into an abnormal structure.

Many proteins are functional only in the quaternary state, multiple aggregate systems composed of two or more polypeptide chains, often forming multienzyme complexes. For example, hemoglobin (65,000 mol wt) is a tetramer consisting of two alpha

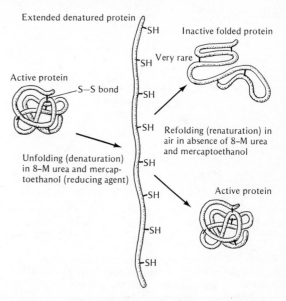

**Figure 7-4** *Denaturation and renaturation of a protein held in a native three-dimensional form by disulfide bonds. As urea dissociates hydrogen bonds and mercaptoethanol reduces disulfide bonds into sulfhydryls (SH), the polypeptide chain unfolds into an extended form. After the denaturing agents are removed, polypeptide chains spontaneously assume their globular form while a few may fold abnormally into inactive forms because of different S—S linkages. (From James D. Watson,* Molecular Biology of the Gene, *Third Edition, copyright © 1976, 1970, 1965 by W.A. Benjamin, Inc., Menlo Park, California.)*

chains and two beta chains (Figure 7-5) held together by the same weak bonding forces that account for intramolecular structure. A functional configuration is achieved only when all four chains assemble into a globular aggregate with the iron-containing heme in a specific position. It is this complex molecule that actively binds and transports oxygen.

When this complex is denatured by urea, the alpha and beta subunits dissociate, and oxygen binding capacity is lost. If the denaturing agent is removed, the subunits automatically reassemble into the native active configuration, and oxygen binding is restored.

A single gene mutation that substitutes a valine residue for a glutamic acid at one position in the beta chain changes it into hemoglobin S, or sickle cell hemoglobin. Its three-dimensional configuration is totally different from normal hemoglobin A, which accounts for the characteristic change in shape of the

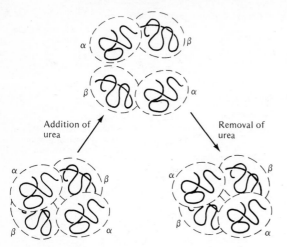

α β

β    α

Addition of
urea

Removal of
urea

α    β

α    β

β    α

β    α

**Figure 7-5**  *Denaturation and renaturation of a more complex protein such as hemoglobin which, in its native state, has a quaternary structure. The addition of urea breaks hydrogen bonds holding the alpha and beta subunits together. These reassociate into a functional hemoglobin after the denaturing agent is removed.*

sickled red blood cell. In this instance, the three-dimensional conformation of a major cellular protein actually alters the shape of the cell. The rules governing molecular organization translate into specific form changes at a cellular level of organization.

Most multiple protein systems behave like hemoglobin and exhibit similar self-assembly properties; they are reversibly denatured and renatured by changes in ionic strength, temperature, pH, and so on. In spite of rotational properties within a molecule and its freedom to assume any one of many configurations, the homeostatic conditions of the cell insure assembly of only one native functional state.

## SELF-ASSEMBLY OF CELL ORGANELLES

Under most physiological conditions, groups of subunits automatically assemble into the functional organelles of a cell. Some cell organelles are composed of identical subunits, but others are polytypic, self-assembled from mixtures of nonidentical subunits. A ribosome, for example, is a complex structure containing 30 to 50 different proteins associated with three different RNA molecules. The size, number, and variety of subunits determine organelle complexity. A subunit may consist of more than one polypeptide

chain. In most cases the information for organelle assembly is within the structure of each reacting subunit, but sometimes exogenous information is required to complete the morphogenetic process. Part of the information for organelle assembly also resides in the cytoplasmic ionic milieu and pH, which determine the final configuration of organelles in the cell.

### Spherical structures made by globular subunits

As a subunit for biological structure, most polypeptide chains randomly fold into globular molecules, which in turn self-assemble into multienzyme systems or complex organelles. Many enzyme proteins are complexes of two or more polypeptide chains (Table 7-1). The space-filling tendencies of these subunits and the types of ordered structures they produce are defined by their number and size; the greater their number the more opportunities for different structural designs.

Small numbers of subunits (less than 10) pack together into a limited number of arrangements. Two, four, and eight units pack together like the cleaving blastomeres in an embryo. Many enzymes are composed of eight globular subunits organized as a cube, with each subunit at a vertex. Additional subunits fill the faces of the cube to form a more compact spherical structure.

**TABLE 7-1   Subunits in Proteins**

| Protein | Molecular Weight | Number of Peptide Chains |
|---|---|---|
| Hemoglobin | 65,000 | 4 |
| Alkaline phosphatase (*E. coli*) | 80,000 | 2 |
| Enolase | 82,000 | 2 |
| Hexokinase | 96,000 | 4 |
| Aldolase | 142,000 | 6 |
| γ-globulins | 150,000 | 4 |
| Catalase | 250,000 | 4 |
| β-galactosidase | 520,000 | 12–16 |
| Glyceraldehyde phosphate dehydrogenase | 145,000 | 4 |
| Lactate dehydrogenase | 126,000 | 4 |
| Glutamate dehydrogenase (beef liver) | 2,000,000 | 40 |

Source: From H. Sund and K. Weber, *Angew. Chem. Int. Ed. Engl.* 5:231–245, 1966.

Many globular subunits (100 or more) spontaneously assemble into spherical shells with subunits closely packed in hexagonal array on the surface (Figure 7-6). These three-dimensional lattices, or *capsids,* have evolved as efficient containers for packing viral genomes; the nucleic acid is folded up within the hollow space of the capsid. When denatured, capsids disaggregate into globular subunits which spontaneously reassemble into the original capsid structure if pH and ionic strength are changed. The same mathematical rules governing construction of the familiar geodesic dome also operate at the molecular level to order subunits into three-dimensional capsids. Given the number of units, we can almost predict their three-dimensional structure.

### Rodlike structures made by globular proteins

With thousands of subunits, self-assembly principles generate new ordering patterns as globular units organize into long cylinders or fibers. Identical subunits may become stacked by repeating the same bonding pattern and automatically form helices. For example, tobacco mosaic virus (TMV) consists of a 300-m$\mu$ "coat" protein cylinder surrounding an RNA core (Figure 7-7). The coat protein consists of 2000 to 2600 globular subunits packed in ordered helical arrays into a cylindrical tube with the helical RNA strand occupying the central core, bound to the inner surface of the subunits.

Infective virus is dissociated into its protein coat and RNA by denaturation, and the isolated coat may be further disaggregated into its globular subunits by alkali at high dilution. Under proper conditions of temperature and pH, the entire mixture of RNA and protein subunits self-assembles into infectious virus. In the absence of RNA, subunits polymerize into helical structures of varying lengths depending on pH and ionic strength. The length of RNA defines the normal size of the viral particle, perhaps by providing nucleotide binding sites that stabilize a specific structure. Subunits do not polymerize individually but seem to add on as collections of many units organized as discs. These interact with the exposed RNA strand and interlock into the growing cylinder. The RNA strand seems to control the pattern of disc addition.

Globular subunits also self-assemble into bacterial flagella (Figure 7-8A). A flagellum is about 20 nm (nanometers, or $10^{-9}$ meters) in diameter and several microns long. Flagella can be sheared from cells and extracted to yield a globular protein, *flagellin.*

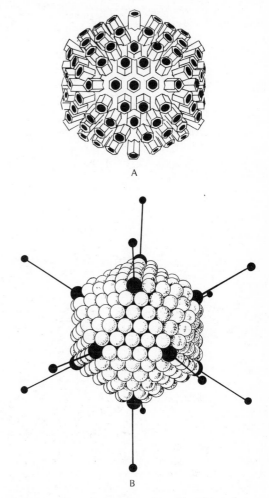

**Figure 7-6** *Model of viral capsids assembled from globular subunits. (A) Model of a viral capsid (rheovirus) consisting of approximately 72 subunits arranged in an icosohedral lattice. The holes in the center of the subunits (capsomeres) are a prominent feature. (From J.T. Finch and A. Klug, J. Mol. Biol. 13:1–12, 1965.) (B) Model of an adenovirus particle based on EM studies. This capsid is also organized as an icosohedron lattice but has fibers extending radially from the pentons. (From R.C. Valentine and H.C. Pereira, J. Mol. Biol. 13:13, 1965.)*

Many thousands of these subunits are arranged in a helical surface lattice around a longitudinal axis to produce a sinusoidal filament.

Flagella disintegrate into subunits in acid, and after neutralization these self-assemble into typical sinusoidal flagella. In most cases assembly is slow un-

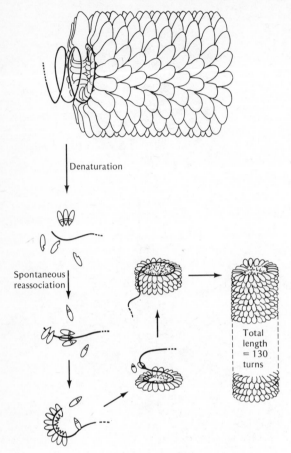

Denaturation

Spontaneous
reassociation

Total
length
= 130
turns

**Figure 7-7**  *Self-assembly of tobacco mosaic virus
particles. The virus consists of a coat protein (shown in
white) surrounding a single molecule of helical RNA
(black helix). When denatured, the protein separates from
the RNA and dissociates into more than 2000 globular
subunits. When denaturing agents are removed, the
subunits self-assemble around the RNA strand,
reconstituting the fully infectious particle. (Based on
H. Fraenkel-Conrat,* Design and Function at the Threshold
of Life. *New York: Academic Press, 1962.)*

subunit structure, produce a different packing arrangement, and alter flagella architecture. An example of such a mutant is *curly,* in which the wavelength is shorter than normal, and *straight,* from a
nonmotile strain, in which flagella lack a coil. Mutant
subunits self-assemble into the typical mutant flagella.
Moreover, mutant molecules dominate the polymerization pattern of mixtures of subunits. A *curly* seed
mixed with normal flagellin subunits reconstitutes
*curly* flagella, and a normal seed with *curly* subunits
also assembles *curly* structures. These results are difficult to explain since we would expect the structure
of the seed to orient the addition of subunits. The fact
that a normal seed fails to order a normal flagellum
with *curly* subunits suggests that the mutant protein
introduces sufficiently large perturbations to change
the packing arrangement.

The structure of a flagellum and its polymerization pattern resemble the behavior of the contractile
protein actin, a ubiquitous molecule that is found in
virtually all cells (see Chapter 8). Actin was originally
extracted from muscle tissue, where it interacts with
another protein, myosin, as part of the contractile ap-

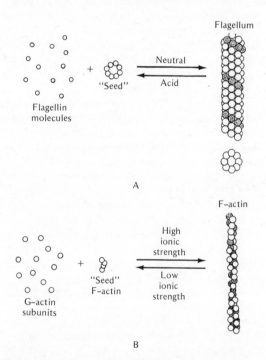

Flagellum

Neutral

Acid

Flagellin
molecules

"Seed"

A

F–actin

High
ionic
strength

Low
ionic
strength

G-actin
subunits

"Seed"
F-actin

B

**Figure 7-8**  *Self-assembly of filamentous structures from
globular subunits. (A) Flagellin-flagella equilibrium. (B)
G-F actin transformation.*

less a "seed," or nucleus of small flagella fragments,
is introduced to initiate polymerization. The sinusoidal pattern is spontaneously generated because the
weak forces between units are maximally stabilized
by a helical, staggered array, which results in a coil.

The wavelength of this coil depends on the three-
dimensional configuration of flagellin subunits, which
in turn is determined by primary amino acid sequences. A mutation in the sequence will change

paratus. Actin filaments (F-actin) are composed of identical globular subunits organized as two helically wound strands (Figure 7-8B). If the ionic strength is lowered, the filaments dissociate into globular subunits, or G-actin, which reversibly polymerize into F-actin filaments of variable lengths if ionic strength is increased. Polymerization of actin is greatly stimulated by seed fragments of F-actin, which suggests that polymerization, once begun, is self-accelerating. G actin contains bound ATP as an essential component in its structure. If ATP is destroyed by ultraviolet irradiation, the molecule denatures. Though polymerization of G- to F-actin is usually accompanied by the splitting of ATP, it is not essential in the process. We will see later that reversible polymerization of actin in cells is fundamental to cell motility and shape changes during morphogenesis (Chapter 8).

Two principles of self-assembly emerge from these observations. (1) The size of assembled structures, particularly of filaments, is variable unless interactions with other size-controlling elements occur. The TMV RNA controls the length of the viral coat by providing specific binding sites that stabilize one length in preference to all others. (2) Self-directed assembly is a slow process in many cases but is accelerated by addition of a nucleating site which exposes an ordered surface for rapid polymerization of subunits. Both such principles probably operate in cells to regulate the assembly of cell organelles.

## Fibrous structures from linear subunits

Fibrous structures are also made by self-assembly of rodlike subunits as in silk or collagen. Some polypeptide chains remain as helical fibers instead of folding into globular molecules. These asymmetric molecules have a periodic repeating structure running their full length and self-assemble into even more complex fibers. Collagen is probably the best example of this organizational pattern.

Collagen is a ubiquitous protein, found in virtually every animal phylum. In some organisms it constitutes one-third of total body protein, serving as a scaffolding within connective tissue. There are several types of collagen, depending on their tissue location. Though collagens in cartilage, skin, and in basement membranes differ with respect to their subunit polypeptide chains, all follow the same design rules in forming a fiber. In the electron microscope collagen fibers appear as banded structures with a typical repeating periodicity (Figure 7-9). This pattern for native collagen comes from the lateral association of many fibrils lined up in register to form bands with a 640-Å repeat pattern. A collagen fibril is composed of a sequence of linear tropocollagen molecules joined end to end. A tropocollagen subunit is a rod, 2800 Å long and 14 Å in diameter, made up of three entwined helical polypeptide chains. One chain has a different amino acid sequence from the other two. Collagens have different combinations of alpha and beta chains. These polypeptide chains are synthesized within a fibroblast (or an epithelial cell) where they spontaneously assemble into tropocollagen molecules. Tropocollagen is secreted into the extracellular space where it in turn spontaneously organizes into huge collagen fibrils attached at first to the cell plasma membrane. The assembly involves an end-to-end alignment as well as lateral association of tropocollagen subunits. Each tropocollagen molecule is polarized, with a head and tail end, and the intrinsic repeat pattern of amino acids favors aggregation so that heads always interact with tails.

Collagen fibrils dissolve in dilute acid into soluble tropocollagen molecules. These reassociate at higher pH into native collagen, with a repeat spacing pattern of 640 Å. Subunits aggregate in staggered parallel array with heads attached to tails, all pointing in the same direction. Many agents affect the assembly process and produce new patterns in vitro. For example, in the presence of ATP, short fibrils reconstitute with a major spacing period of 3000 Å. In this "segment long spacing" (SLS) pattern the monomers are united head to tail in parallel, with all amino acids in register rather than staggered. Still another pattern is the "fiber long spacing" (FLS) in which the subunits are arranged like the SLS type with a 3000-Å repeat pattern except that alternate fibrils run in opposite directions. The intraperiod fine structure is therefore different from the SLS type. Clearly, the type of ordered pattern assumed by collagen subunits is determined by their response to conditions of aggregation.

Bone is normally deposited within a collagen matrix, and its contours as it crystallizes depend on collagen fiber patterns. Bone precursors, the hydroxyapatite crystals, deposit only on collagen fibrils with the native 640-Å periodicity. Here we see that macromolecular patterns generated by self-assembly at one level act as templates for more complex ordering at a next higher level. In organs like skin another type of collagen organizes its fibers into a basement membrane between dermis and epidermis composed of layers of fibrils stacked at right angles to one

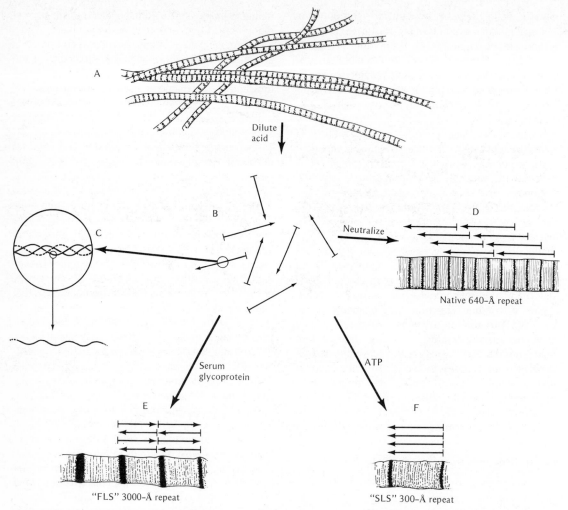

**Figure 7-9**  *Collagen structure and self-assembly. (A) Appearance of mass of collagen fibrils in the electron microscope showing the native axial repeat period of 640 Å. (B) Fibers dissolved in dilute acid yield soluble tropocollagen subunits represented by arrows. (C) A tropocollagen molecule consists of three polypeptide chains entwined about one another in helical configuration. One chain (dotted line) differs from the other two. Each polypeptide chain is itself a helix. (D) A solution of tropocollagen molecules will self-assemble into native collagen fibrils under physiological conditions, but in the presence of ATP, or serum glycoprotein, different fibril structures reconstitute, shown as (E) a fiber long spacing (FLS) pattern and (F) a segment long spacing (SLS) pattern.*

another. How these laceworks come about is still not clear, but we can assume that ordering of one layer of fibrils itself determines the pattern established by each layer that follows. Within this system, an organizational hierarchy develops from prepatterns assembled at lower levels.

There are limits to the diversity of structure that can be built from identical subunits. Though var-

iations appear if conditions of reassociation are changed, such novel configurations are usually nonfunctional. In general, intracellular conditions favor only one of several possible polymerized states.

The assembly of collagen introduces another problem of cellular control of organelle assembly. Evidently conditions within the cell are not favorable for tropocollagen polymerization into collagen fibrils;

this assembly occurs outside the cell, presumably in association with extracellular matrix materials. The cell must be able to package and transport subunits from their sites of synthesis to assembly sites by mechanisms of directed flow that are still mysterious.

### Self-assembly from nonidentical subunits

Most cell organelles are polytypic, composed of lipid or nucleic acid subunits combined with proteins. Self-assembly of nonidentical subunits often results in complex structures—arrays of multienzyme systems organized into lipid membranes as in mitochondria or chloroplasts. These enzyme complexes are more efficient than single enzymes in solution because enzyme reactions are brought into close spatial proximity. Cofactors, precursors, and substrates are rapidly exchanged between reactive sites on adjacent enzyme surfaces.

Many enzymes are not membrane-bound but are organized into different-sized aggregates. A simple mixed aggregate is the enzyme tryptophane synthetase, consisting of two different polypeptide subunits, A and B. The A subunit is a single polypeptide chain, whereas B is composed of two entwined polypeptide chains designated as $B_2$. Each subunit has different enzymatic function: the A enzyme catalyzes the conversion of indole and triose phosphate to indole glycerophosphate; the B enzyme catalyzes the reaction between indole and serine to make tryptophane. When subunits associate into a tryptophane synthetase complex, two molecules of A combine with the $B_2$ unit and acquire an additional catalytic activity, indole glycerophosphate reacting with serine to produce tryptophane. The complete system assembles spontaneously in vitro if A and B subunits are mixed. As subunits aggregate, a conformational change in both molecules creates new reactive sites that recognize additional substrates. Not only does this demonstrate the spontaneity of interaction between different proteins, but also it shows that the aggregate acquires catalytic properties (an epigenetic event?) not found in either subunit. This is of considerable advantage to the cell because the multienzyme complex with its juxtaposition of catalytic surfaces utilizes several substrates more efficiently. Far more substrate is handled in this manner than is possible by enzymes in solution where the reaction rate depends on random collisions between enzyme and substrate. Some complex enzyme systems such as pyruvate dehydrogenase from *E. coli* consist of many

more molecules (Table 7-2). Here three different proteins, each composed of several different polypeptide chains, are organized as a multienzyme system in which enzyme movement is restricted and the intermediates of three separate enzyme reactions do not dissociate. Instead, the active groups of each enzyme are spatially arranged within the complex so that the several reactions are efficiently coupled; the products of one reaction immediately become the substrates of the next reaction in the metabolic sequence.

### The ribosome: a polytypic organelle

As more elements enter into an aggregate, it acquires additional functional properties. Such structures are able to carry out whole sequences of metabolic reactions. The ribosome, for example, originates from stepwise assembly of many heterogenous subunits to make an organelle that synthesizes polypeptide chains. Each step in protein synthesis is controlled by one or several enzymes packed into the ribosome complex.

The activity of the system depends on the stereospecific relationships of its components; a conformational change in one protein or a mutation in its structure can modify function of the entire complex.

Bacterial ribosomes consist of two subunits, a 30S particle and a 50S particle. Neither alone is active in protein synthesis. The 30S particle contains a 16S ribosomal RNA combined with some 20 different proteins (Figure 7-10). The 50S particle is also a nucleoprotein conglomerate of 30 to 40 separate proteins with the 23S and 5S ribosomal RNA. At low magnesium concentration intact ribosomes dissociate into the two subunits. If magnesium concentration is increased, subunits rejoin as an active 70S ribosome, capable of synthesizing polypeptides in vitro. Eukaryote ribosomes, on the other hand, dissociate into 40S and 60S subunits in high potassium plus magnesium and reassociate after potassium is removed. Since ribosome integrity is affected by ion concentration, we assume the two subunits are held together by ionic bonds.

The bacterial system has been resolved in more detail. We will confine the discussion to the 30S subunit although the 50S subunit exhibits similar behavior. The 30S subunit is dissociated into a 23S core and a collection of soluble *split* proteins when centrifuged in a cesium chloride gradient. Split proteins spontaneously reassociate with cores into a functional subunit in less than one minute under physiological

**TABLE 7-2**    **Composition of** *E. coli* **pyruvate dehydrogenase complex**

| Component | Molecular Weight | Number of Molecules | Number of Chains per Molecule |
|---|---|---|---|
| Decarboxylase | 183,000 | 12 | 4 ($\alpha_2\beta_2$) |
| Flavoprotein | 112,000 | 6 | 2 ($\gamma_2$) |
| Transacetylase | 1,000,000 | 1 | 24 |

Source: From L. J. Reed, *The Neurosciences*, ed. G. C. Quarton, T. Melnechuk, and F. O. Schmitt. New York: Rockefeller University Press, 1967, pp. 79–90.

conditions. This pattern of self-assembly from core and split proteins may occur normally in the cell during ribosome biogenesis. No other components are required since the information for correct assembly is in the specific binding sites (hydrophobic ?) in the components.

The 23S core may be further dissociated by treatment with urea and lithium chloride, or phenol, into the free 16S RNA and several core proteins. If urea is removed and the temperature increased to 40°C, the RNA and core proteins reassemble into a 23S core particle which becomes functional in protein synthe-

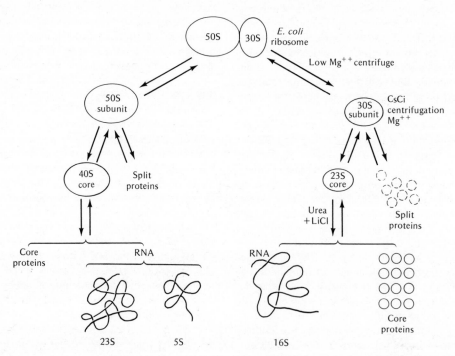

**Figure 7-10**   *Structure and self-assembly of bacterial ribosomes. (A) The* E. coli *ribosome consists of a large 50S subunit and a small 30S particle. These may be separated by centrifugation in low Mg++. Each subunit may, in turn, be dissociated into a 23S core and split proteins. Subsequent dissociation of 23S core with urea and lithium chloride yields 16S RNA and core proteins, of which there are 12. The 30S particle consists of 7 split proteins and 12 core proteins, plus the 16S ribosomal RNA. The reversible arrows indicate the points of self-assembly.*

sis only after recombining with split proteins. The role of 16S ribosomal RNA is critical here; foreign 16S ribosomal RNAs from different bacterial species or from yeast cannot replace *E. coli* RNA in reconstitution of the 23S core. RNAs from closely related species, on the other hand, can substitute in the reaction. Evidently, certain nucleotide sequences are essential to reconstitution, possibly because they furnish complementary binding sites for some ribosomal proteins.

Assembly is temperature-dependent, occurring most rapidly at 40°C but not at 0°C. At low temperatures small intermediate particles are found containing the 16S RNA plus half the ribosomal proteins, the remaining proteins still free in solution. If the temperature is raised to 40°C for a short time, the intermediate particles are activated by some conformational change, and the remaining proteins associate to produce the functional 30S subunit. The assembly proceeds in a stepwise fashion; first 10 to 12 proteins interact with RNA to create an active intermediate with a configuration able to interact with the remaining proteins. A similar stepwise assembly is assumed to take place in the cell.

The role of different protein components in the ribosome can be determined simply by removing proteins from reassociation mixtures. For example, certain split proteins of the 30S subunit are needed for amino acid incorporation and for tRNA and mRNA binding. Other split proteins are dispensable.

Some of the core proteins are probably important in streptomycin sensitivity of ribosomes. Streptomycin inhibits bacterial growth by disrupting ribosome function and protein synthesis. The 30S core proteins from streptomycin-sensitive (wild type) and streptomycin-resistant strains have been systematically reconstituted after eliminating one protein at a time from mixtures with 16S RNA. We learn, at first, that the mutant lesion is not in the RNA of the resistant strain; mutant 16S RNA with sensitive core proteins self-assembles into a sensitive 30S ribosome subunit. On the other hand, omission of a specific protein from reconstitution mixtures containing mutant core proteins results in weakly active but normal-sized 30S particles. Streptomycin resistance depends upon the presence of a specific protein. Since the subunit is of normal size, we can assume that the missing protein plays no part in ribosome structure but has a functional role in translation. Other proteins may be essential to ribosome assembly but dispensable in protein synthesis.

Ribosomes may associate into complex crystalline lattices, but most importantly they can align themselves along membranes of the endoplasmic reticulum where protein synthesis is coupled to transport and secretion of gene products. The association of ribosome with membranes at the organelle level is a visible expression of interactions during membrane formation at the molecular level.

## SHEET SYSTEMS: MEMBRANE ASSEMBLY

A two-dimensional membrane or "cytoplasmic sheet" is the most ubiquitous organelle in all biological systems. Every cell, even the most primitive prokaryote, is bounded by lipid membranes containing an assortment of proteins. Membranes act as diffusion barriers maintaining the integrity of protoplasm, separating a complex, internal aqueous environment from that outside the membrane. This is accomplished by virtue of the special properties of lipid molecules.

Lipids are amphipathic molecules; they are polarized with hydrophobic, fatty acid tails and hydrophilic, glyceride "heads" (Figure 7-11). Consequently, in aqueous solutions pure lipids tend to organize spontaneously into spherical micelles or liposomes, the most thermodynamically stable configuration. A bimolecular layer (bilayer) of lipid molecules surrounds a droplet of water. The bilayer is composed of lipid molecules arranged side by side with nonpolar ends facing internally toward one another away from water and the polar heads facing outwardly toward the aqueous medium. Such artificial liposomes possess the dimensions and staining qualities of natural membranes.

Phospholipids such as lecithins, cephalins, and sphingomyelins are basic constituents of plasma membranes as well as the intracellular membranes of the endoplasmic reticulum, Golgi vesicles, mitochondria, and chloroplasts. The polar head of these lipids is built around a three-carbon glycerol framework to which a phosphate group is attached. In most phospholipids two fatty acid chains of varying length and composition (stearic acid, oleic, palmitic) are combined with two carbons on the glycerol molecule to make up the nonpolar tail. Sometimes additional polar groups are present; for example, choline is attached to the phosphate group in lecithin, or the amino acids serine and threonine may also be present as phosphatidylserine or phosphatidylthreonine.

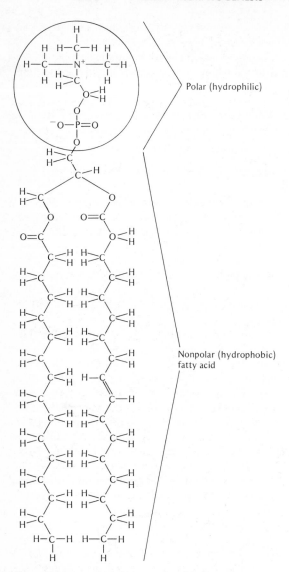

**Figure 7-11** *Amphipathic structure of lipid molecule, phosphatidylcholine, with hydrophilic "head" attached to a twin hydrophobic fatty acid chain.*

Membranes contain a variety of these lipids in varying proportions, with some lipids usually predominating. Mitochondrial membrane lipids consist almost entirely of lecithin and phosphatidylethanolamine but contain no cholesterol, while erythrocyte membranes contain sphingomyelin, cholesterol, and small quantities of lecithin, among others.

All membranes in the electron microscope appear to have the same structure and dimensions: two dense strata, each 20 Å thick, separated by a lightly staining interzone about 35 Å thick. The total "unit membrane" is about 75 Å thick, a structure that varies only slightly from membrane to membrane. These dimensions correspond to that of a lipid bilayer which has led to a unit membrane concept; membranes are lipid bilayers with the dark strata in electron micrographs representing the stained polar heads and the lighter interzone occupied by the hydrophobic tails.

The classic model of membrane structure was based on permeability studies of Davson and Danielli. They proposed a membrane consisting of a lipid bilayer core with proteins aligned on inner and outer surfaces (Figure 7-12A). This resembles the unit membrane model except it does not define the number of lipid monolayers in the core. This model and that of the unit membrane have been modified in the light of recent studies of membrane structure and composition.

Relatively pure membrane preparations from different cells have been analyzed. Only a small proportion of membrane proteins are readily extracted by the usual solvents; these are the "peripheral" proteins seen in the classic membrane model. Most proteins are intimately bound to the lipids, so much so that strong detergents must be used to dissolve the lipids and release the proteins. These are the integral proteins of the membrane. Such tight associations between proteins and lipids are inconsistent with the classic membrane models. It appears that most integral proteins are also amphipathic molecules with polar and nonpolar regions. They are not spread out as monolayers on lipid surfaces but are in a globular configuration, embedded in the lipid bilayer, with hydrophobic groups immersed in the lipids while hydrophilic groups are exposed to the aqueous medium in or outside the cell. These proteins are not covalently linked to the lipids, but appear to be largely independent. Presumably, the proteins or glycoproteins are distributed at random through the membrane at varying depths and in different patterns. Any protein with the correct three-dimensional configuration and distribution of charges (it must be an asymmetric molecule) can occupy a site in the membrane, which probably explains the functional heterogeneity of membrane proteins.

These studies have led to a new model of membrane structure—the fluid mosaic model (Figure

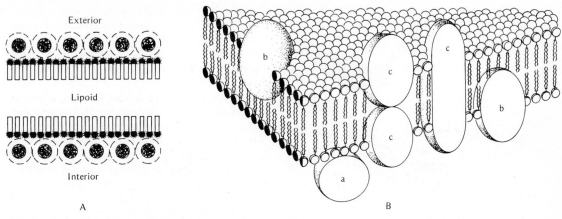

**Figure 7-12** *Models of membrane based on amphipathic structure of lipids. (A) Classic model of membrane as proposed by Davson and Danielli. The outer and inner surfaces of the lipid bilayer are shown coated with protein molecules. (B) Fluid-mosaic model made up of a lipid bilayer with hydrophilic heads facing toward outer and inner surfaces. The bilayer is 45 Å thick and proteins (a, b, c) are embedded in the membrane, in part or all the way through, singly and in pairs. Some proteins (a) are at or near a membrane surface as in the Davson-Danielli model. Bonds between proteins and lipids are probably hydrophobic. (Reprinted with permission from S.J. Singer, Architecture and Topography of Biologic Membranes, Hospital Practice, vol. 8, no. 5, and Cell Membranes: Biochemistry, Cell Biology and Pathology, ed. G. Weissmann and R. Claiborne. New York: HP Publishing Co., Inc., 1975.)*

7-12B). The lipid bilayer is a matrix organized as a uniform two-dimensional layer in which proteins "float" from inner to outer surfaces, or occupy only part of the membrane. Proteins may be single or combined into clusters; only their polar groups are exposed. Such globular particles appear in electron micrographs of membranes prepared by a freeze-etching technique which splits the membrane down the middle of the bilayer and exposes the inner surfaces (Figure 7-13). Randomly arranged 50- to 85-Å particles are attached to inner surfaces, sometimes organized into clusters. These particles disappear after membranes are treated with proteolytic enzymes, which suggests that they are proteins, or contain proteins as an integral part of their structure. Most eukaryote membranes are of this type but differ with respect to number, size, and distribution of these particles. Some membranes such as the myelin membranes surrounding nerve fibers are entirely free of these structures. These membranes are almost pure lipid; they contain little protein and seem to act as barriers to ion flow, a kind of insulation in nerve conduction.

Thompson, working with artificial membranes, proposed that all membranes are formed by self-as-

sembly. He assumed that different lipids act as subunits just as amino acids assemble into proteins. Accordingly, membrane structure is determined by the space-filling properties of lipid molecules and their binding sites. Contrary opinion questions whether self-assembly accounts for membrane formation since they seem to grow from preexisting membranes acting as templates for the addition of new precursors.

Since membranes differ in lipid and protein content, they may also exhibit different mechanisms of assembly. Mitochondrial membranes are probably synthesized within mitochondria; membranes of the endoplasmic reticulum seem to arise as vesicles detached from nuclear membranes while plasma membranes develop from coalescence of Golgi vesicles.

Both self-assembly and template generation are consistent with most observations of membrane formation. For example, myelin sheath membranes that surround vertebrate peripheral nerve fibers grow by a template-generated mechanism. In the electron microscope a myelin sheath consists of a regular series of concentric light-dark bilayers surrounding the central axon (Figure 7-14). These simple membranes are generated from Schwann cells which surround the

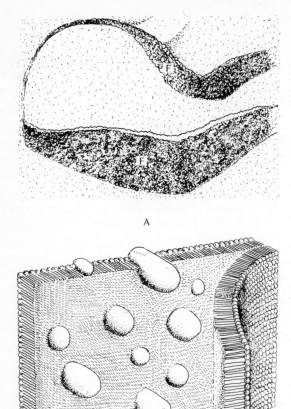

A

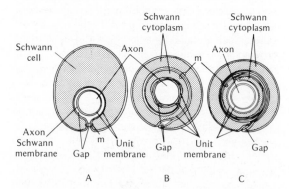

cursors synthesized within Schwann cell cytoplasm migrate to the membrane surface and simply add on to a preexisting membrane. The alignment and insertion of new lipid molecules within the membrane are fixed by the preexisting membrane acting as a template.

On the other hand, a lipid bilayer structure organizes spontaneously in vitro from solutions of purified lipids. When brain phospholipids are allowed to swell on water, they form typical myelin figures, resembling the concentric ring patterns of the myelin sheath. Here bilayer membranes assemble in aqueous media from lipid subunits without any preexisting membrane. These artificial bilayers are approximately 50 to 60 Å thick, display electrical properties of natural membranes, are equally permeable to water, and will exhibit selective permeability to ions. The addition of various substances to these bilayers, such as proteins and antibiotics, produces artificial membranes which resemble natural membranes. For example, an antibiotic such as valinomycin will increase the rate of passage of $K^+$ ions by hundreds of times the normal rate without affecting the passage of $Na^+$ ions.

Artificially prepared liposomes are also of value in studying membrane properties. These minute vesicles consist of a series of concentric, closed

B

**Figure 7-13** *Freeze-etched preparation of a cell membrane. (A) Diagram of a freeze-etched membrane in the electron microscope showing the fracture planes (F) with numerous spherical particles presumed to be protein molecules or subunit aggregates. (B) A schematic diagram showing how the membrane has been fractured to expose the rough inner faces seen above. (Reprinted with permission from S.J. Singer, Architecture and Topography of Biologic Membranes, Hospital Practice, vol. 8, no. 5, and Cell Membranes: Biochemistry, Cell Biology and Pathology, ed. G. Weissmann and R. Claiborne. New York: HP Publishing Co., Inc., 1975.)*

**Figure 7-14** *Mechanism of formation of myelin by Schwann cell. (A) An early stage with Schwann cell associated with an axon and a short mesaxon (m). (B) As Schwann cell rotates about the axon it spins off additional membrane in concentric rings. The mesaxon is elongated into a spiral. (C) A more advanced stage as unit membranes are compacted into a sheath. (Based on J.D. Robertson, Cellular Membranes in Development, ed. M. Locke. New York: Academic Press, 1964, pp. 1–82.)*

axon during early stages of axon outgrowth in the embryo. As the axon grows out in a spiral pattern, the adherent Schwann cell spins out its own membrane and slowly winds it about the axon in a series of tightly applied layers. Presumably, membrane pre-

membranes, each separated from the next by a layer of water and surrounded by water outside. In this way all hydrophobic tails of lipid molecules avoid water. Liposomes have been used to study membrane interactions with anesthetics, which are presumed to act in vivo by dissolving readily in lipid membranes. We see that many properties of natural membranes can be reproduced from lipids that self-assemble into artificial bilayers in aqueous environments.

But living membranes are dynamic structures as we have seen, in constant motion and turnover. New membranes accumulate as cells grow; biological properties of other membranes change with time as membranes differentiate or respond to physiological needs. In addition, most membranes turn over; proteins and lipids are removed and degraded while new precursors are inserted in their place. Most evidence suggests that new membranes are not assembled de novo from protein and lipid subunits but are added to preexisting membranes which expand as precursors are inserted. New proteins synthesized from radioactive amino acids are found interspersed within the old membranes along with unlabeled proteins (see Chapter 3). As cells divide, daughter cells receive a full complement of membrane systems which undergo growth until the next division, to be passed on to subsequent progeny. Meanwhile the molecules within the membrane undergo continuous replacement, which raises some interesting questions of molecular traffic and its control.

How do specific enzyme molecules "know" the membrane with which to associate, and how do they enter the membrane? Since most enzymes are manufactured in the endoplasmic reticulum, they must somehow get out and migrate to other sites in the cell. These molecules must possess some structural affinity for specific binding sites in different membranes. Perhaps by a sequential process the addition of certain proteins insures, because of complementary configurations, that other specific enzymes will become inserted in their turn. In this way functional enzyme arrays may be built up within membranes, in part by self-assembly rules of matching binding sites.

The membrane systems within eukaryote cells seem to be continuous with one another, at least at some stage in the life cycle. Membranes flow from one compartment to another, usually as small vesicles that detach from larger membranes and coalesce. This process of vesicle formation is common at the cell surface and plays an important role in the turnover of the plasma membrane (Chapter 10). But internal membranes also participate in a directed migration of membranes from organelle to organelle as part of cellular transport and secretory processes (Figure 7-15). Secretory proteins are synthesized within the membrane cisternae of the endoplasmic reticulum. Small vesicles and vacuoles bud off cisternae at one pole and migrate to the Golgi region, where they coalesce into a system of flattened cisternae stacked in parallel. Within the Golgi, proteins are processed and packaged for secretion. Small secretory vesicles detach from Golgi lamellae and migrate to the periphery where they fuse with the plasma membrane and discharge their contents to the outside. Thus vesicle membranes are inserted into the plasma membrane.

Membranes differentiate during this migration process. Membranes of the endoplasmic reticulum contain much more protein than lipid (4:1) while most plasma membranes have about equal amounts of each. Proteins, being the major structural component in endoplasmic reticulum membrane, would

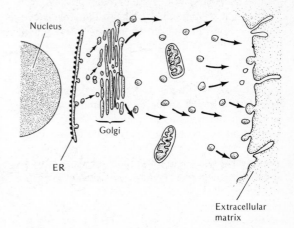

**Figure 7-15**  *Membrane flow pattern as part of Golgi secretory mechanism. Direction of vesicle flow is shown from endoplasmic reticulum (ER) where small primary vesicles are produced. These fuse into the complex lamella system of the Golgi, presumably where secretory products (glycoproteins of the cell surface and extracellular matrix?) are accumulated and processed. Small secretory vesicles bud off from the Golgi and migrate to the cell surface where their membranes fuse with the plasma membranes to discharge their contents. In this manner, new membrane is added to the plasma membrane.*

probably make them more complex and rigid. The same is true for mitochondrial and chloroplast membranes, which means that the model of the fluid membrane as "protein icebergs in a sea of lipid" probably does not apply to these. Outer membranes are thicker than those of the endoplasmic reticulum while Golgi membranes seem to be intermediate, not only in morphology but in chemical constitution, presumably as new constituents are added during processing.

The dynamic flow of these membrane systems and their patterns of differentiation and turnover illustrate that self-assembly is only a first stage in the morphogenesis of organelles. More information is required than that encoded in the structural genes.

## ONTOGENY OF MITOCHONDRIA AND CHLOROPLASTS

The membrane systems of mitochondria and chloroplasts seem to be more stable; they are not part of the membrane flow system described above. It is true that mitochondria undergo dynamic movements in the cell and are often seen (in time-lapse cinemicrography) constricting, extending, and fragmenting, sometimes fusing with other mitochondria.

The mode of membrane formation in these organelles is more highly ordered since membranes are packed with various multienzyme systems involved in energy transformation. Besides, the organization of these organelles is based on information derived from both nuclear and organelle genomes, and we would expect the ontogeny of these organelles to follow a more complicated program than is possible through self-assembly rules. Synthesis and assembly of the products of two genomes have to be coordinated, which, as we have seen, demands efficient communication and precise temporal and spatial regulation. Moreover, we have seen that mitochondrial development is a template-generated mechanism; new mitochondrial components are simply added to old mitochondria; they do not arise de novo (Chapter 3). Nevertheless, we cannot exclude the possibility that components of these organelles such as some multienzyme systems are indeed assembled according to self-assembly rules.

Chloroplasts consist of membrane stacks embedded in a matrix or stroma, surrounded by a typical unit membrane. The intrachloroplast membranes are arranged in parallel lamellae or *thylakoids* which are differentiated in certain regions into thickened membrane discs or grana separated by stroma lamellae (Figure 7-16). Freeze-fracture electron micrographs of the grana and stroma lamellae reveal different-sized particles arranged at random on fracture faces. These may be multienzyme subunits involved in photosynthesis. How does this complicated membranous organelle develop?

If a young seedling is placed in the dark for several days, its leaves turn white; they become etiolated as chloroplast structure changes. Chloroplast stability depends on light, and in continuous darkness

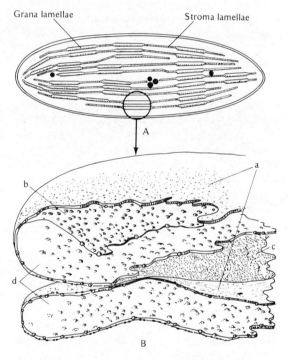

**Figure 7-16** *Membrane structure in chloroplasts of higher plants. (A) Chloroplast showing lamella structure. Lemella discs arranged in stacks are known as grana. Membranes within grana are grana lamellae in contrast with the single or stroma lamellae. (B) An enlarged freeze-fracture view of the grana lamellae shows different sized particles in the fracture faces of the discs, a, b, c, and d. Chlorophyll and the many photosynthetic enzymes are embedded in these membranes.*

chlorophyll, photosynthetic enzymes, and lamellae disappear, resulting in empty saclike vesicles or *etioplasts*. Though these are abnormal structures produced by unbalanced growth, they resemble the immature plastid and have been useful in studying the ontogeny of chloroplast lamellae because development can be controlled easily with light. The predominant structure in an etioplast is a paracrystalline array of tubules, the *prolamellar body* within the matrix. This develops from a rearrangement of vesicles that pinch off lamellae as they disperse in darkness.

Chloroplast development from an etioplast (greening) is a stepwise process that requires light (Figure 7-17). A few seconds of illumination will initiate lamella formation, coupling lipid membrane assembly to synthesis of photosynthetic pigments. Almost instantaneously upon exposure protochlorophyll is converted to chlorophyll, a process that may take place within the prolamellar body. This is followed by a lag in chlorophyll synthesis (either in the dark or light) during which the prolamellar body disperses and its components rearrange into primary layers, possibly as small vesicles fuse into larger sacs. During this phase chlorophyll undergoes a spectral shift in absorption maximum which may reflect its new orientation in the lamellae. With continuous illumination dispersed primary sheets line up into lamellae stacks and form grana. At this time a rapid synthesis of chlorophyll and photosynthetic enzymes begins.

Here too we see that membrane systems are generated from preexisting membranes in a template-generating mechanism. New membranes are added to old ones; old membranes differentiate as new protein or enzyme systems are inserted into the bilayer. Self-directed assembly does not seem to play a major part.

## A PROGRAM OF VIRAL MORPHOGENESIS

Since membrane templates are required to build complex organelles such as chloroplasts, then self-assembly principles must be supplemented with additional genetic information. What kind of recognition and binding rules govern the organization of these organelles? The best system is again provided by virus-infected prokaryote cells. Here complex viruses containing up to 100 different macromolecular

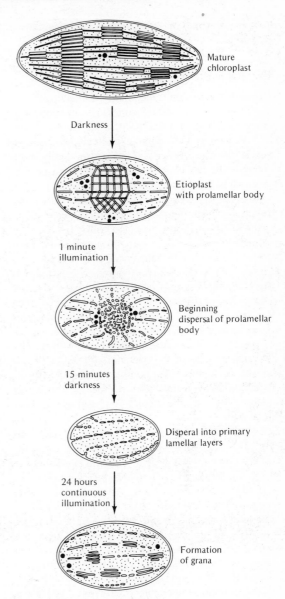

**Figure 7-17**  *Ontogeny of chloroplasts from etioplasts in higher plants. See text for details.*

constituents are constructed within minutes. These viruses are almost as complex as plastids and may serve as model systems for organelle morphogenesis. It turns out that the most incisive tool in analyzing viral morphogenesis is a genetic one; mutants which interfere with specific steps in viral assembly can be studied.

## The viral life cycle

The virus in question is the phage T4 infecting the familiar bacterium *E. coli* (Figure 7-18). In contrast to tobacco mosaic virus with only a few genes (where self-assembly is sufficient to explain viral design), T4 phage has over 100 genes and is structurally more complex. Its DNA is coiled tightly within a protein polyhedral shell making up the phage head. The head is similar to those in many spherical viruses with the characteristic icosahedral symmetry. Connected to the head by a short neck (and collar) is a tail consisting of a contractile sheath surrounding a central core. At one end of the tail, a plate or base is attached from which extend six short spikes and six tail filaments arranged symmetrically around the end plate. These viral organelles are made up of different protein subunits, each playing a specific role in the infective process. The head surrounds and packages the coiled viral genome; the spikes and fibers affix the viral particle to the bacterial cell wall; the tail sheath, neck, and collar complex acts as a syringe, injecting the DNA into host cytoplasm as the sheath contracts, driving the tubular core through the bacterial wall.

The life cycle begins when the virus attaches to the host cell wall and injects its DNA into the cytoplasm. Once inside, the viral genome takes over the host's metabolic machinery and directs it to synthesize viral DNA and proteins. Within a minute, viral proteins are being synthesized by host ribosomes on messages transcribed on viral genes. The viral developmental program begins as these first genes are turned on to synthesize enzymes required for viral DNA synthesis: viral-specific polymerases, nucleotide synthetases, and so on. DNA replication begins within 5 minutes after infection; 3 minutes later, a second set of genes begins to transcribe messages directing the synthesis of structural proteins of head and tail components. Morphogenesis involves the assembly of these subunits into heads and tails, which are later joined together as mature virus. At 13 minutes after infection, the first mature viruses appear, but DNA and protein synthesis continue for an additional 12 minutes until approximately 200 virus particles have matured within the host cytoplasm. Finally, a viral coded enzyme, lysozyme, is synthesized which

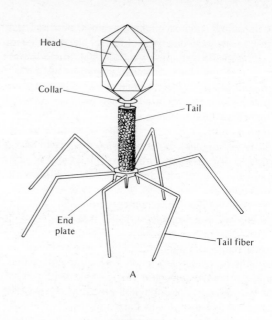

A

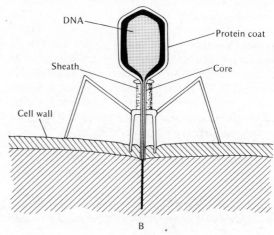

B

**Figure 7-18**   *Structure of T4 bacteriophage. (A) A view of the complex phage with its head, collar, tail, and tail fibers. (B) Cross section of phage attached to a bacterial cell wall at the moment of infection. The phage head consists of a protein coat (icosahedral lattice of globular units) surrounding a central core of DNA. During infection, the DNA is injected through the contracted tail sheath into the bacterial cytoplasm. Some genes are concerned with DNA synthesis, but many other genes code for head, tail, and tail fiber proteins respectively. The boxes indicate the lesions noted in synthesis of phage particles as a result of gene mutations. The lysates of such mutant-infected cells contain the unassembled phage components shown.*

destroys the cell wall and releases infectious particles able to resume the life cycle in other host cells.

At any time during the life cycle, infected cells may be killed, broken open, and examined by electron microscopy. Different states of viral morphogenesis are seen, but because of heterogeneity of the host cell population and the absence of synchrony, the temporal sequence of events is difficult to reconstruct. This problem has been overcome, however, by use of viral mutations which interfere with morphogenesis at different steps in the assembly process. This genetic tool can resolve a rapid series of events into a clearly defined sequence. Moreover, the analysis provides the best indication of the operation and control of genetic instructions in the generation of form at the organelle level.

From the amount of DNA in the virus, it was calculated that over 100 genes control viral morphogenesis. But how can mutations be obtained in a viral genome that is essentially haploid with only one copy of each? A mutation is usually lethal; it modifies viral proteins, making them nonfunctional so that viral development stops at the point where the particular protein is required. The mutant virus cannot be recovered for use under these conditions. It was only when "conditionally lethal" mutations were found in which a mutant protein is functional under some conditions but not others that viral mutations could be studied. For example, many viral mutants are temperature-sensitive and cannot grow if host cells are maintained at high temperatures (restrictive conditions) but produce normal virus at lower temperatures (permissive conditions). Permissive conditions favor growth and permit study of viral genetics; viral development under restrictive conditions is abnormal and stops at various points in the cycle, depending on the mutation. Since all infected cells would be blocked at the same point, electron microscopic examination is made easier, and we can infer what step is controlled by specific genes and even get some clues as to how genes function.

Large numbers of conditional mutants were studied in this fashion and mapped as a circular map of more than 75 genes for viral development (Figure 7-19). Two groups of genes can be distinguished, "early acting" genes controlling enzymes for DNA replication and "late acting" genes governing synthesis of viral structural proteins and their assembly into heads, tails, sheaths, and so on. Mutations in the early acting genes prevent normal DNA replication, because some key enzyme is not formed or the temporal events are disturbed. No viral components are made by these mutants.

More than 40 late genes are characterized by observing which products accumulate within infected cells. Only bits and pieces of virus are formed in cells infected by mutant viruses of the late-acting variety. The absence of certain viral components indicates that the mutated gene codes for the subunit proteins that make up the specific structure. For example, in lysates of cells infected with mutants of genes 3, 5, 6, 7, and 8, only heads and tail filaments are found; tail sheaths and cores are absent, which means these genes control tail protein synthesis. Mutants in the cluster 34–38 produce normal tails and heads but no tail filaments, which means that genes 34–38 direct the synthesis of tail fiber protein subunits. Evidently failure to make one building block does not affect the synthesis of others. Morphogenesis is not a continuous sequential process beginning at one end and gradually adding on additional material. It is more like an assembly-line process with three independent production lines making tails, heads, and filaments. Ultimately these completed units are combined to make a mature viral particle.

One feature of the map is gene clustering; many genes controlling one component are grouped together as if they are functioning as an operon under coordinate control. For instance, the cluster of genes 20–24 controls head formation; mutation in any one of these genes prevents the appearance of normal heads. This, however, does not mean that every gene in the cluster codes for head structural proteins. Gene 23 is primarily responsible for the major subunit of the head, which, incidentally, accounts for 80 percent of the total protein of the virus. Cells infected with mutants of the other genes still contain the subunit protein. If these genes contribute no protein to the head, what is their role in head formation? We learn that these genes are regulatory rather than structural, controlling head shape.

The normal head is a kind of prolate icosahedron and achieves this shape through the interaction of these regulatory or *morphopoietic genes*, as Kellenberger calls them. Gene 66, for example, though not part of the cluster, directs the synthesis of an "elongation factor" without which normal viral heads are not produced. Instead, short heads with 20 percent less DNA appear. Gene 20 produces a "rounding-up factor" that closes off the icosahedral shape. A mutation

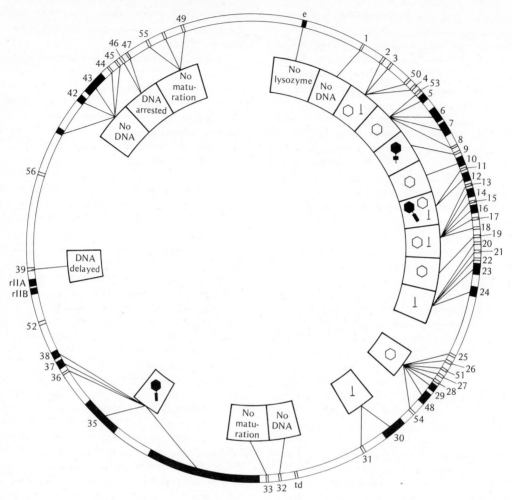

**Figure 7-19**  *Circular map of the viral chromosome showing the position of various head and tail genes. Some genes related to the same function are clustered, though this is not true generally. Early genes code for enzymes involved in DNA synthesis. The boxes indicate the content of cell lysates of mutants at specific loci showing the unassembled phage components. (Based on Building a Bacterial Virus by W.B. Wood and R.S. Edgar. Copyright © 1967 by Scientific American, Inc. All rights reserved.)*

in this gene results in long tubular heads, or "poly-heads," which accumulate within the host cell. This mutation behaves as if the elongation factor is completely uncontrolled. Finally, a mutation in gene 21 induces the formation of abnormal heads that contain little or no DNA. Since gene-23 product can be organized into different structures by products of neighboring genes, the structural information within subunits controlled by gene 23 is inadequate to self-assemble a fully formed head. Additional topological

information is supplied by the morphopoietic genes, 20, 21, 66, and perhaps others.

The same principle seems to be true for genes concerned with tail and tail fiber formation. Though contributing no structural proteins to the mature virus, certain genes are essential to their normal assembly. If all genes code for a specific protein product, what role is played by the products of these assembly genes in the viral production line? Since these products are not detected in mature virus, the genes involved may be regulatory (repressors?), controlling on-off behavior of structural genes, or else they may serve as a kind of scaffolding for viral assembly that is removed after mature virus is formed. They would be analogous perhaps to a conveyor belt in an industrial assembly-line system. They could also act as the "nuts and bolts" or "glue" holding components like heads and tails together, but in such small amounts as to be undetectable in the mature virus.

An important breakthrough in the problem of the action of morphopoietic genes has come from studies in which intact viruses are assembled in vitro from extracts of host cells infected with different viral mutants. The first experiment was an attempt to attach tail fibers to an otherwise completed virus to see whether a mature infectious virus could be produced. It is, in a sense, a complementation experiment done in the test tube rather than in the host cell. The products of two mutant genomes are combined in vitro to see whether they generate intact virus since neither one alone can do this. In the experiment, cells infected with a virus carrying mutations in tail fiber genes 34, 35, 37, and 38 were lysed and the almost complete virus (except for fibers) was released and isolated. Other cells, infected with a mutant of gene 23 defective in head formation (but able to make intact tails and fibers), were also broken open and their contents purified to yield an extract of tails and fibers. When the incomplete viral particle preparation was mixed with the tail fiber extract and incubated in vitro, the amount of infectious virus increased a thousandfold within a short time. Examination of the mixture showed that fully mature virus had been assembled.

Similar incubation mixtures involving combinations of more exacting mutants were studied. For example, mutant extracts were prepared containing tails and free tail fibers but without heads (gene 23), and another, blocked in tail formation, containing heads and free tail fibers but no tails (gene 27). The incubation mixture gave rise to intact virus, which meant that a two-step self-assembly process was taking place. First, the heads attached to tails followed by the attachment of tail fibers to the resulting particle. By carrying out such a complementation test in vitro, Wood and Edgar determined what gene controlled which step in the morphogenetic sequence; they could identify the specific sites of action of "nuts and bolts" in the assembly line.

This method of in vitro complementation also furnished the assembly-line sequence. For instance,

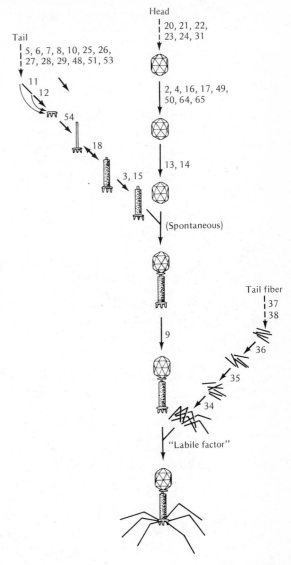

**Figure 7-20** *Assembly-line pathways for viral morphogenesis. Three separate pathways are involved, one each for heads, tails, and tail fibers. Each component is synthesized separately while "assembly" genes put the parts together to form a complete phage. The numbers refer to the specific gene products involved at each step, many of which do not act as structural genes but seem to be required for assembly. The solid arrows show which steps in the pathway can be duplicated in vitro. (From Building a Bacterial Virus by W. B. Wood and R. S. Edgar. Copyright © 1967 by Scientific American, Inc. All rights reserved.)*

genes 15 and 18 both direct tail completion. The analysis showed that in the absence of the product of gene 18 no contractile sheaths were made, but if gene 18 were functional and gene 15 defective, sheath units were unstable. Though they could assemble on the tail core, they would fall away. It was found that the product of gene 15 acts like a "button" at the upper end of the core to stabilize the sheath.

The morphogenetic pathway for virus assembly includes three independent conveyor-belt systems leading respectively to head, tail, and tail fiber formation. These systems eventually meet to assemble mature virus (Figure 7-20). Only after each component is fully mature can a virus assemble. The most important generalization is that morphogenesis of a complex virus calls for more information than is provided by self-assembly. Morphogenetic information from regulatory and morphopoietic genes is needed to complete maturation of major components before other genes direct final viral assembly.

Viral morphogenesis, an ordered sequence of events with each step controlled by coordinate action of structural and assembly genes, serves as a model for the ontogeny of complex eukaryotic organelles such as membranes, mitochondria, cilia, and so on, where many more genes and protein products are involved. Many programs of eukaryote organellogenesis indicate that after the initial phases of self-assembly, the first ordering step of morphogenesis at the macromolecular level, second and third ordering systems are necessary. These have additional information for construction—in the form of scaffolding, glue, nuts, and bolts—to complete a functional structure. It is here in such assembly-line construction processes that the many regulatory genes in eukaryote cells come to play an important role.

## SUMMARY

We see that a principle of biological organization is the utilization of identical subunits to assemble more complex structures at a next higher level. The shape and complexity of a structure depend on the number of interacting subunits in addition to the distribution of reactive sites. Subunits spontaneously interact to form large aggregates, generating symmetries and patterns. The information for these structures is coded in linear sequences of subunit molecules. Self-assembly, operating at the level of molecules and macromolecules, is a primary morphogenetic principle responsible for the generation of form. It automatically creates more complex structures with properties exceeding those of individual subunits. This principle of the "whole being greater than the sum of its parts" is applicable at all levels of biological organization.

Spontaneous assembly into organized structures may explain the origin of mixed macromolecular aggregates such as viruses, bacterial flagella, and even eukaryote organelles. Structure of subunit molecules and the physiological conditions under which they assemble provide all the information for creating ordered systems.

Membrane systems may form spontaneously, but more usually they arise from preexisting membranes, the coalescence of small membrane-bound vesicles. Membrane formation depends on topological information provided by a preexisting membrane, but new membrane properties may be acquired as different proteins are inserted into a growing membrane. As organelles become more complex, additional information is required to generate form. This is particularly true of membrane-rich mitochondria and chloroplasts.

A model for organelle assembly is provided by our new knowledge of viral morphogenesis. Morphogenesis of complex viruses like T4 bacteriophage calls for more genetic information than is found in subunit molecules. Specific genetic information is required to prepare building blocks for assembly into organized structures, but the products of morphopoietic genes, acting as the scaffolding on which construction takes place, are also required for normal morphogenesis. In the following chapter, we will see how preexisting blueprints for pattern operate in the morphogenesis of single cells.

**REFERENCES**    Anfinsen, C. B. 1968. Self-assembly of macromolecular structures. Spontaneous formation of the three-dimensional structure of proteins. *The emergence of order in developing systems*, ed. M. Locke. New York: Academic Press, pp. 1–20.

_____. 1973. Principles that govern the folding of protein chains. *Science* 181:223–230.

Blout, E. R. 1967. Conformation of proteins. *Neurosciences,* ed. C. C. Quarton, T. Melnechick, and F. O. Schmitt. New York: Rockefeller University Press, pp. 57–66.

Boardman, N. K., Linnane, A. W., and Smillie, R. M., eds. 1971. Autonomy and biogenesis of mitochondria and chloroplasts. Amsterdam: North-Holland.

Bretscher, M. S., and Raff, M. C. 1975. Mammalian plasma membranes. *Nature* 258:43–49.

Casjens, S., and King, J. 1975. Virus assembly. *Ann. Rev. Biochem.* 44:555–611.

Caspar, D. L. D. 1966. Design and assembly of organized biological structures. *Molecular architecture in cell physiology,* ed. T. Hayashi and A. G. Szent-Gyorgi. Englewood Cliffs, N.J.: Prentice-Hall, pp. 191–207.

Creighton, T. E., and Tanofsky, C. 1966. Association of the $\alpha$ and $\beta_2$ subunits of the tryptophane synthetase of *E. coli. J. Biol. Chem.* 241:980–990.

Gross, J., Lapiere, C. M., and Tanzer, M. L. 1963. Organization and disorganization of extracellular substance: the collagen system. *Cytodifferentiation and macromolecular synthesis,* ed. M. Locke. New York: Academic Press, pp. 175–202.

Kushner, D. J. 1969. Self-assembly of biological structures. *Bacteriol. Rev.* 33:302–345.

Lebeuvier, G., Nicolaieff, A., and Richards, K. E. 1977. Inside-outside model for self-assembly of tobacco mosaic virus. *Proc. Nat. Acad. Sci. U.S.* 74:149–153.

Nomura, M. 1973. Assembly of bacterial ribosomes. *Science* 179:864–873.

Oosawa, F., Kasai, M., Hatano, S., and Asakura, S. 1966. Polymerization of actin and flagellin. *Principles of biomolecular organization,* ed. G. E. W. Wolstenholme and M. O'Connor. London: Churchill, pp. 273–303.

Platt, J. R. 1961. Properties of large molecules that go beyond the properties of their chemical subgroups. *J. Theoret. Biol.* 1:342–358.

Reinert, J., and Ursprung, H., eds. 1971. *Origin and continuity of cell organelles.* New York: Springer-Verlag.

Sager, R. 1972. *Cytoplasmic genes and organelles.* New York: Academic Press, Chapters 7 and 8.

Singer, S. J., and Nicolson, G. L. 1972. The fluid mosaic model of the structure of cell membranes. *Science* 175:720.

Thompson, T. E. 1964. The properties of bimolecular phospholipid membranes. *Cellular membranes in development,* ed. M. Locke. New York: Academic Press, pp. 83–96.

Von Wettstein, D., Henningsen, K. W., Boynton, J. E., Kannangava, G. C., and Nielsen, O. F. 1971. The genetic control of chloroplast development in barley. Autonomy and biogenesis of mitochondria and chloroplasts, ed. N. K. Boardman, A. W. Linnane, and R. M. Smillie. Amsterdam: North-Holland, pp. 205–223.

Weinstock, M., and LeBlond, C. P. 1974. Formation of collagen. *Fed. Proc.* 33:1205.

Wood, W. B. 1973. Genetic control of bacteriophage T4 morphogenesis. *Genetic mechanisms of development,* ed. F. H. Ruddle. New York: Academic Press, pp. 29–46.

# 8

# Morphogenesis in Single Cells

Cells are distinguished from one another by their shape and organelle arrangement; both these are relatively stable properties and are transmitted faithfully from generation to generation. Regardless of the ebb and flow of cytoplasmic activities, most organelles usually remain fixed in place. But how is the position of an organelle fixed? For example, what factors determine where a flagellum emerges from the cell surface? What distinguishes anterior from posterior in polarized cells? What structures support and what forces change cell shape?

Information flow from gene to protein is not complete until the gene products are assigned to specific organelles in the cell; as we have seen, this process is assisted by morphopoietic genes. Similar regulatory genes direct molecular traffic, the flow of building blocks to sites of assembly.

Organelle assembly and assignment, cell polarity and shape are, in essence, problems of protoplasmic motility. Some cell shapes are unchangeable, defined by rigid walls in plants or exoskeletons in free-living protists. Other cells, such as the amoeba,

change shape continuously. Somatic cells in suspension are generally spherical, but when surrounded by cells in compact tissues they assume hexagonal or pentagonal shapes. Some tissue cells such as the neuron are asymmetrical and maintain their intricate shapes throughout the life of the organism, as if supported by an internal cytoskeleton. For many years such cytoskeletons were postulated to explain form changes in cells. Now we know that microtubules and microfilaments control shape as part of a more general role of controlling all protoplasmic movement.

A cell is often polarized by localization of organelles or metabolic systems within the cytoplasm. The direction of ion flow between cell compartments and between cell and environment is determined by membrane transport systems that pump molecules against concentration gradients. The movement of a single cell is oriented by the position of its flagellum or the direction of ciliary beat. Cells are also polarized by their position within multicellular aggregates which may favor organelle accumulation on "outside" or apical surfaces, as seen in the apical cilia of respiratory epithelium. These aspects of cell architecture, shape, polarity, and organelle location seem to be mutually dependent features of morphogenesis in single cells. Although single-cell morphogenesis appears simple, limited by what cells secrete on their surfaces or assemble within their cytoplasm, we will see that the same ordering forces are involved in eggs and embryos.

## PROKARYOTE MORPHOGENESIS: SPORULATION

Prokaryote cells have a limited morphogenetic repertoire; they assemble flagella or secrete cell walls. Some species sporulate when conditions are unfavorable and secrete a specialized, protective spore wall. Sporulation is a model system for studying gene regulation in development since easily controlled envi-

ronmental signals switch a vegetative cell from a growth-duplication cycle to a program of spore wall synthesis. The trigger for this new morphogenetic program is starvation, a depletion of exogenous nutrient molecules that stops growth. As division genes are switched off, new genes are expressed, and wall precursors are synthesized.

For example, *Bacillus subtilis* is triggered to sporulate when nutrients are exhausted. Cells stop growing and begin to make proteins and enzymes that are absent in the vegetative cell. Sets of sporulating genes (about 100) are activated and a sequence of metabolic adjustments initiate a program of morphological change (Figure 8-1).

Within 1½ hours after sporulation begins, the two chromosomes within the vegetative cell condense into a continuous thread. Cell proteins turn over and enzymes, such as ribonuclease and exoprotease, spill into the medium. These enzymes are es-

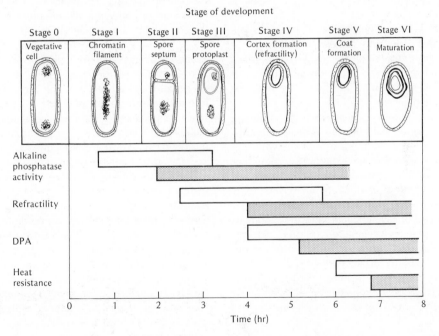

Stage of development

| Stage 0 | Stage I | Stage II | Stage III | Stage IV | Stage V | Stage VI |
|---|---|---|---|---|---|---|
| Vegetative cell | Chromatin filament | Spore septum | Spore protoplast | Cortex formation (refractility) | Coat formation | Maturation |

Alkaline phosphatase activity

Refractility

DPA

Heat resistance

Time (hr)

**Figure 8-1**   *Program of spore morphogenesis in* Bacillus subtilis. *Morphological stages are above and some key biochemical events are illustrated in the diagram below. The white rectangles represent those periods when cells acquire the ability to express the property; the relevant genes are transcribed at these times. Actinomycin D treatment of cells prior to or during this period prevents expression of the biochemical or morphological events to follow. Treatment afterwards has no effect. The tinted rectangles indicate when the event is expressed; for example, an enzyme activity is measurable. DPA refers to dipicolinic acid, a compound essential in spore wall formation and heat resistance. (From J. Mandelstam,* Symp. Soc. Exp. Biol. 25:1–26, 1971.)

sential to sporulation inasmuch as mutants, unable to synthesize any one of them, do not sporulate and are blocked in early stages. Outside the cell, proteases may convert exogenous proteins into amino acids that may be utilized for synthesis of new sporulation-specific proteins.

In the next hour a spore septum forms at one end of the cell with one of the chromosomes drawn into it. The septum enlarges and separates from the mother cell membrane as an independent spore protoplast inside the mother cell wall. Meanwhile, wall materials are deposited between the double membranes of the spore protoplast, the spores becoming visible as refractile areas in the cell. A principal spore wall constituent, dipicolinic acid, which is never seen in the vegetative cell, is synthesized at this stage. Finally, new materials are deposited in the spore coat, and spores mature, losing water. They become resistant to many organic solvents and heat and survive in the complete absence of water. Wall synthesis is the principal phenotypic expression of this developmental program.

Biochemical and morphological programs are interdependent; blocking an early morphological step stops the entire sequence while biochemical mutations prevent most morphological changes. It is only during sporulation that long-lived message molecules are detected.

The transition from a vegetative to a sporulating cell involves sequential changes in gene transcription. Cells are committed to a specific enzyme synthesis at different times in the sporulating cycle. For example, alkaline phosphatase genes are transcribed within the first hour of sporulation, but the enzyme is detectable only one hour later; the message molecules are stable for several hours and can support enzyme synthesis even if gene transcription is blocked. Genes are turned on at one time, but their protein products are translated at later times. Prokaryote sporulation is a simple case of information flow being translated into visible morphogenetic changes in cell shape and structure.

## THE CORTICAL FIELD: ASSIGNMENT OF SURFACE ORGANELLES

Free-living protists are large eukaryote cells with a complicated cell architecture. Their cytoplasms are filled with numerous vacuoles, tubules, fiber systems, and granules of various sizes which permit them to

behave as whole organisms. The most extensive specializations are found in the outer cortical layer located beneath the plasma membrane. In this region complex linear arrays of cilia are anchored, many of which are fused into large bristlelike structures known as *cirri*. The number of ciliary rows, or *kineties*, is usually constant for each species and seems to be regulated by the total size (surface area) of the cell. Many large cilia collect near the mouth and fuse into stiff membranelles as an oral apparatus. What regulates the number and pattern of these organelles when new sets are reconstituted at each division?

Cortical morphogenesis of one large ciliate, *Stentor*, has been studied in some detail. This cone-shaped, pigmented cell is about 1 to 2 mm long when fully extended and is equipped with a contractile system, which, when stimulated, shortens the cell to more than 50 percent of its length. In the cortex, extending from the base to the apex of the cone, are about 100 pigmented stripes of graded widths alternating with rows of cilia. At the broad end of the cone is a whorl of concentrically oriented stripes around the central mouth with a circle of stiff membranelles.

We have already mentioned some of the nuclear events associated with cell division in this species (see Chapter 3). During this same division period each daughter cell receives half the maternal ciliature and must assemble new rows. Besides, the posterior daughter cell must also regenerate an oral apparatus. It is the process of oral morphogenesis that has been studied most thoroughly.

The reconstitution of an oral region that normally occurs during each division can be experimentally reproduced by cutting off the old oral apparatus of a mature cell (Figure 8-2). Shortly after wound closure the new oral primordium arises as a clear rift in the cortex in the region of fine striping. Within this region new rows of cilia are produced, and they align themselves into large bands that fuse into membranelles and gradually coil into a whorl as the anlage migrates anteriorly. The primary event in this process is the multiplication of individual ciliary units in distinct rows within the cortical field of the oral primordium.

How does an oral primordium assume its proper place within the overall cortical pattern? The pigment bands in the cortex are arranged in a gradient of stripe widths over the surface; narrow stripes in the post-oral sector become increasingly wide around the cell. In one sharply defined region, the finest stripes lie next to the widest, and it is always in this region of

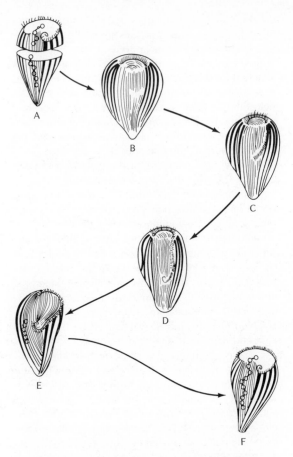

**Figure 8-2** *Regeneration of an oral primordium in Stentor after removing the original one. (A) Section of anterior end. (B) Posterior region closes wound and oral primordium appears as diagonal, clear rift in region of fine striping. (C) Primordium longer. Multiplication of fine stripes and short oral cilia appear. (D) Enlargement of posterior band of anlage where invagination will occur. Primordium now joined with remnant of old membranellar band. (E) Coiling invagination to form gullet. Fine-stripe multiplication posterior to the anlage. (F) Regeneration complete as anlage shifts anteriorly. (Based in part on V. Tartar,* The Biology of Stentor. *Oxford: Pergamon Press, 1960.)*

continues to reproduce as a doublet animal for many cell generations, with the doublet cortical pattern inherited as a stable phenotypic character. The new region of stripe contrast acts as an organizing center at each division and promotes oral regeneration. Nucleate fragments without any cortical regions of stripe contrast produce no oral primordium and eventually die of starvation because of inability to feed. Information for the organization of an oral primordium seems to reside in the cortex and not in the nucleus.

The nearest wide stripe seems to act as an inducer in the narrow-stripe region. Implanting a region of wide stripes into a fine-stripe zone produces two additional loci of stripe contrast in addition to the original primordium site, and three oral primordia appear. Conditions for primordium formation seem to extend over some distance within the cortex and are transmissible from cell to cell. For example, if two *Stentors* are grafted in parabiosis (two whole animals fused together) and the oral region is excised from one, new primordia appear in both animals, one at the cut surface while the other undergoes a reorganizational replacement of an intact organelle. A signal

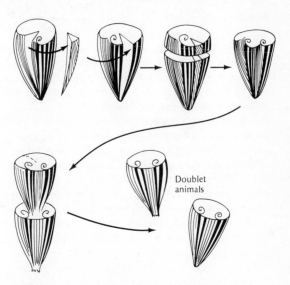

Doublet animals

**Figure 8-3** *Role of cortical striping pattern in regeneration of oral primordia. A sector of thin stripes is grafted into the wide-stripe region and the oral region is removed. An extra oral region is regenerated in the region of stripe contrast and a doublet animal is produced. When this animal undergoes fission, the persistent regions of stripe contrast in the cortex insure that all progeny will continue to form two oral regions after each division.*

contrasting stripes that an oral primordium appears during regeneration. If a region of fine striping is implanted in a wide-stripe area and the cell is induced to reorganize by cutting off the oral region, an oral region regenerates in the usual place while a second oral primordium appears in the new area of stripe contrast (Figure 8-3). A doublet *Stentor* results which

to reorganize extends from a regeneration primordium to the other, nonregenerating cell.

The strength of an inductive signal depends on cell size and is transmitted only when the regenerating cell is larger than the nonregenerating cell. On the other hand, regeneration is inhibited if a large nonregenerating cell (morphostatic) is grafted to a smaller regenerator. Oral induction may depend either upon total cell mass via a diffusible gradient of an inducing factor, or it may depend on the cortex itself as it responds to perturbations in cortical organization. Although we still haven't resolved these alternatives, the fact that formation of a new oral primordium is inhibited by the presence of a preexisting one suggests that the "state of the cortex" exercises control. A new primordium is induced only after the preexisting oral apparatus disappears, either by surgical removal or by reorganization during division. Changes in the physical state of the oral cortex act as a signal to trigger the morphogenetic events. Does this mean that the cortex is autonomous in oral primordium regeneration?

## Factors in cortical regulation

A simple enucleation experiment demonstrates that the cortex does not contain all the information for organelle assembly and position. Surgical removal of the macronucleus prevents primordium formation and oral regeneration. But nuclear information is not required continuously throughout the period of regeneration. If the cell has begun to develop a primordium before enucleation, then regeneration will continue for one or two stages more before the primordium is resorbed. On the other hand, if primordium development has gone beyond the membranellar band stage, enucleation has no effect, and a complete primordium develops. This suggests that nuclear instructions for organelle assembly are synthesized very early, before the membranellar stage, and that assembly and position assignment within the cortex are independent of nuclear control. This agrees with the observation that primordium morphogenesis is blocked by actinomycin D only if cells are treated at early stages of regeneration. After an oral anlage is formed, actinomycin does not inhibit the final stages of morphogenesis. We assume that RNA template instructions for organelle subunit proteins are synthesized at the beginning of morphogenesis for assembly later in the cortex. Apparently, assignment of ciliary units in the cortex and their assembly from precursors

is controlled by a cortical mechanism. What, then, is the nature of a ciliary unit, and how does it develop?

Each kinety is a linear array of repeating ciliary units (Figure 8-4). A unit contains one or two cilia attached to basal granules or kinetosomes embedded in a hexagonal pit, a long fiber arising beside the kinetosome, a blind sac to the right of the fiber, and membranous vesicles. Units communicate by means of the kinetosomal fiber running beneath the surface, thereby coordinating ciliary beat.

The cilium is a cylindrical organelle (axoneme) constructed from bundles of microtubules in a matrix. In cross section it exhibits the familiar nine-plus-two arrangement of paired microtubules. The outer nine doublets are continuous with tubules in the basal granules or kinetosome, which resembles the fine structure of centrioles. In fact, basal granules and centrioles are probably interconvertible. A cross section of a basal granule shows a pinwheel pattern of nine triplet tubules surrounded by an amorphous matrix. These structures are essential to cilia (and flagella) formation and are present at the base of all cilia and flagella (except in prokaryotes).

The position of a new cilium in the cortex seems to be determined by a preexisting ciliary unit. Each new cilium always emerges from a specific location within the old unit, as if the organizing forces for ciliary assembly exist within the old unit. For example, a new kinetosome is found immediately anterior to the old kinetosome, and new fibers always arise on the right side of kineotsomes and extend forward. If new ciliary patterns are copied on old patterns, then cortical instructions are inherited. A doublet *Stentor*, as we have seen, persists in transmitting its doublet phenotype to daughters because it possesses two cortical regions of stripe contrast. The classic experiment of cortical inheritance was performed on *Paramecium* by Beisson and Sonneborn in 1965. They removed a fragment of cortex from one cell and grafted it in reversed orientation into the cortex of another. The pattern of upside-down units within the graft was reproduced faithfully at each division for many generations. The pattern is stable and is inherited autonomously without benefit of genetic instructions. Changing the nuclear or cytoplasmic genome had no effect on inheritance of the inverted pattern, nor did massive exchanges of cytoplasm or influences from normally oriented adjacent units correct the inverted pattern.

Some have suggested that kinetosomes contain a small amount of DNA which codes for ciliary as-

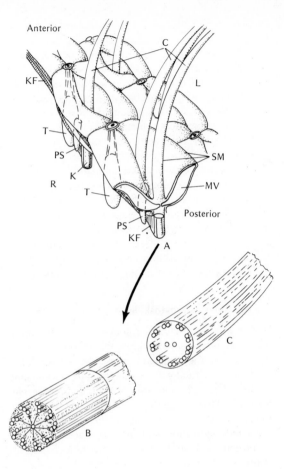

sembly. This is unlikely because the evidence for the existence of kinetosomal DNA is by no means conclusive. The simplest explanation for the inheritance of these ciliary patterns is that cilia structure and orientation depend upon mechanisms of basal body formation and its ability to act as a nucleating site for microtubule assembly in cilia (Figure 8-5).

A basal body usually develops anterior to the proximal end of a preexisting basal body, always with its long axis at right angles to it and separated by a gap. The basal body arises from a sequential addition of microtubules to form first a complete ring of singlets, soon followed by the appearance of doublets,

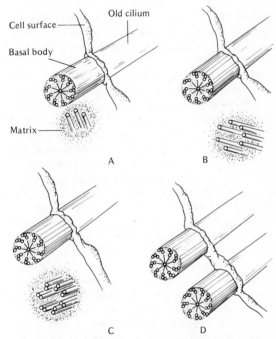

**Figure 8-4** *Diagram of a ciliary unit (Paramecium) and detailed structure of basal body and cilium. (A) Part of two contiguous anterior-posterior rows of cilia with only two repeating ciliary shown in each row. Cilia on left are cut in section to reveal internal structures. R = right; L = left; C = cilium; K = kinetosome; KF = kinetosomal fiber; MV = membranous vesicle; PS = parasomal sac; SM = surface membrane; T = trichocyst. [From T. M. Sonneborn, "Integration of Genetic Material into Systems," in R. Alexander Brink, ed.,* Heritage from Mendel *(Madison: The University of Wisconsin Press; © 1967 by The Regents of the University of Wisconsin), p. 386.] (B) Diagram of a basal body showing proximal end in cross section with pinwheel of 9 triplets (microtubules) in a matrix. Part of cilium axoneme is shown connected to basal body. (C) Part of cilium, cut to show 9-plus-2 arrangement of its doublet microtubules. The 9 outer doublets are continuous with the 9 triplets of the basal body.*

**Figure 8-5** *Development of basal body in cortex of Paramecium. (A) Diagram of part of ciliary unit showing basal body, surface membrane, and emerging cilium. Nearby, in a matrix region, singlet microtubules are forming, their axes at right angles to the existing basal body. (B) A circle of singlet microtubules in a new basal body. A few may now show a doublet structure. (C) A ring of doublets has assembled and is assuming its position beneath the membrane. (D) The basal body is mature, serving as a nucleating site for the assembly of ciliary microtubules. Presumably the membrane expands between old and new basal bodies as the cell grows. (Based on electron micrographs in R. V. Dippell, Proc. Nat. Acad. Sci. U.S. 61:461–468, 1968.)*

then triplets. The microtubules seem to condense out of matrix material adjacent to the old basal body, and possibly connected to it by fine filaments. The triplets may then serve as nucleating centers for the assembly of the outer fibers of the ciliary axoneme. The elongation of a cilium then proceeds by distal addition of microtubular precursors.

The kinetosome is the organizing unit of the cortex. Since it initiates cilia assembly, the preexisting pattern of kinetosomes will in turn determine all new patterns that arise. This, however, is not the only factor since kinetosomes are integrated into an overall cortical pattern. As new kinetosomes appear within an expanding cortical field, they migrate into a new position as if following an underlying "prepattern" beneath or in the cortical gel. Positioning of each kinetosome is first determined by its relationship to a preexisting granule, but its assignment to a row probably results from interactions with other kinetosomes within an expanding cortical field. An all-over cortical pattern seems to direct individual kinety positions.

Some of these instructions may come from the nuclear genome. Frankel studied a recessive mutant in another ciliate, *Euplotes*, which caused a reduction in the number of old and new ciliary units that appear in specific regions of the cortex after division. This suggests that mutant defects may be confined to early stages of cilia development. In addition, the number and orientation of kineties in the mutant are abnormal as if some basal bodies fail to conform to the orienting rules imposed by their neighbors. A nuclear product, other than microtubule precursors, seems to be essential in guiding kinetosome orientation, which means that the cortical mechanism is not strictly autonomous and that some nuclear information sets the rules for all-over body patterns. Nevertheless, the role of the protist cortex in determining surface organelle assignment suggests that it may be a blueprint of patterning information in all cells.

## DETERMINATION OF CELL SHAPE

The shapes of most single-celled protists are species-specific and defined by a tough outer pellicle or rigid skeleton which develops from silicaceous or calcareous deposits. Delicate networks of silicon spicules are deposited within a frothy mass of cytoplasm and grow into intricate designs. In his classic book *Growth and Form* Darcy Thompson attributed the mathematically perfect design of these skeletons to surface

tension forces within the frothy cytoplasm which controlled deposition of their silicaceous precursors. Though somatic cells have no visible skeleton, they maintain a variety of asymmetrical shapes, and for many years cell shape was also based on surface tension forces. Cells were compared to soap bubbles, whose shapes could be explained entirely on physical and mathematical principles. But no matter how mathematically consistent these models were, they were not accurate since surface tensions measured in cells were incapable of maintaining most cell shapes.

Some early cytologists proposed an invisible cytoskeleton, a network of fibrous elements oriented within the clear hyaline cytoplasm, which acts as a scaffolding on which organelles are suspended. No cytoskeleton was ever seen, but the orientation of certain organelles within cells suggests an underlying framework supporting the contours of the cell.

### Microtubules as a cytoskeleton

Within the last few years cytoskeletons, made up of microtubules oriented within the cytoplasm, have been seen in the electron microscope. Microtubules are found in virtually all eukaryote cells at some time in their life cycle and are the principal motile element in cilia and flagella. They make up the bulk of mitotic spindles which initiate chromosome movement and segregation, and they are aligned in parallel bundles in the cytoplasm of asymmetric cells such as neurons. Microtubules are dynamic structures, appearing and disappearing in different regions and at different times in the cell cycle. These rapid changes in microtubule organization account in part for changes in cell shape and for the movement of some cytoplasmic organelles. Changes occur because microtubules exist in an equilibrium state of soluble subunits reversibly self-assembling into tubules wherever needed.

Microtubules are cylindrical rods of indefinite length and 250 Å in diameter (Figure 8-6). The cylinder wall is made up of globular subunits arranged into a left-handed helix with 13 subunits per turn. The subunits are polypeptides of a protein, tubulin, which is usually isolated as a 6S dimer consisting of one $\alpha$ and one $\beta$ subunit, each of molecular weight 54,000 but of different amino acid sequence.

Purified preparations of tubulin have been used to study the equilibrium assembly of microtubules in vitro. Microtubules reversibly polymerize in vitro under conditions of neutral pH and low $Ca^{++}$. At 0°C

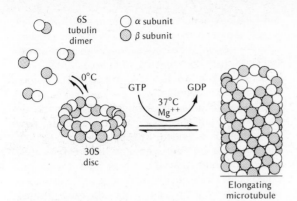

**Figure 8-6** *Self-assembly polymerization equilibrium of tubulin subunits and microtubules. At 0°C, tubulin is in equilibrium between 6S dimer and 30S disc. At neutral pH, low Mg⁺⁺, and accompanied by the splitting of GTP to GDP at 37°C, the dimers polymerize on the 30S nucleating site to lengthen into a microtubule.*

tubulin was found to exist in equilibrium between two states of aggregation, a native 6S dimer and a larger 30S "disc." The disc seems to be an intermediate in assembly since removal of discs by centrifugation prevents tubule assembly in vitro. Restoration of the discs stimulates polymerization.

Polymerization occurs optimally at 37°C; micro-

tubules dissociate when the temperature is lowered to 0°C. The reaction requires low concentrations of $Mg^{++}$ and the splitting of GTP. The role of the 30S discs in the assembly process is not certain inasmuch as other nucleating structures may exist. Elongation of the microtubule is unidirectional as subunits condense at one end. The conditions that promote assembly in vitro seem to be similar to those in vivo. Both require neutral pH, low ionic strength, and both are reversed by low temperature. There are certain differences, however, inasmuch as the equilibrium kinetics in vitro seem to follow a different reaction mechanism than those in vivo. The reason for these differences is not known.

Colchicine inhibits polymerization. For many years cytologists used colchicine to inhibit mitosis without knowing why. We now know that it dissociates microtubules of the spindle apparatus by combining with subunits, thereby driving the equilibrium toward dissociation. A specific assay for the presence of microtubule proteins in tissues is colchicine binding. Radioactive colchicine combines specifically with tubulin subunits in tissue homogenates in direct proportion to the amounts present. Colchicine also affects shape and motility in a variety of cells, which suggests microtubules may be implicated in these activities (Table 8-1). It is therefore used as a probe to

**TABLE 8-1  Cell Activities Sensitive to Colchicine**

| Cell Type | Effect on Cell Activity |
|---|---|
| Spinal ganglion nerve cells | Inhibition of axon elongation; no effect on growth cone activity |
| Chinese hamster cells in culture | Inhibition of continuous spindle fiber formation |
| *Allium* root meristem | Inhibition of spindle formation |
| *Tetrahymena* | Inhibition of macronuclear division and cell growth |
| *Echinosphaerium* | Breakdown of microtubules of retractable axonemes |
| *Stentor* | Inhibition of oral morphogenesis |
| *Nitella* | Inhibition of orderly cell wall formation and sperm morphogenesis (branched flagella produced) |
| Mouse macrophages in culture | Induction of pseudopodlike processes and amoeboid motion in gliding cells; cytoplasmic microtubules disappear |
| Chick embryo (neurula cells) | Inhibits neurulation (formation of neural tube); cells become spherical rather than elongate; inhibition of interkinetic migration of nuclei |
| *Fundulus* melanocytes | Inhibition of pigment granule migration |
| *Ambystoma* neural plate cells | Inhibition of cell elongation |

Source: Based in part on L. Margulis, *Int. Rev. Cytol.* 34:333–361, 1973.

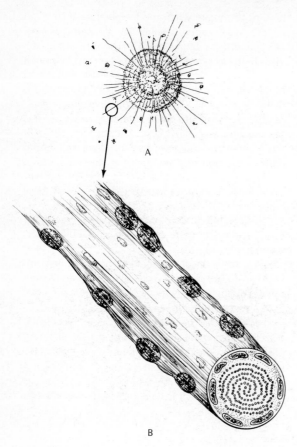

**Figure 8-7** *Microtubular organization of the axopodia of* Echinosphaerium. *(A) A light microscope view of the organism with radially projecting axopodia, the principal feeding organelles. (B) A longitudinal section of an axopodium showing the parallel arrangement of microtubules within the cytoplasm. A cross section of the axopodium shows the concentric arrangement of microtubules. This microtubular core supports the extended axopodium and facilitates the directional flow of cytoplasm.*

cine should dissociate microtubules and disrupt cell shape; removing the drug should reverse the process.

For example, the feeding organelles of the amoebalike protist *Echinosphaerium* consist of many filamentous axopodia radiating from a central body (Figure 8-7). An axopodium with attached prey is drawn into a food vacuole within the cell body by retraction (it seems to "melt" into the cell body), and a new axopodium is extended. Running the full length of the axopodia are cores composed of approximately 500 microtubules arranged in two concentrically oriented coils. These microtubules support the extended axopodium while the streaming cytoplasm surrounding the core captures prey.

Axopodia retract and disappear after application of hydrostatic pressure, low temperature, or colchicine. Simultaneously, the microtubular core dissolves. When hydrostatic pressure is released, microtubules reassemble, axopodia extend, and cytoplasmic streaming resumes. Extension of the axopodium depends on polarized growth of microtubules. Before the axopodium protrudes, an oriented array of microtubules fills the cytoplasm of the central cell body, increasing in length as new microtubules are added to the concentric ring pattern. Extension and retraction of axopodia can be explained by reversible assembly and disassembly of microtubules from a pool of tubulin subunits.

Similar observations have been made in most asymmetric cells; microtubules are usually aligned parallel to the long axis, and when these are dissociated, cells become spherical. These correlations do not prove that microtubules alone control cell shape. Microtubule assembly may be a secondary response to available new space in a cytoplasm expanding for other reasons. The best working model, however, is that cellular asymmetries are preceded by assembly and lengthening of microtubules. As these continue to elongate by condensation of subunits, cytoplasm flows in conformity with its growing skeleton.

### Factors ordering microtubules in cells

Some microtubules appear to float freely in the cytoplasm; others interact laterally and form bundles. In order for cell shape to change, to elongate along a single axis, microtubules must be anchored at some point. Parallel bundles seem to arise as a result of cross-bridge interactions. Lateral arms along a tubule may stabilize bundles of tubules into complex arrays.

determine whether microtubules are essential to certain cellular processes (see Chapter 5).

One way to study cell shape and its dependence on microtubules is to examine the ultrastructure of cells with asymmetrical shapes and expose them to colchicine or low temperature to disrupt microtubules. Changes of cell shape should correlate with the state of microtubule assembly. Addition of colchi-

The axopodial microtubules in *Echinosphaerium* are interconnected by cross bridges into a network of radial spokes extending from a center.

Similar but smaller cross bridges have been seen between spindle microtubules and are implicated in a motile mechanism for chromosome movement. But lateral interactions do not explain how microtubules are aligned in specific orientation in the cell. A nucleating site, a region where microtubule assembly is initiated, is necessary. We have seen that basal granules and centrioles act as initiating sites for microtubules in cilia and spindles respectively. They may also act as initiating centers for cytoplasmic microtubules in nonciliated or nonmitotic cells. Immunofluorescence microscopy has been used to visualize microtubules and other contractile structures in the cell. Inasmuch as tubulin, the principal protein, has been purified, relatively pure antibodies can be made against it, tagged with a fluorescent dye, and used to localize microtubules in cells. Such antibodies applied to mouse fibroblasts in vitro show a brilliant display of fibrous elements when examined, particularly in cells that have attached to the substratum. The microtubules fill the cell, many terminating near the membrane while others seem to follow the cell contours. Many tubules appear to radiate from a polar region near the nucleus. This center exhibits some interesting properties; it consists of tubulin, is usually fixed in the perinuclear area above the nucleus, and it persists in the presence of colchicine or low temperature, both of which cause immediate dissolution of all other tubulin elements. It seems to be a stable, cylindrical structure, approximately 2 to 3 m$\mu$ in length. When colchicine is removed, the microtubules begin to regenerate, elongating from the center out toward the plasma membrane; growth of the microtubule seems to be unidirectional. Since the center is the region where centrioles are located, the best interpretation is that the orienting center is equivalent to the centrosphere region of the cell. It suggests that the centriole may serve as a nucleating site for microtubules in interphase as well as for mitotic cells.

Other cytoplasmic microtubules may be oriented by different types of nucleating centers. For example, as ectoderm cells of the sea urchin blastula begin to elongate, three nucleating centers, or satellites, appear near centrioles from which microtubules radiate and extend parallel to the axis of elongation (Figure 8-8). Though these centers appear in cells undergoing shape changes, it is uncertain whether they are

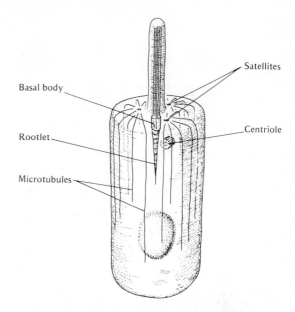

**Figure 8-8**  *Nucleating sites for microtubular orientation in cells. A reconstruction of the epidermal cell based on an EM analysis showing positions of nucleating satellites, microtubules, basal granules, and centrioles. (Based on L. Tilney and J. Goddard, J. Cell Biol. 46:564–575, 1970.)*

sites for microtubule regeneration or for the attraction of ends of microtubules. Be that as it may, the problem of cell shape becomes one of defining those conditions that induce centers for microtubule assembly and stabilize their position.

### Microtubules and protoplasmic movement

Since microtubules make up the core of cilia and flagella and move chromosomes in mitosis, may we assume these structures also participate in other cytoplasmic movements? Flow of protoplasm over the microtubule lattice in the axopodium of *Echinosphaerium* could result from microtubular action but the mechanism of force generation is still mysterious. Surprisingly, after colchicine treatment the collapsed axopodial cytoplasm still exhibits flow of some sort, which suggests that microtubules, at least in this instance, are not the sole driving force.

In fish melanocytes, however, microtubules are involved in aggregation and dispersal of pigment granules (Figure 8-9). These cells in the skin are under neurohumoral control, innervated by a pair of nerves.

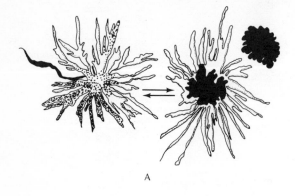

A

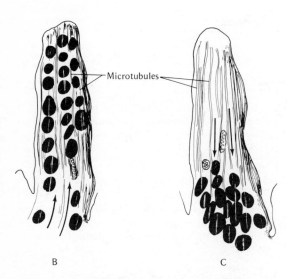

Microtubules

B                              C

**Figure 8-9**   *Migration of melanosomes (pigment
granules) in fish melanocytes. (A) Two melanocytes in the
epidermis. On the left is an expanded melanocyte with
pigment granules distributed throughout the cytoplasmic
arms or processes, as seen in the light microscope. On
the right is a contracted melanocyte; all pigment granules
have accumulated in the center of the cell while the
cytoplasmic processes remain extended. (B) A diagram of
an EM enlargement of an extended cytoplasmic process.
The pigment granules are in linear order surrounded by
microtubules arranged parallel to the long axis of the
process. (C) An EM diagram of a contracted process. The
microtubules remain aligned in the process but all
pigment granules have migrated toward the cell center.
Arrows indicate direction of pigment granule movement.
(Based on D. Bikle, L. G. Tilney, and K. R. Porter,*
Protoplasma *61:322–345, 1966.)*

A variety of environmental changes, including light
and background contrast, signal the nervous system,
which triggers melanocytes to contract or expand
into discrete color patterns in the surface. These have
adaptive value in startle reactions or in concealment
in a specific environment. The cells also respond to a
pituitary hormone, melanocyte-stimulating hormone
(MSH), and expand.

We find that melanocytes always retain their extended form; contraction and expansion reflect aggregation and dispersal of pigment granules within
the cell as they flow to and from the cell center. Some
microtubules are oriented toward the center while
others are aligned in the arms. The cell's extended
shape into irregular arms is probably supported by
its tubular cytoskeleton, and it seems that the reversible condensation and dispersal of pigment granules
is also mediated by these same microtubules. Agents
like colchicine also disperse condensed pigment
granules and prevent their reaggregation. This suggests that regulation of melanocyte behavior stems
from a direct neurohumoral effect on microtubular
function. Again, the mechanism of force generation
is not understood.

The dispersal action of MSH can be mimicked
by cyclic AMP, the second messenger of hormone action. Evidently, MSH acts by increasing the intracellular levels of cyclic AMP. Drugs such as epinephrine,
known to have an antagonistic action by causing pigment condensation, exercise their effect by lowering
the levels of cyclic AMP. Not all melanocytes behave
in this manner (amphibian melanocytes exhibit opposite responses to these drugs), but these studies show
that the reversible and rapid displacement of intracellular granules can be driven by microtubules. We
will see later that pigment granules and other organelles within the cytoplasm of eggs move about, particularly in the first few minutes after fertilization, and
here too microtubules may be involved (see Chapter
11).

Organelle shape may also be affected by microtubules. For example, elongation and condensation
of the macronucleus in *Stentor* during the mitotic cycle is probably controlled by surrounding microtubules (see Chapter 3). Microtubules align themselves
parallel to the elongating macronucleus during the
division cycle. The position of a nucleus in the cytoplasm may also be affected by microtubules embedded in the overlying cortex. When a macronucleus is
transplanted into another *Stentor,* it usually aligns it-

self beneath the cortex as if moved by intracellular fibers. One can imagine a model of cortical control of nuclear position and behavior with microtubules transmitting signals between cortex and nuclear surface. In this way expansion of the cell cortex during growth can be monitored by the nucleus as cues are transmitted via a microtubular network. As discussed in Chapter 5, such signals from the cell surface may coordinate mitosis with cytokinetic events during cell division.

Microtubules exhibit two roles in cell morphogenesis: supportive and motive; they seem to initiate and maintain cell shape, and they generate specific motive forces. Often, they express a dual role simultaneously. In cilia and flagella they combine the supportive role of a skeletal fiber with a dynamic role of motile force generation. Microtubules also exist in more labile states, as in the mitotic spindle, where they are poised in an equilibrium between subunits and polymers, organized to initiate chromosome movements. But not all changes in cell shape are due to microtubule systems. Most protoplasmic movements including many changes in cell shape are generated by contractile machinery based on reversible polymerization of the protein actin.

## MICROFILAMENTS AND PROTOPLASMIC MOVEMENT

Often it is difficult to decide which organelles participate most directly in cell-form changes. More than one organelle may be involved in the various types of motile behavior. Protoplasmic streaming and amoeboid movement are probably the most important. The familiar form changes of a migrating amoeba point up the rapidity with which such processes can alter cell shape. These motile changes seem to depend on contracting microfilaments, rather than on microtubules.

Streaming is seen in large plant cells, like *Nitella,* where streams of cytoplasm circulate files of chloroplasts rapidly around the cell beneath the cell wall. The more fluid interior endoplasm flows against a stationary, semirigid cortex or ectoplasm. At the endoplasm-cortex interface, numerous 50- to 80-Å filaments seem to be oriented in the direction of flow. Microtubules found outside the stream, in association with the plasma membrane, are not involved because colchicine does not inhibit streaming. It

seems that most plant cell microtubules participate in cell wall assembly rather than in cytoplasmic movement.

Cytoplasmic streaming is also found in plasmodia of the acellular slime mold *Physarum.* The plasmodium is an irregular, multinucleated mass of protoplasm made up of a network of veins or channels in which cytoplasm surges in various directions. The walls of the channels are composed of a more rigid ectoplasm surrounding the freely flowing endoplasm. Again, numerous bundles of 50- to 80-Å microfilaments are found in the ectoplasm.

Microfilaments are ubiquitous structures found in all eukaryote cells, even nonmotile ones. Cells do change shape and do exhibit cytokinesis, both of which depend upon the orientation of bundles of microfilaments. Several size classes have been detected, a common 40- to 80-Å class and thicker 100- to 120-Å tonofilaments, which seem to be attached to plasma membrane specializations known as desmosomes (see Chapter 10). In axons even larger neurofilaments, up to 160 Å, have been noted. We have mentioned earlier that the drug cytochalasin B seems to disrupt microfilaments in dividing cells. Because of this property, the drug has been used to determine if microfilaments are essential to various types of protoplasmic movement. For example, low concentrations inhibit movement and block cytokinesis of cultured fibroblasts, inhibit cytoplasmic streaming in *Nitella,* prevent cleavage of many eggs, and alter cell shape. The drug is not specific for microfilaments, however, nor does it disassemble all types of filamentous elements. It inhibits respiratory metabolism, protein synthesis, and may even alter membrane permeability. Though streaming is inhibited in *Nitella,* many microfilaments remain intact in the presence of the drug. Cytochalasin, therefore, cannot be used diagnostically to identify microfilaments as we use colchicine to identify microtubule protein.

The 40- to 80-Å microfilaments are probably contractile. Their chemical properties are almost identical to those of F-actin, a contractile protein from the thin filaments in muscle. Most of our knowledge about contractile proteins has come from studies of vertebrate skeletal muscle, which consists of elongated multinucleated cells filled with parallel myofibrils (Figure 8-10). Muscle fibers are striated because of the periodic packing arrangement of actin and myosin filaments, the two principal muscle proteins. In the electron microscope the myofibril con-

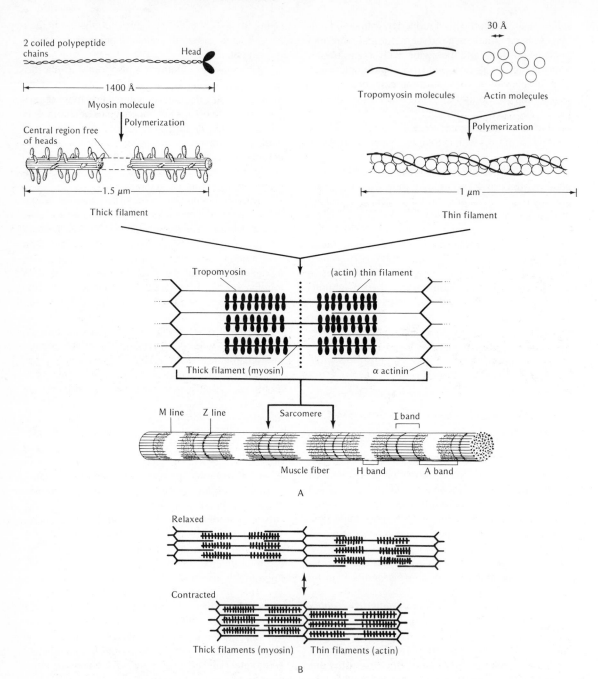

**Figure 8-10**  *(A) Assembly of muscle proteins into myofibrils is a stepwise process from globular subunits to myofibril bundles. (B) Sliding filament model of muscle contraction; actin filaments make and break cross bridges with thick myosin filaments. (From James D. Watson,* Molecular Biology of the Gene, *Third Edition, copyright © 1976, 1970, 1965 by W. A. Benjamin, Inc., Menlo Park, California.)*

sists of a series of distinct light and dark bands, the I and H bands, respectively. The repeating unit of muscle is a sarcomere extending from a dark Z line in the middle of an I band to the next Z line. Z lines contain another muscle protein, $\alpha$-actinin, to which actin filaments attach.

Myofibril assembly takes place within muscle cell cytoplasm in discrete self-assembly steps, very much like the process of viral morphogenesis; several separate assembly lines converge until the final fiber is constructed. Actin filaments polymerize from globular subunits along with another protein, tropomyosin, into helical thin filaments. Myosin, large polarized molecules of several subunit chains, with long tails and short heads, polymerize into thick filaments with many lateral extensions, or cross bridges. The sarcomere is assembled from a spontaneous lateral association of thick and thin filaments, with boundaries set by $\alpha$-actinin attachment sites in the Z line. During contraction the sarcomere shortens because actin filaments slide alongside myosin thick filaments as cross bridges between them are made and broken. The signal for contraction seems to be the release of calcium ions, triggered by a nerve impulse.

The principal elements of this motile apparatus are the contractile proteins actin and myosin, a regulatory protein tropomyosin which may regulate the flux of calcium ions and $\alpha$-actinin, the site of actin filament attachment. These proteins are also found in nonmuscle cells but are not usually organized into visible fiber systems.

Actin may be one of the principal proteins in all eukaryote cells, making up 10 to 20 percent of total cell protein. Surprisingly, this amount is about the same as that found in skeletal muscle, which suggests that it is a basic cellular constituent, perhaps as important as ribosomes. In contrast to muscle, cytoplasmic actin is less stable, with most of it in the form of a precursor pool of soluble G-actin in equilibrium with small amounts of filamentous F-actin. Actin filaments (microfilaments) combine with heavy meromyosin of muscle to form typical arrowhead structures characteristic of the actomyosin complexes in muscle. The rapidity with which actin filaments are reversibly assembled in different cell locations suggests, however, that a different type of regulatory machinery must be involved in nonmuscle cells. Although cytoplasmic actins share many amino acid sequences with muscle actin, they have different properties; they are less efficient in stimulating myo-

sin ATPase activity, the energy-generating reaction in muscle contraction.

Myosin is also widely distributed and is probably present wherever actin is found, but in much smaller concentrations than in muscle (less than 1 percent of total cell protein). This means the ratio of actin to myosin is much higher in nonmuscle cells, another indication that the mechanism of contraction might differ significantly from muscle. Many cytoplasmic myosins resemble muscle myosin and assemble in vitro into thick filaments with cross bridges. Both tropomyosin and $\alpha$-actinin have been isolated from brain and platelets and may, like actin, be a constituent of most eukaryote cells. We see that all the proteins of a contractile apparatus may be present in all eukaryote cells.

If these contractile elements are part of a cell's basic machinery, we want to know where they are found and how they are mobilized in various types of protoplasmic movement. As mentioned earlier, electron micrograph studies reveal microfilaments in gel-like cortical regions, usually beneath plasma membranes. In many cases filaments appear to attach to thickened plaques in the membrane and are most prevalent in actively contracting regions.

Contractile elements may also be visualized in the light microscope by immunofluorescence. Pure muscle proteins from vertebrate muscle are used as antigens for the preparation of fluorescent antibodies. Antibodies prepared against smooth muscle actin, skeletal muscle tropomyosin, and actinin were applied separately and in combination to rat embryo fibroblasts in culture just as they attach and spread out on the substratum. These antibodies react specifically with each protein and reveal its location by bright fluorescence (Figure 8-11). Within 4 hours after plating, patches of bright fluorescent foci can be seen in the cell after exposure to antibody against $\alpha$-actinin. At this same stage, similar bright spots are seen with actin antibody, but in addition numerous fibers radiate from these foci in a regular polygonal lattice. Tropomyosin antibody, in these early stages, stains only diffusely.

A few hours later, after the cell has attached and spread out, a regular polygonal network of $\alpha$-actinin foci and actin bundles covers the entire cell, with bright fibers extending to the clear edges of the cell. At this time tropomyosin antibody stains only the elongated fibers between vertices. A combination of actin and tropomyosin is found in the arms of the

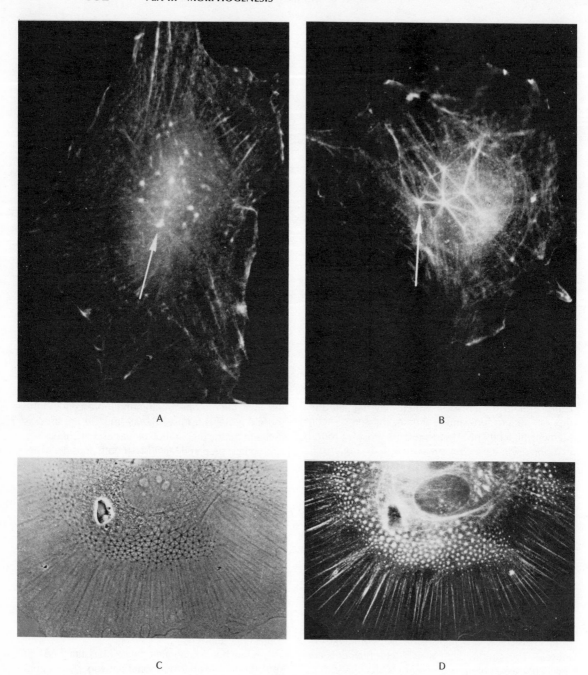

**Figure 8-11**   *Immunofluorescent localization of actin, tropomyosin, and α actinin in rat fibroblasts. (A) A cell treated with antibody against α-actinin 4 hours after plating. Bright fluorescent spots appear in the cell. (B) A similar cell treated with antiactin antibody. In addition to bright vertices, the arms of a polygonal network begin to appear extending from the vertices. (C) Cell seen in indirect immunofluorescence to show α-actinin antibody polygonal patterns. (D) Cell seen in direct immunofluorescence in which the bright vertices are most evident. (From E. Lazarides, J. Cell Biol. 68:202–212, 1976.)*

polygons within the network while actin and $\alpha$-actinin are together at the vertices. Since tropomyosin readily polymerizes with actin, it may function as a cofactor during actin polymerization or may even confer structural stability on the helical fibers within the polygonal lattice. The absence of tropomyosin at the vertices makes it possible for actin to freely interact with $\alpha$-actinin, possibly acting as a nucleating site for the assembly of the complex lattice network. The long fibers that extend to the edges of the cell membrane contain all three proteins, with tropomyosin assuming a periodic pattern along the bundle. The location of the vertices is uncertain; some may be on the plasma membrane, others on internal membranes. Though this regular network of fiber bundles seems to correlate with cell spreading in vitro, it may be an example of the reversible assembly of contractile machinery in different types of cell movement.

Similar fiber systems have been demonstrated in other cells. Movements in the growth cones of axons or the numerous microvilli on the surfaces of cells may also depend on reversible changes in a lattice-work of contractile proteins.

The location of myosin in this system has not yet been determined; not enough is present to be detected easily in most cells. If it acts as it does in muscle, we would expect it to interact with actin fibers. There are sufficient differences between contractile systems in nonmuscle cells to suggest that each type of protoplasmic movement, though employing the same elements, may be regulated by a different mechanism.

Recruitment of this contractile machinery in local regions of cells can account for major changes in cell shape. The formation of the contractile ring in a cleaving egg is an example of a dynamic mobilization of filaments for transient constriction of a dividing cell. The very lability of the system engenders vast versatility for different types of protoplasmic movement, from streaming cytoplasm, intracellular transport of organelles, formation of pseudopodia and cytoplasmic processes, and the movement of cells over substrata.

## CELL POLARITY

Except for spherical cells most tissue cells are polarized; that is, they possess an axis with an apical and basal region. Shape and polarity go hand in hand, since the former often determines the latter. Polarity

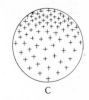

**Figure 8-12**  *Types of cell polarity. (A) Polar field polarity due to the asymmetrical distribution of organelles within or on the cell surface. (B) A structural orientation polarity arising from orientation of organelles within the cell or in the cell cortex. (C) Gradient polarity arises in cells as a consequence of a gradient distribution of organelles' substances or physiological activities.*

is an important property in cell morphogenesis; it may regulate the direction of cell movements and even the pattern of cell division. It also establishes metabolic differentials across the cell and "assigns" organelles to specific locations.

Organizing forces in the cytoplasm may produce one of three different types of polarity (Figure 8-12). Cell organelles may be restricted to one region, as is seen in ciliated borders of epithelial cells lining the respiratory tract. The axis of the cell is visibly defined by organelle accumulation, and cutting this cell into two eliminates its polarity. A second type of polarity is based on a uniform distribution of oriented structures throughout the cytoplasm with the cell axis defined by their orientation. In this second type, polarity is a property of the entire cell rather than of a region. The organelles align themselves in response to forces operating within the cytoplasm, and polarity is retained even after cutting. A muscle cell is a good example; its longitudinal myofibrils possess a polarity similar to that of the surrounding cytoplasm.

A third polarity pattern is based on intracellular gradients either of organelles or of metabolic systems with peaks of activity at one end diminishing gradually toward the other. "Double gradients" may appear, each with a high point of metabolism at one pole diminishing toward the opposite pole. The yolk-pigment double gradient in the frog egg is an example. Yolk is concentrated at the vegetal pole while pigment is highest at the opposite pole.

### A cell polarized by a basal granule

As a first approximation, cell polarity may be based on intrinsic polarity of macromolecules. Most subunit molecules such as amino acids or sugars are asymmetrical and are themselves polarized, in either a D- or L-form. This in turn is translated into a polarity of larger molecules, as in DNA and protein, where sequences are read in only one direction. Many large polymers possess a head and tail as we have seen in Chapter 7. Microtubules also exhibit unidirectional growth during polymerization which is correlated with elongation of surrounding cytoplasm. Microtubules seem to acquire polarity from basal granules or other nucleating sites which determine the pattern of subunit condensation. This polarizing influence from basal granules is also expressed as a pattern of oriented cilia beating synchronously in waves that direct movement of a ciliated cell. In this way polarity at the molecular level is transformed into organismic asymmetries.

For some cells polarity seems to arise de novo. For example, cell polarity appears in *Naeglaria gruberi* when an amoeboid cell transforms into a flagellated cell (Figure 8-13). This organism exists in two alternative motile states, amoeboid or flagellated. As an amoeba it possesses no intrinsic polarity, forming

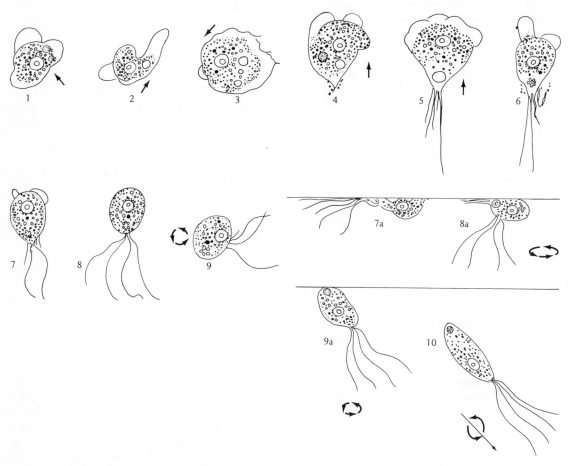

**Figure 8-13**  *Polarity changes during the amoeboid-flagellate transformation of* Naegleria gruberi *in distilled water. The arrows indicate direction of motion. (1–3) The amoeboid form: polarity changes as the direction of pseudopodial motion. (4) Polarized form: pseudopodia directed opposite site of future flagella development. (5, 6) Filiform pseudopodia emerge at posterior pole. (7–10) Development of active flagella. (7a–9a) Appearance of the transformation as seen from the side with the amoeboid form resting on a surface. In stages 4-10 the contractile vacuole remains at the posterior pole. (From E. N. Willmer,* J. Exp. Biol. *33:583, 1956.)*

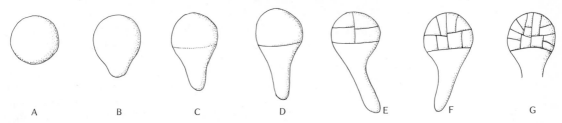

**Figure 8-14**   *Development of the* Fucus *egg. (A) 4 hours after fertilization. (B) 16 hours after fertilization: a rhizoidal bulge has appeared but no cell wall has formed. (C) 18 hours: a wall forms between rhizoid and thallus cell. (D) 26 hours: further elongation of rhizoid and early cell division of thallus cell.*

pseudopodia in all directions. Within minutes after exposure to an inducing signal (low ionic strength) pseudopodial outgrowth becomes polarized; only one pseudopodium appears at one end and exhibits directed movement. Shortly thereafter fine projections appear at the posterior end and transform into a flagellum. This latter region now becomes the anterior, leading end, reversing the polarity of the cell. Polarity of the cell is probably determined by factors controlling the position of the basal granule. In most cells these granules preexist or are derived from centrioles, but in *Naeglaria* no preexisting centrioles or basal granules have been seen. This implies that the basal granule arises de novo by mechanisms still unknown and becomes fixed within the cortex at the posterior end. It is not clear whether the granule forms randomly in any region of the cell and migrates to the posterior, or whether it is assembled in its final position. It does become fixed in the region of an amoeboid cell that is most rigid and under a state of contraction.

### Eggs polarized by transient membrane changes

The plasma membrane may be the primary site for controlling some types of cell polarity since it is first to receive stimuli from the environment. A highly localized signal impinging on the cell surface may induce polarity in a nonpolarized cell. Or if membrane structure is not homogenous but is a mosaic instead, then polarity may arise as a result of differential permeabilities at the cell surface. This would generate ion gradients within the cytoplasm which could direct the movement of organelles to specific sites. Perhaps the best example of this mechanism is found in the eggs of the seaweed, *Fucus*.

*Fucus* is a common brown seaweed found growing on rocks and pilings. It reproduces sexually, forming motile sperm and eggs which are fertilized in sea water. After fertilization the zygote divides unequally into a small rhizoid-producing cell and a larger thallus progenitor (Figure 8-14). The rhizoid develops into a holdfast organ that anchors the plant to a substratum.

Before any sign of division the position of the future rhizoid is seen as a local bulge or outgrowth which lengthens into a tube. Though the point of outgrowth seems to be random in a population of eggs, the event, when it occurs, establishes the future polarity of the developing embryo as well as that of the adult plant.

The position of rhizoid outgrowth can be fixed by a variety of environmental signals; the polarity of outgrowth is induced by gradients in light, temperature, pH, electric fields, and even polarized light (Figure 8-15). Rhizoids grow randomly in different directions when eggs germinate in the dark. If unidirectional light of a specific wave length and intensity is applied, all rhizoids grow away from the source toward the darker region. In an electric field rhizoids grow toward the positive pole, and in polarized light rhizoids grow in the direction of the plane of polarized light. Auxin, a plant hormone, also seems to polarize outgrowth.

Polarity determination seems to be a gradual process in normal germination. Approximately 50 percent of *Fucus* eggs are germinated by 15 hours, but imposition of an environmental factor accelerates this process. Almost 100 percent of the eggs are polarized by light in 8 hours. Polarity is labile within the first 9 hours after fertilization; that is, a variety of orienting influences such as light can fix rhizoid polarity

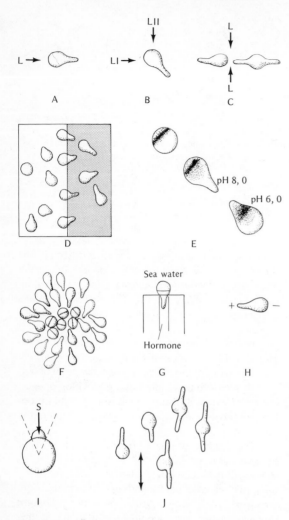

**Figure 8-15** *Environmental factors that polarize Fucus eggs. (A) Light from one side only. (B) Light first in direction LI followed by light from LII. (C) Opposing light beams of equal strength. (D) Partial illumination using half transparent and half opaque background. (E) Centrifuged eggs in seawater at two different pHs. (F) Group effect. (G) Local application of growth hormone (auxin). (H) Polarity in an electric field. (I) Polarization affected by sperm(s) entry. (J) In plane polarized light coming from above and below; the arrow shows the direction of vibration of the electric vector. (From A. Kühn,* Lectures on Developmental Physiology. *New York: Springer-Verlag, 1971.)*

prior to that time. Between 9 and 12 hours, however, eggs become less responsive, and polarity is determined. A point of no return has been reached.

How can so many different stimuli elicit the same response? Is there a common mechanism for all? In a series of studies by Jaffe, it has become clear that localized changes in membrane permeability to ions may be the initial polarizing event, perhaps the common denominator, in establishing polarity.

Shortly before an egg is polarized it sets up a minute electric current that passes through the cytoplasm. To measure these currents Jaffe placed a series of eggs in a capillary tube after exposure to a polarizing light source (Figure 8-16). As rhizoids emerged, measurable currents were detected at the ends of the capillary because all eggs were in series and all currents were polarized in the same direction. As 90 percent of the eggs germinated, a transcellular current of approximately $6 \times 10^{-11}$ amp passed through each egg and persisted even if the polarizing light source was removed. As soon as egg polarity is determined, the current appears and is self-generated.

An interpretation of current flow suggests that positive ions flow into the expanding rhizoidal end, driving current through the cell toward the opposite end. The membrane at the growing end seems to have undergone a depolarization; the normal negative resting potential of the egg becomes positive. Such depolarizations in other cells usually mean that positive ions are entering the cell as potassium ions (the principal intracellular cation) leave.

The nature of the ion involved was also determined. The future rhizoidal and thallus ends of polarized *Pelvitia* eggs (a related seaweed) were independently exposed to $^{45}Ca^{++}$ to compare rates of calcium uptake and efflux. The evidence indicated that there is a directed flow of calcium ions into the cell; they enter at the rhizoidal end and leave at high rates at the thallus end. This transcellular calcium current diminishes as the period of photosensitivity passes and germination begins. For a brief period a significant current is generated across the cell, possibly as a result of membrane leakage at the rhizoid end while a calcium pump is activated at the opposite end to secrete calcium. Such changes in membrane properties could be induced by a redistribution of $Ca^{++}$ channels and pump sites within the fluid membrane in response to the external polarizing influence.

What effect could such calcium currents have on the polarity of rhizoid outgrowth? An electric field of sufficient charge density could affect the distribution of negatively charged particles or vesicles within the cytoplasm. If this occurs, then these organelles

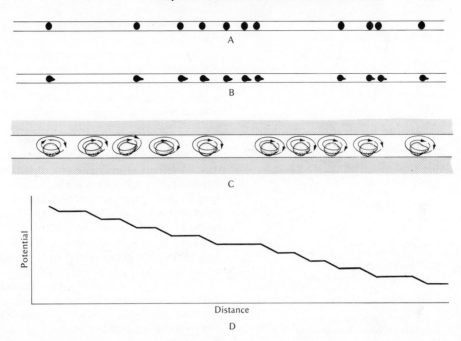

**Figure 8-16**   *Transcellular electric currents induce polarity of Fucus eggs. (A) Eggs in a capillary tube before germination, placed in series so as to facilitate measurement of exceedingly small transcellular currents. (B) Same eggs 26 hours after fertilization. All rhizoids polarized as consequence of self-generated transcellular current. (C) Schematic view of inferred current pattern in a tube. A transcellular current is set up in each egg before outgrowth, in which positive ions enter each embryo at rhizoidal end, flow through the cytoplasm and out the quiescent thallus end, returning through the medium to complete the loop. (D) Graph of inferred change of potential along the tube. (From L. F. Jaffe,* Adv. Morphogen. *7:295–328, 1968.)*

would tend to migrate in the direction of $Ca^{++}$ influx, the more positive growing end. In fact Golgi vesicles, carrying cell wall precursors, do accumulate at the growing tip. Since growth depends on cell wall elongation, we would expect wall precursors to migrate to the region. Quatrano has demonstrated the accumulation of sulfated polysaccharides, or fucoidins, in the presumptive rhizoid end a few hours before the appearance of an outgrowth. Fucoidins are essential components of the cell wall, and these appear to accumulate preferentially in the expanding wall of the rhizoid. According to Quatrano, as Golgi vesicles accumulate membrane-bound polysaccharides, they are sulfated by enzymes which may increase the net negative charges of the vesicle membranes. In response to the calcium current, negatively charged Golgi vesicles transmigrate the cytoplasm and accumulate at the end that is destined to become the rhizoid.

Calcium itself may be important in wall formation. We know that microtubules are involved in

plant cell wall deposition, although it is not clear how they act. Inasmuch as calcium ions affect microtubule assembly, calcium gradient of fixed and free ions, with a high point at the rhizoid end, might affect microtubule assembly and allow a free flow of wall precursors to the growing tip.

### Differential membrane permeability and polarized movement in Paramecium

Many cells such as *Paramecium* are morphologically polarized into an anterior and posterior end and exhibit directed motility. Movement in this cell is ciliary, and its direction depends upon the direction of ciliary beat, coordinated by a network of fibers between kinetosomes. During forward feeding movement all cilia stroke from front to back in waves, starting from the anterior end. If an animal contacts a mechanical object or noxious stimulus, it exhibits an avoidance reaction; the ciliary beat is immediately reversed, and the cell backs off to resume a new for-

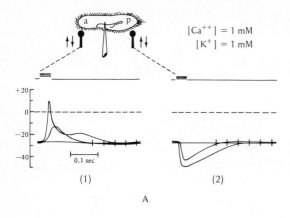

$[Ca^{++}] = 1$ mM

$[K^+] = 1$ mM

0.1 sec

(1)                    (2)

A

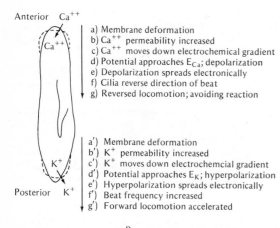

Anterior   $Ca^{++}$

$Ca^{++}$

a) Membrane deformation
b) $Ca^{++}$ permeability increased
c) $Ca^{++}$ moves down electrochemical gradient
d) Potential approaches $E_{Ca}$; depolarization
e) Depolarization spreads electronically
f) Cilia reverse direction of beat
g) Reversed locomotion; avoiding reaction

$K^+$

a′) Membrane deformation
b′) $K^+$ permeability increased
c′) $K^+$ moves down electrochemical gradient
d′) Potential approaches $E_K$; hyperpolarization
e′) Hyperpolarization spreads electronically
f′) Beat frequency increased
g′) Forward locomotion accelerated

Posterior   $K^+$

B

**Figure 8-17**  *Polarity in Paramecium. (A) Diagram shows an immobilized specimen impaled by an intracellular recording electrode. Mechanical stimuli were applied to either the anterior or the posterior regions of the cell. (1) Anterior receptor potentials elicited by three intensities of mechanical stimulation of the anterior end. (2) Posterior receptor potentials elicited by mechanical stimulation of the posterior end. Deflections in the upper trace show the duration and relative intensity of pulses activating the piezoelectric crystal. (B) Model linking mechanical stimuli to motor response patterns. Anterior and posterior ends are top and bottom, respectively. Stimulation of forward end results in the stepwise sequence of events leading to the avoidance reaction. Mechanical stimulation at the posterior end leads to forward motion. The polarity of the cell and the directed motor patterns stem from the fact that the membrane is differentially permeable to Ca and K ions at each end. (From Y. Naitoh and R. Eckert,* Science *164:963–965, 1969. Copyright © by The American Association for the Advancement of Science.)*

ward pathway until it avoids the obstacle. How do these rapid changes in ciliary beat come about?

Polarized movement has been studied by Naitoh and Eckert who measured changes in intracellular electrical potentials while mechanically stimulating anterior and posterior ends (Figure 8-17). An electrode placed in any region of the cell shows a negative resting potential of −30 mV in relation to the external environment. If the anterior end is mechanically stimulated, the membrane rapidly depolarizes as if a net flow of positive ions enters the cell. This depolarization is accompanied by the reversed ciliary beat, or avoidance reaction. Stimulation at the posterior end elicits an entirely different reaction, a hyperpolarization (the interior of the cell becomes more negative) as if positive ions diffuse out of the cell, and this is accompanied by a forward ciliary beat.

Exposure of cells to calcium or potassium solutions of different concentrations produces different potential changes in response to mechanical stimuli. The response suggests that mechanical stimulation at the anterior end induces a transient increase in calcium permeability whereas stimulation at the posterior end is accompanied by increased permeability and efflux of potassium. Opening of calcium channels results in an influx of calcium ions which increases the intraciliary calcium concentrations by increments of $10^{-6}$ M for each millivolt of calcium response and causes reversed beating. $Ca^{++}$ ions seem to regulate direction and frequency of the ciliary contractile mechanism. This can be shown by treating cells with a detergent to increase membrane permeability to all ions. Such paramecia swim forward in the presence of ATP and $Mg^{++}$ ions, but if $Ca^{++}$ concentration is increased above $10^{-6}$ M, ciliary beat is reversed, and the animals swim in reverse. Lowering calcium concentration forces the animals to move forward again.

Several behavioral mutants have been obtained which exhibit no avoidance reaction. Intracellular recordings of the "Pawn" mutant indicate that they fail to depolarize in response to the stimulus; the influx of calcium ions does not occur which suggests the membrane is altered by the mutation. This was further verified by showing that when these mutants are treated with detergent to allow free access of all ions, they will respond to calcium ion concentrations above $10^{-6}$ M, with reversed ciliary beat in the presence of ATP and $Mg^{++}$. It is evident that these mutants, like the wild type, possess a normal contractile mechanism responsive to calcium ion regulation but have a defective membrane which is unable to control calcium permeability. The study shows that membrane potential changes, correlated with the influx of calcium, can regulate a motile mechanism. Like the influx of Ca++ in *Fucus* eggs, such changes can have profound effects on cell growth and morphogenesis as well.

## SUMMARY

At the cellular level of morphogenesis two new phenomena appear: spatial location of organelles within cell architecture and cell shape and polarity. The assignment of organelles to their proper location depends on the cell cortex. The positioning of protozoan ciliary units into patterns is a cortical property that is independent of nuclear instructions. The nucleus provides the information only for organelle building blocks, while assembly and spatial assignment are properties of cytoplasm and cortex, respectively. Moreover, the cortex may serve as a site of attachment for elements within the cytoplasm, like the mitotic spindle, and in so doing segregate organelles.

Until recently the control of shape and form of cells has been obscure. We now know that the microtubule is organized as a cytoskeleton whose orientation is directly correlated with the asymmetrical forms of cells. Dissolution of microtubules is followed by loss of cell shape. Microtubules act as the backbone of such ordered structures as flagella and nerve axons; together with microfilaments, microtubules control cytoplasmic streaming, cell motility, and changes in cell shape. By changing cell shape these organelles underlie cell morphogenesis.

Microtubules assume ordered arrays through self-assembly, but extrinsic factors also operate as microtubules converge on cortical nucleating centers.

Cell polarity is a function of cell shape. It is manifested as a polar distribution of organelles, as oriented filaments, or as a system of metabolic gradients across the cell. Differential permeabilities of the cell membrane set up directed ion flow or electrical fields across the cell which displace organelles. An initially unpolarized cell may become polarized and align itself parallel to the orienting forces in an environmental field.

Finally, cells in groups acquire polarity of position. The properties of insideness and outsideness in a mass of cells create metabolic differentials which affect cell behavior as tissues are organized.

**REFERENCES**

Beisson, J., and Sonneborn, T. M. 1965. Cytoplasmic inheritance of the organization of the cell cortex in *Paramecium aurelia*. *Proc. Nat. Acad. Sci. U.S.* 53:275–282.

Bikle, D., Tilney, L. G., and Porter, K. R. 1966. Microtubules and pigment migration in the melanophores of *Fundulus heteroclitus*. *Protoplasma* 61:322–345.

Borisy, G. G., and Taylor, E. W. 1967. The mechanism of action of colchicine. Binding of colchicine-³H to cellular protein. *J. Cell. Biol.* 34:325–533.

Fulton, C. 1971. Centrioles. *Origin and continuity of cell organelles,* ed. J. Reinert and H. Ursprung. New York: Springer-Verlag, pp. 170–221.

Goldman, R. Lazarides, E., Pollack, R., and Weber, K. 1975. The distribution of actin in non-muscle cells. *Exp. Cell Res.* 90:333.

Gould, R. R., and Borisy, G. G. 1977. The pericentriolar material in Chinese hamster ovary cells nucleates microtubule formation. *J. Cell Biol.* 73:601–615.

Jaffe, L. F. 1968. Localization in the developing *Fucus* egg and the general role of localizing currents. *Adv. Morphogen.* 7:295–328.

———. 1969. On the centripetal course of development, the *Fucus* egg and self-electrophoresis. *Communication in development,* ed. A. Lang. New York: Academic Press, pp. 83–111.

Jaffe, L. F., Robinson, K. R., and Nuccitelli, R. 1975. Calcium currents and gradients as a localizing mechanism. *Developmental biology,* ed. D. McMahon and C. F. Fox. *ICN-UCLA Symp. Mol. Cell. Biol.* Vol. 2. Menlo Park, Calif.: Benjamin, pp. 135–147.

Johnson, K. A., and Borisy, G. G. 1975. The equilibrium assembly of microtubules in vitro. *Molecules and cell movement,* ed. S. Inoué and R. E. Stephens. New York: Raven Press, pp. 119–139.

Kung, C., and Naitoh, Y. 1973. Calcium-induced ciliary reversal in the extracted models of "Pawn," a behavorial mutant of *Paramecium*. *Science* 197:195–196.

Lazarides, E. 1976. Actin, α-actinin, and tropomyosin interaction in the structural organization of actin filaments in non-muscle cells. *J. Cell Biol.* 68:202.

Lwoff, A. 1950. *Problems of morphogenesis in ciliates.* New York: Wiley.

Naitoh, Y., and Eckert, R. 1969. Ionic mechanisms controlling behavioral responses of *Paramecium* to mechanical stimulation. *Science* 164:963–965.

Ng, S. F., and Frankel, J. 1977. 180° rotation of ciliary rows and its morphogenetic complications in *Tetrahymena pyriformis*. *Proc. Nat. Acad. Sci. U.S.* 74:115–119.

Olmsted, J. B., and Borisy, G. G. 1973. Microtubules. *Ann. Rev. Genet.* 8:411–470.

Osborn, M., and Weber, K. 1976. Cytoplasmic microtubules in tissue culture cells appear to grow from an organizing structure towards the plasma membrane. *Proc. Nat. Acad. Sci. U.S.* 173:867–871.

Pollard, T. D., and Weihing, R. R. 1974. Actin and myosin and cell movement. *Crit. Rev. Biochem.* 2:1–65.

Quatrano, R. S., and Crayton, M. A. 1973. Sulfation of fucoidan in *Fucus* embryos. I. Possible role in localization. *Develop. Biol.* 30:29–41.

Robinson, K. R., and Jaffe, L. F. 1975. Polarizing Fucoid eggs drive a calcium current through themselves. *Science* 187:70–72.

Snyder, J. A., and McIntosh, J. R. 1976. Biochemistry and physiology of microtubules. *Ann. Rev. Biochem.* 45:699–720.

Sonneborn, T. 1970. Determination, development and inheritance of structures in cell cortex. *Symp. Int. Soc. Cell Biol.* Vol. 9. New York: Academic Press, pp. 1–14.

Tartar, V. 1961. *Biology of Stentor.* Elmsford, N.Y.: Pergamon.

Tilney, L. G. 1975. The role of actin in nonmuscle cell motility. *Molecules and cell movement,* ed. S. Inoué and R. E. Stephens. New York: Raven Press, pp. 339–386.

Tilney, L. G., and Goddard, J. 1970. Nucleating sites for the assembly of cytoplasmic microtubules in the ectodermal cells of blastulae of *Arbacia punctulata. J. Cell Biol.* 46:564–575.

Tilney, L. G., and Porter, K. R. 1965. Studies on the microtubules in Heliozoa. I. Fine structure of *Actiosphaerium* with particular reference to axial rod structure. *Protoplasma* 60:317–344.

Wessels, N. K., Spooner, B. S., Ash, J. F., Bradley, M. D., Luduena, M. A., Taylor, E. L., Wrenn, J. T., and Yamada, K. M. 1971. Microfilaments in cellular and developmental processes. *Science* 171:135–143.

# 9

# Acquisition of Multicellularity

Large organisms have many more cells than small ones. The evolution of large size has been possible because of multicellularity. Cell populations become integrated into two-dimensional sheets through mutual surface contact and acquire new opportunities for morphogenetic expression. This transition represents a quantum jump wherein cells become subunits for the construction of new organizational hierarchies. More developmental information is available to such systems. For example, the position of a cell in a sheet and its contact relations with neighboring cells is a spatial property that does not exist in isolated cells but is important in a tissue. Metabolic gradients arise in multicellular aggregates, and some cells, physiologically more active than others, become pacemakers, or centers of morphogenetic activity. Such asymmetries lead to regional differences in tissues that transform two-dimensional sheets into three-dimensional organs.

Cells in tissues are in close contact. Their membranes conjoin or are separated by narrow matrix-filled spaces. Accordingly, cells acquire a "social sense" and communicate with one another by exchanging informational molecules. Such cell interactions promote the formation of different cell phenotypes, thereby increasing organismic complexity. Because they share a common extracellular matrix, interacting cell populations may cooperate with each other in carrying out specific functions. Wherever cell membranes abut, specialized morphological adaptations may develop to facilitate exchanges of chemical signals. Cells acquire specialized functions, "lose" their independence, and behave as members of a population.

Since these new tissue qualities are responsible for complex organismic development, we will first see how morphogenetic potential increases when primitive systems organize into multicellular aggregates. They do this prior to asexual reproduction, that is, when they sporulate.

## WAYS OF ACHIEVING MULTICELLULARITY

Three different routes to multicellularity are possible: (1) partition of multinucleated protoplasms or *syncytia* into tissues, (2) cementing dividing cells together with extracellular matrix materials, and (3) aggregation of cells into communities. In the first, primitive syncytia develop from single cells by increasing the number of nuclei in a growing cytoplasm without division. Nuclei are then partitioned into cells by membranes.

In the second pathway, cells remain together after division by secretion of intercellular "glue." The third requires that cells come together at random or in response to an aggregating signal to fuse into a multicellular mass. In all cases extracellular secretions and adhesive substances are important. In fact, the route followed often depends upon the physical properties of a matrix. For example, cellulose walls of plant cells are crystalline extracellular secretions which cement dividing cells into rigid multicellular patterns. Plant growth and morphogenesis are almost completely dependent upon patterns of cell division and wall synthesis. The principal route to multicellularity in plants is therefore the matrix-conditioned pathway.

Most animal cells secrete soft mucopolysaccharide matrices that accumulate between cells and keep them together. Multicellularity is matrix-dependent, but semifluid properties of the matrix permit cell movement. Hence cell motility and changes of cell shape emerge as major factors in animal morphogenesis. Contacts with neighbors are continuously made and broken as cells change shape and position. Accordingly, animal cell populations exhibit greater versatility in morphogenetic expression.

Motile cells can exploit the aggregation pathway to multicellularity. Independent cells may aggregate at some point in a life cycle when "summoned" by diffusible chemical signals. As cells aggregate, they adhere into functional tissues, probably because they secrete extracellular "glues" and establish specialized membrane adhesive sites.

Almost all extracellular matrices of plants, animals, and prokaryotes consist of systems of fibers (cellulose in plants, collagen in animals) embedded in a ground substance made up of glycoproteins. These proteins are basic constitutents of virtually all cell surfaces and extracellular mucoid materials and are the principal cementing substances for the construction of multicellular systems.

## MULTICELLULAR MORPHOGENESIS AND REPRODUCTION

Most multicellular organisms begin their life cycle as a single cell (a zygote or spore) which, by mitotic activity, develops into a multicellular state. The life of such organisms, therefore, may be depicted as a series of periodic cycles between the single-cell and the multicellular stage and back again to the single-cell stage during reproduction. The single cell is best suited for reproduction and dispersal whereas the multicellular condition is adapted for growth and development.

From the point of view of species survival, sexual gametes and asexual spores must of necessity be haploid single cells. In addition, dispersal via motile or even wind-borne spores is facilitated by small size and large numbers, particularly if the vegetative state is large and sessile, as is true for many primitive forms. The evolution of multicellularity in primitive organisms is coupled to morphogenetic programs that segregate reproductive from vegetative cells in fruiting bodies adapted to disperse unicellular reproductive products.

Many algae and fungi possess a syncytial or coenocytic organization of long filaments with numerous nuclei. This organization arises because no cross walls partition replicating nuclei into separate compartments at division, as if the ability to synthesize cross walls is inhibited. For example, certain fungi *(Mucor)* are dimorphic; that is, they exist in two morphological states depending upon conditions of growth, either as a spherical yeastlike cell or as a multinucleated filament resembling a mold. The growth pattern depends upon $CO_2$ concentration; the yeastlike growth is displayed in high $CO_2$, which seems to activate cross-wall synthesis to segregate each daughter cell. In low $CO_2$ no cell wall develops between division products, and the multinucleated condition appears. In these primitive systems morphogenetic transitions during the life cycle are generally coupled to the biochemical regulation of cell-wall synthesis and assembly.

### Physarum: A simple coenocyte

*Physarum polycephalum*, an acellular slime mold, is an irregular mass of protoplasm filled with channels of streaming cytoplasm (Figure 9-1). Although no cross walls are present, it does manage to grow to a large size—of the order of many centimeters. These amorphous masses of protoplasm can carry out complex morphogenetic programs independent of cellular organization and produce fruiting bodies adapted for gamete dispersal and fertilization.

The life cycle begins with germination of a haploid spore. The tough, protective wall dissolves, and two naked myxamoebae emerge and transform into flagellated swarm cells (like *Naeglaria*). After maturing for several hours two gametes fuse to form a diploid zygote. The zygote loses its flagella and transforms into a plasmodium by cytoplasmic growth and synchronous nuclear division without cross-wall formation. The plasmodium enters vegetative growth and flows about feeding on decaying logs. Later, nutritional factors and light interact to trigger morphogenesis of reproductive structures. Several days of starvation are required before a plasmodium becomes competent to respond to a light signal.

The light signal activates a train of biochemical reactions (induction) underlying morphogenesis. As old enzymes disappear and new ones are synthesized, a point of no return is reached (determination), and the plasmodium is irreversibly committed to sporulate.

During sporulation protoplasmic strands flow into a thick continuous layer which cleaves into many small nodules. These slowly elongate into pillars, perpendicular to the substratum, with pigment concentrated in the central axis. A swollen sporangium differentiates at the apex, the center filling with a foam-like cytoplasm and a network of threads containing calcium carbonate. Meanwhile, meiosis occurs in sporangial nuclei to produce haploid spore nuclei. At this point enzymes required for cellulose biosynthesis appear; walls are synthesized around each nucleus, and mature spores differentiate. Presumably the genes regulating wall synthesis are activated only at this stage in the life cycle.

## Switch points in a life cycle

In several species of nonfilamentous aquatic fungi the life cycle is streamlined so that a vegetative growing stage is reduced and compressed into the reproductive phase. One species, *Blastocladiella emersonii*, has been studied because it is easily triggered to follow one of two different morphogenetic programs (Figure 9-2). The life cycle begins with a motile spore with a large flagellum and a distinctive nuclear cap filled with ribosomes. The store of ribosomes is utilized for subsequent morphogenesis which begins

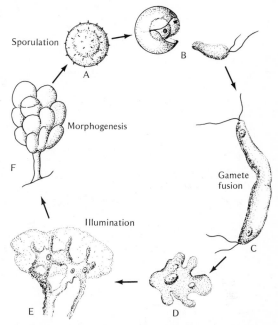

**Figure 9-1**  *Life cycle of the plasmodial slime mold Physarum polycephalum. A spore (A) germinates (B) into haploid motile gametes which fuse (C) to form a zygote (D). This develops into the multinucleated plasmodium with its network of shuttling cytoplasm (E). Specific illumination of the plasmodium stimulates the morphogenetic program and the plasmodium develops numerous sporangia (F) in which nuclei collect and organize into spores (undergo meiosis?) to repeat the cycle.*

when the spore stops swimming, retracts its flagellum, and rounds up. The cell soon germinates; the nuclear cap breaks up and releases ribosomes. This is followed by an unequal division (like *Fucus*) into a small rhizoid and a larger thallus cell. A brief period of exponential growth ensues, accompanied by nuclear division without cross-wall formation. A coenocytic vegetative phase grows directly into a spherical sporangium. At the end of growth thousands of nuclei surrounded by walls are released, as zoospores, to resume the cycle. Exponential growth of vegetative and reproductive phases occurs simultaneously.

After germination the two-celled stage follows either one of two morphogenetic pathways, the "ordinary colorless" (OC) pathway just described or the resistant spore (RS) pathway to production of a thick-walled, pigmented sporangium which becomes dormant. The trigger that determines the path-

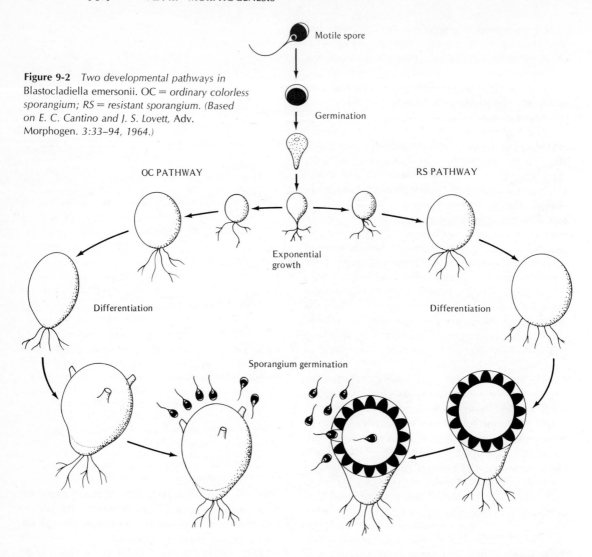

**Figure 9-2**   *Two developmental pathways in* Blastocladiella emersonii. *OC = ordinary colorless sporangium; RS = resistant sporangium. (Based on E. C. Cantino and J. S. Lovett,* Adv. Morphogen. *3:33–94, 1964.)*

way taken is high $CO_2$, which favors the RC program. As we can see, this is similar to *Mucor* morphogenesis, which is also controlled by $CO_2$.

Growth in the OC pathway is rapid with polysaccharide synthesis occurring more rapidly than protein synthesis. Mitosis takes place about once an hour and is probably synchronous in all nuclei. Only after growth stops do cell walls appear to cut up the cytoplasm into uninucleated fragments. Each spore differentiates with a nuclear cap and flagellum and emerges when the sporangium germinates. The entire cycle is completed within 2 to 3 days.

The RC program is basically the same in the

early stage of exponential growth, but the $CO_2$ switch alters the overall metabolic state so that certain biosynthetic pathways (pigment and carotene synthesis) are activated while others are repressed. The metabolic action of $CO_2$ is not understood; some believe it alters key enzymes in the tricarboxylic cycle, but this is by no means certain. The overall growth rate is less than that of an OC cell, but polysaccharide synthesis is greatly stimulated, presumably in preparation for resistant wall assembly. After 36 hours of growth several thousand nuclei have been produced, and a cross wall appears to separate the large sporangium from the rhizoidal mass below. The sporangium with its protective wall gradually becomes dormant, and it can survive a dry state for years. In both cases the multinucleated state switches to a multicellular state at the termination of the growth period; in the OC cell it occurs immediately, but it is delayed in the RC pathway perhaps until germination of the dormant sporangium.

During exponential growth one pathway can be switched to the other by changing $CO_2$ concentration. For example, the RC program is not fully determined until 36 hours, when growth ceases and a small cross wall separates the sporangium from the rhizoid. If $CO_2$ is removed at this point, the RC program continues. RC cells taken from a $CO_2$ environment before this point, even at 28 to 32 hours, are still able to switch to the OC program. When placed in a medium without $CO_2$, the biochemical machinery committed to pigment, carotene, and cell wall synthesis is turned off, and within a few hours the cell behaves as an OC sporangium and liberates zoospores. The point of no return at 36 hours seems to correlate with the appearance of the cross wall, but the actual mechanism of determination is not known.

A critical period of only a few hours (between 32 and 36 hours) marks the transition from a labile state to the fully committed RC program. It is not clear what mechanisms are involved. Several mutants unable to respond to the $CO_2$ signal have appeared spontaneously, but these cannot be used effectively, since *Blastocladiella* has no sexual cycle. A transient period when the cell is poised between two morphogenetic alternatives seems to be a common condition in many developmental systems.

## Polarity and morphogenesis

In *Fucus* and *Blastocladiella* we have seen that a differential division, resulting from intrinsic cell polarity, segregates a rhizoid from a thallus cell with each displaying a different morphogenetic program. The large size attained by some of the coenocytic algae and fungi also creates polarizing situations, perhaps because diffusion gradients are stabilized in such systems. Environmental stimuli such as light or gravity exert differential effects on different parts of the organism and induce polarity. Such polarizations are particularly evident in filamentous species in which cytoplasm at opposite ends of a filament respond independently to environmental signals.

*Bryopsis* is a filamentous green alga organized

as a syncytium (Figure 9-3). At the basal end of a filament are rhizoids which attach to a substratum. The stalk branches symmetrically into featherlike fronds filled with nuclei and chloroplasts. Light and temperature stimulate gametangia formation in the branches, and nuclei divide rapidly into large numbers. They are surrounded by walls to emerge as haploid, motile gametes which fuse into zygotes. These undergo meiosis before germinating into a crawling filament that again develops shoots and rhizoids.

Coenocyte polarity, expressed as shoot-rhizoid differentiation, results from environmental gradients. For example, shoot filaments respond negatively to gravity and positively to light. If a shoot is inverted, the branches and vertical stalk begin to grow upwards to assume a normal position (see Figure 9-3). Furthermore, polarity can be reversed. If an apical filament is inverted in sand, the old rhizoids shrivel, and a new branch emerges from this end to become an apical, branching region while new rhizoids grow from the original apical end. The chloroplasts, moving in streams in the cytoplasm, collect primarily in the end exposed to the light, irrespective of its former function. These streaming movements can be followed by staining with a vital dye; if the apex is stained before inversion, the dye soon finds its way into the exposed rhizoid end which now becomes apical. Conversely, if the rhizoid end is stained before inversion, the dye also streams to the apical end now destined to form rhizoids. The apical and basal ends exchange cytoplasms, which suggests that localizations of specific cytoplasms at different ends may be responsible for polarity. A stronger cytoplasmic polarity arises if the length of a coenocytic filament is increased. Presumably greater lengths favor more extreme metabolic differentials between apical and basal ends. Diffusion through a corridor of cytoplasm is unable to eliminate these differences, and polarized morphogenetic expression results; the filament is "regionalized." Morphogenetic changes in response to exogenous influences are stabilized into apical and basal structures.

## Coenocytic filaments as building blocks

The vegetative structure in fungi is a branching, threadlike filament or hypha organized into networks known as *mycelia*. Filaments are generally multinucleate and seem to have unlimited growth capacity. Elongation occurs at the tips, the weakest points in

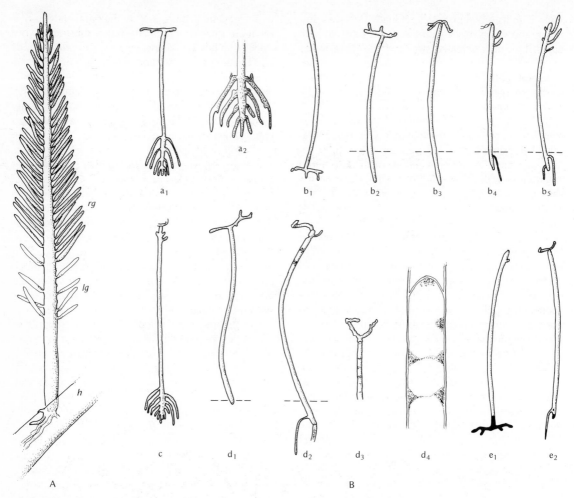

**Figure 9-3** *Polarity in Bryopsis. (A) A mature plant with plumules, rg, that have begun to produce gametes. Some empty gametangia, lg, are shown as well as the rhizoidal attachment at h. (From F. Oltmanns, Morphologie und Biologie der Algen. Jena: Fischer, 1922.) (B) Inversion experiments. ($a_1$–$a_2$) Diagram of plant with closer view of inverted plumule end bearing emerging stemlike structures at the base of each. ($b_1$–$b_5$) Stem inverted in substratum. Original rhizoid end transforms into an apical end with plumules while the apical end develops as a rhizoid. The stippling reflects chromatophore concentration which shifts toward the new apical end. ($c$–$e_2$) Vital stain illustrating the migration of cytoplasm. If the apical cytoplasm is stained before inversion, it is soon found in the original basal end ($c$, $d_1$–$d_4$), and if the rhizoid is stained before inversion ($e_1$), some stain is found in the new basal end, indicating that apical and rhizoidal cytoplasms are displaced to opposite poles during inversion. (From A. Kühn, Lectures on Developmental Physiology. New York: Springer-Verlag, 1972.)*

their cellulose-chitin cell walls, as Golgi vesicles bearing wall precursors migrate into the tip. Environmental factors such as light, nutrition, and temperature may polarize or switch growth patterns from vegetative to reproductive modes. Hyphae grow hori-

zontally on a substratum and in response to cues extend vertically into the air as sporangia or gametangia with thickened pigmented walls, filled with spores or gametes.

In more advanced fungi, the *Basidiomycetes*

(mushrooms), the hyphal filament itself is a morphogenetic unit, packing together into large three-dimensional structures. Hyphae fuse at growing tips or come together laterally as bundles or thick strands and even solid tissues. Growth of such bundles seems to be directed by diffusible factors which promote strand interaction and coordinated outgrowth. Hyphal strands are stimulated to grow as a group by a leading strand while other hyphae are recruited from behind (Figure 9-4). The bundles thicken as more "following" strands enter into the growing

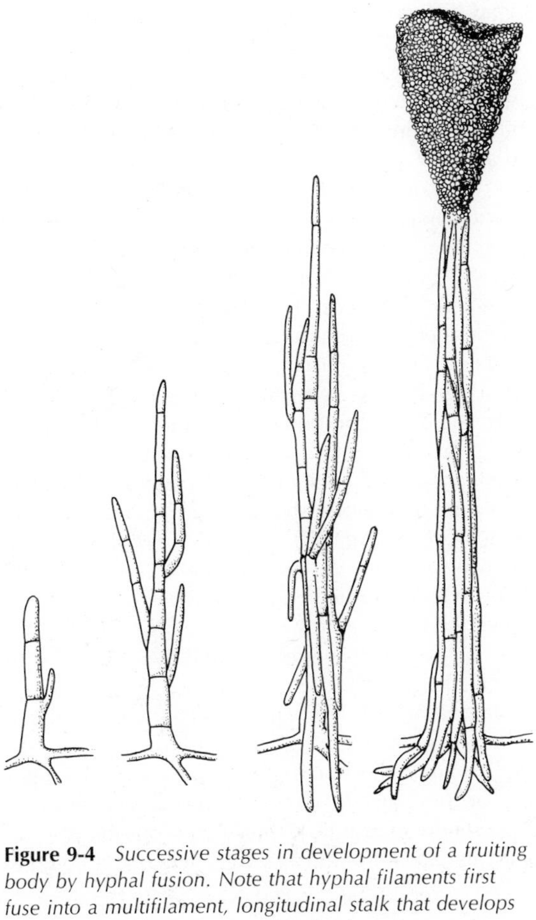

**Figure 9-4** *Successive stages in development of a fruiting body by hyphal fusion. Note that hyphal filaments first fuse into a multifilament, longitudinal stalk that develops into a terminal sporangium. This simple pattern evolves into the more elaborate fruiting programs of the mushrooms.* [*From John Tyler Bonner,* Morphogenesis: An Essay on Development *(copyright © 1952 by Princeton University Press): Fig. 37, p. 98. Reprinted by permission of Princeton University Press.*]

complex. Metabolic gradients arise within strand bundles, and hyphae differentiate into morphological zones; certain regions of high growth activity become pacemakers for morphogenesis and fruiting-body initiation.

The mushroom, like most fungi, starts its life as a spore which germinates into vegetative hyphal filaments. These grow and branch in the ground or in rotten wood into a large mycelial net of strands and filament bundles. Within the mycelial mat a discrete region of filament bundles, the *rhizomorph,* develops with apico-basal polarity. Its apex becomes the principal growing region dominating all morphogenetic activity. Because of its preferred position, it tends to serve as a "sink" for nutrients which are translocated from neighboring hyphae. Accordingly, the rhizomorph grows five times more rapidly than the remaining mycelial strands and becomes delineated into a stalk and cap, shooting into the air to expose gills and spore-forming surfaces of the typical mushroom. Its shape is probably controlled by the dynamics of filament fusion. An overriding regulatory factor transcends cell wall boundaries and organizes independent filaments into functional patterns. Morphogenesis occurs because growth patterns in the mycelium are unequal, polarized by unidirectional environmental factors.

The fleshy mushroom is the culmination of morphogenesis of interacting hyphal filaments. Again a multicellular organization is unnecessary to carry out developmental programs; whole filaments can organize complex fruiting structures.

## GLYCOPROTEINS AND MULTICELLULARITY

Perhaps the most important adaptation for multicellularity is the secretion of extracellular matrices or glues that hold cells together after division or when they aggregate. These matrices may be hard or soft and accumulate between cells where they modulate cell interactions. Though cells may communicate via specialized membrane junctions, they usually exchange materials and regulatory signals through matrix-filled intercellular spaces. Such interactions are bound to influence a cell's morphogenetic behavior, and they probably account for the social behavior of cells. They may be responsible for the specificity of such interactions as cell adhesion, recognition, and fusion (see Chapter 10). When cells remain together in aggregates, materials normally secreted from the cell surface are entrapped within intercellular spaces, thereby creating a new microenvironment. The principal constituent of these extracellular materials are glycoproteins, various fibrils, water, and salts.

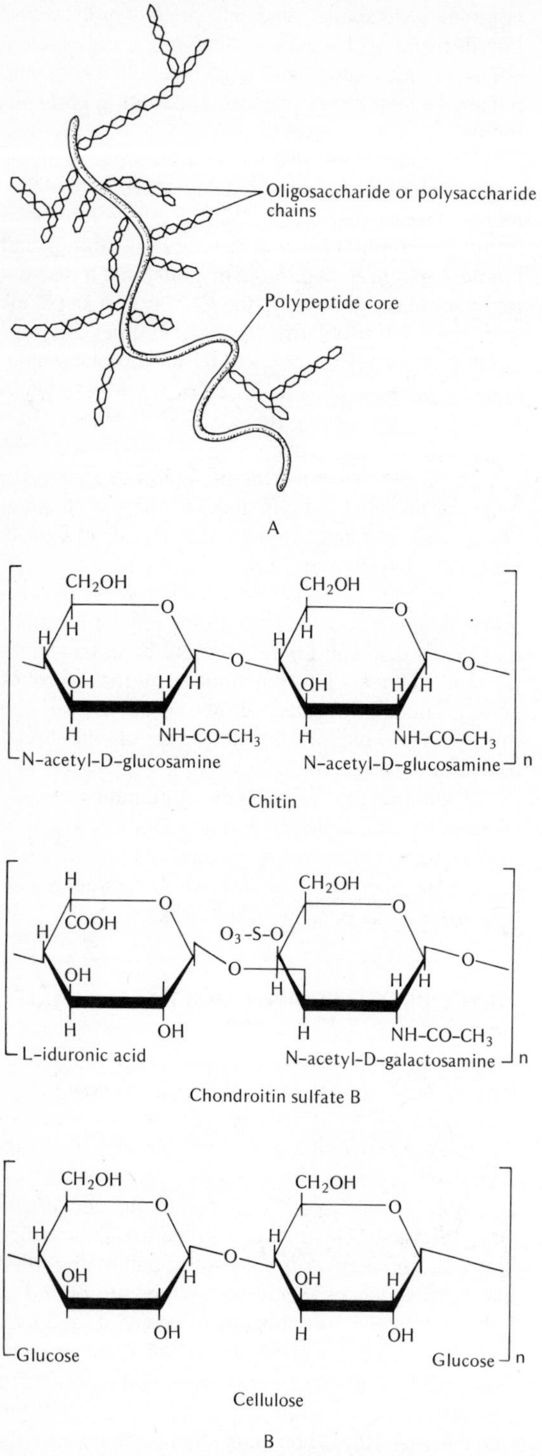

Glycoproteins are proteins to which carbohydrates are covalently attached at various points of the amino acid chain (Figure 9-5). They consist of two classes, true glycoproteins containing typically small oligosaccharides of 4 to 15 sugars and mucopolysaccharides with long-chain polysaccharides containing thousands of sugars. The former are found as integral components of cell membranes, extending from the lipid bilayer to the outside. They also include many enzymes such as ribonuclease, antibodies, hormones (chorionic gonadotropin), and ABO and MN blood group substances on erythrocyte membranes. These glycoproteins may contain 1 to 60 percent carbohydrate by weight.

Mucopolysaccharides, or glycosaminoglycans (GAG), may contain up to 85 to 90 percent by weight of polysaccharide chains attached to small peptides. These substances are the matrix materials that fill the intercellular spaces. This class of glycoproteins includes cellulose, chitin, chondroitin sulfates, keratin sulfates, and hyaluronic acid. They are major constituents of plant cell walls, of mucinlike secretions (saliva), jellies surrounding eggs, the capsules of bacteria, the slimes around algal filaments, and of such well-formed tissue matrices as cartilage and bone. They are ubiquitous, show a wide heterogeneity and specificity, and display an assortment of biological activities. In this chapter we are concerned principally with their role in morphogenesis of simple multicellular communities.

In plant cells the bulk of the GAG matrix is organized as a cellulose cell wall consisting of a crystalline array of long-chain, unbranched polysaccharides combined with a matrix of pectins. This tightly knit collection of fibers accounts for the rigidity of the wall and its highly regular patterns. The deposition of cellulose precursors outside the cell and its assembly into the wall lattice is a complex

**Figure 9-5** *Structure of glycoprotein. (A) Generalized structure of glycoprotein with protein core to which are attached numerous oligosaccharides or even long-chain polysaccharides. (B) Several glycosaminoglycans showing basic repeat unit in these structures.*

process that depends on packaging and transport of wall precursors in Golgi vesicles to the cell surface and such orienting factors that are imposed by microtubular arrays near the wall. Plant cell growth is essentially cell wall expansion, and cell division is usually followed by the synthesis of a new cell wall between daughter cells. This tends to keep all cells together in patterns that depend upon the orientation of cell division planes.

By contrast, most animal matrices are soft, containing fewer fibrils (usually one of the collagen types) loosely arranged in a GAG matrix of chondroitin sulfates or hyaluronic acid in various proportions. Different tissues are surrounded by matrices of different composition, usually varying in proportion of fibrils to GAG components. Sometimes collagen fibrils are ordered into layers of orthogonal lattices in basement membranes that cover most epithelia. These structures are somewhat more stable than most matrices which seem to undergo compositional changes with

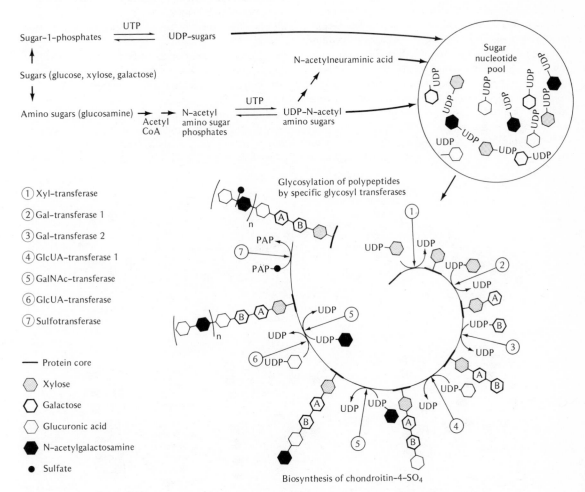

**Figure 9-6**  *Generalized pathway for the synthesis of glycoproteins. A pool of nucleotide sugars is derived from reactions of sugar phosphates (glucose-1-phosphate) with uridine or guanosine triphosphates (UTP, GTP). In addition, sugars are converted into amino sugars (glucosamine) which are acetylated into N-acetyl amino sugar phosphates. These are converted into nucleotide derivatives in combination with UTP. The synthesis of chondroitin sulfate takes place in the endoplasmic reticulum and Golgi vesicles where a polypeptide core is sequentially glycosylated by specific glycosyltransferases. At each enzymatic step a new sugar nucleotide is added to a growing polysaccharide chain.*

time. Such changes, particularly in developing systems, may play an important role in controlling growth and morphogenetic behavior of cells in multicellular communities.

Most GAG components in both plants and animals are products derived from the enzymatic assembly of sugar nucleotides on protein cores. These polysaccharide subunits are synthesized from phosphorylated or amino sugars combined with uridine diphosphate (Figure 9-6) through reaction sequences that are common to all cells.

The protein and carbohydrate moieties of various GAGs are made independently. Protein synthesis can be inhibited without affecting polysaccharide synthesis. The protein core is probably first made on polysomes in the endoplasmic reticulum. As it passes down the lumen of the endoplasmic reticulum, membrane-bound enzymes add carbohydrate by stepwise addition, one sugar at a time. The first sugars are attached to the amino acid chain of serine, asparagine, or hydroxyproline residues, and all subsequent sugars are added to the linearly growing carbohydrate

chain. An activated sugar nucleotide such as UDP galactose acts as a sugar donor, and the acceptor is the growing polysaccharide chain on the protein. UDP is released after the sugar is added to the oligosaccharide. Since different sugars are added in sequence, the specificity of sugar addition is determined by glycosyltransferase enzymes in the membranes of the endoplasmic reticulum. Many different glycosyltransferases must work together to synthesize a glycoprotein. Glycosyltransferases are heterogeneous, and different cells may have different populations of these enzymes, which means that cells synthesize cell-specific or tissue-specific glues (see Chapter 10).

After passage through the endoplasmic reticulum the glycoprotein reaches Golgi vesicles where it is concentrated, packaged, and prepared for secretion (Figure 9-7). The secretion of material involves migration of Golgi vesicles to the surface, fusion of the Golgi vesicle membrane with the outer plasma membrane, and spilling of their contents to the outside (see Chapter 7). This seems to be the mechanism for

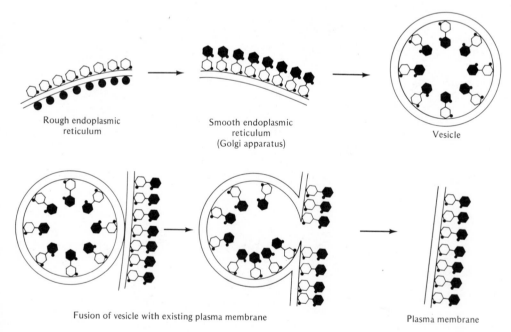

Rough endoplasmic
reticulum

Smooth endoplasmic
reticulum
(Golgi apparatus)

Vesicle

Fusion of vesicle with existing plasma membrane

Plasma membrane

**Figure 9-7**  *Mechanism of plasma membrane formation and exteriorization of glycoproteins. The saccharide rings are shown as open and closed hexagonals; the dark circles are ribosomes of the endoplasmic reticulum. In this scheme, the protein core synthesized in the ER is glycosylated by the sequential addition of UDP-sugars while attached to the membranes of the Golgi. These are pinched off as vesicles which migrate to the cell surface and fuse, thereby adding additional membrane to the outer surface and exposing the now completed glycoprotein moieties. (Based on a model proposed by H. Hirano et al.,* Proc. Nat. Acad. Sci. U.S. *69:2945–2949, 1970.)*

the synthesis and turnover of the plasma membrane itself. It is also utilized for deposition of cellulose precursors during cell wall synthesis and may be a common mechanism for the secretion of most extracellular matrices.

## Rigid walls and cell division patterns

Form in multicellular plants depends upon the orientation of cell division planes in successive divisions. This determines where a new cell wall develops. The simplest case, of course, is the long filament in which all division planes are oriented parallel to one another in the same plane. At each division new cellulose walls are laid down between cells to keep them together as an elongating filament as they grow.

This pattern of division and wall formation may have been established ages ago when the first single-celled motile alga evolved into a primitive filament. For a filament to form, some mechanism must have evolved to keep cells together after division. One mechanism is suggested by division patterns in a group of unicellular green algae, *Chlamydomonas* (Figure 9-8). In this species a single cell divides into two or four motile cells within the original mother cell wall, each secreting its own wall. These cells break out and swim off. Sometimes a single cell secretes a coat of jelly and divides haphazardly into an irregular group of cells in a mass of jelly. This may be regarded as an incipient multicellular stage because the jelly keeps the cells in a loose aggregate. Two related species, *Palmella* and *Geminella,* have both lost the free-swimming flagellated stage and exist only in the multicellular condition surrounded by jelly. *Geminella* division patterns are not oriented at random but are parallel to one another, resulting in a chain of cells, a primitive filament embedded in jelly. We see here a stepwise progression from a unicellular condition to a multicellular filament.

If cross walls did not separate after each division, a true filament, characteristic of most green algae, would result (Figure 9-9). The important point is that multicellularity is brought about by a combination of factors: synthesis of common wall partitions between cells after division made possible, in part, by a jelly matrix holding cells together. In fact most filamentous green algae are surrounded by a mucopolysaccharide gelatinous matrix.

The combination of matrix secretion and cell division contributes to diverse morphologic patterns

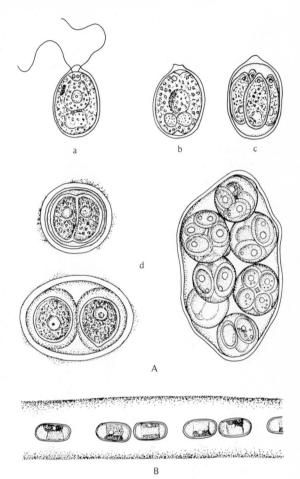

**Figure 9-8** *Multicellularity achieved through extracellular matrices. (A) Life cycle of* Chlamydomonas. *(a,b,c) A single individual undergoes normal mitotic division within the original cell wall. Each daughter cell produces its own cell wall and breaks free. (d) Stages in the development of the palmella stage in which many individuals are held together within a secreted extracellular matrix. (From* The Plant Kingdom *by William H. Brown, copyright © 1935 by William H. Brown. Used by permission of the publisher, Ginn and Company (Xerox Corporation). (B) A portion of a Geminella filament showing individual cells embedded within a gelatinous sheath. (From G. M. Smith,* Fresh Water Algae of the U.S. *Copyright © 1933, 1950 by McGraw-Hill Book Company. Used with permission of McGraw-Hill Book Company.)*

in algae. Evolution from simple filaments to complex three-dimensional tissues can be followed. Simple

**Figure 9-9** *Multicellular shapes and patterns of the algae can be derived from simple rules governing the direction of cleavage planes between each division during growth of organisms surrounded by rigid cellulose cell walls. Changes in the direction of cleavage planes between divisions, in the frequency of division, its duration, and the distribution of division activity within the organism, will produce a variety of filamentous, sheetlike, and solid patterns. [From John Tyler Bonner, Size and Cycle: An Essay on the Structure of Biology (copyright © 1965 by Princeton University Press), Fig. 4, p. 37. Reprinted by permission of Princeton University Press.]*

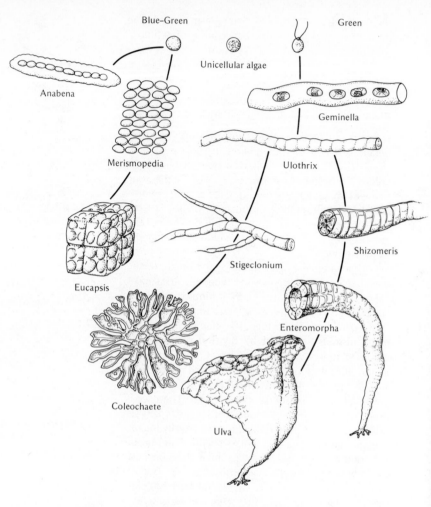

filament organization in the blue-green algae *Anabena* (Figure 9-9) depends on the confining jelly sheath and the parallel orientation of the spindle axes at each division. As long as the spindle axes are parallel, only simple filaments are generated. A filament is converted into gelatinous sheets, however, when the plane of cell division changes. A two-dimensional sheet of another blue-green algae, *Merismeopedia*, acquires its symmetry because division is synchronous, spindle axes are regular, all in one plane, and each successive division is at right angles to the previous one. A sheetlike colony grows in the surrounding jelly with spatial patterning dependent upon the orientation of successive division planes. The adhesive jelly keeps cells together in a loose colony and preserves the division history of each cell.

A three-dimensional pattern is the next step; blue-green algae form perfect boxes or cubes, like *Eucapsis*. Here successive planes of division occur at right angles to one another, occupying all three dimensions.

Modifications of the basic filament plan are introduced in the more complex green algae, and new symmetries arise. The usual life cycle starts with a motile or nonmotile spore which germinates by successive divisions into a small filament. For many species this simple filament is retained into the adult form, but in others, planes of cell division reorient during filament development, and a thicker sheet appears, as in the two-layered fronds typical of many seaweeds like *Ulva*. At some point in development a new set of division planes appears parallel to the surface to form a second layer; all subsequent divisions in both layers proceed perpendicular to the main axis of the plant. The shape of some of the larger green algae is determined primarily by this pattern of cell division. On the other hand, if spindle axes continuously orient parallel to the surface of the plant, a hollow area will form within the center of the filament to produce a cylinder from a two-dimensional sheet. Variations of this cell-division pattern may generate symmetrical branching.

Branching is frequent and regular in green algae, the branch point usually arising where the spindle axis in one dividing cell is at some angle from the main axis of the filament. Immediately following this aberrant division each division of the divergent daughter cell is parallel to the previous one, forming the branch filament. The important point is that aberrant divisions occur at regular intervals along the main axis of the plant, always followed by periods of straight filamentous growth. Secondary branches may appear, but these too show parallel filamentous growth before any new branches are made. The "decision" to branch is made at regular intervals, as if cells are being counted along the filament. Since each species shows a typical branching pattern, we assume branching is under genetic control. Since microtubules are responsible for orienting the spindle axis in plant cells, a shift in division planes may be induced by a change in microtubule orientation. Such reorientations could be signaled by diffusible factors coming from other cells within the filament, a "supercellular" signal controlling branching.

The best hypothesis to explain branching is the presence of a physiological gradient along the filament. Cells recognize their position in the gradient, responding at those points that favor a shift in spindle orientation. Presumably such physiological gradients are created at the first division of a zygote or spore as is *Fucus* (see Chapter 8). Two attached cells convert the initial polarity of a zygote into a gradient system because the first division is usually differential, with a smaller apical cell retaining its high mitotic potential while the large cell begins to elongate and mature. An apical region remains as "pacemaker," a peak of growth activity, which grades off along the

filament to older, more slowly proliferating cells. Rates of photosynthesis or cell wall elongation may form several linear gradients along the filament, some with opposite polarities. Each cell is physiologically unique, exhibiting rates of metabolic activity slightly different from those of its immediate neighbors. At periodic intervals in the filament, a critical combination of gradient properties sufficiently perturbs a cell to alter its division pattern. The periodicity may result from natural physiological oscillations, transmitted through the filament by pulses of chemical signals along the gradient. We see that an "organismic" factor determining a cell's position in the filament may specify its ultimate fate, a view first expressed by Vochting as early as 1877.

### Soft matrices and bacterial morphogenesis

What cellulose walls do for plant cells, soft mucopolysaccharide jellies do for bacteria and animal cells. Most animal cells secrete extracellular materials that cement cells into cohesive tissues. Matrix secretions, by bringing cells into a common extracellular environment, translate individual cell properties into those of a population; the multicellular group now behaves as an organism. Even among primitive bacteria, a jelly matrix converts a group of single cells into a multicellular system which exhibits reproductive behavior.

For example, among the myxobacteria, commonly known as the slime bacteria (Figure 9-10), large amounts of a mucopolysaccharide slimy matrix are secreted by numerous individual cells. Embedded in this matrix, a loose collection of independent cells carries out a program of multicellular morphogenesis.

*Chondromyces* is a species of myxobacteria found in animal dung. The first visible sign of the organism is a collection of yellow, tree-shaped fruiting bodies on the surface, containing oval cysts. The life cycle begins when the cysts germinate, and hundreds of individual rodlike bacteria emerge in a gelatinous matrix. Individual cells are motile, gliding slowly in a slimy secretion that adheres to the substratum. They possess no flagella or other visible locomotory organs. The rods do not move independently but have a tendency to aggregate and stream together in the matrix. Growth and cell division con-

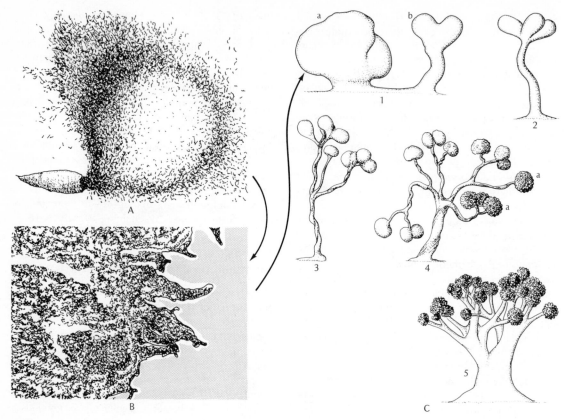

**Figure 9-10** *Life cycle of myxobacteria (*Coprinus *sp.). (A) Germinating cyst spilling out thousands of rodlike bacteria. These soon become embedded in a slimy secretion which migrates over a substratum. (B) A leading edge of such a gelatinous mass containing thousands of multiplying bacteria. At the end of a short migratory stage and in response to environmental cues (nutrition, humidity, light), the mass enters the morphogenetic phase and produces fruiting bodies. (C) Sequence of fruiting-body production. Culmination of fruiting involves the emergence of vertical fruiting bodies in which the bacteria collect into numerous cysts at the ends. (From R. Thaxter, Bot. Gaz. 17:389–406, 1892.)*

tinue in these streams, and the mass increases in size, moving about on the surface as a jellylike clump. Movement of the mass is random and uncontrolled, surging forward in one direction and then returning in the opposite direction. Since growth continues within the mass, cell density increases, which may be an important factor in regulating the rate of movement of the mass and its state of expansion.

Just before the vegetative mass begins to fruit, movement patterns change. The rods aggregate into groups exhibiting oriented movements; the clump now behaves as a cohesive morphological unit

moving "purposely." The trigger for this change is not known but may be nutritional, such as a depletion of an amino acid. The "goal-directed" behavior is dependent on the matrix itself. Bonner showed that a slimy secretion, left as an adherent trail by one motile mass, orients the movement of subsequent masses of rods by a mechanism of contact guidance. For instance, if a small fragment is separated from a larger migrating mass and displaced on the agar surface, it migrates and fuses with the larger mass by following the gelatinous tracks laid down previously rather than by moving directly toward the large aggregate (Figure 9-11).

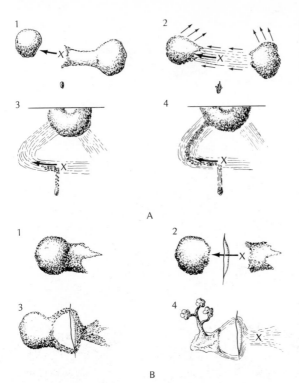

**Figure 9-11**  *Directed migration and coordinated behavior of myxobacterial masses. (A) Experiment showing how one mass or a small population of rods follows the musciladinous tracks laid down by other masses. (1) A center is moved from position X. In the region there is also a small ungerminated cyst below. (2) Cells in right-hand center move to one on the left while both begin to move upwards. The cyst has germinated. (3) Both centers fuse while cells from the cyst move toward the center, not by the most direct route but by following the tracks laid down by the major centers. (B) Experiment showing how aggregating rods avoid an obstacle to join a center. (1) The original population. (2) The center has been moved from X and a groove has been cut in the agar to separate the remaining adherent rods from the center. (3) The cells migrate to the center by going around the groove. (4) A normal fruiting body develops. [From John Tyler Bonner,* Morphogenesis: An Essay on Development *(copyright © 1952 by Princeton University Press): Fig. 62, p. 171, and Fig. 63, p. 172. Reprinted by permission of Princeton University Press.]*

Other factors besides contact guidance may also control movement. In another experiment Bonner separated a small mass from a larger one and cut a groove in the agar between them. The small mass returned to the larger mass, but as it approached the groove, it moved around it toward the central mass again, as if some chemotactic stimulus were directing cell movement since no gelatinous track had been laid down previously.

Evidently some social factor not found in individual cells organizes a group into a coherent unit. Only after completing directed movement can the cell mass transform into a fruiting structure.

When the clumps reach maturity, they round off into a compact knob which elongates off the substratum. Its apical end flares open into a funnel-shaped body wherein rods concentrate, all pointing in the direction of movement. The stalk consists largely of a hardening gelatinous matrix. The apical mass becomes lobed and indents into spherical cysts, creating a treelike structure.

This sequence of events is integrated by properties of the mucilage material enveloping the cell population. New morphogenetic principles emerge when many independent cells share a common extracellular matrix.

## Matrix materials and pattern formation

The matrix may define morphological patterns; that is, its mode of secretion determines the shape and symmetry of a cell colony. In some colonial Protozoa like *Zoothamnion* the branching pattern depends on an extracellular jelly secreted by each cell of the colony. Growth begins with a ciliated spore cell coming to rest on a surface and exuding an adhesive substance directed toward the substratum (Figure 9-12). The cell rises vertically by continuous secretion of a column of adhesive jelly. Shortly thereafter the cell divides into two daughters. Each new cell after division immediately begins to secrete its own column of jelly. Cell division establishes a branch point in the colony, and each division increases the bifurcations in the jelly matrix. A symmetrical branching colony emerges as each new cell generates its own adhesive stalk.

A pathway to multicellularity is cell replication within an extracellular matrix. The population becomes polarized because its increased size prevents free diffusion of substances. Chemical gradients are stabilized within the population and different regions exhibit autonomous morphogenetic behavior.

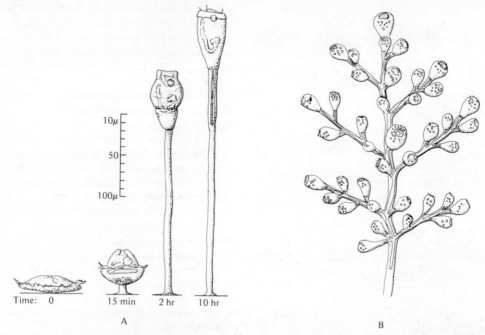

**Figure 9-12** *Extracellular matrices and the shape of Protist colonies. (A) Stages in development of a colony of Zoothamnium alternans, a colonial ciliate like Vorticella. Elevation from the surface is accomplished by secretion of a column of extracellular material. Within 10 hours an individual cell is fully extended and a "muscle" fiber has formed within the stem. (Based on E. Fauré-Fremiet,* Biol. Bull. *58:28–51, 1930.) (B) A fully grown colony of Zoothamnium arbuscula showing the branching pattern derived from the growth of extracellular matrix columns from each new individual. (From L. H. Hyman,* The Invertebrates—Protozoa through Ctenophora. *Copyright © 1940 by McGraw-Hill Book Company. Used with permission of McGraw-Hill Book Company.)*

Because of these polarities, regions also exhibit differential responses to exogenous stimuli.

## CHEMOTACTIC-INDUCED AGGREGATION AND MORPHOGENESIS IN CELLULAR SLIME MOLDS

Another route to multicellularity is the aggregation of independent, motile cells into integrated communities. Though cell aggregation occurs spontaneously in the life cycle of some primitive groups of organisms, it is not a widespread phenomenon.

Two different aggregation pathways may be followed. The first involves chemotaxis, whereby a specific chemical stimulus summons cells to an aggregating center. In the second, random cell collisions in highly dense populations may lead to aggregation. For successful chemotactic aggregation,

cells must be motile, but more importantly they must respond to aggregating signals by directed movement. Furthermore, and this is true for aggregation via random collision as well, motile cells must adhere and remain together after they make contact. To do so requires adhesive surfaces or the secretion of extracellular cements, or both.

The life cycle of the cellular slime molds displays almost an ideal system for analyzing the consequences of achieving multicellularity through chemotactic-induced aggregation of independent cells. Each of the requirements seem to be met by sequential changes in the nature of cell surfaces.

### Life cycle of a slime mold

*Dictyostelium discoideum* is a cellular slime mold which exists in two functional states in its life cycle: a population of unicellular, vegetative amoebae and a multicellular pseudoplasmodium that develops into a fruiting body. All growth, DNA synthesis, and cell division are restricted to the vegetative phase, while morphogenesis, pattern formation, and cell differentiation occur during the multicellular phase. Development is separate from growth and is expressed only by the multicellular phase.

The life cycle of *D. discoideum* starts with a haploid spore shaped as a small capsule (Figure 9-13). On moist agar each spore splits during germination to liberate a single-celled amoeba which feeds on bacteria. The amoebae grow and divide repeatedly by binary fission, with generation times of 3 hours, yielding thousands of independent cells. Usually after the food supply is exhausted (or possibly as a result of accumulation of a cell division inhibitor in the medium), aggregation begins. At different sites in the agar plate a few amoebae begin to move toward a central point. Cells migrate independently and are incorporated into thickened, compact heaps or aggregating centers. Mobility is directional; pseudopodia orient toward the centers, the source of the chemotactic signal. Cells may aggregate at some distance from a center, forming streams of aggregating cells which in turn emit signals to attract other amoebae. In this species signaling spreads from the central mass to the cellular streams, acting as relay stations in aggregation.

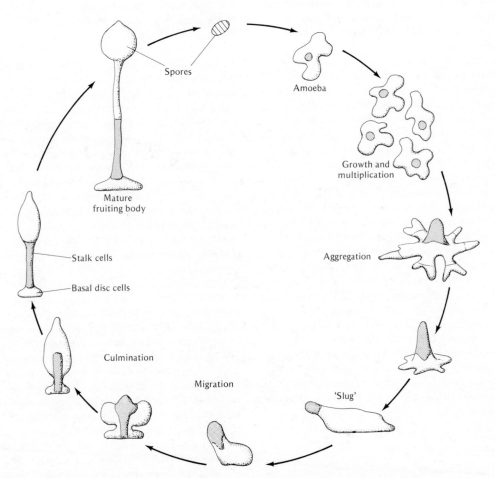

**Figure 9-13**  *Life cycle of the slime mold* Dictyostelium discoideum. *(From James D. Watson,* Molecular Biology of the Gene, *Third Edition, copyright © 1976, 1970, 1965 by W. A. Benjamin, Inc., Menlo Park, California.)*

An aggregate of *D. discoideum* may contain anywhere from a few hundred to hundreds of thousands of cells, depending upon the original density of the culture. Little cell division takes place during or after aggregation, which implies that size and shape of all multicellular structures depends on the original number of cells in the aggregate.

The aggregate soon elongates into a cigar-shaped slug, varying in length from 1.1 to 2 mm. The slug or *grex* glides over the substratum, leaving a slimy secretion or track behind, just like myxobacterial aggregates previously described. Movement may result from coordinated amoeboid activity of cells at the lower surface of the slug, moving along an exudate. The slug possesses an anterior-posterior polarity and responds to environmental stimuli such as light and temperature. It may migrate for a period of several hours before migration ceases and fruiting begins.

Soon after a slug forms, it contains two cell populations: anterior and posterior, or prestalk and prespore cells, respectively. The larger anterior cells, with large nuclei, develop into the stalk of the mature fruiting body, while the posterior cells become spores. Genetically identical cells differentiate into two distinct cell populations, perhaps because of the positions they occupy in a multicellular community. On the other hand, prespore and prestalk cells may have acquired their "biases" earlier during aggregation, and they sort out within the developing slug to their respective positions. In either case we see that cell position may determine cell fate.

During fruiting or culmination the slug extends into the air as an elongating papillalike structure. The large anterior prestalk cells become vacuolated and are displaced into the center of the mass. During elongation prestalk cells stream up on the outer surface to the apex and become trapped in the central core as stalk cells. The prespore cell mass rises simultaneously with the increase in height of the mature stalk. Prespore cells differentiate into mature spores at the midpoint of culmination, which ceases when all prestalk cells have been converted into mature stalk.

## Chemotaxis and aggregation

Many questions have been asked of this system. For example, if all amoebae arising from a single spore are genetically identical, what establishes the ag-

gregating centers, and how are cells attracted to them? Bonner's experiments in 1947 showed that chemotaxis was involved. For example, an aggregating center on a coverslip will stimulate amoebae on a neighboring coverslip to migrate toward the edge closest to the aggregating center; diffusion of a chemical stimulus from the aggregating center across the gap initiates directed movement. Likewise, if a piece of agar, which formerly contained an aggregate, is placed among vegetative amoebae, the amoebae stream toward the agar, collect there, and transform into a fruiting body. Even completed slugs and incipient fruiting bodies continue to secrete the chemotactic signal. If a slug is placed among populations of vegetative amoebae, they migrate toward the mass, most amoebae collecting at the apical end of the slug, which suggests that it is the source of the chemotactic signal.

The chemotactic substance was originally called *acrasin*, after the group *Acrasiales* in which its existence was demonstrated. We now know it is the simple nucleotide sugar cyclic AMP, the "second messenger" of hormone action (see Chapter 6). At low concentrations of cyclic AMP amoebae begin to aggregate. This substance acts on the amoeba cell surface, polarizing the direction of pseudopodial activity, thereby increasing the rate of directed movement. Normally, as amoebae move, pseudopodia seem to go out in all directions. In the presence of cyclic AMP, however, pseudopodia are directed toward the highest concentrations. Microcapillary applications of cyclic AMP to different points on the amoeba surface induce immediate projection of pseudopodia.

*D. discoideum* is one of only a few species that produce substantial amounts of cyclic AMP and employ it as an aggregating signal in development (Figure 9-14). Cyclic AMP is secreted at the beginning of aggregation and increases rapidly, reaching a peak at the end of aggregation after slugs have formed. This means that the cells possess the complete enzyme system for generating cyclic AMP and regulating its intracellular as well as extracellular concentration. In this organism both adenylcyclase and phosphodiesterase are membrane-bound.

An aggregation center arises from one vegetative cell, called the *founder cell*, which differs only slightly from its neighbors except for its active secretion of cyclic AMP. Any cell may become a founder cell and secrete cyclic AMP. There is a certain probability that some cells will acquire this active state; the larger the population, the greater the number of founder cells and the greater the number of aggregating centers.

Steep gradients of cyclic AMP are immediately established around founder cells. Founder cells also secrete phosphodiesterase, which transforms cyclic AMP to 5'-AMP near the cell surface. The transformation of cyclic AMP outside the cell always insures that its concentration is highest nearest the cell surface. Phosphodiesterase activity is itself governed by a protein inhibitor secreted by amoebae before aggregation. This system of enzymes, inhibitors, and chemotactic signals can be put into a model to explain aggregation. Cyclic AMP is assumed to direct the motility of vegetative amoebae.

During the growth phase phosphodiesterase activity increases, reaching a peak 1 to 2 hours before the end of growth. Accordingly, not enough cyclic AMP is around to initiate aggregation. At this point inhibitor secretion increases in some cells (the inhibitor may be calcium ion, which is known to inhibit diesterases), and the diesterase is inactivated. Cyclic AMP accumulates and is secreted by founder cells; centers are established as surrounding amoebae migrate toward founders. Pulses of cyclic AMP are generated by the centers, and sensitive cells respond by moving toward their neighbors into aggregating streams. Such cells also begin to generate pulses of cyclic AMP as the stream acquires a signal relay function and the chemotactic signal is amplified. It appears that the optimal signals for aggregation are pulses of cyclic AMP rather than steady concentrations. Signal pulses tend to propagate as waves of chemotactic activity over the aggregating population, and one sees surges of cell movement followed by short inactive pauses.

Aggregation is genetically controlled. Aggregate-less mutants have arisen spontaneously and are transmitted asexually through many generations of amoebae. Some mutants fail to aggregate because they synthesize low levels of cyclic AMP and do not produce aggregating centers. Others fail to respond to the chemotactic signal emanating from the centers; still others move abnormally.

An example of a nonresponding mutant is AGGR 39. Aggregation occurs in wild type because movement of vegetative amoebae is polarized by cyclic AMP; pseudopodial formation is restricted to one pole of the cell. On the other hand, AGGR 39

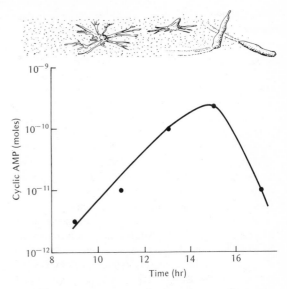

**Figure 9-14** *Changes in the amount of cyclic AMP secreted by aggregating amoebae. The peak activity is reached at the completion of aggregation, after amoebae have streamed into the aggregation center. It then falls off rapidly as the migrating slug moves off. The slug continues to produce cyclic AMP in a polar manner with the high point at the anterior end. (From Hormones in Social Amoebae and Mammals by J. T. Bonner. Copyright © 1969 by Scientific American, Inc. All rights reserved.)*

amoebae continuously form pseudopodia all over their surface; that is, their movements are not oriented in the presence of cyclic AMP. Being unresponsive to the chemotactic signal, they are incapable of polarization and directed migration.

Before aggregation occurs, cells must become responsive to the cyclic AMP signal and they must acquire adhesive properties to bind them into the aggregate. Both involve changes in the structure of cell surfaces.

Vegetative amoebae move actively during feeding and make frequent contacts without being contact-inhibited. During multiplicative growth cells do not cohere, nor do they respond to cyclic AMP. Growing cells require several hours to mature into aggregation-competent cells, and in that period between cessation of feeding and initiation of aggregation a change occurs at the cell membrane: cells become responsive to cyclic AMP and acquire cohesive properties. In both cases the surfaces seem to differentiate speicific  cyclic AMP receptor sites and contact sites, each regulated by cyclic AMP.

Perhaps the most important acquisition is the ability to bind cyclic AMP. It has been shown that cyclic AMP acts on the cell surface and not intracellularly as a second messenger. Radioactively labeled cyclic AMP binds specifically to amoebae surfaces toward the end of the growth period. Competence to bind cyclic AMP seems to depend upon a membrane-bound receptor; the binding kinetics resemble a hormone-receptor interaction (see Chapter 6). Besides, slime mold species that do not aggregate in response to cyclic AMP do not bind cyclic AMP, presumably because they lack the receptor.

The nature of this receptor is not known since its presence has been demonstrated only indirectly through binding studies. Some recent experiments suggest it may be a phosphorylated protein. For example, the addition of ATP to a suspension of preaggregative amoebae accelerates aggregation by 3 to 5 hours. During this same period there is also a marked stimulation of cyclic AMP binding to cell surfaces, which suggests that ATP is activating receptors. Two major proteins are phosphorylated at this time, which raises the possibility that ATP phosphorylation of receptors activates binding of cyclic AMP, thereby promoting aggregation.

It is not clear how cyclic AMP polarizes the motile behavior of amoebae. Cells move by amoeboid motion, a process of ectoplasmic gel contraction mediated, probably, by actin microfilaments. Most evidence suggests that the cell surface in amoeboid movement is polarized, inasmuch as new membrane is made at the pseudopodial tip by a process of exocytosis (the fusion of Golgi vesicles with the plasma membrane), while at the posterior, contracting end the reverse process of endocytosis is taking place whereby small vesicles are pinched off from the membrane to enter the cytoplasm. Both these processes are known to be affected by cyclic AMP and intracellular calcium in other cell systems. It may be exercising a similar effect on membrane motility in aggregating amoebae.

## Adhesive changes in the cell surface

The second event that takes place in the surface of preaggregative amoebae is the acquisition of cohesive properties, the capability of making stable connections with other amoebae during aggregation. As we have seen, subtle changes occur in the surface during growth. For example, young amoebae treated with the chelating agent EDTA, which binds calcium ions, will not aggregate. Older amoebae, similarly treated, will aggregate, presumably independently of divalent cations. This suggests that during the maturation phase before aggregation new molecular assemblies for adhesion are synthesized at the surface, a synthesis requiring $Ca^{++}$ and $Mg^{++}$ ions. Once these appear, cell cohesion occurs even in their absence.

What distinguishes the surfaces of old from those of young amoebae? Are new chemical constituents inserted in the surface during maturation, or do membranes reorganize and expose preexisting adhesive sites? These questions have not yet been satisfactorily resolved. Membranes of old amoebae contain glycoproteins not found in young cells. These substances may be part of the adhesive mechanism, but their presence in the surface does not preclude membrane reorganization (a reshuffling of membrane components into new patterns) as the underlying mechanism for maturation of adhesive sites.

There are indications that new surface antigens appear in aggregating cells. Antibodies made against aggregation-competent cells will inhibit cohesion of these cells even if the antibody molecule is made univalent by fragmentation. In other words, this one-armed antibody combines with an antigenic site on such cells and prevents cohesion without affecting cell motility or response to the chemotactic signal.

The surface of aggregation-competent cells is also structurally different from that of vegetative amoebae. A freeze-etch study of the plasma membrane of growing and aggregation-competent amoebae reveals that fewer but larger particles appear in the membranes of the latter. Large particles are about 100 Å in diameter compared to the smaller 60- to 70-Å particles. Moreover, vegetative amoebae treated with cyclic AMP do develop the larger membrane particles. Either smaller elements floating in the lipid bilayer fuse into larger units, or new units are inserted when an amoeba is competent to aggregate. Evidently, the surface of an aggregating amoeba is both structurally and functionally different from that of a feeding amoeba. Some specialized membrane components (receptors or contact sites?) play a role in integrating aggregating cells into a multicellular community.

## New morphogenetic information created by multicellularity

Once a slug is formed, a multicellular system is created with new developmental information. It exhibits a stable anterior-posterior polarity composed of two different cell populations: prespore cells posteriorly (about 80 percent of the population) and a smaller population of prestalk cells anteriorly. Cells in this system can "sense" that they are in contact with adjacent cells or when they have been disaggregated. They can also determine their position along the longitudinal axis and sense when that position changes. Finally, they can monitor the progress of the aggregate along a morphogenetic pathway.

Cells may move independently within the slug as it glides over the substratum, but they usually retain their relative position until culmination and fruiting. This positional stability is seen when vitally stained slug fragments are combined with colorless pieces (Figure 9-15). The remodeled slug wanders about, a sharp border marking each cell group during fruiting; there are no exchanges between stained and unstained portions.

Front and back regions of the migrating slug exhibit different behavior. For example, movement is polarized with the apical end always advancing first. It controls the rate and direction of migration. If a slug is cut in half and the posterior piece rotated 180° before the fragments are joined, the original polarity is retained in each half, and the posterior fragment separates, wanders off, and establishes a second fruit-

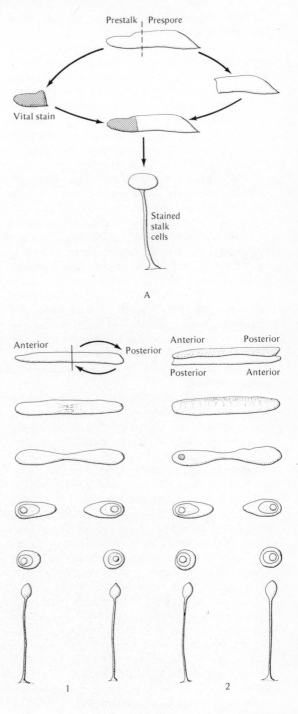

**Figure 9-15** *Polarity as a consequence of multicellularity in slime molds. (A) Spatial position of prespore and prestalk cells is fixed within the slug; no migration of cells between anterior and posterior ends occurs. A vitally stained anterior fragment grafted to a posterior fragment retains its position throughout slug migration and fruiting; all stained cells are found in the stalk. (B) Polarity of fragments is fixed. (1) Fusion of a rotated posterior fragment with an anterior fragment leads to a slug with a sharp wound surface in the center. In a short while the slug separates into two, each subsequently reorganizing into a normal but smaller fruiting body. (2) Fusion of two slugs in reversed polarity gives an identical result; the anterior ends of each seem to dominate the direction of migration and migrate off in opposite directions. (Based on J. T. Bonner, Biol. Bull 99:143–151, 1950.)*

ing body. Posterior fragments may be induced to fruit immediately, without further migration, forming almost normal fruiting bodies. Anterior pieces, on the other hand, form abnormal fruiting bodies consisting almost entirely of stalk cells. Timing seems to be important here since front pieces fruit in half the time under these conditions. If these are allowed to migrate for 24 hours, they will fruit normally and produce a normal population of spores. These fragments gradually regulate proper proportions of prespore to prestalk cells without any cell division.

How is polarity established? Inasmuch as differential cyclic AMP production persists in the slug, with the apical end secreting more than the posterior, it is postulated that polarity results from a gradient of cyclic AMP synthesis within the slug. The tip, a morphologically distinct nipple, seems to control morphogenesis and may do so by releasing periodic pulses of cyclic AMP. But this hypothesis fails to explain polarity of artificially prepared aggregates. Spherical aggregates derived from agitated cell suspensions remain spherical in liquid but become polarized into migrating slugs or directly into fruiting bodies when set out on a solid substratum. Polarity in this case may be induced by exposure to a substratum-air interface.

Now we ask; what accounts for determination of prestalk and prespore cells in the slug? By the late aggregation stage, even before a slug develops, prespore cells can be distinguished from prestalk cells. In the electron microscope the cells that become prespore cells contain large organelles called *prespore vacuoles* (PV) while presumptive prestalk cells lack them. These structures are not seen in vegetative amoebae; they appear only after aggregation. This suggests that mutual cell contact or interaction may be necessary for differentiation. Cyclic AMP alone does not induce PV formation in vegetative amoebae, nor does it stimulate PV formation in aggregating cells that have been dispersed into a single cell suspension. Such cells do possess the large membrane particles previously described, but they show no evidence of PV formation, except for those few cells in contact with other cells. Cell contact seems to be essential for the differentiation of PV. Contact, however, is a necessary but not a sufficient condition for PV formation; the large membrane particles are also important. If these particles are removed by concanavalin A, no prespore cells differentiate. It appears that large particles on opposing cell surfaces must interact for prespore cells to differentiate.

Inasmuch as Con A also affects the adenylcyclase system in cell membranes (see Chapter 5), the mechanism of Con A action here is not clear. In any case these contact interactions, seen only in the multicellular state, are essential for early differentiation and polarization of the migrating slug. Contact between complementary sites on adjacent cells may be important in cell interactions and may explain the ability of cells to sense their position in the population and recognize their neighbors.

We see that topological relationships dominate multicellular morphogenesis; small metabolic differences in local regions polarize the behavior of a cell mass. Some type of cell communication must explain the ability of pseudoplasmodia to regulate fruiting by controlling the proportions of prespore to prestalk cells. For instance, isolated slug fragments develop, after migration, into a complete fruiting body, adjusting cell number to give the normal proportion of one-third stalk cells to two-thirds spore cells. Even in extremely small mutant fruiting bodies, consisting of 12 to 15 cells, identical proportions of stalk cells to spores are maintained. All regulation takes place without mitotic activity. Evidently the position of a cell in the slug does not irreversibly determine its fate. Groups of cells are capable of readjusting their developmental behavior as conditions change. We do not understand how regulation occurs, but new spatial relationships set up in small fragments may reprogram cell fate.

Cyclic AMP may also be a diffusible signal for cell communication within the multicellular population. At low cyclic AMP concentrations and in the presence of inhibitors of protein synthesis, vegetative amoebae transform directly into stalk cells. Apparently they do not need new protein synthesis to differentiate as stalk cells, which suggests cyclic AMP is acting as an inducer. This is consistent with the fact that the apical end of the slug (the prestalk cell end) continues to secrete large amounts of cyclic AMP.

Stalk cells are also induced by exposure to lithium ions which promotes calcium uptake by cells. The prestalk region of the slug accumulates calcium ions; it is the high point of a calcium gradient along the longitudinal axis of the slug. Again, we see the synergistic action of cyclic AMP and calcium as regulatory molecules.

As culmination begins, biochemical differences between prespore and prestalk cells become more pronounced. Stalk and spore cells differentiate because two separate but coordinated biochemical programs are turned on. Cellulose is the primary chemical product of both programs; it is not synthesized in vegetative amoebae but appears only when the slug undergoes morphogenesis. This means that new enzymes for cellulose synthesis must be synthesized or preexisting enzymes must be activated. Indeed, as the intact slug culminates in fruiting, new proteins appear and old ones disappear. These new proteins include enzymes necessary for cellulose and mucopolysaccharide biosynthesis.

The sequence of gene-controlled events underlying fruiting is a developmental program expressed only after the multicellular population has become integrated (Figure 9-16). Evidently one of the consequences of multicellularity is the turning on of specific sets of genes for mucopolysaccharide synthesis. Some enzymes are synthesized early during aggregation (enzymes 1, 2, 3) while others appear late after

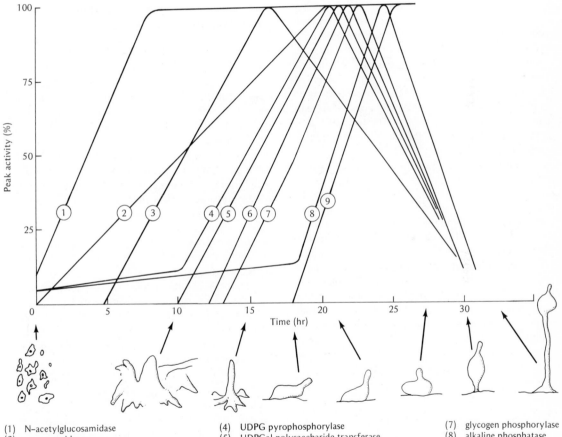

(1) N–acetylglucosamidase
(2) α–mannosidase
(3) trehalose–6–phosphate synthetase

(4) UDPG pyrophosphorylase
(5) UDPGal polysaccharide transferase
(6) UDPGal epimerase

(7) glycogen phosphorylase
(8) alkaline phosphatase
(9) β–glucosidase–2

**Figure 9-16** *Program of biochemical differentiation during development of* Dictyostelium discoideum. *The morphological stages are shown beneath the graph. All activities have been normalized to 100 percent. Note that some enzymes are found in vegetative amoebae at low concentrations (1 and 2, 4 and 8) while others appear during aggregation (3) The key enzymes involved in cell wall synthesis make their appearance after aggregation is completed (5,6,7,9). In most cases, enzymes reach a peak of activity, usually preceding a morphogenetic event, and then decline, either as a result of release into the medium or cessation of synthesis. (Based on J. M. Ashworth, Symp. Soc. Exp. Biol. 25:27–50, 1971.)*

the slug has developed (enzyme 9). Toward the end of aggregation, during plasmodium migration, small amounts of mucopolysaccharides are detectable in the slug. More accumulates as culmination begins at 20 to 25 hours, and a rapid synthesis is noted, reaching a peak when the spore capsule appears. Mucopolysaccharide accumulation coincides with differentiation of spore and stalk cell walls.

The appearance and disappearance of enzymes points to differential gene expression during a specific developmental program. Genes transcribe for a while and then stop when feedback signals turn them off. This pattern seems to be true for a number of different enzymes involved in polysaccharide metabolism. The pattern of gene regulation in these multicellular aggregates resembles the pattern underlying sporulation in prokaryotic bacteria. In both cases accurate timing of gene action is essential for normal morphogenesis. For instance, in some abnormal fruiting mutants stalk and spore cells differentiate at an accelerated rate because the critical enzyme appears many hours earlier than normal. The timing of gene expression is disturbed in the mutant and abnormal fruiting bodies develop.

Slime mold morphogenesis demonstrates, within a brief life cycle, a transition from an independent, unicellular state to the multicellular condition involving the creation of new systems properties that depend on surface membrane changes. Contact interactions within a cell population induce physiological differences in cell phenotypes, and new genetic programs are turned on. Though cells occupy different positions within the mass, their developmental pathways are coordinated through cell communication via membrane-bound transmitter and receptor molecules. The major rules governing morphogenesis of multicellular systems that relate to cell recognition, adhesion, and communication have been established in these primitive organisms.

## AGGREGATION IN SPONGES

As opposed to chemotaxis, aggregation of other systems results from random collisions between migrating cells and is dependent on cell density, cell motility, and adhesion. This pattern of random aggregation has been studied primarily in artificially dissociated tissues and was first observed in sponges by H. V. Wilson in 1912.

A sponge is a loosely organized association of cells surrounding a calcareous or siliceous skeleton (Figure 9-17). Several different cell types (flagellated choanocytes, spicule-forming cells, and so on), embedded in a matrix, are integrated into a primitive multicellular organism, intermediate between the colonial protozoa and the higher metazoa.

H. V. Wilson was one of the first to cut up pieces of the adult red sponge *Microciona* and squeeze them through fine bolting silk into seawater. This breaks up the sponge and produces a suspension of cell fragments, individual cells, and small cell clusters. Cells adhere to the substratum and migrate actively until they come into contact with one another, whereupon they form small clusters. Such clusters continue to move about actively, sweeping up other cells and clusters with which they come in contact. As a result, aggregates fuse into a few large masses. In a few days they flatten and differentiate spicules and canals with flagellated chambers aligned with choanocytes. Within a period of weeks elaborate branching occurs, an opening or osculum appears, and a small sponge develops.

This process of reconstitution resembles slime mold aggregation only in part. First, no chemotactic signal brings the cells together. As the cells migrate on the substratum, they meet at random and adhere immediately, forming large aggregates. An important point, however, is that in contrast to slime mold amoebae, sponge cells are adhesive from the very beginning, and simple contact is sufficient to form a cohesive aggregate.

Sponges exhibit species specificity in aggregation. In mixed cell suspensions of the red species *Microciona* with the blue-pigmented species *Haliclona*, only cells of the same species adhere to form morphogenetically competent aggregates. Initially mixed aggregates do form, but these cannot complete morphogenesis. Cells of the same species sort out into independent aggregates because they form more stable adhesive contacts with one another than they do with cells of the opposite species, and only these species-specific aggregates undergo morphogenesis.

Some idea of the nature of this adhesive property has come from the experiments of Humphreys and Moscona (Figure 9-18). They found that mechanically separated sponge cells reconstitute normally and at any temperature. On the other hand, if sponge fragments are dissociated chemically, that is, with calcium- and magnesium-free seawater, then reconsti-

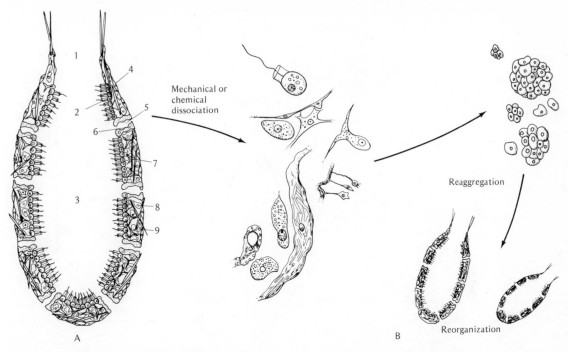

**Figure 9-17** *Sponge cells and reaggregation. (A) Cross section of a simple asconoid sponge to show the arrangement of tissue layers. (1) Osculum through which water passes. (2) Layer of choanocytes or collar cells. (3) Spongocoel or feeding cavity. (4) Epidermis. (5) Pore through porocyte; water entrance channel. (6) Porocyte. (7) Mesenchyme. (8) Amoebocyte. (9) Spicule, a portion of the sponge skeleton. (B) When sponges are squeezed through bolting cloth or dissociated chemically in Ca- and Mg-free seawater, the tissues separate into a suspension of individual cells and clumps of cells, as shown. These motile cells aggregate at random into larger motile clumps which in turn increase in size as more cells are randomly collected into the aggregates. Within a short time, the aggregates reorganize into small sponges. (Based on L. H. Hyman,* The Invertebrates—Protozoa through Ctenophora. *Copyright © 1940 by McGraw-Hill Book Company. Used with permission of McGraw-Hill Book Company.)*

tution takes place only after calcium and magnesium are restored to the medium at room temperature, but they do not reconstitute at 5°C. Apparently chemical dissociation removes an aggregation-promoting factor that reappears only as cells reaggregate at higher temperature. Mechanical disruption, on the other hand, does not eliminate this important factor.

The aggregation factor dislodged by chemical treatment is in the supernatant fluid. When this fluid is restored to chemically dissociated cells at 5°C, cell reaggregation occurs. Significantly, supernatant fluid from chemically dissociated *Microciona* promotes only *Microciona* reaggregation without any effect on *Haliclona*. Moreover, in mixtures of chemically dissociated cells, *Microciona* cells aggregate and reconstitute masses only in the presence of their own supernatant fluid, segregating from the surrounding

*Haliclona* cells, which do not aggregate. Cell recognition of specific cell types during aggregation seems to reside in an isolatable chemical entity.

The factor stimulating aggregation seems to be a surface component, most likely a glycoprotein. An active aggregation promoting factor has been isolated from supernatant fluids obtained from suspensions of dissociated *Microciona partheia*. This partially purified factor consists of glycoprotein (about 50 percent protein and 50 percent carbohydrate) and appears in the electron microscope as a large complex molecule made up of a 800-Å circle of fibers to which are attached numerous extended arms about 1100 Å long. The factor is degraded if calcium is removed, and in high calcium concentrations it tends to form gels. It specifically promotes aggregation of *Microciona* cells but not *Haliclona*.

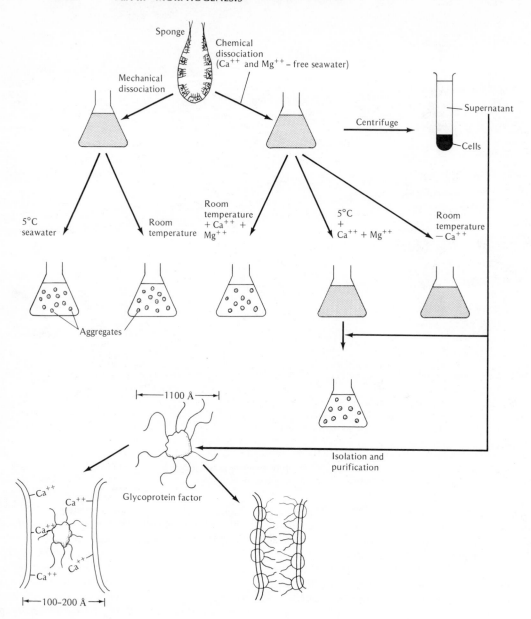

**Figure 9-18**  *Differences between mechanically and chemically dissociated sponge cells. A sponge dissociated mechanically through silk bolting cloth leads to a cell suspension which reaggregates at all temperatures including 5°C. Chemically dissociated sponge cells (Ca++- and Mg++-free seawater) aggregate only at room temperature and not at 5°C when restored to a medium containing Ca and Mg. They will aggregate at 5°C if the supernatant fluid from chemically dissociated cells is added. Isolation and purification of the factor in the supernatant fluid yields a glycoprotein shown in the center, about 800 Å in diameter and having numerous arms. This factor may act as an extracellular ligand between cells, possibly binding at Ca++ sites in the neighboring cell membranes, or it may be a constituent of the cell membrane itself.*

So far the role of calcium in this system is not understood, but any model for cellular adhesion and multicellularity in sponges must be based on a three-component system: an extracellular ligand or cementing substance (the glycoprotein factor cited in the experiments above); a binding site on the cell surface, perhaps a protein which specifically interacts with the factor; and divalent cations. When all three components are present, specific adhesions can occur and aggregation results. Removal of any one prevents adhesion. At high temperature chemically dissociated cells regenerate the membrane ligand and the extracellular cement. At low temperatures, they can do neither but will reconstitute if factor is added since this is the only component removed by chemical dissociation. The divalent cation may serve as a covalent bridge between extracellular factor and binding site on the cell surface.

Intimate cell contacts established within the aggregate form an integrated system with new properties, and a regular developmental program unfolds. As in the slime mold, spatial-temporal factors take over cell fate. Polarity is established, inside-outside gradients are formed, and cells respond to their position by following different developmental pathways to generate a small sponge.

Multicellularity in this experimental system is acquired through random collision of cells followed by specific cell adhesions. Cell aggregation depends on species-specific binding surfaces either in the membranes themselves or between membranes in the form of intercellular cements. Similar adhesive phenomena may account for morphogenesis of cell populations in higher organisms.

## INTEGRATION IN SIMPLE CELLULAR COMMUNITIES

One prerequisite for multicellularity is cell communication. Whether by membrane contacts or through extracellular matrices, cells do exchange developmental information. This accounts for "organismic" properties as distinct from cellular properties. The multicellular mass now behaves as a whole, particularly in reproduction.

We have seen a low level of integration in the simple myxobacteria, in which a mucilaginous secretion surrounding individual rods transforms them into a multicellular fruiting body. We do not know how cells communicate here, but the matrix clearly plays an important part. In more complex cellular slime molds and sponges spatial factors polarize cell populations, and different cell phenotypes differentiate. Communication between different cells opens up new avenues of morphogenetic interaction. How is cell communication mediated, and how does it integrate independent cell behavior into that of a multicellular community?

When morphogenetic behavior in one region harmonizes with cell activity elsewhere, then purposeful, whole organismic behavior is seen. Cell communication implies that cells produce and respond to informational signals; their surfaces are organized to send and receive. Communication can take place at a distance via diffusion of chemicals or hormones, as in chemotactic aggregation of cellular slime molds, or it may even take place within extracellular matrices. More often, after cells form cohesive aggregates, they may communicate directly at specialized points within their membranes. These membrane junctions will be mentioned briefly here and described in more detail in Chapter 10.

Specializations of the membrane, called *gap junctions*, allow free ion flow between adjacent cells, and their presence is assumed if electric current can pass from one cell interior to another. A narrow 20- to 30-Å gap between adjacent membranes marks the site of these junctions in the electron microscope. Current flow is facilitated by these specialized low-resistance channels, which means that ions and even larger molecules (under 1000 mol wt) can pass. Small regulatory molecules such as cyclic AMP could diffuse rapidly within a cell population and even establish gradients as in the slime mold slug.

Only cells that cohere strongly establish this type of junctional communication. After sponge cells are dissociated, for example, they show no ionic communication; their membranes exhibit low permeability. If the cells are brought into contact experimentally, they form a low-resistance junction which facilitates current flow, and within minutes cells establish ionic communication. This occurs when cells are brought together at random, suggesting that gap junctions develop anywhere and are not permanent features of the cell membrane.

The nature of the extracellular matrix surrounding a junction also controls the effectiveness of cell communication. When a *Haliclona* cell is brought into contact with a *Microciona* cell, no ionic communication occurs. These cells do not form stable cohesions, possibly because they do not use the same cementing substance. Chemically dissociated cells of a single species will not form junctions unless its specific aggregation factor is added to the medium. When the correct factor is furnished, good insulation is formed around the junction, and the cells communicate. If the insulation is allowed to leak calcium ions, however, communication breaks down.

Inasmuch as cells communicate through gap junctions, they can acquire information about the size of the population and their own position within the population. Both, we have seen, are essential factors of morphogenetic patterning in the slime molds. A cell population, behaving as a junctional system, has a finite volume, bounded at the periphery by its nonpermeable membranes and extracellular insulation. Thus a diffusible chemical signal in such a system is volume-dependent and can be equated with a change in concentration of a diffusible substance. Any cell within the population can determine population size by sensing the decrease in concentration of the signal as it is diluted by an increasing cell mass.

Information for position is also a function of a finite volume; the local concentration of a substance depends upon the position of a cell emitting and sensing a signal. In this way local concentrations sensed at different times can be used to determine position. A centrally located cell in a chain of cells will detect an earlier decline in concentration of the signal than will a peripheral cell. If both are producing equal bursts of signals at different times, the concentration falls more rapidly in the central cell. If a pacemaker cell appears, then position is determined by sensing the time and concentration of the signal coming from the pacemaker.

These two important parameters of a cell population determine polarity in large aggregates and contribute to pattern formation (see Chapter 16). The important point here is that the formation of multicellular aggregates leads to functional integration of cells and the formation of goal-directed, whole-organism behavior. The transition from independently acting cells to a cell community at the tissue level of organization is made possible, in part, by junctional communication over broad expanses of tissue.

## SEXUAL REPRODUCTION AND CELL SPECIALIZATION

One of the consequences of multicellularity is that cells specialize; they can carry out specific functions and thereby utilize cell machinery and resources more efficiently. Cell specialization, however, relies on cell communication, not only locally between adjacent cells but over greater distances — between "outside" and "inside" cells. The behavior of one group of cells must coordinate with another, and this is usually accomplished by diffusible chemical signals or hormones.

In the evolution of multicellularity, differentiation of reproductive from somatic cells was probably the first step in cell specialization. A mechanism must have evolved to segregate these two populations at a certain time in the life cycle so that reproductive cells could detach from the parental body and establish a new individual.

In one group of colonial green algae, the volvocales, we can follow the evolution of reproductive from somatic functions. The colonies of most species are composed of a finite number of cells, and once formed, no new cell divisions take place until a reproductive phase is initiated. Such integrated colonies, or *coenobia,* are usually organized as hollow spheres of cells embedded in a jelly matrix. In the most primitive species individual flagellated cells are held together haphazardly by fine protoplasmic threads and exhibit limited intercellular communication.

The simplest coenobium is *Gonium,* a flat plate of four central and twelve peripheral cells embedded in a matrix (Figure 9-19). Only the peripheral cells are flagellated, an indication of an inside-outside polarity. In this species all cells behave alike in asexual and sexual reproduction; each is able to develop into a smaller colony.

*Eudorina* shows a slightly more complex arrangement, a hollow sphere of 32 flagellated cells embedded at the periphery of a matrix. Four anterior cells divide more slowly than the others; they remain vegetative or behave asexually and form only daughter coenobia. All others exhibit both asexual and sexual reproduction. One of the first evolutionary changes from a colony to an organism is the loss by a few cells of sexual reproductive potentialities.

In a related species, *Pleodorina,* with 128 cells, half the cells in the anterior portion of the coenobium are purely vegetative, having lost both sexual and asexual reproductive powers. These anterior cells are smaller than reproductive cells and cannot divide even if isolated in a nutrient medium. Only posterior cells retain proliferative potential and develop asexually into daughter coenobia within the body of the mother.

Examples of the most complex colonial organization are the large spherical colonies of *Volvox* containing 1500 to 20,000 small flagellated cells. The life cycle of this group is controlled by hormones that coordinate asexual and sexual reproductive phases. Most importantly, however, these organisms demonstrate that factors controlling cell division and cell size are critical in determining reproductive versus somatic cell differentiation.

Most cells of *Volvox carteri* are vegetative, and only a few large cells—no more than 16—are reproductive gonidia. Gonidia are located in the periphery at the posterior half of the spheroid just beneath the vegetative cells which are in direct communication by a network of fine protoplasmic strands. All cells including gonidia are haploid.

A gonidium cell begins to develop asexually within the parental body while the parent is still young (Figure 9-20). The cell enlarges and then divides synchronously, very much like a spirally cleaving egg into 8- and 16-cell stages (see Chapter 11). At the 32-cell stage differentiation of vegetative from asexual reproductive gonidial cells occurs, as 16 cells at the anterior half undergo an unequal division, the larger cell becoming the gonidial initial and the smaller a somatic cell initial. Meanwhile, all posterior cells divide equally. All 16 gonidial initials stop dividing at this stage. Somatic cells continue to divide asynchronously, becoming smaller with each division until a hollow ball forms with a finite cell number (about 3000). These somatic cells then lose the capacity to divide. Gonidia which have not divided appear larger and more distant from each other; they always retain their original position within the anterior half of the embryo. At this point the sphere undergoes a dramatic morphological change as it turns inside out on itself (gastrulates?), exposing polarized somatic cells to the outside where they now develop flagella.

Daughter colonies rotate freely within the parent colony. The parent soon dies, releasing 16 young colonies to begin independent existence, each with 16 gonidial cells. They grow by secretion of matrix,

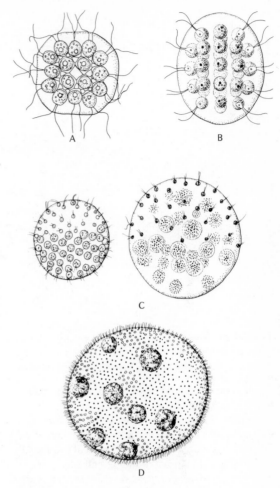

**Figure 9-19** *Simple and complex volvocales. (A) Typical colony of* Gonium pectorale. *(B) Mature colony of* Eudorina elegans. *(C) Mature colony of* Pleodorina californica. *Individual on right is undergoing asexual colony formation. (D) Asexual spheroid of* Volvox *containing 8 large vacuolate gonidia. (From G. M. Smith,* Fresh Water Algae of the U.S. *Copyright © 1933, 1950 by McGraw-Hill Book Company. Used with permission of McGraw-Hill Book Company.)*

rather than by cell enlargement or cell division, filling the space between cells with jelly until large spheroids are produced. As they grow, a new cycle of gonidial development begins within each daughter colony.

A sexual phase may be hormonally induced in *Volvox*. Male and female strains are separate, each developing asexually from a gonidium. Sexual dif-

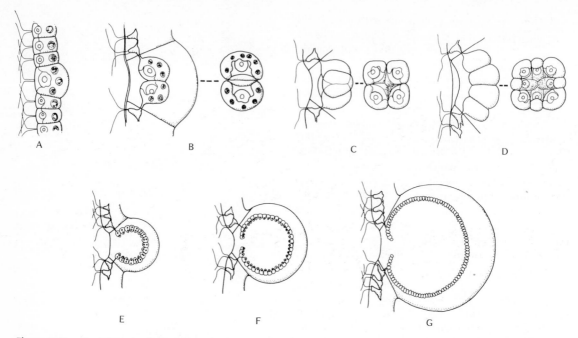

**Figure 9-20** *Gonidium development in* Volvox. *(A) Gonidium and somatic cells of a young colony. (B) Side and surface views of gonidium at 2-cell stage. (C, D) The same of 4- and 8-cell stages. (E, F, G) Sections of the developing embryo before inversion. (From M. A. Pocock,* Ann. S. African Museum *16:523–646, 1933.)*

ferentiation is induced by exposing a strain (male or female) to a "hormone" derived from sexually mature males. All gonidia in male coenobia differentiate into male spheroids when exposed to the "hormone." The early cleavage divisions are identical to asexually developing gonidia, but at the 32-cell stage, in contrast to asexual development, no differential divisions occur; all cells continue to divide equally until 256 cells are produced. In response to the hormone the division program is altered. It is the last division resulting in 512 cells that is unequal, and small somatic cells segregate from large sex cells in a ratio of one somatic cell to one androgonial cell. Within 24 hours the large androgonial cells undergo rapid cell division to form packets of 64 to 128 biflagellate sperm. The male spheroid (smaller than an asexual spheroid) then escapes from the parent.

The inducer substance also signals female strains to develop sexually. Gonidia in females develop into egg-bearing spheroids. Cleavage begins in the usual fashion until the 32-cell stage, and like males, the next division is equal, yielding 64 cells. The differentiating cell division occurs at the following division when 32 anterior cells and some posterior cells divide unequally, yielding 35 to 45 egg initials. These larger cells stop replicating while the remaining somatic cells develop into a typical spheroid, which is released from the mother colony. Again we see that differentiation of somatic from reproductive cells occurs after a differential cell division; the large cell in each case is the reproductive cell precursor.

Within a few hours after male spheroids are released, sperm packets are shed, penetrate the female spheroid, and dissociate into individual naked biflagellate sperm which fertilize the eggs. Within 3 to 4 days the zygotes, still enclosed in the female spheroid, develop pigment, secrete a thick wall, and become dormant. On germination the zygote undergoes meiosis, producing only one haploid biflagellate zoospore which swims for a short time and then undergoes cleavage to form a small germling individual with eight gonidia.

Developmental mutants have been obtained which interfere with sexual differentiation. Some mutants (R-1) have fewer androgonia—a ratio of six or more somatic cells to one androgonium, rather than a one-to-one ratio. The mutant gene changes the time of the unequal division to an earlier division, and fewer androgonia are produced. In a sense this is a mutation of a gene regulating the division behavior of cells, a case of a gene expressed phenotypically at the organismic level although it is manifested in individual cells.

The division behavior of other mutants is consistent with the hypothesis that cell size is a factor in the differentiation of reproductive from somatic cells. These mutants show premature cessation of cell division during embryonic development; they complete only 5 to 7 cleavage divisions instead of the normal 10 to 12 seen in wild type. The cleavages seem normal but stop prematurely to yield small colonies with only 64 to 128 cells. Differentiation of gonidial cells occurs after an unequal division, but proportionately many more gonidia develop; sometimes as many as 50 out of 64 cells are gonidia.

Perhaps the best hypothesis to explain the overproduction of reproductive cells in the mutant is that its limited division program produces many abnormally large cells which differentiate as reproductive rather than somatic cells. Cleavage cells in the mutant are 8 m$\mu$ rather than a normal 3 to 3.5 m$\mu$ at the end of division. And in the wild type only large cells derived from unequal divisions differentiate as asexual gonidia or as sexual reproductive cells. In mutants most somatic cells are large enough to differentiate into reproductive cells. Though other factors unrelated to cell mass may account for differentiation, it seems to be the simplest hypothesis.

These experiments show how genetic factors may control the evolution of colonial organization to multicellularity. Sets of regulatory genes controlling cell division seem to underlie all phases of morphogenesis in these forms. Asexual embryos differ from sexual ones only with respect to the time at which a critical unequal division occurs. More importantly, the only difference between males and females is the timing of differential division. The division is a determinative event since each daughter cell follows a different developmental program. Even the number of divisions is important since this determines the final size and may determine a cell's fate as somatic or reproductive. We will see again and again, in different morphogenetic systems, including those in higher organisms, that the timing of differential divisions is at the basis of cytodifferentiation. Perhaps the first step in the evolution of an organism from a colony is the ability to control the timing and location of these divisions in the life cycle.

## SUMMARY

We have considered three pathways to multicellularity: (1) conversion of a multinuclear syncytium into cellular compartments, (2) cementing of cells into organized groups by matrix secretions, and (3) aggregation of motile single cells into multicellular masses by chemotactic signals or random collisions. Once cell groups form, they are organized into a hierarchical system with new morphogenetic properties. Filaments or sheets of cells have spatial and gradient properties not found in single cells. The opportunities for morphogenesis increase as simple cell populations exploit these properties by transforming cell sheets and filaments into complex fruiting bodies.

The secretion of polysaccharide matrices seems to be one activity common to all pathways to multicellularity. All organisms including prokaryotes share metabolic pathways for polysaccharide matrix synthesis. Besides cementing cells into functional aggregates, the matrix itself plays a direct role in morphogenesis. The pattern of colonies depends on the manner of matrix secretion as well as specificity of some cell adhesions and contact behavior. Acquisition of metabolic systems for synthesizing extracellular polysaccharides is probably one of the most important steps in the evolution of cell social behavior.

Cells, as tissues, develop new rules of morphogenetic patterning. They acquire space-recognition properties and can change position and shape, thereby transforming a cell sheet into a tube or solid mass. The matrix also acts as a vehicle for intercellular communication. Additional morphogenetic possibilities come from cell interactions as they exchange regulatory molecules over short distances in a matrix. Cells also develop more intimate relationships through direct membrane contacts or junctional communication. As a result, the independent behavior of cells is coordinated into community behavior. Cell groups function as a whole during reproduction as they transform into specialized structures for spore or gamete production and release. Regulatory molecules are exchanged between cells that direct morphogenetic processes. Simple cell colonies become integrated organisms when they reproduce, and segregation of reproductive from somatic cells seems to depend upon previous division history. Differential divisions segregate large cells which retain reproductive potential from smaller somatic cells which lose mitotic potential. As organisms become larger, more complex tissues specialize to adapt to new physiological demands. Morphogenesis of cells into tissues and organs in higher organisms also utilizes matrix interactions, contact and adhesive behavior, cell motility, and shape changes. The next chapter will consider the level at which cells shape organs.

**REFERENCES**

Aldrich, H. C., and Gregg, J. H. 1973. Unit membrane structural changes following cell association in *Dictyostelium*. *Exp. Cell Res.* 81:407–412.

Bonner, J. 1952. *Morphogenesis*. Princeton, N.J.: Princeton University Press.

————. 1969. Slime mold aggregation. *Sci. Am.* 22:79.

Dworkin, M. 1973. Cell-cell interactions in the myxobacteria. *Symp. Soc. Gen. Microbiol.* Vol. 23. New York: Cambridge University Press, pp. 125–142.

Gerisch, G. 1968. Cell aggregation and differentiation in *Dictyostelium*. *Current topics in developmental biology*, ed. A. A. Moscona and A. Monroy. Vol. 3. New York: Academic Press, pp. 157–197.

Gerisch, G., Malchow, D., Huegen, A., Nanjundiah, V., Roos, W., and Wick, U. 1975. Cyclic AMP reception and cell recognition in *Dictyostelium discoideum*. *Developmental biology*, ed. D. McMahon and C. F. Fox. *ICN-UCLA Symp. Mol. Cell. Biol.* Vol. 2. Menlo Park, Calif.: Benjamin, pp. 76–88.

Henkart, P., Humphreys, S., and Humphreys, T. 1973. Characterization of a sponge aggregation factor; a unique proteoglycan complex. *Biochemistry* 12:3045–3050.

Humphreys, T. 1963. Chemical dissolution and in vitro reconstruction of sponge cell adhesions. Isolation and functional demonstration of the components involved. *Develop. Biol.* 8:27–47.

Konijn, T. M. 1972. Cyclic AMP and cell aggregation in the cellular slime mold. *Acta Protozool.* 11:137–143.

Loewenstein, W. R. 1968. Communication through cell junctions. Implications in growth control and differentiation. *Emergence of order in developing systems,* ed. M. Locke. New York: Academic Press, pp. 151–183.

Loomis, W. F. 1975. *Dictyostelium discoideum*. A developmental system. New York: Academic Press.

Matto, J. M., and Konijn, T. M. 1975. Enhanced cell aggregation in *Dictyostelium discoideum* by ATP activation of cyclic AMP receptors. *Develop. Biol.* 47:233–235.

Pall, M. L. 1975. Mutants of *Volvox* showing premature cessation of division: Evidence for a relationship between cell size and reproductive cell differentiation. *Developmental biology*, ed. D. McMahon and C. F. Fox. *ICN-UCLA Symp. Mol. Cell Biol.* Vol. 2. Menlo Park, Calif.: Benjamin, pp. 148–156.

Sampson, J. 1976. Cell patterning in migrating slugs of *Dictyostelium discoideum. J. Embryol. Exp. Morphol.* 36:663–668.

Sauer, H. W. 1973. Differentiation in *Physarum polycephalus. Microbial differentiation. Symp. Soc. Gen. Microbiol.* Vol. 23. New York: Cambridge University Press, pp. 375–405.

Sheridan, J. D. 1973. Functional evaluation of low resistance junctions: Influence of cell shape and size. *Am. Zool.* 13:1119–1128.

Starr, R. C. 1970. Control of differentiation in *Volvox. Changing syntheses in development,* ed. M. N. Runner. New York: Academic Press, pp. 59–100.

Truesdell, L. C., and Cantino, E. C. 1971. The induction and early events of germination in the zoospore of *Blastocladiella emersonii. Current topics in developmental biology,* ed. A. A. Moscona and A. Monroy. Vol. 6. New York: Academic Press, pp. 1–44.

Wilson, H. V. 1910. Development of sponges from dissociated tissue cells. *Bull. Bur. Fisheries* 30:1–30.

Yu, N. Y., and Gregg, J. H. 1975. Cell contact mediated differentiation in *Dictyostelium. Develop. Biol.* 47:310–318.

# Cell Surface in Morphogenesis

We have seen that specificity of cell adhesion is essential to multicellularity. Only cells of the same species cohere into morphogenetically active aggregates, whereas mixtures of cells from different species sort out according to species type before morphogenesis can begin. Cells recognize their neighbors; though they may contact and adhere temporarily to cells of a different species, they "prefer" to interact with their own kind.

In complex morphogenetic systems diversity of cell types arises during embryonic development; tissue specificity of adhesion and recognition emerge as the important feature of tissue and organ construction. Extensive cell rearrangements, migrations, and associations are required for tissue assembly. Cells born in one region detach and migrate independently to other target areas where they reaggregate selectively into tissue and organ primordia. They follow prescribed routes and "home in" on specific targets. Such perambulations call for cell surfaces that can recognize migration paths and tar-

get sites. As we have seen, the cell surface is important in cell division as well as morphogenesis. In both processes specific membrane components such as receptors, enzyme systems, or ion channels are involved. The surface is a complicated lipid bilayer in which many different protein and glycoprotein species float. Many of these proteins have regulatory functions; besides, their number and distribution in the bilayer seem to change continuously as the fluid membrane turns over. How does this dynamic structure acquire the ability to organize stable tissue fabrics? In this chapter we will examine the properties of the cell surface which seem to be essential in tissue morphogenesis. These include cell adhesion, cell recognition, and cell communication.

## THE NATURE OF THE CELL SURFACE

It is difficult to state precisely where a cell surface begins and ends. It is a complex structure consisting

223

**Figure 10-1**  *The outer surface of several different cells. (I) The amoeba cell surface with its lipoprotein membrane (M) cut in cross section, lying in a vertical plane normal to the page. It shows the trilaminate unit membrane appearance as seen in the electron microscope. Cytoplasm is represented by (C), and glycocalyx (G) extends to the right, attached to the membrane as a juxtamembranal layer. (II) Neurospora cell surface. The membrane (M) is highly folded, capable of movements in which the glycocalyx (G), or cell wall (W), does not participate. The cytoplasm (C) is seen under the membrane. (III) Internal and external surfaces of a vertebrate blood capillary endothelial cell. Cytoplasm in light stippling. Upper extracellular space E is the lumenal surface, separated from the cytoplasm by the membrane (M) which is invaginated into numerous pinocytotic vesicles on both surfaces (K, L, V). Glycocalyx (G) or basement membrane (basal lamina B) in dense strippling on both surfaces, thinner on lumenal surface than below. In basal lamina below are collagen fibrils. (IV) Surface of the erythrocyte. The membrane (M) separates cytoplasm (C) from extracellular space (E). The membrane has the typical trilaminate unit membrane appearance but no glycocalyx component is seen on the outer surface in the electron microscope, although it is known to exist because polysaccharides and sialic acids are detected by chemical and immunological methods. (From H. S. Bennett, Handbook of Molecular Cytology, ed. A. Lima-de-Faria. New York: American Elsevier, 1969, pp. 1261–1293.)*

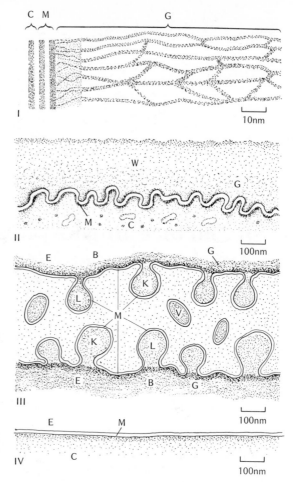

of more than the plasma membrane with its integral and peripheral proteins. It includes the extracellular materials secreted outside or adherent to the membrane (about 100 Å beyond the outer leaflet) that make up the *glycocalyx* and may also include subjacent microfilaments and microtubules attached to proteins floating in the bilayer (Figure 10-1). Outer surfaces show considerable variation, from the relatively clean red cell surface to the fuzzy surface of an amoeba, a diversity attributable to the structure of the glycocalyx. Glycoproteins are the principal constitutents of the glycocalyx. Even the simple erythrocyte surface is covered with glycoproteins (glycophorin) that emerge from the lipid bilayer and extend some distance to the exterior (Figure 10-2). The carbohydrates (sialic acid) in this molecule are attached like branches of a tree to a protein core that is partly embedded in lipid and partly in the cell interior, where it may attach to contractile molecules beneath the membrane. In more complex glyco-

calyces the glycoprotein "trees" may be so numerous as to form a "forest."

The outer membrane is not a static barrier separating the cell contents from the environment but is in active motion, thrown up into folds, microvilli, and ruffles. Different cells display different microvillar patterns. The surface folds fuse with one another, forming minute membrane vesicles that incorporate the surrounding fluid environment into vacuoles. The cell seems to be "drinking" its environment, a process known as *pinocytosis or endocytosis*. Large macromolecules such as proteins gain entry in this way; that is, substances are incorporated into a cell even though the membrane itself is impermeable to them. The movement of pinocytotic vesicles from the surface toward the interior is the

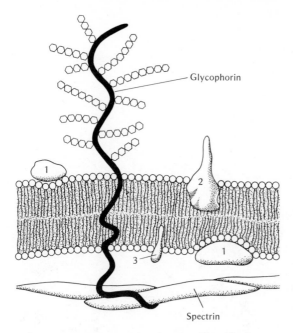

**Figure 10-2** *Surface of an erythrocyte. Glycophorin extends from interior to some distance from exterior. Other proteins (1, 2, 3) in membrane also shown. Attachment of glycophorin to spectrin is speculative. Spectrin is presumed to have contractile properties. (Reprinted with permission from V. T. Marchesi, The Structure and Orientation of a Membrane Protein, Hospital Practice, vol. 8, no. 6, and Cell Membranes: Biochemistry, Cell Biology and Pathology, ed. G. Weissmann and R. Claiborne. New York: HP Publishing Co., Inc., 1975.)*

reverse of secretion, or *exocytosis,* in which cytoplasmic vesicles containing secretory products (or membrane precursors) migrate to the plasma membrane and fuse with it to discharge their contents. We have here a two-way traffic of macromolecular transport and membrane turnover.

## "Protein icebergs in a sea of lipids"

The structure of the membrane has been described as "protein icebergs in a sea of lipids." This implies that lipids under physiological conditions exhibit fluid properties while the proteins are free to float in the plane of the membrane. Fluid properties of the lipid bilayer can be detected with fluorescent antibodies.

Antibodies prepared against human and mouse

cell surface antigens were tagged with dyes that fluoresce either red or green in the fluorescence microscope. Human cells treated with antihuman antibody appeared to be covered with a red fluorescence whereas antibody-treated mouse cells appeared green (Figure 10-3). Using the somatic cell fusion technique (see Chapter 3), mouse and human cell hybrids were made. Minutes after fusion, hybrid cells treated with both antibodies exhibited half red and green fluorescence, marking human and mouse cell membranes, respectively. Within 40 minutes at 37°C, however, the antigenic components of the membranes were completely intermixed; red and green fluorescence was evenly distributed over the entire surface. Apparently mouse and human cell membranes flowed together. The rate of intermixing was unaffected by inhibitors of protein synthesis or energy production; only low temperatures delayed reshuffling of membrane components. Intermixing is due to displacement of membrane antigens in a viscous lipid layer rather than synthesis and insertion of newly formed membrane components.

A fluid membrane also explains the polarized flow of other surface components such as antigen receptors in lymphocyte membranes. Lymphocytes are immune competent cells possessing immunoglobulin antibodies in their membranes. Antibodies can be made against these membrane-bound immunoglobulins; the antibodies can then be tagged with a fluorescent dye and made to react with the lymphocyte surface. Initially the surface is uniformly coated with fluorescent antibodies, but within 30 minutes these antibodies collect in bright patches and migrate to one pole of the cell as a bright fluorescent cap which is incorporated into the cell by pinocytosis. Cap formation does not occur at 0°C, a result to be expected if membrane fluidity is diminished at low temperatures. Proteins in the membrane are mobile under physiological conditions. Since lymphocytes treated with cytochalasin B do not exhibit capping, some have proposed that actinlike microfilaments beneath the cell membrane are attached to proteins floating in the lipid bilayer. Contraction of these filaments could displace proteins within the membrane and change their distribution on the surface. This conclusion is doubtful, however, because cytochalasin has many other effects on cell surface behavior besides disrupting microfilaments.

We have seen in Chapter 5 that Con A also af-

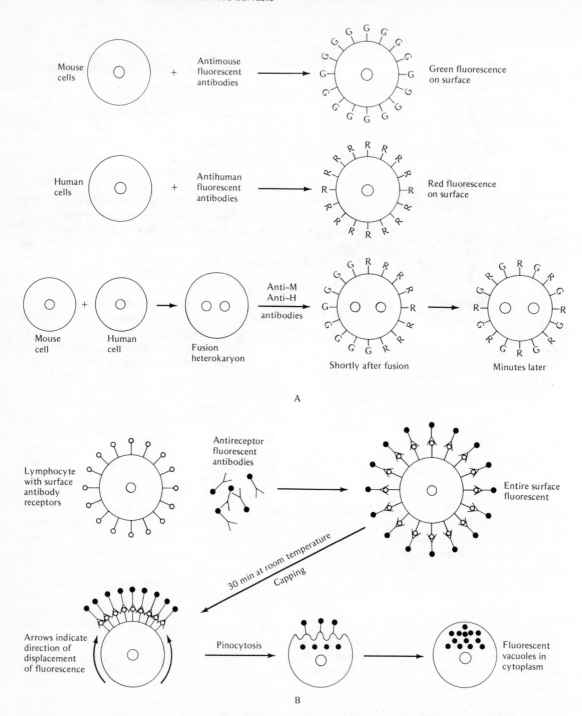

**Figure 10-3**  *Two experiments demonstrating fluid nature of plasma membrane. (A) Frye-Eddidin experiment on cell fusion. (B) Capping experiment. See text for explanation.*

fects membrane fluidity; it interferes with patch and cap formation in lymphocytes. The effect of Con A is inhibited by low temperatures and, more specifically, by colchicine. Since colchicine does not combine with Con A or displace it from lymphocyte surfaces, its principal site of action must be on microtubules. According to Edelman, many of these affects on membrane fluidity can be explained by a model of the cell surface that includes a scaffolding of microtubules beneath the membrane attached to proteins in the bilayer (Figure 10-4). According to this model, integral proteins are either bound or free, with bound proteins attached to the microtubule scaffolding. Presumably only free proteins are able to move in the bilayer; bound proteins can move only after they are released from microtubules, which occurs when microtubules disassemble. Thus membrane receptors are displaced and exhibit capping because microtubules, in a dynamic equilibrium between assembly and disassembly, are triggered to disassemble. Con A inhibits membrane mobility because it attaches to its surface receptors. Being a tetrameric molecule, it interacts with itself and produces extensive crosslinking over the surface. This stabilizes the underlying tubules so that they are less likely to dissociate. Since colchicine and low temperature dissociate tubules, they will release the receptors which are then free to move in the membrane. This model attempts to relate these fluid properties of the cell surface and its modulation by microtubules (and microfilaments) to many processes including cell division, cell interaction, and cell differentiation (see Chapter 5).

## Intercellular spaces

The membranes on cells in tissues do not abut on adjacent membranes but are usually separated by 100- to 200-Å spaces filled with an amorphous staining material.

Is this gap real, or is it an artifact of fixation, as many claim? Cells shrink after fixation and gaps may arise as extracellular material is released. But shrinkage cannot be the sole explanation for an extracellular space, since in the same electron micrograph, a 100-Å gap region may lie adjacent to regions where membranes are in direct contact between two cells. Furthermore, these spaces are present in living tissues. Large electron-dense molecules such as ferritin, 100 Å in diameter, infiltrate spaces between tissue cells if injected before tissues are fixed. No infiltration

Various types of receptors

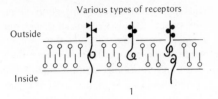

Equilibrium states of receptors

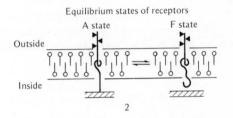

Immobilization of receptors

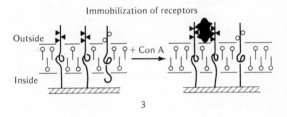

Colchicine effects

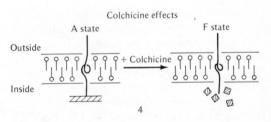

**Figure 10-4**  *Cell surface interactions with concanavalin A and colchicine in terms of a fluid mosaic model. (1) Various protein receptors in the lipid bilayer, some interacting with cytoplasmic elements inside directly or indirectly. (2) These latter proteins interact with colchicine binding proteins (tubulins?) in an equilibrium consisting of an anchored state (A) and a free state (F). (3) Cross linking of some receptors by Con A may alter the structural configuration of colchicine binding proteins and thereby change the A–F equilibria of other receptors; they are less easily displaced in the membrane. (4) Colchicine dissociates microtubules into subunits and also alters the A-F equilibria in the direction of the F state. (From G. Edelman,* Cell Surface in Development, *ed. A. A. Moscona. New York: Wiley, 1974.)*

occurs if tissues are fixed before the injection. Since large molecules penetrate only living tissues, we conclude that the gap is real and not a fixation artifact.

In the electron microscope the gap is not empty but is filled with an amorphous staining material, a common mucopolysaccharide matrix shared by adjacent cells. It is unlikely that this is a fixation artifact since many gaps are filled with long fibers, structures that would not be expected to leave cells during fixation. The reality of an extracellular matrix is important. If it did not exist, cells could not establish firm contacts between their surfaces when separated by 100- to 200-Å gaps.

### Membrane also a charged surface

The outer surfaces of most cells behave as if they have a *surface potential,* or a net negative charge. Because of their charged surfaces cells placed in an electric field will migrate at a rate proportional to the density of charge on the surface. The net surface potential is an average of the positive and negative charges distributed over the surface, and its magnitude varies in different cells. Since most cells have a net negative charge, they migrate toward the positive pole. This technique of cell electrophoresis permits comparisons of surface-charge densities on different cells.

Surface-charge density is not intrinsic to the cell, since cells are bathed in media containing hydroxyl and hydrogen ions plus many positively charged salt ions. These ions adsorb to the exposed charges on the cell surface and generate an electrostatic field, grading off in intensity from a high point near the cell membrane to zero at some distance from the surface. Thus cell mobility in an electric field is dependent upon medium pH and ionic strength in addition to surface potential.

Since all cells migrate to the anode, the net surface charge is negative. Therefore they would tend to repulse one another as they come together. They must overcome this electrostatic repulsive force in order to adhere.

The net charge density of the surface is determined largely by the outermost constituents of the plasma membrane. Accordingly, enzymatic removal of certain components (polysaccharides, proteins) or addition of reagents that react specifically with certain groups such as amino ($NH_3^+$) or carboxyl ($COO^-$) groups will alter cell mobility in an electric field.

Most surface potential is generated by negatively charged carboxyl groups. For example, after red cells are treated with trypsin, the net surface charge decreases because negatively charged COOH groups are eliminated. Glycoproteins, however, seem to contribute most to surface potential since the enzyme neuraminidase, an enzyme that liberates sialic acid from glycoproteins, greatly reduces charge density at the surface. Enzymatic release of sialic acid residues from human erythrocytes decreases net negative charge by 80 percent. We can assume that any factors which alter surface-charge density will also affect cell contact and adhesion.

## CELL ADHESION

Some simple physical models of particle interaction are relevant to cell surface behavior in adhesion. If we treat cells as charged spheres that behave like flocculating colloidal particles, then the repulsive and attractive forces that govern particle flocculation might also apply to cell adhesion.

Two similarly charged spherical particles generate a certain potential energy of repulsion, a force that varies with distance; the closer the particles the greater the force. Such particles will adhere when their potential energy through Brownian movement overcomes the repulsive force. We find, however, that particles larger than 2000 Å would not have enough potential energy to come closer than 60 Å, and particles the size of cells would experience even greater difficulties in overcoming the repulsion barrier. Hence Brownian movement could not be important in cell adhesion.

The attractive forces that overcome this barrier and bring cells into intimate contact at distances less than 40 Å may be the so-called weak Van der Waals–London forces that arise from interactions between dipolar atoms. Any dipolar atom such as water exhibits electromagnetic oscillations which generate weak attractive forces. Cell surfaces made up of similar atoms might expose many dipolar groups and establish weak electromagnetic attractions. These forces act best over short ranges, but long-range attractive forces may also arise.

Attractive and repulsive forces each vary with distance but in different ways. As a result, the total energy of interaction between two particles will also vary with distance, as shown in Figure 10-5. There-

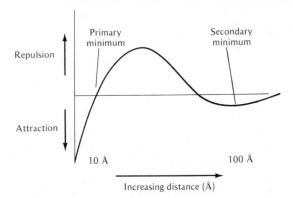

**Figure 10-5**   *Energy of interaction between particles as a function of distance. The secondary minimum at approximately 100 Å is a region where weak attractive forces are generated, perhaps sufficient to promote adhesion. (From A. S. C. Curtis, The Cell Surface. London: Logos Press, 1967.)*

fore the energy barrier to contact lies between two regions of attraction, a short range, or *primary minimum,* and a long range of weak attractive forces, or *secondary minimum.* Attractive forces are maximal at these distances. Anything that reduces repulsive forces, particularly in the region of the primary minimum where attractive forces are high, should promote close cell contact. The existence of a secondary minimum suggests that cells at some distance might develop weak attractive forces.

Can we determine which distances between cells permit effective contact on the basis of these formulations? Probably not, because we cannot define the true outer boundary of a cell. A glycocalyx surrounds the outer layer of the plasma membrane, and a field of adsorbed ions from the medium extends from the cell periphery as a diffuse zone. In addition, cells are not true spheres but are covered with numerous fine microvilli which may interdigitate with those of opposing cells before their large surfaces interact. All factors considered, it can be shown that repulsive and attractive forces are balanced at a distance of between 3 and 6 Å (the primary minimum). Accordingly, cells should make the most stable contacts at or below these distances.

Curtis proposed that cells may also make stable attractions in the region of the secondary minimum. Adhesion here is more labile because of the gap between surfaces and the weaker attractive forces. Both types of adhesion, however, are nonspecific. Any two particles having the same dielectric constants, immersed in the same medium, will adhere if repulsive forces are overcome. At physiological conditions Curtis calculated that cells would generate stabilization energies large enough to keep them together at the 100- to 200-Å gap seen between tissue cells. The existence of extracellular matrices and the fact that cells are subjected to the potential energies of Brownian movement opposing attractive forces make adhesion in the secondary minimum theoretically more difficult, however.

A glycoprotein layer up to 100 to 200 Å thick on the cell surface minimizes the significance of the gap between cells. Initial contacts could be made between glycoproteins extending at distances up to hundreds of ångströms from the surface by mechanisms unrelated to forces of repulsion and attraction.

## Microvilli and cell adhesion

A physical treatment of cells as colloidal particles fails to take many factors into account, such as contractile properties of the cell surface, the ordering properties of water around protein molecules, the existence of extracellular matrices, the role of calcium in decreasing surface-charge densities, and the microvillar nature of most cell surfaces. Cells are not inanimate spherical particles; their surfaces are thrown into numerous fine cytoplasmic extensions, folds, and ruffles. These are the first to "reach out" and make contact with other cells, and the forces operating at the tips of microvilli differ from those operating between smooth spheres. The large radius of curvature at fine microvillar tips can account for attractive forces far in excess of those calculated for spherical surfaces. In fact it is likely that all free surfaces of cells are covered with these cytoplasmic extensions that are filled with a network of fine microfilaments (see Chapter 8). The rapidity with which these structures appear on the surface and change shape suggests their activity is controlled by a contractile system.

## Surface potential and cell adhesion

As cells come together, we would expect their negative surfaces to repulse one another and prevent contact. Moreover, cells with high net surface potentials should be less adhesive. Is there any evidence that surface potential does in fact affect adhesive behavior?

One study carried out by Katsuma Dan as early as 1936 showed that the adhesiveness of echinoderm eggs to glass is indirectly related to surface-charge potential; as net negative surface charge increases, eggs become less adhesive. In addition, tumor cells with high surface potential are less adhesive than normal cells, a property that may account for tumor cell invasiveness and metastasis (Chapter 21).

A reduction in surface potential may increase cell adhesiveness. If the high negative charge on tumor cell surfaces is reduced by addition of positively charged substances such as polylysine, the cells behave as a cohesive, organized tissue, and when combined with normal tissues in vitro, they do not invade them.

Presumably removal of anionic charged groups of glycoproteins from the cell surface should reduce net negative charge and increase cell adhesiveness. Enzyme removal of sialic acid residues from the surface of different cells has led to conflicting results. For example, adhesiveness does increase after blood platelets are treated with neuraminidase. On the other hand, trypsin-disaggregated chick neural retinal cells do not become more adhesive after surface negativity is reduced.

Removal of positively charged ions ($Ca^{++}$ and $Mg^{++}$), decrease in ionic strength, or a rise in pH all increase surface-charge potential and reduce cell adhesion. This is consistent with the hypothesis, and tissues exposed to such treatments usually dissociate into independent cells. The calcium chelating agent, EDTA, is an effective reagent in dissociating tissues. But all cations do not have the same effect on cell adhesion, in spite of having equivalent charge properties. Calcium is better than magnesium, while both are more effective than barium or strontium. This means that reduction of surface negativity by Ca ions is not its primary role in adhesion.

The adhesive properties of embryonic cells change during development, and sometimes these correlate with changes in surface-charge density and morphogenetic behavior. For instance, amphibian embryonic cells become more adhesive during gastrulation; this correlates with a reduction in surface negativity as seen by electrophoretic increased mobility. At least in some developing systems surface-charge density is consistent with a cell's adhesive behavior, although no causal relationships have as yet been demonstrated.

Cells adhere to a wide range of nonliving substrates such as glass, plastic, and cellulose nitrate, some of which may be uncharged or even negatively charged at physiological pH. This nonspecific binding suggests that high surface potentials on opposing surfaces are overcome during adhesion. Cell-glass adhesions may be qualitatively different from cell-cell adhesions, although this is uncertain since most media contain proteins which coat the glass substratum and interact with cell surfaces. Both types of adhesion are calcium-dependent, are decreased by enzymatic treatment, and share other properties consistent with the hypothesis that cell-to-glass and cell-to-cell adhesion may be based on similar protein-protein interactions.

The fluidity of the cell surface may also relate to its adhesive behavior. As we have seen, mobility of the surface and of proteins embedded in the lipid bilayer may depend upon a system of contractile microfilaments suspended on a scaffolding of microtubules within the cell cortex. Surface fluidity could be affected by calcium ions which may control contractile proteins such as actin and myosin. Could calcium's role in surface adhesion derive from its direct role in the contractile mechanism of the cell surface?

Several years ago Jones proposed a model of cell adhesion based on the contractile state of the cell surface. According to this model, cells are least adhesive when their surfaces are contracted and most adhesive when surfaces are relaxed. The surfaces of motile cells such as an amoeba or a fibroblast are temporarily adhesive at those points in contact with the substratum and least adhesive elsewhere. Jones finds that motile cells detach from the substratum in the presence of $10^{-3}$ M ATP, a metabolite that promotes contraction, while they become more adhesive if ADP is added instead. High ADP concentration characterizes the relaxed state in muscle. This hypothesis suggests that adhesion and surface motility are two sides of the same coin, and later on we will see that both adhesion and motility are implicated in the aggregation and sorting out of cells from mixtures of different cell types.

Surface motility and adhesion may not be related as the above model implies, but membrane fluidity could displace adhesive sites, thereby affecting contact behavior. Once repulsive forces are overcome and contact is made, other factors may stabilize the interaction. Mucopolysaccharide matrices may accumulate between cell surfaces to "glue" them together, or other specialized junctions may be made.

## Specialized contact sites

Cells do make adhesive contacts with one another, often to the point of membrane fusion (Figure 10-6). Evidently, repulsive forces between cells can be overcome. Where surfaces of adjacent cells are in close proximity, their membranes organize into specialized structures that seem to bind one cell to another. A cell makes relatively few firm contacts with its neighbors; most of its surface is in "contact" across the 100- to 200-Å gap. But these few contacts may be adequate to cement cells into a tissue fabric.

One specialized contact is the *tight junction* or *zonula occludens* in which the outermost leaflets of the membrane bilayer of two cells fuse; in the electron microscope we see three dense lines, the middle one thicker than the outer two. The space between cells is obliterated at the junction, although on either side the usual 100- to 200-Å intercellular gap is seen.

Tight junctions are found in epithelial tissues and are most common in situations where there are sharp impermeable barriers between two compartments. Tight junctions are usually located on the lateral borders of cells near the apical surfaces, completely encircling each cell like a "seam." Fusion of membranes into a tight junction requires a high degree of structural specificity, at least at the local region of fusion. When they are found in epithelia, tight junctions seem to act mostly as insulators, preventing ion flow between interior cavities and the exterior milieu. Cohesiveness of cells in an epithelium also depends on tight junctions maintaining close side-to-side association in a two-dimensional plane. In this way the motile activity of individual cells may be translated into spreading properties of a multicellular sheet.

Other membrane specializations seem to be more directly involved in cell adhesion. These include *desmosomes* that appear as deeply staining plaques at discrete points in the membrane between opposing cells. In a desmosome, outer leaflets of the unit membranes are thicker and more highly differentiated than neighboring regions, with fibrous and tubular elements extending into the adjacent cytoplasm. A gap of 200 to 250 Å between cells is filled

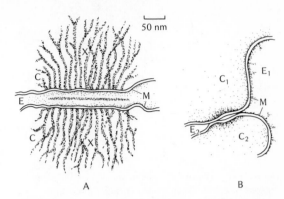

**Figure 10-6**   *Specialized cell junctions. (A) A desmosome with intracellular densities. Filaments (X) on either side extend into cytoplasm. M = membrane; C = cytoplasm; E = extracellular space. (B) A tight junction showing the regions of membrane contact as three dark lines. Two tight junctions are separated by a gap junction in this instance. (From H. S. Bennett,* Handbook of Molecular Cytology, *ed. A. Lima-de-Faria. New York: American Elsevier, 1969, pp. 1261–1293.)*

with an amorphous dense matrix. Desmosomes vary in structure and are commonly found in compact tissues like heart muscle and epithelium. In heart tissue, for example, they form the intercallated discs that join muscle cells end to end. The filamentous elements associated with some desmosomes have the same dimensions as contractile microfilaments, again pointing up the possible relationship between specific adhesive sites and an underlying contractile system.

Desmosomes and tight junctions are uncommon in invertebrate tissues. In their place are *septate desmosomes* with similar adhesive functions. These structures consist of a series of ladderlike septa, extending between the outer leaflet of unit membranes across the 100- to 200-Å gap. There are no filaments or intracellular tubules, nor is the gap filled with an intercellular deposit.

Desmosomes gradually appear in embryonic tissues as specializations of the cell surface. In the early chick embryo, for example, we find no desmosomes but we do see tight junctions. The tissues are easily dissociated at these stages, but in older embryos, when tissues are more highly organized and discrete, desmosomes are found at numerous points in cell membranes, and the tissues are much more difficult to

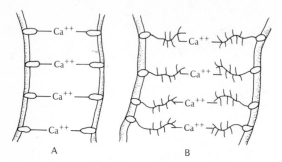

**Figure 10-7**    *Role of Ca++ in cell adhesion. (A) Calcium bridge hypothesis. Calcium ions covalently bound to anionic groups on proteins (COOH) embedded in membranes of opposing cells. (B) Calcium bridge–glycoprotein interactions. This differs from A in that calcium ions are covalently bound to polysaccharide moieties extending from glycoproteins in cell membrane.*

disaggregate. After chick embryo tissues are disaggregated by trypsin, desmosomes seem to disappear, but as reaggregation begins, fully formed desmosomes reappear between contacting membranes and become more common in fully organized aggregates. The high correlation between tightness of adhesive contact and high incidence of desmosomes suggests that these, above all other junctions, play a key role in cell adhesion.

Desmosomes are persistent structures; they do not disappear after cell membranes are isolated. Only the extracellular matrix of the desmosome is susceptible to chemical agents. In the absence of calcium or after trypsin digestion, extracellular matrix disappears and desmosomes cleave, forming half desmosomes that persist in the dissociated cells. Such structures may act as sites for local adhesion during cell reaggregation.

Desmosomes disappear during embryogenesis whenever cells detach from an epithelial sheet. For example, cells of the neural epithelium lose desmosomes before withdrawing from the periphery to migrate as neuroblasts. Embryonic cells lose desmosomes when they become motile. The integrity of an epithelial sheet seems to depend on the number and distribution of these specialized contact points.

## Role of calcium in cell adhesion

Since Herbst first separated sea urchin embryos into blastomeres by calcium-magnesium-free seawater in 1883, tissues of all kinds have been dissociated by removing divalent cations. When calcium is restored to the medium, cells reassociate and adhere as an aggregate. What role is calcium playing in keeping cells together?

One hypothesis assumes that calcium acts as a "bridge" between negatively charged anionic groups (carboxyl) projecting from the proteins or mucopolysaccharides on the outer surface of two opposing cells (Figure 10-7). One cell recognizes and adheres to another by accurate point-to-point matching of patterns of anionic groups. Different cells adhere less strongly because their respective surface patterns of carboxyl groups, being dissimilar, make fewer calcium bridges. Cells with matching surface patterns would account for stronger bonding. A calcium bridge hypothesis requires that cell surfaces be closer than 10 Å, one order of magnitude smaller than the 100- to 200-Å gap seen in the electron microscope. Even if we assume the large gap is an artifact, the concept of a patterned distribution of binding sites requiring point-to-point fit is difficult to reconcile with a fluid-mosaic model of the membrane and the random manner in which adhesive contacts are made. A fluid membrane could not sustain a definite pattern of anionic groups on the surface. If, on the other hand, the extracellular space is filled with a glycoprotein cement, calcium may be binding to and stabilizing the cement at each cell surface.

Our understanding of the role of divalent cations in adhesion is still unsatisfactory. Other roles for calcium have been proposed. It may be essential to the structure of the membrane itself, or it may reduce the net negative repulsive charge on the cell surface, permitting Van der Waals attractive forces to bring cell surfaces together at the primary minimum.

## Cell aggregation as a measure of adhesion

To understand the importance of cell adhesion in morphogenesis, we should determine the adhesive forces between cells and how they change. Attempts to study adhesive properties by measuring the forces required to separate cells from one another or from substrata have been unsatisfactory. Such experiments are difficult to interpret because cell adhesion depends on such factors as the amount of cell surface involved in adhesion, the conditions of measurement, as well as intrinsic binding forces between cells. Accordingly, several indirect methods have been de-

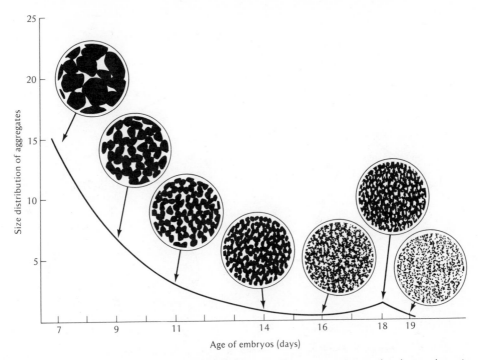

**Figure 10-8**  *Effect of embryonic age on size distribution of retina cell aggregates. Twenty-four-hour cultures in standard medium rotated 70 rpm. Young retinas seem to be most adhesive in that they form the largest aggregates. As developmental age increases, retinal cells form smaller aggregates until late stages when they fail to form any aggregates. (From A. A. Moscona, J. Cell Comp. Physiol. Suppl. 60:65–80, 1962.)*

veloped to compare adhesive properties of different cells. One of these is cell aggregation.

Studies of sponge cell aggregation showed that specific glycoproteins are involved in cell adhesion. Similar techniques have been applied to studies of adhesive behavior of embryonic cells. Tissues removed from chick embryos are dissociated by incubation in trypsin for varying lengths of time. Mildly trypsinized cell suspensions, when washed and restored to a normal medium, reaggregate and within a few days redifferentiate into their tissue of origin: embryonic cartilage cells form cartilage, heart cells develop beating heart, and kidney cells organize into tubules.

Such dissociation-reaggregation systems have been used to compare adhesive properties of cells from different tissues. Moscona, using a technique of controlled aggregation in rotating shaker flasks, was able to eliminate extraneous factors that affected aggregation (such as cell motility) so that rate of aggregation and size of aggregates could serve as measures of cell adhesiveness; cells forming large aggregates were assumed to be strongly adhesive, whereas few, small aggregates were indicative of weak adhesion. Moscona observed that chick retinal cells exhibit a declining ability to aggregate with increasing embryonic age (Figure 10-8). This was interpreted as a failure of older, more specialized cells to reconstitute adhesive sites after trypsin dissociation.

Moscona also found tnat aggregation was temperature-sensitive; at low temperatures (15°C) only minute clusters appeared, whereas these same cells formed typical large aggregates at 38°C. Temperature sensitivity suggested that cells must metabolize and synthesize surface adhesive factors.

Aggregation was also sensitive to inhibitors of protein synthesis, but the cells were not damaged by the inhibitor; they reaggregated normally after it was washed out. The initial phases of aggregation were unaffected, but after a few hours aggregation stopped. Again, this implied that synthesis of a surface ligand was necessary for aggregation.

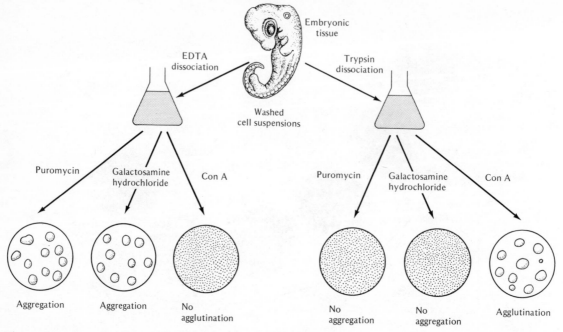

**Figure 10-9** *Aggregation behavior of embryonic tissues depends upon methods of dissociation. A comparison of trypsin- and EDTA-dissociated chick tissues under different aggregation conditions. In the presence of puromycin and galactosamine hydrochloride only EDTA-dissociated cells will aggregate while trypsin treatment alters cell surfaces so that they agglutinate in the presence of concanavalin A.*

This conclusion was somewhat premature inasmuch as trypsin damages cell surfaces and proteins escape from cell interiors. The membranes of such cells must be repaired for cells to survive, and repair does require protein synthesis, a temperature-dependent process. The relative insensitivity of early stages to the inhibitors suggests that some binding sites persist on cell surfaces after trypsin treatment, but those important for aggregation are synthesized later.

The aggregative behavior of cells does depend upon the dissociation procedure (Figure 10-9). Tissues treated with EDTA in calcium- and magnesium-free media are dissociated by removing cations without releasing most surface proteins. Accordingly, EDTA-treated embryonic cells exhibit different aggregation behavior. They reaggregate in the presence of a protein synthesis inhibitor, puromycin, suggesting either that these membranes are in less need of repair than trypsin-dissociated cells or that EDTA does not strip off adhesive sites. Another difference is that trypsin-dissociated cells will agglutinate after exposure to Con A whereas EDTA-dissociated cells

will not. Evidently, trypsin exposes concanavalin receptor sites whereas EDTA does not.

It appears that cell aggregation is a function of the dissociation procedure and not necessarily a measure of intrinsic adhesive properties. Nevertheless, once membranes are repaired, we can assume that aggregative behavior does tell us something about the nature of cell surfaces. In fact, aggregation, an indirect method of measuring cell adhesion in different cells, suggests that differences in the aggregative performance of a tissue correlate with the relative adhesiveness of its constituent cells. The aggregation procedure demonstrates that embryonic cells behave as if they recognize their immediate cell neighbors. Like sponge cells, they prefer to aggregate with cells of their own kind, and they will sort out from cell mixtures to form discrete tissue-specific aggregates.

**Tissue specificity of cell aggregation**

We have seen that aggregation in primitive systems is species-specific and that surface glycoproteins may be critical in the process. These systems have served as models for selective cell adhesion in more complicated developmental systems. According to Moscona's cell-ligand hypothesis, selective adhesion and embryonic cell recognition are brought about by specific cell surface components or surface ligands that link adjacent cells into cohesive, morphogenetically competent aggregates. Specificity of interaction resides in the chemical structure and configuration of these surface molecules, which may be in the membrane or in the matrix. Embryonic cells differ with respect to their surface ligands or to their topographical pattern in or on the surface. Accordingly, cells with chemically similar ligands may display different affinities because of differences in distribution of ligands over the surface.

The hypothesis also suggests that, like slime mold amoebae, surface patterns of ligands undergo maturation during development. Presumably as cell surfaces mature, newly synthesized ligands are inserted into the membrane. Such ligand patterns confer changing specificities on cells during the course of development. In early embryonic stages they permit cells to distinguish broad tissue categories such as ectoderm from endoderm, but as development proceeds, cells acquire tissue-specific ligands that permit them to distinguish one tissue from another. Finally, when fully specialized, surfaces acquire cell-specific ligands that distinguish different cell types within a tissue. Only cells with complementary ligands recognize one another and cohere.

The hypothesis is not very different from an antigen-antibody concept of cell aggregation specificity that was proposed independently by Albert Tyler and Paul Weiss in the 1940s. Protein-protein interactions between two cell surfaces behaved like antigen-antibody reactions. The groups on one cell surface interacted sterically with complementary groups on the opposing cell surface. Only cells with similar sites recognize one another, and the chemical nature and pattern of distribution of these sites on the surface determine the strength and specificity of adhesion.

The presence of surface ligands has been shown by preparing antibodies against whole cells and demonstrating that they specifically inhibit aggregation of cell suspensions. One possible interpretation is that antibodies inhibit aggregation by blocking adhesive sites on the surface.

According to Moscona, glycoproteins in the cell surface are ligands involved in adhesion of embryonic cells. A supernatant fluid from cultures of embryonic chick retinal cells promotes the aggregation of chick neural retina but has no effect on other embryonic tissues. Retinal cells aggregate into morphogenetically active masses which differentiate into retinal tissues. The aggregation-enhancing factor does not accumulate at low temperatures nor is it produced by retinal cells in which protein synthesis is inhibited, suggesting that the factor is synthesized and released into the medium. Only trypsinized cells respond to this retinal aggregation factor; mechanically dissociated cells are insensitive. Presumably, parts of the cell surface must first be exposed by trypsin before the factor can act. If cells are incubated for two hours in the absence of factor, they no longer respond. Trypsin-exposed sites are no longer accessible, perhaps as a consequence of some repair mechanism. Note that this initial period of "repair" corresponds to the period insensitive to protein synthesis inhibitors, which suggests that aggregation is a more complicated process than implied by the hypothesis, since a cell surface must be properly organized to interact with the ligand. Ligands are taken up from the medium by retinal cells and seem to be inserted in (or on) the cell surface since antibodies prepared against purified factor specifically inhibit retinal cell aggregation.

The retinal aggregation factor has been isolated and purified as a glycoprotein. It is only one of a few cases of a pure ligand in embryonic cell aggregation, although other factors have been found. Specific glycoprotein factors promote aggregation of cells from sea urchin embryos, and others seem to enhance aggregation of cells from the embryonic mouse cerebrum. Such extracellular factors may be more widely distributed, but we still do not know how they function in cell adhesion or aggregation, partly because different dissociation procedures have variable (and unknown) effects on cell surfaces and damage membranes to different degrees.

Edelman's group has studied the mechanism of action of such aggregation factors. They isolated an aggregation factor from chick retina tissues known as *cell association factors* (CAF) which promotes aggregation of retinal cells. Fluorescent antibodies prepared against purified CAF react with the cell surfaces of several embryonic tissues including retina, brain, liver, kidney, and heart, but retina cells seemed to bind more than the others. CAF localization on the surfaces of many different tissue cells suggests lack of specificity, but some kind of specificity could be demonstrated if conditions for the assay were changed.

Fine nylon fibers were coated with the antibody to CAF and placed in trypsinized suspensions of cells from different tissues. The cells bind to the fiber, and the number of cells bound per fiber length is a function of the affinity between cell surface ligands and the antibody. Brain, retina, and liver cell suspensions at different ages were prepared from chick embryos, and their binding to anti-CAF-coated nylon fibers was compared under standard conditions (Figure 10-10). Three tissue-specific binding profiles were obtained.

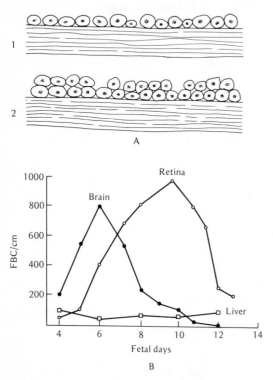

**Figure 10-10**  *Nylon fiber technique for studying surface interactions. (A) Appearance of nylon fiber in two different assays. (1) Assay of cell affinity for antibody-treated nylon fiber. (2) Cell-cell adhesion assay. Number of cells attached to standard length of fiber, FBC/cm is basis for assay. (B) Binding of embryonic chick cells as function of age to fiber coated with anti-CAF from chick retina. (From U.S. Rutishauser et al., "Expression and Behavior of Cell Surface Receptors in Development," in Volume II ICN-UCLA Symposia on Molecular and Cellular Biology: Developmental Biology, copyright © 1975 by W. A. Benjamin, Inc., Menlo Park, California.)*

At 4 days all tissues showed low levels of binding, but by 5 and 6 days brain cells showed a greater affinity for the antibody than retina cells. The retina was much slower; it reached its peak of binding at 10 days when affinity of brain cells had declined. Liver cells showed very little change in their low binding affinities during development. It is interesting that retina binding also showed a sharp decline between 10 and 14 days. These results are consistent with the hypothesis that properties of cell surfaces do change during development, and they indicate that specificity is relative, depending upon the age of the tissue.

A more direct assay of cell-cell binding, using the nylon fiber, correlated binding of cells to anti-CAF with binding between cells taken from embryos at different ages. Here an anti-CAF-coated nylon fiber was first covered with cells of one tissue. The cell-coated fiber was then placed in a suspension of cells from the same tissue or from different tissues at different ages, and the number of cell-cell adhesions was counted directly in the microscope. In this way a direct measure of cell-cell binding was obtained. The data showed that binding of cells to anti-CAF correlated with their binding to one another. For example, binding between two 9-day retina cells, two 6-day brain cells, or between 9-day retina and 6-day brain was highest. On the other hand, 6-day retina, 13-day retina, or 9-day brain cells bound poorly to each other. Pretreatment of cells with anti-CAF before carrying out the binding assay completely eliminated cell-cell binding, which indicates that CAF sites on the cell surface are involved. The fact that the CAF ligands are found on all cells suggests that specificity of binding is not a result of different tissues synthesizing tissue-specific ligands. Rather, it appears that different tissues may share the same ligands, and specificity arises because their number and topographic distribution differ at different developmental stages. The mobile properties of the cell surface, particularly as it displaces receptors and other membrane components, may also mobilize binding ligands.

The interaction of CAM on the surface of one cell with that of another has been examined. CAM seems to bind to the surface as an inactive, large-molecular-weight glycoprotein (approximately 240,000 mol wt), or pro-CAM state. Such molecules are not adhesive. Adhesive activation depends upon a surface-modulated event, the splitting of a fragment from pro-CAM by a proteolytic enzyme that seems to be associated with retinal tissues. The smaller CAM fragment remaining in the surface (about 150,000 mol wt) is converted into an adhesive binding site, and cells with exposed CAM fragments adhere. Antibodies made to the cleavage fragments of pro-CAM interfere with short-term binding of chick embryonic retina cells.

As we have seen, retina and brain cells exhibit differential binding to anti-CAM antibodies with advancing age; the number of CAM receptors increases, but this is followed by a decline in responsiveness. It could be shown that the number of cells possessing pro-CAM increases with age and attains a plateau rather than declines. Accordingly, it appears that the surface of embryonic cells may undergo an epigenetic modulation by proteolytic activation of surface components only at specific developmental stages, perhaps at those times when adhesive interactions are important to morphogenesis.

In a unifying model Edelman has attempted to correlate the adhesive phenomena described here with the fluid properties of the cell membrane which modulate cell division and motility. He proposes that the membrane coordinates all signals that control growth, cell motility, and cell interactions. Surface modulation of receptors may signal specific metabolic reactions essential to various cell properties. Thus adhesive interactions between two cells may cause receptors to redistribute and signal differentiation. An example of this was seen in the slime mold differentiation of prespore cells. More elaborate surface mechanisms may have evolved in embryonic systems to regulate morphogenesis.

### Steps in cell adhesion

No single hypothesis explains all adhesive phenomena. Many observations need to be synthesized into a coherent model of cell adhesion. Visible changes in surfaces of aggregating cells as seen in the scanning electron microscope, coupled to what is known about chemical changes at the surface, may serve as a first step in constructing such models. For example, after an initial repair phase to recover from trypsin treatment (possible restoration of surface adhesive sites) most cells in suspension are spherical, covered by numerous short microvilli. The first cell contacts are made between the tips of microvilli projecting from motile surfaces. Because of the large radius of curvature at the tips of microvilli, very high energies of attraction are generated, enough to overcome the re-

pulsive forces of negatively charged surfaces. Adhesions between matching surface ligands stabilize the membrane, which continues to expand over neighboring membrane surfaces. Such wider areas of membrane apposition may stimulate the formation of tight junctions, or desmosomes. Cells begin to line up in small chainlike aggregates, held together at only a few points initially, but these contact areas increase as adjacent membranes are "pulled in" on either side so that the glycoproteins extending from the two surfaces overlap a 200- to 400-Å space. Because of close apposition of membranes, mucopolysaccharides begin to accumulate between extracellular spaces. Such matrix interactions may further stabilize the membranes and diminish their inherent motility. As cells persist in these stable, quiescent associations, additional intercellular matrix materials are secreted and accumulate within the tissue spaces, filling the gap to its 100- to 200-Å dimension. At this point desmosomes and tight junctions have matured (repaired?) to keep cells together in more specific associations. Such aggregates may organize and differentiate into distinct tissue arrangements.

The above is only a speculative mechanism for the events of cell aggregation. Other combinations of events may explain the facts as well.

## SORTING OUT OF EMBRYONIC TISSUES AND CELL RECOGNITION

According to the cell ligand hypothesis for specific adhesion and embryonic cell recognition, embryonic cells should experience progressive changes in adhesion to correlate with their morphogenetic behavior. These changes should be seen as differences in aggregation behavior which should correlate with changes either in the specific surface ligand or, more likely, in their number and distribution. One aspect of behavior that can be measured is sorting out from mixed cell suspensions, a property that has been equated operationally with cell recognition. Embryonic tissues do show distinct sorting out properties either when combined with one another as fragments or as cell suspensions. Embryonic cells recognize their own kind and make preferred associations with one another to produce morphogenetically active aggregates. Cell recognition, like adhesion, is a property of the cell surface. In fact sorting out and specific cell adhesion may depend upon the same surface

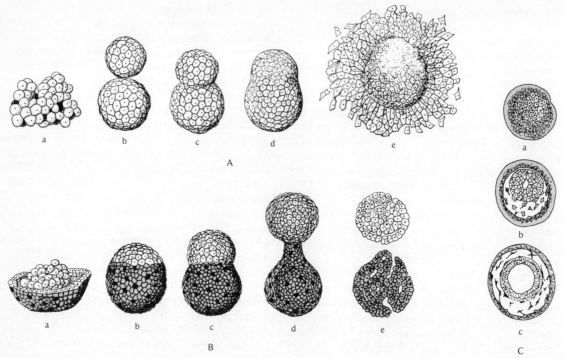

**Figure 10-11** *Combination of different amphibian embryonic tissues. (A) Combination of two fragments of endoderm. (a) Cells as they appear shortly after isolation. (b, c, d) Fusion of fragments after they round up. (e) Attachment to substratum and outward migration of cells. (B) Combination of ectoderm and endoderm. (a) A cup of ectodermal tissue is filled with endoderm. (b) Tissues round up and fuse initially. (c,d) Later endoderm segregates from ectoderm. (e) Cross section of separated tissues. (C) Combination of ectoderm, endoderm, and some mesoderm. (a) After rounding up and fusion. (b,c) Cross section of aggregate showing mesoderm as loose mesenchymal cells in a cavity surrounding the endoderm which has formed a lumen. Both are surrounded by pigmented ectoderm, a spatial arrangement that is also found in the intact embryo. (From J. Holtfreter, Arch. f. Exp. Zellforsch. 23:169–209, 1939.)*

ligands. Examples of sorting behavior come from amphibian and chick embryos.

### Sorting out of amphibian cells

Cell recognition appears early in amphibian development. Three distinct germ layers arise: ectoderm, mesoderm, and endoderm. Each occupies a definite position in the embryo: the ectoderm is on the outside, surrounding a layer of mesoderm, which in turn envelops the endoderm (see Chapter 12). They differ in morphogenetic behavior and developmental fate, in part because of their position in the embryo and their specific recognition behavior. Recognition is displayed when tissue fragments from different germ layers are combined in vitro. For example, two endoderm pieces fuse intimately, whereas endoderm combined with ectoderm forms no stable union; the two tissues separate (Figure 10-11). A cohesive fusion mixture is obtained only if a piece of mesoderm is interposed between ectoderm and endoderm. The ectoderm spreads over both, but each tissue segregates from the others, taking up a position within the mass comparable to its position in the intact embryo, with mesoderm surrounding the internal endoderm.

As the embryo ages, parts of the same germ layer acquire different tissue affinities and sort out from each other when fused or mixed as cell suspensions. Sorting from cell mixtures is easily followed with amphibian embryonic tissue because germ layers are distinguishable by cell size and pigmentation; for example, endoderm cells are large and full of yolk, whereas the smaller ectoderm cells are heavily pigmented.

Two derivatives of the primary ectoderm, epidermis and neural ectoderm, sort out from one another in cell mixtures, even before they are recognizably different in the embryo (Figure 10-12). When dissociated cells from each tissue are mixed, they reaggregate at random. Later, cells move over each other, making and breaking surface contacts with like and unlike cells and gradually sorting out with homologous cells aggregating into small clumps. A stable equilibrium is achieved when pigmented epidermis preferentially segregates at the periphery surrounding the interior neural cells organized as a vesicle. These recognition specificities are acquired within a matter of hours because ectodermal cells do not sort out before this period.

Summarizing the many combinations that were

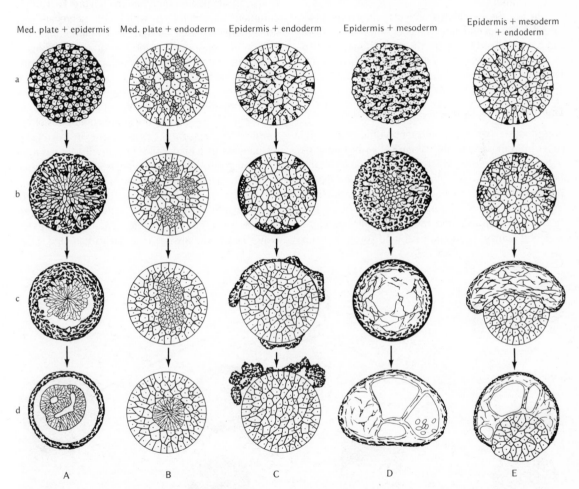

**Figure 10-12** *Sorting out from mixed cell aggregates of amphibian embryonic tissues. (A) Randomly arranged epidermis (black) and medullary plate cells (white) move in opposite directions and sort out into epidermis and neural tube. (B) Medullary plate cells (small) mixed with endoderm segregate into a central mass of neural tissue surrounded by endoderm. (C) As with tissue fragments, recombined ectodermal and endodermal cells first form fusion mixture but sort out and separate from each other. (D) Epidermis and mesodermal cells form random mixture that sorts out into a thin layer of epidermis surrounding coelomic cavities, mesenchyme, and blood cells, all mesodermal derivatives. (E) A mixture of cells from all three layers leads to segregation of outer ectoderm surrounding mesenchymal derivatives and a ball of endoderm. (From P. Townes and J. Holtfreter, J. Exp. Zool. 128:53–120, 1955.)*

studied, it was found that (1) cells of each germ layer share a generalized adhesive property, which explains aggregation into random mixtures; (2) before cells sort out, they first acquire a germ-layer specificity; (3) the final position taken up by a cell within a mixture corresponds to its normal position in the embryo; and (4) in later development regional recognition specificities develop within each germ layer—derivatives of the same germ layer sort out from one another.

Just as the hypothesis predicts, embryonic cells pass through several different recognition states, from more generalized to the specific, which may explain their morphogenetic properties in the embryo at different developmental stages.

## Embryonic chick tissues

Sorting-out experiments with embryonic chick cells differ from those with amphibian cells since they involve older tissues in which most cells have already begun to acquire specialized function. Each tissue, presumably, has already acquired tissue-specific recognition properties. A variety of cell mixtures has been used, such as retina and heart, limb-bud cartilage and kidney, heart and liver, and so on. In each case cells initially form random mixtures; that is, heterologous cells adhere in mixed aggregates. Within a few days, however, cells sort out into an equilibrium state with each tissue occupying a preferred position in the aggregate, one tissue usually enveloping another. The tissues are identifiable because they differentiate; cartilage cells secrete a matrix, heart cells develop myofibrils, and kidney cells form tubules.

Several basic sorting-out patterns were obtained from various tissue combinations (Figure 10-13): (1) complete segregation, in which cells segregate into nonadherent masses, as is the case of amphibian ectoderm and endoderm at the end of aggregation; (2) incomplete segregation, in which cells sort into distinct adherent masses in partial contact; (3) a pattern in which one tissue totally envelops another, with the latter continuous or discontinuous, distributed as small clusters throughout the larger continuous phase (most chick segregation patterns are in this class); (4) patterns in which cell types are distributed nonrandomly, but neither cell type seems to occupy an internal or external position.

Under standardized conditions, a hierarchy of tissues based on sorting-out patterns could be arranged. Liver cells segregated externally to heart, while heart was always external to cartilage. Liver

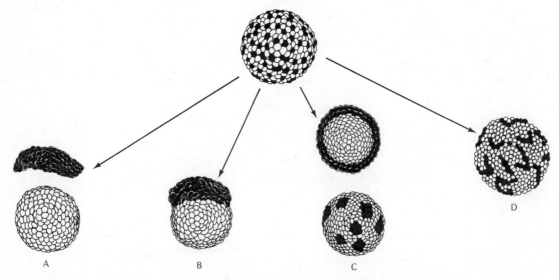

**Figure 10-13**   *Generalized sorting-out patterns exhibited by a mixture of embryonic cells. (A) Complete segregation. (B) Partial segregation; tissues still adherent. (C) Inside-outside sorting out. Upper is continuous with outside tissue completely surrounding inner mass of tissue; lower is discontinuous, inside cells segregated into several discrete masses. (D) Random segregation; no one tissue has preferred inside-outside arrangement.*

also segregated externally to cartilage and to retina. In the hierarchy of tissues those at the top of the list usually segregate externally to those immediately below, with the last tissues on the list segregating internally to all others.

Sorting out is independent of cell interactions since the position of a tissue in the hierarchy is independent of the tissue with which it is combined. Sorting out reflects intrinsic adhesive properties of cell surfaces; heart cells "prefer" to adhere to heart cells, liver to liver, and so on. The most adhesive occupy inside positions in all combinations.

As we have noted, trypsin dissociation produces membrane lesions, which means the sorting-out hierarchy may be artifactual, that is, sorting-out behavior of a tissue may be due to differences in its sensitivity to enzyme damage rather than to intrinsic recognition specificity. The position of a tissue in the hierarchy, however, coincides with its inside-outside relationships in the embryo, suggesting that sorting out depends on the same cell properties that account for morphogenetic behavior in vivo. Moreover, Steinberg found that undamaged tissue fragments, when fused with one another, sort out like trypsin-dissociated cell mixtures and give similar inside-outside hierarchies. Though the kinetics of sorting out of fused fragments is different from that of cell mixtures, the final equilibrium pattern is the same. Since tissue fragments are not enzymatically treated, we conclude that sorting-out patterns of cell mixtures do reflect an intrinsic cell surface property and are not in vitro artifacts.

The tissue hierarchy is not rigid; reversal of positions has been observed, particularly between tissues close together in the hierarchy such as liver and heart. Depending on conditions, liver or heart may occupy the inside position. Two factors are involved in position reversal: (1) the relative proportion of cells of each tissue type and (2) the age of the tissue. When dissociated heart cells are in the majority, heart envelops liver; when liver cells are in the majority, liver envelops heart. These results differ from those obtained with fused tissue fragments. In all fragment fusion experiments liver always envelops heart. How is this difference explained?

Since trypsin dissociation alters surface properties, membranes must be repaired before cells can adhere. The conditions under which this repair process occurs affect sorting out. For example, if fragments of heart and liver are each cultured for 2½

days in vitro before fusion, they sort out with liver enveloping heart. But if heart is cultured only ½ day and liver 2 days, then heart envelops liver. Heart cells are more adhesive than liver in the first few hours after trypsin treatment. This temporal factor suggests that in vitro surface properties of heart tissue change during membrane repair and reverse relative to liver so that two completely different sorting-out patterns can be obtained.

We see that sorting out patterns are complicated by age of tissue, method of dissociation, conditions of reaggregation, and length of time in vitro. These factors have not been satisfactorily resolved for all tissue combinations. Nevertheless, this has not prevented speculation about the meaning of sorting out and its bearing on cell-specific adhesion and recognition.

Most models explain selective segregation and cell recognition on the basis of cell adhesiveness rather than cell motility; that is, cells sort out because they possess tissue-specific adhesive properties. Cell motility, as we will see, may be equally important, particularly if adhesion and surface contractility are based on the same mechanism.

## Glycoproteins and sorting out

We have already alluded to Moscona's cell ligand hypothesis to explain adhesive specificity; it may also account for sorting out. In this model, each tissue type secretes a tissue-specific ligand or extracellular matrix that forms stable adhesions between cell surfaces.

Specificity could reside in glycoproteins in the surface, since these are sufficiently diverse chemically. As we have seen, a few glycoproteins have been isolated that promote specific adhesion.

Carbohydrates at the cell surface are necessary for cell recognition in vitro. For instance, suspensions of mouse teratoma cells aggregate in a complete medium with amino acids, vitamins, and so on, but they do not aggregate in a balanced salt solution. By a process of stepwise addition of constituents to the salt solution, it was found that only the amino acid L-glutamine promoted aggregation. Moreover, the sugars D-glucosamine and D-mannosamine could substitute for glutamine in this system, which indicated that the metabolic pathways responsible for synthesizing amino sugars, the basic carbohydrate constituents of glycoproteins, are involved in cell adhesion. The synthesis of adhesive sites seems to de-

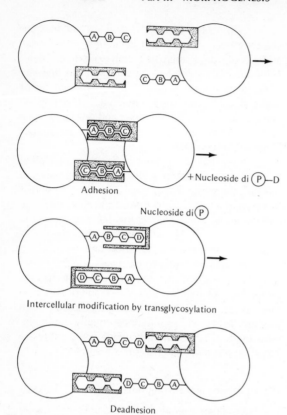

Adhesion

+ Nucleoside di (P)—D

Nucleoside di (P)

Intercellular modification by transglycosylation

Deadhesion

**Figure 10-14**   *Model of cell adhesion based on gly-cosyltransferase enzymatic activity. Cells adhere because of enzyme substrate interactions at the respective surfaces. The glycosyltransferases on one surface combine with the appropriate sugar nucleotide acceptors on the opposing cell surface (A—B—C—D). Deadhesion results if the carbohydrate acceptor site is altered by trans-glycosylation. This weakens the enzyme substrate and the cells separate. (From R. B. Kemp et al., Progress in Surface and Membrane Science, ed. J. F. Danielli, M. D. Rosenberg, and D. A. Cadenhead, vol. 7. New York: Academic Press, 1973, pp. 271–318.)*

pend upon a supply of specific sugars for glycoprotein synthesis.

We have already learned that such sugars are the sites for attachment of concanavalin A, the plant lectin. Con A sites on cell membranes, however, are not involved in sorting out. If Con A receptor sites on chick embryonic cells are covered with univalent Con A, they aggregate normally and will also sort out from cell mixtures into the same tissue hierarchies obtained from untreated cells. The univalent molecule, unable to cross link, does not inhibit membrane mobility. On the other hand, tetrameric Con A, the native molecule, does make cell surfaces less mobile, and sorting out is inhibited, perhaps because adhesive sites are not free to move. Con A, by stabilizing the surface, may reduce adhesiveness by interfering with formation of microvilli, the sites of initial cell contact.

Another hypothesis implicating carbohydrates as bridging agents in specific recognition has been proposed by Roseman and Roth. The model is based on the interaction of enzymes on one cell surface with their substrates on opposing cell surfaces, very much like a lock and key interaction (Figure 10-14). The substrates, in this instance, are surface carbohydrates, and the enzymes are glycosyltransferases that polymerize polysaccharides from sugar nucleotide precursors (see Chapter 8). There are multiple glycosyltransferases, each one specific for a different polysaccharide receptor and substrate, and different tissues would possess different surface enzymes. Surface recognition, according to this hypothesis, is based on the specificity of enzyme-substrate interactions between sites on opposing membranes. The association of enzyme on one cell with its specific substrate on another cell accounts for adhesion. Enzymatic transfer of the donor sugar dissociates the enzyme-substrate complex, whereupon cells separate.

In support of this hypothesis, it was found that removal of a terminal galactoside residue from the surfaces of chick embryo neural retinal cells altered their adhesive properties. Specificity of aggregation was lost; many more retinal cells attached to liver cells after treatment than before. Intact retinal cells were also found to have glycosyltransferases on their surfaces. When activity of these enzymes was inhibited, specific adhesion of retinal cells was also affected, which suggests that the enzyme may play a role in cell adhesion. The key to this hypothesis is the demonstration of different glycosyltransferases on the surfaces of different cells. So far this has not been done.

### Timing: an adhesion hypothesis

Of many hypotheses explaining sorting out, only Curtis' timing hypothesis acknowledges membrane damage after tissue dissociation and assumes that cell adjustment to the in vitro environment requires repair of membrane lesions. During recovery cells migrate at random, colliding with one another, making temporary adhesive contacts. It is during this period that adhesive properties can exhibit changes such as position reversal. Curtis proposed that dissociated cells in vitro take different lengths of time to reestablish adhesive sites. He assumed that all cells have identical binding properties and that sorting out in mixed aggregates occurs because some cells acquire their adhesive capabilities earlier than others. According to this view, the first cells to become adhesive migrate to the surface where shearing activity is minimal as cells move against one another. At the surface they form a firm adhesive boundary which attracts and entraps only cells of the same tissue type (homotypic). This process of entrapment tends to herd the heterotypic (unlike) cells in the aggregate toward the center; the latter, of course, acquire adhesive properties much later.

Tests of this hypothesis have shown that one can indeed change inside-outside relationships of tissues in mixed aggregates by varying dissociation times. For example, in the combination of amphibian ectoderm and endoderm, ectoderm invariably is found on the outer boundary surrounding endoderm. If, however, endoderm is dissociated several hours before ectoderm to permit early repair, then endodermal cells are usually found at the outer boundary with the ectodermal cells herded within the aggregate. Experiments with other vertebrate tissues (liver and heart) show that inside-outside relationships can be changed by manipulating the time of dissociation of each tissue type. A major criticism of this model is that it explains tissue spatial patterns in the intact organism on the basis of a response to artificial induced dissociation damage. It is difficult to see how repair of membrane damage could account for in vivo patterns. The hypothesis is further weakened by the behavior of intact tissue fragments. As we indicated, intact tissues sort out in patterns identical to those shown by cell suspensions. Since intact tissues are relatively undamaged by trypsin dissociation, the timing hypothesis cannot account for the sorting-out patterns of fused tissue masses. Nonetheless, Curtis' hypothesis correctly emphasizes timing and adjustment to in vitro conditions as key factors in cell aggregation and sorting out.

## Differential adhesiveness

According to a hypothesis proposed by M. Steinberg, selectivity in sorting out may depend on differential adhesiveness of tissue cells. This property may arise because cells from different tissues possess qualitatively different adhesive sites in their surfaces, or all cells may share the same sites which may differ in terms of number or spatial distribution. Cells making strong adhesions have many more sites than those making weak associations. Irrespective of the precise mechanism of adhesion, a differential adhesion hypothesis can account for sorting-out patterns if the adhesive affinities between cells are treated as a simple thermodynamic system. Cells are pictured as randomly colliding balls with different adhesive strengths, and cell mixtures are assumed to resemble multiple-phase systems of immiscible liquids, with cells replacing molecules. We know that after thorough mixing of the liquids, equilibrium occurs when the two liquids are separated by a phase boundary. This is caused by differences in the cohesive forces between molecules. Likewise, equilibrium is achieved in cell mixtures after cells have sorted out, presumably into a spatial arrangement in which cell surfaces exhibit minimal surface tension. This too arises because the adhesive forces between cells differ. Cells with strong recognition tendencies have greater cohesive affinities and make permanent contacts on collision, whereas other cells cohere less firmly. The more cohesive cells will tend to sort out from the less cohesive into a separate "phase."

This model predicts that in a mixture of randomly colliding cells, weak adhesions between cells will be replaced by stronger ones (Figure 10-15). In mixtures of A and B cells, with A more adhesive than B, A cells will tend to associate selectively with one another. As cells move about, they make random contacts: A-A or A-B. Since A-A contacts are more cohesive than B-B contacts, A cells will continually replace heterotypic A-B contacts with isotypic A-A contacts until all A cells cohere. The A cells will tend to form larger clusters, collecting internally with most of their surfaces in contact. The less cohesive B cells, excluded from the tightly bound A clusters, will be displaced to the periphery. Equilibrium is reached when the more cohesive A cells, as an internal phase, are enveloped by the less cohesive B cells, or external phase.

A most important prediction of the model is that

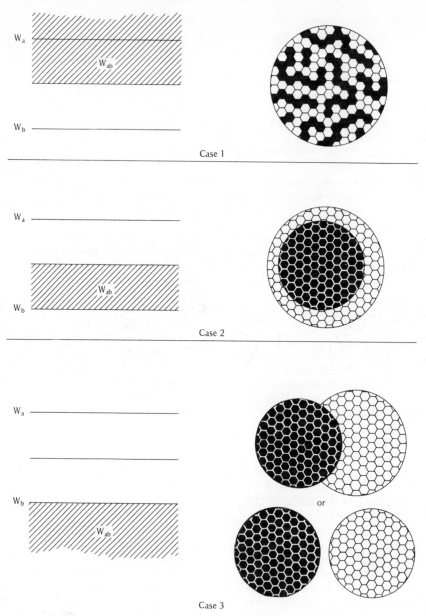

**Figure 10-15**   *Steinberg model of sorting out based on differential adhesion between two cell types in a binary mixture. The adhesive strength between cells is denoted as $W_a$ (A-A cell interaction), $W_b$ (B-B cell interaction), and $W_{ab}$ (A-B adhesive interactions). In case 1, the value for $W_{ab}$ is equal to or exceeds the average adhesive strength $W_a$ and $W_b$, a condition that favors intermixing of A and B cells. In case 2, if the strength $W_{ab}$ equals or exceeds $W_b$ but is less than $W_a$, then the more cohesive A-A cells will tend to collect in the center surrounded by the less coherent B-B cells which surround them, the situation described in the text. In case 3, $W_{ab}$ falls below $W_b$ and at equilibrium the two tissues separate in varying degrees. (From M. Steinberg, Cellular Membranes in Development, ed. M, Locke. New York: Academic Press, 1964, pp. 321–366.)*

the relative positions (internal or external) taken up by tissues during sorting within a binary (two-phase) cell mixture would be hierarchical. That is, if tissue A sorted internal to tissue B, and tissue B was internal to tissue C, then it could be predicted that in a mixture of A and C cells, tissue A would be internal to tissue C. The observations were in full accord with these predictions. For example, in mixtures of chick embryonic limb-bud cartilage with heart, heart enveloped cartilage at equilibrium. In another mixture a 5-day heart ventricle was enveloped by embryonic liver, and in a mixture of limb-bud cartilage and liver, cartilage was the internal tissue, enveloped by liver. The hierarchy in this case, in order of decreasing cohesiveness, would be cartilage, heart, and liver.

The model's confirmation depends on our ability to measure adhesive forces between cells. An indirect method of measuring cell adhesion correlates the centrifugal force required to deform a tissue aggregate with its relative cohesiveness. The experiments assume that a centrifugal force tends to flatten a tissue fragment but that this force is resisted by a tissue's cohesive properties. The greater the tissue's cohesion the less it is distorted by the centrifugal force and the more spherical it remains. Therefore the shape of a tissue after centrifugation is a measure of its strength of cohesion. In this way a hierarchy of tissue cohesiveness can be made, and it turns out that the hierarchy obtained by centrifugation agrees with the one obtained from sorting-out experiments. These results tend to support the differential adhesion hypothesis.

One difficulty with this method for measuring adhesion is that it does not actually measure the adhesive forces between cells; instead, it measures a tissue property called tissue cohesiveness. The tissue property, however, is complicated by many extraneous parameters in addition to cell adhesion: the amount of extracellular matrix, the relative proportion of different cell types in the aggregate, the appearance of a basement membrane on the surface of the aggregate, and even the contractile properties of cells.

Another observation that seems to be inconsistent with Steinberg's differential adhesion model is that reported by S. Roth and J. Weston in their studies of cell adhesion using the collecting aggregate method. A collecting aggregate is prepared by allowing dissociated cells of a specific tissue to reaggregate into a large compact mass. These aggregates are then mixed in rotary shaker flasks with isotypic (same cell type) or heterotypic (different cell type) suspensions of marked cells whose nuclei have been labeled with radioactive phosphorus or thymidine. The mixtures are incubated for varying periods, and the number of labeled cells adherent to the aggregate is determined by measuring the radioactivity.

Here too specificity of adhesion is shown. Isotypic cells make stronger adhesions with isotypic aggregates than they do heterotypically; that is, more labeled retina cells attach to retina aggregates than to liver or heart aggregates and vice versa. All isotypic adhesions are stronger than all heterotypic adhesions; A-A and B-B cells adhere more strongly than A-B. This is not predicted by the differential adhesion hypothesis since heterotypic adhesions should be intermediate; that is, AA > AB > BB. In these experiments no heterotypic interaction was stronger than either isotypic combination.

Steinberg argues that the collecting aggregate system differs significantly from sorting out. The latter takes several days compared with the few hours required in the former procedure. Besides, cell motility and displacement occur during sorting out, and these are not important in the collecting aggregate method. Thus the spatial relationships achieved in sorting out are a function of more than the initial adhesiveness of tissues. The collecting aggregate system is a kinetic assay that measures only the initial rates of adhesive interaction, which, as we know, are somewhat artifactual since membranes are undergoing repair. This means the collecting aggregate method is not a real test of the differential adhesion hypothesis.

The evidence of a wide distribution of identical surface ligands in different embryonic tissues is consistent with the hypothesis, however. It suggests that ligands shared by all cells may be differentially mobilized in the cell surface to organize as adhesive sites. Cells sharing similar ligand patterns would form more stable associations and sort out from cell mixtures. The problem of sorting out then becomes one of defining the factors in the cell surface that promote changes in the distribution and topography of surface ligands.

Though the differential adhesion hypothesis is a quantitative model it does not exclude other mechanisms of adhesion between cells. Because of the fluid nature of the cell surface, similar sites may be displaced into qualitatively different configurations on the surface and generate adhesive affinities only between cells bearing complementary surface patterns.

### Cell movement and sorting out

Any model of sorting out assumes that cells in an aggregate are capable of movement, presumably by adhering to the surfaces and extracellular matrices of neighboring cells. In all previous models random movement is assumed and is not a factor in sorting out. On the other hand, sorting out may result if cells exhibit directed movement in mixed aggregates. A simple chemotactic model, proposed originally by Holtfreter in 1939, can account for some equilibrium situations achieved in cell mixtures.

According to this hypothesis, each tissue produces a specific chemical within the mixed aggregate, luring homologous cells toward the center of highest concentration. Cells moving about at random tend to home in on these attractive centers. Another kind of chemotactic mechanism could consist of a gradient of a nonspecific metabolite such as carbon dioxide within the aggregate; the highest concentration at the center would attract certain responsive cells. Other cells more sensitive to high oxygen concentrations would be attracted toward the periphery to establish the outside-inside relationship.

Though chemotaxis is important in slime mold aggregation (Chapter 9), there is no evidence for similar chemotactic mechanisms in embryonic tissues, nor do metabolic gradients play an important part. If, for example, a carbon dioxide gradient was responsible for sorting out of embryonic tissues, the inside tissue should occupy and maintain a central position during sorting out, because $CO_2$ concentration would always be highest in the center. This does not always occur; in fact innermost tissues are more often eccentrically positioned in the aggregate. Furthermore, directed cell movement is not seen in aggregates. In time-lapse films of aggregates, cell movements are more random than oriented.

Most theories assume that for cell movement to occur, parts of the cell surface must be anchored to a substratum. For example, an amoeba moving over a glass substratum is in contact at only a few points on its ventral surface. The ruffled membranes of fibroblasts are the motor force, but the cell body is attached to the substratum at many points. Movement

is possible because the force generated by the ruffled membrane pulls the cell along as it makes and breaks adhesive contacts with the surface.

As a result, the strength of adhesion between a cell surface and its substratum affects the rate of cell movement. The greater its adhesiveness the less likely it can move freely. This means that the physical or chemical nature of the substratum will affect the rate of cell movement. It can be shown, for example, that fibroblasts move more rapidly over collagen-coated surfaces than they do over glass or plastic. Cells from different tissues may "prefer" some substrata over others and exhibit differential motility when migrating on identical substrata. Sorting out, therefore, may depend upon differential motility within mixed aggregates, perhaps as a consequence of differences in relative adhesion to one another or between cell surface and extracellular matrix within the aggregate. The more motile cells might tend to displace the less motile cells toward the interior and envelop them, forming the typical sorting-out patterns.

The movement of fibroblasts on artificial substrata has been studied. A cell tends to attach to a substrate and flatten out; the degree of flattening is directly related to the extent of adhesion to the substrate. When attached, fibroblasts begin to move, sending out their ruffled membranes as waves of undulations at the anterior end.

Differential substrate adhesion also determines the direction of cell movements. One may prepare artificial surfaces coated with substances varying in adhesive affinities for cells. Cells move in one direction along these surface gradients, generally toward regions of greater adhesiveness. Polarity of cell movement seems to be based on the degree of adhesiveness between the cell surface and substratum.

Cells move at random over a surface, usually away from the central area of maximum cell concentration, but exhibit oriented movements if placed on an organized substratum. They also synthesize and secrete matrix materials which cover the exposed surfaces. Most cells, therefore, move over a macromolecular substratum, a layer of adsorbed molecules called a ground mat or microexudate. If the ground mat is itself oriented by lines of tension within the macromolecular layer, we find that cell movement is polarized along these external force lines.

For example, cells placed in plasma clots orient their movement and pseudopodia according to the direction of fibers within the clot. This phenomenon, called *contact guidance,* may explain the directed migration of certain cell populations in the embryo such as nerve cell axons. Cells progress more rapidly over oriented fibers in plasma clots than within the clot itself, which suggests that rates of cell movement may be influenced by mechanical tensions within a substratum. Such oriented movements may arise within an aggregate and affect sorting out behavior.

If cell surface motility is inhibited by cytochalasin B, then both aggregation and sorting out are affected. For example, aggregation of embryo chick heart cells is enhanced by the agent. On the other hand, heart-liver cell mixtures do not sort out; the aggregates fail to round up, and cells remain as random mixtures. The drug interferes with motile behavior and produces abnormal morphogenesis. Since the cells do aggregate, we can assume that generalized adhesiveness is unchanged by the reagent, but specific adhesiveness may be altered if membrane architecture has been disturbed. Within a few minutes after treatment with cytochalasin B, cultured cells retract peripheral processes and round up, assuming a smooth surface. Microvilli reappear shortly after the reagent is removed, and cells resume normal movement.

## Contact inhibition

Another mechanism that polarizes cell movement is contact inhibition. Leo Loeb in 1921 noticed that cells in cultured explants tended to migrate away from tissue centers toward the periphery. To explain this, he proposed that surface movement stops at the site of contact when two cells meet but continues at other noncontact points. As a result, cells tend to move away from one another. This observation was not pursued until 1953 when Abercrombie and Heaysman observed that ruffled membranes of embryonic chick heart fibroblasts discontinued movement whenever they contacted another cell surface; the membrane surface became *contact-inhibited,* and movement in another direction was initiated. As cells made more contacts with their neighbors, their motility declined. If fibroblasts growing out from two explants came into contact at the advancing edge of the culture, very few cells overlapped. The surfaces of the leading cells stopped their ruffled activity, and the cells established what appeared to be a monolayer and did not move, whereas a cell with one free edge continued motility. If a monolayer was torn, cells at

the torn edge, released from contact inhibition, immediately migrated over the glass surface.

When cells are contact-inhibited, they adhere at the contact points since cells snap apart when contacts are broken. Except for tumor cells, most somatic cells in vitro exhibit contact inhibition. Tumor cells tend to migrate over one another or over normal cells after contact and may even penetrate spaces between cells (see Chapter 21). Instead of forming monolayers, they overlap and pile up on one another, forming multilayered foci.

Contact inhibition of membrane ruffling has been followed in detail by time-lapse cinephotography. Ruffling always stops whenever membranes make adhesive contact. Analysis of ruffling behavior and rate of motility indicates that cell movement is independent of ruffling. In fact the results suggest that ruffles are themselves a by-product of movement. The leading edge of the membrane may be the principal organ of locomotion.

The extent and direction of tissue migration may be controlled by contact-inhibition phenomena, particularly of tissue explants in vitro. Expansion of tissue into a circular perimeter around an explant may be explained on this basis. Similarly, the directed expansion of epithelium over a wound surface may depend on motile behavior of non-contact-inhibited cells at the free edge, pulling a contact-inhibited cell sheet behind them.

Two hypotheses for contact inhibition have been proposed: (1) the motor drive of cells is inhibited when surfaces make contact, and (2) cells exhibit differential adhesive behavior. According to the latter, cells develop stronger adhesive contacts with the substratum than with an opposing cell. Consequently, they cannot move over the cell and stop. Presumably their motile machinery is still working, but cells "prefer" not to overlap another cell since its surface is less adhesive. The front end of the cell, its motile surface, makes stronger adhesions with the substratum and pulls the posterior portion off the surface into a new position. It could be shown that mouse fibroblasts placed on a small adhesive island surrounded by a less adhesive substratum do not move off. They are, in a sense, contact-inhibited because of differential adhesiveness and not because motile activity is inhibited. Contact inhibition based on differential cell adhesion resembles Steinberg's hypothesis for sorting out. According to this view, movement of cells in confluent monolayers is not inhibited by con-

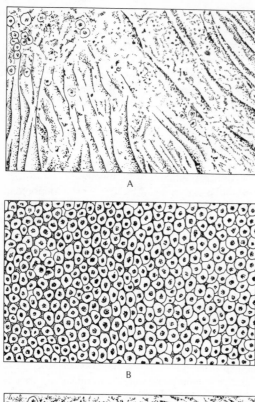

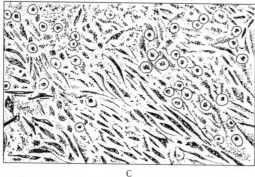

**Figure 10-16**  *Effect of cyclic AMP on contact inhibition. (A) A culture of mouse embryonic fibroblasts. Medium changed every 24 hours and photographed at 97 hours when cells were contact-inhibited, lined up in parallel arrays. (B) Mouse L cell (a tumor cell line) in culture after 120 hours. Medium changed every 24 hours. Non-contact-inhibited. Spherical cells piled up over one another in dense masses. (C) A similar culture of L cells but grown in the presence of the cyclic AMP derivative N-butyl-cAMP. Cells are contact-inhibited and have assumed the growth pattern of normal fibroblasts. (From G. S. Johnson, R. M. Friedman, and I. Pastan, Proc. Nat. Acad. Sci. U.S. 68:425–429, 1971.)*

tact, but their tendency to overlap other cells is inhibited.

Adhesion and motility may be the opposite faces of the same coin. That is, a highly motile surface tends to be nonadhesive, while a quiescent surface is more likely to make stable adhesive contacts. In fact motile, non-contact-inhibited cells may become more quiescent by treatment with cyclic AMP. Non-contact-inhibited tumor cells migrate over one another, but in the presence of cyclic AMP they become contact-inhibited and align themselves in parallel arrays (Figure 10-16). This transformation is blocked by colchicine, which suggests that microtubules are involved in both contact inhibition and the orderly association of cells into monolayers. Cyclic AMP also inhibits motility of other cells such as leucocytes, which differs from its effect on slime mold amoebae. Since the enzyme system that synthesizes cyclic AMP is located in the plasma membrane, a surface change that activates adenylcyclase (or inactivates the phosphodiesterase) may be a first step in contact inhibition.

The fact that sorting out does occur in mixed aggregates means that cells move within the aggregate; they are not contact-inhibited but tend to migrate over one another. It is uncertain whether contact inhibition plays any role in modulating the sorting-out pattern. Cells move at different rates within the aggregate, and as small aggregates develop, these too continue to move, albeit more slowly than single cells. As more cells are recruited into aggregates, their size increases, and their rates of movement decline. Such declining motility patterns, rather than contact inhibition, may account in part for the final biphasic pattern at equilibrium.

### Cell homing behavior

The directed movement of cells to specific organ sites during embryogenesis is another example of cell recognition behavior. Cells must pick up cues along the route in order to "home in" on their destination or identify their destination when they arrive. The routes taken by some peripatetic cells are highly specific, and as soon as they reach their goal, they stop migrating. Such recognition behavior is displayed by a variety of embryonic and adult cell types such as primordial germ cells, presumptive pigment cells, and lymphocytes. In this section we will consider homing-in behavior of lymphocytes and postpone discussion of the others to Chapters 11 and 12, respectively.

Lymphocytes tend to home in on specific organs such as lymph nodes and spleen as they percolate through the blood stream. They attach to capillary walls, perhaps by an adhesive mechanism, and migrate into tissue spaces. Lymphocytes leave the circulation in the postcapillary venules of the lymph node, passing between, or possibly through, cells in venule walls. They then leave the lymph nodes directly through the lymphatic circulation and enter the blood stream to be transported throughout the body. They return to the lymph node to repeat the cycle. A single lymphocyte repeats such a cycle hundreds of times, suggesting that it has a specific affinity for the walls of the lymph nodes. In addition, lymphocytes manufactured in the thymus gland, a key organ in the immune system, also tend to home in, colonizing spleen and lymph nodes. It appears as if lymphocytes recognize various organs of the immune system and localize there as they pass through the circulation. If labeled lymphocytes are treated with enzymes that digest membrane carbohydrate, fewer cells accumulate in the spleen while more accumulate in the liver. The addition of sugars such as N-acetylglucosamine reverses the effect of the enzyme. This suggests that modification of cell surface glycoproteins alters the homing mechanisms of these cells by disturbing their recognition properties.

In the case of lymphocyte homing, specific recognition sites on cell surfaces may be involved, since trypsin also abolishes lymphocyte localization in lymph nodes, presumably by digesting surface constituents. Later, these cells begin to reappear in the lymph nodes as trypsin-damaged membranes are renewed and recognition sites are reinserted in the surface.

Homing of blood-borne cells may involve a combination of specific entrapment within tissues coupled to directed migration through capillary walls, possibly in response to local chemical cues. Migratory cell populations such as embryonic neural crest are more active than blood-borne cells and migrate directly through tissue spaces, often over great distances. A discussion of these cells is postponed to Chapter 12.

## THE CELL SURFACE AND CELL COMMUNICATION

One important consequence of multicellularity is that cells communicate directly with their neighbors, not only via diffusible molecules across intercellular spaces but more directly across specialized membrane junctions between adjacent cells. These gap junctions have been mentioned briefly in Chapter 8. They are intermediate between tight junctions and desmosomes and seem to be primarily concerned with cell communication rather than adhesion.

At a gap junction the space between opposing cell surfaces is 20 to 40 Å instead of the usual 200 Å. In the electron microscope these structures are complex and consist of a series of fine channels or pores about 15 Å in diameter that pass through the abutting membranes (Figure 10-17). These channels are large enough to allow the passage of ions and small molecules, which suggests that the cytoplasms of these cells are continuous. Large macromolecules and organelles are not exchanged. If a dye or other marker substance is injected into one cell, it is seen to diffuse into neighboring cells. The channels are large enough to allow passage of dyes up to 1000 mol wt, which means that the cytoplasm of adjoining cells acts as a syncytium, sharing regulatory molecules that may even synchronize metabolic activity.

Gap junctions were first detected electrically by passing a current through one cell and picking up the somewhat attenuated signal in adjacent cells. In salivary gland epithelia of *Drosophila* larvae Loewenstein showed that a current introduced at one end of a chain of cells could be detected 10 or 15 cells away, which suggests that all cells are coupled in ion flow. A fluorescent dye injected into a cell diffuses rapidly through many cells in the epithelium. These electric junctions have been detected in epithelia of many different tissues and seem to be a primitive but widespread means of direct cellular communication. Wherever cells are electrically coupled, gap junctions are seen in the electron microscope between opposing membranes; in tissues where no electrical coupling is detected, no gap junctions are present.

These channels are labile and appear and disappear when cells come into contact or are separated. The assembly of these junctions is rapid and can occur at any place in the cell membrane. Two cells brought into contact in vitro will establish junctional communication within 10 minutes and will self-seal as soon as they are separated. Surprisingly, the cells do not have to be from the same tissue nor from the same species for coupling to occur. Possibly the ease of inducing junctional communication in cultured cells may correlate with the fact that most cells have dedifferentiated into a common cell type.

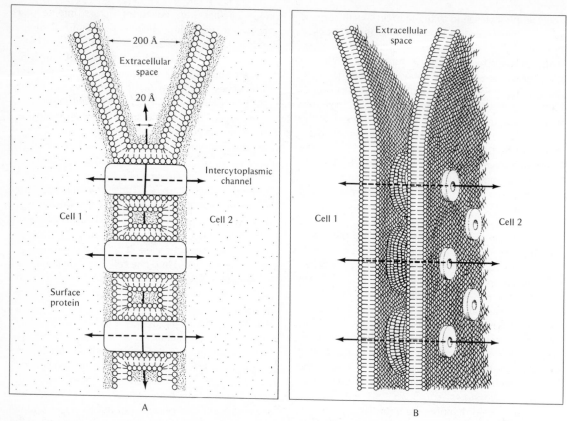

**Figure 10-17** *Structure of the gap junction. The molecular architecture of the gap junction as seen in cross section (A) and three dimensions (B) based on electron microscopic and freeze-fracture analyses. The surface proteins on the inner and outer surfaces, indicated in A in fine stippling, are not shown in B. The gap junction permits passage of ions and small molecules through channels that form a mosaic of circular units. An extracellular channel (arrow in A) allows passage of substances through the extracellular space without entering cells. The lipid bilayer of cell membranes is simplified. (Reprinted with permission from G. D. Pappas, Junctions between Cells, Hospital Practice, vol. 8, no. 8, and Cell Membranes: Biochemistry, Cell Biology and Pathology, ed. G. Weissmann and R. Claiborne. New York: HP Publishing Co., Inc., 1975.)*

It is unlikely that coupling occurs between differentiated cells in vivo.

Regulation of junctional communication depends upon calcium ion concentration (Figure 10-18). The internal free calcium concentration in most cells is less than $10^{-6}$ M compared with an external concentration of $10^{-3}$. A calcium pump in the membrane maintains this differential by pumping calcium out of the cell, and if this internal concentration increases, communication between cells is effectively blocked. Injections of calcium into cells or membrane damage which allows calcium leakage will seal channels and

inhibit ion flow. Similar uncoupling effects are observed if the membrane calcium pump which pumps calcium out of the cell or the mitochondrial calcium pump which accumulates calcium is inhibited by a drug. As soon as these pumps are activated again, calcium concentration is restored to its low levels, and the cells are electrically coupled.

Such widespread coupling of cells in embryonic tissues may coordinate growth and morphogenetic activities of cell populations. By sharing regulatory molecules, sheets of cells can behave as a community and display new patterns of sheet behavior rather than cell behavior. According to this hypothesis, movements of small populations of cells are communicated to other cells as waves of activity from a pacemaker region. Many small molecules, including ions, have regulatory functions, even to the extent of regulating gene expression. Such regulatory signals could pass through a tissue, sequentially activating specific sets of genes that translate into proliferative activity or cell differentiation. And as we have mentioned earlier, such short-range communication may

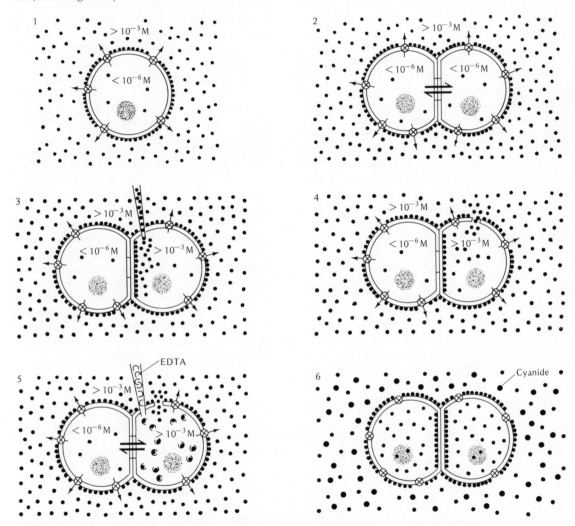

**Figure 10-18** *Role of calcium in cell coupling and cellular communication. (1) In a normal cell, Ca is held at very low levels compared to the outside concentration by an active calcium pump. (2) When such cells are brought into contact they establish gap junctions and are actively coupled. (3, 4) If the interior concentration of calcium is increased either by direct injunction or by lesions in the membrane, calcium ions accumulate across the junction, block the channels, and uncouple the cells. (5) Injection of the calcium chelating agent EDTA will reverse the blockage by soaking up the calcium, and communication is reestablished. Uncoupling is also produced (6) by stopping the calcium pump with a metabolic inhibitor, cyanide. (Reprinted with permission from W. R. Loewenstein, Cellular Communication by Permeable Membrane Junctions, vol. 9, no. 11, Hospital Practice, and Cell Membranes: Biochemistry, Cell Biology and Pathology, ed. G. Weissmann and R. Claiborne. New York: HP Publishing Co., Inc., 1975.)*

assist a cell in sensing the size of a population and its relative position in that population. Gap junctions appear in embryonic tissues at different developmental periods and seem to play an important communication role during morphogenesis.

## SUMMARY

The cellular properties of adhesion and recognition are fundamental to morphogenesis, and both phenomena show varying degrees of cell and tissue specificity. Cells "prefer" to associate with like cells more than with unlike cells. Recognition and adhesive specificity account for the ordering of multicellular populations into organized tissues and organ systems.

Both properties are difficult to measure quantitatively. They overlap; adhesive specificities may account for cell recognition. Cells with strong binding affinities tend to adhere more readily than cells which make weak bonds. As a result, mixed tissues sort out into stable configurations with one tissue enveloping another.

All cells seem to adhere by a nonspecific adhesive mechanism, but functional association depends on more precise recognition. Recognition may entail membrane "organelles," such as tight junctions or desmosomes, that join opposing cell surfaces. In addition, spaces between cells are filled with a matrix, a cement binding cells together into strong associations. Specificity of cell associations in such a case depends on the cementing material rather than on the cell membrane. Specific interactions do not require that cells of common origin share identical binding sites. All somatic cells may use one binding mechanism with specificity residing in the number and distribution of surface binding sites. Cells with more sites make stronger adhesions and will therefore tend to segregate from those with fewer sites.

Adhesive interactions explain sorting out and cell homing behavior only in part; differential motility is also important. Motile behavior depends upon surface contacts with other cells and with the substratum. Here too recognition is implicated, since motility of normal cells is inhibited when surfaces make contact. Such behavior may play a more important role in controlling motility of cells organized as two-dimensional sheets in vitro than it does in vivo. Cells in vivo display some movement in three-dimensional aggregates.

The cell surface also participates directly in cell communication in multicellular systems. Gap junctions between cells facilitate the flow of ions and regulatory molecules which may integrate individual cells into a cohesive sheet with new morphogenetic properties. The transient nature of these elements and the frequency with which they appear in embryonic tissues suggest that they are part of a primitive communication system that may regulate spatial relationships of cells during morphogenesis.

**REFERENCES**     Abercrombie, M. 1970. Contact inhibition in tissue culture. *In Vitro* 6:128–142.

Albertini, D. F., and Anderson, E. 1977. Microtubule and microfilament rearrangements during capping of Con A receptors on cultured ovarian granulosa cells. *J. Cell Biol.* 73:111–127.

Cook, G. M. W., and Stoddart, R. W. 1973. *Surface carbohydrates of the eukaryotic cell.* New York: Academic Press.

Curtis, A. S. G. 1967. *The cell surface: Its molecular role in morphogenesis.* London: Logos Press.

————. 1973. Cell adhesion. *Progress in biophysics and molecular biology,* ed. J. A. V. Butler and D. Noble. Oxford: Pergamon Press, pp. 315–380.

Edelman, G. M. 1976. Surface modulation in cell recognition and cell growth. *Science* 192:218–226.

Edelman, G. M., Yahara, I., and Wang, J. L. 1973. Receptor mobility and receptor-cytoplasmic interactions in lymphocytes. *Proc. Nat. Acad. Sci. U.S.* 70:1442–1447.

Frye, L. D., and Edidin, M. 1970. The rapid intermixing of cell surface antigens after formation of mouse-human heterokaryons. *J. Cell. Sci.* 7:319–335.

Furshpan, E. J., and Potter, D. D. 1968. Low-resistance junctions between cells in embryos and tissue culture. *Current topics in developmental biology,* ed. A. A. Moscona and A. Monroy. Vol. 3. New York: Academic Press, pp. 95–127.

Lilien, J. 1968. Specific enhancement of cell aggregation in vitro. *Develop. Biol.* 17:657–678.

Loewenstein, W. R. 1973. Membrane junctions in growth and differentiation. *Fed. Proc.* 32:60.

Moscona, A. A. 1962. Analysis of cell recombination in experimental synthesis of tissues in vitro. *J. Cell Comp. Physiol.* 60 (Suppl. 1):65–80.

Moscona, A. A., ed. 1974. *The cell surface in development.* New York: Wiley.

Nicolson, G. L. 1974. The interaction of lectins with animal cell surfaces. *Int. Rev. Cytol.* 39:89–190.

Phillips, H. M., and Steinberg, M. S. 1969. Equilibrium measurements of embryonic chick cell adhesiveness. I. Shape equilibrium in centrifugal fields. *Proc. Nat. Acad. Sci. U.S.* 64:121–127.

Roseman, S. 1974. Complex carbohydrates and intercellular adhesion. *Biology and chemistry of eukaryotic cell surfaces,* ed. E. Y. C. Lee and E. E. Smith. New York: Academic Press.

Roth, S. 1968. Studies on intercellular adhesion selectivity. *Develop. Biol.* 18:602–631.

Roth, S., and Weston, J. 1967. The measurement of intercellular adhesion. *Proc. Nat. Acad. Sci. U.S.* 58:974–980.

Rutishauser, U. S., Yang, D. C. H., Thiery, J. P., and Edelman, G. M. 1975. Expression and behavior of cell surface receptors in development. *Developmental biology,* eds. D. McMahon and C. F. Fox. *ICN-UCLA Symp. Mol. Cell. Biol.* Vol. 2. Menlo Park, Calif.: Benjamin, pp. 3–18.

Staehelin, L. A. 1974. Structure and function of intercellular junctions. *Int. Rev. Cytol.* 39:191–283.

Steinberg, M. S. 1964. The problem of adhesive selectivity in cellular interactions. *Cellular membranes in development,* ed. M. Locke. New York: Academic Press, pp. 321–366.

_____. 1970. Does differential adhesion govern self-assembly process in histogenesis? Equilibrium configurations and the emergence of a hierarchy among populations of embryonic cells. *J. Exp. Zool.* 173:395–434.

Steinberg, M. S., and Garrod, D. R. 1975. Observations on the sorting out of embryonic cells in monolayer culture. *J. Cell Sci.* 18:385–403.

Taylor, R. B., Duffus, W. P. H., Raff, M. C., and DePetris, S. 1971. Redistribution and pinocytosis of lymphocyte surface immunoglobulin molecules induced by antiimmunoglobin antibody. *Nature* 233:225–229.

Townes, P. L., and Holtfreter, J. 1955. Directed movement and selective adhesion of embryonic amphibian cells. *J. Exp. Zool.* 128:53–120.

Tyler, A. 1947. An autoantibody concept of cell structure, growth, and differentiation. *Growth* 10:6–7.

Weiss, P. 1958. Cell contact. *Int. Rev. Cytol.* 7:391–423.

Weissmann, G., and Claiborne, R., eds. 1975. *Cell membranes, biochemistry, cell biology and pathology.* New York: HP Publishing.

# The Egg as a Developmental System

The egg is a unique cell; it is the cell fated to segregate from the organism and develop into a new individual. Larger than most somatic cells, the egg derives its developmental potential not only from a munificent endowment of organelles but also from the organization of its outer cortex and plasma membrane. Moreover, an egg contains special cytoplasmic substances, *morphogens,* that are responsible for morphogenetic expression. The essence of an egg, therefore, resides in its store of informational molecules and their functional organization. E. B. Wilson (*The Cells in Development and Heredity,* 3d ed, New York: Macmillan, 1925), impressed by these remarkable cells, stated:

> That a single cell can carry the total heritage of the complex adult, that it can in the course of a few days or weeks give rise to a mollusc, or a man, is one of the great marvels of nature. In attempting to attack the problems here in-

volved, we must from the outset hold fast to the fact that the specific formative energy of the germ is not impressed on it from without, but is somehow determined by an internal organization inherent in the egg and handed on intact from one generation to another by cell division. Precisely what this organization is we do not know. We do know that it is a heritage somehow perpetuated by cell division and that development is only a further extension of the processes that have been going on since life began. The dramatic aspect of development is due to the apparent simplicity of its starting point, the suddenness of its onset, the prodigious speed with which it goes forward, and the complexity of the results achieved within a span so brief.

Developmental programming, however, is only a secondary role for eggs; they are primarily gametes designed for sexual recombination. In most cases, an egg is a meiotic product and transmits a species' genetic endowment to the next generation. In primitive algae and fungi the adult organism is haploid; gametes are usually isogamous. Male and female gametes are morphologically and physiologically identical, except for differences in mating type. The zygotes obtained from fertilization usually become dormant and undergo meiosis before germinating into haploid spores which develop into the adult filament. The arrested meiosis of the zygote persists over the dormant period until germination is initiated.

*Oogamy*, or gamete differentiation into a large egg and a small motile sperm, may have evolved as the egg acquired developmental potential, perhaps in a manner exemplified by *Volvox* (see Chapter 9). In this species developmental potential resides in the asexually reproducing gonidia, the large reproductive cells. Retaining proliferative potential, they undergo regular cleavage (like most eggs) to produce a multicellular sheet organized as a hollow ball. At this point, the hollow ball inverts, a morphogenetic change that resembles gastrulation (see Chapter 12). On the other hand, *Volvox* gametes, though differentiated into a large egg and a small motile sperm, are true sexual recombinants and do not exhibit a developmental program. Instead, a zygote becomes dormant and completes meiosis before it germinates into a large zoospore. The haploid zoospore then develops, like a gonidium, into an individual coenobium. During the course of evolution the egg, because of its large size, may have taken over the developmental role of the gonidium; its large size gives it the proliferative potential to produce a multicellular population. In addition, at some point in evolution, meiosis was suppressed in the zygote and a diploid state emerged to evolve as the more dominant phase of the life cycle in the advanced algae and, of course, in the evolution of plants and animals.

An egg's primary developmental expression is rapid proliferation. Most eggs contain a reserve of cell division precursors that sustain an early period of synchronous division called cleavage. These divisions are rapid (cell cycles are short) and in contrast to all other proliferating cells, cleaving cells do not grow between divisions. Instead, cell size becomes progressively smaller with each division until the small size of the adult cell is achieved. The cleavage process produces populations of cells organized as two-dimensional sheets, with new morphogenetic properties.

The special properties of eggs depend upon two factors, their developmental history (or mode of origin) and their growth in the ovary when they store developmental information.

## ORIGIN OF EGGS: GERM CELL DETERMINATION

In many species, ranging from nematode worms to mammals, certain cells are precociously singled out as progenitors of all sex cells. Usually, during the first few divisions of the zygote, a few cells, the primordial germ cells, are segregated from the rest and invade a developing gonad where in the mature adult they differentiate into gametes. The factors responsible for early germ cell determination are localized in egg cytoplasm as special morphogenetic substances or *germinal cytoplasms*. We can say that an egg begins when these germinal cytoplasms interact with nuclei of presumptive germ cells. Their ultimate fate as gametes, months or even years later, is determined at that time.

It was August Weismann, in 1892, who focused attention on germ cell origins with his theory of "the continuity of the germplasm." He followed the fate of the *pole cells* in early insect embryos and observed that these future germ cells migrated to the gonads. According to his theory all multicellular organisms contain a special material, the *germplasm*, which faithfully transmits the complete hereditary makeup of the species from one generation to the next. Early in development, certain cells, the future germ line, segregate from all others, the somatic cell line or *soma*. Weismann believed that only the germ line retained the full chromosome complement of the species while cells in the somatic line lost nuclear determinants during the course of development. The body, or soma, was viewed as the temporary custodian of a continuous line of germ cells which pass on their hereditary material from one generation to another.

Mechanisms of germ cell determination vary; some seem to correlate with changes in chromosome number while others are more subtle and do not display any visible nuclear changes.

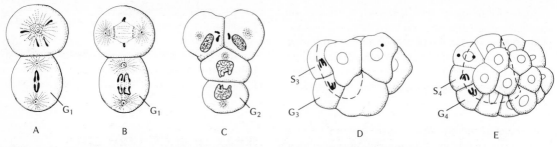

**Figure 11-1**  *Chromosome diminution and germ cell determination in Ascaris megalocephala. (A–C) Chromosome diminution in second cleavage division. Diminuted chromosomes in upper cell (B) appear in cytoplasm after division (C). $G_1$ cell divides normally in plane perpendicular to animal-vegetal axis. In third division all cells except $G_2$ cell undergo diminution. (D) Seven-cell stage. (E) Twenty-four-cell stage. In each case only one cell undergoes normal division, the G or germ-line cell, while all others undergo diminution. $G_3$, $G_4$, $S_3$, and $S_4$ are germ and soma derivatives of $G_2$.*

## Ascaris: chromosome diminution

Boveri's analysis of *Ascaris* development supported Weissmann's theory. This nematode species has only two chromosomes which replicate normally in the first zygotic division into a dorsal and ventral cell (Figure 11-1). In the second division the dorsal cell undergoes chromosome diminution; that is, the heterochromatic ends of each chromosome fragment into small bits and are lost in the cytoplasm while central portions are transmitted to daughter cells. The ventral cell, on the other hand, divides into two cells, each with complete chromosomes. Later, in the third division, one of these cells undergoes diminution while the other retains its full chromosome set. Thus at the eight-cell stage only one cell, the primordial germ cell, retains the full chromosome complement. It continues to divide, always producing one cell with the full chromosome set and another with a diminished set. Each time, only the heterochromatic ends of the chromosomes are lost. All cells with diminuted chromosomes become somatic cells of the adult worm whereas the cell with two undiminished chromosomes is the progenitor of the germ line and its progeny invade the gonad and become gametes. The germ line in *Ascaris* retains the complete genetic endowment of the species and passes it on to the next generation.

Boveri showed that a factor in egg cytoplasm prevents chromosome diminution in the germ cell line. The presumptive germ cells receive this factor at first cleavage since it is localized in the vegetal or lower half of egg cytoplasm. If cytoplasmic constituents are displaced by centrifugation so that both cells

receive the factor, then no diminution is observed. Moreover, if both nuclei are displaced into the animal-half region of the egg before completion of the two-cell stage, they are deprived of the vegetal factor and exhibit diminution. If the factor is destroyed by ultraviolet irradiation, then all nuclei undergo diminution. This UV-sensitive material is segregated to the germ cell line during the first few cleavage divisions. In its presence chromosomes replicate and segregate normally; in its absence, they fragment and undergo diminution. The factor seems to inhibit a cytoplasmic system preventing normal chromosomal replication and segregation during mitosis.

Why does the germ line retain the full genome complement? Analyses of germ cell determination in dipteran insects suggest a role for heterochromatin in gamete differentiation.

## Chromosome elimination in gall midges

In the eggs of gall midges *(Wachtiella)* a region of cytoplasm at the posterior pole of the egg is distinguishable from the rest (Figure 11-2). This polar plasm, a deeply staining granular cytoplasm dispersed in a reticular framework, determines the germ cell line.

During the first three divisions, nuclei divide synchronously in the central yolk as a syncytium to produce 8 nuclei. One of these nuclei enters the polar plasm which pinches off as a discrete cell, a pole cell. At the next division a wave of mitosis passes through the 7 nuclei in the center which proceed to eliminate most chromosomes; approximately 30 chromosomes (mostly heterochromatic) out of a total of 40-plus fail to separate at anaphase and form a dense elimination mass in the cytoplasm. Each of the 14 nuclei resulting from this division has 8 chromosomes, the normal number for male and female somatic tissues, respectively. The pole cell, on the other hand, divides later into two cells, each with a full chromosome set of 40-plus. As cleavage continues, somatic nuclei become smaller but the large pole cells are easily identified. At the end of cleavage, when more than 1000 nuclei have appeared in the central yolk, somatic nuclei migrate to the egg cortex, forming a layer of cells at the periphery of the blastoderm. The primordial germ cells at the pole later invaginate and migrate to the developing gonad.

If the polar plasm is separated by a fine ligature from cleavage nuclei in the center, then all eight nuclei eliminate chromosomes in the next division. Moreover, displacement of granular plasm from the posterior pole into the central yolk area protects any nucleus with which it interacts from chromosome elimination at the fourth division. Evidently polar plasm prevents chromosome elimination. At the fourth division, egg cytoplasm signals chromosome elimination in all nuclei except those protected by polar plasm, but the nature of this factor and its mode of action are obscure. One possibility is that kinetochores of eliminated chromosomes do not function normally since these chromosomes do not migrate in the mitotic spindle and are subsequently lost to the cytoplasm. Polar plasm may protect chromosomes by acting at the kinetochore region.

A full chromosome set, however, is not essential for all aspects of germ cell behavior. For instance, germ cell migration to the gonads is independent of the 30 additional chromosomes. If a ligature segregating polar plasm is released after all nuclei have eliminated chromosomes, then any nucleus with eight chromosomes entering the polar plasm enlarges and divides independently of all somatic nuclei. It behaves like a primordial germ cell and migrates into the developing gonad. This means that 30 elimination chromosomes are unnecessary for primordial germ cell migration, but polar plasm is essential. These embryos develop into normal adults. Testes are normal in males with viable sperm (only eight chromosomes) but ovaries are rudimentary; oogenesis is completely inhibited and no mature eggs develop. It appears that germ cell chromosomes are

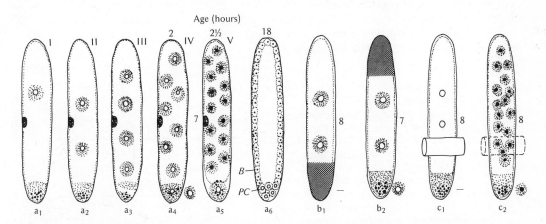

**Figure 11-2**   *A series of experiments showing the role of polar plasm in the gall midge* Wachtiella persicariae. *(From I. Geyer-Duszynska, J. Exp. Zool. 141:391–441, 1959.) The a₁–a₆ series illustrates the events during normal cleavage. Polar plasm is indicated by stippling and granules at lower end. First three mitotic divisions (I, II, III) are normal (a₁–a₄). One nucleus enters polar plasm. At fourth mitosis, all nuclei except pole nuclei eliminate 30-odd chromosomes (shown in black; white before elimination). In a₆, B indicates blastoderm and PC pole cells. In b₁–b₂, cauterization of polar plasm results in chromosome elimination in all eight nuclei at fourth mitosis, whereas only seven are affected if the opposite end is cauterized. Diagrams c₁–c₂ illustrate a ligature experiment. Segregation of polar plasm leads to elimination in all eight nuclei. In c₂, hair is removed after fourth mitosis; all eight nuclei have eliminated chromosomes including any that enter polar plasm.*

indispensable for oogenesis but have no role in spermatogenesis. Recently it has been shown that the 30 extra chromosomes do transcribe RNA during oocyte maturation.

## Polar plasm in Drosophila

The polar plasm of the *Drosophila* egg has been studied with the electron microscope. In *Drosophila*, chromosome elimination does not occur in somatic nuclei. Large pole cells appear in an early cleavage at the posterior pole, divide asynchronously, and migrate to the gonads in later development. Polar plasm undergoes characteristic changes through several cleavage divisions. Polar granules are meshes of fibrous elements about 0.5 to 1 $\mu$ in diameter, rich in RNA and often attached to mitochondria. After fertilization, they break down and associate with the nuclear envelope of any nucleus that enters the pole. At this stage, chromatin becomes more compact in discrete regions and nucleoli appear. But we do not know how granules affect nuclear function. One effect they have is to induce asynchrony of pole cell mitosis; its cell cycle lengthens in contrast to somatic cells.

Perhaps the most direct test of polar plasm as a germ cell determinant is to inject polar plasm into somatic cells and change their developmental fate in the direction of germ cell formation. Just such a test has been carried out with *Drosophila* by Illmansee and Mahowald (Figure 11-3).

**Figure 11-3** *Experiment demonstrating the determination of germ cells by transplantation of polar plasm in* Drosophila. *Polar plasm from a wild type fly is withdrawn in a micropipette and injected into the anterior pole of eggs of mutant flies (mwh e; minute, white eye, and ebony). The anterior cells formed in this region during cleavage and blastoderm formation are removed and transplanted into the posterior pole of cleaving embryos of another mutant, y w sn³ (yellow, white, and singed). All these host embryos developed into mutant y w sn³ adults and were mated to other y w sn³ flies. Of the 92 fertile host flies mated, 88 flies gave only y w sn³ progeny, indicating they contained no germ cells derived from the transplanted cells. Four flies, however, gave the progeny shown. The high percentages of wild type progeny obtained from these flies indicated that some of the germ cells came from the transplanted cells and must have had the mwh e genotype to account for the appearance of the wild type flies. (From K. Illmansee and Λ. Mahowald, Proc. Nat. Acad. Sci. U.S. 71:1016, 1974.)*

In normal development polar plasm is found in the posterior region of the *Drosophila* egg, and after the seventh synchronous mitotic division, a few nuclei enter the posterior region to emerge as spherical pole cells. These divide asynchronously to produce about 55 pole cells while the remaining somatic nuclei complete three more synchronous divisions and migrate to the periphery to form a blastoderm.

At an early four-nuclei stage polar plasm was sucked out of the posterior region of wild type flies and injected into the anterior region of mutant flies homozygous for multiple wing hair and ebony body

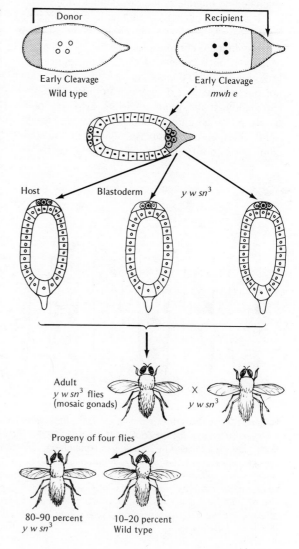

(mwh e). The recipient eggs continued to develop until the blastoderm stage, at which time groups of spherical-appearing cells at the anterior end were sucked out with a micropipette (normally no cells in the anterior region give rise to any germ cells or gonad derivatives) and transplanted into the posterior region of host embryos of the mutant yellow, white, and singed y w sn³. When these flies developed into adults they were mated with yellow-white-singed flies y w sn³; the phenotype and genotype of their offspring were then analyzed. If polar plasm induced anterior somatic cells to become germ cells, then some of the host y w sn³ flies receiving a cell transplant should have mosaic gonads, that is, gametes of yellow-white-singed genotype and gametes of the multiple wing hair ebony genotype. Since y w sn³ genes are on the X chromosome and mwh e mutant genes are on the third chromosome, then among the progeny of a mosaic fly we should find a certain frequency of wild type offspring. Of 125 fertile host flies receiving cell transfers, four turned out to be mosaic; that is, they produced wild type and yellow-white-singed offspring. Clearly anterior somatic cells of mwh e had been transformed to primordial germ cells after receiving polar plasm from a wild type fly.

Histological examination of recipient embryos showed polar granules in some anterior cells, a strong argument in support of the hypothesis that polar granules are specifically involved in determining the germ cell fate of a cell nucleus.

## Germ cell determination in vertebrates

Primordial germ cell determination in vertebrates also correlates with a precocious segregation of germ from soma. As in insects, segregation of primordial germ cells is dependent upon a special egg cytoplasm. In frog eggs, this unique cytoplasm (germinal cytoplasm) stains differently from the surrounding vegetal yolk and can be followed during early cleavage. The cells bearing germ plasm are larger than most, but their mitotic behavior is no different from that of somatic cells, at least in early stages. If the vegetal pole of the frog egg is irradiated with ultraviolet light shortly after fertilization, a large proportion of the animals that develop are sterile; they lack functional gonads. A cytoplasmic component, sensitive to UV and necessary for normal gamete development, is segregated during the first cleavage divisions into presumptive germ cells.

Later, large primordial germ cells carrying the stainable germinal plasm can be localized in the vegetal "belly" region of the embryo (Figure 11-4). In young tadpoles they can be seen migrating in single file along the border of the intestine. Their identity as germ cells has been shown by the following experiment. A belly region from an embryo of one Xenopus strain is replaced with a block of ventral tissue from another strain bearing a nucleolar marker gene. Cells bearing this gene have only one nucleolus rather than two and can be easily identified in the embryo. The embryos develop into animals whose gonads contain gametes genetically identical to the donor strain and not the host; that is, they carry the nucleolar marker. Primordial germ cells in the graft populate the host gonad and generate gametes characteristic of the donor strain. Most importantly, when these animals are mated their progeny resembles the strain from which the graft was derived, and not that of the host.

If germinal plasm is localized at the vegetal pole, then it should be possible to transfer it from egg to egg. D. Smith found that the UV effect could be corrected by injecting normal egg cytoplasm from the vegetal pole into irradiated eggs which do not yield any germ cells. Twenty-three to forty-seven percent of irradiated eggs receiving injections of vegetal cytoplasm developed into tadpoles with germ cells and fully formed gonads. The number of germ cells restored seemed to correlate with the amount of vegetal cytoplasm injected; moreover, no germ cells appeared if cytoplasm from other egg regions was injected. Clearly, the cytoplasmic factors for germ cell determination are localized in the vegetal pole of the egg.

A clue as to the nature of the UV-sensitive material was obtained by an action spectrum analysis; that is, eggs were irradiated at several different UV wavelengths and the extent of inhibition of germ cell formation was measured for each wavelength. Peak inhibition was obtained at 2537 Å, the absorption peak for nucleic acids. Since the vegetal pole region does stain positively for RNA, the active germ plasm may be RNA, a ribonucleoprotein, or even DNA. An electron micrographic study of the vegetal pole reveals small clusters of dense granules that resemble polar granules in insect eggs. These stain for RNA and do migrate into the cleaving blastomeres as predicted by the UV studies. The fact that germ cell determination depends on similar organelles in egg cytoplasm

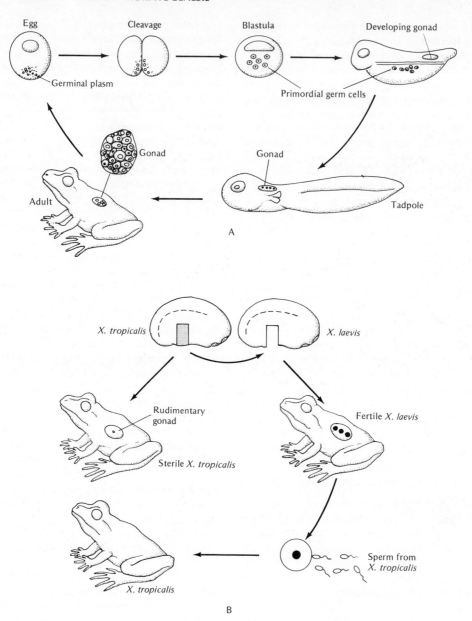

**Figure 11-4**  *Continuity of germ plasm in the frog. (A) Germ plasm, localized in the vegetal pole of eggs, enters several blastomeres during cleavage. These become primordial germ cells which migrate in the wall of the larval intestine to the developing gonad. They invade the gonad and multiply as oogonia or spermatogonia, respectively. Maturation is completed in  the adult, and during oogenesis additional germ plasm is synthesized and stored in the oocytes. (B) Experiment to demonstrate the presence of primordial germ cells in the ventral region of the early embryo. A block of ventral tissue from an early neurula of Xenopus tropicalis is grafted into a X. laevis neurula of the same stage. The donor is allowed to heal and metamorphose into an adult. This animal contains rudimentary gonads with no gametes. The X. laevis host metamorphoses into a fertile adult which (in the case shown here) produces eggs that are genetically those of X. tropicalis, as demonstrated by mating with a X. tropicalis male.*

in such unrelated groups as insects and frogs suggests a common underlying mechanism.

A precocious segregation of the germ line from the somatic line is also seen in birds and mammals, including humans. Primordial germ cells can be identified very early in the development of mammals. Long before the gonads develop, 100 primordial germ cells may be seen in the endodermal yolk sac epithelium. They are usually large cells that show a high level of alkaline phosphatase activity. The cells migrate up the dorsal gut mesentery into the germinal ridge (future gonad) and multiply as they migrate until some 5000 cells ultimately invade the gonads.

## Germ versus soma

Among insects and vertebrates, special germinal plasms are segregated early into progenitors of the germ line. But all germ cells do not arise in this manner nor is there always an early segregation of germ from soma. There is no sharp distinction between germ and somatic cell lines in plants. Isolated meristematic cells from root tips, shoot apices, and buds develop into new plants. Even somatic cells of carrot roots develop into complete plants with flowers and fruits when cultured in vitro (see Chapter 17). Meristematic tissues that normally give rise to flowers with sexual reproductive structures also develop into vegetative shoots. Clearly germ and soma are interconvertible in plants.

Precocious differentiation of germ cells is not an important developmental feature in many invertebrate animals. There is no separate germ line in coelenterates and flatworms. Gonads and gametes may arise from various somatic tissues. In coelenterates, gonads develop from either the outer ectoderm or inner endoderm lining the gastrointestinal cavity. Interstitial cells in these tissue layers proliferate and differentiate into gametes. These undifferentiated cells normally replenish the stinging cells or nematoblasts on the tentacles, but changes in temperature or nutrition stimulate these cells to transform into gametes. Here we see that the same progenitor cell can give rise to germ or soma, depending on environmental conditions.

There is a reversible relationship between germ and soma. But this depends upon the animal group. If gonads are excised in insects and vertebrates, sterility results. All potential germ cells are sequestered in the gonads, and no other cells differentiate into germ cells after gonad removal. In many invertebrates, however, excision of the gonads and neighboring tissues does not sterilize the animal since new functional gonads may regenerate from adjacent somatic tissues. The reason for precocious segregation of germ from soma in insects and vertebrates is not understood.

Germinal plasm probably triggers specific nuclear responses during the course of embryogenesis. Even in lower invertebrates where the dichotomy between germ and soma is not sharp, presumptive germ cells are unique; only they enter meiosis, differentiate, and store developmental information. The potential to undergo meiosis distinguishes germ from soma and it is likely that the extra chromatin in the germ line of *Ascaris* and gall midges carries genes that regulate meiosis.

## OOGENESIS: THE ORGANIZATION OF DEVELOPMENTAL INFORMATION

An egg acquires its developmental potential during oogenesis, a period of growth and maturation in the ovary. The trigger for oogenesis is the onset of meiosis which begins after the egg has already completed a long developmental history.

Primordial germ cells arise early in the vertebrate embryo and begin to migrate to the site of the future gonad, sometimes over considerable distances, proliferating along the way. After germ cells invade the gonad, it completes development, as if gonadal differentiation depends upon factors produced by germ cells. In fact, gonadal sex (that is, whether a testis or ovary develops) seems to correlate with the site of germ cell invasion, at least in humans. In males, primordial germ cells invade the central medullary regions of the immature gonad which develops into a testis. Female germ cells invade the cortex of the immature gonad and it is this outer region that develops into an ovary at the expense of the medulla. The development of the testis is dependent upon testicular-inducing substances that are probably coded by genes in the Y chromosome of gonadal cells. In the absence of the Y chromosome in females, no testicular inducing substances are produced and the gonad develops as an ovary.

Although determination of primordial germ cells occurs early in vertebrate embryogenesis, differentiation into functional gametes, or *gametogenesis,* is delayed for a long period; in humans it is delayed for years, until reproductive maturity, or puberty. Meiosis marks a basic transition in gamete development. Once a gonocyte enters meiosis, it is committed to gamete differentiation. Meiosis is the scaffolding on which the events of oogenesis hang. Accordingly, oogenesis is divided into a premeiotic or proliferative phase and a meiotic phase. The latter is subdivided into early meiotic events interrupted by a later vegetative phase of oocyte growth, yolk synthesis, and maturation. The pattern for amphibian oogenesis is shown in Figure 11-5.

### The premeiotic phase

The premeiotic phase is a period of oogonial proliferation within the ovary, usually during embryonic, larval, or fetal life. This occurs in all vertebrates, but its timing differs in major groups. In mammals and birds, primordial germ cells proliferate into a limited number of oogonia; the premeiotic phase is completed once in the life of an organism, long before sexual maturity. No proliferating oogonia are seen in an adult ovary. In fish and amphibians, the premeiotic proliferative phase is seasonal throughout the adult life of the organism. New oogonia are replenished each mating season from a reserve population of mitotically active gonocytes in mature ovaries.

Not all oogonia mature into functional eggs since many do not survive into the adult. Surviving oogonia usually begin meiosis in larval and fetal stages and when they do, they become *primary oocytes.* This marks the transition between premeiotic and meiotic phases.

Egg development in vertebrates is a wasteful process. Most oogonia fail to mature and degenerate. Such wholesale attrition of oogonia attends gonadal development in human females. Counts made early in development indicate that approximately 1700 primordial germ cells migrate and invade the indifferent female gonad at one month of fetal development. These proliferate actively so that at two months of gestation, the gonad contains approximately 600,000 oogonia, and by the fifth month, when the gonad assumes its female characteristics, there are 7 million. Mitosis usually ceases at this point, and many oogonia enter the first meiotic prophase and become oocytes. The remaining oogonia undergo

atresia so that at birth only 2 million oocytes remain, many still degenerating. After birth there is a continuing decline so that by puberty only 300,000 oocytes remain and these too continue to undergo atresia. Each month, several oocytes are stimulated to mature, but only one usually completes meiosis and ovulates; all others degenerate. This means that during the roughly 40 years of reproductive life in a woman (until menopause), only 400 to 500 eggs will have been ovulated. All others degenerate since no oocytes can be detected in the ovaries of menopausal women.

The reasons for reproductive wastage are not understood. Similar wastage occurs in most other vertebrates although not to the same extent. There may be a selective advantage in that it insures that some mature eggs are always available for fertilization. In humans, some mature eggs will have been in arrested meiosis for 40 or 50 years, which may explain the higher incidence of congenital abnormalities among children from older women. Delayed ovulation may lead to overly ripe eggs that are more susceptible to chromosomal aberrations (see Chapter 20).

### The meiotic phase

Before oogonia enter meiosis, their chromosomes replicate as if preparing for another mitotic division. The amount of DNA doubles from the diploid 2C value to a 4C or tetraploid value. Shortly thereafter, prophase begins and replicated chromosomes condense, but in contrast to mitosis, prophase is considerably protracted, lasting for months and even years. The arrested prophase is the most distinguishing feature of oocyte meiosis. The oocyte grows and acquires its developmental information during this phase. The dual functions of an egg, its role as a sexual recombinant and its morphogenetic potential, are acquired during this long prophase. Because of its duration, prophase may be arbitrarily divided into five distinct stages (see Figure 11-5).

### Leptotene

This phase is similar to an early prophase in mitosis. The chromosomes appear as long, thin threads without definite structure. Each has already duplicated and consists of two chromatids. At the end of this stage, a constant number and size of small chromo-

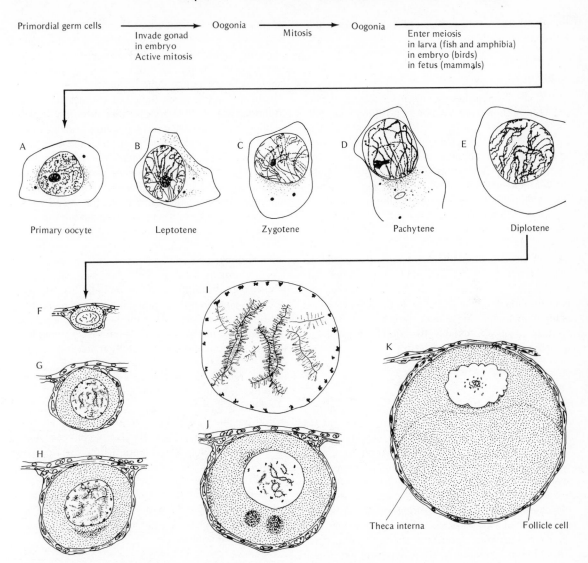

**Figure 11-5** *Oogenesis and meiosis in vertebrates. (A–E) Early meiotic phases in the cat oocyte. (A) Primary oocyte at the beginning of meiosis. Nucleus still in interphase at time of DNA and chromosome replication. (B–E) Meiotic prophase, the leptotene to diplotene. Meiosis is usually arrested in diplotene for many months or years (12 years in humans). (F–K) The vegetative growth phase in the amphibian oocyte. (F) An early diplotene stage comparable to (E) above. The oocyte is surrounded by a follicle cell layer and another layer of epithelial cells, the theca interna. (G) The diplotene nucleus enlarges as lampbrush chromosomes appear, each joined at places where crossovers occurred earlier. (H–J) Later stages of nuclear enlargement into the germinal vesicle. (I) An isolated germinal vesicle from a growing oocyte showing the structure of lampbrush chromosomes. Nucleoli are numerous (about 1000 per germinal vesicle) and are localized at the periphery beneath the nuclear membrane. (K) A late stage at the end of growth. The germinal vesicle moves to the animal pole and its membrane begins to shrink. The chromosomes condense in the center of the germinal vesicle in preparation for the first meiotic division and extrusion of the first polar body.*

meric granules may be seen dispersed along the chromatid threads.

## Zygotene

This stage distinguishes meiosis from mitosis. While the chromosomes are still long and threadlike, homologous chromosomes, the identical paternal-maternal homologs, pair with one another along their long axes at several places in the chromosome. This pairing process, known as *synapsis,* continues until the homologous chromosomes are aligned along their full length. A special structure, the *synaptinemal complex,* is seen in the paired homologs, or *bivalents,* as a ribbonlike element made up of two parallel, dense rods separated by a uniform 1000- to 2000-Å gap. A median longitudinal element in the gap is connected to the rods by lateral fibers. Synaptinemal complexes have been seen primarily in spermatogenesis and consist largely of protein. Each homolog is lined up on one side of the complex, held in place by its elements. Presumably, the complex is involved in pairing.

## Pachytene

This is the terminal phase of pairing. Chromosomes are visibly thicker and more closely paired. The number of bivalents equals the haploid number of the species. For example, a human oocyte at this stage would have 23 bivalents. It is during pairing, or the synaptic period of pachytene, that crossing over between homologous chromosomes occurs. Presumably stresses and strains of pairing cause breaks in chromosomal DNA, but these are immediately repaired by enzymes during the shortening process. An additional synthesis of DNA occurs at this time, as the ends of broken strands are rejoined by insertion of a few nucleotides. Since chromatids are physically intimate at this stage, their broken free ends are repaired at random, leading to chromatid exchange and crossing over between homologous chromosomes, the principal genetic recombination mechanism of meiosis.

## Diplotene

The consequences of crossing over become evident during diplotene. Homologous chromosomes begin to separate from one another. The chromatid structure of each homolog is now seen with each bivalent consisting of four chromatids. Inasmuch as crossover points prevent homologs from separating, these are seen as chiasmata along the bivalents.

In spermatogenesis, the chromosomes continue to shorten during diplotene and become visibly coiled as they are packaged for segregation and the first meiotic metaphase. In most oocytes, however, instead of coiling into a compact chromosome the paired, crossed-over chromosomes become more diffuse, longer, and threadlike, transforming into a typical "lampbrush" structure. Chromatid threads unwind at many points along the axis and are thrown out into numerous lateral loops. Meanwhile, the nucleus enlarges into a germinal vesicle. Prophase stops abruptly at this point, usually for long periods of time. In humans, for example, a diplotene stage is established by the fifth month of prenatal life and all oocytes persist in this condition until sexual maturity, 12 to 15 years later. At puberty small groups of oocytes resume meiosis in response to periodic hormonal stimulation; one usually matures to complete meiosis during ovulation each month. Many oocytes remain in the diplotene stage for 50 years until the human female stops producing eggs.

During diplotene, oocyte growth and differentiation begin. The oocyte genome and maternal genomes in surrounding accessory cells actively synthesize developmental information for storage in egg cytoplasm. This ushers in the vegetative phase of yolk, glycogen, and lipid synthesis as diverse organelles accumulate in the cytoplasm. The developmental organization of the egg is established, a process that may take several months under continuous hormonal control.

## Diakinesis

After the growth phase is completed, meiosis resumes in most eggs, usually in response to hormonal stimulation. Threadlike chromosomes within the germinal vesicle coil into compact elements; bivalents assume a more rounded shape with homologs joined to each other at their terminal ends. The shortened, compacted bivalents prepare to enter the first meiotic metaphase. The germinal vesicle breaks down, releasing its contents into the cytoplasm, and the metaphase of the first meiotic division begins.

Eggs need not complete meiosis before they are released from the ovary. They may be ready for fertilization at various stages in meiosis as long as vegetative growth is completed. Depending on species, meiosis may be completed within the ovary, in the oviduct, or outside the body before or even after fertilization.

In metaphase, bivalents move into the center of the spindle, each with two undivided kinetochores, and homologous chromosomes separate at anaphase. Another recombination event occurs here since paternal and maternal sets segregate at random. An anaphase group consists of a haploid set of chromosomes (each composed of two chromatids joined at the kinetochore) instead of a diploid number of chromosomes as in mitosis. A daughter nucleus after the first metaphase contains a 2C amount of DNA with each chromosome made up of two chromatids. A telophase stage ensues as relaxed coiled chromosomes are restored to a resting or interphase nucleus. The two cells are now *secondary oocytes* but differ drastically in size. During division, the spindle migrates up near the surface of the animal pole and a small cell, or *polar body,* is separated from the larger egg. All developmental information is retained in the large cell. An equal division would essentially destroy the developmental potential of the egg since each daughter would receive only half the information.

Again, the time of first polar body formation varies with species; it may occur within the ovary or oviduct, or may occur outside after the egg is laid or even after it is fertilized. The meiotic process is completed, however, when the 2C amount of DNA is reduced to the haploid C amount through the next or *second meiotic metaphase.* This is accomplished by a true mitotic metaphase, in which kinetochores divide and chromatids segregate. A second polar body is formed and the mature egg results. The first meiotic division is referred to as *reductional,* since this reduces the number of chromosomes to the haploid number, while the second is the *equational* division where no change in chromosome number occurs but the DNA level becomes haploid.

Cytoplasmic growth occurs during the long period of suppressed meiosis, in the diplotene stage, when chromosomes are in a diffuse lampbrush condition, "opened out" and actively expressing the oocyte genome. As oocyte cytoplasm differentiates, metabolically active chromosomes synthesize quantities of gene products that are organized into the morphogenetic substratum of the egg.

## Vegetative growth phase: storage of developmental information

The vegetative phase is the period of oocyte growth when stores of gene products accumulate in the cytoplasm. These products are coded in the oocyte genome and in the genome of accessory maternal cells surrounding the oocyte.

The relationship between developing oocyte and maternal tissues is variable (Figure 11-6). Oocytes may develop as independent cells, floating freely in body fluids, usually after detaching from the coelomic epithelium. They absorb nutrients directly from body fluids, a situation found in many annelid worms. In most instances oocytes develop in a localized organ, an ovary. They may mature independently of surrounding ovarian cells or may be directly associated with follicle and/or nurse cells. Most vertebrate eggs are surrounded by follicle cells derived from somatic cells of the gonad; nurse cells are typical of insects and arise from the same stem cells as oocytes, the primordial germ cells. In ovaries pole cell progenitors give rise to two cell populations, oocytes and nurse cells. Among the more primitive insects (grasshoppers), oocytes are surrounded only by follicle cells. In most advanced insects (flies), both follicle cells and nurse cells are found, the latter as a cluster at the anterior end of each oocyte, alternating along the length of the ovary. In some beetles, nurse cells form a subterminal nutritive chamber attached to the developing oocyte.

Among oocytes developing independently of nurse cells, the oocyte germinal vesicle serves as the principal source of developmental information. It swells to a volume several hundred times that of the original oogonial nucleus, and its newly extended lampbrush chromosomes become metabolically active. By contrast, oocytes surrounded by nurse cells generally lack a germinal vesicle or lampbrush chromosomes. Instead, nurse cell nuclei enlarge, become metabolically active, and synthesize gene products for storage in oocyte cytoplasm.

Another sign of genomic activity is the appearance of functional nucleoli within the germinal vesicle. Species with small, yolk-poor eggs—for example, sponges, hydroids, and echinoderms—usually

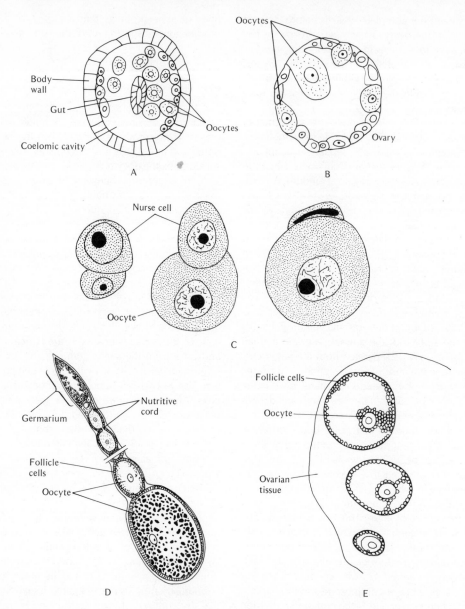

**Figure 11-6**   *Some examples of oocyte relations with accessory cells. (A) Diagrammatic representation of diffuse pattern in some annelids where oocytes bud off from the lining of the coelomic epithelium and grow within the coelomic space, deriving nutrients from the coelomic fluid. (B) An echinoderm ovary in which oocytes grow within the ovarian epithelium, usually free of any accessory cells. They grow into the lumen of the ovary and usually fill it when mature. (C) An oocyte–nurse cell arrangement in the annelid Ophryotrocha. The large single nurse cell contributes its nutritive materials to the growing oocyte until the nurse cell degenerates. (D) An example of a telotrophic ovary in insects in which a single ovariole (as shown) consists of a germarium where oocytes develop attached to the cells in the germarium by nutritive cords which funnel substances to be stored during oogenesis. In this instance each oocyte is surrounded by a single layer of follicle cells. (E) A mammalian ovary showing the developing follicles made up of several layers of follicle cells surrounding a cavity in which the oocyte grows.*

contain one large metabolically active nucleolus, while those with yolk-rich eggs, such as amphibians, may contain thousands of uniform nucleoli distributed at the germinal vesicle periphery.

Let us first examine the informational contribution of the oocyte nucleus itself, that is, the products of lampbrush chromosomes and nucleoli. Then we will look into the role played by nurse and follicle cells in preparing the oocyte for its developmental program.

### The lampbrush stage: informational RNA from the oocyte genome

Lampbrush chromosomes are found in oocytes of diverse groups, including humans, but their behavior has been most thoroughly studied in amphibian oocytes. In the frog, *Xenopus,* diplotene begins as chromosomes elongate into the lampbrush condition. *Vitellogenesis,* the period of yolk synthesis and storage, also begins; cytoplasm becomes opaque as yolk platelets accumulate. The height of growth and metabolic activity, however, is reached later when lampbrush chomosomes are in their most extended, active condition.

The lampbrush stage, though it persists for 4 to 8 months in *Xenopus,* is a transient stage. While the oocyte is still growing and yolk is deposited, chromosomes lose their lateral loops and begin to condense, a process that is completed in mature oocytes. The germinal vesicle enlarges to accompany cytoplasmic growth, but the chromosomes contract to their ultimate compact configuration in preparation for meiotic metaphase.

Lampbrush chromosomes are homologous bivalents attached at one or more points where crossovers have occurred (see Figure 11-5). The long axis of each chromosome, made up of two chromatids, consists of a series of chromomeres, tight coils of the principal axial thread (Figure 11-7). Lampbrush loops are regions where chromomeres have uncoiled and extend laterally and symmetrically on either side. Most loops are small, but several large loops are easily identified because of their distinctive morphology. All chromomeres, interchromomeric segments, and loop axes are composed of DNA.

Some large loops are as long as a complete bacterial chromosome with over 1000 genes, and most of the uniform smaller loops are many times larger than the length of DNA that would code for an aver-

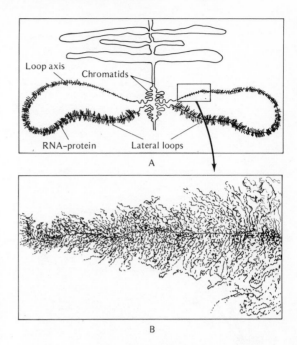

**Figure 11-7**   *Lampbrush chromosomes in newt* (Triturus) *oocytes. (A) Postulated structure of a pair of lampbrush chromosome loops. Paired chromatids are thrown into lateral loops. These are made up of RNA-protein fibers extending from a DNA axial filament. (From J. G. Gall, Cytodifferentiation and Macromolecular Synthesis, ed., M. Locke. New York: Academic Press, 1963, pp. 119–143.) (B) EM micrograph of a portion of a loop showing the RNA-protein fibers of increasing length extending from the DNA axis. These RNA strands are in the process of transcription from left to right. RNA polymerase molecules are visible at the base of each fibril. (Based on O. L. Miller et al., Cold Spring Harbor Symp. Quant. Biol. 35:505–512, 1970.)*

age-sized protein such as the hemoglobin beta chain. We can say, therefore, that a loop must represent more than one gene site. Some propose that a loop consists of many copies of the same gene arranged in tandem.

Approximately 20,000 loops are found in the whole chromosome set of the salamander, *Triturus,* which, according to a calculation, is only 5 percent of the total genomic DNA length in a nucleus; the remaining 95 percent must be tightly coiled in chromomeres. This suggests that only 5 percent of the genome is "opened out" and available for transcription at any one time. Does this mean that only 5 per-

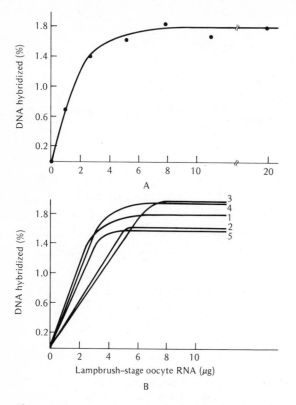

**Figure 11-8**   *Saturation curves of Xenopus DNA with lampbrush-stage oocyte $^{32}$P-RNA. These curves primarily measure the expression of redundant gene sequences. (A) A single representative saturation curve. (B) Five different curves with oocytes from the same animal to show the degree of variation. Each sample contained 5 μg $^{3}$H-DNA and the amounts of $^{32}$P-RNA shown. (From M. Crippa, E. H. Davidson, and A. E. Mirsky, Proc. Nat. Acad. Sci. U.S. 57:885, 1967.)*

cent of the total oocyte genome is actually expressed during the lampbrush stage?

All loops actively synthesize RNA. Radioactive precursors of RNA are incorporated initially into loop regions, which means loops are sites of gene transcription. Although RNA seems to be synthesized along the main axis, this could be an artifact resulting from the collapse of small loops against the axis during fixation. It is possible, therefore, that loops are the only sites of RNA synthesis. Protein is also synthesized, or, more likely, accumulated along the loop axis. A few giant asymmetrical loops in *Triturus* show polarized RNA synthesis; their thin regions incorpor-

ate labeled precursors earliest. Precursor incorporation moves progressively around the loop to the thickened portion until the loop is completely labeled after 14 days. It appears as if an RNA-protein product accumulates at the thick end of the loop as newly synthesized material is transported along the axis. To explain polarized transcription, Gall and Callan proposed that the loop axis is in motion, moving along with its RNA product spinning out from the thin end of the chromomere and rolling in at the thick end. They suggested that all loops operate in this way. Although only 5 percent of the genome is in loop form at any one time, the long duration of the lampbrush stage would permit most chromomeres to spin out for expression at some time, which meant that 100 percent of the genome was expressed during the lampbrush stage. If this hypothesis was true, every maternal gene would be expressed in the egg.

How accurate was this model? In other words, what proportion of the lampbrush genome is actually transcribed as informational RNA? An answer was obtained by DNA-RNA hybridization experiments using rapidly labeled RNA obtained from lampbrush-stage oocytes (Figure 11-8). From these experiments, we learned what proportion of the redundant and single-copy genomes are transcribed during the lampbrush stage. The saturation curves for hybridization indicated that approximately 3 percent of the total redundant genome and 1.5 percent of the single-copy genome are transcribed as relatively stable gene products. Approximately 50 to 100 times more RNA is transcribed from redundant genes than from single-copy genes.

The total value of 4.5 percent is very close to the calculated 5 percent value for the proportion of the genome in the form of loops. Inasmuch as the same 5 percent is made throughout the lampbrush stage, the Gall and Callan hypothesis of a moving genome with 100 percent genomic expression was inaccurate.

Evidently only a small proportion of redundant and nonredundant genes is transcribed in the oocyte. The oocyte genome transcribes no more information than most somatic cells in spite of its larger size and lampbrush condition. The sequential labeling pattern seen in giant loops may reflect a packaging process where RNA is progressively complexed with protein as it is synthesized. Electron microscopic examination of spread lampbrush chromosome loops has shown the genes in the act of transcription. RNA polymerase molecules are attached to the principal axis of the loop from which fibrils of increasing length extend. The main axis of the chromosome is at the end with shorter fibrils; this is where transcription begins. As transcription continues along the DNA strand, the fibrils of RNA lengthen. Presumably protein associates with these fibrils as they form and the ribonucleoprotein product is released.

Later, we will find that these lampbrush messages are not used during oogenesis but may be "delayed action" ribonucleoprotein messages stored in the cytoplasm in a nonfunctional state to be translated in embryogenesis (see Chapter 13).

## Synthesis and storage of ribosomes

No other cell, even the most metabolically active liver cells, can match the oocyte in its rate of ribosome production and storage. The ribosome reserve sustains the embryo through most of its development until it begins to synthesize its own ribosomes, after hatching.

During the early stages of amphibian oogenesis, before the lampbrush stage, RNA synthesis is one of the principal synthetic activities. Most of the RNA made during this period is transfer 4S RNA and 5S ribosomal RNA; very little 28S and 18S RNA are synthesized at this time. These species of RNA begin to appear in large amounts only after the ribosomal genes have been amplified in the many thousand nucleoli.

The nucleolus is the principal organelle for ribosome biosynthesis and assembly, and most oocytes contain many active nucleoli. When single nucleoli are present, they are usually large and densely staining, much larger than any somatic cell nucleolus. More often, thousands of small active nucleoli (for example, 1400 in *Xenopus* oocytes) are dispersed throughout the enlarged germinal vesicle. In either case, nucleolar function is considerably amplified in the growing oocyte since ribosomal RNA is actively synthesized throughout the lampbrush stage.

The gigantic nucleoli and/or the multiple nucleolar systems in many oocytes provide the best case of redundancy of a gene locus (see Chapter 2). Genes coding for ribosomal RNA are copied thousands of times within oocyte nucleoli, on templates in the nucleolar organizer region.

Before the lampbrush stage in *Xenopus* oocytes, an amplification mechanism selectively replicates this chromosome region, releasing thousands of DNA copies into the surrounding nucleoplasm where they, in turn, organize into typical nucleoli. A thousand nucleoli, each carrying many copies of ribosomal cistrons, greatly enhance an oocyte's capability for ribosome synthesis. Of the total DNA extracted from mature *Xenopus* oocytes, approximately 75 percent is derived from nucleoli with the remaining 25 percent localized in meiotic chromosomes. Most oocyte DNA codes for ribosomal RNA; there is several hundred times more DNA complementary to ribosomal RNA than is found in any somatic cell DNA. This DNA, presumably pure ribosomal RNA genes, can be isolated and used as a template for in vitro RNA synthesis. It yields a product similar, but not identical, to ribosomal RNA, presumably because the in vitro system lacks regulatory factors that start and stop transcription at the right places. Here we see the isolation of a single gene capable of transcription in vitro.

Ribosomal RNA genes from osmotically shocked nucleoli have been caught in the act of transcription, or, as Oscar Miller put it, "in flagrento transcripto" (Figure 11-9). A fibrous region unwound from nucleolar cores consists of a fine circular fiber, coated periodically with matrix-covered segments varying from 8 to about 1000. Each matrix unit consists of about 100 fibrils attached by one end to the core axis, running in graded lengths of short to long; that is, the core axis is polarized. The matrix is ribonucleoprotein and the core is DNA; only the matrix region incorporates labeled RNA precursors. Where the matrix fibril is attached to the core, we find a minute granule, the RNA polymerase enzyme molecule transcribing its way along the DNA template. Presumably the polymerase attaches at the initiating site of the gene and migrates along the axis in a polarized manner, transcribing a 45S ribosomal RNA precursor segment (Chapter 2). The graded lengths of fibrils are different stages in synthesis of this precursor. Polymerase granules are not found on intergene segments (spacer regions) which, as we know, are not transcribed. Millions of polymerase molecules simultaneously synthesize 45S ribosomal RNA precursors which are processed in nucleoli into 28S and 18S RNA molecules.

Gene amplification, through selective replication of ribosomal RNA genes, probably reduces the time required for the synthesis of all ribosomal RNA stored in the oocyte. If no amplification occurred, it would take approximately $1.7 \times 10^5$ days to synthe-

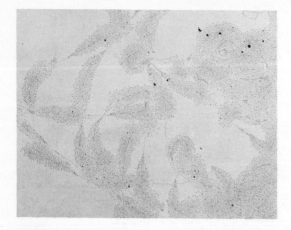

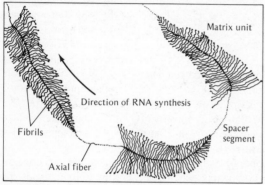

**Figure 11-9** *Active ribosomal RNA genes spill out of an osmotically shocked nucleolar core and are photographed in the electron microscope. An interpretation is shown in the diagram below. The long axial fiber, in the form of a circle, is composed of DNA and protein. The DNA segment within each matrix unit is a gene for ribosomal RNA (18S and 28S), each separated by a spacer segment which is not transcribed. The lateral fibrils are ribosomal RNA complexed to proteins in the process of transcription, the shortest fibrils marking the site at which transcription begins. See text. (From O. L. Miller et al., Cold Spring Harbor Symp. Quant. Biol. 35:505–512, 1970.)*

size the total amount of oocyte ribosomal RNA, assuming a rate of synthesis identical to that found in somatic tissues. Replication of thousands of additional template sites for ribosomal RNA reduces the time to 3 to 6 months, the duration actually required by the oocyte to complete this phase of its maturation.

Although most ribosomes are used in embryogenesis, there is evidence of ribosome turnover, even in mature oocytes. Ribosomal RNA continues to be synthesized in mature oocytes, at lower rates than during the lampbrush period, and some ribosomal RNA is synthesized in mitochondria. Contrary to earlier views, ribosomal RNA synthesis does not shut down completely in mature eggs.

## Informational contributions from accessory cells

In species lacking a lampbrush stage, oocytes receive informational molecules from surrounding follicle and nurse cells. The metabolism of both accessory cells is coupled to oocyte growth. The intimacy is so great that we often find direct cytoplasmic connections between nurse cells and/or follicle cells and the developing oocyte. The principal roles of these accessory cells are to (1) supply oocyte nutrient reserves during its growth, (2) act as a selective barrier between the vascular system and the oocyte, transporting essential molecules (yolk precursors) into oocyte cytoplasm, and (3) synthesize accessory egg membranes.

Nurse cells, found predominantly in insects, are progeny from oogonial divisions in the ovary. At some point a differential division separates cells destined to become oocytes from their sisters which develop into nurse cells. Sometimes this division correlates with visible differences between oocyte and nurse cell nuclei. In one classic example, the water beetle *Dytiscus*, a large Feulgen (DNA) positive mass in an oogonial nucleus is segregated into only one of the two cells following an oogonial division. This process continues into subsequent divisions, so that only one cell out of a total of 16 retains this extra DNA. This cell is destined to become the oocyte, whereas the others, without the Feulgen positive body, become nurse cells. In *Dytiscus,* this extra DNA represents a mass of ribosomal RNA genes, a mechanism of gene amplification similar to nucleolar amplification in amphibian oocytes.

*Drosophila* oocytes lack lampbrush chromosomes and do not synthesize RNA. Nurse cells are the principal source of maternal genetic information and ribosomes. Amplification mechanisms for RNA synthesis are found in nurse cell genomes which become highly polytenic (giant chromosomes) and metabolically active in RNA synthesis. The volume of a nurse cell nucleus, nucleoli, and cytoplasm in *Drosophila* doubles once every 4 to 5 hours during the period of oocyte vitellogenesis.

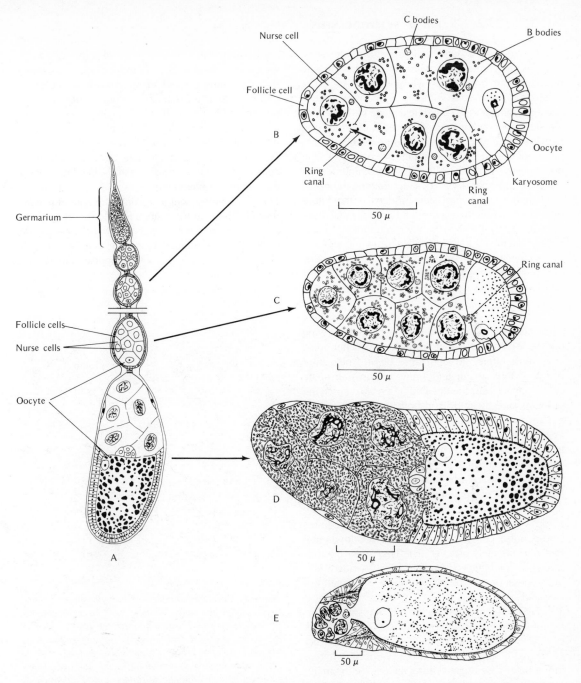

**Figure 11-10** *Nurse cells and oogenesis in* Drosophila. *(A) A single ovariole showing sections of the vitellarium. In the germarium, nurse cell–oocyte complexes are formed with 15 nurse cells and a single oocyte connected through ring canals. Various cytoplasmic inclusions pass from nurse cells into the oocyte during growth. (B) An early ovarian chamber. (C) A later chamber when yolk synthesis begins. Yolk = black dots. (D) A larger ovarian chamber with lipid droplets (open circles) entering oocyte. The yolk spheres have coalesced in oocyte cytoplasm. (E) A last stage ovarian chamber in which the oocyte has almost reached its maximum volume. The nurse cells, reduced in size, form an anterior cap above the oocyte. (From R. C. King,* Ovarian Development in Drosophila melanogaster. *New York: Academic Press, 1970.)*

As *Drosophila* nurse cells develop, they retain cytoplasmic connections with the oocyte, forming ring canals or bridges. Oocyte and nurse cells develop from stem cells within a germarium, a constricted chamber in the ovariole (Figure 11-10). A single stem cell undergoes a differential division; one daughter remains as a stem cell while the other completes four divisions to form a cluster of 16 cells, an oocyte plus 15 nurse cells, all intercommunicating via cytoplasmic bridges. The cell retaining the maximum number of cytoplasmic bridges is the future oocyte, one of the two cells with four ring canal connections. Only these two show synaptinemal complexes, but the future oocyte is the only one with access to the fluid of a canal joining two egg chambers. The presumptive oocyte usually appears at the first division while all subsequent divisions produce nurse cells.

*Drosophila* ovaries are *polytrophic;* that is, only a few nurse cells connect directly with each growing oocyte, which means that products from other nurse cells must funnel through these before they enter oocyte cytoplasm. In *telotrophic* ovaries, all nurse cells connect with the developing oocyte, emptying nutrients directly into it. In both systems, informational macromolecules (RNA messages) from nurse cell nuclei enter the oocyte via cytoplasmic bridges. In some cases, DNA from nurse cell nuclei is also extruded directly into the cytoplasm of the oocyte. For example, in some hemipterans (mealybugs), large DNA masses of coalesced nurse cell nuclei pass directly into oocyte cytoplasm through cytoplasmic cords. The oocyte cytoplasm in this species acquires whole nurse cell genomes, whose developmental role is completely obscure. The DNA may be conserved as a functional template for subsequent embryogenesis or may be broken down into small nucleotides.

Two different amplification mechanisms for storing informational molecules in oocytes have evolved. On the one hand, oocyte nucleus and nucleoli become highly specialized for RNA synthesis, whereas in the other case nurse cell nuclei assume the major burden for informational macromolecule synthesis and pump these into the growing oocyte through cytoplasmic bridges.

Follicle cells on the other hand, play a different role. They are organized as a single layer of epithelial cells surrounding the developing oocyte (Figure 11-11) but do not perform the same functions as nurse cells. Though they may synthesize some substances for oocyte storage, they act more as a selective barrier, mediating transfer of materials from the blood into oocyte cytoplasm. The connections between oocyte and follicle cells are intimate. Fine cytoplasmic processes from the follicle cell interdigitate with thousands of microvilli of the oocyte surface.

Some mammalian follicle cells, the *granulosa cells,* mature into endocrine glands. In early stages of ovarian development they grow with and envelop the oocyte. Development stops until the animal reaches reproductive maturity, when follicle-stimulating hormone (FSH) from the pituitary gland stimulates oocyte maturation and follicle cell mitotic activity. The follicle enlarges as additional layers appear, and a lumen

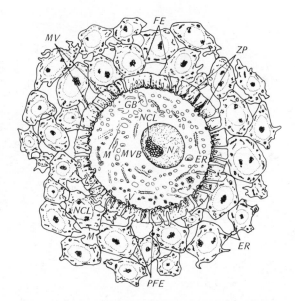

**Figure 11-11** *Oocyte–follicle cell relationship in guinea pig as seen in the electron microscope. Extensions from the follicle cells (PFE) extend through the zona pellucida (ZP) to the oocyte surface, where desmosomes form. Microvilli (MV) on the oocyte surface extend into the zona, intermingling with the follicle cell extensions. The small pits beneath the oocyte surface and small vesicles (V) just below the surface probably represent micropinocytotic activity of the oocyte surface. The mitochondria appear at M, Golgi bodies at GB, endoplasmic reticulum at ER, and multivesicular bodies at MVB. Also shown: NCL = nucleolus; N = nucleus; FE = follicle cell epithelium. (From E. Anderson and H. W. Beams, J. Ultrastruct. Res, 3:432–446, 1960.)*

develops, filling with fluid secreted by follicle cells. Another hormone signal from the pituitary, luteinstimulating hormone (LSH), induces ovulation and the follicle ruptures. The remaining follicle cells synthesize and secrete large amounts of the pregnancy hormone, progesterone, becoming, in essence, an endocrine gland, the corpus luteum that functions through early pregnancy.

Amphibian follicle cells contribute little to the informational content of the egg. For example, primordial germ cells from embryos of *Xenopus tropicalis* can be transplanted into embryos of another species, *Xenopus laevis*. Each species differs significantly in a number of phenotypic traits, such as skin pigment patterns. The germ cells migrate to and populate the host's developing gonad. In such combinations, the mature oocytes are derived from the donor species, but they are surrounded by follicle cells from host gonadal tissue. Do oocytes interacting with foreign follicle cells receive developmental information from the host genome? If they do, then host females should produce eggs that develop into embryos or larvae exhibiting host as well as donor traits. In all instances donor eggs always developed donor phenotypic traits, with no contribution from host follicle cells. In contrast to nurse cells which make significant contributions to the informational content of the oocyte, follicle cells contribute little developmental information.

Informational molecules in oocyte cytoplasm have multiple origins. The oocyte nuclear genome itself produces a major portion, particularly the nucleic acid components of the egg. Nurse cell genomes also contribute to the store of nucleic acid templates. Follicle cells may facilitate the passage of other nutritive substances such as yolk, synthesized elsewhere in the female body and transported to the oocyte for storage. Yolk, the principal protein of the egg, is not actively synthesized by the oocyte, in most cases. Instead, the oocyte is often a passive recipient of yolk precursors which are assembled into yolk granules within oocyte cytoplasm.

## The synthesis and deposition of yolk

Yolk, the major nutritional reserve of the oocyte, accumulates during the vegetative growth phase and accounts for the rapid increase in size of the maturing egg. In *Drosophila*, for example, oocyte volume increases 100,000 times in 3 days of yolk deposition. It is a period marked by vigorous synthetic activity in certain organs of the vitellogenic or yolk-synthesizing female. Oocyte growth often dominates the physiological state of the female organism; her hormonal system is principally committed to synthesis, transport, and deposition of yolk in the oocyte.

Yolk makes up more than 90 percent of oocyte dry weight in some vertebrate species. In the amphibian oocyte it is localized in platelets of variable size and crystalline internal structure (Figure 11-12). Two major proteins predominate, one rich in lipid, known as *lipovitellin*, and a smaller phosphoprotein called *phosvitin*.

Yolk proteins are synthesized in the amphibian liver and transported to the oocyte via the circulation. A yolk protein precursor, a lipophosphoprotein called *vitellogenin*, is detected in the serum of vitellogenic

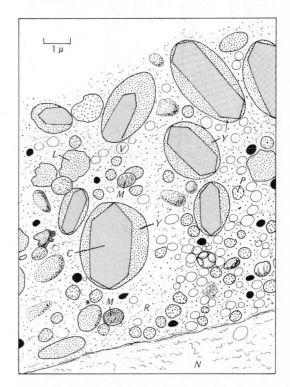

**Figure 11-12** *Yolk platelets in the amphibian oocyte. Numerous yolk platelets fill oocyte cytoplasm, each with a crystalline core surrounded by a granular cortex. N = nucleus; R = ribosomes; P = pigment granule; M = mitochondria; L = lipid droplet; V = vesicle; Y = yolk platelet; C = crystalline core. (Based on S. Karasaki, J. Cell Biol. 18:135, 1970.)*

females during the period of oocyte growth. Its synthesis is under hormone control since males injected with estrogen can be induced to synthesize this serum precursor.

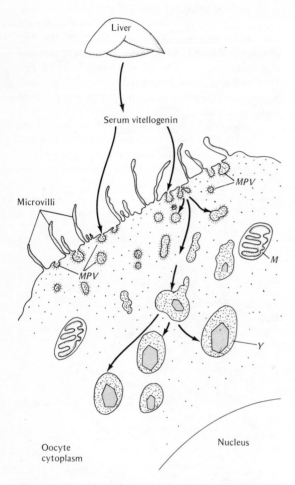

**Figure 11-13**  *Formation of yolk platelets in the amphibian oocyte. In a vitellogenic female, serum vitellogenin is synthesized in the liver and secreted into the circulation. In the ovary, developing oocytes are undergoing active micropinocytosis, presumably under the influence of a hormone. Coated vesicles appear at the surface and selectively incorporate the serum vitellogenin into vesicles which enter the cytoplasm beneath the surface. The vesicles coalesce into larger vesicles in which lipovitellin and phosvitin are formed and the characteristic yolk platelet with a crystalline core appears. MPV = micropinocytotic vesicles; M = mitochondria; Y = yolk platelet.*

Serum vitellogenin disappears rapidly in vitellogenic females because it is taken up by the ovaries. If radioactively labeled vitellogenin is injected into a female along with other labeled serum proteins, it is incorporated into the ovary about 50 times more rapidly than the others. Selective incorporation of yolk protein precursors is not understood, but it may relate to the pattern of micropinocytosis. Micropinocytotic behavior of the oocyte membrane is a selective process, in contrast to pinocytosis of most cells which seems to be a nonselective drinking of surrounding fluids. The vesicles in micropinocytosis are much smaller, usually lined with small granules, and these appear in great numbers near the surface of oocytes showing rapid yolk deposition. Presumably a hormonal signal, pituitary gonadotropin, acts directly on the cell surface and stimulates micropinocytotic uptake of serum vitellogenin. The appearance of these vesicles on the surface of growing oocytes may be the first stage in yolk platelet formation (Figure 11-13). Since vitellogenin is not detected in mature yolk, it must be a precursor of phosvitin and lipovitellin. During platelet formation, the precursor is converted to yolk proteins by proteolysis. The yolk platelet begins at the cell periphery as a small droplet and gradually increases in size as it migrates into the interior where it crystallizes into a definite pattern. In *Rana pipiens* oocytes, crystalline yolk platelets are sometimes seen within the mitochondrial matrix, suggesting a direct role in yolk platelet formation. This is, however, not a general mechanism of yolk formation; it is found only in Ranidae and represents only 1 to 2 percent of the total yolk produced. Most yolk platelets are assembled within the cytoplasm as micropinocytotic vesicles enlarge with crystalline yolk.

Micropinocytotic uptake of yolk precursors from extracellular fluids is a common mechanism of yolk accumulation in insects. In *Cecropia* moths, yolk synthesis takes place in fat bodies in pupal life; later, in the adult, yolk precursors are transported to the gonad where they are taken up by growing oocytes. Not all oocytes are passive in yolk synthesis, however. In the guppy, for example, yolk is synthesized within the oocyte endoplasmic reticulum and is stored in Golgi vesicles as yolk platelets.

In addition to yolk, other nutritive substances such as glycogen, lipids, and oil droplets are stored in the egg. Glycogen granules are probably one of the most important energy reserves and in the frog egg, they are more numerous than ribosomes. In addition, various kinds of fat droplets and lipids accumulate, either localizing at the periphery of the oocyte or distributing uniformly throughout the cytoplasm. As a rule, lipids made elsewhere in the organism are stored as droplets in oocyte cytoplasm prior to yolk platelet synthesis. Carbohydrates and lipids act as nutrient reserves for energy production during embryogenesis. A lipid store supplies the phospholipid precursors for plasma membrane synthesis during cleavage, when new cells are rapidly produced.

Included among the morphogenetically important molecules in the egg are tubulin, the subunits for microtubule formation, actin and presumably myosin, the components of a cytoplasmic contractile system, and a store of histones. These molecules may be distributed randomly throughout egg cytoplasm, perhaps in the form of soluble precursor pools, available for recruitment in specific regions of the egg during the early stages of development. The egg undergoes a series of rapid cell divisions after fertilization and must make spindles, cleavage furrows, and chromosomes. Apparently all the essential precursors for division are stored in the egg cytoplasm during oogenesis.

## Oocyte membranous organelles

The membranous organelles of the oocyte, mitochondria, endoplasmic reticulum, and Golgi apparatus make up an essential portion of the egg's developmental organization. Though some of these organelles are necessary for oocyte metabolic activity, the bulk is stockpiled in oocyte cytoplasm to sustain the developing embryo to hatching.

Mitochondria are the most numerous of all membranous organelles. The structure of oocyte mitochondria is typical of that found in most somatic cells; in some oocytes, they are more spherical, smaller, and have fewer cristae than mitochondria in surrounding follicle cells or in adult tissues.

Large numbers of mitochondria collect in the oocyte, more than are needed for respiratory metabolism. There are approximately $2 \times 10^6$ mitochondria in a mature *Rana pipiens* oocyte compared with the 200 to 1000 found in most somatic cells. In very young oocytes, they are localized at the periphery of nuclei, but later, as the oocyte grows, they are dispersed randomly in the cytoplasm. During the period of maximum yolk deposition, mitochondria are often seen at the periphery of the cytoplasm in areas of active micropinocytosis, a process requiring large amounts of energy.

Mitochondria formed during oocyte growth probably arise from autonomous replication of their DNA. In fact, in many mature oocytes, mitochondria DNA makes up the bulk of total DNA; for example, up to 1000 haploid equivalents in *Xenopus* (Table 11-1). In some oocytes, mitochondria come from nurse cells, migrating down nutritive cords into oocyte cytoplasm. Some eggs may accumulate a heterogeneous population of mitochondria with different genetic backgrounds.

Metabolically, oocyte mitochondria resemble those found in somatic cells; they carry on oxidative phosphorylation. Yet the heterogeneity of oocyte mi-

**TABLE 11-1  Content and Distribution of DNA in Eggs and Oocytes**

| Species | DNA Content | | Distribution of DNA, % | | |
|---|---|---|---|---|---|
| | Total Picograms | Haploid Equivalents | Percent Nuclear | Percent Recovered in Mitochondria | Mitochondrial (%) Estimated from Physical Properties |
| *Xenopus laevis* (eggs) | 3100 | 1000 | About 1 | 78 | About 100 |
| *Rana pipiens* (eggs) | 4500 | 600 | About 1 | 70 | About 100 |
| *Lytechinus pictus* (eggs) | 8.3 | 9.5 | About 12 | 55 | 80 to 90 |
| *Urechis caupo* (oocytes) | 10 | 10 | 40 | 60 | 60 |

Source: From I. Dawid, *Oogenesis,* ed. J. Biggers and A. Scheutz,© 1970, University Park Press, Baltimore.

tochondria in terms of size and morphology may reflect a biochemical heterogeneity; that is, some mitochondria may be more active than others. It has even been suggested that egg mitochondria are metabolically immature and during the course of embryogenesis become mature, assembling batteries of mitochondrial enzymes that might be characteristic of each tissue type. For instance, sea urchin egg mitochondria lack cytochrome c but acquire it during development. Though there may be changes in mitochondrial enzymes, the total number of mitochondria remains constant throughout development to hatching in both sea urchins and frogs; there is no evidence of mitochondrial DNA replication during development. Mitochondria stored in the egg are parceled out to the cells of the developing embryo until the number of mitochondria per cell is equal to that found in adult tissues.

Most oocytes possess a smooth endoplasmic reticulum of small vesicles dispersed evenly through the cytoplasm. Membranes of the endoplasmic reticulum seem to peel off from the nuclear envelope in some oocytes while in others endoplasmic reticulum membranes close to the nuclear membrane form an array filled with annulae, or pores, and are known as *annulate lamellae*. These appear initially as small blebs of the nuclear envelope that align into lamellae stacks. These bodies contain RNA but their function is not understood. Though the relatively simple, smooth endoplasmic reticulum found in most oocytes suggests that they are not actively synthesizing proteins, they probably are. The simplicity of the endoplasmic reticulum indicates that proteins are being synthesized for storage rather than for secretion, the usual function of the rough endoplasmic reticulum in most secretory tissues, such as pancreas.

A Golgi apparatus is also present in most oocytes, performing a variety of secretory, storage, and transport functions. The stacked lamellae vesicles are a typical feature of young oocytes in the process of vitellogenesis, implicating them in yolk synthesis. On the other hand, Golgi might be concerned with the elaboration of membranes of the endoplasmic reticulum. In some oocytes, Golgi first subdivides into multiple aggregates of vesicles and tubular elements which migrate to the periphery where they become uniformly distributed in the cortical cytoplasm. The vesicles fill with a dense material, fuse, and coalesce into large granules that collect in the cortical region underneath the plasma membrane (Figure 11-14). These cortical granules consist of a fibrous core surrounded by an amorphous matrix, bounded by a

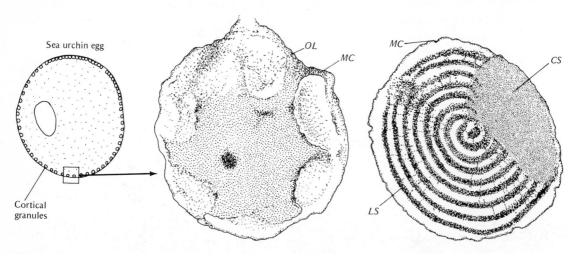

**Figure 11-14**  *Fine structure of cortical granules in the sea urchin egg. Cortical granules are seen beneath the membrane of the mature egg, and in the electron microscope they consist of a membrane (MC) surrounding a central compact structure (CS) surrounded by a more granular mass. The cortical granule membrane is close to the membrane of the egg (OL). On the right is a section of cortical granules in a late oocyte. These are in the process of maturation and consist of the compact structure associated with lamellar units (LS). The latter are associated with each other by fine filaments. (From E. Anderson, J. Cell Biol. 37:515, 1968.)*

membrane. The cortical granule is common in mollusks, echinoderms, and anuran eggs.

Where found, most cortical granules seem to participate in the fertilization reaction. In sea urchin eggs, for example, they are lined up beneath the plasma membrane of the unfertilized egg but disappear almost immediately on fertilization. The granules cannot be displaced by centrifugation: they seem to be fixed in the cortical gel. The material in the cortical granule is not homogenous but consists of several glycoproteins that associate with the fertilization membrane and perivitelline fluids surrounding the egg.

There is no consistent pattern for the behavior of cortical granules. Some break down at fertilization but in other species they may remain intact or coalesce before fertilization into a homogenous jelly layer outside the egg surface. Some even persist into cleavage stages. We cannot generalize on the role of cortical granules in fertilization other than to say that in some eggs they serve a specialized function and must be included as an essential component of the developmental organization of the egg. We will discuss their role in fertilization in a later section.

## The egg cortex

Developmental information is also stored in the egg cortex, that is, the plasma membrane and underlying cortical gel (see Chapter 9). Except for those eggs with cortical granules, the cortex is not morphologically different from the more fluid cytoplasm below with its mitochondria, ribosomes, and glycogen granules. In the electron microscope, however, the cortex of some eggs appears as a dense layer consisting of small granules and microfilaments. During embryogenesis, organelles such as basal bodies appear in the cortical gel of embryonic cells.

Mitochondria, yolk granules, and lipid droplets localize in the peripheral cytoplasm of immature oocytes, but eventually disperse into the cytoplasm as the oocyte grows. Later, when yolk synthesis is at its peak, the cortex and plasma membrane are thrown up into numerous folds or microvilli which interdigitate with similar large projections extending from follicle cells. The region becomes active in micropinocytosis with vesicles localizing in microvilli and in the adjacent cortical region. The most distinctive feature of the egg cortex is the network of microfilaments that fill the microvilli extending from the surface. The microfilaments may be contractile proteins (actin?), which would account for the micropinocytotic activity of the surface.

During the period of maximum oocyte growth, the plasma membrane takes in large macromolecules for storage in the cytoplasm. Because of an asymmetrical position of the oocyte in relation to the ovary and its blood supply, substances collect more rapidly at one end of an oocyte than at the other and establish polarity gradients. Accordingly, organelles are not uniformly arranged in oocyte cytoplasm and asymmetries arise. Vitellogenin is incorporated more actively at the animal-pole surface of the amphibian egg than at the vegetal pole. Yolk platelets, however, are distributed as a gradient in most eggs with highest concentration in the vegetal half. Pigment granules are more numerous in the animal-half cortex of the egg or confined to specific cortical regions at the equator. These localizations seem to depend upon orienting properties of the cortical gel as well as on the permeability of the plasma membrane.

The cortex in the egg of the pond snail *Limnaea* seems to localize special subcortical cytoplasms. These patches of cytoplasm with unique staining properties form an asymmetric pattern over the oocyte surface. The pattern corresponds exactly to the configuration of follicle cells surrounding the oocyte in the gonad and persists beneath the cortex as the egg passes down the oviduct, as if held there by attractive forces in the cortex. According to Raven, the cortex acquires this property after interacting with follicle cells. These cytoplasms may be followed in dividing eggs, where they maintain a constant subcortical position in spite of cortical displacements caused by cell division. Although the subcortical cytoplasms coalesce and re-form during cleavage, they show a precise pattern of segregation into the individual blastomeres. If the eggs are centrifuged during cleavage, the subcortical cytoplasms are displaced according to their densities in the stratified cytoplasm, but after a few hours they return to their original sites underneath the cortex. Does this imply that the future symmetry of the embryo is specified by a cortical blueprint beneath the egg plasma membrane?

## Polarity and symmetry

From the above, we see that asymmetries in cytoplasmic organization generate egg polarity. Factors responsible for polarity differ for each egg. Some eggs (amphibian) contain a graded distribution of yolk, the bulk localized in the lower vegetal pole of the egg. These eggs usually orient themselves with respect to gravity, with animal pole uppermost. Egg polarity also resides in the cortex, sometimes as a variable thickness of cortex in different regions but more frequently as a gradient of pigment granules entrapped in the cortical gel. At the beginning of oogenesis, the nucleus starts out in a central position. Polarity arises as the oocyte matures and the nucleus moves into its eccentric position in the animal half, displaced largely by the asymmetric accumulation of yolk.

The position of the oocyte in the ovary and the arrangement of surrounding nurse and follicle cells often directs egg polarity. In echinoderms, for example, the animal pole is always situated at the free end of the oocyte with the portion attached to the ovary wall becoming the vegetal half. Materials coming from nurse cells establish a unidirectional flow. One region of the oocyte receives the materials first as they pour through connecting nutritive cords and this tends to polarize an egg.

The initial arrangement of organelles is not fixed; reshuffling may occur as the egg matures. In teleosts, for example, yolk granules appear first at the vegetal pole of the egg. During the last phase of oogenesis, a bipolar differentiation occurs rapidly as egg substances rearrange themselves. The germinal vesicle moves toward the periphery at the animal pole while the cytoplasm, interspersed between yolk platelets, shifts toward the same pole beneath the cortical layer. In this way, a thick cap of cytoplasm forms at the animal pole which will give rise to the embryo proper while yolk and lipid droplets accumulate below the clear cytoplasm.

Bilateral symmetry of the embryo is also preformed in the organization of some eggs; that is, a bilateral distribution of organelles in the oocyte corresponds to the future dorsoventral axis of the embryo, a situation found in the snail egg.

The origin of symmetry in mammalian embryos is still unresolved. In fact, it is now believed that the future plane of symmetry of the vertebrate embryo, the left-right axis, is not preformed in the mature egg. Rather, it appears later, either at fertilization or during cleavage.

The significance of egg polarity lies in the fact that this property is often translated into the major axes in the embryo. For example, the animal-vegetal polarity of the amphibian egg becomes the anterior-posterior axis of the embryo, with the head appearing at the animal pole and the anus developing near the vegetal pole. The same applies to the axes of symmetry of the embryo; that is, dorsal-ventral and left-right. These, too, are often "preformed" in the egg, sometimes as a result of an asymmetrical distribution of organelles or pigmented cytoplasms. In many cases egg polarity correlates with the position of the centriole. Its position determines the site of the asters in a developing spindle, and accordingly will determine the axes of cleavage.

Centrioles also serve as nucleating sites for cytoplasmic microtubules which possess an intrinsic polarity of assembly, lengthening in a unidirectional manner. This may organize the major axes of a cell, and, if a large population of cells is polarized along the same axis, this, too, will translate into axial symmetries of the embryo.

Centrioles or basal granules also initiate cilia and flagella which develop in specific cortical regions; they establish an apical-basal axis in an embryo with ciliated apical surfaces exposed to the outside. These initial egg polarities, acquired during oogenesis, act as blueprints for the future development of the embryo.

Although many eggs start out with an initial symmetry, it need not be maintained in the embryo. Future axes of symmetry of the embryo often arise after the organizational changes following fertilization. Organelles are reshuffled through the egg by a process of cytoplasmic streaming known as *ooplasmic segregation,* probably the most important axis-determining event in many invertebrate eggs.

In other species, polarity seems to be fixed in cortical structures and ooplasmic segregation does not have any effect on future embryonic polarity or symmetry. As long as the egg cortex remains intact, the original polarity and symmetry of the unfertilized egg are retained. Though organelles are displaced by cytoplasmic movements after fertilization or experimentally shifted by centrifugation, they seem to return to their original position within hours. For example, fat droplets appear first at the vegetal side of the *Limnaea* oocyte, forming a dense cap. If ovarian oocytes are stratified by centrifugation, the droplets are driven centripetally, but within six hours after centrifugation they return to their original location at the vegetal pole, perhaps attracted there by the cortical field.

No generalizations can be made about the origin

of egg polarity. Polarity is conferred from the outside by the egg's position in the ovary, by its relationship with nutritive cells, and by its attachment to body wall and ovary epithelium. Except for polarity within the rigid cortical gel and/or plasma membrane, it is a labile feature of egg organization which may be modified by ooplasmic segregation at fertilization or cleavage. Not every egg shows a rigid cortical prepattern, which means that the origin of the polar axes in eggs and embryos is unique for each group of organisms.

## Meiotic maturation

Though a full complement of organelles and nutritive substances is packaged into egg cytoplasm by the end of oogenesis, other more subtle events are completed before the egg matures and is ready for fertilization. Meiotic arrest is broken; meiosis resumes as the germinal vesicle breaks down. Before this occurs, however, the prophase lampbrush chromosomes condense into a diplotene state in preparation for the first meiotic metaphase and polar body formation. In most species, these final events of egg maturation are controlled by hormones.

Germinal vesicle breakdown marks the end of the vegetative phase of oogenesis and initiation of meiotic metaphase. The time of its occurrence varies with species. In the sea urchin, germinal vesicle breakdown and polar body formation are completed within the ovary before eggs are shed. In related species such as the starfish, eggs are shed into the sea before germinal vesicle breakdown while in many worms germinal vesicle breakdown is induced only after sperm entry.

Hormonal factors regulate oocyte maturation, ovulation, and shedding in invertebrates and vertebrates (Figure 11-15). For example, at the end of starfish oogenesis, *Asterias* oocytes are still in the germinal vesicle stage. Ovulation is induced by a gonad-stimulating hormone secreted from the radial nerve, and eggs are shed. Hormone secretion is probably regulated by some environmental stimulus such as temperature or light. The radial nerve factor turns out to be a small peptide (2100 mol wt) that can be isolated directly from nervous tissue or from the coelomic fluid of naturally spawning females. It induces gamete shedding when injected into the coelomic cavity of ripe animals. The gonad-stimulating hormone, however, is not directly responsible for spawning or for germinal vesicle breakdown. Instead, it acts

directly on ovarian tissue, which is stimulated to secrete another substance, 1-methyladenine. Germinal vesicles do not break down in oocytes isolated in calcium-free sea water, but when 1-methyladenine is added, germinal vesicle breakdown is triggered and meiosis is completed. This substance is a principal meiotic inducing substance for other starfish species and is synthesized within ovarian tissue in response to the gonad-stimulating hormone. If ovarian fragments are treated with radial nerve factor for 1 to 3 hours, the supernatant can then be added to isolated oocytes and induce germinal vesicle breakdown within 30 minutes.

1-Methyladenine seems to act on the oocyte surface rather than on the germinal vesicle. Germinal vesicles break down only when oocytes are immersed in 1-methyladenine and not when it is injected into the cytoplasm, even at 10 times the concentration. The events leading to germinal vesicle breakdown are triggered by some change at the oocyte surface.

A similar sequence of events is seen in frog oocytes. Here, too, maturation is programmed by a hormonal signal, presumably pituitary gonadotrophin. This acts directly on follicle cells which are induced to synthesize and release a steroid hormone resembling progesterone. The hormone then acts on the oocyte surface which triggers germinal vesicle breakdown, completion of the first meiotic metaphase, and polar body formation.

Full size frog oocytes (about 1 mm in diameter) will respond to exogenous progesterone in vitro, even when dissected from their follicle membranes. The germinal vesicle breaks down and the first polar body appears. In general, oocytes smaller than this (0.8 mm) do not respond, presumably because they do not have steroid receptors on their surfaces. These probably develop during growth to full size. The hormone signal on the surface induces a cytoplasmic change and the release of a maturation-inducing factor (MIF). Cytoplasm taken from oocytes treated with progesterone will induce maturation when injected into untreated oocytes. It is MIF which induces germinal vesicle breakdown and the subsequent events of meiosis. The nature of this factor is unknown but it is not species-specific since cytoplasms from one species induce maturation when injected into oocytes of foreign species. In distantly related groups of organisms, echinoderms and amphibians, a similar hormonal mechanism controls the final stages of egg maturation.

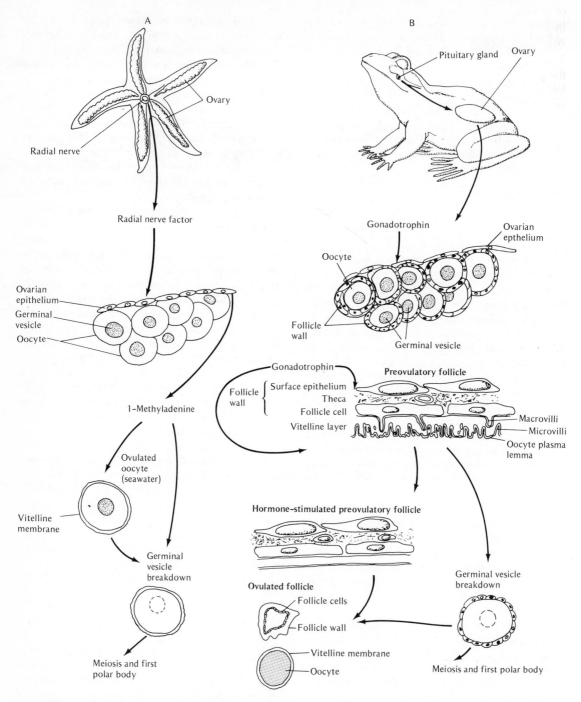

**Figure 11-15**  *Comparison of meiotic maturation in starfish and frog. (A) Starfish. The principal organs are the radial nerve and the ovaries in each arm. A radial nerve factor (gonad-stimulating hormone) is secreted into the body cavity where it diffuses to the ovary and stimulates ovarian epithelium to secrete 1-methyladenine. This, in turn, acts on the oocytes*

## Accessory egg membranes and jelly coats

Oocyte maturation is not complete until additional structures are produced outside the egg membrane proper. For most eggs, a gelatinous or mucoid outer envelope is usually secreted around the entire egg as a protective coat or capsule that also plays an important role in fertilization. The principal accessory structure is the vitelline membrane found close to the plasma membrane of the mature egg (Figure 11-16). This membrane, the first penetrated by sperm, becomes the fertilization membrane.

It is likely that the vitelline membrane has a dual origin in most cases inasmuch as the oocyte and adjacent follicle cells each contribute components to its formation. The development of this bipartite structure of the vitelline membrane has been studied in amphibian oocytes.

Immature oocytes show no sharp demarcation or space between the plasma membrane and the membrane of the follicular epithelium cells. As the oocyte grows, its surface is thrown up into minute microvillus folds and projections. A mucoidal matrix is secreted adjacent to the membrane, filling up a space between follicle epithelium and oocyte. The gap becomes increasingly wide as granular material collects in spaces between microvilli and the macrovilli projections from the follicular epithelium. A different, more amorphous matrix material is now secreted by follicular cells into the same space, forming a bipartite structure of fibrous elements embedded in an amorphous matrix.

All circulating nutrients, such as yolk precursors, appear to pass through matrix-filled intercellular spaces between follicle cells to growing oocytes. The entire matrix becomes the vitelline membrane at ovulation, as microvilli diminish in size and the oocyte leaves the follicular epithelium.

Hence the vitelline membrane is really mis-

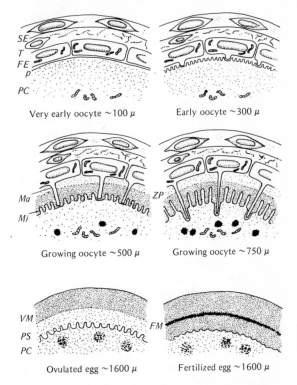

Very early oocyte ~100 μ          Early oocyte ~300 μ

Growing oocyte ~500 μ          Growing oocyte ~750 μ

Ovulated egg ~1600 μ          Fertilized egg ~1600 μ

**Figure 11-16**   *Development of the vitelline layer in the amphibian oocyte. SE = surface epithelium; T = theca; FE = follicular epithelium; p = oocyte plasma membrane; PC = peripheral cytoplasm; VM = vitelline layer; FM = fertilization membrane; PS = perivitelline space; Ma = macrovilli from follicle cells; Mi = microvilli from oocyte surface; ZP = zona pellucida. (Based on S. Wischnitzer, Adv. Morphogen. 5:131–179, 1966.)*

named; it is not a membrane but a mucoprotein extracellular matrix that is both fibrous and amorphous. It is much thicker and stronger than the fine egg plasma membrane and has tensile and elastic properties.

---

to induce both ovulation and germinal vesicle breakdown, followed by the first meiotic division and formation of the first polar body. (B) Frog. The principal organs are the anterior pituitary and the ovary. Gonadotrophin secreted by the pituitary stimulates two components, the oocyte itself and the follicle wall surrounding the oocyte. In the electron microscope, the follicle wall is seen to consist of a layer of follicle cells with microvilli interdigitating the microvilli emerging from the oocyte surface. These are separated by the vitelline layer. Outside the follicle cell layer is a thecal layer consisting of cells and extracellular material with numerous (contractile?) microfilaments. The hormone induces retraction of microvilli and presumably contraction of the thecal layer to induce ovulation. It also induces interactions between follicle cells and oocyte surface, including the appearance of a hormone that induces germinal vesicle breakdown followed by the first meiotic division and polar body formation. The egg is ovulated in the second meiotic metaphase stage.

After ovulation, the egg passes into an oviduct, usually from the body cavity. Leading into the oviduct may be shell glands, jelly-producing glands, or albumin-producing structures which secrete materials around the egg as it moves down the passage. In the frog, epithelial cells of the oviduct secrete three discrete concentric rings of jelly around the egg. In the chick, the white or albumin and the shell are added to the egg by secretions from accessory glands.

The jellies surrounding most eggs consist of complex mucopolysaccharides similar to extracellular matrices that surround most somatic cells. As we will see later, these jellies play an important part in the fertilization reaction. They are the first egg components contacted by sperm. For many aquatic eggs, the jelly serves to maintain egg buoyancy and support the egg as it floats in water. The jellies are also used for the attachment of eggs to solid objects during egg laying.

Many eggs have an additional structure: an outer *chorion* applied to the egg as it completes its maturation. The chorion is a tough, thick, protective layer which, in insects, is designed to reduce water loss and protect these terrestrial eggs from mechanical damage. A small opening, a *micropyle,* is usually present at the animal pole where sperm enter.

### Storage of developmental information

In many species, presumptive eggs are segregated precociously from somatic cells early in cleavage as primordial germ cells, a process mediated by special cytoplasms localized in the zygote. Cells acquiring this special plasm are committed to a different division program from somatic cells and tend to migrate to the developing gonad where they differentiate as gametes.

These cells retain the full genetic endowment of the species and engage in meiosis, a process that correlates with the events of egg differentiation. Meiosis is arrested, often for many years, while the oocyte stores all its developmental information. During oogenesis, the egg genome along with accessory cell genomes synthesizes and stores genetic information in egg cytoplasm. Oocyte nucleoli also become active and synthesize a rich supply of ribosomes. These are stored in egg cytoplasm along with nutritive and morphogenetic materials and organelles such as glycogen, lipid, yolk, mitochondria, tubulin, actin, and histones. Oocyte organization is entirely a product of the maternal genome.

Yolk is a unique protein stored almost exclusively in eggs. Though synthesized elsewhere in the female body, yolk accumulates as large crystalline aggregates or platelets in oocyte cytoplasm and acts as a source of precursors for embryonic protein and lipid synthesis.

Contrary to the conditions in most cells, the egg surface and cortex also seem to be specialized. A large portion of the developmental blueprint is laid down in the cortex and the plasma membrane.

Eggs, then, are large cells with high developmental potential because they accumulate and assemble nutritive and informational reserves, often distributed in a polarized manner. This organization seems to be primed for rapid proliferation into two-dimensional multicellular sheets which, in turn, transform into an embryo. Before an egg can play out this program, however, it must be activated by fertilization.

## FERTILIZATION AND EGG ACTIVATION

Fertilization, besides its obvious role in sexual recombination, triggers the developmental program in a rapid burst of metabolic activity. In a sense, this moment of activation "saves" the life of the egg, precariously balanced between quick extinction or growth and morphogenesis. Because of their enormous size and disproportionate nucleocytoplasmic volumes, mature unfertilized eggs have a limited life after ovulation, often a matter of a few hours. They "ripen" or age, gradually losing developmental information as their nuclei and cytoplasm become disorganized. If overripe eggs are fertilized, they develop abnormalities, including chromosomal aberrations whose magnitude is directly related to the egg's age before fertilization (see Chapter 20). Many known congenital abnormalities in humans have been attributed to the development of overripe eggs. Ripening induces irreversible changes in egg organization, and unless an egg is rescued by fertilization, it soon dies. Hence fertilization accomplishes three purposes: (1) sexual recombination, (2) activation of the developmental program, and (3) sustaining egg survival.

Fertilization may occur at different times during egg maturation (Table 11-2). Many marine eggs are fertilized in the germinal vesicle stage, an act which stimulates completion of oocyte meiosis. Others are fertilized after the egg completes meiosis and forms two polar bodies. A full range of intermediate condi-

**TABLE 11-2  Stage of Egg Maturation at Which Sperm Penetration Occurs in Different Animals**

| Young Primary Oocyte | Fully Grown Primary Oocyte | First Metaphase | Second Metaphase | Female Pronucleus |
|---|---|---|---|---|
| Brachycoelium | Ascaris | Aphryotrocha | Amphioxus | Coelenterates |
| Dinophilus | Dicyema | Cerebratulus | Most mammals | Echinoids |
| Histriobdella | Dog and fox | Chaetopterus | Siredon | |
| Otomesostoma | Myzostoma | Dentalium | Rana | |
| Peripatopsis | Nereis | Many insects | Xenopus | |
| Saccocirrus | Spisula | Pectinaria | | |
| | Thalassema | | | |
| | Asterias | | | |

Source: From C. R. Austin, *Fertilization*, © 1965, p. 87. Reprinted by permission of Prentice-Hall, Inc., Englewood Cliffs, New Jersey.

tions exists with eggs fertilized at different times between the germinal vesicle stage and the terminal stage of meiosis.

No generalizations can be made about the metabolic or physiological state of an egg at the moment of fertilization. Eggs fertilized in the germinal vesicle stage are still metabolically active, on the verge of completing two meiotic divisions. Others, having completed meiosis, become metabolically depressed before the egg is laid, a condition which may contribute to the ripening phenomenon. In such eggs, the vigorous metabolism of oogenesis with high rates of respiration and macromolecular synthesis diminishes at the end of oocyte growth. In many mature eggs, nucleic acid and protein synthesis are depressed, unable even to sustain the normal maintenance requirements of the cell. Even when protein synthesis continues at rates comparable with those in the oocyte, the unfertilized egg still has a limited life.

Egg activation occurs at sperm attachment. Within seconds, egg structure is reorganized, particularly within the cortex, and the developmental program begins.

## Sperm organization

Before we discuss sperm-egg interactions, let us describe the structure of the sperm cell. This cell is specialized for the delivery of packaged genetic material for sexual recombination. Usually it is a motile cell equipped with a strong flagellum to propel the cell in an aqueous medium. It is devoid of most cytoplasmic organelles that normally sustain cell viability, such as ribosomes and endoplasmic reticulum. Furthermore, its nucleus is essentially inactive; its DNA is tightly packaged into an almost crystalline structure within the sperm head, reduced to minimum volume for effective transport. Accordingly, the life of a sperm cell is short; it depends upon metabolic substrates provided within accompanying fluids (seminal fluid) for its metabolic energy. Within the male, sperm survive for long periods and may even survive in the female genital tract if kept in special storage organs or spermatheca. Once shed into dilute media, however, metabolic substrates are rapidly exhausted as sperm swim and they soon die. Hundreds of millions of sperm cells are usually produced to insure that a few successfully fertilize eggs.

Although observers have noted considerable variation in structure, sperm follow a common morphological pattern. Nonflagellated, amoeboid sperm are found in such groups as nematodes, spiders, and crustacea. Most sperm are flagellated tripartite structures consisting of head, midpiece, and motile tail. This specialized cell develops from a haploid spermatid, one of four meiotic products of a spermatocyte. The streamlined architecture of a sperm cell results during spermatid differentiation as all excess cytoplasm is sloughed off while a tail emerges.

The situation in human spermatogenesis is seen in Figure 11-17. After the second meiotic division, the Golgi vesicles and lamellae accumulate large amounts of mucopolysaccharides. The centrioles at the opposite periphery of the cell generate a set of microtubules, growing out as an axial filament, the future flagellum. The pair of centrioles differentiates

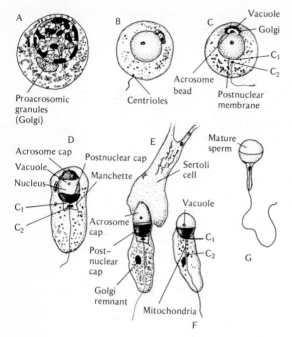

A
Proacrosomic
granules
(Golgi)

B
Centrioles

C
Vacuole
Golgi
$C_1$
$C_2$
Acrosome
bead
Postnuclear
membrane

D
Acrosome cap
Vacuole
Nucleus
$C_1$
$C_2$
Postnuclear cap
Manchette

E
Mature
sperm
Sertoli
cell

Acrosome
cap
Post-
nuclear
cap
Golgi
remnant
Vacuole
$C_1$
$C_2$
G
Mitochondria
F

**Figure 11-17**  *Human spermatogenesis, $C_1$ and $C_2$ indicate centrioles. (Based on O. E. Nelsen,* Comparative Embryology of the Vertebrates. © *1953 by Blakiston Company, Inc. Used with permission of McGraw-Hill Book Company.)*

into a basal body and a ring centriole, the latter migrating along the lengthening microtubules. The Golgi accumulation fuses into a large *acroblast* which adheres to the nuclear membrane and transforms into an *acrosome*. Meanwhile, chromosomes and DNA are packaged into a tight crystalline array in the nucleus.

As the flagellum lengthens, cytoplasm elongates with it, carrying the mitochondria into the region surrounding the growing microtubules. Note that throughout this process the plasma membrane retains its peripheral position around the entire cell, including the anterior head with acrosome, as well as the lengthening tail. Extension of the tail is probably due to the assembly of microtubules into a characteristic nine-plus-two array of alpha and beta tubules, an elaboration of the pattern preexisting within the centriole "cartwheel" from which these seem to organize. Mitochondria accumulate behind the nucleus to make up, with the ring centriole, the midpiece of the mature sperm. Golgi remnants along with any excess

cytoplasm surrounding the tail are sloughed off. The sperm cell is now ready for a motile life, an energy-transducing system (mitochondria within the midpiece) prepared to drive the flexible tubular rods that make up the flagellum core.

Variations are found in acrosomes, nuclei, midpieces, and flagella, but all sperm have similar metabolic activity. They are principally catabolic, usually capable of anaerobic and aerobic oxidation of such carbohydrate substrates as fructose, glucose, lactate, and pyruvate in sperm fluids. The principal business of a sperm cell is to swim to an egg, make contact with it, and deliver its genetic material. Swimming is accomplished by sinuous waves of the flagellum, generated at the basal granules. It should be noted, however, that sperm motility is not always the principal means of sperm transport. In mammals, for example, fluid flow within the female genital tract coupled to uterine contractions probably is responsible for transporting sperm to the proper site.

### Interactions between sperm and egg

Gamete fusion is followed by fertilization, the union of two different haploid genomes into a single diploid zygote. Sperm-egg interactions call for cellular recognition (Chapter 10), a property residing in outer membranes and extracellular matrices. In prokaryotes and protists, where gametes are morphologically identical, mating specificity has evolved to insure fusion of genetically different cells. Plus and minus mating types have evolved within a species. In this way, asexual reproducing cells within the same mating type are prevented from fusing with one another. For example, in many bacteria and unicellular algae such as *Chlamydomonas*, conjugation is accomplished only between opposite mating types; only cells with different surface specificities (plus and minus) interact. Mating-type mechanisms have evolved to prevent self-fertilization and promote outcrossing, thereby enhancing species' genetic variability. Specificity of mating type is a genetically determined trait segregating as a simple Mendelian gene. Presumably, these genes specify some physical-chemical aspect of the outer plasma membrane that insures recognition specificity.

Mating-type substances are at the cell surface, in the plasma membrane, in appendages of the surface such as cilia or flagella, or in extracellular matrices. In yeast, for example, surface glycoproteins of different mating types will interact and agglutinate the cells, presumably in a manner analogous to an antibody-antigen reaction. The cilia of two opposite mating types of *Paramecium* are involved in conjugation, an interaction resembling agglutination. Cells of one mating type are agglutinated only if treated with cilia isolated from an opposite mating type. In this organism, mating-type genes code for ciliary mucoproteins that control cell recognition during conjugation.

Sperm-egg interactions are not fundamentally different from cell fusion reactions in yeast and protozoa. They, too, involve surface recognition between membranes and jelly matrices of the egg and those surrounding the sperm head, and glycoproteins also play a major role in recognition. During the course of multicellular evolution, the properties for membrane fusion have been restricted to only a few cell lines, usually those destined to become gametes. Presumably as gametes differentiate, mating-type genes are expressed and their products are incorporated into outer membranes and extracellular secretions.

Recognition mechanisms in fertilization rely on cell contact. In most cases, chemotaxis is not involved; gametes are not attracted to one another over great distances but interact only after random collision. On the other hand, substances acting like hormone attractants are found in fungi, lower plants, and some coelenterates. They are produced by both gamete types, operate at a distance at low concentrations, and are clearly essential to the fertilization reaction.

### Fertilization in plants: chemotactic systems

As early as 1884, it was recognized that fertilization in plants involved a chemotactic phenomenon. The sperm of liverworts, mosses, and ferns react chemotactically, swarming about the small openings leading into the narrow necks of female archegonia, as if attracted to substances leaking out of eggs at the base. In the water fern *Marsilea*, sperm are entrapped in a gelatinous matrix extruded from the opening of the archegonium. Once embedded in the jelly, they swim about at random until attracted to a funnel-like opening where they enter as if following a chemical track. Some attracting substances have been identified; for example, such simple compounds as sucrose or L-malic acid seem to work in mosses and ferns.

Two kinds of attractive systems have been studied, one involving motile gametes (chemotactic) and one showing directed growth or chemotropism. A specific hormone, *sirenin*, attracts sperm in the water mold *Allomyces* at concentrations as low as $10^{-10}$ M. By metabolizing the hormone, sperm create a concentration gradient and swim toward the female gametes where hormone concentration is usually highest.

Hormonal interactions are more complex in chemotropic phenomena where gametes send out fertilization tubes. In the water mold *Achlya*, the male antheridial hypha grows toward the oogonial wall. A transverse wall forms, cutting off a small antheridial gamete from which a fertilization tube develops to penetrate the oogonial wall and grow to the eggs. This directed growth of the antheridial hypha is induced by diffusible substances coming from the oogonium. Growing antheridial hyphae, in turn, induce oogonial development by secreting male hormone. Several hormones seem to be involved in a sequence of feedback interactions between the two gametes. Since certain amino acid mixtures induce directed outgrowth of male hyphae, no specific hormones can be implicated; hyphae could be attracted to nutrient-rich regions surrounding the oogonium.

Pollen tubes of higher plants also show directed growth toward ovary tissues or ovules. If pollen grains are placed a short distance from a clump of ovules on the surface of nutrient agar, they all tend to grow out toward the individual ovules. A hormone may not be essential inasmuch as a crystal of calcium carbonate, placed at a distance from pollen grains, also induces directed outgrowth.

Specificity of hormone action in plant reproduction is questionable since cross-fertilization can occur; for example, sperm from several species are attracted to the same oogonial site. Even pollen from one species can grow into the ovary and fertilize eggs of another. Comparative study of sex hormone systems in plants is still only in its infancy, but these systems should furnish more information about chemotactic mechanisms and the specificity of cell surface interactions.

### Fertilizin: recognition substances in animal fertilization

The sessile life style of many coelenterates may require a chemotactic mechanism for fertilization. Though chemotaxis is implicated in coelenterate fertilization, it is rare in most other groups. If sperm are not attracted to eggs, how is fertilization brought about and, more importantly, what determines its specificity?

Factors that mediate sperm-egg interactions even before they make contact were identified in 1914 by F. R. Lillie, who proposed the first theory of the physiology of fertilization. He observed that egg water (sea water surrounding unfertilized sea urchin eggs) agglutinated sperm and activated their motility. The reaction was specific since sperm from related species were unaffected. The factor, called *fertilizin*, came from the egg jelly coat which slowly dissolved as eggs stood in sea water. Lillie postulated that fertilizin was continuously secreted by the egg before fertilization, but it is now clear that fertilizin is a constituent of both jelly coats and egg membranes.

For sperm to be agglutinated by fertilizin, they must have receptor sites on their surface. These receptor sites, or "antifertilizins," interact with a bivalent (two-armed) fertilizin molecule. Neighboring sperm cells fuse because bivalent fertilizin molecules form a bridge between sperm cell surfaces. The interaction between the sperm and the egg resembles an antigen-antibody reaction; fertilizin in egg membranes may behave as an antibody combining with antigenic sites on the sperm surface.

The bivalent nature of the fertilizin molecule is illustrated by the reversibility of sperm agglutination. Agglutinated sperm dissociate and swim off shortly after agglutination. These sperm are now insensitive to fertilizin and cannot be agglutinated again. Apparently separation breaks the fertilizin molecule, leaving one combining site attached to the sperm surface which blocks all further interactions with fertilizin.

The fertilizin molecule has been partially characterized as a mucopolysaccharide rich in sialic acid residues with a molecular weight of approximately 82,000 or more. It seems to be a homogeneous molecule in that only one major constituent is usually extracted from the jelly coat. Such mucoproteins, typical of matrix materials secreted by somatic cells, are important in surface recognition (see Chapter 10). In this case, they probably insure the initial entrapment of randomly swimming sperm in jelly surrounding the egg. Jelly coats, vitelline membranes, and even egg plasma membranes share identical mucoproteins, which means that the specificity of interaction between sperm and egg persists as the sperm makes its way into the egg cytoplasm from the surrounding medium.

Similar sperm agglutination reactions have been found in other species, including mammals. In amphibians, fertilization is impossible without materials dissolved from egg jelly coats; that is, eggs, denuded of their jelly, are not fertilizable but become so when encapsulated in jelly taken from another egg.

The agglutination reaction per se probably has no direct role in fertilization except for the entrapment phenomenon mentioned earlier. It denotes, however, a molecular matching between egg mucopolysaccharide secretions and complementary molecules on sperm surfaces. An important effect of egg water is on sperm motility; sperm are motile for longer periods in its presence. Furthermore, aged sperm may be reactivated by the addition of egg water, but it is uncertain if sperm activation correlates with agglutinating properties of egg water. There is no evidence that sperm motility is enhanced by purified preparations of fertilizin, for example.

In addition to increased sperm motility, egg water also transforms sperm structure, an essential step in sperm activation. Sperm must undergo certain physiological and structural changes to prepare them for the fertilization act. This is particularly important in the mammalian egg where the phenomenon of sperm activation is known as *capacitation*. Before they can fertilize eggs, mammalian sperm first interact with female genital tract secretions, then with the secretions emanating from envelopes surrounding the mammalian egg. Capacitation changes in sperm structure are subtle and often difficult to detect. One visible structural reorganization, however, that does take place in egg water or after contact with egg jelly is the *acrosome reaction*.

**The acrosome reaction**

The acrosome, the most anterior structure of the sperm head, is the vesicle or granule sitting atop the sperm nucleus (Figure 11-18). In the electron microscope, its organization is more complex: a central granule embedded in a clear vesicle, surrounded by a separate membrane. In some sperm, a rodlike element is seen below the acrosomal granule, usually embedded in an indentation in the anterior portion of the nucleus. The entire organelle is derived from the Golgi apparatus in the differentiating spermatid, and mucopolysaccharides, among other proteins, are its major constituent.

On exposure to egg water or jelly, a sperm acrosome "explodes." The sperm plasma membrane at

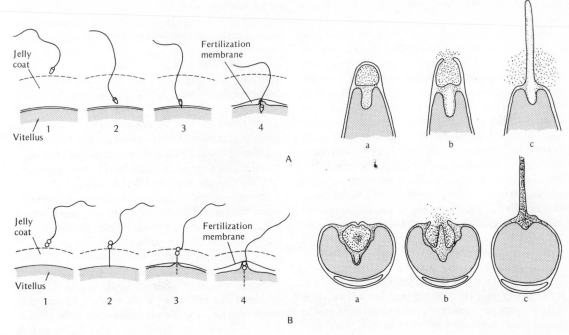

**Figure 11-18** *Acrosome reaction in two echinoderm species. (A) Reaction in the sea urchin,* Arbacia. *On the left is the reaction seen in the light microscope. (1) Approach of the spermatozoon. (2) Passage of sperm through jelly coat. (3) Acrosome reaction initiated by contact with vitelline membrane and projection of filament into vitelline surface. (4) Absorption of sperm and formation of fertilization cone and membrane. On the right is the acrosome reaction seen in the electron microscope. (a) Intact acrosome. (b) Opening of acrosome after fusion between plasma membrane and outer acrosome membrane; release of lytic agent. (c) Projection of filament and reduction of material of acrosome. (B) Reaction in starfish,* Asterias. *On the left is the sequence in the light microscope. (1) Approach of sperm cell. (2) Acrosome reaction provoked by contact with jelly surface and projection of long filament through jelly to vitelline surface. (3, 4) Absorption of spermatozoon; formation of vitelline cone and membrane. On the right is the sequence in the electron microscope. (a) Intact acrosome. (b, c) Opening of the acrosome by fusion between plasma membrane and outer acrosomal membrane; release of lytic agent; projection of filament. (From C. R. Austin,* Fertilization, © 1965. *Reprinted by permission of Prentice-Hall, Inc., Englewood Cliffs, New Jersey. Drawn from electron micrographs published by J. C. Dan, Exp. Cell Res. 19:13, 1960.)*

the tip of the acrosomal vesicle dissolves and fuses with the acrosomal membrane to form an opening through which the acrosomal vesicle contents are released. These emerge as an expanding tubule which may be almost as long as the entire spermatozoon. In mammalian eggs, the acrosomal reaction is initiated before sperm contacts the outer egg envelope or *zona pellucida*, while it is still migrating in the genital tract. It is important to note that the membrane surrounding the acrosomal tubule is derived from an inner membrane of the acrosomal vesicle. It is this membrane that makes first contact with egg jelly, vitelline membrane, and plasma membrane, in that

order. The acrosomal tubule penetrates the jelly, reaching the egg surface before the sperm head. The first membrane contacting the egg surface is a mosaic consisting of the acrosomal membrane fused with the sperm plasma membrane.

The acrosome reaction was observed more than 100 years ago, when Fol in 1870 first described fertilization in the starfish egg, but it is only within the past few years that its significance has been realized. Fertilization cannot occur unless the acrosome is triggered. If whole sperm are injected with a micropipette directly into sea urchin egg cytoplasm, the egg does not fertilize nor does it develop. Fertilization is a cell surface phenomenon; that is, unless sperm and egg membranes undergo a sequence of interactions before fusion, fertilization does not occur.

What role, then, do fertilizin and antifertilizin play in this mutual activation process? Since egg water triggers explosion of the acrosomal filament, one suggestion is that fertilizin, combining with antifertilizin sites on the acrosome membrane, changes its permeability to ions. There is no evidence supporting this hypothesis since other factors in egg water besides fertilizin may be responsible for the acrosome reaction. As indicated before, specificity of fertilization suggests an immunological reaction between complementary receptor sites in sperm and egg membranes. Specificity, however, is not absolute. We know that eggs of one species of echinoderm can be fertilized by sperm of distantly related species to make species hybrids. A high degree of specificity of antigen-antibody cannot apply to sperm-egg interactions. The fact that antibodies made against fertilizin or egg jellies do interfere with fertilization does not implicate these materials in the fertilization mechanism itself. Antibodies attached to membrane surfaces may, by steric hindrance, interfere with other structures at the egg surface essential to the fertilization reaction.

Complementary receptor sites on cell surfaces seem to account for mating-type reactions in lower organisms and might also explain membrane fusion of sperm and egg. Fertilizin-antifertilizin reactions express molecular complementarity in two opposing cell surfaces. The recognition properties of the egg membrane are amplified by the vitelline membrane and jelly envelopes carrying identical receptor mucoproteins. This facilitates entrapment of sperm as they move randomly about the egg. Motile sperm in the vicinity of the egg are "captured" by swollen jelly envelopes with numerous exposed reactive sites. The explosion of acrosomal filaments also increases the frequency of random contact between sperm and egg jelly. Once these contacts are made, the ensuing attachment of acrosomal filament to the egg surface brings the sperm into the egg.

All acrosomes and acrosome reactions are not identical. There is considerable variation in the mechanism of the reaction and the mode of sperm entry. In *Limulus* sperm, for example, the acrosome is an elaborate coiled filamentous structure that contains actin, the contractile protein, as a principal component. Discharge of the acrosome activates the

actin filament, which like a contracted coil extends rapidly to make contact with the sperm surface. The acrosome filament then contracts, pulling the sperm into the egg surface. Two other proteins are associated with actin, one of which is α-actinin. Apparently these proteins keep actin organized in a paracrystalline state in the sperm and the factor stimulating the acrosome reaction causes a rearrangement of these filaments so that they now extend in parallel bundles in the acrosome filament. In another species, *Thyone* (a sea cucumber), the acrosome consists largely of actin held in a nonpolymerized monomeric state. The acrosome reaction in this instance involves changes in the state of polymerization: monomeric actin precursors polymerize into a long filament which contracts and again pulls the sperm toward the egg surface. The identification of actin as the principal dynamic molecule in these acrosome reactions suggests that it may also be a constituent of most other acrosome reactions. It is to be noted that myosin or tropomyosin is not found in these systems, which means that contractile mechanisms unrelated to the sliding filament model in muscle must be implicated.

### Attachment and penetration of sperm

To reach the egg surface, sperm must penetrate the outer envelopes including the vitelline membrane. Some sperm release a lytic substance from the acrosomal vesicle which digests the egg jelly. True membrane lysins have been obtained from sperm of the common mussel, *Mytilus edulis*. These lytic substances also digest holes in vitelline membranes, probably by an enzymatic process.

The lytic substance in mammalian eggs is an enzyme, hyaluronidase. Mammalian eggs are surrounded by a thick zona pellucida covered by follicle cells of the cumulus oophorus, and the sperm must penetrate this cell layer as well as the zona. To do this, the sperm head releases the enzyme hyaluronidase (located in the acrosome), which specifically digests the intracellular cement holding the cumulus cells together and also dissolves the jelly of the pellucida, both of which contain large concentrations of hyaluronic acid. The capacitation reaction of mammalian sperm is correlated with the acrosome reaction and the release of hyaluronidase. Many mammalian sperm spend several hours in genital tract secretions before they are capacitated; until then,

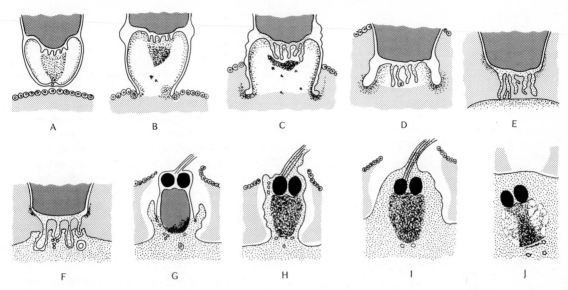

**Figure 11-19** *Fertilization in the annelid worm* Hydroides. *(A–C) Stages in the acrosome reaction which occurs when sperm contacts vitelline surface. Lysin released and part of vitelline membrane is dissolved. Acrosome consists of several short filaments that penetrate vitelline layer. (D) Accommodation of projections of acrosome membrane in invagination of vitelline surface soon followed by fusion of this membrane with the egg plasma membrane. (E–J) Passage of the sperm head into the vitellus. The egg cytoplasm flows up under the sperm plasma membrane and around sperm head, midpiece, and tail, forming the fertilization cone. Membrane fusion has occurred at G. (Based on electron micrographs from L. H. Colwin and A. L. Colwin,* J. Biophysic. Biochem. Cytol. *10:255–274, 1961.)*

they are unable to penetrate the zona and fertilize the egg.

How does the sperm penetrate the egg after the acrosome filament contacts the egg surface? Fusion and sperm entry in a worm, *Hydroides*, have been followed in the electron microscope. The sequence of events is shown in Figure 11-19.

After a portion of the vitelline membrane has been pierced, the membrane of the acrosome fuses with the plasma membrane of the egg. As we will see, once this contact is made, an activation reaction instantaneously passes over the egg surface. The activated surface at the point of attachment elevates, emerging as a small cytoplasmic protuberance or fertilization cone. The egg is not the passive recipient of a penetrating sperm; rather, it actively participates in sperm entry. The cytoplasm of the fertilization cone probably attaches to the acrosomal filament, crawling up the filament, and pulls it into the cytoplasm. Movements of the sperm flagellum are not involved in ingression of the filament and sperm head; sperm enters as if engulfed by egg cytoplasm. Fusion of the egg membrane with the outer sperm membrane cre-

ates a cytoplasmic bridge through which the naked sperm nucleus enters the egg cytoplasm.

### The egg activation reaction

The minute fertilization cone emerging at the site of the acrosome filament is part of an overall activation of egg membrane and cortex. The change that occurs at the point of membrane fusion is transmitted as a wave or impulse over the egg surface and subjacent cortex. The surface undergoes a change in ion permeability, accompanied by altered electrical properties of the membrane. Electrical potential changes at the moment of fertilization resemble changes occurring in a nerve when an impulse passes over it. In the egg, however, the time constant is much longer than it is in a neuron. In the starfish *Asterias*, for example, the inside of the egg has a membrane potential of −30 to −60 mV, and at fertilization the potential decreases about 5 mV, a change lasting about 1 minute (Figure 11-20). Larger drops of membrane potential are seen in the eggs of the fish and toad, and the magnitude of the potential change varies with the ion con-

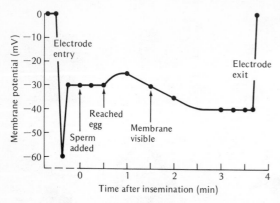

**Figure 11-20**  *Changes in membrane potential of the starfish (Asterias) egg at fertilization. The membrane potential difference of the unfertilized egg is initially 60 mV (inside negative) as soon as the electrode penetrates, but this soon stabilizes at −30 mV. Within 30 seconds after sperm addition, the potential decreases by about 5 mV, to be followed later by an increase in potential to −40 mV, where it remains steady. (From A. Tyler, A. Monroy, C. Y. Kao, and H. Grundfest,* Biol. Bull. *111:153–177, 1956.)*

tent of the medium. If more sodium is added, the potential drop is reduced while the addition of potassium increases the change by 10 to 15 mV. The response to these ions suggests that the primary activation event is a change in ion permeability at the site of sperm contact. Sodium, potassium, and calcium diffuse rapidly into the frog egg, reaching a peak within a minute after sperm attachment. These ions are probably redistributed between nuclear and cytoplasmic compartments after fertilization, and the proportion of free to bound ions also changes. The permeability change is then propagated as a wave over the membrane within a matter of seconds.

Of all the ions involved, calcium seems to play a major role. In the absence of calcium, fertilization does not occur. Sperm viability, motility, and fertilizing capacity are also diminished in calcium-free media. Furthermore, the acrosome reaction is calcium-dependent; without it, no acrosome filament explodes. Egg stability also depends upon calcium; in its absence, changes in viscosity and rigidity of the cortex are observed.

Changes in the distribution of calcium within the egg are noted at fertilization. Calcium seems to be bound to the plasma membrane and cortex, but after

membrane fusion, calcium is released, becoming freely diffusible within the cytoplasm, moving from one compartment to another. This "calcium activation" event may change protein configuration in membrane and cortex, with consequent changes in functional properties.

Metabolic activation of the egg is similar to other excitation phenomena, where membrane-bound calcium also plays a principal role. In nerve excitation, we assume that resting membranes bind calcium that is released during transmission of the nerve impulse. A calcium-dependent adenosine triphosphatase in the membrane acts like a calcium pump, moving ions across the membrane and pumping unbound calcium out of the neuron. The similarity between fertilization and excitatory phenomena suggests that egg membranes behave in a comparable manner.

Calcium may be involved in maintaining protein configuration; any change in its association with proteins changes protein function. Conformational changes of membrane proteins may account for increased permeability after fertilization. Finally, calcium exchanges across membranes in excitatory phenomena are coupled to cyclic AMP, a regulatory molecule that activates many enzymes, including protein kinases. Protein kinases are enzymes that phosphorylate proteins such as histones and may also phosphorylate microtubular proteins that function in cell division. Since histones have a structural and possibly a regulatory role in chromosome function, we can see that activation of histones and microtubule proteins, among others, might be essential to further development of the egg. Activation is also accompanied by enhanced phosphate metabolism. We find that the incorporation of radioactive phosphate is markedly stimulated immediately after sea urchin eggs are fertilized. Unfertilized sea urchin eggs exposed to labeled phosphate rapidly reach ionic saturation, but immediately after fertilization the rate of transfer of these molecules across the membrane is increased.

### The cortical reaction

Egg membrane activation is intimately tied to changes in the egg cortex. In sea urchin eggs, where cortical granules break down at fertilization, granule and egg membranes fuse to form a mosaic. Such an altered membrane may contribute to the change in permeability. If cortical granule breakdown is inhibited, then the permeability changes seen in the sea urchin egg do not occur. It is tempting to speculate that cortical granule membranes in this egg are already differentiated as special transport systems. When fused into the plasma membrane, they are activated, altering membrane permeability. If this were so, the cortical reaction would be the principal activation event at fertilization, at least in those eggs with cortical granules. Let us examine this reaction in more detail.

Within seconds after membrane contact, changes occur in the egg cortex. The fine microvilli of the egg membrane are withdrawn and the egg surface becomes smooth. The dense granular layer of the cortex becomes a homogeneous region of oriented microfilaments adjacent to the plasma membrane. These changes signal the altered physiological state of the cortex.

Cortical granule transformation in sea urchin eggs entails several independent changes in the granule and egg membrane (Figure 11-21). The process is initiated at the site of sperm attachment. First, cortical granule membranes fuse with the egg plasma membrane, thereby opening the cortical granule to the outside. The dark lamella-like contents of the cortical granule are explosively released and attach to the expanding vitelline membrane, already rising off the egg surface. The clear matrix of the cortical granules spills into the space between the elevating fertilization membrane and the plasma membrane proper. The matrix remains attached to the outer membrane of the egg and that, plus the lighter material, makes up the hyaline layer of the egg, now penetrated by surface microvilli. Between the hyaline layer and the fertilization membrane is a fluid-filled space known as the *perivitelline space*. Thus cortical granule contents contribute to the fertilization membrane, the mucoprotein material of the hyaline layer adjacent to the egg surface, and also furnish materials to the fluid-filled perivitelline space.

Blocking cortical granule breakdown interferes with certain events of fertilization. If eggs are drawn into capillary tubes and inseminated at one end, the capillary walls prevent propagation of the cortical stimulus over the egg surface and only one surface shows a cortical reaction. Only this region possesses an elevated fertilization membrane, while cortical granules persist at the opposite side. In spite of this

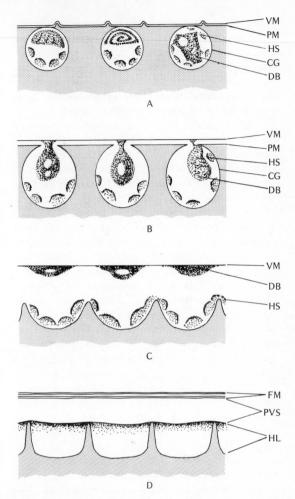

**Figure 11-21** *Cortical granule breakdown and formation of the fertilization membrane (FM) in the sea urchin egg. (A) Surface of the unfertilized egg showing vitelline membrane (VM), egg plasma membrane (PM), cortical granules (CG) composed of central core of dark bodies (DB) plus some hemispheric globules near the cortical granule membrane. (B) Explosion of the cortical granules, extrusion of dark bodies, fusion of cortical granule membrane with plasma membrane, and beginning of lifting of the vitelline membrane. (C) The dark bodies join the vitelline membrane while hemispheric globules (HS) adhere to the plasma membrane, building up an extracellular matrix over the surface which becomes the hyaline layer (HL), as seen in part D. (D) Dark bodies fuse with vitelline membrane, forming the fertilization membrane which is separated from the egg surface by a perivitelline space (PVS) (Based on Y. Endo, Exp. Cell Res. 25:383–397, 1961.)*

polarization, some fertilization events continue, even in the region where cortical granule breakdown does not occur. Pronuclei form but they do not fuse, nor does the egg cleave. Although the normal sequence of fertilization events is blocked by preventing cortical granule breakdown, some events such as pronuclear formation are independent of it.

The sequence of cortical granule fusions with the egg membrane resembles an enzymatic activation traveling as a wave over the egg surface from the point of sperm attachment. It is, in sea urchin eggs, propagated over the surface in 20 to 30 seconds. One hypothesis for impulse spread is that a self-propagating chain reaction is initiated by enzymes formed at the site of sperm attachment, which diffuse over the egg surface and initiate cortical granule breakdown. For example, minutes after fertilization, lipases and proteases become active, and these may digest cortical granule membranes to initiate breakdown. Another enzyme activated at fertilization is a calcium-dependent adenosine triphosphatase, which could explain the requirement for calcium at fertilization. Splitting the ATP may furnish energy for the explosive release of the granules, or for contractions of the egg surface.

There are conflicting opinions as to the significance of the cortical granule reaction at fertilization. In some eggs, such as the barnacle *Barnea*, cortical granules explode independently of fertilization. In the annelid worm *Chaetopterus*, they do not break down at fertilization but are retained during early cleavage. Activation of the egg cortex at fertilization may not be linked to cortical granule breakdown. Apparently cortical granules have various roles in the egg, depending on species. In fact, cortical granule breakdown may not contribute to egg activation but may participate only in formation of the fertilization membrane and hyaline layer. According to this view, cortical granule breakdown is, itself, an effect of a more fundamental activation of the egg cortex rather than a cause.

What is the nature of this propagated impulse that activates the egg? The events occur within seconds after sperm attachment, too rapid to resolve the primary trigger. We assume the trigger is "pulled" at the plasma membrane, most likely as a change in ion permeability. This may activate all subsequent metabolic events since changes in Na+ and Ca++ ion distribution within cellular compartments may affect

protein configuration, enzyme kinetics, and metabolism (see Chapter 13).

## The block to polyspermy

In most species, successful fertilization occurs when one and only one sperm enters the egg. When more than one enters, a genetic imbalance is created and development is abnormal. In 1902, Boveri noted that dispermic sea urchin eggs develop into bizarre forms. He could show that the first division of the egg was abnormal because a multipolar spindle formed. The triploid chromosome complement was randomly distributed to three cells at the first division, and subsequent cells were aneuploid and developed abnormally. How, then, is polyspermy prevented at fertilization?

When the vitelline membrane lifts off and transforms into a fertilization membrane, it acts as a mechanical barrier to the penetration of other sperm. Presumably the configuration of mucoprotein in the membrane is altered so that sperm membranes cannot fuse with or penetrate through the barrier.

The block to polyspermy, however, is more than a lifting of the fertilization membrane; it is intrinsic to the egg surface itself. At the moment of sperm attachment, before the membrane lifts, there are many sperm close to or attached to the egg surface (vitelline membrane), yet only one penetrates. The block may involve a rapid transformation of egg surface receptors that prevents attachment of more than one sperm.

The nature of the block to polyspermy varies from one species to another. In certain crustaceans, many sperm simultaneously attach to the egg surface and induce the elevation of fertilization cones, each in contact with a sperm head. The first cone surrounding a sperm and withdrawing it into the cytoplasm prevents the others from engulfing their sperm. In some sea urchins, the block depends upon the barrier formed by the fertilization membrane. Removing the membrane permits additional sperm to attach and penetrate. In others, however, the block resides in the egg cortex because membrane removal is not followed by penetration of other sperm. The block to polyspermy is removed in such eggs by placing them in calcium-free medium, where they can be refertilized. Since the block may be experimentally removed, the surface of the fertilized egg does not develop an irreversible block to polyspermy.

Some idea of the block to polyspermy in sea urchin eggs has come from studies of fertilization in *Strongylocentrotus*, with the scanning electron microscope (Figure 11-22). Many sperm attach to the vitelline membrane within a 25-second period after insemination. The vitelline membrane is thrown up into numerous regularly spaced arrays of microvillar projections, and sperm acrosome filaments seem to attach to these. At 20 to 25 seconds or so, a fertilizing sperm enters and the cortical granule breakdown reaction is initiated. As the wave of cortical granule breakdown passes over the egg surface, most of the remaining sperm are released from the vitelline surface, which itself is undergoing a change as it lifts off. The arrangement of microvilli alters so that the surface now appears smoother.

Attachment-detachment of sperm correlates with cortical granule breakdown and release of a protease that is recoverable from the medium. The protease will alter the vitelline surface of unfertilized eggs in a manner identical to that occurring during fertilization. It has been proposed that the protease hardens the vitelline membrane and makes it refractive to sperm attachment. Simultaneously, it destroys the attachment of accessory sperm sites in the vitelline surface, and sperm are released.

The block to polyspermy in some sea urchin eggs involves a two-step process, "fast and slow." The fast block is associated with electrical changes in the egg surface and is completed in less than 3 seconds after sperm attachment. As we have seen, the slow block is associated with formation of the fertilization membrane. A comparison of the electrical properties of monospermic and polyspermic eggs indicates that in the fast block, polyspermy is prevented by a rapid depolarization of the egg surface, from $-60$ mV to a potential greater than zero, about 5 to 10 mV. Eggs showing a negative potential after insemination (about $-30$ to $-10$ mV) invariably behave as polyspermic, dividing into three or four cells at the first division. Only eggs having achieved a positive potential within seconds after insemination turn out to be monospermic.

Fertilization can be prevented by imposing a current so that the resting potential of the unfertilized egg is greater than 5 mV; no sperm fertilize and no fertilization membrane forms, although many sperm attach to the egg surface. If the current is turned off, the membrane potential becomes negative and fertilization occurs. It seems that the fast block to polyspermy involves a membrane change, possibly associated with the early influx of ions and membrane depolarization.

The machinery for the block to polyspermy seems to be incorporated into the egg late in maturation. An immature sea urchin oocyte becomes polyspermic if exposed to sperm, but the mature, unfertilized egg does not. It is interesting that only the unfertilized egg cortex is birefringent; that of the oocyte and fertilized egg is not. This suggests that a specific molecular arrangement is necessary for a defense mechanism to operate. If this structural orientation at the surface of unfertilized eggs is destroyed by trypsin, the block mechanism is obliterated and such eggs become highly polyspermic.

Ascidian eggs rely on the chorion or tough outer membrane to prevent multiple sperm entry; in mammals, the zona pellucida prevents polyspermy. In the latter case, penetration of one sperm changes the organization of the zona so that no other sperm can penetrate. The only generalization we can make is that when the fertilization membrane barrier has elevated from the surface of the egg, the major block to polyspermy has been produced. As long as the membrane is present, no new sperm can penetrate the egg surface.

Polyspermy is a normal occurrence in many species of insects and salamanders. Several sperm enter the cytoplasm but only one succeeds in fusing with the egg nucleus, while the others degenerate.

## Parthenogenesis

Eggs activated artificially respond with the cortical reaction, membrane elevation, and changes in egg metabolic activity; some even develop into abnormal larval forms. Such development without sperm is called artificial parthenogenesis. In many marine eggs, simply changing the tonicity of the sea water is enough to bring about the activation reaction. Unfertilized sea urchin eggs placed in hypertonic sea water or butyric acid will activate and even begin to cleave, producing an irregularly shaped blastula. Some of these eggs go on and form a ciliated larva. Pricking a frog egg with a needle is enough to activate it; the fertilization membrane lifts, the egg rotates within the perivitelline space, and the second polar body is formed. The egg does not cleave, however,

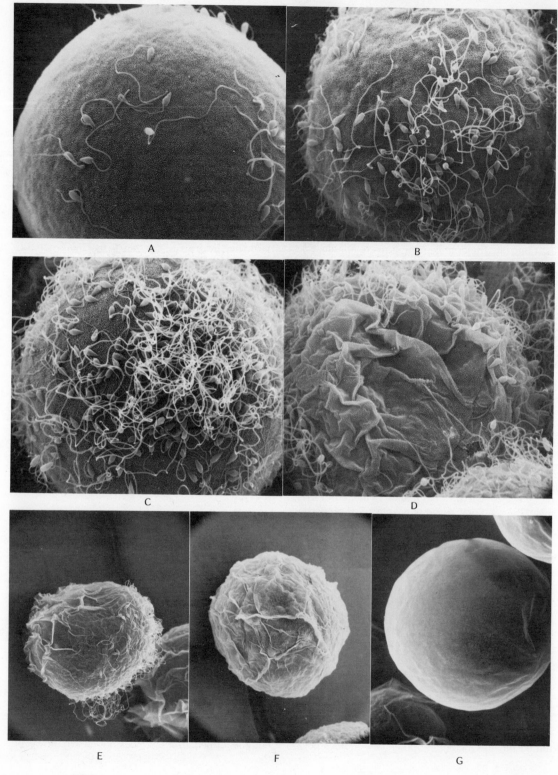

A

B

C

D

E

F

G

294

because a second factor contributed by the sperm is lacking. This factor seems to be the basal granule of the midpiece, which acts as a centriole in egg cytoplasm to set up the first cleavage spindle. Development can be initiated artificially by introducing this second factor, usually by dipping the needle in freshly drawn frog blood. Presumably, when erythrocyte nuclei are introduced into the egg, they act as cleavage-initiating centers. About 1 to 5 percent of eggs activated in this way develop as haploid embryos. The frequency of artificial parthenogenesis can be increased to 25 percent by injecting granule fractions derived from adult tissues such as mitochondria, microsomes, or even ribosomes. Such fractions may contain centrioles serving as the principal activating factor. Though we can increase activation with cytoplasmic fractions, the problem of the second factor and its role in the amphibian egg is still uncertain.

The diversity of physical-chemical agents inducing parthenogenesis obscures the underlying activation mechanism. Ultraviolet irradiation, temperature, salinity, and pH are but a few of the activating factors. The one common feature is that all agents first act on the egg plasma membrane, which, as we have seen, behaves like a coiled spring ready to be triggered into action at the slightest stimulus. Once the egg is activated, a developmental program can unfold in the absence of sperm.

Certain compounds, called *calcium ionophores*, can activate sea urchin and frog eggs. An ionophore facilitates the release and flow of bound ions, particularly calcium. Apparently the ionophore induces the release of bound intracellular calcium in the sea urchin egg, and this is followed immediately by the action potential of fertilization, cortical granule breakdown, membrane elevation, and stimulation of metabolism. The activated egg does not cleave, however. Parthenogenetic agents may act by releasing calcium.

Parthenogenesis can occur accidentally in many organisms, usually by suppression of polar body for-

mation in meiosis. It has been reported in turkeys but no natural occurrence has been found in mammals. Artificial parthenogenesis can be induced in some mammalian eggs, and haploid embryos have been obtained which do not survive very long.

Parthenogenesis is a natural occurrence in the life cycle of bees and aphids (Figure 11-23). Development is apomictic; that is, no sperm nucleus enters the egg, so that there is no mixing of genomes. The female worker bee is derived from normally fertilized eggs, whereas males arise parthenogenetically and start out life from haploid eggs. Only the germ line remains haploid, since all somatic tissues, as is true for many insects, become highly polyploid or polytenic during larval development. The male produces normal gametes because meiosis does not occur to prevent further reduction of the chromosome set. Each spermatogonium undergoes one mitotic division without any crossing over and produces two haploid sperm. Meiosis in females is normal.

This system of *haplodiploidy* (haploid parthenogenetic males, diploid females) is not as common as systems in which diploid females arise parthenogenetically as in flatworms, rotifers, and arthropods. In complete parthenogenesis, every individual arises from an unfertilized egg; the species consists entirely of females, and males appear rarely, if at all. Sexuality is lost in these groups; the entire adult life is devoted to feeding and reproduction. The system's advantages lie in the population's high reproductive capability, since each individual is a female, producing more offspring. On the other hand, the system loses the advantages offered by sexual recombination, which means the only source of species variation is mutation. Another system known as cyclic parthenogenesis, found in rotifers and aphids, retains the advantages of sexuality. These groups benefit from both methods of reproduction since parthenogenesis and bisexuality are found in an animal's life cycle. Aphids, for example, increase rapidly during the parthenogenetic portion of the cycle. In spring, eggs hatch

**Figure 11-22**   *Sequence of sperm-egg interactions in the sea urchin* Strongylocentrotus purpuratus *as seen in the scanning electron microscope. (A) One second after insemination. (B) Five seconds after. (C) Fifteen seconds after. (D) Thirty seconds after. Although many sperm seem to attach at first, only one has successfully activated the egg, seen at the arrow in (D). The cortical reaction has progressed from this point as a circular wave marked by the circular area from which sperm have detached. (E) Forty-five seconds after. Cortical granule breakdown has covered half the egg. (F) Fifty-five seconds after. Cortical reaction complete; all but fertilizing sperm have detached. (G) Three minutes after. A hardened fertilization membrane showing projections on the surface. (From M. J. Tegner and D. Epel,* Science *179:685–688, 1973. Copyright © 1973 by the American Association for the Advancement of Science.)*

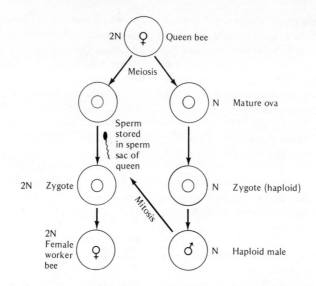

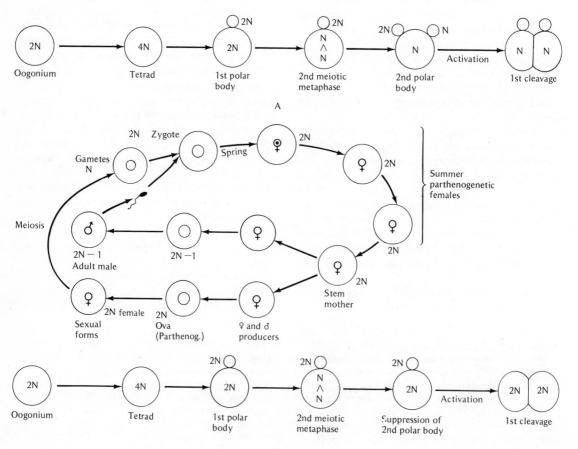

A

B

as females, which invade plants and reproduce many parthenogenetic offspring through the summer months. Gene recombination is achieved in the single sexual generation at the end of the summer. Eggs from some winged females become sexual individuals which mate. From these fertilized eggs, female nymphs emerge in the spring the following year, to repeat the cycle.

Parthenogenetic diploid females may arise in several ways. Commonly, during oogenesis, the first meiotic division is normal, and a first polar body forms. The second meiotic division is suppressed, and the second polar body nucleus fuses with the egg nucleus to produce the diploid. In other cases, meiosis is normal but the diploid condition is restored early in cleavage by fusion of cleavage nuclei in pairs. Finally, in some groups such as the brine shrimp *Artemia*, meiosis is entirely suppressed in the female, the maturation divisions of the egg being entirely mitotic or equational in character.

If sperm are not involved, how is a parthenogenetic egg activated? The precise mechanism is not known in all species, but in the wasp *Habrobracon*, activation seems to result from a mechanical stimulus imparted to the egg as it is squeezed through the ovipositor during laying. Eggs protected from this compression process do not develop. Since so many artificial agents stimulate development, we can assume that comparable activation mechanisms must have evolved in natural parthenogenesis. Changes in ion concentration or osmolarity as the egg is laid might be sufficient to activate. Since cases of parthenogenesis arise from an altered meiotic mechanism, activation could result after the polar body nucleus fuses with the egg nucleus. This act in itself, resembling fusion of egg and sperm nuclei, could stimulate development. Finally, the evolution of parthenogenesis could have eliminated the need for an activation signal. As eggs mature in the ovary, they might develop immediately without going into a metabolically arrested state.

Except for its genetic and second-factor contributions, the importance of sperm in embryogenesis is relatively minor. The egg alone is the repository of developmental information and can express this potential when stimulated either by sperm or by some artificial agent.

## Ooplasmic segregation: the determination of embryonic axes of symmetry

One of the principal results of activation is reorganization of egg cytoplasm. This, in turn, may change the original polarity of the egg and affect the future embryonic axes. Many cytoplasmic organelles flow about the egg after fertilization to take up new positions, often within a matter of seconds after sperm entry. This cytoplasmic restructuring, known as ooplasmic segregation, creates new axes of polarity and symmetry. Morphogenetically active materials are redistributed in the cytoplasm and come to occupy specific regions. As a result, some morphogenetic substances are segregated into certain cells during the first few cleavage divisions. Sorting out morphogenetic substances in this way may determine cell fate early in development just as it does in the germ line.

In 1905 Conklin carried out the classic study on the ascidian *Styela* (Figure 11-24). The ripe unfertilized egg of this species is laid in the germinal vesicle stage and is full of gray yolk concentrated in the center. Surrounding the yolk is a peripheral layer of protoplasm in which are embedded yellow pigment granules and many minute mitochondria. A region of clear cytoplasm surrounds the germinal vesicle at the animal pole. As soon as the egg enters seawater, it begins to mature, the germinal vesicle breaks down, and the first metaphase spindle appears at the animal

**Figure 11-23** *Parthenogenetic life cycles in bees and aphids. (A) Haploid parthenogenesis in bees. The diploid queen bee lays haploid eggs after meiosis. Most are fertilized by stored sperm as they pass out and develop as diploid female workers. At certain times some eggs emerge unfertilized; these develop as haploid males which mate with another queen on a nuptial flight, and she starts the cycle again in a new hive. The males produce sperm by mitosis rather than meiosis. The mechanism of haploid parthenogenesis is shown below in which oocytes complete a normal meiosis but the haploid egg is activated without sperm and develops as a haploid male. (B) Diploid parthenogenesis in aphids. Females emerging in the spring produce several generations of females by diploid parthenogenesis resulting from suppression of first or second polar body (shown below). At summer's end some females produce sexual males and females by diploid parthenogenesis, males differing from females in lacking one sex chromosome. These, in turn, produce haploid gametes through normal meiosis which fuse to form diploid zygotes that emerge again in the spring as parthenogenetic females.*

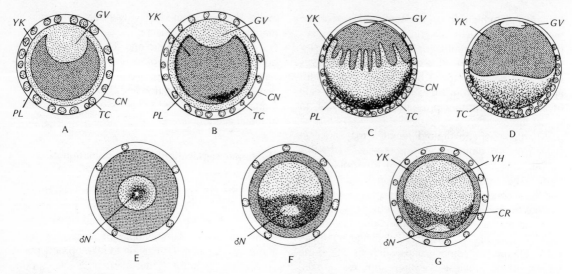

**Figure 11-24**  *Ooplasmic segregation after fertilization in the egg of Styela partita. (A) Unfertilized egg before germinal vesicle (GV) fades, showing central mass of gray yolk (YK), peripheral layer (PL) of yellow cytoplasm, surrounding test cells (TC) and chorion (CN). (B) Germinal vesicle breaks down and its contents plus clear cytoplasm surrounding it spread peripherally. (C) Five minutes after fertilization. The peripheral cytoplasm streams toward the lower pole, the site of sperm entry, thereby exposing the underlying yolk mass in the upper hemisphere. The test cells also collect at the lower hemisphere. (D) Later stage. Yellow and clear cytoplasm collect in the lower half of the egg with a cap of clear cytoplasm appearing over the yellow cytoplasm below. (E–G) Successive stages drawn at intervals of 5 minutes as seen from the vegetal pole. (E) The area of yellow protoplasm is smallest and the sperm nucleus (N) is a small clear area near its center. (F) and (G) are stages in the spreading of yellow protoplasm until it covers nearly the whole of the lower hemisphere (yellow hemisphere, YH). At the same time the sperm nucleus and aster move toward one side of the yellow cap and a crescent (CR) begins to form at this side. (From E. G. Conklin, J. Acad. Sci. Philadelphia 13:1, 1905.)*

pole. The egg remains in this state until fertilized. After sperm entry, the first and second polar bodies form; within 5 minutes, ooplasmic segregation takes place as the clear cytoplasm and the peripheral layer of yellow cytoplasm together stream violently toward the vegetal pole, forming a cap. They separate, with the clear cytoplasm spread out over the yellow plasm below. Meanwhile, gray yolk granules congregate near the animal half, at the maturation spindle. The sperm, now a pronucleus, penetrates the vegetal region, migrates through the clear cytoplasm toward the animal hemisphere, and shifts toward the future posterior side of the egg. In so doing, it drags the clear and yellow plasms toward the posterior side near the equator, creating a clear layer of cytoplasm above a yellow crescent on one side of the egg, the future dorsal-ventral plane of the embryo. The gray yolk granules are likewise displaced into the vegetal-pole region. This is the state of cytoplasmic localization before the first division. It displays an asymmetri-

cal distribution of cytoplasms that will be irreversibly fixed after the first division, with each cell receiving different populations of egg organelles.

Ooplasmic segregation in the amphibian egg involves two components, the cortex and the adjacent cytoplasm. Segregation results from contraction of the cortical gel over the stationary yolk mass. The egg of *Rana temporaria* consists of a pigmented cortical region of variable thickness extending from the animal half and surrounding a central yolk mass (Figure 11-25). Attachment of sperm to the egg surface initiates a surface reaction with cortical granule breakdown. The site of sperm entry is marked by a tail of pigment granules drawn in from the surface as the sperm penetrates the deeper cytoplasmic layers. When the vitelline membrane lifts, the egg rotates freely in the perivitelline space so that the animal half lies uppermost with the heavier vegetal half below. Within 1 to 2 hours after fertilization, the cortex contracts over the passive yolk mass in the center, toward the point of sperm entry. This produces the so-called *gray crescent*, a slightly pigmented region just below the equator, visible within 2 hours after fertilization. An originally pigmented region becomes "gray" because cortical pigment has shifted out. The gray crescent now marks the future dorsal side of the embryo and fixes the dorsal-ventral axis. The point of sperm entry corresponds to the future plane of bilateral symmetry of the embryo; that is, it usually coincides with the plane of first cleavage and occupies the ventral side of the embryo.

Before its appearance, the site of the crescent

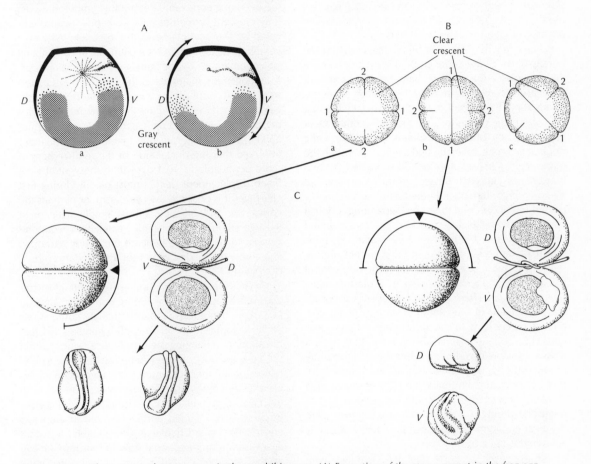

**Figure 11-25** *The gray (or clear) crescent in the amphibian egg. (A) Formation of the gray crescent in the frog egg (Rana temporaria) after fertilization. (a) One to two hours after insemination, before the gray crescent has formed. The path taken by the sperm is marked by a trail of pigment through the cytoplasm. The two pronuclei are forming in the animal hemisphere. The yolk mass in the center is shown as a crescent in the vegetal hemisphere. The pigmented cortex of the animal half extends below the equator in a decreasing concentration gradient. (b) After formation of the gray crescent. The arrows show the direction of contraction of the pigmented cortex, exposing a more pigment-free gray crescent in the dorsal (D) side, opposite the site of sperm entry. (B) First and second cleavage divisions in salamander (Triton) egg showing relationship between cleavage furrows and clear crescent (light stippling). (a) Vegetal view showing first cleavage passing through the clear crescent. In (c) and (d) the first cleavage does not pass equally through the crescent, but second cleavage does. (C) Spemann's ligature experiments. If eggs of the first type (a) at the end of first cleavage are tied off with a ligature as shown, each blastomere develops into a normal embryo. The crescent with the arrow indicates the dorsal side of the embryo and the extent of the clear crescent. On the right-hand side, an egg shown in (b) above is ligatured, and in this case the ventral (V) half rather than the dorsal half develops into an embryo, a result Spemann did not obtain. (From A. Dollander, Arch. Biol. 61:1, 1950.)*

and the future embryonic axis may be changed by rotating the egg. If the egg is rotated before sperm entry,
the action of the sperm in fixing embryonic symmetry is overcome. Apparently artificial rotation of the
egg within its membrane mimics the cortical changes
appearing after fertilization, and the plane of bilateral
symmetry of the embryo (for example, first cleavage)
usually develops perpendicular to the axis of rotation,
passing through the egg's polar axis.

Spemann in 1901 was first to suggest that the
principal axis of the future embryo is organized from
the gray crescent area. The salamander egg does not
form a gray crescent; a "clear crescent" arises instead, about 85 percent of the time. Usually the first
cleavage plane passes through the center of the clear
crescent; each blastomere retains a portion of the
original crescent. Spemann tied a ligature through the
plane of this cleavage furrow and separated the two
blastomeres and each developed into a normal but
dwarf embryo. Infrequently the first cleavage furrow
passed in the opposite direction, separating a dorsal
blastomere with the clear crescent area from a ventral
blastomere. After blastomeres were separated, most
dorsal blastomeres, containing the crescent region,
developed into normal embryos whereas ventral
blastomeres stopped developing in a blastula stage
(Dauerblastula). These results, however, are not always obtained; dorsal-ventral symmetry is not correlated precisely with the distribution of the clear
crescent. Dollander in 1950 showed that ventral
halves also develop normally while dorsal halves may
produce the typical Dauerblastula. He pointed out
that Spemann did not mark the clear crescent area
before ligaturing eggs and could not accurately correlate the position of the first cleavage furrow with
the clear crescent. Hence this original observation by
Spemann is open to question.

In frog eggs, the embryonic axis correlates with
the position of the gray crescent region. The neural
tube with brain and spinal cord develops from this
region. Accordingly, some have proposed that the
gray crescent cortex or the subjacent cytoplasm may
contain morphogenetic materials essential to formation of the head-tail organization of the embryo. This
concept was based on experiments by Schultz as
early as 1894. He showed that inversion of fertilized eggs produced double embryos. Yolk granules
at the vegetal pole streamed en masse toward the animal half of an inverted egg (Figure 11-26). If inverted
between fertilization and first cleavage and returned

to a normal position, eggs developed normally. On
the other hand, if an egg was inverted at the beginning of first cleavage or during the two-cell stage,
double embryos developed; two embryonic axes
were produced.

Changing the distribution of yolk in relation to
the overlying cortex initiates another embryonic axis;
gray crescent properties have somehow been duplicated by shifting yolk beneath the cortex. Wherever
yellow yolk is adjacent to pigment, an organizing
site comparable to the gray crescent may appear.
Morphogenetic materials in the gray crescent region
may shift in response to a gravitational field and localize elsewhere. Changing the distribution of cytoplasms may also affect the egg cortex or even permeability of the egg plasma membrane. Dollander
observed that the dorsal side of the fertilized egg is
more permeable to vital dyes than the ventral side.
Since displacement occurs most readily during first
cleavage, it is possible the cytoplasm, or perhaps the
cortex, is most labile at this time. Simple inversion of
the egg stimulates ooplasmic segregation; yolk and
cytoplasm redistribute beneath the overlying cortex
and a new organizing center for a secondary embryo
is created.

What forces move cytoplasms or attract organelles into specific regions during ooplasmic segregation? One hypothesis is that particulate components
of the cytoplasm are electrically charged, and fertilization, by changing the charge distribution within
the cell, induces charged particles to move to one
region or another. In this sense, segregation of organelles is comparable to polarity determination in
the *Fucus* egg, which results from intracellular electrophoresis of charged particles within an imposed
electric field (see Chapter 8). Presumably, this field
arises from a unidirectional flow of ions into the egg
at the moment of fertilization, when membrane
permeability changes.

Another hypothesis, first proposed by Raven in
1960, is that a "cortical morphogenetic field," a
combination of membrane-cortical properties, "directs" the displacement of cytoplasmic substances
during ooplasmic segregation. The localizing forces
may indeed emanate from the cortex since it seems
to orient microtubules and microfilaments, the organelles of cell motility and cytoplasmic flow. Microtubule assembly and microfilament contractility are
both sensitive to changes in calcium ion concentration, one of the principal ion changes that occurs at

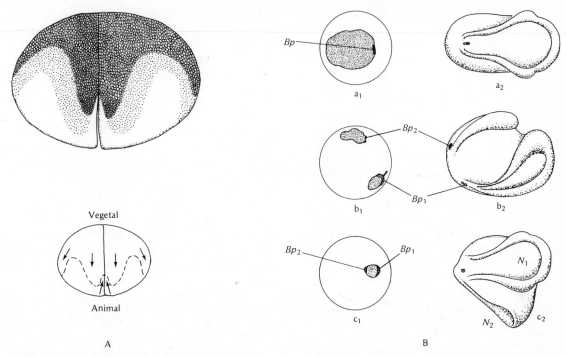

**Figure 11-26** *Inversion of a frog egg at first cleavage. (A) Flowing of the vegetal yolk through an inverted egg with a diagram below showing the direction taken by yolk toward the animal pole. (From J. Pasteels, Adv. Morphogen. 3:363–388, 1964.) (B) Three different results of an inversion experiment. On the left are the gastrula stages obtained showing the position of the invaginating dorsal lip (Bp). The secondary yolk regions are stippled. In $a_1$, the original dorsal lip appeared and only a single neural plate developed at $a_2$. In $b_1$ two separate dorsal lips appeared and two neural axes developed at different sides of the embryo ($b_2$). In $c_1$, two dorsal lips developed close to one another and a joint double neural axis developed, $N_1$ and $N_2$. (From A. Dalcq, L'Oeuf et Son Dynamisme Organisateur. Paris: Albin Michel, 1941.)*

fertilization. In addition, both types of organelles are also responsive to changes in concentration of cyclic AMP, another regulatory molecule that probably fluctuates at fertilization. The flux of tubules and filaments after fertilization may orient cytoplasmic flow just as they displace pigment granules in fish melanocytes (Chapter 8) or induce axoplasm flow in neurons. Localization of specific cytoplasmic substances may reflect the fact that these are preferentially associated with microfilaments or microtubules, or both. The directed flow of cytoplasms could occur because microtubules and microfilaments are anchored in specific regions of the cortex so that a combination of cortical contractions with microtubular displacements could account for ooplasmic segregation.

In many eggs, ooplasmic segregation does not occur immediately after fertilization. Instead, it is a progressive process beginning at fertilization but

completed after the first few divisions. Cytoplasmic localization is achieved by coupling ooplasmic segregation to cleavage, a situation found in the egg of an annelid worm, *Nereis*.

The unfertilized egg contains a large central germinal vesicle surrounded by a double row of oil droplets. After activation, large alveolarlike granules in the cortical layer are extruded and swell into a thick jelly. The germinal vesicle ruptures and slow movements of cytoplasmic constituents begin. Most mitochondria aggregate into a small subcortical ring below the animal pole, and these are incorporated into only one of the two cells at the first division. Cytoplasmic movements continue well into the second division, and by the four-cell stage, ooplasmic segregation is complete with clear protoplasm in the animal half and the oil droplets and yolk spheres localized in the vegetal hemisphere. In this egg, cytoplasmic reorganization is stabilized by the cleavage furrowing process.

Timing of ooplasmic segregation determines the cytoplasmic constituents in each cleavage blastomere. If cytoplasmic localization occurs early, immediately after fertilization as in *Styela*, then first cleavage produces qualitative differences in each blastomere as each receives different morphogenetic material. If it occurs late, after several cleavage divisions as in *Nereis*, then qualitative differences in blastomere cytoplasms are delayed. In this case, the cleavage process itself determines the localization of special cytoplasms. Finally, there are eggs, such as that of the sea urchin, which exhibit no visible segregation except for a cortical shift in distribution of pigment granules. In this species, however, cytoplasmic localizations are completed precociously during oogenesis; the developmental blueprint is programmed before fertilization. The division process alone partitions egg organization into cell populations with different developmental fates.

## NUCLEOCYTOPLASMIC INTERACTIONS DURING EGG MATURATION AND FERTILIZATION

In Chapters 3 and 8 we saw how nucleocytoplasmic interactions control single-cell morphogenesis. Similar programmed sequences between nucleus and cytoplasm mediate the events of egg maturation and fertilization. As the oocyte passes from oogenesis to fertilization, we detect changes in cytoplasmic states which, in turn, affect nuclear and cytoplasmic patterns of macromolecule synthesis.

Most evidence points to the cytoplasm as the site of gene regulation; a different "cytoplasmic state" at each stage triggers specific nuclear responses. These "states" are revealed by transplanting embryonic or adult somatic nuclei into amphibian oocytes and eggs at different stages in the maturation sequence. Irrespective of its activity at the time, an injected nucleus is reprogrammed in response to cytoplasmic signaling circuits characterizing each state.

For example, if a nucleus is taken from an embryo with low levels of ribosomal RNA synthesis and transplanted into young oocytes actively synthesizing ribosomal RNA, the nucleus swells, takes up cytoplasmic proteins, and begins RNA synthesis. DNA synthesis is shut down in such nuclei as ribosomal

RNA synthesis is initiated. If, however, the same nucleus is transplanted into a mature oocyte after germinal vesicle rupture, it does not synthesize RNA; instead the injected nucleus condenses into a compact mass, and, in some cases, small chromosomes may appear. The nucleus assumes a state comparable to that of the condensed chromosomes released after breakdown of the germinal vesicle. Finally, a nucleus actively synthesizing ribosomal RNA taken from late embryos and injected into an unfertilized egg stops synthesizing RNA and begins replicating DNA within 1½ hours. RNA synthesis of adult brain nuclei is inhibited after they are injected into fertilized egg cytoplasm; they are switched on to DNA synthesis. The cytoplasm of a fertilized egg is in a replicating state, and the abrupt switch to DNA synthesis reflects genome reprogramming by the cytoplasm. A brain-cell nucleus, destined never again to replicate its DNA, is induced to do so in egg cytoplasm.

After mitotically inactive nuclei from adult frog brain or red cells are injected into unfertilized egg cytoplasm, they increase enormously in volume, just like pronuclei before synthesizing DNA. This probably results from an uptake of protein. $^{125}$I-labeled proteins such as histones and serum albumin are taken up by nuclei and concentrated after injection into oocyte cytoplasm. Large proteins, up to 60,000 mol wt, can diffuse across the nuclear membrane. On the other hand, proteins do not leave the nucleus in oocytes. The exchange is one-way.

Activated egg cytoplasm probably owes its "proliferative state" to a store of cell division precursors, enzymes, and, above all, regulatory molecules. Cytoplasm progressively acquires its active state during oogenesis. We find, for example, that immature oocytes and mature oocytes with intact germinal vesicles do not stimulate DNA synthesis in injected adult brain nuclei. Although oocytes contain the full complement of stored precursors of a mature egg, they still lack activating factors for DNA synthesis. Only in the final stages of maturation, when the germinal vesicle ruptures after hormone stimulation, does the cytoplasm become "conditioned" to initiate DNA replication. Gurdon proposed that one of the factors contributed by the germinal vesicle is DNA polymerase, or a substance that activates a masked form of the enzyme already present in the cytoplasm. Purified DNA from a variety of sources, when injected into unfertilized eggs, acts as a template for DNA synthesis.

It has always been assumed that the contents of germinal vesicles are essential for both fertilization and cleavage. The assumption is based on classic experiments on starfish eggs which showed that oocytes with intact germinal vesicles or oocytes from which intact germinal vesicles had been removed could not be fertilized. An enucleated egg fragment, on the other hand, prepared after germinal vesicle breakdown, could be fertilized.

The activation reaction of the starfish egg is independent of germinal vesicle content but is dependent upon a surface (or cytoplasmic) change induced by the ovulating hormone, 1-methyladenine. Isolated oocytes from which germinal vesicles have been removed fail to elevate fertilization membranes on insemination, but a similar group of oocytes treated with 1-methyladenine for 30 minutes raise fertilization membranes after insemination. Egg fertilizability seems to depend upon surface changes induced by hormone treatment.

Breakdown of the germinal vesicle releases a large volume of nucleoplasm, rich in proteins. These substances in the cytoplasm contribute to subtle maturation events essential for cleavage but not for fertilization. This role of the germinal vesicle was demonstrated by Costello in 1946 in *Nereis* eggs (Figure 11-27). Immature eggs shed in seawater are still in the germinal vesicle state. The germinal vesicle breaks down after activation, emptying its contents into the cytoplasm, and the chromosomes resume meiosis to form two polar bodies. Although sperm attach to the egg surface, their penetration into egg cytoplasm is delayed almost 1 hour until maturation is completed. Such eggs may be centrifuged before or after germinal vesicle breakdown and cut into halves. Only halves containing germinal vesicle contents are able to cleave. Apparently the breakdown of the germinal vesicle releases cleavage-initiating substances into the cytoplasm without which development is not possible.

The nature of the germinal vesicle contribution to cleavage is still unknown. One possibility is that the germinal vesicle produces precursors for the mitotic spindle, perhaps as microtubule subunits. Smith and Ecker explored this possibility in the frog oocyte *(Rana pipiens).* Here, too, an oocyte can be activated in the absence of germinal vesicle contents if it has been exposed to the hormone progesterone. Such eggs, though activated, do not cleave. On the other hand, the eggs cleave if the germinal vesicle has

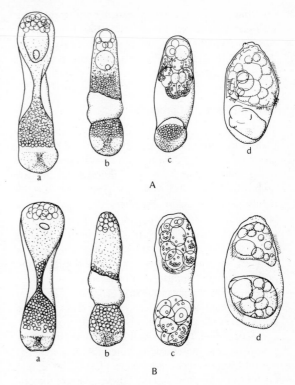

**Figure 11-27**   *Role of germinal vesicle contents in cleavage of Nereis eggs. (A) Centrifuged egg before germinal vesicle breakdown. (a) Shortly after centrifugation showing germinal vesicle and oil droplets displaced toward centripetal pole while heavier yolk is at the centrifugal half. (b) After cutting the two halves and insemination. Both activated but only the centripetal half with the germinal vesicle cleaves (c) and develops into a ciliated larva (d). (B) Egg centrifuged after germinal vesicle breakdown. The egg chromosomes are displaced toward the centripetal half. (a) Shortly after centrifuging. (b) After cutting and insemination. (c, d) In this instance both halves cleave and may develop into abnormal ciliated larvae. (Based on D. Costello, J. Morphol. 66:101, 1940.)*

broken down or if the contents of the germinal vesicle are injected into the cytoplasm.

They found that colchicine, the agent that binds microtubular proteins, did not preferentially bind less protein in enucleated oocytes than in normally nucleated oocytes. In fact, protein from isolated germinal vesicles bound little, if any, of the labeled colchicine. Though this means that microtubule protein is not contributed by the germinal vesicle, it does not rule out the possibility that a regulatory factor necessary for microtubule assembly is involved. Another hypothesis, as we have noted, is that the germinal vesicle contributes factors necessary for DNA replication such as DNA polymerase.

### Syngamy: fusion of pronuclei

The true act of fertilization occurs when male and female nuclei fuse before first cleavage and genomes are combined. The ultrastructural changes accompanying these events have been followed in the sea urchin by Longo and Anderson. They find that the tightly packaged sperm chromatin begins to unravel after entering egg cytoplasm. As chromatin disperses, the nuclear envelope forms from a coalescence of small vesicles collecting near the nuclear surface. The chromatin becomes more diffuse at the periphery until it appears as an entangled network of filaments. The nuclear envelope expands as a typical double membrane with its characteristic nuclear pores. The enlarged nucleus is the male pronucleus.

Meanwhile, the proximal centriole of the sperm establishes the sperm aster as microtubular elements assemble and extend into the cytoplasm. The female nucleus also enlarges, taking up cytoplasmic materials. While both nuclei enlarge into pronuclei, DNA synthesis begins in the unraveled chromosomes. The pronuclei move in the cytoplasm toward one another, probably propelled by microtubules appearing in the cytoplasm. At the completion of DNA synthesis pronuclei membranes make contact and fuse into a single large zygote nucleus. Shortly thereafter, the chromosomes condense and the metaphase of the first division begins.

### CLEAVAGE: THE PRODUCTION OF MORPHOGENETICALLY ACTIVE CELL SHEETS

At fertilization, the egg is activated, initiating DNA replication and cell division. The original organization of the egg is partitioned into a multitude of small cells that ultimately make up the body of the embryo and the adult organism. This partitioning process, known as cleavage, is, as we have just seen, coupled to another redistribution process, ooplasmic segregation. Both result in a quantitative and often qualitative segregation of the egg's cytoplasm into cell populations organized as two-dimensional sheets. It is these multicellular epithelia that carry out morphogenetic programs.

Cleavage is a period of rapid, synchronous cell division as hundreds and sometimes thousands of cells are produced within a matter of hours. Cell generation times are, in some instances, shorter than those in bacteria. Cells become smaller with each cleavage, and the abnormal nucleocytoplasmic ratio of the egg is rapidly restored to that of somatic cells. In the sea urchin *Paracentrotus,* for example, as cytoplasmic volume decreases, nuclear volume also diminishes from an original 2000 $\mu^3$ to less than 100 $\mu^3$ by the time 512 cells develop. Cleavage is an epigenetic process, and new properties of multicellularity emerge with each division; that is, cells are in contact along their surfaces and communicate across matrix-filled spaces. The cleaving embryo is controlled by surface interactions between emerging cell populations. Cleaving cells communicate through gap junctions which means they may be exchanging ions and small regulatory molecules.

Rates and patterns of cleavage are important developmental properties. In primitive forms such as coelenterates and sponges, cleavage is often random and disorganized, unrelated to cytoplasmic localization. In most other eggs, however, cleavage displays a precise pattern and rhythm, segregating egg cytoplasms into specific cells that occupy fixed positions in the embryo. In these forms, cleavage pattern seems to determine cell fate in the future embryo.

As we examine cleavage patterns in different organisms, we want to know whether the ultimate fate of a cell population depends upon its cleavage history. Just as germinal plasms determine the fate of germ cells, do specific cytoplasms segregate into somatic cells to determine their fate? If the fate of a cell depends on its cleavage history, then we speak of cleavage as *determinate.* Cleavage and cell differentiation proceed concurrently. If, however, cleavage is unrelated to the fate of cells, then it is *indeterminate,* independent of cell differentiation. All cleavage patterns tend to partition different cytoplasms of the egg, and it is only the time at which cell fate is determined that distinguishes one from the other. Cell fate is fixed early in determinate cleavage; in indeterminate cleavage it may be fixed only a few divisions later.

## Irregular cleavage

Before we review the more precise cleavage patterns characteristic of most organisms, let us look briefly into a simple case of indeterminate cleavage found in the lower metazoa. In many coelenterates, cleavage is irregular; the direction of cleavage planes or the partitioning of egg organization seems irrelevant to the future development of the embryo.

The development of the egg of *Tubularia,* a colonial hydroid, illustrates this pattern (Figure 11-28). Eggs develop while attached to the reproductive hydranth (gonophore) in the colony. Cleavage is irregular with distorted furrows and division planes appearing at random. Crowding of eggs in the gonophore increases this irregularity. No rules seem to be followed. Cleavage may even result in polynucleate cells since furrows do not cut off independent cells. The newly formed blastomeres are disorganized and continue to proliferate asynchronously until a blastula, a mass of cells with a small internal cavity, is formed. This mass gradually differentiates into a larva with fine extended arms surrounding an oral region. If blastomeres are separated at early stages, each will cleave irregularly and develop into a normal embryo. The eggs of some coelenterates behave like a disorganized cytoplasmic mass, producing a large number of cells which develop into a simple larva. Apparently the egg contains no special cytoplasms to be assigned to certain cell populations. Since larval organization is simple, it can be achieved without any precise localization of cytoplasmic morphogens.

## General features of regular cleavage

As metazoa evolved, their eggs became more highly organized and developmental information localized in discrete cytoplasmic regions. Complex larvae were produced. They survived for longer periods and had to be equipped with many different cell types to carry out specialized functions. The cleavage

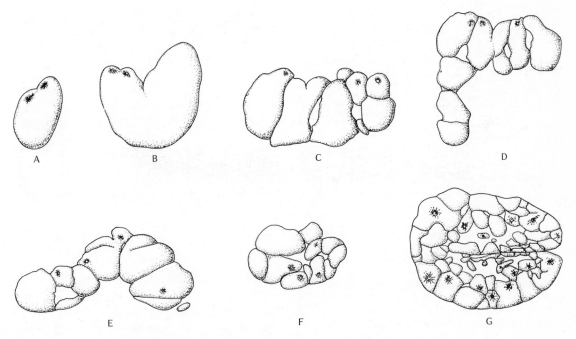

**Figure 11-28** *Patterns of irregular cleavage in the hydroid* Tubularia. *(A–B) Beginning of first cleavage. (C–F) Various views of early cleavage showing the random pattern of cleavage furrows. (G) A cross section of a blastula stage. This entire process takes place within the gonangium. (From W. Hargitt,* Bull. Museum Comp. Zool. Harvard *53:159–212, 1909.)*

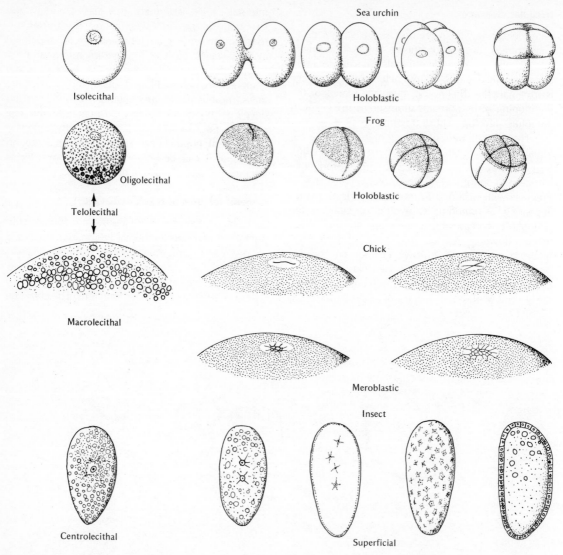

**Figure 11-29**  *Cleavage patterns as a function of yolk content. Isolecithal and some oligolecithal eggs such as that of the frog exhibit a holoblastic cleavage pattern. Macrolecithal eggs of chick, squid, and fish exhibit a meroblastic unequal cleavage of the small amount of cytoplasm sitting on top of a large yolk mass. The centrolecithal eggs of insects exhibit superficial cleavage (a misnomer, since the cytoplasm does not cleave until the end of nuclear divisions in the central yolk mass). Cross sections of insect cleavage are shown.*

process acquired precision in partitioning the egg's developmental potentialities into specific cells. Cleavage became a factor in determining cell fate in the embryo by defining cell position as well as its cytoplasmic constituents.

One feature of egg organization affecting cleav-

age patterns is the amount of stored yolk (Figure 11-29). Small eggs (sea urchin or human) with a small amount of yolk uniformly distributed through the cytoplasms are known as *hololecithal (isolecithal)* eggs. Cleavage in these eggs is total or *holoblastic*, cutting through the egg from one pole to the other. In most holoblastic eggs, the cleavage furrow begins simultaneously and equally at animal and vegetal poles.

As the amount of yolk increases, it tends to localize at the vegetal pole, as in the frog egg. Cleavage is still holoblastic, but the cleavage furrow first appears at the animal pole and gradually cuts through the egg to the vegetal pole.

Holoblastic cleavage may be equal or unequal. In most eggs it is equal, with each daughter blastomere equivalent in size, one-half the volume of the original egg. Many annelid and mollusk eggs, however, exhibit unequal cleavage; one blastomere is much larger than the other, retaining most egg cytoplasm and yolk. This visible difference in cytoplasmic volume correlates with the future fate of these cells in the embryo.

Large yolky eggs like that of the bird, fish, or squid are *telolecithal* eggs. The yolk mass makes up the bulk of cell volume, pushing cytoplasm into a thin peripheral layer either at the surface or at one pole. Cleavage in these eggs is partial or *meroblastic*. The cleavage furrow does not cut through the egg but stops at the border between the upper cytoplasmic mass and the large yolk below. This cleavage pattern is also called *discoidal* because the embryo proper develops as a small disc-shaped mass on top of the yolk. Only peripheral cytoplasm is partitioned into embryonic cells, which develop as a two-dimensional sheet surrounding the yolk. Cells closest to the yolk are usually organized as a syncytium, a cytoplasmic layer with many nuclei. This layer extracts yolk nutrients for the developing embryo.

Most insect eggs display a completely different pattern. Yolk in these *centrolecithal* eggs is concentrated in the center, surrounded by a thin layer of cytoplasm. The egg nucleus in the center of the yolk is enveloped in a small island of cytoplasm communicating with the periphery via fine cytoplasmic strands. Cleavage nuclei replicate independently within the yolk without benefit of cleavage furrows until hundreds or thousands of nuclei form in the central yolk mass. At a common signal, these "free" nuclei migrate to the peripheral cytoplasm where they are rapidly partitioned into a cellular layer by the appearance of membranes. This pattern of cleavage is known as *superficial cleavage*, since the primary cell layers are formed at the egg surface.

In general, eggs with regular cleavage patterns follow the same basic rules. The first cleavage is usually in the plane of the major axis of the embryo, passing from the animal to the vegetal pole. The second cleavage spindle is always oriented at right angles to the first, passing through the same polar axis. Four blastomeres are formed, occupying separate quadrants side by side. Still following the same rule, the third cleavage furrow, usually appearing synchronously in all four blastomeres, is perpendicular to the first two and is horizontal and parallel to the equator. This separates four animal-half blastomeres from four blastomeres in the vegetal half.

All patterns of cleavage are simple variations of this theme, following the rule that the spindle (1) appears in the region of maximum protoplasmic mass and (2) is oriented parallel to the longest axis of protoplasm. The division furrow cuts this axis transversely. Though the first three cleavages of most eggs follow this rule, subsequent cleavages are more complicated. Changes in mitotic rhythm and orientation of cleavage furrows, as well as the inequalities of division, distort the patterns.

The right-angle orientation of successive cleavage planes may be attributed to the centriole, an organelle that orients the spindle in the longest protoplasmic axis. As young centrioles always orient at right angles to old centrioles, they establish a new spindle axis perpendicular to the old spindle axis set by the mother centriole.

Three types of regular cleavage patterns, radial, bilateral, and spiral, will be described before we explore their relationship to the determination of cell fate.

### Radial cleavage

The radial cleavage pattern found typically in echinoderm eggs follows the rule most closely. During the first few cleavage divisions, blastomeres become tightly arranged in a radially symmetrical pattern about a single axis passing through the animal-vegetal poles of the egg (Figure 11-30). Any plane passing through this axis divides the cleaving embryo into symmetrical halves. The first two divisions are total, equal, and at right angles to one another. At this point, variations arise. In the sea cucumber *Synaptia,* the third division is also equal, the four animal-half cells lying directly over the blastomeres in the vegetal half. When viewed from either pole, the blastomeres have assumed a distinct radial symmetry. The right-angle cleavage pattern continues in a regular alternation between vertical and horizontal divisions, but

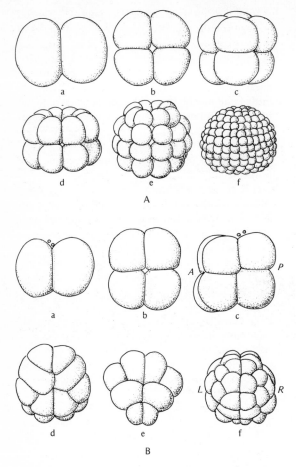

**Figure 11-30** *Radial and bilateral cleavage compared.
(A) Radial cleavage of the sea cucumber Synaptia. Note
that each division plane results in equal division and
stacking of blastomeres in equal rows, one above the
other. Any number of axes passed through the animal-
vegetal poles will give rise to symmetrical halves. (B)
Bilateral cleavage of an ascidian egg. (a) Side view.
(b) View from the vegetal pole. (c) Left side view showing
anterior (A) and posterior (P) ends of the embryo. (d) A
dorsal and (e) a ventral view of the same stage showing
the bilateral symmetry of the embryo. (f) A later stage
from the dorsal side showing the left-right (L, R)
symmetry of the embryo.*

planes are perpendicular to those in the vegetal half,
where division is horizontal and unequal. This results in a 16-cell stage composed of three tiers of
cells; a tier of eight animal-half cells sitting on four
large vegetal-half cells below which is a cluster of
four very small cells or micromeres. In spite of these
unequal divisions, the radial pattern persists since
cells are arranged symmetrically around an axis passing through the animal-vegetal poles. Each species of
echinoderm shows variations in this pattern, particularly after the third division.

### Bilateral cleavage

A major modification of the radial pattern is bilateral
cleavage, illustrated by the tunicate egg *Clavelina*
(Figure 11-30). Here anterior-posterior and right-left
axes of the future embryo are established by the first
two cleavages. In contrast to the radial plan, only one
plane passing through the major axis will yield any
symmetry. These cleavages are again perpendicular
to one another, passing through the animal and vegetal poles. The first cleavage is equal while the second
cleavage is unequal, forming two large and two small
blastomeres. Anterior and posterior regions of the
future embryo, as well as right and left sides, can be
identified at the four-cell stage. The first cleavage
plane marks the right-left symmetry of the future embryo, while the second separates anterior from posterior.

We see that bilateral cleavage patterns correlate
with the future planes of symmetry of the embryo,
larva, and adult. Even in meroblastic eggs such as
that of the squid, adult symmetries are determined
early in the first few cleavages. In the squid egg,
cleavage occurs only in a very small cytoplasmic
layer at the animal pole; the first cleavage furrow coincides with the future median plane of the embryo.
Other blastomeres are arranged in bilateral symmetry
around that axis.

Bilateral cleavage patterns often appear late in
embryos which start with radial or spiral cleavage.
We will see that most division patterns are often a
combination of two types.

### Spiral cleavage

The most striking modification of the basic radial plan
is spiral cleavage, characteristic of several major
phyla, the annelids, the mollusks, the flatworms, and

only in this species is the radial pattern followed to
the letter. In the sea urchin, the regular alternation of
horizontal and vertical cleavages does not occur (see
Figure 11-38). The fourth cleavage in the upper animal-half tier of cells is vertical and equal, but the

the nemerteans. Instead of being parallel or perpendicular to the egg axis or equator, cleavage spindles are oriented obliquely, usually at a 60-degree angle. As a result, blastomeres rotate away from the radial plane at each division and do not lie in properly arranged tiers, one directly above the other (Figure 11-31). They become stacked into an alternating spiral pattern, difficult to trace because the cells are compressed into a three-dimensional configuration of minimal surface area like a cluster of soap bubbles.

The oblique orientation of these spindles is not seen in the first division, which passes through the animal-vegetal pole in the usual fashion although it itself may be slightly oblique. The orientation manifests itself at the second division — perpendicular to the first and also slightly oblique, but pointing in the opposite direction. This results in four blastomeres oriented precisely with respect to one another — so much so that each blastomere may be identified and its progeny followed well into the late embryo. Two of the diagonally opposed blastomeres line up toward the animal pole slightly above the other two. The two blastomeres at the upper half of the cluster are in direct contact at the animal pole, whereas the two lower blastomeres contact each other at the vegetal pole. It is customary to identify the two upper cells as

A and C, respectively, whereas the two lower cells are B and D. Since polar bodies always occupy the animal pole, identification of these cells is usually quite simple.

The third cleavage clearly sets off the spiral pattern from all others. As in the radial plan, it is a horizontal cleavage, but the spindles are all oblique since they are perpendicular to previous spindles. Third-cleavage furrows cut off animal-hemisphere from vegetal-hemisphere cells in a tilted plane not parallel to the equator. The daughter cells in the animal half are shifted in a clockwise or a counterclockwise direction with respect to lower vegetal-half cells. In most cases, this division is unequal; the four animal-half cells, or *micromeres*, are smaller than the four vegetal-half cells, or *macromeres*. When seen from the side, a spiral pattern is evident; micromeres are above and to the left of the macromeres, fitted into the depressions between macromeres and forming a "zig-zag" boundary between the two layers. Since the divisions are synchronous at this stage, the four micromeres appear as a group or *quartet*, twisted in a clockwise or counterclockwise direction with respect to the major egg axis. The shift in the clockwise direction is known as a right-handed or *dexiotropic* cleavage, while the shift in the counterclockwise di-

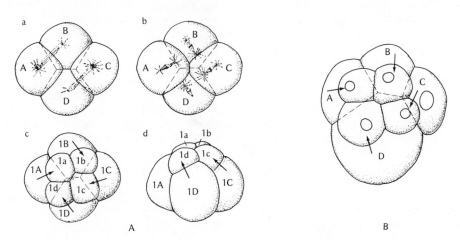

**Figure 11-31**  *Spiral cleavage patterns in mollusk eggs. (A) Cleavage in the mollusk* Trochus. *First cleavage is the usual holoblastic type with the cleavage furrow passing through the animal-vegetal poles. (a) The end of second cleavage as seen from the vegetal pole forming the A, B, C, and D blastomeres. (b) The beginning of third cleavage (animal-pole view). Spindles at right angles to one another but shifted obliquely in each blastomere. (c) The eight-cell stage as seen from the animal pole. (d) The same stage seen from the side. (From E. Korschelt,* Vergleichende Entwicklungsgeschichte der Tiere. *Jena: G. Fisher, 1936.) (B) An eight-cell stage (animal pole) of the bivalve* Unio. *In this case the cleavage is unequal, giving rise to a large D blastomere. Arrows indicate direction of formation of micromere at third cleavage. (From W. E. Kellicott,* A General Textbook of Embryology. *New York: Henry Holt, 1914.)*

rection is known as a left-handed, *leiotropic* or *sinistral* cleavage.

In most spirally cleaving eggs, the division planes alternate at each division; the first is right-handed, the second left-handed, the third right-handed, and so on, an alternating pattern continuing to the blastula stage. Micromeres as well as macromeres divide in this pattern.

After the first quartet of micromeres is produced at the third cleavage, each succeeding macromere division is also unequal, forming a small micromere and a larger macromere. Inasmuch as macromeres divide synchronously, additional quartets of micromeres are produced in succession. Three and sometimes four tiers of micromeres are stacked one beneath the other. Meanwhile, micromeres also continue to divide. After the sixth or seventh cleavage, most animal-hemisphere cells are micromeres and their derivatives, while the vegetal-half cells will often have four large macromeres plus some of their derivatives. All cells are compressed into a tight sphere, held together by a hyaline layer beneath the vitelline membrane. Cells assume a position providing minimal surface contact, usually hexagonal-faced spheres with membranes meeting along a line at angles of 120 degrees.

The rhythm and pattern of spiral cleavage are more rigid than those in other types of cleavage, as if each cell carves out a specific region of egg cytoplasm. More importantly, the number of cells that make up the embryo and larva, when organ systems appear, is quite small, in some cases also fixed. Accordingly, cells are easily identified, and their division history or cell lineage may be traced into the swimming larvae. By analyzing cell lineages, we can study the relationship between cleavage pattern and the fate of cells in the embryo.

## Cell lineage

The precise nature of spiral cleavage makes it possible to trace a cell lineage for each larval organ. We find that spirally cleaving eggs exhibit determinate cleavage; that is, differentiation of each cleavage product and its progeny is dependent upon its prior division history.

An example of a cell lineage study is given for *Nereis,* a classic work done by Wilson in 1892 (Figure 11-32). The first four cleavages are synchronous in this species. The first cleavage is unequal, giving

rise to a small AB blastomere and a large CD. Sixty minutes later, the second cleavage furrow produces two equal-sized A,B blastomeres, while the larger CD cell divides into a smaller C and a larger D cell. Cleavage is spiral, alternating dexiotropic to leiotropic with each successive division. At the third cleavage, the first quartet of micromeres separates from four large vegetal macromeres. This quartet is designated 1a–1d, marking the micromere derivative of each A–D macromere. With each succeeding macromere division, the second through fourth micromere quartets are formed, identified as 2a–2d through 4a–4d, respectively. It should be noted that the third cleavage division, in producing the first micromere quartet, segregates yolk and oil droplets into the macromeres, which remain there until the end of cleavage.

Division becomes asynchronous at the fifth cleavage; the 32-cell stage is reached in steps as cell groups divide at different times. The first micromere quartet at the animal pole gives rise to the most apical structures in the larva. It divides into four central cells at the animal pole (designated $1a^1$–$1d^1$) and four trochoblast cells, $1a^2$–$1d^2$. The latter are the progenitors of the band of powerful cilia encircling the larva, known as the *prototroch*. This band develops after only two more divisions, when 12 of the 16 cells produced immediately differentiate cilia and never divide again. The apical tuft of the larva, made up of large cilia extending from the former animal pole, is derived from the four central cells $1a^1$ and $1d^1$. These divide unequally into four small cells at the animal pole, designated $1a^{11}$–$1d^{11}$ and four larger cells $1a^{12}$–$1d^{12}$. The former immediately differentiate into apical tuft cells with long cilia, while the latter contribute to larval epidermis. In this way, each cell can be followed into the larval organs such as gut, eye spots, ciliary bands, and coelomic sacs.

The cleavage patterns and lineages in three major phyla—flatworms, annelids, and mollusks—are similar. The latter two develop a trochophore larva with a common body plan, a sign of close phylogenetic relationship. In each, larval ectoderm comes from the first three micromere quartets, the remaining quartets contributing to mesodermal derivatives while the endoderm comes from the macromeres. The ciliated trochophore bands in these larvae come from the $1a^2$–$1d^2$ cells just described. Two micromere derivatives, the 2d cell of the second quartet and

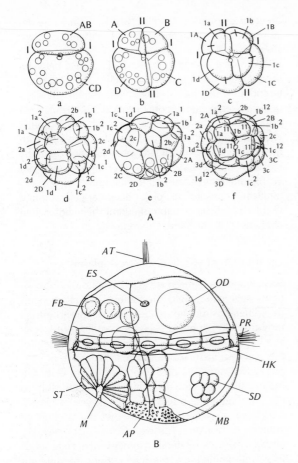

**Figure 11-32** *Cell lineage pattern in* Nereis. *(A) Normal cleavage in* Nereis *tracing the cell lineage. Circles in parts a, b, and e are oil droplets. (a) First division is unequal, yielding a large CD and a small AB blastomere, seen from the animal pole. (b) Four-cell stage from the animal pole, showing large D blastomere. (c) Polar view of eight-cell stage. The first micromere quartet (1a–1d) on top of macromeres 1A–1D. (d) Polar view of sixteen-cell stage. Second quartet of micromeres 2a–2d below first quartet which has also divided to form 1a¹–1d¹ and 1a²–1d² cells. Macromeres 2A–2D also indicated. (e) Sixteen-cell stage from the right side. (f) Polar view of twenty-two-cell stage. Only some of the cells have divided since division is no longer synchronous. The macromeres 2C and 2D divided into third quartet derivatives 3c–3d, but macromeres 2A and 2B are delayed. Likewise none of the second micromere quartet have divided as yet. (From E. B. Wilson, J. Morphol. 6:361–480, 1892.) (B) Trochophore larva of* Nereis *(diagrammatic) about 20 hours after fertilization, viewed from left side. AT = apical tuft; ES = eye spot; FB = frontal body; HK = head kidney; M = mouth; MB = mesoblast band; OD = oil droplet in endodermal cell; AP = anal pigment area; PR = prototroch; SD = derivatives of the first somatoblast; ST = stomadeum. (After D. P. Costello, J. Exp. Zool. 100:19–66, 1945.)*

the 4d cell of the fourth (both derivatives of the D macromere), the so-called first and second *somatoblasts,* respectively, assume a special role in the development of larval structures in both phyla. Both give rise to such mesodermal derivatives as heart, muscle bands, and ventral nerve cord, as well as other key tissues. The development of each phylum diverges from this common plan later on, as unique structural adaptations appear, such as shell glands in mollusk larvae and segmented body plans in annelids.

Each blastomere carries out a prescribed number of divisions before it differentiates into a larval structure. Both its position and division program seem to be set at the beginning of development, at the first two cleavages. If any of the four macromeres is separated after second cleavage, each divides exactly as it would have in the intact embryo, both in terms of number of divisions and position of progeny.

Cleavage partitions egg organization according to a predetermined plan, and special cytoplasms and morphogenetic substances are precociously segregated into different cell lines, as we will see. More than in radially or bilaterally cleaving eggs, minor perturbations of spirally cleaving eggs upset the developmental fate of blastomeres, and abnormal embryos result. Though the former are less sensitive to experimental changes, here, too, cleavage soon becomes determinate, and disturbances in rate or pattern may also upset development. The spiral pattern is a rigid cleavage program in which blastomere fate may be set as early as the first division. For this reason it has been called *mosaic;* the progeny of each blastomere contributes one small piece of the "jigsaw puzzle" of the larval body plan. Loss of an early blastomere or a shift in its position results in abnormal larvae. By contrast, development of echinoderms and vertebrates is called *regulative* because removing some cells during cleavage, or altering cleavage planes, does not affect development; the remaining cells can regulate into a normal larva. We will see, however, that this distinction between mosaic and regulative eggs is arbitrary; they differ only with respect to the timing of determinate divisions. In mosaic eggs these divisions occur earlier, sometimes at the first division. Otherwise they do not differ in developmental plan.

## CLEAVAGE AND CYTOPLASMIC LOCALIZATION

In many spirally cleaving eggs, two features stand out: (1) each division seems to segregate egg cytoplasms into different cell lines, and (2) there is an early shift from synchronous to asynchronous division. Most interest has focused on the first of these because of its obvious implications in nucleocytoplasmic interactions and control of gene expression. On the other hand, mitotic rhythms in early development have been noted for many years, but only recently has their relevance to cell determination been recognized. We now believe that temporal aspects of cleavage along with the number of genome replications may be important in genomic expression. Before we look into cleavage synchrony, we will first see how localization of egg cytoplasms affects nuclear function and cell fate.

The original heterogeneities of the fertilized egg are broken up by cleavage into regionally distinct cell populations with different kinds and amounts of cytoplasms. Most yolk in the frog egg, for example, is localized in vegetal-hemisphere cells, whereas pigment granules concentrate in animal-half cells. We have already seen that special cytoplasms such as the polar plasm of insect eggs are segregated into the germ line by cleavage. Likewise, mitochondria are segregated unequally; some cells acquire more than others. While cytoplasm is being partitioned, each cell receives a genetically identical nucleus. Very early in development, however, cell fate is determined; different cell populations or phenotypic clones are established. Does cleavage determine cell fate by distributing different cytoplasms into different regions? If so, how? Do genetically identical nuclei express different developmental information because they come to occupy qualitatively (or quantitatively?)

different cytoplasm? In other words, what is the developmental significance of such differential divisions?

We have seen that the basic difference between determinate and indeterminate cleavage patterns is temporal—the time at which differential divisions occur to segregate egg organization into distinct cell populations. Timing is different in each species, but it is a differential division that usually correlates with the determination of cell fate. We mean two things by the term differential division: (1) daughter cells are visibly different because of the amount or type of cytoplasmic constituents, or (2) daughter cells, though visibly identical, exhibit different behavior when challenged in the same environment.

### Localization of visibly distinct cytoplasms

Sometimes we can see specific cytoplasmic components segregating into blastomeres which develop into different structures in the embryo. This is particularly true in pigmented eggs such as those of *Styela*. We have already seen how ooplasmic segregation in this egg results in a complete rearrangement of various cytoplasmic regions. The question is, does the organization of cytoplasms bear any relationship to the cleavage pattern and subsequent fate of blastomeres in the embryo? Again, we owe the cell lineage analysis to Conklin. The egg, before first cleavage, already exhibits bilateral symmetry and consists of five well-defined cytoplasmic regions of different color and content (Figure 11-33). At the posterior surface are the clear plasm and the yellow crescent, the gray crescent is in the anterior part, gray yolk is localized in the vegetal pole, and a clear yolk is found in the animal half.

Cleavage partitions this organization into different cell groups, easily identified because each contains different cytoplasms. The first cleavage bisects the egg through the animal-vegetal pole, equally through all layers, forming the AB and CD blastomeres. This division marks the future plane of bilateral symmetry of the embryo, separating it into right and left halves. Each blastomere receives an equivalent sample of all cytoplasmic layers. The second cleavage also passes through the animal-vegetal pole, perpendicular to the first. This division is a differential one since it separates the anterior two cells (A and D) with gray crescent material from the posterior two cells (B and C) with yellow crescent cytoplasm.

The third cleavage is equatorial, separating four small animal-half cells from the larger vegetal cells. The division in the posterior side is differential. It slants somewhat and segregates most of the yellow crescent material into the two lower, vegetal-half cells. At this eight-cell stage, the various cytoplasms have been sorted out into different blastomeres, a process that is amplified in subsequent divisions as each of the blas-

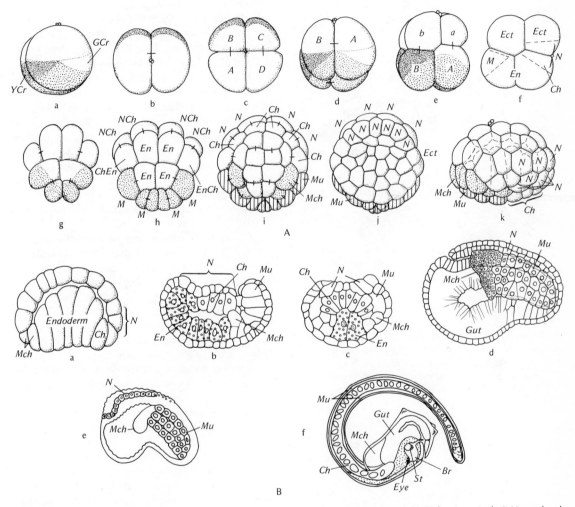

**Figure 11-33** *Cytoplasmic localization during cleavage of the Styela egg. (a, d, e, f, k) Right view. (g, h, i) Vegetal pole view. (b, c, j) Animal pole view. (a–e) First to third cleavage. Y Cr = yellow crescent; G Cr = gray crescent. (f) The presumptive cytoplasmic materials are shown localized after the third division into blastomeres destined to become specific larval structures. Ch = chorda; Ect = ectoderm; En = endoderm; M = mesoderm. Note that yellow-crescent cytoplasm is localized in presumptive mesoderm while clear cytoplasm is in N, the presumptive nervous system. (g) Sixteen-cell stage. (h) Thirty-two-cell stage; presumptive mesoderm stippled. (i, j) Sixty-four-cell stage. (k) Seventy-six- to one-hundred-twelve-cell stage. Mch = mesenchyme; Mu = muscle cells. (Based on E. G. Conklin, J. Acad. Sci. Philadelphia 13:1–119, 1905.) (B) Later stages showing fate of special cytoplasms in the larva. (a) Early gastrula. (b, d) Longitudinal view of early larva showing neural tissue (N), muscle (Mu), and mesenchyme (Mch). (c) Cross section of (b). (e) Early larva. (f) Mature larva showing musculature of tail, brain (Br), eye, and statocyst (St) derived from neural tissue.*

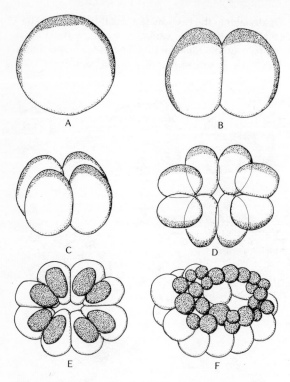

**Figure 11-34**  *Cytoplasmic localization during cleavage of the* Beroë *egg. These embryos have been stained for the Nadi reaction which indicates the presence of mitochondria. The mitochondrial-rich cytoplasm stains positively and is seen localized at first in the cortex as a cortical cap at the animal pole of first and second cleavages (A–C). It is segregated to all blastomeres at the third division (D), but in subsequent divisions, the cytoplasm localizes almost exclusively in the eight micromeres (E). These continue to divide (F) and eventually give rise to the eight rows of ciliated comb plates in the larva. (Based on G. Reverberi,* Experimental Embryology of Marine and Fresh Water Invertebrates, *ed. G. Reverberi. Amsterdam: North-Holland, 1971.)*

tomeres divides into smaller cells. The original cytoplasmic layers of the fertilized egg have become different cell groups which eventually develop into different larval organs.

By tracing the cell lineage of each blastomere, Conklin showed that the posterior cells containing the deep yellow crescent protoplasm give rise to all the larval muscles in the tail, the anterior cells with light gray cytoplasm become the notochord and neu-

ral tube of the larva, the gray yolky vegetal cells become the endoderm of the gut, and the clear cytoplasm in the animal-half cells becomes larval ectoderm and sense organs. After fertilization the egg becomes a mosaic of organ-forming substances that are segregated into groups of cells by cleavage. Cytoplasmic localization in this egg is critical to normal development; displacement of cytoplasmic layers by centrifugation before first cleavage disturbs development and abnormal embryos result.

In some cases, precocious segregation of special cytoplasms can be related to specific organelles such as mitochondria. The sorting out of mitochondria during cleavage of the egg of the ctenophore *Beroë* seems to be correlated with the development of ciliated comb plates, the locomotory organs of this group of animals. In the dark-field microscope, the outer ectoplasm of the *Beroë* egg appears emerald green, in contrast to the dark endoplasm and yolk. This light region is enriched in mitochondria (Figure 11-34).

During first cleavage, the green cytoplasm localizes as a cap in the animal half of the first two blastomeres. The four-cell stage is a loose arrangement of two small and two large blastomeres with green cytoplasm at the pole. Segregation continues to the 16-cell stage as eight small micromeres filled with the emerald green cytoplasm pinch off from the larger macromeres. The subsequent fate of these blastomeres is clear; each gives rise to one of the eight ciliated bands of the larva. The localization of mitochondria in these cells insures that ciliated cells are provided with the bulk of the egg's store of energy-transducing organelles.

If a small blastomere containing green cytoplasm is isolated, it develops into a dwarf, an abnormal larva with a single ciliated band. Isolation of a large cell lacking green ectoplasm produces an abnormal larva without a ciliated band. Possession of mitochondria-enriched cytoplasm is essential to development of a ciliated band and it can form independently of any interactions with neighboring cells.

### Polar lobes

Sometimes there is no pigment or yolk to mark the morphogenetic regions of an egg. The existence of special cytoplasmic regions that determine cell fate is revealed by unusual cleavage behavior. For example, in spirally cleaving eggs of *Ilyanassa* (gastropod snail), the first division is unequal because of polar lobe formation.

*Ilyanassa* eggs are fertilized internally before laying and develop synchronously in a fluid-filled capsule (Figure 11-35). As the mitotic apparatus for first cleavage appears, the vegetal region of the egg balloons out to form a large cytoplasmic lobe equal in size to the two constricting blastomeres. At this *trefoil* stage, the polar lobe is connected to one of the two blastomeres, the future CD cell, by a fine cytoplasmic bridge. The lobe contains most vegetal cytoplasm and yolk. As the first cleavage is completed, the contents of the lobe flow into one blastomere to form a large CD cell attached to a small AB blasto-

mere. At the beginning of the next division, a second polar lobe emerges from the CD blastomere, and its contents flow into the D macromere, making it the largest of all four cells.

Subsequent development is typically spiral and determinate. The characteristic veliger larva of gastropods develops in 4 to 5 days and consists of a shell, stomach-intestine, foot, otocyst (sense organ), and a velum, a series of ciliated whorls at the head end serving for motility and food gathering. As in most determinately cleaving eggs, each blastomere, as part of a mosaic, follows a predetermined division program to produce specific larval structures.

What is the significance of the polar lobe? A simple surgical experiment tells us it is essential to development (Figure 11-36). With a fine glass needle, the lobe may be removed at the trefoil stage. The remaining lobeless embryo develops abnormally. All cells at the four-cell stage are equal in size, and the subsequent cleavage pattern of the D macromere is altered. Normally, its cleavage behavior is different from that of the others, with an independent rhythm and pattern. In the lobeless embryo, all cells of the D quadrant cleave like all others. The larva of a lobeless embryo is incomplete, a monster with numerous large velar cilia, pigment, and an endodermal mass. Such key organs as shell, foot, otocyst, eyes, or beating heart, the primary derivatives of the D quadrant, are missing. The abnormal larva swims about for 9 to 10 days and eventually dies.

Removing the lobe has two effects: (1) the cleavage pattern of the D quadrant is altered and (2) all larval derivatives of the D quadrant are eliminated. Evidently the lobe contains factors that regulate cleavage, along with information for differentiation of several larval organs. Lobe contents seem to be excluded from the cleavage process to insure their localization in the D blastomere. The first division in this species is determinate since the fate of the AB blastomere is different from that of the CD. An isolated AB blastomere develops like a lobeless embryo. The isolated CD blastomere produces an almost normal larva with shell, foot, heart, and occasionally one or two eyes. The most notable defect is the velum, which is reduced, suggesting that the AB blastomere contributes most of the cells that make up the anterior portion of the larva, including the eyes.

With each successive division, different organ-forming materials within lobe cytoplasm are progressively segregated into blastomeres of the D quad-

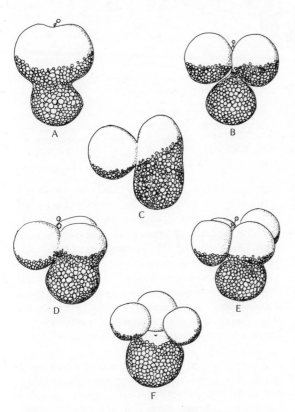

**Figure 11-35**   *Polar lobe formation during cleavage of* Ilyanassa *eggs. (A) Before first cleavage a lobe appears at the vegetal pole. (B) Trefoil stage; lobe fully extended. (C) Lobe contents flow into presumptive CD blastomere as cleavage is completed. (D–F) Polar lobe formation in second division and its resorption into the D blastomere. (From T. H. Morgan,* Experimental Embryology. *New York: Columbia University Press, 1927.)*

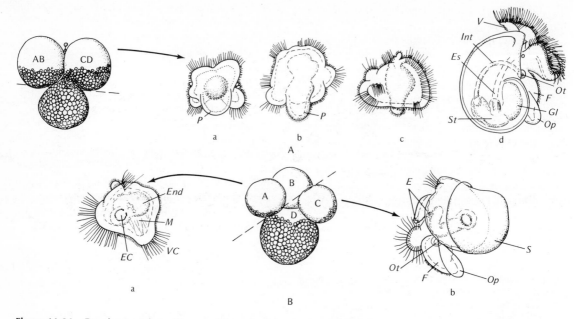

**Figure 11-36**  *Developmental significance of the polar lobe. (A) Removal of polar lobe (D) at the trefoil stage gives rise to ciliated larvae shown in (a–c). (d) A normal veliger larva.* Int = *intestine;* Es = *esophagus;* V = *velum plus velar cilia;* Op = *operculum;* F = *foot;* Ot = *statocyst;* Gl = *shell gland;* St = *stomach. (B) Separation of AB from CD blastomeres gives rise to different larvae.* VC = *velar cilia;* End = *endoderm;* EC = *enteric cavity;* M = *muscles;* S = *shell;* E = *eye.*

rant. By destroying the D macromere after successive divisions, Clement showed that each micromere, 1d, 2d, 3d, and 4d, receives different derivatives of polar lobe cytoplasm. Since the D macromere is always the largest cell of the embryo, it is easy to identify and puncture with a glass needle. The division at which the operation is performed determines which D-quadrant micromere remains. For example, if the D cell is destroyed at the four-cell stage, the remaining ABC embryo develops as a lobeless embryo; it lacks all D cell derivatives. If the D cell is punctured after formation of the first quartet of micromeres, the surviving embryo is made up of all ABC derivatives plus the 1d micromere; and if the D cell is destroyed after the second micromere quartet forms, the embryo has all derivatives of the original ABC plus a 1d and a 2d cell. In this way, Clement could create ABC embryos with different derivatives of the D quadrant and determine the contribution of 1d, 2d, 3d, and 4d micromeres to the larva.

Retention of the 1d cell by the ABC blastomeres does not improve development; they still behave as a lobeless embryo, suggesting that no morphogenetic

materials in the lobe are segregated from the D cell at this division. Addition of the 2d cell, however, results in shell gland formation; but it is the 3d cell, after the third quartet appears, that contributes most polar lobe contents. The resulting larva (ABC + 1d, 2d, and 3d) contains velum, eyes, and well-developed shell gland. Sometimes a foot is present, but heart and intestines are still absent. Addition of the 4d cell gives a completely normal embryo except for its smaller size. The remaining 4D macromere is relatively unimportant, functioning as a nutritive cell without contributing to any larval organ.

Except for the 1d cell, each micromere of the D quadrant acquires a specific cytoplasm during cleavage, and it in turn contributes to the development of a specific larval organ. These cells also receive materials that mediate communication between cells of one quadrant and those of another. We have already pointed out that a lobeless larva or an ABC larva lacks eyes; eyes develop only if the 2d and 3d cells are added. This suggests that eye differentiation depends upon lobe materials apportioned to the 2d and 3d micromeres. Clement demonstrated, however, that eye development depends upon interactions between these D derivatives and micromeres of the A and C quadrants. If, at the eight-cell stage, micromeres 1a and 1c are punctured, an eyeless larva is produced in spite of the fact that it contains all polar lobe contents. Eyes do not form unless 1a and 1c are present, yet we also know that eyes do not form unless the 2d and 3d cells are also present with their supply of morphogenetic materials. The conclusion from these experiments is that eyes form when the 1a and 1c cells interact with the 2d and 3d cells. The polar lobe presumably contains eye-inducing substances which change the fate of the 1a and 1c cells and transform them into eye progenitors. Because of their spatial relationships, these micromeres interact as long as lobe contents are present.

We tend to think of the polar lobe as a "bag" of morphogenetic materials parceled out into D-quadrant cells. We should note that in addition to its contents, the "walls of the bag," the membrane and cortex of the lobe region, are also partitioned at each division. Besides, all shape changes associated with polar lobe formation are properties of the vegetal cortex of the egg. If the fertilized egg is cut into animal and vegetal fragments, only the vegetal half forms polar lobes at each of the first two divisions. The animal half cleaves into equal-sized blastomeres. Moreover, a lobe isolated at the trefoil stage continues to contract and expand, forming a second lobe at exactly the time a second lobe forms in situ. The contractile properties associated with lobe formation are intrinsic to the vegetal cortex, possibly because of localized contractile microfilaments.

If the cortex is vital to lobe formation, does it also have informational content and furnish morphogenetic materials to the embryo? Clement tried to answer this question by centrifuging eggs to displace lobe contents. The eggs fragment into animal and vegetal halves in the centrifugal field (Figure 11-37) because the heavier yolk is concentrated in the vegetal region. This results in "light" yolk-free animal halves and heavy, yolky (and presumably polar-lobe rich) vegetal halves. One can also reverse egg orientation within the centrifugal field to obtain yolk-filled animal halves and light, yolk-free (also free of polar-lobe cytoplasm) vegetal fragments. A study comparing development of such fragments with different cortices revealed that only vegetal fragments (with or without yolk) form polar lobes, which is consistent with the fact that this property resides in the vegetal

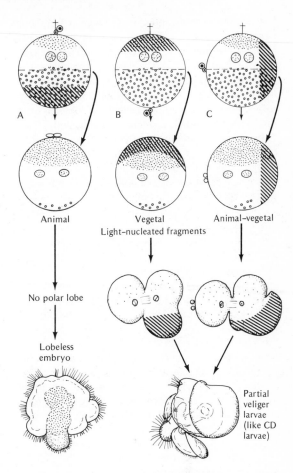

**Figure 11-37**  *Role of polar lobe cortex in* Ilyanassa *development. Centrifugation of eggs leads to three patterns of stratification with respect to egg axis and polar lobe region. Direction of arrow indicates direction of centrifugal force. (A) Clear nucleated area in animal half (animal pole marked by polar bodies). Stippling = oil cap; small circles = yolk granules; diagonal lines = region of polar lobe; dashed line = zone of fragmentation. (B) Clear nucleated area appears in vegetal half. (C) Clear nucleated fragment forms midway between animal and vegetal halves if egg is oriented 90 degrees with respect to direction of centrifugal field. The development of each nucleated half is followed. Only the vegetal and animal-vegetal fragments, those containing the original polar lobe region, will form a polar lobe; these will develop into partial larvae with eyes, shell gland, and so on. Fragments without the lobe region develop as lobeless embryos. (Topmost figure based on A. C. Clement, Develop. Biol. 17:165–186, 1968.)*

cortex. Most significantly, those vegetal fragments containing the same cytoplasms as animal fragments, produce 10 times more larvae with polar-lobe derivatives such as eyes, heart and shell, than do animal halves. Centrifugation does not separate lobe-forming capabilities from their morphogenetic role. Irrespective of cytoplasmic contents, if lobes appear, the larvae that usually result possess lobe derivatives. Though these experiments point to the polar-lobe cortex as a site of developmental information, they do not rule out the possibility that a clear lobe cytoplasm, or cortical components such as microfilaments, difficult to displace by centrifugation, remain in vegetal fragments to account for their improved developmental performance.

So far, biochemical characterization of lobe contents has not uncovered any qualitative differences in cytoplasm. The amounts and kinds of RNA and protein in the lobe are only quantitatively different from those found in the AB blastomere. Rates of RNA synthesis in lobeless and normal embryos are different, but the differences are not great enough to explain the vast differences in developmental capacity between normal and lobeless embryos. Examining the lobe with the electron microscope also reveals no specialized structures; its organelles resemble those in other blastomeres. Because of its large size, however, the lobe supplies a disproportionate amount of organelles to the D quadrant. But this alone could not account for the developmental contribution of this cytoplasmic region. One is left with the hypothesis that the lobe cytoplasm and cortex localize unidentified morphogenetic materials (microfilaments, or specific surface configurations) that are sorted out during cleavage.

It is easy to see why early embryologists identified the determinate cleavage pattern as a "mosaic work." As soon as a cell appears, its developmental fate is fitted precisely into the overall structural plan of the embryo. A functional larva is produced because the organization of the egg is so fixed and the cleavage pattern so rigid that organ-forming materials are accurately sorted out into each cell. With few exceptions, the blastomeres do not interact during this process. Each cell seems to play out its role independently of all others. But this is not the case in indeterminately cleaving embryos. In these, egg organization is partitioned into spatially distinct cell populations which must interact to produce normal development, a situation seen in the sea urchin embryo.

## Morphogenetic gradients and blastomere interaction: the sea urchin egg

Many eggs, such as the sea urchin egg, have no visibly different cytoplasms or polar lobes, nor do they display elaborate ooplasmic segregation. The egg is homogeneous, filled with yolk and other nutrients, and usually cleaves indeterminately. Does this mean that no prelocalizations exist in these eggs, that cleavage plays no role in determining cell fate?

If organ-forming substances are prelocalized in the egg, as in *Beroë*, then parts of the egg should exhibit a different developmental potential. Eggs may be cut before and after fertilization, and the developmental fate can be followed. If cleavage does partition egg components, then fragments of the egg should exhibit development comparable to cleavage products derived from the same regions. To what extent is this true for the sea urchin egg?

We have already mentioned the early cleavage pattern in the sea urchin. The first two cleavages are equal and meridional (Figure 11-38). The third cleavage is equatorial and cuts the four cells in two equal groups of four animal-half and four vegetal-half cells. The fourth cleavage is the most distinctive. In the animal half, cleavage furrows are perpendicular to the previous third cleavage and produce eight equal-sized cells known as *mesomeres*. The four vegetal cells divide equatorially and unequally into four very small micromeres cut off at the vegetal pole from four large macromeres. The 16-cell stage, then, consists of eight mesomeres at the animal half, four large macromeres in the vegetal half, and four small micromeres clustered at the vegetal pole. Again, when the 16-cell stage divides into 32 cells, the divisions in both halves are different. In the animal half, mesomeres divide equatorially, resulting in two tiers of eight cells each, known as $An_1$ and $An_2$. The four large macromeres in the vegetal half divide in a meridional plane and give rise to a single tier of eight macromere cells. The four micromeres divide into a cluster of eight smaller micromeres, still at the vegetal pole.

Except for the micromeres which divide asynchronously, the remaining divisions of both mesomeres and macromeres are synchronous. When the next division occurs, all cells divide equatorially; two tiers of animal-half cells yield four tiers, each of eight cells, making a total of 32 cells, and the vegetal-half macromeres divide equatorially, producing two tiers each of eight cells ($Veg_1$ and $Veg_2$), while the eight micromeres divide into a cluster of 16 small cells at the vegetal pole. All divisions subsequently become asynchronous as the early blastula forms with its central blastocoele cavity surrounded by a layer of ciliated epithelial cells. The embryo becomes motile, rotating within its membrane. The later blastula hatches with an apical tuft at the animal pole, a cluster of long cilia exceeding the length of most body cilia. This larva subsequently develops into the mature *pluteus* larva with ciliated arms, skeleton, mouth, digestive tract, and anus.

A cell lineage for many larval structures has been established. All mesomere cells ($An_1$ and $An_2$) develop into the animal half of the blastula, while macromeres and micromeres contribute to the vegetal half. The micromeres give rise to the *primary mesenchyme*, which differentiates into skeletal elements. Cells derived from $Veg_2$ invaginate during gastrulation and become intestine, secondary mesenchyme, and coelomic sacs. Progeny of $Veg_1$ macromeres form the ventral or anal side of the pluteus larva, while blastomeres of the $An_2$ tier produce the epithelia in the oral and dorsal side of the pluteus. $An_1$ blastomeres also contribute to oral and dorsal sides of the pluteus as well as produce the apical plate with stiff cilia. Even in this so-called regulative embryo, the position of each cell during cleavage seems to determine a cell's fate.

We can explore the organization of the unfertilized egg by cutting it in half with a fine glass needle. Production of egg fragments is called *merogony* and each fragment is a *merogone* (Figure 11-39). When eggs are cut along the animal-vegetal axis, parallel to the first cleavage plane, and fertilized, two normal but dwarf larvae develop. If, however, an egg is cut along the equator into animal and vegetal halves and inseminated, development of each is entirely different. Cleavage of animal halves is abnormal; all divisions are equal, giving rise to equivalent mesomere cells. No micromeres form at the fourth cleavage, and no primary mesenchyme appears. Such animal halves always develop into blastulae with long apical tuft cilia covering the entire surface. An oral field may develop, and a spherical Dauerblastula is formed without gut, skeleton, or arms.

The vegetal halves fare better. They exhibit regular cleavage, forming micromeres at the fourth division. Though blastulae lack the long apical tuft cilia, subsequent development results in plutei ranging from almost normal to spherical with reduced arms.

Gradations in development are obtained with larger fragments. Thus if eggs are fragmented above

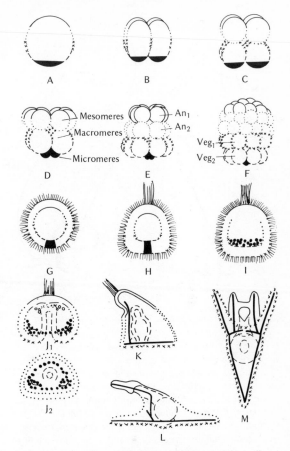

**Figure 11-38** *Cleavage in the sea urchin and the cell lineage. Outline of the fertilized egg divided into regions: solid line = animal pole; dotted lines = region to the equator below animal fifth; crosses = region just below equator; dashes = next fifth of vegetal half; black region = vegetal pole. (A–F) During cleavage these regions are partitioned and by the 16-cell stage, (D), mesomeres, macromeres, and micromeres can be identified. At (E) the animal half has divided into two tiers of cells: $An_1$ and $An_2$. At the 64-cell stage, (F), each tier of cells has partitioned out the respective regions of the original egg. The macromeres have divided into two tiers: $Veg_1$ and $Veg_2$. The subsequent fate of these tiers of cells is seen in larval development, the micromere region becoming the skeletal elements of the pluteus larva, the macromeres giving rise to the stomach intestine, and the mesomeres giving rise to the outer ectoderm, the apical tuft with cilia. (G) Blastula. (H–I) Mesenchyme blastula. ($J_1$) Gastrula. ($J_2$) Ventral view of gastrula. (K) Prism stage. (L, M) Pluteus side view and dorsal view. (Based on S. Horstadius, Biol. Rev. 14:132–179, 1939. With permission of Cambridge University Press.)*

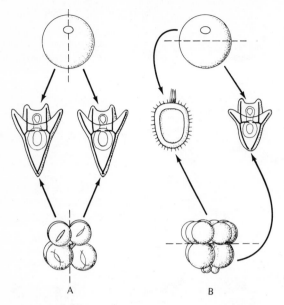

the lower hemisphere. This continues at the next division when four small micromeres are partitioned off, carrying with them the region richest in vegetal-half factors. For normal development, animal- and vegetal-half cells interact to produce the embryo.

The behavior of isolated blastomeres reflects this segregation. Blastomeres isolated after the first and second divisions develop into normal but dwarf plutei. Clearly, the first two cleavages of the sea urchin egg are indeterminate, but the third equatorial cleavage is determinate; it is a differential division separating animalizing from vegetalizing potentialities. Four animal-half cells isolated as a group develop exactly as an animal-half fragment, forming Dauerblastulae with numerous apical tuft cilia; such larvae are called *animalized* larvae. The vegetal group of four produces nearly normal plutei, called *vegetalized* larvae; they lack apical tuft cilia. Segregation continues in the next division, when micromeres appear at the 16-cell stage. Eight isolated animal-half mesomeres develop into an apical tuft blastula, but four large macromeres plus micromeres produce vegetalized larvae. On the other hand, meridional halves at this stage always develop normally since they retain a normal balance of animal-vegetal factors.

At the 64-cell stage, Horstadius isolated and recombined blastomeres. His results showed that as long as a balance between vegetal and animalizing tendencies is maintained, development is almost normal, irrespective of the specific cell combinations (Figure 11-40). For instance, an animal half in isolation develops in a typical animalized fashion, but addition of different kinds and numbers of vegetal-half cells alters the developmental pattern. A vegetal gradient can be demonstrated by combining different vegetal-half blastomeres with the animal half. If only the first tier of macromeres ($Veg_1$) is combined with an animal half, apical tuft cilia are confined to the animal pole, and a swimming blastula with a stomadeum and a ciliated band appears, resembling a larva. It develops no further, however. A partial vegetalizing influence provided by the eight macromeres shifts the animal pattern toward a normal larva.

If the animal half is combined with the furthermost vegetal-half layer, $Veg_2$, then a small abnormal pluteus with arms, stomadeum, gut, and skeletal elements is formed. On the other hand, an almost normal pluteus larva is produced when an animal half is combined with one tier of vegetal-half cells plus four

**Figure 11-39** *Merogone experiments on the sea urchin embryo. (A) Cutting the egg along the animal-vegetal axis (in a meridonal plane) gives rise to two halves, each of which, when fertilized, will give rise to an almost normal pluteus larva. Cleavage partitions the egg into equivalent halves along the animal-vegetal axis inasmuch as separation of an 8-cell stage along the same axis will give two halves that will develop normally to a pluteus (bottom). (B) Section of an egg along the equatorial plane separates animal from vegetal half and each of these will exhibit a different developmental pattern. The animal half develops into an abnormal blastula (as shown) with large apical cilia and does not develop into a larva, whereas the vegetal half develops into a larva. Separation of a 16-cell embryo gives the same result.*

the equator, development of vegetal halves is almost normal; likewise, large animal-half fragments that include more of the vegetal-half region produce more normal-looking larvae with apical tufts. The unfertilized egg seems to be polarized, with animal-half and vegetal-half determinants distributed gradient fashion along the animal-vegetal axis. For development to occur, fragments must contain a balance of animal and vegetal factors. Cleavage does exactly what the experimenter does with his needle; it partitions a preexisting polarity into separate blastomeres. The first two cleavages produce four blastomeres with equivalent proportions of animal-vegetal factors, but at the third equatorial cleavage, animal-half factors are segregated from vegetal substances in micromeres. As more vegetal cells are added, development becomes more normal, but macromeres are not too important since an animal half combined with just four micromeres is enough to produce a small pluteus with almost intact structures. Micromeres have the strongest vegetalizing properties, with $Veg_2$ and $Veg_1$ macromeres possessing decreasing potentialities. In the intact embryo, mesomeres never develop into endoderm, but in combination with micromeres they are induced to invaginate and become the endoderm of the archenteron. Hence development of a normal pluteus is dependent not on absolute but on relative amounts of animal-half and vegetal-half material and the interaction between the two. Cells communicate perhaps by exchanging

vegetalizing substances and thereby regulate into a normal larva. Substances from micromeres may act by repressing animalizing tendencies within the egg.

Animal factors that repress vegetal development are also present as a gradient, with maximum activity at the animal pole. An $An_1$ tier of cells completely suppresses vegetal-inducing tendencies of one micromere in combination with macromeres. Under these conditions, four micromeres are needed to attain proper balance and produce a normal larva. An $An_2$ tier, however, has less of a repressing tendency since two micromeres can produce a normal larva while four micromeres are excessive, producing a more vegetalized larva.

A double gradient concept of the organization of

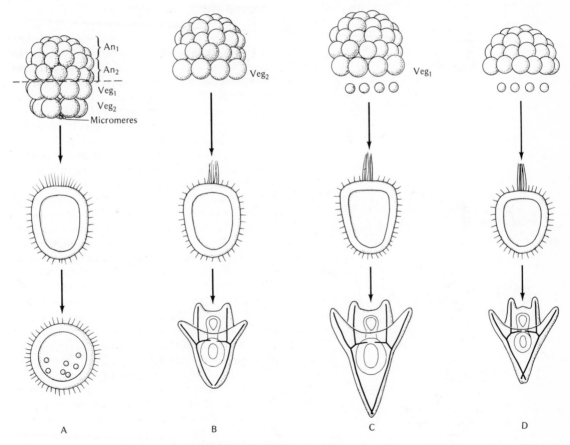

**Figure 11-40** *Differentiation of animal-half fragments from a 64-cell stage with different combinations of vegetal-half derivatives. (A) Appearance of 64-cell stage and developmental pattern of isolated animal half, $An_1$, $An_2$. (B) Isolated animal half plus $Veg_2$ tier of cells. (C) Animal half plus $Veg_1$ and four micromeres. (D) Animal half plus four micromeres. (Based on S. Horstadius, Pubbl. Staz. Zool. Napoli 14:251, 1935.)*

the sea urchin egg, first proposed by Boveri in 1901 and later elaborated by Runnstrom and Horstadius, is the best model proposed to explain these results (Figure 11-41). Each developmental potentiality is distributed as a gradient along the animal-vegetal axis of the fertilized sea urchin egg. The potentiality for micromere formation is restricted to the vegetal hemisphere with its maximum at the pole, grading off progressively toward the animal hemisphere. About halfway between the vegetal pole and the equator, micromere-forming capacity ceases. Likewise, the potential for archenteron formation and spicule development also resides in the vegetal half. On the other hand, apical tuft formation and the development of ectodermal structures such as arms are confined to the animal half, again as a gradient starting at the animal pole. The developmen-

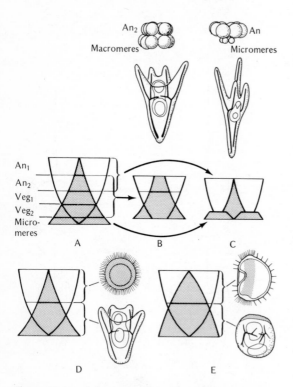

**Figure 11-41**  *Animal-vegetal double-gradient concept to explain sea urchin development. (A) Pattern as it appears in the entire embryo. (B) Gradient pattern in An₂ plus macromeres. (C) An₁ and An₂ plus micromeres. (D, E) Isolated animal and vegetal halves; two variants with different balance between animal and vegetal gradients. (From A. Kühn,* Lectures on Developmental Physiology. *New York: Springer-Verlag, 1971.)*

tal expression of each gradient in isolation seems to be an exaggeration of tendencies in the egg which are correctly balanced in normal development. These two developmental gradients, prelocalized in the unfertilized egg, interact during cleavage. Cutting the egg along the gradient perturbs the gradient system and results in animalized or vegetalized larvae, depending on the ratio of animal to vegetal factors remaining in the fragment. Likewise, cleavage partitions the gradient system into tiers of blastomeres with different developmental tendencies. Because cells communicate, the balanced gradient system is maintained and normal development occurs.

What in the egg is responsible for this morphogenetic gradient? The simplest hypothesis is that each gradient is based on a distribution of distinct animalizing and vegetalizing substances. One way to test this hypothesis is to rearrange the egg cytoplasm by centrifugation to see if it affects development. The cytoplasmic contents of the unfertilized sea urchin (*Arbacia*) egg may be stratified into layers; a lipid layer of fat droplets at the centripetal pole, a hyaline layer of clear cytoplasm containing the nucleus near the lipid layer, a third mitochondrial layer, a dense fourth layer of yolk, and finally a small cap of pigment accumulating at the centrifugal pole (Figure 11-42). The axis of stratification is oriented at random with respect to the original animal-vegetal axis. Because of this new cytoplasmic organization, cleavage is abnormal. The first cleavage plane is usually perpendicular to the axis of stratification; the second is parallel to it but perpendicular to first cleavage, and the third cleavage is perpendicular to stratification and to all previous cleavages. In comparison to normal cleavage, in this case second and third cleavage planes are interchanged. Blastomeres at the four-cell stage contain different cytoplasms than they would normally. Two blastomeres are filled with clear cytoplasm with no yolk, whereas the other two are rich in yolk. Significantly, when micromeres appear they always emerge at the original vegetal pole, irrespective of the axis of stratification. In spite of profound alterations in cleavage pattern and cytoplasmic distribution, some normal plutei develop. Blastomeres with abnormal proportions of yolk, mitochondria, pigment, lipid droplets, and so on, can regulate into normal larvae. We can only conclude that the double gradient system does not reside in a graded distribution of organelles easily displaced by centrifugation. This leaves the gel-like cortex and the outer plasma membrane as possible sites for gradient lo-

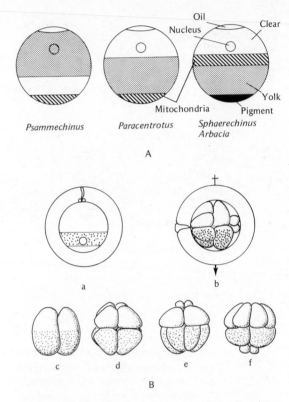

Figure 11-42 *Stratification of the sea urchin egg and its effect on cleavage and distribution of cytoplasmic organelles. (A) Three patterns of stratification in four different species of sea urchin egg. (From H. G. Callan, Biochem. Biophys. Acta 3:92–102, 1949.) (B) Micromere formation in relation to stratification in centrifuged eggs. (a) Normal egg with jelly coat and micropyle with two polar bodies marking animal pole. Nucleus seen beyond pigment ring. (b) Stratified 16-cell stage. Arrow indicates direction of centrifugal force. Animal pole on left. Note that micromeres form from original vegetal pole. (c) Two-cell stage of stratified egg. Cleavage plane is perpendicular to stratification. (d) Eight-cell stage of centrifuged egg. (e) Sixteen-cell stage of centrifuged egg in which previous animal pole corresponds to centrifugal pole. Note micromeres appear at original vegetal pole. (f) Sixteen-cell stage in which previous animal pole corresponds to centripetal pole. (From A. Kühn, Lectures on Developmental Physiology. New York: Springer-Verlag, 1971.)*

calization along with some subjacent cytoplasm not displaced by low-speed centrifugation.

Cortical (and membrane) components are segregated at higher centrifugal forces. When this is done, development is abnormal. Stratified eggs are stretched into dumbbell shapes and ultimately are torn in half: the light centripetal half (white half) contains lipid, egg nucleus, most clear cytoplasm, mitochondria, and a small amount of yolk; the lower heavy or red half contains most of the yolk and pigment and few mitochondria (Figure 11-43).

Fertilized light centripetal halves cleave at the normal rate but in an irregular pattern. No micromeres appear, and development generally stops at the Dauerblastulae stage. Some abnormal plutei develop, however. Yolky halves also cleave abnormally, but occasionally small, heavily pigmented abnormal plutei are produced. We see here a resemblance to the merogony experiment, in which half eggs exhibit different developmental capacities (see Figure 11-39).

These centrifugation experiments indicate that cortical and membrane components are crucial to development. When these latter elements are torn apart by centrifugation, it is rare that development of halves is normal. Evidently maintenance of normal cortical patterns is more essential to development than is the spatial arrangement of cytoplasms.

In support of this argument, we note that up to 50 percent of the egg cytoplasm may be sucked out with a micropipette without affecting development. A dwarf pluteus larva develops with normally proportioned organ systems. Though animalizing and vegetalizing factors are removed in equal proportions when cytoplasm is sucked out, gradients do not seem to be dependent upon specific cytoplasmic substances but reside in elements not easily displaced, such as the egg cortex and membrane or cytoplasms adherent to these structures.

### Nature of gradient system

It has been suggested that the animal-vegetal gradient is physiological rather than morphological — a graded distribution of metabolic activities such as respiration or protein synthesis. According to this view, the high points of metabolism are at the animal and vegetal poles. Respiratory gradients have been revealed by the application of vital dyes which change color when oxidized or reduced. The vital dye Janus green stains sea urchin mitochondria, which are uniformly green under anaerobiosis. The dye acts in place of oxygen as hydrogen acceptor, and when it is reduced it changes color from light green to red, then becomes colorless (Figure 11-44). The color change does not take place simultaneously

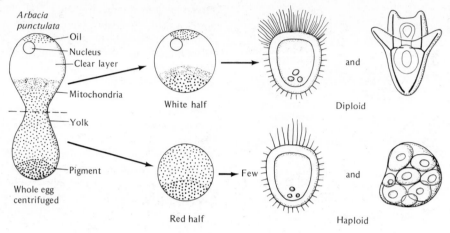

**Figure 11-43** *Stratification of sea urchin egg (Arbacia punctulata) and separation into fragments by centrifugation. Egg is separated into a light white half and a heavy red half which contains most of the yolk and pigment. Fertilization of these will result in the development of Dauerblastulae and abnormal plutei, as shown. (After E. B. Harvey, Biol. Bull. 62:155–164, 1932.)*

in a whole embryo but follows a gradient pattern, appearing first at the vegetal pole and spreading progressively over the vegetal hemisphere until it reaches the equator. At this point, the animal pole begins to change to red, establishing another gradient of color change which expands over the animal hemisphere toward the equator until the entire embryo is bleached. So far, this dye reduction gradient is the only metabolic system which correlates with the double morphogenetic gradient. Isolated animal-half embryos only display a dye reduction gradient beginning at the animal pole; the vegetal gradient is suppressed, while vegetalized embryos do not show any animal pole reduction gradient. The bleaching sweeps up from the vegetal pole until the embryo is colorless.

Since Janus green combines with mitochondria, some have assumed that a dye reduction gradient is really based on a gradient of mitochondria in the embryo. There are, in fact, some claims of more mitochondria in animal-half cells than in vegetal-half regions, which, if true, would explain why dye remains oxidized for a longer period in animal halves. Indeed, if a mitochondrial gradient is the basis for the morphogenetic gradient, then displacing mitochondria by centrifugation should have an effect on development; but, as we have seen, this does not happen. The basis for dye reduction is poorly understood. It may equally involve differences in membrane permeability to ions, or to oxygen.

Morphogenetic gradients are labile and are responsive to many physical-chemical agents which affect metabolism. Various chemicals have been used to alter the gradient system in the sea urchin embryo. One such agent is the lithium ion. In 1892, Herbst showed that sea urchin eggs immersed in low concentrations of lithium salts developed in a typical vegetalized fashion. In lithium-treated embryos, primary mesenchyme cells migrate precociously toward the animal pole. Rather than invaginating into the blastocoele cavity, the ventral region everts externally, forming a monster with an everted gut and no epidermis, a structure known as an exogastrula.

The ion has so strong a vegetalizing influence that it overcomes the normal animalizing tendencies of the animal-half embryo in isolation. Animal halves developing in the presence of lithium ions gastrulate and form endodermal structures; in some cases, typical plutei may be seen. Lithium ion seems to behave like the vegetalizing factor since it has the same inductive effect as the micromeres; it transforms animal developmental patterns into a normal-appearing larva by enhancing the vegetal gradient. Here we have an agent that has a specific effect on the morphogenetic gradient, but it is not useful in analyzing the metabolic basis of the physiological gradient.

Unfortunately, lithium has too many metabolic

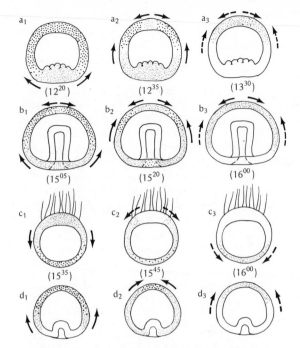

$a_1$ $(12^{20})$   $a_2$ $(12^{35})$   $a_3$ $(13^{30})$

$b_1$ $(15^{05})$   $b_2$ $(15^{20})$   $b_3$ $(16^{00})$

$c_1$ $(15^{35})$   $c_2$ $(15^{45})$   $c_3$ $(16^{00})$

$d_1$   $d_2$   $d_3$

**Figure 11-44** *Dye-reduction gradients in* Paracentrotus lividus *embryos. Janus green is bleached by reduction first to an intermediate diethyl safranin, a blue color, designated as large dots, and then a red color, shown as small stippling. The series $a_1$–$d_3$ shows time after fertilization when embryos were placed in stain. Changes occur in left-to-right sequence in each row. a = late blastula; b = gastrula; c = animal half of an embryo; d = vegetal half of the same embryo. Note that the dye-reduction gradients traverse the embryo in opposite directions in the latter two. (After S. Horstadius, J. Exp. Zool. 120:421–436, 1952.)*

effects; its action is nonspecific. It alters protein configuration, thereby changing solubility and function. It is not surprising, therefore, that many metabolic systems such as protein synthesis and respiration are inhibited by lithium ions. Since lithium also combines with nucleic acids, it will interfere with DNA replication and cell division. Finally, and perhaps most importantly, lithium interferes with membrane permeability by replacing sodium in the sodium-potassium pump. Since the plasma membrane is the first to interact with lithium ions, its permeability will change and ion transport through various cellular compartments will be altered.

Other substances have animalizing effects. Eggs immersed in low concentrations of sodium thiocy-

anate or trypsin or exposed briefly to low concentrations of metal ions such as zinc or mercury develop in an animalized fashion, forming a blastula with numerous apical tuft cilia and very much reduced intestine. The agents capable of animalization are diverse and seemingly unrelated. The one feature they have in common is the ease with which they combine with proteins. These poisons are preferentially absorbed in the vegetal region of the egg, perhaps reflecting a differential permeability. This means that vegetalizing tendencies are destroyed, creating an imbalance that promotes animalized development.

The action of these agents has furnished little information about the physiological basis of gradient systems in eggs. Unraveling this problem is difficult because gradients may reside in egg cortices and membranes, structures that have been inaccessible to analysis until recently. We have seen that many experiments point to the membrane and cortex as sites of developmental information. Animal and vegetal gradients may arise from differences in membrane permeability at the two poles, or differences in the nature of membrane-bound enzyme systems. The plasma membrane could be a mosaic, with a graded distribution of properties from one pole to another. This alone could account for morphogenetic and even physiological differences of animal and vegetal halves.

### The egg cortex as a morphogenetic blueprint

Some suggest that the cortex with its associated microtubules and microfilaments, by controlling the displacement of membrane receptors and enzyme systems, modulates metabolism in growth, division, and cell surface interactions (see Chapters 8 and 10). We saw in Chapter 8 that the protist cell cortex is involved in the assignment of ciliary position during morphogenesis, and in the experiments just described the egg plasma membrane and underlying cortex seem to be sites for the localization of morphogenetic information. Centrifugation and merogone experiments on polar lobes in *Ilyanassa* and the sea urchin egg show this only indirectly, but Arnold's studies on the squid egg show more directly that the cortex may be organized as a blueprint of developmental information, arranged as a mosaic of preprogrammed regions that develop into specific adult organs.

The squid egg is large, about 1.6 mm in diameter, with a thin layer of peripheral cytoplasm surrounding a large yolk mass in the center. Polar bodies appear after fertilization and most peripheral cytoplasm sweeps up to the animal pole to form a blastodisc, very similar to ooplasmic segregation in the teleost egg. This leaves a thin layer of cytoplasm surrounding the yolk, a region comparable to the cortex of most other eggs since it consists of a plasma membrane and a subjacent cortical layer containing only a few cytoplasmic granules and mitochondria. Cleavage is discoidal and incomplete, and a small population of blastomeres (the blastoderm) appears at the animal pole sitting on a syncytial yolk epithelium surrounding the yolk mass. As cleavage continues the blastoderm progressively expands toward the vegetal pole and covers the yolk epithelium. The embryonic organs such as eyes, mantle, and arms emerge in the animal half while blastoderm cells covering the vegetal half develop into a yolk sac.

To study the role of the cortex Arnold used an ultraviolet microbeam to irradiate small regions of the cortex in the noncellulated (that is, before the blastoderm migrates over the yolk) egg. Only the plasma membrane, cortex, and subjacent cytoplasm in the yolk epithelium are affected; no nuclei are present. After irradiation, cleavage is normal and the cellulated blastoderm covers the yolk surface including the site of the UV lesion. Such embryos usually develop abnormally; they lack the specific organs or tissues normally developing at the irradiated site. In this case, destruction of a cortical region of the egg, well before it has been covered with cells, is sufficient to prevent development of specific organs.

Similar results are obtained if parts of the noncellulated cortex are constricted by a ligature and segregated from the rest of the embryo, or if they are treated with small agar blocks soaked in cytochalasin B, the drug known to inhibit cytokinesis in many eggs. Topical application of cytochalasin B (2.0 $\mu$g/ml) for only 10 minutes is enough to inhibit normal morphogenesis of a specific organ-forming area even though cleavage is normal and the treated region is covered by the blastoderm. The specific action of cytochalasin (or of UV irradiation) is not known, although it has been implicated as an inhibitor of contractile microfilaments in cleaving eggs and some motile cells. In the squid egg, the drug also inhibits the fusion of cytoplasmic vesicles with the plasma membrane, a process that seems to be a normal concomitant of fertilization. Cytochalasin may alter the

motile machinery in local cortical sites, or it may prevent the insertion of vesicle membranes into plasma membranes at these sites. In any event, local lesions in the egg cortex and membrane, induced by either UV irradiation or cytochalasin treatment, prevent cells that occupy such regions from carrying out a morphogenetic program. This implies that regionally specific morphogenetic information is localized in the egg cortex (or membrane) and is somehow transferred to blastoderm cells after they arrive. These observations on the squid egg are consistent with the more classic experiments mentioned previously that point to the egg surface rather than cytoplasmic organelles as the principal site of developmental information in the egg. Similar results have been obtained on insect eggs which reveal a comparable pattern of developmental mosaicism.

## Mosaic and regulative eggs: the insect egg

A distinction has often been drawn between determinately and indeterminately cleaving eggs. Determinate eggs exhibit mosaic development. Cells precociously determined by a differential division develop into a specific portion of the larval body. The embryo is constructed as a mosaic, each cell fitting into its predetermined location in the larva. Indeterminately cleaving eggs, on the other hand, are regulative because parts of the embryo, when isolated, reorganize and develop into whole rather than partial larvae. Some cleavage patterns may be modified without affecting development since determination occurs later.

There are, however, no real differences between these two developmental patterns other than the timing of differential divisions. In some, the first division is differential, whereas in others it is the third equatorial division that separates animalizing from vegetalizing tendencies. We find, therefore, a complete spectrum of developmental patterns, from the highly regulative coelenterate eggs to the extreme mosaic patterns in some insects. Most developmental systems, like that of the sea urchin, exhibit both regulative and mosaic phases; they differ only in the time at which the transition from one to the other occurs.

Within the same insect order, mosaic and regulative species are found. Because of the egg's unusual cleavage pattern, the transition between the regulative and the mosaic phases is easily identified. Cleavage of insect eggs, as we have seen, involves replica-

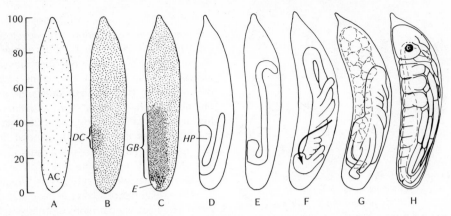

**Figure 11-45**   Platycnemis pennipes. *Normal embryonic development. (A) Cleavage; each dot is a cleavage nucleus. (B) Onset of germ-band formation. (C) Germ band before infolding. (D–E) Infolding. (F) Formation of segments and appendages in folded embryo. (G) Outfolding and twisting of embryo, yolk cleavage complete. (H) Shortly before hatching. AC = activation center; DC = site of the differentiation center; E = origin of infolding; GB = germ band; HP = head process. (From A. Kühn, Lectures on Developmental Physiology. New York: Springer-Verlag, 1971.)*

tion of free nuclei in yolk until a certain number of divisions are completed, after which nuclei migrate to the peripheral cytoplasm to form the blastoderm, a cell sheet out of which the embryo is constructed (Figure 11-45). Cleavage is separated from the beginning of embryo formation. For each species, we may ask, when does the mosaic or determined phase begin? Are larval or adult organ systems determined before or during cleavage, when a blastoderm forms?

To answer this question, small regions of the egg are segregated by a ligature or destroyed by cauterization or UV irradiation at different times in development. If the procedure is carried out before or during cleavage, nuclei cannot enter the cauterized region at blastoderm formation. If partial embryos result, the egg is defined as mosaic, but if a normal embryo is produced, the egg is regulative.

By these operational criteria, most mosaic eggs are found among Coleoptera, Diptera, and Lepidoptera. In the housefly, for example, if portions of the egg cortex are cauterized before cleavage, partial embryos result. The larval structure lost corresponds to the one normally produced by cells that occupy this region when the blastoderm forms. Damage to different cytoplasmic regions results in loss of different larval structures, as if the egg periphery contains organ-forming materials arranged as a two-dimensional mosaic. In these strictly mosaic species, it seems that genetically equivalent cleavage nuclei are

committed to follow a developmental program after interacting with peripheral cytoplasms.

The egg of the dragonfly *Platycnemis* is a more regulative system, but only to a degree. Destruction of anterior portions of the egg before cleavage has no effect on development. If the egg is ligatured across the middle at the four-nucleus stage, the posterior half of the egg develops into a small but complete embryo. In this case, the posterior region with only half of the egg organization regulates into a normal embryo. If the posterior end is ligatured from the rest of the egg before cleavage nuclei have had time to interact with posterior cytoplasm, no embryo appears from the large anterior fragment (Figure 11-46). The peripheral cytoplasm at the posterior pole is essential for normal development. If the ligature is loosely tied so that some cleavage nuclei enter, or if the ligature is applied after nuclei have entered and interacted with the posterior cytoplasm, then an embryo develops. The posterior region, called the activation center, is essential to most insect development; an embryo cannot form unless cleavage nuclei interact with it. After this interaction, new morphogenetic substances are produced which diffuse toward the anterior end of the egg, progressively determining other blastoderm regions. Up to this point the egg is regulative; afterward it becomes mosaic. At an early stage of blastoderm formation, cauterization of a small region at the posterior end ($\frac{1}{9}$ of the egg length) results in

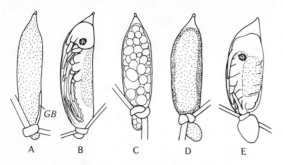

**Figure 11-46**   *Demonstration of the posterior activation center and its role in programming cleavage nuclei. (A, B) Ligation close to the posterior pole at an early cleavage stage (activation center still accessible to cleavage nuclei) does not interfere with normal embryo formation. (C, D) Ligation a little further forward (at same stage) segregates activation center from cleavage nuclei and germ-band (GB) formation is prevented although yolk cleavage takes place and a blastoderm appears. No embryo develops; this time it behaves as a mosaic. (E) Ligation at a later stage in the same region as (C) and (D), after cleavage nuclei have interacted with activation center cytoplasm, will give rise to an embryo lacking only in a posterior abdominal region, showing some regulation, but also exhibiting a mosaic quality inasmuch as the abdomen is lacking. (After F. Seidel, Verh. Deut. Zool. Ges. 38:291–336, 1936.)*

normal development. At a later stage, a larger region of the posterior end (¹/₇ of egg length) can be cauterized without affecting development, indicating that the activation center has had more time to influence the rest of the blastoderm. After the entire blastoderm is determined as a mosaic, even minor damage to an anterior portion results in developmental abnormalities.

Some strictly mosaic eggs show regulative properties. The silkworm *Bombyx mori* is a mosaic form; cauterizing cortical regions early in cleavage causes partial embryo formation. The embryo is determined before cleavage nuclei arrive at the periphery. An examination of these partial embryos, however, shows that the mosaicism is not complete. Occasionally, a fully formed intestine develops, although its inductive site in the cortex has been destroyed, or integument appears along the border zone between injured and uninjured parts, suggesting that there is regulative capacity in some regions. The egg of *Drosophila*, also considered strongly mosaic, is capable of normal development after being punctured

by a glass needle before fertilization. A partial loss of cytoplasm does not interfere with development.

The terms "regulative" and "mosaic" refer to two extremes of a continuous spectrum of developmental patterns based on the timing of embryonic determination. There is no unique component of egg organization that accounts for the difference in timing; it seems to be a property of the total developmental plan of the species. In general, insect eggs with a generous amount of periplasm form large mosaic embryos. Determination takes place earlier than in those species with small eggs and small embryos.

Two features of cleavage may determine cell fate. First, cleavage partitions morphogenetic materials so that genetically identical cleavage nuclei are assigned to different cytoplasms. Accordingly, qualitatively different nucleocytoplasmic interactions may produce cell clones, each committed to a different phenotypic pathway. Second, the timing of this event varies with species and may correlate with the transition from synchronous to asynchronous cleavage patterns. It is here that the temporal component of cleavage affects cell fate.

## SYNCHRONOUS AND ASYNCHRONOUS DIVISION

The first few cleavage divisions of most eggs, except those showing irregular cleavage, are synchronous; every blastomere replicates its DNA and progresses through each phase of the cell cycle at the same time. In a sense, cell cycle time or the rhythm of cell division acts as a clock measuring off the pace at which the developmental program unfolds. Initially all cleaving blastomeres seem to possess the same cell division clock and are synchronized, but soon local cell populations establish their own mitotic rhythms. Morphogenetic events acquire regional properties, attuned to the division cycles of discrete cell populations.

The length of cleavage synchrony varies in different species. Some maintain synchrony for long periods, possibly up to 15 cell cycles in the frog embryo; in others, it is brief, only 2 to 3 cycles in spirally cleaving forms. In the sea urchin, division synchrony persists for three divisions but is lost at the fourth division as micromeres slow down and acquire their own mitotic rhythm while macromeres and mesomeres divide synchronously for 6 to 8 more cell cycles. Likewise, pole cell nuclei in insect eggs are delayed at the fourth division while all other cleavage nuclei replicate synchronously for 7 to 8 additional cycles.

Division synchrony is not found in all eggs. In nematodes, rotifers, and mammals, it does not extend beyond the first division. Each daughter cell at the two-cell stage divides to its own rhythm, an asynchrony that becomes more pronounced in subsequent divisions. Early division asynchrony may correlate with precocious determination of blastomere fate. The best example is in *Ascaris*, where the first division is differential, separating animal and vegetal hemispheres. Each then proceeds to divide independently, and, as we have seen, the animal half cell, the progenitor of the somatic cell line, undergoes chromosomal diminution while the vegetal cell completes normal mitosis and gives rise to the germ line.

A similar correlation between division asynchrony and determination exists in other embryos, although the relationship is not as apparent, since synchrony persists for many cell cycles and the cell population is large when determination occurs. Nevertheless, this period is often marked by transition to asynchronous mitotic patterns with different regions dividing at different rates. A mitotic gradient is gradually established over the embryo, and mitotic waves pass periodically from one pole to the other.

In the sea urchin *Paracentrotus*, for example, the vegetal region acts as the pacemaker. Synchrony persists for six cycles, except for the micromeres which establish their own mitotic rhythm at the fourth cleavage. At the seventh cleavage, mitosis begins first in the vegetal macromeres sitting directly over the micromeres and spreads toward the animal hemisphere. This gradient prevails through the next two cycles, but gradually disappears at the tenth division, when over 1000 cells are present. Gastrulation, the first sign of morphogenesis, usually begins at this time, and at the vegetal pole. In insects the transition occurs at the 9th division and the gradient progresses anterior to posterior (Figure 11-47).

Transition to asynchrony occurs earlier in spirally cleaving eggs, at the third division in some cases. In eggs showing unequal cleavage, the large D blastomere often begins third cleavage precociously. Later, a mitotic gradient may appear as a wave passing from vegetal to animal pole, and at the fifth or sixth cleavage it reverses. Cells within a quadrant, particularly the D quadrant, will often divide together at rates different from those in other quadrants. Though an early onset of asynchrony correlates with precocious cell determination in these embryos, the precise relationship is obscure.

Determination may depend on the amount of time available for genome expression during the cell cycle. In synchronous division, cell cycles are very short, 15 minutes to 1 hour. Most of this period is devoted to DNA replication, in S and in mitosis. $G_1$ and $G_2$ phases are either absent or brief. Consequently, time for genome expression is limited. The cell cycle remains short and constant during the synchronous phase and lengthens after synchrony is lost and mitotic gradients are established. Though all phases lengthen, the $G_1$ phase exhibits the greatest change, and it is possible that some cells are committed to their fate at this time.

Asynchrony is characterized by visible signs of nuclear activity. For instance, nucleoli appear in frog embryos at the asynchronous blastula stage of several thousand cells. Moreover, transcriptive activity of cell nuclei also becomes evident; RNA is synthesized. By contrast, in mouse eggs with somewhat longer cell cycles and no obvious synchronous phase, nucleoli appear early at the four-cell stage. Specific cells that display independent mitotic rhythms, such as insect pole cells, sea urchin micromeres, or the derivatives of the D macromere in spirally cleaving eggs, seem to be determined precociously and assume an important morphogenetic role in the embryo.

Two factors may contribute to division synchrony during cleavage: (1) a large store of mitotic precursors in the egg and (2) junctional communication between blastomeres that transmit pacemaker signals coordinating division cycles. We know that eggs contain large stores of ribo- and deoxyribonucleotides and the enzyme systems for their synthesis and interconversions. Some eggs, such as the brine shrimp *Artemia*, have rich supplies of strange nucleotides such as guanine tetranucleotide, which may serve as a nucleotide precursor store. These nucleotide reserves decrease during the synchronous phase and reach low levels at the time asynchrony begins. In addition, eggs also contain reserves of tubulin, actin, and histones, other essential molecules for a rapidly proliferating cell population. The blastomeres need not divert the machinery of protein synthesis to manufacture these substances.

Cleaving cells also communicate with each other across gap junctions. Early blastomeres of the newt are in electrical communication; a stimulating

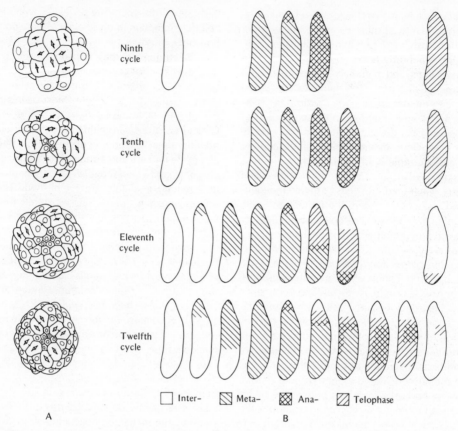

Ninth cycle

Tenth cycle

Eleventh cycle

Twelfth cycle

☐ Inter-    ▨ Meta-    ▧ Ana-    ▨ Telophase

A                                                    B

**Figure 11-47** *Synchronous cleavage and mitotic gradients. (A) Mosaic cleavage pattern in the embryo of the mollusk* Trochus *embryo. Note the synchronous division of corresponding cells in each quadrant. (B) Mitotic gradients during blastoderm stages in the development of the insect* Calliphora erythrocephalla. *In the ninth cycle, synchrony is almost complete, but in the tenth cycle asynchrony begins and a gradient of mitosis appears with the beginning of anaphase. By the eleventh cycle, cells enter metaphase in gradient fashion, and with subsequent cycles the gradient becomes more pronounced. (From I. Agrell,* Synchrony in Cell Division and Growth, *ed. E. Zeuthen. New York: Interscience, 1964.)*

microelectrode inserted in one blastomere generates a voltage that can be recorded in distant cells. Likewise, cells of the discoidally cleaving squid, chick, and fish embryos are in intimate ionic exchange. In the squid, junctional communication between embryo and yolk persists until larval stages, after which it disappears. These specialized junctions for cell communication may facilitate the flow of regulatory molecules that promote synchronous division behavior. Perhaps synchrony is lost when the cell population becomes too large and diffusion rates cannot keep up with rates of signal initiation from pacemaker regions. In addition, junctional contacts are dynamic and change; some disappear and others are

made. This may explain the emergence of regional properties as the egg is partitioned, since small local populations may retain junctional communication and lose contact with distant cells. Subdivision into several semiautonomous but overlapping populations may delay signaling; extracellular spaces must be traversed as cell distances increase and junctional contacts are lost. Mitotic gradients could appear at this time, and morphogenesis might begin shortly thereafter as discrete cell groups acquire different properties.

Mitotic gradients are probably the first all-over pattern in the embryo reflecting regional differentiation. These gradients may amplify polarities preexisting in the egg. Because of temporal differences in the cell cycle, determination may follow a mitotic gradient. Vegetal regions, perhaps the first to experience a changing cell cycle, usually initiate the first morphogenetic events. Later, as tissue and organs develop, the appearance of local mitotic gradients also seems to correlate with patterns of cell differentiation (see Chapter 17).

Summarizing, we see that cleavage does more than propagate multicellularity; it also determines the fate of cells in the embryo. By segregating morphogenetic materials, it creates different cell lineages with unique developmental programs. The timing of these events varies with each species, but it tends to occur when rapid division synchrony shifts to an asynchronous, lengthening cell cycle.

The accuracy of DNA replication and mitosis insures that the original zygote genome is reproduced faithfully; each cell receives identical genetic information. Yet early in the mosaic egg and later in the regulative egg, cell clones with potentialities for different phenotypic expression appear. We now ask, does partitioning of different cytoplasmic materials elicit different modes of nuclear expression? That is, do nuclei change as they sort out into different cytoplasms?

## CLEAVAGE AND NUCLEAR DIFFERENTIATION

When the problem of cleavage and its effect on cell fate was first raised at the turn of the century, it was not evident that all cleavage nuclei are genetically identical. The uncertainty stemmed from such observations as chromosome diminution in *Ascaris*, where cleavage nuclei are visibly different. In fact, August Weismann at the end of the nineteenth century proposed a theory of "idioplasm" to account for the origin of different cell phenotypes during embryogenesis. According to this hypothesis, development was explained by a nuclear segregation process. At each cleavage division, after germ separates from soma, different nuclear determinants (idioplasm) are released into somatic cell cytoplasm, specifying the fate of each blastomere and its descendants. Sorting out of nuclear factors continues at every cleavage division, presumably until all cell lines arise. Each cell receives a different constellation of idioplasm determinants and thereby experiences a different de-

velopmental fate. The theory predicts that all cleavage nuclei are qualitatively different, a kind of nuclear differentiation accompanying partition of egg cytoplasms. Primordial germ cells, however, differ from somatic cells; they retain a complete set of determinants and are thereby equipped to transmit whole germ plasm to the next generation.

According to the model, even first cleavage is differential; that is, each daughter blastomere differs since its nucleus releases a different set of determinants. The early experiments of Wilhelm Roux (1895) supported the model. He destroyed one cell of the two-cell stage in the newt embryo with a needle (Figure 11-48). The surviving cell continued to develop, producing a partial embryo (right or left side) with one-half of the organ components of a complete embryo. This suggested that half the nuclear determinants had already been segregated at the first division. Meanwhile, Hans Driesche demonstrated that isolated blastomeres of the two- and four-cell sea urchin embryo produced normal larvae, which meant that each blastomere was the same and no nuclear segregation had taken place. Driesche stressed the strong regulative properties of embryonic cells. How could these differences be resolved?

It was later realized that Roux's experiment was crude because the dead blastomere attached to the surviving one mechanically interfered with the latter's development. The key experiment came in 1901 when Spemann showed that if the blastomeres are actually separated at the two-cell stage by a ligature, each develops into a completely normal but dwarf embryo. But this experiment only ruled out Weismann's notion that the nuclei were different at the two-cell stage. What about later stages of cleavage? Was it not possible that segregation of determinants did in fact occur later on?

Spemann devised a clever experiment to evaluate the equivalence of cleavage nuclei. Using a fine hair as a ligature, he constricted the fertilized egg of the newt into two halves just as it began to cleave. Only a partial constriction was made to retain a small channel of cytoplasm between the two blastomeres, but the zygote nucleus was restricted to one of the two blastomeres, and only this one cleaved. The enucleated blastomere, though still attached, did not cleave. At the 8- or 16-cell stage, one of the daughter nuclei in the cleaving half migrated through the cytoplasmic bridge into the uncleaved blastomere. Now that it had a nucleus, it was able to divide

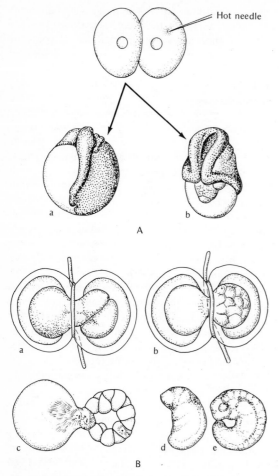

Hot needle

A

B

**Figure 11-48** *Roux and Spemann experiments showing developmental capacity of early cleavage blastomeres in the amphibian egg. (A) Roux's experiment on frog embryo. (a) Half embryo obtained after injury to one of the first two blastomeres. (b) So-called anterior embryo, after injury to one blastomere at two-cell stage in which cleavage was at right angles to median plane. This embryo may be interpreted as a whole embryo in which lateral lips of the blastopore have failed to close because of material in injured half. (B) Spemann experiment on newt embryo demonstrating equivalence of cleavage nuclei. (a) Cleavage beginning in nucleated half. (b) A nucleus has penetrated the uncleaved half and a cell boundary appears between this and the cleaving half. At this point the ligature is tightened. (c) A cross section of (b). (d) Embryo from half which received a cleavage nucleus. (e) Embryo from half which contained the original zygote nucleus. (From H. Spemann,* Embryonic Development and Induction. *New Haven: Yale University Press. Copyright © 1938 by Yale University Press.)*

normally, and at that moment the ligature was tied tightly to separate both halves into two independent entities. Each half continued to develop, one slightly delayed, until two completely normal dwarf embryos appeared. According to the Weismann hypothesis, the half that cleaved first, with the original zygote nucleus, should have retained all the original nuclear material since the other blastomere received only $1/16$ or even $1/32$ of the total nuclear material, having received its nucleus after the fourth or fifth division. Clearly, if nuclei were segregating determinants at each division, then a nucleus of a 16-cell stage should have a developmental potential different from that of a zygote nucleus. Since both halves developed normally, 16-cell-stage nuclei were unchanged and identical to the zygote nucleus, thereby disproving Weismann's hypothesis. Because of the precision of mitosis, each cleavage nucleus receives an identical copy of the zygote genome. If early cleavage does have an effect on cell fate, it does not change developmental properties of nuclei.

In recent years the experiment has been extended to nuclei from later embryonic stages by the technique of nuclear transplantation. Conceivably, stable nuclear changes may occur at the end of cleavage, in the blastula, when nucleoli appear, cell cycles lengthen, and asynchronous division begins.

**Nuclear transplantation**

Only after Briggs and King developed a nuclear transplantation procedure for the frog embryo was it possible to test nuclei from later stages. The procedure is simple; blastula nuclei are transplanted into enucleated, activated eggs, and their capacity to support a complete developmental program is assessed. An intact egg organization with all its developmental potential is provided; if a nucleus has been irreversibly altered in development, it should not be capable of supporting normal development.

The procedure begins with the newly laid unfertilized egg of *Rana pipiens* (Figure 11-49). After the egg is activated by pricking it with a glass needle, a fertilization membrane is elevated and the female nucleus completes maturation, forming the second polar body. The site at which the second polar body forms, containing the entire spindle apparatus, can be easily plucked out with a glass needle since it appears close to the surface. The activated, enucleated egg now serves as the recipient of donor nuclei. If nothing is done to the egg, it soon dies and cytolyzes. If, how-

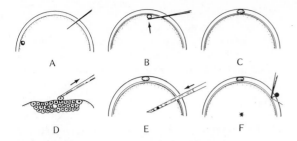

**Figure 11-49** *Technique of transplanting nuclei from embryonic cells into activated enucleated eggs of* Rana pipiens. *(A–C) Steps in preparing enucleated eggs. The egg is activated with a glass needle (A) and enucleated when polar pit appears on surface of egg marking site of second meiotic metaphase of female nucleus. The enucleation leaves an exovate on the surface beneath the vitelline membrane. (D–F) Transplanting blastula nuclei. In (D), a donor cell from a blastula partially dissociated in Ca- and Mg-free salt solution containing EDTA is drawn into a micropipette; the cell is broken because the pipette diameter is smaller than that of the cell. The broken cell is injected (E) and the nucleus is released into the egg cytoplasm. As pipette is withdrawn, a small canal of cytoplasm is drawn out of the surface coat and this is cut with a needle to prevent leakage. [From Robert Briggs and Thomas J. King, "Specificity of Nuclear Function in Embryonic Development," in* Biological Specificity and Growth, *ed. Elmer G. Butler (copyright © 1955 by Princeton University Press), figure 2, p. 216. Reprinted by permission of Princeton University Press.]*

ever, a nucleus is introduced, the egg's fate is entirely different. A donor cell from any region of the blastula is broken as it is sucked into a micropipette to release its nucleus and this broken cell is injected into recipient egg cytoplasm. More than 80 percent of injected blastula nuclei induce normal cleavage to the blastula stage, and approximately 50 to 80 percent of these blastulae develop into normal tadpoles which can be reared to fertile adults. All blastula cell nuclei behave like zygote nuclei; they support a complete developmental program even after 12 to 15 cell generations, during which they have been exposed to different cytoplasms. We conclude, therefore, that there is no genetic change in nuclei during cleavage.

This is not surprising in the case of the indeterminately cleaving frog egg, since blastula cells exhibit regulative capacity; meridional halves of the blastula will develop into normal dwarf larvae. Likewise, in the regulative egg of the dragonfly *Platynemius,* destruction of a single cleavage nucleus with

UV before it migrates to the egg cortex has no effect on development. These cleavage nuclei are equivalent before they interact with the periplasm of the egg cortex.

In determinately cleaving eggs, however, we might expect precocious nuclear changes associated with the early determination of cell fate. Nuclei have been successfully transferred in mosaically developing insects (bees and *Drosophila*). Single wild type nuclei taken from cleavage, preblastoderm, and syncytial blastoderm stages of *Drosophila* were injected into unfertilized eggs from genetically marked strains. Most eggs developed into abnormal embryos which were identical to abnormal control embryos obtained simply by puncturing the egg. This means the abnormal embryos could be attributed to damage produced by the injection procedure and not to abnormal nuclei. Tissues from such nuclear transplant embryos or larvae, if cultured in a host larval abdomen for awhile, did produce some adult wild type structures when larvae metamorphosed. We can conclude that nuclei from late cleavage stages of mosaic embryos are not genetically modified; they have the capacity to produce a variety of larval and adult cell phenotypes after transplantation.

Cleavage creates cell sheets made up of genetically identical cells. If cells acquire any phenotypic commitment at this time, it is not stabilized as a nuclear genetic change. Multicellularity, however, brings with it entirely new systems properties (gradients, inside-outsideness, cell communication, to mention only a few), and cell sheets exploit these to initiate morphogenesis.

## THE MORULA AND BLASTULA: THE FIRST CELL SHEETS

As we have seen, the size of an egg (that is, its yolk content) determines the cleavage pattern. Yolk also determines how newly forming blastomeres arrange into a cell sheet. Mechanical forces exerted by the surrounding fertilization membrane and hyaline layers, as well as the contact relationships established between blastomeres, also contribute to the shape of a cleaving cell population.

Generally, individual blastomeres are spherical except where they come into contact. Probably because of surface tension, they cluster into a spherical group with a minimal amount of surface area exposed. Hololecithal eggs usually form a population

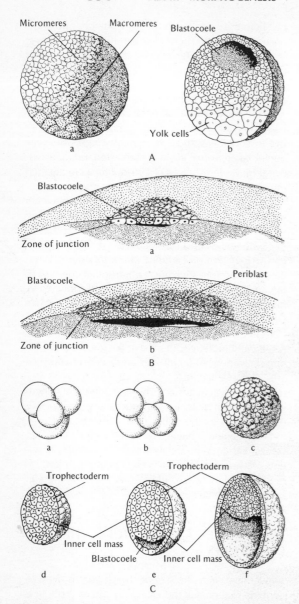

**Figure 11-50** *Blastula formation in (A) frog, (B) chick, and (C) mammal. (Based on A. F. Huettner,* Fundamentals of Comparative Embryology of the Vertebrates. *New York: Macmillan. Copyright © 1941 by Macmillan Publishing Co., renewed in 1969 by Mary A. Huettner.)*

of equal-sized cells in a tight cluster known as a *morula,* the Latin word for mulberry, which they resemble. This term, applied to most early cleavage stages, refers to a cluster of blastomeres tightly

packed within the fertilization membrane; the cells have not organized into a tissue sheet as yet. A morula in telolecithal eggs is composed of a small flattened disc of cells grouped at the animal pole. No morula is found in the centrolecithal eggs of insects.

During cleavage, cells tend to move toward the periphery of the cluster, leaving a small fluid-filled space in the center, known as the *segmentation cavity.* By the end of cleavage, the first cell sheets are organized as a blastula, usually a sphere of cells surrounding the cavity, or *blastocoele.* The segmentation cavity in such *coeloblastulae* varies with species, depending on the amount of yolk and the cleavage pattern. In many spirally cleaving eggs, no cavity appears; the smaller micromeres accumulate as a cluster of cells over the larger vegetally placed macromeres. At the end of cleavage, this results in a solid blastula known as a *stereoblastula.*

The blastula cavity is surrounded by a single layered sheet of small tightly packed cells forming an epithelium, or *blastoderm.* This pattern is typical of most echinoderm embryos. In the more yolky egg of the frog, the blastocoele is displaced toward the animal half with large yolk-laden cells forming a multilayered platform underneath. With more yolk (for example, the chick or fish embryo), the blastula is known as a *discoblastula* because it appears at the animal pole as a small multilayered flat disc separated from the yolk by a narrow segmentation cavity. The superficially cleaving eggs of insects produce a *periblastula* without a blastocoele cavity as nuclei collect in the peripheral layer. In a sense, a cavity is present but filled with yolk from the beginning of cleavage.

Blastula formation in the mammalian embryo is distinctive. Cleavage is regular, a typical morula is formed (Figure 11-50), and a small cavity appears inside the dividing cells, becoming progressively larger. The cluster of cells differentiates into two distinct groups, an epithelial-like layer surrounding the expanding cavity and an inner cell mass displaced to one pole of the sphere. Though these eggs have no yolk, they form a discoidal blastula just as in a chick embryo. The inner cell mass spreads inside the cavity as a flat disc, resembling the chick blastoderm. This stage of mammalian development is known as the *blastocyst,* and it is at this time that the embryo implants into the uterine wall. The epithelial layer or *trophoblast* surrounding the blastocyst cavity is not part of the embryo proper but eventually becomes part of the placenta or extraembryonic membranes as the embryo implants.

In a blastula, specialized cell adhesions appear in the most peripheral layer of cells and a true cell sheet is formed. As cells cleave, they are integrated into a physiological continuum by tight junctions and cytoplasmic bridges. Tight junctions appear at apical ends of epithelial cells and organize a coherent sheet. The epithelial layer surrounds the developing embryo and controls the exchange of materials between the environment and the blastocoele.

How does this first embryonic cell sheet, the blastoderm, develop? This question has been studied in the hollow-sphered blastula of the echinoderm embryo, where the hyaline layer at the cell periphery plays an important part in blastula formation. The hyaline layer is the extracellular matrix of the cleaving embryo. One hypothesis suggests that the hydrostatic pressure within the blastocoele cavity forces the cleaving cells to the periphery. According to this view, fine cytoplasmic processes penetrate the enveloping hyaline layer, creating a cohesive expanding surface. As cell number increases and cells become smaller, they secrete fluids into extracellular spaces and these fluids accumulate in a central cavity. Hydrostatic pressure builds up in the cavity, impinges on outermost cells adhering to the hyaline layer, and generates a hollow sphere.

Another hypothesis of blastula formation stresses the role of the hyaline layer in keeping cells together. As cells cleave, they round up and reduce their point of mutual contact. Their outer surfaces, however, are still embedded in the hyaline layer by fine protoplasmic projections and, as a result, a tangential force is exerted with each cleavage division that tends to expand the peripheral cell layer as central regions fill with blastocoele fluid. Since cleavages are synchronous, this produces rhythmic tensions, and divisions are translated into pulsations of contraction and expansion. After a division, tension is reduced and cell contact is increased. The volume of the blastula is temporarily diminished but is expanded as new cells are added in the peripheral layer at the next cleavage. Because of the radial cleavage pattern and the attachment of cells to the hyaline layer, an overall radial symmetry is imparted to the blastula.

The blastocoele, with its changing fluid environment, becomes part of new gradient systems developing within the embryo. A spherical cell sheet separates an external fluid environment from a different internal one in the blastocoele. The spatial property of "inside-outsideness" is now a reality as epithelial cells become polarized. The outer surfaces of cells of the peripheral layer are thrown into fine protoplasmic filaments embedded in a hyaline layer. Smooth inner surfaces are exposed to a blastocoele cavity of different ionic composition from the surrounding fluid environment. Cells are polarized, basal and apical ends exposed to different ionic environments which may, in turn, lead to differential permeability with inner surfaces generally more permeable than outer surfaces.

One of the first structural signs of this newly acquired polarity is the emergence of cilia. Most blastulae of marine embryos develop cilia and begin to swim about within the fertilization membrane. These motile organelles emerge on the outer surfaces of epithelial cells because basal bodies, from which cilia develop, localize in the cortex, beneath the plasma membrane. Even contacts between cells acquire polarity. Specialized contact points such as gap and tight junctions are not distributed uniformly over the cell surface but are found only at lateral surfaces, at the more distal regions of the cell. The free proximal surfaces lack these membrane modifications.

Individual cell properties are coordinated within sheets, which now acquire morphogenetic potential. The original polarity and gradients within the egg, along with the contractility of the cortex, are translated into properties of a multicellular sheet. Superimposed on these are additional systems properties of inside-outsideness, intercellular communication, and sheet mobility. Ion diffusion gradients across the blastocoele cavity and through cell sheets generate patterns of electrical potential differences which, in turn, may initiate morphogenetic gradients over the embryo. For example, the mitotic gradients that arise at the end of the synchronous phase could reflect changes in the ion flow pattern throughout the embryo, with groups of cells "identifying" their position across gap junctions. In this manner a sheet develops regional properties.

The discoidally cleaving fish egg produces a blastoderm on the yolk surface. At the periphery of this cleaving mass, in the periblast, cells are continuous, with a thick cytoplasmic layer surrounding the yolk. Distinct regions begin to appear in the blastula, with the outermost cell layer forming a cohesive ectodermal sheet, extending over a loose mass of cells at the animal pole. The ectodermal sheet is continuous with the syncytial cytoplasm surrounding the yolk. The blastocoele cavity is a space in the inner cell mass above the yolk. Three distinct regions can be identified: the expanding layer, the inner cell mass, or future embryo, and the peripheral syncytium fused with yolk cytoplasm. The expanding layer behaves as a cohesive, integrated sheet because tight junctions are found exclusively in this layer. These cells are in junctional communication as ionic currents flow rapidly through the expanding layer. The inner cell mass, loosely organized at first, becomes more compact as development proceeds and as the blastocoele cavity is obliterated. Ionic communication occurs between these cells but not as rapidly as in the expanding layer because junctional contacts are fewer and not as intimate. No tight junctions appear in this tis-sue, nor do membranes of neighboring cells come closer than the usual 100- to 200-Å gap. These cells are in poor communication with the yolk mass below since currents generated in the yolk do not migrate into the inner cell mass. Even before morphogenesis begins, a regionalization is apparent, and without knowing anything else, we can almost predict that the fates of these cell sheets will be different.

Although cells at the blastula stage seem to be developmentally equivalent, we assume that regional differences become amplified in time and are gradually translated into regionally specific patterns of morphogenetic behavior. In the next chapters we will see that cell sheets do become semiautonomous, following separate developmental pathways as organ systems are fashioned in the embryo.

**REFERENCES**

Agrell, I. 1964. *Synchrony in cell division and growth*, ed. E. Zeuthen. New York: Academic Press, pp. 5–27.

Arnold, J. M., and Williams-Arnold, L. D. 1976. The egg cortex problem as seen through the squid eye. *Am. Zool.* 16:421–446.

Bennett, M. V. L., and Trinkhaus, J. P. 1970. Electrical coupling between cells by way of extracellular space and specialized junctions. *J. Cell Biol.* 44:592–610.

Briggs, R., and King, T. J. 1952. Transplantation of living nuclei from blastula cells into enucleated frogs' eggs. *Proc. Nat. Acad. Sci. U.S.* 38:455–463.

Brown, D. D., and Dawid, I. B. 1968. Specific gene amplification in oocytes. *Science* 160:272.

Callan, H. G., and Lloyd, L. 1960. Lampbrush chromosomes. *New approaches in cell biology*, ed. P. M. B. Walker. New York: Academic Press, pp. 23–46.

Clement, A. C. 1971. Ilyanassa. *Experimental embryology of marine and fresh-water invertebrates*, ed. G. Reverberi. Amsterdam: North-Holland, pp. 188–214.

Counce, S. J. 1973. The causal analysis of insect embryogenesis. *Developmental systems: Insects*, ed. C. H. Waddington and S. Counce-Niklas. Vol. 2. New York: Academic Press, pp. 1–156.

Czihak, G. 1971. Echinoids. *Experimental embryology of marine and fresh-water invertebrates*, ed. C. Reverberi. Amsterdam: North-Holland, pp. 363–506.

Dan, J. C. 1970. Morphogenetic aspects of acrosome formation and reaction. *Adv. Morphogen.* 8:1–39.

Davidson, E. 1977. *Gene activity in early development.* 2d ed. New York: Academic Press.

Eddy, E. M. 1975. Germ plasm and the differentiation of the germ cell line. *Int. Rev. Cytol.* 43:229–281.

Geyer-Duszynska, I. 1959. Experimental research on chromosome elimination in Cecidomyidae (Diptera). *J. Exp. Zool.* 141:391–488.

Gurdon, J. B., and Woodland, H. R. 1968. The cytoplasmic control of nuclear activity in animal development. *Biol. Rev.* 43:233–262.

Horstadius, S. 1949. Experimental researches on the developmental physiology of the sea urchin. *Pubbl. Staz. Zool. Napoli*, Suppl. 21:131–172.

Illmansee, K., and Mahowald, A. 1974. Transplantation of polar plasm in *Drosophila. Proc. Nat. Acad. Sci. U.S.* 71:1016.

Mahowald, A. P. 1971. Origin and continuity of polar granules. *Origin and continuity of cell organelles*, ed. J. Reinert and H. Ursprung. New York: Springer-Verlag, pp. 159–169.

Metz, C. B., and Monroy, A. 1967, 1969. *Fertilization.* 2 vols. New York: Academic Press.

Morgan, T. H. 1927. *Experimental embryology*. New York: Columbia University Press.

Pasteels, J. J. 1964. The morphogenetic role of the cortex of the amphibian egg. *Adv. Morphogen.* 3:363–388.

Rappaport, R. 1975. Establishment and organization of the cleavage mechanism. *Molecules and cell movement*, ed. S. Inoué and R. E. Stephens. New York: Raven Press, pp. 287–303.

Raven, C. P. 1961. *Oogenesis: The storage of developmental information*. Elmsford, N.Y.: Pergamon.

Ridgway, E. B., Gilkey, J. C., and Jaffe, L. F. 1977. Free calcium increases explosively in activating medaka eggs. *Proc. Nat. Acad. Sci. U.S.* 74:623–627.

Roux, W. 1964. Contributions to the developmental mechanics of the embryo. *Foundations of experimental embryology*, ed. B. H. Willier and J. M. Oppenheimer. Englewood Cliffs, N. J.: Prentice-Hall.

Smith, L. D. 1966. The role of a "germinal plasm" in the formation of primordial germ cells in *Rana pipiens*. *Develop. Biol.* 14:330–347.

Tilney, L. G. 1975. Actin filaments in the acrosomal reaction of *Limulus* sperm. Motion generated by alterations in the packing of the filaments. *J. Cell Biol.* 64:289–310.

Wilson, E. B. 1892. The cell lineage of *Nereis*. *J. Morphol.* 6:361–480.

Woodland, H. R., and Adamson, E. D. 1977. The synthesis and storage of histones during the oogenesis of *Xenopus laevis*. *Develop. Biol.* 57:118–135.

# Organogenesis: Behavior of Cell Sheets

Most multicellular organisms are constructed from two-dimensional cell sheets. By cell proliferation, fertilized eggs transform into two-dimensional epithelia, which thicken, fold, invaginate into tubes and vesicles, and shape themselves into an organism.

As we have seen, eggs cleave rapidly into a blastula, an epithelial sheet surrounding a central cavity. This sheet then transforms into three primary germ layers of the embryo in the Bilateria (flatworms to vertebrates) and two germ layers in the Radiata (sponges and coelenterates).

New sheets arise from blastula epithelium by a program of tissue displacements known as gastrulation. The blastula epithelium expands or invaginates, or both, so that organ-forming areas that were outside are now inside. The outermost sheet, the ectoderm, surrounds a middle mesodermal sheet which covers an innermost endoderm. By the end of gastrulation, a single blastula sheet has become three concentric cylinders making up the body plan of the organism. Each germ layer derivative, occupying a different position in the embryo, is committed to develop into

a different set of organ systems. The basic body plan of most Bilateria is that of a tube within a tube—an inner digestive tube surrounded by the outer body wall, separated by a body cavity or *coelem*. The acquisition of an embryonic body plan, therefore, involves transformation of two-dimensional sheets into tubes.

A cell sheet may be defined as a monolayer of cells displaying integrated behavior. Though it may consist of more than one cell layer, the several layers behave as one. Sheet cohesiveness depends on cell communication mediated by gap and tight junctions, extracellular matrices, and desmosome connections. All these develop during cleavage; as a result, sheet behavior arises epigenetically as cell motility is translated into an expanding or invaginating epithelium. Most movement takes place on substrata, such as a basement membrane, or the extracellular matrix between sheets. Some substrata may favor directed migration. Guided by surface features of the ground mat, the leading edge of a sheet expands in directions of preferred adhesive contacts, usually outward into

338

unoccupied areas, covering all available surfaces. Marginal edge cells display vigorous activity at their free surfaces; they exhibit ruffling movements while interior cells are dragged along passively. If the sheet is torn, cells at the free edges develop ruffled membranes and the sheet expands to cover the wound. When leading edges of two epithelial sheets come into contact, they show a form of contact inhibition. Ruffled membrane activity at the marginal zone ceases and sheets fuse into a continuous monolayer.

Some sheets tend to spread more readily than others. If two sheets are combined, the one with greater spreading tendencies envelops the other, which may explain the inside-outside relationships of tissues within the embryo (see Chapter 10). Ectoderm, as we have seen, has the greatest spreading tendencies; it always envelops all others and occupies an outside position. In their turn, endoderm and mesoderm sheets tend to invaginate or spread inside into cavities beneath the ectoderm.

Morphogenetically active sheets arise early in development, during gastrulation, usually at the end of the asynchronous cleavage phase. An epithelial sheet of the blastula transforms, by a series of inward displacements, into three primary germ layers.

## GASTRULATION: THE ORIGIN OF CELL SHEETS

A blastula is an equilibrium stage, poised between a synchronously dividing cell population and a morphogenetically active sheet. The spherical sheet of the blastula is not homogenous; it may consist of small cells in the animal hemisphere and larger ones in the vegetal half. Gastrulation movements are affected by the size of the blastocoele cavity, the thickness of the epithelium, and the relative sizes of cells in the epithelium.

Two basic movement patterns are involved in gastrulation, *epiboly* and *emboly* (Figure 12-1). The first is a natural tendency of epithelial sheets to expand into free areas over a substratum; the word is derived from the Greek, meaning a ''throwing on'' or ''extending upon.'' Epibolic sheets are usually outside, or ectodermal, and tend to envelop and surround inner sheets. Sometimes sheets expand toward a single point on the surface, a movement known as *convergence*.

A second type of movement, usually opposite to

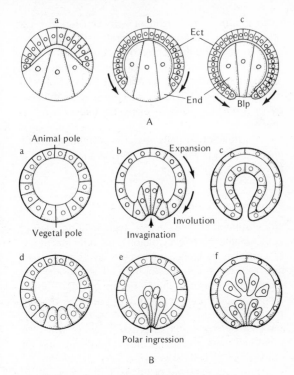

**Figure 12-1** *Types of gastrulation movements. (A) Diagram showing gastrulation by epibolic expansion of ectoderm (Ect) over large endodermal cells (End), forming a blastopore (Blp). (B) More complex embolic movements during gastrulation. Heavy line indicates the original (contractile?) cortex of the egg. In (a) through (c) the invaginating cells retain contact with the original cortex and pull it into the embryo as the blastopore forms. This is accomplished by a combination of epibolic expansion, involution, and invagination. In (d) through (f), during ingression, attachment to the cortex is broken.*

epiboly, is emboly. Tissues fold in or invaginate; that is, a sheet migrates inside a cell aggregate or beneath another layer because of local changes in cell shape or motility. Embolic movements include *invagination, involution,* and *ingression,* movements displacing future endoderm and mesoderm into the interior of the embryo. Invagination, or the sinking in of a layer of cells, is similar to punching in a soft tennis ball. Involution is the turning in of an expanding layer at a specific point. Here a surface layer of cells turns inside itself and continues to expand along the inner surface of the superficial layer. In ingression, small groups of independent cells separate from a primary epithelial layer and migrate into the blastocoele cavity or other embryonic spaces as a new layer. *Delamination* is a type of ingression that forms an inside sheet by mass separation of a layer of cells from an epithelium. This is sometimes accomplished by division of cells in a plane parallel to a tissue space.

Gastrulation usually includes combinations of embolic and epibolic movements, as illustrated in a few species.

### Epiboly: the stereogastrula

Since stereoblastulae lack a segmentation cavity, epiboly rather than invagination is the principal mode of gastrulation. Upper micromere layers proliferate and migrate as an enveloping cell sheet over the larger macromeres. Sometimes this epibolic movement coincides with ingression of the larger macromeres toward the center. The ultimate result is the same, a *stereogastrula* with two layers of cells: an inner cell mass, the future meso-endoderm, and an outer sheet of cells developing into all ectodermal structures. An example of this pattern is seen in the developing gastropod snail *Crepidula* (Figure 12-2).

Epibolic expansion of a micromere sheet over the macromeres below suggests some change in mo-

tility or adhesive relationships or both. Ectoderm glides over macromere surfaces until it fuses at the vegetal pole, closing up the free surface available for further expansion. Before fusion, however, expanding ectoderm forms a circular outline at the vegetal pole known as the *blastopore,* through which macromeres are seen. In some species the blastopore is completely covered by epibolic ectoderm, whereas in others it remains open, becoming the future mouth of the larva. Meanwhile, proliferating endodermal cells, the second germ layer, by a combination of embolic movements and displacements create a cavity, the *archenteron* or future larval digestive tract.

Most stereogastrulae develop in spirally cleaving forms, and in general, the mesoderm arises in a similar manner in all. Cell lineage analysis reveals two types of mesoderm; a larval mesenchyme derived from ectoderm, and a true mesoderm arising from endoderm. The mesenchyme comes from certain ectodermal cells of the second and third micromere quartets released as free cells into the blastocoele cavity. In most trochophore larvae, these mesenchyme cells differentiate into larval muscle. Eventually this muscle degenerates during larval metamorphosis at which time the true adult mesoderm emerges from the 4d cell of the fourth quartet, often called the *mesentoblast.* The 4d cell slips into the blastocoele near the blastopore and divides into two cells, left and right *teloblasts,* which lie as a pair in front of the future larval anus. Each then divides into a band of cells, the mesodermal bands, expanding anteriorly on either side of the larval body. These mesodermal bands become the segmentally arranged muscle somites in adult annelids, but break up into free mesenchyme cells in mollusk larvae. As these larvae metamorphose, mesenchyme reaggregates into a pair of cell masses; in each cell mass a coelomic cavity appears, and these fuse into a single large cavity, the pericardium. Adult heart, blood vessels, kidney, and gonads develop from this tissue in both annelids and mollusks.

### The teleost egg: yolk covered by an expanding sheet

The best example of epibolic expansion of an epithelial layer is seen in teleost embryos. The organization of the teleost blastula has been described in the previous chapter. The caplike blastodisc overlays the yolk mass below. Between blastocoele and yolk is a thin syncytial layer of multinucleated protoplasm,

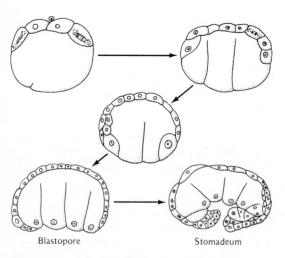

Blastopore                    Stomadeum

**Figure 12-2** *Epibolic expansion to form a stereogastrula in* Crepidula fornucata, *a gastropod snail. Note that small micromeres expand over large macromeres.*

the periblast, connecting the edges of the blasto-dermal cap as a rim of cytoplasm surrounding the yolk. The blastula consists of a cohesive layer of ec-todermal cells, the epiblast, below which is a loosely organized hypoblast layer. The epiblast is multilay-ered, composed of a tightly knit upper layer of non-embryonic cells, the enveloping layer, and an inner cell mass which develops into the embryo proper.

At the beginning of gastrulation, the edges of the blastoderm thicken into a *germ ring* (Figure 12-3). Its most posterior portion develops into the *embryonic shield,* the principal site of invagination in the em-bryo. The hypoblast initiates gastrulation by invaginat-ing as a tongue of migrating tissue anteriorly toward the future cephalic end of the embryonic shield. Meanwhile, the expanding layer spreads radially in all directions over the surface of the yolk, progres-sively covering it from the animal pole and finally fus-ing at the vegetal pole as a concentric ring or blasto-pore. At the same time, the embryonic axis is estab-

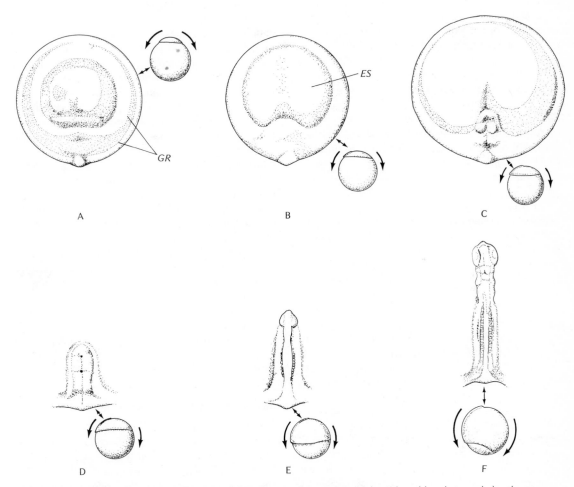

**Figure 12-3**   *Epiboly and early development of the teleost embryo* (Salmo iredeus) *from blastula to early head process stage, showing a dorsal view of the isolated blastoderm at each stage accompanied by a lateral view of the entire embryo. (A) Blastula. (B) Advanced gastrula. (C) Young neurula. (D–F) Epibolic expansion. (D) One-quarter epiboly. (E) One-half epiboly. (G) Three-quarters epiboly. GR = germ ring; ES = embryonic shield. Arrows indicate direction of expansion of embryo over yolk. (Based on C. Devillers,* Adv. Morphogen. *1:379–428, 1961; from J. Pasteels,* Arch. Biol. *47:205, 1936.)*

lished by the invaginating hypoblast. The germ ring at the margin is attached to the periblast which is the advancing edge of the expansion.

How are epibolic movements of the expanding layer explained? At one time, expansion was attributed to the contractile periblast and yolk surface layer. It was assumed that contractions of the yolk surface layer pulled the edges of the blastoderm down over the yolk, but this is only part of the story. Though periblast precedes the blastoderm as a motile marginal zone during epiboly, it is not the primary motive force. If blastoderm is severed from the periblast and thereby separated from the yolk surface layer, the former continues independent epibolic movements. A blastoderm has its own inherent tendencies to spread, because the motility of independent cells is integrated by tight junctions into that of an expanding sheet (Chapter 11).

Between the blastula and gastrula stages, cell adhesiveness increases. This promotes migration of the expanding layer on a periblast substratum; a blastoderm isolated from its periblast and placed on other substrata rounds up and does not spread; it "prefers" a periblast surface, presumably because it is more adhesive.

Periblast expansion is better understood. The periblast is continuous with the yolk surface layer and spreads epibolically, flowing into the yolk cortex until it thickens. This expansion tendency of the periblast is, apparently, preprogrammed in the unfertilized egg cortex. We find that the cortex of unfertilized *Fundulus* eggs is contractile and elastic before cleavage. If a small hole is made in the egg, it is quickly sealed by radial contractions of the surrounding coat toward the center. A similar hole in the yolk surface layer of a blastula is repaired by cells expanding into the free space in a radial fashion. In other words, as blastomeres cleave, they acquire intrinsic spreading properties of the egg cortex, which means that the potentialities for sheet behavior preexist within the egg surface even before fertilization. Contractile microfilaments in the egg cortex may be involved. These may persist in all peripheral cells during cleavage as a continuous contractile system extending from cell to cell. This, plus cohesiveness derived from tight junctions, gives the sheet expansion tendencies.

Periblast expansion over the yolk mass is polarized because the sheet can flow only into the free space over the yolk, toward the vegetal pole. The animal-pole area is covered by the expanding layer and embryo. Polarity of expansion is probably aided by the fact that the cells within the expanding layer are contact-inhibited, except those at the marginal edge, the most active region. The result is complete envelopment of the yolk by epibolic expansion of a cell sheet.

## Invagination: Amphioxus

The most direct way to make two layers from a hollow ball is by invagination, that is, indenting the sphere at one point. Gastrulation in the protochordate *Amphioxus* illustrates how invagination is initiated by changes in cell shape. First, the blastula begins to flatten as the large ventral cells become columnar and then wedge-shaped; they elongate and widen on their inner surfaces (Figure 12-4). The vegetal pole buckles in as a result, forming a wide cuplike depression. The ventral layer expands into the blastocoele cavity toward the animal pole until it has obliterated the cavity space. The cavity of the cup, the future archenteron, communicates with the exterior through the blastopore. The outside ectoderm and inside endoderm are continuous at the lips of the blastopore. Invagination of endoderm is assisted by epibolic expansion of the ectodermal sheet in the animal half, which closes the lips of the blastopore and lengthens the embryo in an anterior-posterior direction. The basic body plan of a tube within a tube emerges later when the mouth develops as a minute invagination at the anterior and breaks into the archenteron.

Most endodermal invaginations are not as simple as in *Amphioxus*. Several embolic movements such as ingression and invagination combine to produce most echinoderm gastrulae.

## Invagination and ingression: echinoderms

The dynamics of sea urchin gastrulation rely on cooperative interactions between cellular and sheet properties. A period of cell ingression precedes the first tissue movements of gastrulation. Ingression begins in the swimming blastula, a ciliated hollow ball. Its small blastocoele enlarges as cells proliferate and fluid accumulates (Figure 12-5). The first signs of morphogenesis appear at the vegetal pole as cells change shape from elongate wedges to cylinders; the ventral region thickens into a plate. This is ac-

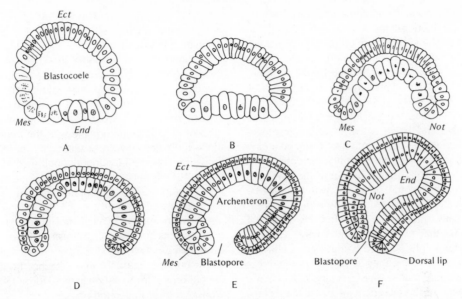

**Figure 12-4**   *(A–F) Stages in gastrulation in* Amphioxus. *Ect = ectoderm; Mes = mesoderm; Not = notochord; End = endoderm. (Based on E. G. Conklin, J. Morphol. 54:69–118, 1932.)*

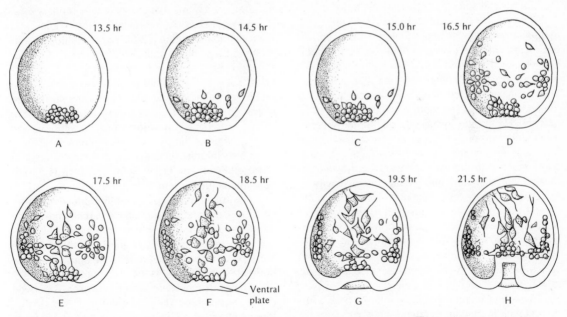

**Figure 12-5**   *Gastrulation in the sea urchin embryo. The sequence of changes in the sea urchin* Clypeaster japonicus. *Time after fertilization is indicated at the top right. (A–D) Immigration and movements of primary mesenchyme cells into the blastocoele cavity. (E) Secondary mesenchyme cells appear (larger stippled cells) and migrate in cavity toward roof of the blastocoele. (F) Continued ingression of secondary mesenchyme, thickening of ventral plate, and first signs of invagination. Primary mesenchyme cells clustering into aggregates prior to spicule formation. (G) Beginning of spicule formation, invagination of ventral plate well on its way. (H) Secondary mesenchyme cells attached to roof of blastocoele "pull" invaginating ventral plate into blastocoele cavity. (From K. Okazaki,* Embryologia *5:283–320, 1960.)*

companied by a shift in microtubules from the outer cortex to the interior cytoplasm, oriented parallel to the cell axis. The vegetal pole of the embryo flattens as vegetal-pole cells become more motile. While distal-lateral surfaces are stabilized by desmosomes and tight junctions and are virtually immobilized by microvilli embedded in the hyaline layer, inner cell surfaces exposed to the blastocoele cavity become less adhesive and begin active cytoplasmic movements. Pseudopodia bulge into the blastocoele cavity, pulsating actively, and motility increases. This surface activity is translated into movement as cells detach from the ventral plate and wander into the blastocoele cavity. Once inside, these *primary mesenchyme* cells round up and develop fine filopodial pseudopodia that extend into the cavity. Microtubules are randomly oriented within these cells, except in filopodia where they lie parallel to the direction of extension. Ingression of primary mesenchyme and formation of the mesenchyme blastula are completed several hours before gastrulation begins. The primary mesenchyme cells are destined to differentiate into skeletal spicules of the pluteus larva.

Meanwhile, adjacent cells in the ventral epithelium experience similar surface changes but do not escape into the cavity, perhaps because of their greater lateral adhesiveness. Presumably these cells retain strong lateral connections via tight junctions. They fill in the gaps left by escaping mesenchyme and the region flattens even more. Soon these cells elongate, buckle in, and invaginate, pulling the sheet in as a unit because of lateral cell adhesions. Since their inner borders are more motile than their outer surfaces attached to the hyaline layer, the direction of invagination is inward, into the blastocoele cavity.

This proclivity for invagination is intrinsic to the ventral plate. Isolated ventral plate displays similar behavior, forming small invaginations, whereas animal halves do not. Motility is an intrinsic property of vegetal-pole cells, a tendency, as we have seen, which preexists in the fertilized egg since only vegetal fragments of an egg gastrulate.

A second ingression phase follows or is concurrent with invagination. After a slight pause, the tip of the invading sheet begins to show renewed pulsatory activity. New mesenchyme cells (the secondary mesenchyme), larger than primary mesenchyme cells, squeeze from the sheet and crowd into the blastocoele cavity. This may be considered the ingression phase. Fine filopodia are spun out into the cavity, their tips extending as far as the inner wall of the ani-

mal half. Attached as they are by filopodia, secondary mesenchyme cells exert a tension on the invaginating tip, pulling it further into the blastocoele.

Contraction of filopodia in secondary mesenchyme probably contributes to invagination of the archenteron in the second and final phase of gastrulation. We know filopodia exert tension because the surface is pulled in wherever they attach. Besides, rates of invagination correlate with the frequency of filopodial extensions from secondary mesenchyme. Sometimes shortening filopodia pull secondary mesenchyme cells out of the advancing tip into the cavity. Finally, suppression of filopodial activity stops invagination.

Primary mesenchyme is not observed in other echinoderms (for example, starfish). The ventral plate invaginates halfway into the blastocoele cavity before any secondary mesenchyme detaches from the archenteron tip. Invagination in these forms takes place without a prior ingression phase.

Since asymmetric shapes of primary mesenchyme correlate with aligned microtubules in the cytoplasm, we assume that filopodia in secondary mesenchyme also contain oriented microtubules and microfilaments (Figure 12-6). Agents inhibiting microtubule assembly, such as colchicine and hydrostatic pressure, or those preventing filament contractions, such as cytochalasin, do indeed inhibit sea urchin gastrulation. Actinlike microfilaments may be responsible for cell movements and changes of shape. Moreover, any disturbance in cell adhesion or properties of the hyaline layer (removal of $Ca^{++}$ ions) also inhibits invagination. No single factor is responsible for sea urchin gastrulation; it is a multiple interacting system.

Gastrulation in the sea urchin produces an inner endodermal sheet surrounding an archenteron. The mouth breaks through where archenteron tip contacts the inner wall of the animal half, while the blastopore becomes the future larval anus. The basic body plan of the larva is thereby established.

The third mesoderm sheet, however, is derived entirely from secondary mesenchyme. Recall that these cells emerge from the invaginating archenteron tip and wander into the blastocoele cavity. They proliferate in the blastocoele, forming a large population. When the archenteron tip contacts the animal half, mesenchyme cells aggregate into two masses on either side and develop into a slender tube, the coelom, running along each side of the developing esophagus and stomach. The walls of the coelom

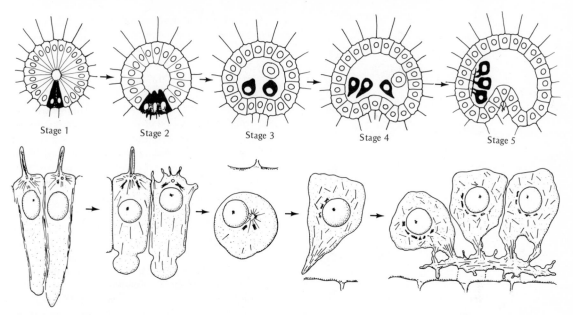

**Figure 12-6**   *Role of microtubules in control of cell shape changes in primary mesenchyme cells during ingression. In the upper drawings are primary mesenchyme cells (black) exhibiting shape changes at different stages of ingression: stage 1 = early blastula; stage 2 = late blastula; stage 3 = first primary mesenchyme; stage 4 = exploratory migrations; stage 5 = formation of cable syncytium before spicule formation. In the lower half the distribution of microtubules at each stage is shown. These organelles orient parallel to the asymmetrical axes of the cell except when the cell is spherical at stage 3, when they seem to radiate from a central nucleating site. (From I. R. Gibbins, L. G. Tilney, and K. R. Porter, Develop. Biol. 18:523–529, 1968.)*

make up the mesodermal sheet. These constrict and enlarge anteriorly, forming the *hydrocoele* chamber, the future adult water vascular system. Metamorphosis of the larva involves transformations of the hydrocoele into the adult body plan, while most other larval structures degenerate (see Chapter 18).

## Invagination combined with involution: the amphibian gastrula

As in echinoderms, local changes in cell adhesion and motility initiate invagination in amphibian blastulae. Epibolic spreading of an epidermal sheet and involution, or the rolling in of tissue at the primary invagination site, contribute to invagination. The structure of the blastula, with its multilayered sheet, large yolk-filled cells in the ventral half, and eccentrically placed blastocoele, also affects the pattern of tissue displacements.

Much of what we know about amphibian gastrulation comes from the pioneering vital dye marking

experiments of Vogt and the transplantation experiments of Johann Holtfreter in the early decades of this century. Vital dye marking experiments actually permit us to follow the movements of tissues as they expand and invaginate. In addition, the technique allows us to determine the ultimate fate of a region as it develops in the embryo.

The vital dye technique was originally introduced by Vogt to prepare fate maps of the amphibian embryo. In this method, small regions of the surface of a blastula or early gastrula are stained with vital dyes such as Nile blue sulfate, and the stained tissues are followed into later embryonic stages (Figure 12-7). Only those cells receiving the dye retain the stain; it usually does not diffuse into neighboring cells. Hence it serves as an effective marker for the morphogenetic behavior of small cell groups.

The movements of gastrulation are followed by staining different regions of the early blastula. For example, a series of vital dye marks in the animal half of a urodele blastula are displaced as the

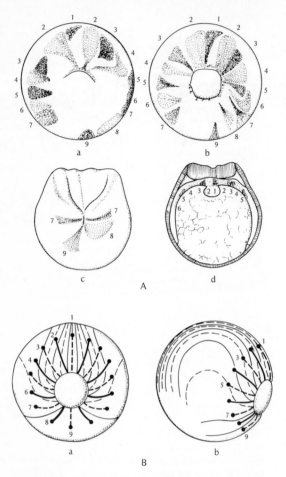

**Figure 12-7** *Gastrulation in amphibians. Epiboly and involution as seen by vital dye-marking experiments. Numbers indicate position of dye marks during tissue movement. (A) Gastrulation in* Triturus alpestris *after marking the marginal zone of an early gastrula. (a) Dorsal lip stage. (b) Completed blastopore with involution of overlying tissues at the lips of the blastopore. (c) Closure almost complete with formation of neural plate and folds. (d) Cross section of embryo showing stained regions in roof of archenteron. (B) Summary diagram of movements of marginal zone tissue during gastrulation. (From W. Vogt,* Wilhelm Roux' Arch. Entwicklungsmech. Organ. *120:384–706, 1929.)*

surface undergoes epibolic expansion. The marks tend to converge toward a central region below the equator, the site of the future dorsal lip. Ultimately, as the dorsal lip forms and expands, the dye marks involute over the dorsal lip region, disappearing into the blastocoele cavity. By a judicious application of

vital dyes and analysis of the displacement of these tissues into the later embryo, Vogt was able to construct a fate map of the blastula and early gastrula tissues. The prospective fate of a region is the tissue or organ it becomes in the fully developed embryo or larva.

A fate map for the urodele blastula shows the spherical sheet divided up into many presumptive organ-forming regions (Figure 12-8). First, we see that the polarity of the fertilized egg is retained in the blastula. The animal and vegetal poles occupy their respective positions, with the animal half corresponding to the anterior end of the embryo, the vegetal half, the posterior. Presumptive organ-forming areas are arranged with respect to the animal-vegetal axis. A future epidermal layer covers part of the animal half, extending well into the equator. Its prospective fate is head and body epidermis. The other portion of the animal half is the presumptive neural layer, fated to develop into the nervous system of the embryo. Within these two areas are the future sense organs. The marginal zone at the equator and just below it, the site of the original gray crescent, is the future notochordal and mesodermal area, which develops as the somites and the vertebral column. The remaining vegetal half of the blastula includes all the presumptive endodermal derivatives, namely, the gut, the visceral pouches, and the pharynx. In this way, a spherical sheet is regionalized in terms of developmental potential.

The dynamic movements and tissue displacements during gastrulation have been examined in various anuran and urodele embryos. Although the basic plans are similar, there are sufficient differences between the two to suggest that we cannot safely extrapolate from one species and assume it is the pattern for all. Each species has evolved special modifications of a common plan of gastrulation to adapt to the peculiarities and rhythms of its own developmental processes.

The pattern to be described is more typical of anurans than of urodeles because it is simpler. The late anuran blastula consists of a multilayered ectodermal sheet surrounding a blastocoele cavity (Figure 12-9). Large yolk-laden cells form a solid mass at the floor of the blastocoele, while smaller animal-half cells organize into a tightly knit epithelium at the roof. The fate map differs significantly from the urodele map in that the presumptive mesoderm is already inside the embryo as an inner layer beneath the ectoderm. Only the future notochordal region is in

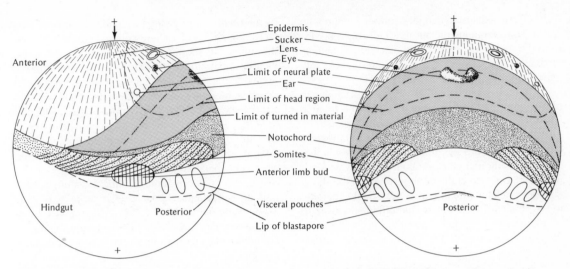

**Figure 12-8** *Fate map of the late blastula or early gastrula of a urodele, based on dye-marking experiments. On the left is a view of the left side and on the right is a view from the posterior or dorsal lip side. Crosses mark animal-vegetal axes. (From W. Vogt,* Wilhelm Roux' Arch. Entwicklungsmech. Organ. *120:384–706, 1929.)*

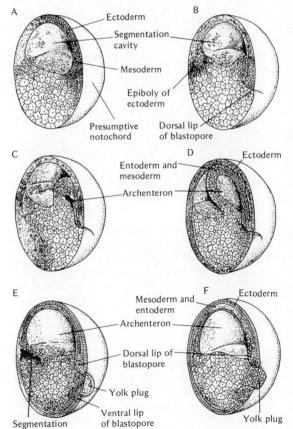

the outer layer in the marginal region. This, along with endoderm, is displaced into the interior by gastrulation, where it assumes its place in the embryo body plan.

Certain cortical events precede gastrulation. In the late blastula, cortical tension in the vegetal pole increases and it contracts, moreso on the future dorsal side. This displaces, by epibolic expansion, the smaller pigmented animal-half cells in the marginal zone well below the equator. These initial changes are probably assisted by microfilaments. Cells in the dorsal marginal zone are smaller than ventral marginal cells, a factor that probably reduces cortical tension in this region. This leads to localized changes in cell surface behavior, particularly in the region of the future dorsal lip of the blastopore.

**Figure 12-9** *Gastrulation of the frog embryo as seen in hemisected embryos. (A) Late blastula; note that ectoderm consists of several cell layers, the innermost of which is presumptive mesoderm. (B) Early gastrula and dorsal lip formation; beginning invagination of endoderm and mesoderm including presumptive chorda region. (C–E) Continued invagination, formation of archenteron, and gradual obliteration of blastocoele cavity. (F) Yolk plug stage and invagination of ventral lip of blastopore. (From A. F. Huettner,* Fundamentals of Comparative Embryology of the Vertebrates. *Copyright © 1941 by Macmillan Publishing Co., renewed in 1969 by Mary A. Huettner.)*

Epibolic expansion cannot continue indefinitely; spatial considerations dictate that cells must be displaced somewhere, and the best place turns out to be the blastocoele cavity. It is the only "free space" into which epithelia may expand.

The first visible sign of gastrulation is a sinking in of a few presumptive endoderm cells at the original gray crescent or dorsal side of the embryo (Figure 12-10). Like ventral plate cells in the echinoderm embryo, these cells become mobile and less adhesive, stretching into elongate bottle shapes, wider at inner surfaces than at the periphery. Again the shape change correlates with microtubules oriented parallel to the long axis. Contracting microfilaments seem to transform cuboidal epithelial cells into bottle cells, their narrow distal tips attached to the outer surface by microvilli interdigitating with neighboring cell surfaces (Figure 12-11). These cells eventually sink into the interior as motility of the wide inner portion increases and a groove, the dorsal lip of the blastopore, appears. Pigment granules collect in the fixed distal tip of each bottle cell as it invaginates, thereby marking the site of the dorsal lip. Adjacent cells, attached to bottle cells by tight junctions, seem to be drawn in over the dorsal lip by the more motile bottle cells in the interior. As a result, the depression lengthens into a crescent-shaped pigmented slit.

Bottle cell formation, a critical feature of amphibian gastrulation, is an intrinsic property of cells at the marginal zone and is not controlled by any exogenous agents. Isolated fragments of invaginating endoderm placed on a substratum of yolky endoderm

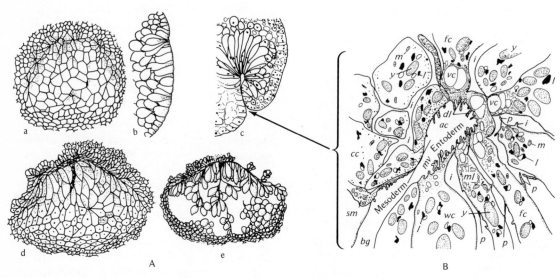

**Figure 12-10** *Bottle cell formation in the amphibian gastrula. (A) Light microscope views of the surface or cross section of dorsal lip region. (a, b, d, e) Ambystoma. (c) Triturus. The onset of dorsal lip formation is shown in (a) and (b). Only pigment accumulation is noted on some cells in the yolk region. In (b) the stippling of some bottle cells marks the site of pigment accumulation. In (c), a longitudinal section at the invaginating dorsal lip shows many bottle cells surrounding the beginning archenteron cavity. In (d), a surface view, note shape of cells above and below lip. In (e) the surface is shown after removal of the region of ingression ventral to the dorsal lip; note the many bottle cells attached at the site of the dorsal lip. (a, b, d, e, based on J. Holtfreter, J. Exp. Zool. 94:261–318, 1943; c, after W. Vogt, Wilhelm Roux' Arch. Entwicklungsmech. Organ. 120:384, 1929.) (B) Electron microscope view (parasaggital section) of bottle cell region of the anuran Hyla at a stage comparable to (c) above. The groove, bg, has the appearance of a "U" with the base of the "U" being the innermost extension of the groove. Long-neck bottle cells (fc), wedge cells (wc), and cuboidal cells (cc) line the groove. The distal surfaces of these invaginating cells contains a dense layer (dl) and vesicles (vc). Note the intercellular spaces (i) bridged by pseudopodial-like processes between cells (p), the archenteron (ac), the membranous layer (ml) with microfilaments, yolk platelets (y), mitochondria (m), lipid droplets (l), and microvilli (mv). The dense layer contains many microfilaments. A specialized mesodermal cell with long cytoplasmic finger (sm) is also seen. (From P. Baker, J. Cell Biol. 24:95–116, 1965.)*

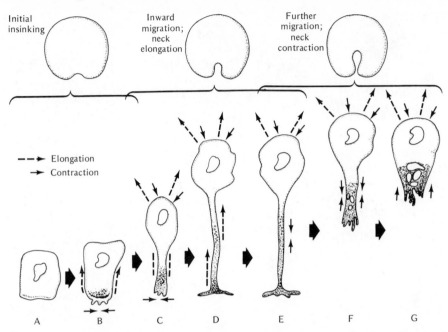

**Figure 12-11** *Proposed scheme for transformations of invaginating bottle cells during gastrulation based on hypothesis of contraction and expansion of dense layer (presumably containing contractile microfilaments). The end of the cell neck remains securely adhered to adjacent cells. (A) The cell on the surface before gastrulation. (B) The distal surface contracts causing initial sinking in of blastoporal pit. (C–D) The proximal cell surface begins to migrate inward, deepening the groove; the distal surface continues to contract and the neck elongates. (E–G) The neck contracts after it is maximally elongated. This shortening, together with the pull exerted by the migrating inner ends, furthers invagination and pulls adjacent cells into the groove. (From P. Baker, J. Cell Biol. 24:95–116, 1965.)*

in vitro will invaginate in a matter of hours, just as they do in situ (Figure 12-12). Cells elongate into bottle shapes, their narrow necks attached to the surface, and drag neighboring cells into a depression resembling a dorsal lip. If inner endoderm cells are used instead, they too invade the substratum but do not form bottle cells and do not generate a blastoporal groove.

According to Holtfreter, the so-called surface coat of the egg plays a major role in invagination. He claims, "The dynamic tendencies of the embryo can be traced back to properties of the unfertilized egg." The nature of the coat in the egg is not clear. No specialized outer extracellular layer is seen in the electron microscope except for clear hyaline material. Surface coat properties are probably due to the cortical region of the egg, which contains contractile microfilaments. As in the teleost egg after cleavage, cells at the surface are joined into a cohesive contractile layer by lateral membrane junctions (tight junctions or desmosomes). This layer acts as a unit; it

has spreading tendencies and communicates the pull of invaginating bottle cells to other cells in the region. An isolated piece of superficial or "coated" ectoderm shows its contractile nature by curling inwardly or spreading if placed on a substratum of endoderm.

The entire animal half of the blastula undergoes epibolic expansion. The dorsal lip enlarges and becomes crescent-shaped because animal-half cells stretch, converge toward the blastopore, and involute over the lip, spreading underneath the superficial layer. Epiboly is most noticeable in the dorsal region but gradually extends laterally and ventrally as the dorsal lip enlarges. More cells invaginate at a ventral blastopore lip, and the expanding blastopore becomes a circular opening. The blastopore is slowly reduced to a yolk plug stage while epibolic movements sweep superficial layers of cells into the interior. Inside, the invading chordamesoderm pushes and pulls yolky endoderm forward to line the cavity of the archenteron. The roof of the archenteron comes from

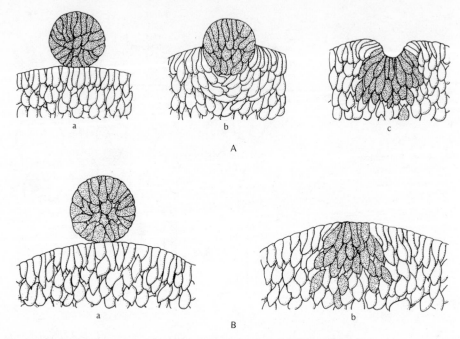

**Figure 12-12** *Invagination behavior of isolated tissue fragments. (A) A graft of blastoporal cells partly covered by the "surface coat" (the contractile cortical layer of the amphibian embryo) sinks into an endodermal substratum and forms a blastoporal groove, pulling in adjacent cells at the surface. (B) An aggregate of uncoated endoderm is incorporated into an endodermal substratum but forms no groove. (From J. Holtfreter, J. Exp. Zool. 95:171–212, 1944.)*

endoderm pushed in by the chordamesoderm, plus any superficial ectoderm that has invaginated over the dorsal lip. The floor of the archenteron consists of yolk-laden cells pulled forward from the blastocoele floor.

Not all spreading tendencies are derived from the surface coat. Endoderm also spreads; it too is an expanding epithelial sheet migrating anteriorly inside the embryo. Nevertheless, "coated" tissues (the most superficial layer with the contractile cortical region) tend to spread more than inner uncoated tissue layers.

Endoderm seems to have a dual property; it invaginates and spreads. Only a small superficial layer of endoderm surrounding the archenteron is capable of expansion; the remainder, made up of large yolk-filled cells, is more cohesive and passive, usually carried along by other expanding sheets. Spreading tendencies of endoderm change. Young gastrula endoderm spreads less than older neural endoderm, indicating that motile and adhesive properties change with age. Isolated blastula and early gastrula cells are less adhesive than cells from later stages.

Ectoderm has a greater spreading tendency than endoderm. If superficial ectoderm is removed from the blastoporal region of a late gastrula, the internal endoderm and neighboring ectoderm tend to spread over the denuded area. Because of its greater spreading tendency, ectoderm covers most of the area, preventing further overgrowth by the endoderm.

Though all sheets develop from blastula epithelium, each acquires tissue-specific spreading and adhesive properties as it assumes a new position in the embryo. When a given sheet is combined with fragments of other sheets, they sort out, each tissue occupying a position corresponding to the hierarchy of inside-outside relatedness in the intact embryo (see Chapter 10). This recognition property appears late, toward the end of gastrulation.

Coated ectoderm (the sheet containing cortical microfilaments) at mid to late gastrula stages spreads over the surface of the endoderm and envelops it. Its collection of filaments gives it greater motile properties, and its outer nonadhesive surface prevents it from sinking into the substratum. Presumptive neural tissue, with an exposed surface coat, behaves identically; as long as the coat contacts the substratum, it will not sink in, but if a strip of neural tissue is allowed to curl up, exposing its inner adhesive surfaces, then it sinks into endoderm. On the other hand, uncoated ectoderm immediately sinks into endoderm, or is enveloped by it. After gastrulation, surface recognition properties change. A piece of neural tissue or subsurface epidermis placed on an endodermal substratum is rejected.

The mesodermal sheet has a complicated origin. According to Vogt's fate maps, urodele presumptive mesoderm and notochord lie superficially at the marginal zone as a wide crescent in the blastula, just above the endoderm (Figure 12-13). Vogt's maps for anurans were not as well-documented, but it turns out that anuran mesoderm has a different orgin. Presumptive mesoderm in an early anuran gastrula occupies an inner position just beneath the most superficial ectoderm layer. Accordingly, all anuran

mesoderm is derived from internal cells that do not invaginate. The combined endo-mesodermal sheet advances as a unit under the ectoderm, and the mesodermal sheet separates, probably by delamination, at the end of gastrulation. Some urodele mesoderm is also inside but most occupies a peripheral position, invaginates behind the endoderm cells, involuting over the dorsal and lateral lips of the blastopore. When the ventral lip appears later, the remaining mesoderm involutes and expands into the interior. Inside, mesoderm and endoderm are at first a composite multilayered sheet; the layer adjacent to the archenteron is endoderm, and that above is the presumptive chordamesoderm. Mesoderm detaches itself early from endoderm, and both sheets expand and invaginate separately underneath the ectoderm. The result, however, is the same in both; three separate layers encircle the yolk mass and archenteron cavity. Inside, most mesoderm concentrates in the dorsal midline of the embryo, forming a thick layer, whereas lateral and

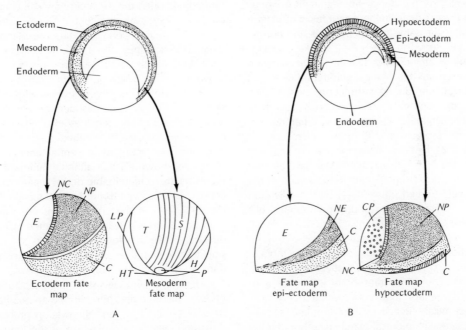

**Figure 12-13**   *Origin of mesoderm and fate maps of germ layers in early amphibian gastrulae. (A) Urodeles. (B) Anurans. Note in both cases that the mesoderm, even at the early gastrula stage, is a layer beneath the ectoderm. The anuran ectoderm consists of two layers, an epi-ectoderm and a hypoectoderm, whereas the urodele ectoderm is only one layer. This affects the subsequent fate of the layers. In both, only the chorda is derived from the invagination of ectoderm, more so in urodeles than in anurans. The fate map for anuran mesoderm is probably similar to that shown for urodeles. E = epidermis; NC = neural crest; NP = neural plate; C = chorda; LP = lateral plate; T = tail; S = somites; HT = heart; H = head; NE = neural ependyma; CP = cranial placodes, P = pronephric primordium. (Based on S. Lovtrup, Acta Zool. 47:209–276, 1966.)*

ventral regions remain thin. It is thickest over the roof of the archenteron but absent at the most anterior end where the archenteron tip joins with the stomadeum to form the mouth.

The morphogenetic center for gastrulation movements seems to reside in cells at the dorsal lip. If this region is removed from an early anuran gastrula, the remaining tissues continue developing but do not invaginate normally. Instead, marginal zone tissues extend over the underlying yolky endoderm, forming lateral blastoporal lips that develop into an abnormal ring embryo. The invagination center (dorsal lip) must be in direct physical continuity with the lateral marginal zone tissue for normal invagination to occur. If continuity is interrupted by implantation of strips of ventral ectoderm, the dorsal lip region behaves independently and extends into a cylindrical outgrowth while the lateral marginal zone extends over the yolk to invaginate into wide blastoporal lips and a ring embryo. On the other hand, Cooke showed no effect of removing the dorsal lip region in *Xenopus* and *Bombina* gastrulae. In these anurans, gastrulation continued at the same rate as controls and normal embryos developed in most cases. Cooke interprets these results as an indication that the dorsal lip region does not exert any mechanical guiding influence on surrounding tissues during gastrulation; the tendency to invaginate seems to be an autonomous property of cells in the marginal zone region. The difference between these results is difficult to reconcile.

How does the dorsal lip region acquire its ability to initiate the site of invagination? We have already seen that the dorsal lip appears in the region of the gray crescent formed immediately after fertilization (Chapter 11). For reasons still unknown, bottle cells appear here first and initiate invagination. The dorsal marginal region may have a higher rate of oxygen consumption, possibly because of greater membrane permeability. Enzymes such as adenosine triphosphatase, involved in the transport of sodium and potassium ions, are also more active on the dorsal side. This higher metabolic activity may stimulate the contractile system within dorsal marginal cells and bring about changes of shape typical of bottle cells.

The potential to form a dorsal lip is not confined to the dorsal marginal zone since any region of the egg surface can be induced to invaginate. We have seen that inversion of eggs at first cleavage will induce two dorsal lips in the animal half. Inversion displaces yolk and possibly subcortical cytoplasms that may contain factors that stimulate dorsal lip invagination. A local constellation of factors (microfilaments or yolk) transforms epithelial cells into invaginating bottle cells.

These factors seem to reside in yolky endoderm cells on the dorsal side of the vegetal half of the embryo which possesses invagination tendencies. Experiments have shown that four animal-half blastomeres isolated from the eight-cell stage of the salamander embryo develop into a small spherical cellular mass that does not gastrulate. The vegetal cells, on the other hand, organize into a spherical embryo that forms a dorsal lip and in many cases develops into an embryo. Invagination tendencies reside in the vegetal half of the embryo.

In follow-up experiments carried out in Nieuwkoop's laboratory, isolated animal-half fragments of early *Ambystoma* blastulae also develop into undifferentiated spherical masses of cells (Figure 12-14). No blastoporal lips appear, nor is any mesoderm formed, although the fragment may still contain part of the former gray crescent region. At the same stage, an isolated yolky vegetal core without marginal zone develops into a solid mass of endoderm cells but does not gastrulate. When the two are combined, however, a blastopore often forms at the border, and abnormal embryos develop with almost 20 percent of their tissues mesodermal. Only in combination is a dorsal lip center established. Most mesoderm seems to come from animal-half cells, presumably after interacting with ventral endoderm. It is significant, however, that although animal-half fragments retain an identifiable dorsal region (gray crescent), the site at which a dorsal lip appears is set by the endoderm. If the combinations are made with both dorsal sides of each half oriented in the same direction, then the dorsal lip appears at the dorsal side. If, however, they are oriented in opposite directions at 180 degrees, then the most frequent site for the blastopore correlates with the orientation of the endodermal core and not with the animal half, with its gray crescent region. Nieuwkoop proposes that the dorsal-ventral organization of the future mesoderm is determined by endoderm. He believes the original dorsal-ventral axis set at fertilization preferentially affects the endoderm, which later determines the site in the blastula surface where a dorsal lip will appear. The nature of this organizing influence is unknown, but this asymmetry is important in all subsequent morphogenesis, as we will see. It defines a

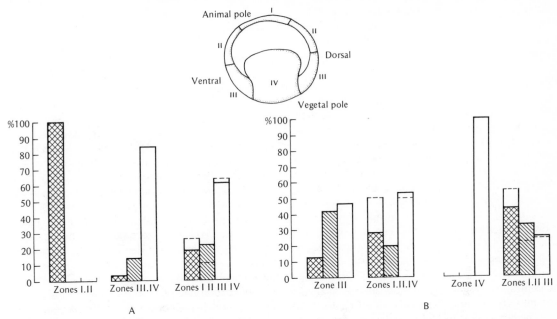

**Figure 12-14** *Origin of mesoderm in urodeles. Above is a diagram of a late blastula stage of* Ambystoma mexicanum *divided into explant regions. Zones I and II are animal-half regions above the floor of the endodermal yolk mass in the center. (See Figure 12-13 for fate maps.) Zone III is the entire marginal zone below the equator (region of dorsal-lip invagination), and zone IV is the endodermal yolk mass including the vegetal pole. (A) Zones I and II are isolated from zones III and IV and cultured as an explant in Holtfreter's solution. The histograms indicate the percentages of different germ-layer derivatives in the isolates after development as determined by comparing relative areas occupied by these tissues in cross sections. Crosshatching = ectodermal derivatives; oblique = mesoderm; white = endodermal. (B) Additional isolates showing the importance of region IV in determining the development of mesodermal tissues from the animal half. Dotted lines indicate calculated quantitative composition of sum of corresponding isolates. (From P. D. Nieuwkoop,* Wilhelm Roux' Arch. Entwicklungsmech. Organ. *162:341, 1969.)*

morphogenetic field with a dominant center in dorsal endoderm, near the marginal zone. According to Waddington, such regions should be called "individuation fields," an operational term which identifies an initially formless region that becomes capable of developing into a specific organ. In this case, the dorsal lip region is the center of an epithelial field that becomes the primary axis of the amphibian embryo.

### Cell properties changed by gastrulation

Gastrulation rearranges all organ-forming areas of the blastula into three sheets, each with different inside-outside relationships and recognition properties. Isolated amphibian blastula cells are similar morphologically and are not very adhesive. They do not attach readily to a substratum and are amoeboid,

forming broad lobopodia. After gastrulation, new cell types appear in each germ layer with different surface properties. Cells become more adhesive; gastrula ectoderm cells now attach firmly to a glass substratum and flatten out. Still other cells appear (bottle cells) with fine filopodia forming ruffled membranes as they glide along the substratum. The surfaces of gastrula cells change; their charge densities are greater than those of blastula cells, which means gastrula cells migrate more rapidly in electric fields. Invaginating chordamesoderm cells exhibit the greatest change in charge density, which may correlate with the fact that negatively charged sulfated mucoproteins, such as chondroitin sulfate, are actively synthesized in these cells during invagination.

Lovtrup suggests that the primary cell transformations at the gastrula stage involve changes at the cell surface; an amoebocytelike blastula cell becomes a more motile bottle cell with filopodia and ruffled membranes. These latter cells, the progenitors of mesenchyme, organize as a morphogenetic center. The amoebocyte also transforms into another cell type, an adhesive cell or *colonocyte*, which assembles into cohesive epithelial sheets, possibly through some type of junctional specialization such as tight junctions. The principal genes acting during gastrulation may involve those coding for cell surface components, either protein constituents of the membrane (adhesive sites, receptors?), extracellular matrices adherent to the surface, or contractile elements in the surface that contribute to cell and tissue movement. Electron micrographs of invaginating tissues show numerous specialized junctions (desmosomes?) which may integrate cells into tissue sheets. Such surface changes could account for tissue displacements and the acquisition of specific recognition properties that result in sorting out and regionalization into discrete functional areas (see Chapter 10).

Let us now look at chick gastrulation where similar morphogenetic processes are adapted to a two-dimensional blastoderm sitting on a large yolk mass. In this case, ingression rather than invagination is the principal morphogenetic factor in sheet formation.

## Chick gastrulation: ingression of endoderm and mesoderm

The chick egg is fertilized as it passes down the oviduct and begins to cleave before shell membrane formation. Cleavage and blastulation are completed in the uterus up to the time of laying. Because of its cleavage pattern, a chick blastoderm develops as a disc between the vitelline membrane and the large yolk mass below (Figure 12-15). The egg is laid in the blastula stage, a central yolk-free cellular blastoderm, the *area pellucida,* with a blastocoele and a more peripheral *area opaca* adherent to the yolk, organized as a syncytium, or *periblast.* An *embryonic shield* is visible in the posterior half of the area pellucida, where cells have accumulated. This posterior thickening has important developmental properties, as we will see. Before incubation, the embryo is at the unincubated blastoderm, or prestreak, stage. During incubation, the embryo develops from the area pellucida while area opaca tissue digests yolk.

The embryo is itself composed of two sheets, a cohesive superficial layer, the *epiblast* or ectoderm, and an underlying loose collection of cells in the blastocoele, the *hypoblast.* Hypoblast cells arise from cleaving inner cells, whereas superficial cells join

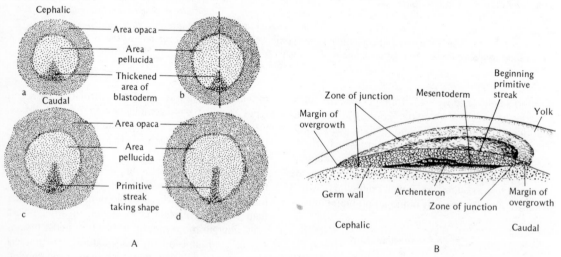

**Figure 12-15**  *Early stages of gastrulation in the chick embryo during formation of the primitive streak. (A) Top view of blastoderm. (a) Three to four hours incubation. (b) Five to six hours incubation. (c) Seven to eight hours incubation. (d) Ten to twelve hours incubation. (Based on N. Spratt, J. Exp. Zool. 103, 259–304, 1946.) (B) Hemisection of one stage showing thickened area marking the primitive streak, the detachment of cells (ingression?) from the anterior end of this region to form mesentoderm. (From A. F. Huettner,* Fundamentals of Comparative Embryology of the Vertebrates. *Copyright © 1941 by Macmillan Publishing Co., renewed in 1969 by Mary A. Huettner.)*

into an epiblast epithelium. The posterior hypoblast cells organize into a thin squamous epithelium beneath the epiblast, a process that gradually extends anteriorly.

A fate map as determined by dye marking experiments of the chick blastoderm at this stage indicates that the epiblast is the source of skin epidermis, neural tissue, future notochord, somites, and lateral mesoderm tissue. The epiblast also contains the presumptive endoderm. The fate of the hypoblast is more difficult to analyze by marking techniques because it is covered with epiblast. When an inverted blastoderm is cultured in vitro with hypoblast uppermost, marking experiments show that some endoderm and notochord are derived from hypoblast, but most of it develops into yolk sac membranes.

Each cell sheet expands on a different substratum whose nature seems to affect the rate and direction of sheet migration. For example, isolated pieces of epiblast do not expand on an agar base, but only do so on the inner surface of a vitelline membrane. Presumably this surface is organized to support epiblast adhesion and spreading. The hypoblast, on the other hand, will not adhere or spread on the vitelline membrane but will adhere to inner surfaces of the epiblast.

These two cell sheets (principally the epiblast) are the source of cells for germ layers and all formative movements shaping the embryo. Though morphogenetic movements have been studied with several cell marking techniques, the problem of endoderm origin in the chick embryo is still controversial, and several different models have been proposed.

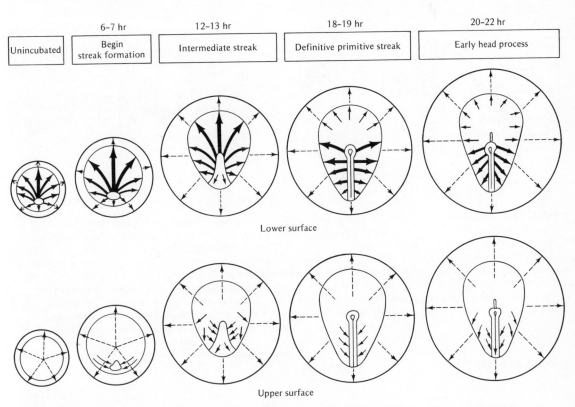

**Figure 12-16** *Dye-marking experiments to illustrate morphogenetic movements of lower surface of blastoderm in relation to upper surface (epiblast) before and during primitive streak formation. The movements of these layers are shown in relation to general growth expansion (dotted arrows) of the entire blastoderm. The thickened arrows from the anterior portion of the primitive streak mark the ingressing cells of presumptive endoderm and mesoderm as well as the movements of the hypoblast. These ingressing cells are derived from the epiblast as they converge toward the primitive streak and descend into the tissues below. (From N. Spratt and H. Hass,* J. Exp. Zool. *144:139–157, 1960.)*

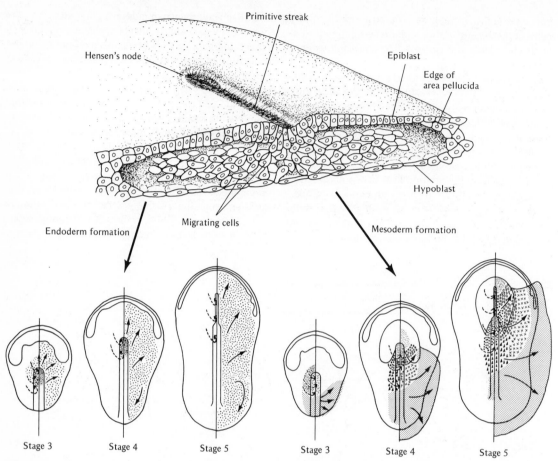

**Figure 12-17**   *Morphogenetic movements of ingressing cells from the primitive streak to form endoderm and mesoderm. Upper diagram is a hemisected view of the primitive streak showing the invaginating and ingressing cells from the epiblast entering the segmentation cavity and spreading out to form the endoderm and mesoderm. Below, on the left the pattern of endoderm formation is shown, based on dye-marking experiments. The dashed arrows indicate the invaginating movements of the epiblast cells as they enter the primitive streak area. For sake of clarity only, invagination is shown on the left side while the expanding movements of the underlying endoderm (stippled area) are shown with solid arrows. Below, right, the expansion of the mesodermal layer and its gradual regionalization into presumptive mesodermal regions is shown. Fine stippling = chordal or prechordal area; gross stippling = axial mesoderm; open circles = cephalic mesenchyme; full circles = somatic mesoderm; open squares = presumptive amniocardiac vesicles; full squares = lateral mesoderm; crosses = extraembryonic mesoderm. (Upper figure based on B. I. Balinsky, An Introduction to Embryology. Philadelphia: Saunders, 1975. Lower figures based on G. Nicolet, J. Embryol. Exp. Morphol. 23:79–108, 1970.)*

Classic studies of tissue movements relied on vital dye marking experiments, or marking with carbon particles. In more recent studies, cells and tissues are labeled with radioactive thymidine incorporated into nuclei. Grafts of marked cells are then transplanted into unlabeled embryos, and the migrations of cells and tissues can be followed by radioautography.

The *primitive streak* appears after 6 to 7 hours of incubation. Epiblast spreads laterally and posteriorly as cells at its leading edges develop long pseudopodial extensions on the substratum (Figure 12-16). Sheet cohesiveness is initially maintained by tight junctions, but more stable junctions (desmosomes) appear in later stages when the epiblast stretches as a single-layer continuum. The epiblast converges toward the posterior border of the pellucid areas and forms a conical thickening, as cells pile up in a morphogenetic center known as the initial primitive streak. This region initiates the definitive embolic movements of gastrulation.

The primitive streak begins to elongate anteriorly as more epiblast cells accumulate in the streak region. After 12 to 13 hours of incubation, it has attained an intermediate stage, extending from the posterior margin to the center of the pellucid area. At this time, a longitudinal groove appears in the middle of the streak as epiblast cells become flask-shaped and invaginate into the blastocoele (Figure 12-17). Their basal ends are highly motile, whereas their apical ends remain fixed in the surface by junctional contacts. In this way, they pull neighboring cells into the groove, migrating in as a cohesive sheet. The streak area is equivalent to the blastoporal lip in the frog gastrula because endoderm and mesoderm invaginate at this point. Ingressing cells detach from the thickened inner layers in the groove and migrate into the blastocoele cavity between epiblast and hypoblast. The less adhesive cells immigrate in as mesenchyme and form a loose network. Where they make contact, they communicate by gap and tight junctions. Immigrating cells are replaced by more epiblast cells converging toward the streak area. These immigrating *mesentoblast* cells, the progenitors of both endoderm and mesoderm, also make contact with and migrate on hypoblast and epiblast surfaces. Once these invading cells establish junctional contacts among themselves, they begin to move as an organized sheet.

Meanwhile, the hypoblast expands anteriorly on the inner surface of the epiblast. Some hypoblast expands into the area opaca to become the extraembryonic endoderm, while other hypoblast cells attach to the mesentoblast and are carried along by the latter's migration. The combined effect is a rolling in of epiblast cells over the lips of the groove. The expansion pressures in the epiblast force the groove to become narrow and extended anteriorly. Meanwhile, the most anterior portion of the groove thickens into a depressed knot of cells known as *Hensen's node*. The

streak is now sharply delineated in the midline of the pear-shaped area pellucida, extending to three-fourths the length with primitive groove and primitive pit. This stage is known as the *definitive primitive streak* and is usually completed after 18 to 19 hours of incubation. Elongation of the primitive streak stops when no more epiblast cells are available to replenish the streak area. The streak then begins to shrink; the anterior end with Hensen's node recedes posteriorly. While Hensen's node recedes, presumptive notochordal cells in the node are released, migrating downward and anteriorly to extend the notochord along the principal embryonic axis. Lateral sheets of mesoderm are also laid down backwards like a carpet, as the primitive streak shortens and neurulation begins with the appearance of neural folds (Figure 12-18).

Initial elongation of the streak, followed by its recession and ultimate disappearance, resembles the morphogenetic movement of blastopore lips in the amphibian embryo, transposed from a three-dimensional sphere to a two-dimensional disc. As in the amphibian, a remnant of the primitive streak (blasto-

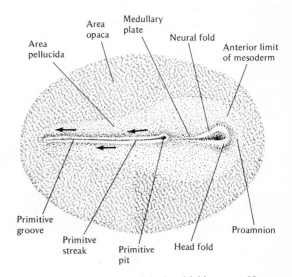

**Figure 12-18** *Formation of the head-fold stage at 19 hours of incubation. The primitive streak is regressing at this time (arrows) as the embryo anterior to Hensen's node begins to grow out. (From A. F. Huettner, Fundamentals of Comparative Embryology of the Vertebrates. Copyright © 1941 by Macmillan Publishing Co., renewed in 1969 by Mary A. Huettner.)*

pore?) is associated with the anus or cloaca in the chick. After the streak disappears, continuity between the three layers is also lost, and they become independent expanding sheets. The endoderm has a dual origin. The original hypoblast at the floor of the blastocoele makes a minicontribution to the gut; most endodermal tissue is derived from cells that migrated in at the primitive streak to become the roof and walls of the developing archenteron.

Spratt and Haas suggest that cell division in the posterior margin of the area pellucida contributes to endoderm formation. Their claim that pressure exerted by this rapidly proliferating zone or dominant growth center in the lower cell surface forces cells away from the mitotic center is based on experiments with unincubated blastoderms isolated in vitro. If a blastoderm is inverted so that the epiblast is in contact with an agar surface, the movement of exposed cells of the hypoblast layer may be followed by marking with carmine particles. Such movements display a fan-shaped pattern, as illustrated in Figure 12-16, and continue well into primitive streak stages. According to the hypothesis, a high mitotic rate in the growth center orients the movement of newly formed cell populations. The significance of proliferation in these morphogenetic movements is in ques-

tion because the posterior region of the pellucid area and later the primitive streak itself do not have higher mitotic activity than the rest of the blastoderm. No doubt cell proliferation is an important source of migrating cells, but sheet expansion does not depend upon localized regions of cell division activity.

Nevertheless, the posterior pellucid zone does have organizing properties; it seems to determine polarity of the embryonic axis (Figure 12-19). If the pellucid area of the blastoderm is rotated 180 degrees within the outer marginal border, the embryo that develops is oriented in the opposite direction. If, however, a central square within the pellucid area is cut out, without including any posterior tissue, and rotated, then the embryo assumes its normal orientation. Spratt and Haas believe that cells from the posterior region migrate out rapidly and dominate other regions by preventing their outgrowth.

The significance of the posterior region to normal morphogenesis is still not understood. The conditions of in vitro culture may set up artificial movements within the blastoderm; the blastoderm is lying on its back, so to speak, with its "belly" uppermost, free to expand. The organizing ability of a center is also canceled out when combined with other centers. It does not always behave like an individuation

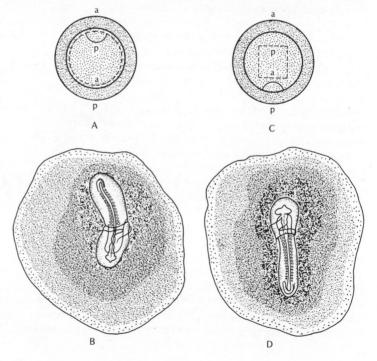

**Figure 12-19** *Evidence for the "dominant growth center" in the posterior of the unincubated early blastoderm. The center is indicated by the semicircular line in (A) and (C). (A) The pellucid area is cut from the area opaca and rotated 180 degrees so that posterior (p) and anterior (a) ends are inverted. (B) The resulting embryo develops with reversed polarity. (C) The central portion of the pellucid area is cut out and rotated 180 degrees, leaving the growth center in its original place. (D) The resulting embryo develops with normal polarity. (From N. Spratt and H. Haas, J. Exp. Zool. 147:57–94; 271–293, 1961.)*

field since a combination of three blastoderms with their dominant posterior centers abutting produces a complete spectrum of results, from no embryo to a system with three axes facing each other. If cell division is ruled out, the interactions observed must be explained by factors that organize the movement of expanding sheets.

All gastrulation movements are restricted to a very small area of the animal pole of this egg since sheet expansion is limited by attachment of the area opaca to the yolk. The marginal edge of the area opaca shows no tendency to spread over the yolk, as in the fish, but expands only gradually.

Gastrulation is completed at about 20 to 22 hours of incubation, at which point the blastoderm is in the so-called head process stage. The midline area of notochordal tissue has developed into a rigid rod, anterior to the receding primitive streak. As the streak regresses posteriorly, the embryo develops anterior to it, the axis of the notochordal rod behaving as the scaffolding around which organ systems develop. A fraction later (23 to 25 hours), primitive streak and Hensen's node regress even more, and a *head fold* develops in the epiblast region over the most anterior portion of the notochord and establishes the basic body plan.

By the end of gastrulation, three separate epithelial sheets have been established: a superficial ectoderm, a middle mesoderm, and an endodermal sheet attached to the yolk below. All organ-forming tissues have assumed their proper position and orientation for construction of the embryo, and fate maps made at this time indicate how the various regions are arranged within the three germ layers. By a series of local changes in cell and tissue motility, these cell sheets undergo morphogenesis into their respective embryonic organs.

The machinery for these morphogenetic changes resides in a contractile system probably composed of actin-myosin-like microfilaments oriented within motile cells (see Chapter 8). Contraction of these filaments may lead to shape changes or may be directly involved in movement of cells over substrata. The behavior of cell sheets is probably due to adhesions between cells via specialized junctions such as desmosomes. The latter may be part of a motile system since microfilaments attach to desmosome thickenings and extend into the cytoplasm. These may be anchoring sites for contraction. In addition, the scaffolding on which cells change shape is probably made up of oriented microtubules, a dynamic equilibrium system of assembly and disassembly of tubulin subunits (see Chapter 10). Individual motile properties of cells are thereby transduced throughout an entire sheet and translated into integrated tissue movements during organ formation.

## NEURULATION: TUBULAR NERVOUS SYSTEM FORMED BY AN OUTSIDE SHEET

Nervous systems are most highly developed in arthropods and chordates, and they exhibit similar patterns of morphogenesis. The nerve cord in each arises from a thickened and folded ectodermal sheet, ventrally in arthropod embryos and dorsally in chordates. Though the final outcome is different, neural tube formation in these two phyla entails similar morphogenetic processes. Development of the amphibian nervous system illustrates how a two-dimensional sheet transforms into a tube, the first organ system that is established in the vertebrate embryo.

### Neurulation in amphibians

The starting point for tubulation is the late gastrula, after the three primary sheets have developed. The process in the spherical gastrula of amphibians is typical of most vertebrates. Although modifications have evolved in disclike embryos of birds and mammals, neurulation in all forms follows a common plan. Some details of the process differ in anurans and urodeles, but these differences are only minor.

The anterior-posterior axis of the amphibian embryo (frog) is blocked out in the late gastrula, with the blastopore occupying the most posterior region (Figure 12-20). A dorsal-ventral axis becomes evident as neurulation begins, when the superficial ectoderm thickens in the dorsal region into a flattened neural plate. This is the medullary plate stage.

Subsequently, neural folds appear at the lateral margins of the plate, increasing in thickness as they slowly rise (Figure 12-21). A longitudinal depression, the neural groove, appears in the center of the plate and deepens as the folds rise and roll in toward one another. Meanwhile, the notochord below the neural ectoderm elongates and increases in diameter, thereby lengthening the embryo along the anterior-posterior axis. Notochordal cells become highly vacuolated and enlarge, which accounts for the rapid

**Figure 12-20** *Early neurulation in the frog embryo. (A) Partial section of late gastrula or yolk plug stage. (B) Partial section of medullary plate stage of beginning neurula; ectoderm is thickening dorsally over notochord to establish the medullary plate. (C) Partial section of a beginning neural fold stage; the neural folds appear on either side of the midline. (From A. F. Huettner,* Fundamentals of Comparative Embryology of the Vertebrates. *Copyright © 1941 by Macmillan Publishing Co., renewed in 1969 by Mary A. Huettner.)*

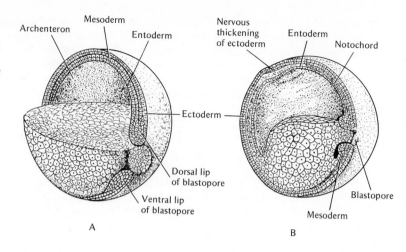

A                                                    B

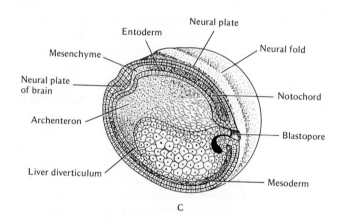

C

lengthening of the embryo during this period. The folds join into a hollow cylinder, the neural tube, which soon detaches from the superficial ectoderm.

The displacement of cells in the presumptive neural ectoderm has been studied by staining them with vital dyes and following them with time-lapse microcinephotography (Figure 12-22). These studies indicate that regions of the sheet expand in different directions at different rates. The most anterior median cells of the plate are displaced most rapidly in an anterior direction while more lateral cells sweep up toward the midline. Groups of cells in the posterior migrate more slowly; in fact, those in the posterior midline seem to behave as a "dead point" and remain relatively fixed. These coordinated movements tend to lengthen the embryo in an anterior-posterior

direction with a resultant change in surface area. Cells do not migrate independently and do not change contact relationships with their neighbors. The sheet behaves as a whole; each portion changes to conform to displacements occurring in other regions.

Undoubtedly these displacements involve directed expansion of different portions of the sheet. But how? Clearly, the low levels of mitotic activity in the neural plate cannot contribute significantly to the morphogenetic movements. What seems to be most important is the degree of cell elongation in different regions of the plate. The plate thickens, particularly at the lateral folds when cells elongate from a low columnar to a tall columnar epithelium (Figure 12-23). Displacements of cells seem to occur because cells change shape rather than migrate. The folds move toward one another as cells in the middle plate area become wedge-shaped; their apical ends become narrow while basal ends remain wide, and they assume a bottle shape, which favors rolling up of the plate. Eventually the folds join, first in the midtrunk region and finally in the anterior, or future brain. Fusion of folds and separation from the overlying ectoderm produce a neural tube beneath the body ectoderm.

What forces drive this dynamic morphogenetic process? Though this question is still unanswered, some of the mechanics of neurulation have been worked out at the ultrastructural level in anurans and urodeles. The primary events rely on cell shape changes, with cell adhesion and motility also playing an important part.

Because the neural plate in urodeles consists of one cell layer, the events of neurulation at the ultrastructural level are easier to describe; but in essence the phenomena observed here are similar in anurans *(Xenopus)* in which multiple layers participate.

Basically, neurulation depends upon forces generated within local cell populations of the neural plate by contractions of microfilaments on a labile microtubular cytoskeleton (Figure 12-24). Within the cytoplasm are microtubules and 50- to 70-Å diameter microfilaments, the former distributed at random while the latter are distributed in small bundles, usually near the cell periphery, just inside the membranes and at lateral borders. As neurulation begins, neural ectoderm cells elongate, a process that coincides with an increased number and orientation of bundles of microtubules along the longitudinal axis of the cell. Many apical microfilaments seem to attach to desmosomes that join cells into cohesive tissues. Meanwhile, in these same cells, apical microfilaments, 50 to 70 Å in diameter, increase in number and assemble into circumferential bundles around the apex parallel to the cell surface. These filaments attach to both sides of desmosomes in adjoining cells and produce an interconnected network over the neural plate. As neurulation continues, apical microfilament bundles thicken, a change coincident with the narrowing of the cell apex and formation of the so-called bottle cell. The synchrony of these two events suggests that microfilaments act as purse strings, contracting into a narrow ring at the apex.

Presumptive epidermis cells, which do not ac-

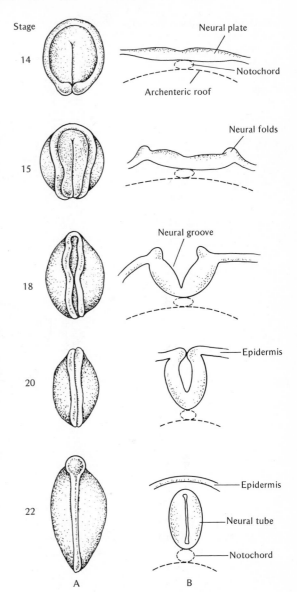

**Figure 12-21** *Formation of neural tube in the urodele Ambystoma. (A) How embryo appears at different stages (dorsal view). (B) Transverse sections of the neural tube region showing thickening of neural plate, appearance of folds, and fusion of folds to form neural tube. (From Early Morphogenesis and Pattern Formation in the Central Nervous System by Bengt Kallen in* Organogenesis, *edited by Robert L. DeHaan and Heinrich Ursprung. Copyright © 1965 by Holt, Rinehart and Winston. Reprinted by permission of Holt, Rinehart and Winston.)*

**Figure 12-22** *Analysis of neurulation in the urodele Ambystoma. A coordinate grid is superimposed over the dorsal surface of the embryos to facilitate tracking of pigmented cells by time-lapse cinematography. The transformation of the coordinate grid corresponds to the apparent displacement of cells over the surface during early stages of neurulation (stages 13–15). Subsequent analysis of these cell displacements indicates that differential changes in cell shape and not cell migration are responsible for the changes in the medullary plate. (From M. B. Burnside and A. G. Jacobson,* Develop. Biol. *18:537–552, 1968.)*

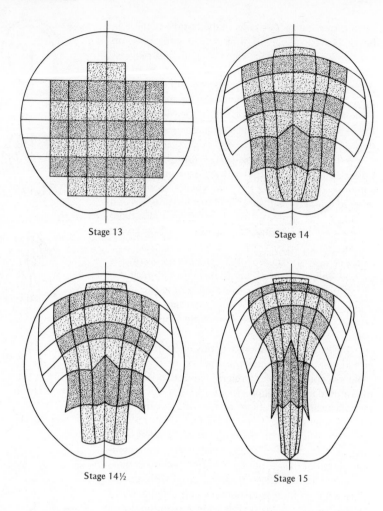

Stage 13          Stage 14

Stage 14½          Stage 15

tively participate in the rolling-up movements, show a different pattern of shape changes. They flatten, retaining a low columnar form throughout the process with microtubules oriented at random in the cell. Only a few thin 50- to 70-Å filaments are found in these cells; instead, thicker microfilaments 70 to 100 Å in diameter appear. These may have a structural rather than a contractile role, inasmuch as they attach to desmosomes across the cell to form an interconnecting network. Cells of the superficial ectoderm flatten into a squamous organization, a change coordinated by a network of desmosome contacts to which microfilaments or microtubules, or both, are attached.

If *Xenopus* gastrulae are treated with vinblastin sulfate or heavy water (agents known to interfere with microtubule assembly), presumptive neural cells become spherical and neurulation is completely inhibited. Microtubular integrity is essential to neurulation.

Movements within the neural plate are autonomous. Isolated pieces of salamander neural plate display similar movements and form neural tubes just as they do in the intact embryo. Folding is intrinsic to the cells at the edge of the plate. Extirpation of the neural plate between the folds has no apparent effect on neurulation because folds are brought together by wound healing movements instead. Likewise, extirpation of folds has no effect; as long as viable neural tissue is present, a neural tube will form. These shape changes occur in neural ectoderm as long as the underlying mesoderm is present.

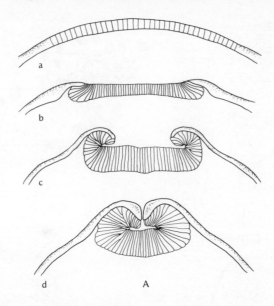

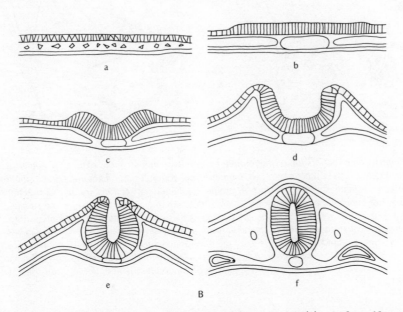

**Figure 12-23**   *Shape changes in neural ectoderm cells during neurulation. (A) Urodeles. (a) Stage-13 preplate presumptive neural ectoderm. (b) Stage-14 neural plate; note lengthening of plate cells results in thickening compared to lateral ectoderm. (c) In stage 16, neural folds move mediad. Elongation continues in center but at folds, wedge-shaped cells appear. (d) In stage 19, neural folds have met and superficial ectoderm is fusing to cover neural tube. Note large wedge-shaped cells. (B) Shape changes seen in chick neurulation are similar to those in urodeles. (a) Preplate blastoderm. (b) Elongating cells produce neural plate. (c) Early neural folds; note beginning of some wedge-shaped cells. (d) Early neural groove. (e) Late neural groove; wedge-shaped cells are numerous in base of groove. (f) Neural tube stage. (From P. Karfunkel, Int. Rev. Cytol. 38:245–272, 1974.)*

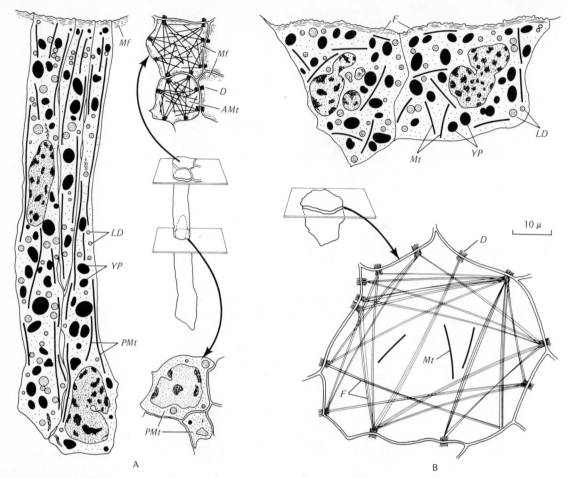

**Figure 12-24** *Comparison of orientation of microtubules and microfilaments in an elongating neural plate cell with a flattening epidermal cell. (A) Neural ectoderm cell. Numerous microtubules are aligned parallel to the long axis of the cell (PMt = paraxial microtubules). These are seen in profile in longitudinal sections and in cross sections in transverse sections of the cell. Just beneath the apical surface, microtubules are aligned parallel to the free surface in a layer across the cell apex (AMt = apical microtubules). In the same plane as these latter tubules, microfilaments (Mf) (50–70 Å) are organized as a circumferential bundle around the cell apex in a purse-string fashion. YP = yolk platelets; LD = lipid droplet; D = desmosome. (B) Epidermal cell. Beneath the free surface thick (70–100 Å) filaments (F) are arranged in discrete bundles which seem to be continuous in some places with desmosomal filaments and may span the cell from desmosome to desmosome (D). Microtubules (Mt) appear to be randomly oriented. (From B. Burnside, Develop. Biol. 26:416–441, 1971.)*

## SEGMENTATION AND ENLARGEMENT OF THE TUBE: BRAIN FORMATION

The vertebrate neural tube is a hollow cylinder running the full length of the embryo, larger in the anterior or cephalic end than it is in the posterior trunk. Even during neurulation, the future brain region is preindicated by an enlarged, broader groove and thicker, deeper folds. Specific regions of the future nervous system are determined at a neural fold stage; before this stage, the tissues are still labile. Parts of the presumptive spinal cord are interchangeable. Specification or *regionalization* of the nervous system, we will see, depends on the position of presumptive neural cells in relation to the chordamesoderm below.

After the tube is formed, the wall of the future brain constricts into several distinct compartments or *neuromeres,* best seen in chick and mammalian embryos (Figure 12-25). Many constrictions appear at first, but these diminish to three basic regions. The most anterior cavity or *prosencephalon* is the largest, ballooning out toward the overlying head ectoderm early in neurulation to become the rudiment of the forebrain. Behind it is another large cavity made up of several neuromeres which together form the *mesencephalon* or midbrain. Five smaller cavities posterior to the midbrain fuse to make up the hindbrain or *rhombencephalon,* continuous with the spinal cord. The brain develops within the superficial ectoderm known as a head fold, the most anterior portion of the chick embryo, while the remaining neural tube develops into the spinal cord. How does this tube acquire its axial organization and what is re-

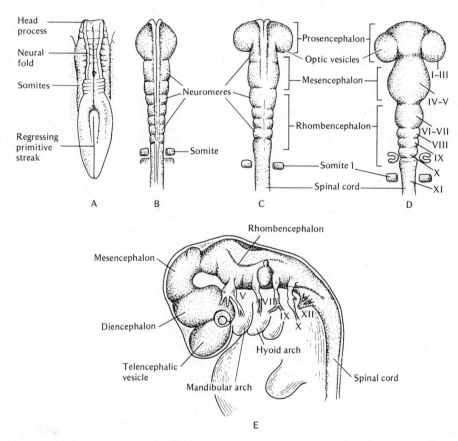

**Figure 12-25** *Development of the vertebrate brain regions in the chick embryo (A–D) and the mammalian embryo (E). (A) Dorsal view of embryo with 4 somites, about 24 hours of incubation. Neural folds have not closed over to form tube. (B) Dorsal view of brain of embryo with only 7 somites. Neural tube has closed and brain regions have been segmented into neuromere constrictions. Most anterior segment or prosencephalon is expanded as optic vesicles begin to grow out. (C) Dorsal view of brain of embryo with 11 pairs of somites. (D) Dorsal view of brain of embryo with 14 somites showing sites or origin of various cranial nerves. (E) Lateral view of mammalian embryo (at a later stage than D) showing flexion of head and brain and arrangement of different regions. Disencephalon and telencephalon are derived from the prosencephalon by constriction. Roman numerals mark cranial nerves. (A–D based on O. E. Nelsen,* Comparative Embryology of the Vertebrates. *© 1953 by Blakiston Company, Inc. Used with permission of McGraw-Hill Book Company.)*

sponsible for its regionalization into brain and spinal cord?

## Fate maps: the problem of determination

Vital dye-marking experiments indicate how tissues are folded into the principal body axis of the embryo. We find that anterior-posterior polarity is established by the direction of chordamesoderm invagination during gastrulation. This germ layer involutes over the dorsal lip and expands anteriorly beneath the future head of the embryo. The prospective fate of various regions of the neural plate can also be mapped by staining with vital dyes. Large dye marks map the position of future brain regions while smaller marks furnish a detailed map of sense organs and sensory or motor areas of the brain.

Mapping techniques are helpful in following the movements and ultimate position of local tissues, but they tell us nothing about tissue developmental potentialities during morphogenesis. During gastrulation and neurulation tissues become irreversibly committed to their prospective fate; that is, they are determined. Determination is defined operationally (see Chapter 17). A tissue is removed from its normal relationships in the embryo and challenged by a different environment; either it is transplanted to a foreign region in another embryo, or it is explanted to a neutral environment in vitro. If a fragment of tissue develops into its prospective fate (as determined by vital dye experiments), irrespective of its new surroundings, then we say the tissue is determined; it has acquired a developmental bias that is unaffected by exogenous influences. On the other hand, if a tissue does not differentiate into its prospective fate but develops in accordance with its new environment, then the tissue fragment is undetermined; it is still labile and responsive to morphogenetic stimuli. Using these techniques of transplantation and explantation we can evaluate the developmental potential of parts of germ layers during gastrulation and neurulation.

Grafting a tissue fragment into another region of the same embryo is defined as an *autoplastic transplantation* whereas exchanges between embryos of the same species are called *homoplastic* transplantations. Instead of using tissue of the same species, where host cannot be distinguished from graft, researchers often use embryos of different species in which host and donor are distinguished by pigmen-

tation, nuclear staining, or cell size. In such cases, tissues exchanged between different species, but within the same genus, are known as *heteroplastic* transplantations; exchanges between different genera—for example, frogs and salamanders—are called *xenoplastic* transplantations.

Since transplantation is one of the principal approaches in studying the fate of embryonic tissues, we should point out an immunological problem that may arise in these experiments. Vertebrates possess an immunological system that recognizes foreign proteins, either as soluble molecules in the circulation or as carried in by cells in tissue or organ grafts (see Chapter 18). Hence grafts of tissues between juveniles or adults of the same species are usually rejected because grafted cells are recognized as foreign by the host's immunological mechanisms. But grafts of embryonic tissues do survive in most instances. One reason is that embryos and young larvae do not have a functional immune mechanism at the time of grafting. The immune system, like all other systems, must mature, a process of varying length in each species. In most amphibians, it becomes fully mature in late swimming tadpoles. Any tissue graft heals in and grows normally in foreign embryo hosts (even those from distant genera) because the host immune system is immature and unable to respond. In most cases the results of the experiment are obtained before the host immune system matures and rejects the foreign tissue.

Sometimes, rejection can be prevented if a large graft is used; the large amounts of tissue antigens tend to swamp out the host immune system, which then becomes tolerant to the foreign tissue. Such foreign grafts may survive well into the adult. The closer the relationship between donor and host, the greater the tendency for such grafts to be accepted. Xenoplastic transplantations usually result in the earliest graft rejection, with heteroplastic and homoplastic transplantations following, in that order.

## Determination in blastula and early gastrula

We know from fate maps that the presumptive ectodermal area of the late amphibian blastula becomes larval skin. Nevertheless, the transplantation test proves that its prospective fate is labile; it is not rigidly determined but responds to influences from the surrounding tissue environment. If presumptive ectoderm is transplanted into the prospective neural

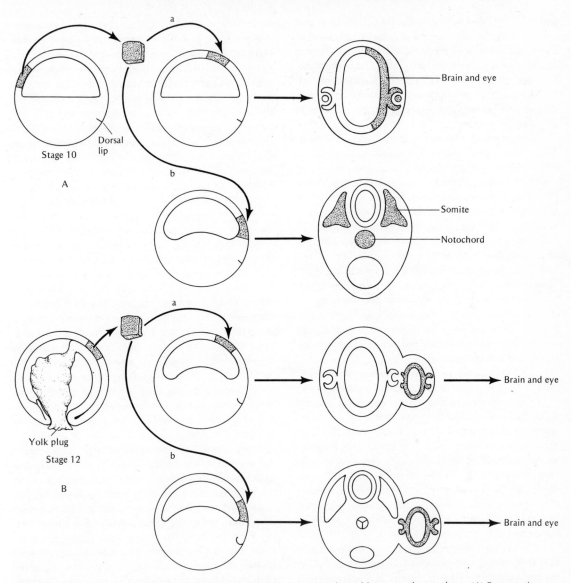

**Figure 12-26**   *Prospective fate and prospective potency of early gastrula and late gastrula ectoderm. (A) Presumptive epidermis of the early amphibian gastrula (stage 10) is grafted onto the presumptive neural ectoderm region of an early gastrula (a) or the chorda region (b). In subsequent development the graft integrates with the tissues of the region and develops accordingly, as brain and eye in the first instance and as notochord and somites in the second. (B) The presumptive neural ectoderm from a late gastrula (stage 12) is grafted into similar regions of early (or late) gastrula, but in this instance the graft develops only into its prospective fate, neural and eye tissues irrespective of the region of the embryo. In this instance its prospective potency is the same as its prospective fate; the tissue is determined.*

region, it develops as neural tissue and becomes incorporated into the newly forming brain (Figure 12-26). If placed in the lateral marginal zone of a gastrula, a mesodermal region, it will be converted into mesodermal derivatives. Finally, if inserted in a dorsal lip region, it can reorganize into notochord, somites, and even neural tissue. Pieces of the mesodermal marginal zone are equally labile; a presumptive mesodermal region can develop into a variety of ectodermal structures, including skin and neural tube. In other words, the blastula sheet is undetermined; its wide range of developmental behavior, known as prospective potency, far exceeds prospective fate of the blastula ectodermal sheet.

The situation in the early gastrula is very similar. The classic cases involve heteroplastic transplantations between newt species *Triturus cristatus* and *Triturus taeniatus*. *T. cristatus* embryos are completely devoid of pigment, whereas *T. taeniatus* cells contain black pigment granules. In this way graft and host tissues can be distinguished. At the beginning of gastrulation, just as the dorsal lip appears, transplantation of presumptive *T. taeniatus* epidermis and neural ectoderm to the ventral regions of a *T. cristatus* gastrula always results in a "regionwise" developmental pattern; that is, each piece develops according to the region in which it was transplanted. Like blastula epithelium, these tissues at the beginning of gastrulation are still labile and undetermined; prospective potency still exceeds prospective fate.

Within a matter of hours, however, this potentiality is lost. In a mid to late gastrula, transplantation of a piece of presumptive neural ectoderm into an epidermal area leads to development of brain or spinal cord instead of skin. Transplants of late gastrula epidermis into presumptive brain areas develop as skin and not as brain, although they may invaginate along with it. At this stage, as we have seen (Chapter 10), presumptive neural tissue now recognizes epidermis as "foreign" and sinks into it without forming a cohesive union; their surface properties have changed.

During gastrulation, tissues change from a labile to a determined, autonomous state and develop independently of their surroundings; their prospective significance is the same as their prospective fate. Do these changes occur intrinsically, or do they depend upon exogenous influences from surrounding tissues?

To answer this question, J. Holtfreter devised a simple salt solution that would support development of tissue fragments in vitro, completely isolated from neighboring tissues and other host influences. Initially, Holtfreter observed that early amphibian gastrulae explanted in hypertonic saline solutions gastrulated abnormally. Instead of invaginating and migrating into the blastocoele cavity, endoderm and

mesoderm migrated out into the medium away from the overlying ectoderm; that is, the embryo exogastrulated. A wrinkled mass of presumptive ectodermal tissue completely separated from the invaginating endomesoderm. Both continued to develop, but only the endomesoderm differentiated into identifiable derivatives such as muscle, notochord, and gill. The presumptive epidermis and neural ectoderm did not differentiate. Though endoderm and mesoderm could differentiate autonomously, ectodermal derivatives seemed to be dependent upon interactions with neighboring tissues. Evidently ectodermal tissues did not become determined under these conditions.

When fragments of presumptive epidermis from an early urodele gastrula are isolated in isotonic saline, they exhibit minimal differentiation; either they round up into spherical, pigmented masses, or they spread on a glass substratum as a simple epithelium (Figure 12-27). No characteristic skin tissues differentiate. Likewise, no brain tissues differentiate from isolated presumptive neural ectoderm, which suggests that they, too, are undetermined. Nevertheless, these same tissues display a wide range of developmental capacities when transplanted into another embryo. Are in vitro conditions unfavorable for differentiation, or do fragments of ectoderm require specific conditions furnished only by neighboring tissues in a host embryo that cannot be supplied in vitro?

We know that the saline solution can support differentiation since other tissues from the early gastrula do self-differentiate when explanted. A dorsal lip fragment from an early gastrula develops into notochord, somites, and even some neural tissues when cultured in the same solution. Such fragments are heterogenous, containing cells from each germ layer. Development of these isolated tissues exceeds their normal behavior in situ. Marginal tissues of the gastrula behave as a diffuse individuation field in which prospective significance exceeds prospective fate. These results suggest that a saline solution can support differentiation if a tissue is already determined.

The diversity of tissues developing from isolated dorsal lips seems to depend on total tissue mass. If varying numbers of dorsal lip fragments are fused into single explants and cultured in vitro, their developmental potentialities correlate directly with the total tissue mass. A single fragment develops into a limited amount of notochord and somites with

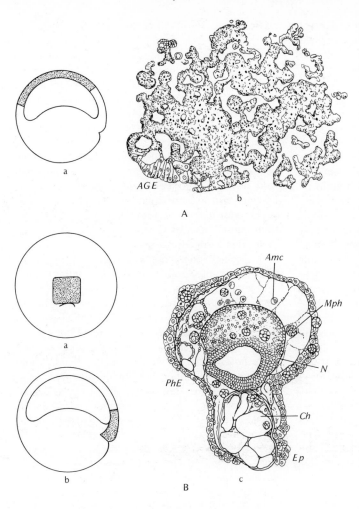

**Figure 12-27** *Comparison of developmental behavior of explants of early gastrula presumptive neural ectoderm* (Rana esculenta) *with dorsal lip when explanted into a salt solution (Holtfreter's solution). (A) Presumptive ectoderm explanted is shown stippled in (a), and in (b) a loose mass of atypical ectoderm develops with some adhesive gland epithelium (AGE). (B) Stippled regions in (a) and (b) indicate two views of dorsal lip. In (c) are seen some of the tissues derived. PhE = sac of pharyngeal epithelium; Ep = epidermis; Ch = chorda; N = neural vesicle; connective tissue contains amoebocytes (Amc) and macrophages (Mph).* (From A. Kühn, Lectures in Developmental Physiology. *New York: Springer-Verlag, 1971; based on J. Holtfreter,* Wilhelm Roux' Arch. Entwicklungsmech. Organ. *138:657–738, 1938.)*

some neural tissue, but as the number of fragments increases, more neural, ectodermal, and even endodermal derivatives are obtained, until almost perfectly formed embryos with complete axial organizations are seen after 10 fragments are combined. The potentialities of this region seem to be far greater than expected. On the other hand, from the vital dye fate maps increasing tissue mass means more contaminating ectoderm and endoderm cells in the explant, and more ectodermal derivatives should develop. In contrast to presumptive ectoderm, dorsal lip regions are determined early and self-differentiate in vitro.

The dorsal marginal zone (as we have already pointed out) acquires this property after interacting with the endodermal region. In fact, it acquires the property to self-differentiate notochord and muscle,

very early, during cleavage. Nakamura and his students have studied the fate of isolated blastomeres taken from the dorsal marginal zone at different stages of cleavage in newts and in anurans (Xenopus). They find that blastomeres taken from the 32-cell stage and grown in a salt medium do not differentiate any notochordal or muscle tissue. At the 64- to 128-cell stage, a small percentage of the isolated blastomeres will form some notochord and myotome, but it is only later when the embryo contains about 100 cells that isolated blastomeres are able to differentiate muscle. It seems that the dorsal marginal zone does not have the property to self-differentiate into notochord shortly after fertilization but must acquire this property after interacting with yolky vegetal-half cells in the future endoderm. Once having acquired its self-differentiation tendencies in early cleavage, the dorsal marginal zone begins to behave as a strong morphogenetic center. Differentiation of other early gastrula tissues (including endoderm) now becomes dependent upon interactions with the dorsal marginal zone.

This is seen in the behavior of explants of presumptive epidermis and neural ectoderm taken from late gastrula embryos. At this stage, most ectoderm has been underlain by and interacted with invaginating endomesoderm. Isolates of presumptive epidermis still behave as before; they develop into an irregular, atypical epidermis. We know they have been determined because explants isolated with a small amount of attached mesenchyme do develop into well-differentiated epidermis. This determination is labile; transplanted ectoderm still reacts to stimuli from different embryonic regions but will only differentiate into various sense organ components (lens, olfactory and auditory placodes). By the end of gastrulation, presumptive epidermis has acquired self-differentiation capabilities and develops only as epidermis.

Presumptive neural ectoderm from late gastrula is more determined. Isolates in saline now differentiate into portions of the nervous system; only modified brain vesicles develop, including some pigment cells. The capacity of self-differentiation is limited, but as development continues into later stages (early neurula), isolates of presumptive neural ectoderm do self-differentiate into neural structures. It appears that the self-differentiating properties of the presumptive neural ectoderm are stabilized in stages: (1) late gastrula, when poorly differentiated brain vesicles and pigment cells develop, (2) early neurula, with the appearance of some forebrain and eye structures, and (3) mid to late neurula, when neural ectoderm is fully determined and virtually all regions of the nervous system will differentiate.

In summary, during gastrulation the developmental potential of a germ layer diminishes; a labile, responsive ectodermal sheet is progressively restricted to its prospective fate. The behavior of presumptive ectoderm suggests that it loses developmental potential as it acquires the ability to self-differentiate. The process of determination in the vertebrate embryo is accompanied by restriction and stabilization of specific developmental programs in local regions of a cell sheet.

## Regionalization of the nervous system

We return to our question of how a simple neural tube is transformed into a nervous system with brain, sense organs, and spinal cord. This is the problem of regionalization. From the moment of closure of the blastopore at the end of gastrulation, different parts of the future neural axis of the embryo are determined. At first, early medullary plate, well before neural folds thicken, is committed in a general way to develop into neural tissue. Though not fully regionalized, its anterior-posterior axis has been established. Fragments of anterior neural ectoderm transplanted to posterior regions of a host fuse and cooperate to form normal nervous systems, as long as they retain their original anterior-posterior polarity. Brain-spinal cord specification has not occurred at early medullary plate stages; they are interchangeable at this time.

If large fragments of anterior neural plate are rotated 180 degrees in place (with or without the underlying mesoderm), they retain their original anterior-posterior polarity and differentiate into brain regions in reversed orientation, irrespective of their position in the host (Figure 12-28). On the other hand, changing lateral-medial orientation by transplanting anterior plate fragments between embryos results in normal development. Anterior-posterior polarity of the neural axis is established before medio-lateral polarity and is an intrinsic property of the neural ectoderm at an early neural plate stage, before folds appear. Later, after folds appear, lateral-medial fate is stabilized and determined. Neural ectoderm acquires polarity and develops autonomously, irrespective of the orientation of the mesoderm substratum.

If small neural plate fragments are rotated instead, even at neural fold stages, they develop in conformity with the surrounding region. Small grafts develop into normally polarized neural elements after rotation. Presumably size affects polarity; large areas of the neural ectoderm seem to be determined, whereas smaller cell populations within such areas continue to respond to new axial cues. The intrinsic polarity of small fragments may be swamped out within the polarity of a larger surrounding tissue.

All presumptive neural ectoderm of an early neural plate stage is pluripotent; any portion can develop into brain and spinal-caudal structures. Potency soon declines, however, as each region is committed to a more limited developmental expression. Anterior explants of late neurulae (late plates and early neural tube stage) develop into brain, while more posterior regions develop only as spinal cord. The sheet acquires a mosaic property, with each region destined to produce specific neural structures.

Wherein does regional specificity arise, in the responding ectoderm or in the chordamesoderm substratum? We have already seen that once determined, neural ectoderm is autonomous and develops independently of the underlying mesoderm. But is this true at all stages?

In Chapter 11 we learned that the future neural axis of the amphibian embryo is fixed at fertilization when the gray crescent forms. This region, the future site of dorsal lip formation, becomes a morphogenetic center during cleavage. In the following sections, we will see how the dorsal lip region is involved in neural axis formation.

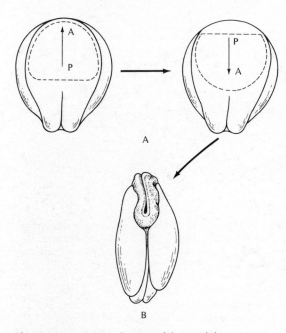

**Figure 12-28** *Regionalization of the amphibian nervous system. Determination of the anterior-posterior axis. (A) Rotation of a large portion of the medullary plate at an early neurula stage in* Ambystoma. *The plate between the neural folds in a stage-14 neurula is rotated 180 degrees. (B) At the end of neurulation, the graft has produced a neural tube in reversed orientation. (Based on F. C. Roach,* J. Exp. Zool. *99:53–77, 1945.)*

## Neural induction

The ability of the amphibian dorsal lip region to organize a neural axis was first demonstrated in Spemann's laboratory in 1924. Its properties were discovered in experiments designed to test the developmental potency of gastrula tissues. We have already seen that fragments of presumptive epidermis and neural ectoderm exchanged between two early gastrulae *(Triturus)* develop normally according to the site. Developmental potency of the dorsal lip was studied in a similar manner. In the original experiment by Hilde Mangold, an associate of Spemann, a piece of *T. cristatus* dorsal lip was transplanted near the lateral lips of a blastopore of a *T. taeniatus* host gastrula in contact with presumptive ventral ectoderm (Figure 12-29). The less pigmented graft tissue was incorporated into the host embryo, where it continued to develop into mesodermal organs, notochord and somites. The host developed a normal neural axis but at the same time, in the lateral region of the graft, a secondary neural plate appeared containing neural folds derived from host ectoderm which developed into a secondary embryo. The notochord of the secondary embryo consisted of graft cells, but the somites contained both graft and host tissues. The neural tube contained only a small number of graft cells; most of its cells and those of the kidney tubules, endoderm of the gut, and some sensory structures came from host ventral ectoderm. *T. taeniatus* ventral ectoderm tissues had been altered by interaction with the chordamesoderm graft and had differentiated as neural tissue rather than as skin.

This phenomenon of primary embryonic induction, where the developmental fate of ectoderm is altered by interactions with dorsal lip mesoderm, won the Nobel Prize for Spemann in 1926. It led to a feverish period of activity in many laboratories since it was the first real break in efforts to understand how the embryo developed. Later, it was shown that inducing potentiality is limited to the dorsal lip region. Other fragments of a gastrula fail to induce a neural tube; only presumptive mesoderm is active. Spemann introduced the term "organizer" to describe the developmental potency of the dorsal blastoporal lip. It has a dual property: (1) self-organizing into an embryonic axis with notochord and somites and (2) inducing overlying ectoderm to organize into a neural tube.

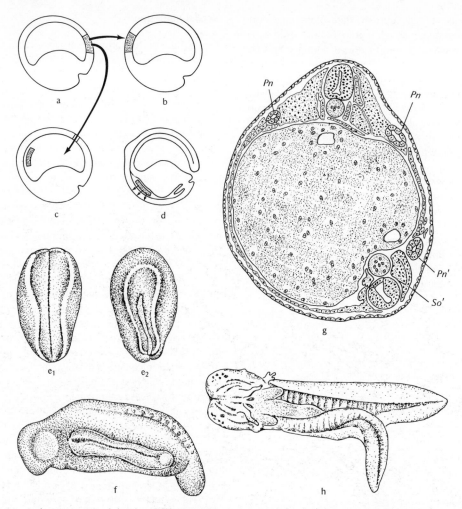

**Figure 12-29**  *Inductive action of the dorsal lip. Transplantation of the dorsal lip (a) into ventral ectoderm (b) or into the blastocoele (c). Position of implant next to ectoderm at the end of gastrulation is shown in (d). A dorsal view of primary neural plate (Triturus taeniatus) is given in (e₁), whereas (e₂) gives the ventral view of secondary neural plate with a pale, longitudinal, stretched, implanted strip of T. cristatus tissue. In (f) is the secondary embryonic axis with ear vesicles at front of neural tube, somite rows, and tail bud (most tissues are derived from host T. taeniatus). (g) is a cross section through (f) showing secondary embryonic axis at lower right with chorda, induced somite, and neural tube. In (h) an embryo with well-developed secondary twin is shown. Graft tissue is light compared to darker host. (Pn = pronephric tubules, which are entirely of the host; Pn' = pronephric tubules of secondary embryo; So' = somite of secondary embryo.) (From A. Kühn,* Lectures on Developmental Physiology. *New York: Springer-Verlag, 1971; based on H. Spemann and H. Mangold,* Wilhelm Roux' Arch. Entwicklungsmech. Organ. *100:599–638, 1924.)*

Most early gastrula ectoderm responds to the organizer by developing into neural tissue, but as gastrulation proceeds, ectodermal responsiveness or competence to form neural structures fades. *Competence*, a term coined by Waddington in 1932, refers to the ability of a tissue to exhibit a specific morphogenetic change in response to an inducing stimulus. As an operational definition, it does not describe any recognizable physiological state. In neural induction, competence is defined as the ability of ventral ectoderm to respond to an inducing signal and develop into neural tissues. It is studied by implanting inducers (dorsal lip) into ectoderm of progressively older embryos or by using the ectoderm fold procedure. Pieces of ventral ectoderm from progressively older gastrula stages are folded and transplanted to chordamesoderm or brain regions of a host neurula (Figure 12-30). Although exposed to the same inductive stimulus, these "aging" pieces of ectoderm are less responsive when taken from older embryos. The youngest fragments differentiate anterior brain structures with sense organs (archencephalic) while fragments from late gastrulae give only limited neural response from more posterior (deuterencephalic) brain structures and eventually form only epidermis. Neural competence is a temporal property appearing in ectoderm only after blastulation, reaching a peak in the mid-gastrula and disappearing by the end of gastrulation. Competence may reflect only the other side of the coin of determination; once a cell sheet is committed, it loses competence and its opportunity to carry out other programs is reduced.

Two cell sheets, invaginating chordamesoderm and overlying ectoderm, interact, and as a result the neural axis of the embryo is determined. As presumptive notochord invaginates, it glides anteriorly and laterally along inner surfaces of the ectoderm, separated by a progressively changing extracellular matrix of sulfated mucoproteins. The ectoderm undergoes a spatio-temporal pattern of determination and becomes specified into different regions of the nervous system. Neural induction has been demonstrated in most vertebrate embryos, and in each case chordamesoderm is the inducer.

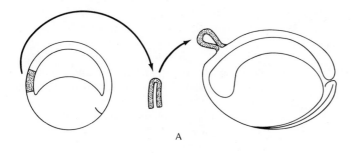

A

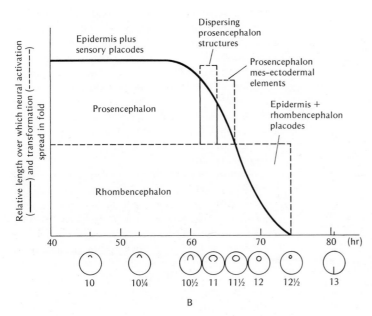

B

Figure 12-30  *Test of ectodermal competence (Ambystoma) by the ectodermal fold technique of Nieuwkoop. (A) Operative procedure. A graft of ventral ectoderm from embryos at different stages of gastrulation is removed, folded into a sandwich, and transplanted over the rhombencephalic region of an early neurula. The neural tissue has strong inductive capabilities, and after some development the graft is examined for the extent of neural induction. (B) A composite curve based on many such experiments showing how competence for neural induction diminishes during gastrulation. Note that by the end of gastrulation, the ventral ectoderm is no longer competent to respond to the strong neural inducing influence of the neural tube. Stages of gastrulation are shown below. Forebrain competence diminishes sooner than hindbrain competence. (From P. D. Nieuwkoop, Biological Organization, Cellular and Subcellular, ed. C. Waddington. Elmsford, N.Y.: Pergamon, 1959, pp. 212–230.)*

Spemann's early experiments indicated that regional properties of the neural ectoderm result from interactions with chordamesoderm located in the roof of the archenteron. If different portions of a neurula archenteron roof are transplanted into host gastrulae, we see that inducing ability is itself regionalized at this stage (Figure 12-31). Anterior archenteron roof usually induces forebrain structures such as telencephalon and eyes, a phenomenon called *archencephalic induction*. Posterior fragments, on the other hand, always induce more spinal cord and trunk mesodermal structures, or *spino-caudal inductions*. Regionality of the neural plate is a property of the underlying chordamesoderm and seems to be acquired as mesoderm invaginates during gastrulation.

These regional differences are acquired gradu-

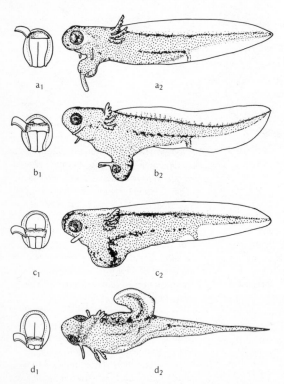

$a_1$    $a_2$

$b_1$    $b_2$

$c_1$    $c_2$

$d_1$    $d_2$

**Figure 12-31**  *Regional specific induction by implantation of different regions of the archenteric roof of early* Triturus *neurula ($a_1$–$d_1$) into early gastrulae. ($a_2$) Head with balancers. ($b_2$) Head with eyes and forebrain. ($c_2$) Posterior part of head with mid-hindbrain and otic vesicles. ($d_2$) Trunk and tail induction. (Based on O. Mangold,* Naturwissenschaften *21:761–770, 1933.)*

ally; blastoporal lip tissue itself undergoes changes in inductive properties during gastrulation. Because of the pattern of chordamesoderm invagination in urodele gastrulae, the first tissue to invaginate at the dorsal lip, the prechordal plate, migrates furthest and ultimately occupies the most anterior portion of the archenteron roof. Later, as more chordamesoderm rolls over the blastoporal lip, more posterior portions of chordamesoderm occupy the lip region. These correspond to trunk-inducing tissue of the medullary plate stage, and finally, toward the end of gastrulation, tissue at the dorsal lip is occupied only by presumptive tail mesoderm which behaves as a tail (spino-caudal) inducer. Thus the inducing power of dorsal lip regions changes during gastrulation as mesoderm sheets involute. Initially, in the early gastrula the dorsal lip consists mostly of prechordal plate and possesses both archencephalic and some spinocaudal inductive properties. In the late yolk plug gastrula, dorsal lip has no archencephalic inductive potential; it induces spino-caudal structures. Meanwhile, the invaginated prechordal plate loses its spino-caudal inducing properties, and as part of the archenteron roof in the head, it induces only archencephalic structures. By the time invagination is complete and the archenteron formed, the most anterior chordamesoderm induces forebrain while the remainder induces trunk and tail structures.

These temporal changes are intrinsic to the invaginating tissue and do not depend on its displacement during gastrulation or on any other influences of the host environment. We know this because isolated dorsal lips show similar changes in inductive capacity during a short period of culture. As dorsal lips mature in isolation, specific inductive properties qualitatively change. Dorsal lips cultured for short periods before implantation into early gastrulae induce mostly archencephalic structures. If they are cultured for longer periods, they can only induce spino-caudal structures.

Another temporal factor, specifically, the length of time an ectodermal region is exposed to invaginating chordamesoderm, also affects its developmental expression. To understand this temporal quality, imagine invagination as a gradual folding of one edge of a sheet on itself and moving underneath toward the other edge (Figure 12-32). Point X on the upper layer will have been exposed to the invaginating lower layer for a longer period of time than point Y at the far edge. The duration of interaction between an-

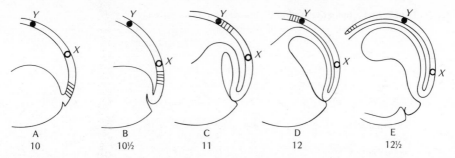

**Figure 12-32** *Timing as a factor in neural induction. (A–E) Changes during gastrulation as chordamesoderm invaginates underneath presumptive neural ectoderm and glides anteriorly. A more posterior point X interacts with the invaginating tissue for a much longer time than an anterior point Y. (Diagonal lines in neural ectoderm indicate region of highest mitotic activity at the stage shown.)*

terior ectoderm and underlying chordamesoderm is much shorter than it is for posterior ectoderm. Besides, posterior ectoderm is first to interact with prechordal plate while it still possesses spino-caudal inductive properties and continues to receive these same signals as more mesoderm involutes. Anterior ectoderm has only received archencephalic signals by the time neurulation begins. To complicate matters even further, competence of presumptive neural ectoderm changes during invagination so that some regions become less competent to respond to archencephalic or spino-caudal inductions. We see that regional determination is a complex phenomenon involving interactions between chordamesoderm, with maturing inductive properties, and ectoderm which becomes less competent. These temporal changes are also seen in the changing patterns of mitotic activity during invagination. As measured by thymidine incorporation into DNA, cells in the ectoderm directly above the advancing tongue of chordamesoderm synthesize DNA more actively than surrounding cells. Accordingly a wave of proliferation passes over the presumptive neural ectoderm as the tip of chordamesoderm advances anteriorly into the embryo. It appears as if local populations of ectoderm directly above the chordamesoderm are induced to complete a division program before they become determined as neural tissue. This wave of mitotic activity in neural ectoderm polarizes the sheet along a posterior-anterior gradient, an axial property that is retained by neural ectoderm in later stages.

Determination of an epithelial sheet depends upon interactions between two adjacent tissues that have established an intimate relationship during mor-

phogenetic displacements. These tissues communicate, the inducing tissue somehow furnishing conditions that evoke the prospective fate of a responding tissue. Inductive interaction is the principal paradigm of embryonic determination. Most vertebrate organ systems emerge as a consequence of such tissue interactions.

### Analysis of primary induction

Spemann's experiments and the "organizer concept" that emerged in the 1920s and 1930s seemed to provide the key to embryonic organization. Though much information has accumulated since then, the mechanism of primary induction still eludes us. Interactions between chordamesoderm and overlying neural ectoderm are probably the most complex of all inductive systems and the least accessible to analysis. Today the problem elicits only a modest effort as interest centers on simpler secondary inductive systems involved in organ determination (Chapter 17).

Over the years, however, a conceptual framework on which to hang our ideas about primary induction has emerged. It is this framework that will be reconstructed in the following sections as we attempt to answer some major questions about primary induction. These same questions are applicable to virtually all inductive interactions.

*Question 1:* Is primary induction specific for the tissues involved, and if so, wherein does specificity reside — in inducing or in responding tissues?

Up until now we have seen that only the dorsal lip can induce, but we know from some early experiments by Spemann that inductive potential is not re-

**TABLE 12-1    Inductive Action of Some Heterologous Tissues[a]**

| Tissue | Type of Induction | | |
| --- | --- | --- | --- |
|  | Archencephalic | Deuterencephalic | Spino-caudal |
| Guinea pig |  |  |  |
| Liver | ++ | + | − |
| Heart muscle | ++ | ++ | + |
| Kidney | (+) | ++ | +++ |
| Bone marrow | − | − | +++ |
| Mouse |  |  |  |
| Adult liver | +++ | ++ | + |
| Adult kidney | − | ++ | +++ |
| Embryonic liver | +++ | + | (+) |
| Embryonic kidney | ++ | +++ | + |
| Rat |  |  |  |
| Retina | − | + | ++ |
| Bone marrow | − | (+) | +++ |
| Leukemic bone marrow | − | − | − |
| Frog |  |  |  |
| Retina | + | − | − |

[a] Inductive action tested in implantation experiments.
Source: Modified from table in L. Saxen and S. Toivonen, *Primary Embryonic Induction.*
London: Logos Press, 1962. Reprinted by permission of Prentice-Hall, Inc. Englewood Cliffs, New Jersey.

stricted to the dorsal blastoporal tissue. Tissues from late embryos also possess inductive potentialities. Pieces of neural plate and neural tube, for example, when implanted into the blastocoele cavity of an early gastrula, induce neural tissue to the same extent as dorsal lip. Spemann called this process *homeogenetic induction;* as ectoderm becomes neural plate, it acquires regional inductive properties. As we have seen, anterior plate regions possess archencephalic tendencies, whereas posterior pieces of the plate tend to elicit a greater proportion of trunk and tail inductions. The regional inductive properties of the underlying chordamesoderm are somehow transferred to overlying neural ectoderm.

Adult tissues, even those from foreign species such as mouse liver and kidney, also exhibit strong inductive capacities. In effect, practically all adult tissues tested, from invertebrates to man, could induce some degree of neuralization in gastrula ectoderm. These results suggest that (1) the inducing signal is not specific, or (2) some intrinsic tendency toward neuralization is triggered in the responding ectoderm by a variety of tissue substances, or (3) all

adult tissues retain the same inducing signal found in the chordamesoderm.

A further complication arose when it was found that different tissues possessed regional specificity of induction. Some, like mouse liver, were strong archencephalic inducers, producing brain and sense organs; others, such as rat bone marrow, were strong spinocaudal inducers. Moreover, the specificity of induction seemed to change with the age of the inducing tissue and its physiological state. Embryonic mouse kidney produced proportionately more archencephalic inductions than did adult kidney, which induced more spino-caudal structures (Table 12-1). This suggests that more than one inducing signal exists in adult tissues.

In general, the primary inductive signal lacks specificity except for those archencephalic and spinocaudal inductions noted above. Later in development, when eye, ear, and other sense organs are induced, some specificity is noted; only specific tissues can induce. On the other hand, specificity does reside in the reacting ectoderm; its morphogenetic display is unchanged regardless of the inducing stimulus.

This is best seen in xenoplastic inductions—transplantations of inducers between different amphibian species. For example, newt embryos develop balancers on either side of the head, ectodermal organs that arise after a secondary induction by underlying mesoderm. In frogs, head ectoderm develops a pair of adhesive structures or suckers instead. If competent frog ectoderm is transplanted to the head region of a newt gastrula, it responds to the newt inductive signal by forming adhesive organs. In the reciprocal transplantation, the newt ectoderm responds to the frog inductive signal by forming balancers. In all cases, specificity of ectoderm responses is in accordance with the ectoderm's genetic background and not with the quality of the host inducing signal. A common inducing signal may exist in all amphibian species, but the specific nature of the response is genetically programmed in the responding ectoderm.

*Question 2:* Does inductive capability depend upon viable, intact cells? Since a variety of embryonic and adult tissues have inductive capability, we now ask whether this is a property of living cells. Early in the analysis of induction, however, it was shown that dead organizer tissue can induce as well as the living implant. Pieces of alcohol-killed blastopore lip implanted into a gastrula produced typical inductions. Once it was realized that intact tissues are unnecessary for induction, the next step was to test extracts and subcellular fractions from embryonic and adult tissues. These also displayed a full range of inductive potentialities. Cell constituents, in addition to intact cells, may be active, suggesting that the inducer is a chemical substance passing from inductor to responding ectoderm.

*Question 3:* If the inducer is a substance produced by the inducing tissue, is it transferred via diffusion or by direct cell contact?

Two hypotheses have been proposed for the mechanism of interaction between inducer and responding tissues: the diffusion hypothesis originally proposed by Mangold in 1932 and the cell contact model proposed by P. Weiss (1950). The diffusion model invokes passage of inducing substances across intercellular spaces from inducer to responding ectoderm. The cell contact model requires close membrane contact for induction; new "molecular configurations" are induced at the membrane surfaces of responding ectoderm cells which, in turn, trigger the events of differentiation in the cytoplasm

below. According to this model, no chemical need be transferred as long as adjacent cell surfaces interact.

Most experiments to test the diffusion hypothesis have given ambiguous results. They rely on the insertion of membrane barriers between inducer and responding tissues. On the one hand, if impermeable membranes such as cellophane are inserted between chordamesoderm and ectoderm in vivo, induction is inhibited. Though this shows that diffusion is necessary for induction, it does not eliminate the need for direct cell contact. On the other hand, if permeable membrane filters are used in vitro, equivocal results are also obtained. Saxen, in 1961, did show induction in 50 percent of the cases, if *Triturus* ectoderm is separated from dorsal lip tissue by filters of known porosity and thickness (Figure 12-33). When thick filters are used, induction is inhibited. Presumably, the inductive stimulus diffuses over minimal distances, but the difficulty with the experiment is that fine cytoplasmic processes may penetrate the pores from both tissues and establish contact within the filter. Electron microscopic examination of these filters indicated that cytoplasmic processes did not make contact within the filter. The pores in these filters, however, are tortuous channels and it is difficult to identify or follow cytoplasmic processes under these conditions. Toivonen's group has reinvestigated this problem using filters prepared by shooting neutron beams through the material. This produces virtually straight-lined pores that can be easily seen in the electron microscope. In the experiments, gastrula ectoderm was separated from dorsal lip tissue by a 25 $\mu$m thick filter with varying porosities, down to 0.1 $\mu$m. The tissues were allowed to interact for a short period, just enough to induce. The ectoderm was then removed and cultured separately. In each case induction was obtained and when the filter was examined, no cytoplasmic processes could be seen. This would suggest that the initial inductive signal does not require cytoplasmic contact; that diffusion is sufficient. This recent experiment is perhaps the best support for the diffusion hypothesis for primary induction.

Cell contact was noticed in later stages of induction, presumably during the period of neural plate regionalization. When neural plate was separated from an inducer, and allowed to interact for a comparable length of time, it was almost impossible to remove the tissue without tearing it from the filter.

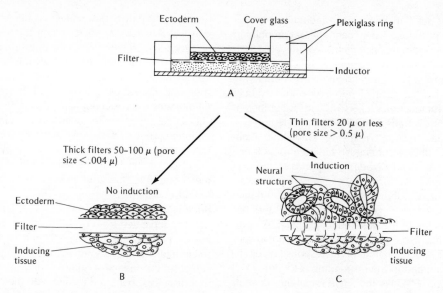

**Figure 12-33** *Transfilter experiments in an analysis of neural induction. (A) Chamber for transfilter experiment showing position of ectoderm, filter, and underlying inducing tissue (or material). The entire assembly is placed in culture medium for hours (or days) and the tissue and filter are examined histologically (or electron microscopically) for evidence of neural differentiation. (B) Thick filters with no pores or extremely small pore diameters are used. Under these conditions, no induction is observed. (C) Thinner filters with larger pores permit an exchange of materials. In the light microscope the filter is filled with matrix materials and stains positively for mucopolysaccharides. In the electron microscope cytoplasmic processes are seen in the filter but not in contact. In such situations induction occurs.*

EM examination of the filters showed many cyto-plasmic processes growing across the filter from the mesodermal tissue. Apparently regional inductive in-teractions require more intimate cell contact than the primary trigger that determines neural tissue.

Another approach is to culture competent ecto-derm in media containing pieces of inducer tissue or in extracts of the inducer. Induction under these con-ditions would occur without benefit of cell contact. Indeed, pieces of ectoderm or even free ectodermal cells do respond to various substances in the cul-ture media with specific patterns of neural and meso-dermal differentiation. But these experiments are also inconclusive, since ectoderm will neuralize in salt solutions without inducer present; that is, ectoderm is capable of *autoneuralization.*

Transfer of radioactive materials from inducer to the responding tissue has also been followed, but here, too, the results are ambiguous. Although ecto-dermal tissue does incorporate large amounts of labeled molecules from mesoderm, their nature is not certain, nor is it clear that the transfer has anything to do with induction.

Perhaps both diffusion and cell contact are in-volved in primary induction in vivo. Interacting tis-sues may exchange molecules across a matrix-filled extracellular space (100 to 200 Å). Mutual contact with the outer cellular matrix may play an important organizing role in the process. In addition, molecules may be exchanged across contact points (gap junc-tions?) between inducer and responding tissues, wher-ever they make contact.

*Question 4:* Does the inductive signal carry in-formation? Is it instructive?

Once the idea took hold that the inductive signal is a substance transferred from inducer to responding ectoderm, the nature of the signal and its mode of action loomed as important questions. Two models of action were proposed: (1) the inductive signal itself carries information and instructs the responding pluri-potent ectoderm to differentiate in a neural direction, or (2) the signal is a noninstructive trigger that elicits developmental expression preprogrammed within the ectoderm. The existence of specific archencephalic and spino-caudal inducers supports the view that in-ducer molecules themselves carry information. In other cases, however, nonspecificity of the inductive signal seems to be the rule and tends to support the second hypothesis. Efforts to isolate and characterize the inducer have revealed that diverse substances — acids, proteins, nucleic acids, and even such inert substances as kaolin — will induce neural tissue in experimental situations. Obviously, nonbiological substances could not be instructive.

Inasmuch as so many unrelated substances had inductive potency, Waddington, Needham, and Brachet, as early as 1936, proposed that inducers act by releasing the true "evocator" already present in the ectoderm. According to them, induction is not the instructive event but is rather an "evocation"; the natural inducer preexisting in an inactive state in the ectoderm is unmasked by any number of inducing signals. Induction according to this model is a relay mechanism. This explains the lack of inductive specificity and focuses on the responding system as the key tissue in induction. We have already noted that the ectodermal response is genetically programmed and unrelated to the inductive stimulus.

Barth observed that *Ambystoma* ectoderm, isolated in a salt solution in the absence of any inducer, yielded a high percentage of neural structures. Later, Holtfreter showed that long exposure to a salt solution or short treatment with urea, or acid, led to cytolysis of newt ectoderm followed by induction, its magnitude correlating with the amount of cytolysis. This phenomenon, called autoneuralization, is consistent with the evocation hypothesis; that is, any agent which nonspecifically damages presumptive neural ectoderm seems to release an inducer and elicit a response (Figure 12-34).

According to the model, a masked inducer is activated in partially damaged cells by sublethal cytolysis and triggers neural differentiation. This hypothesis offers no valid experimental approaches; that is, identification of the natural inducer is impossible since any manipulation of ectoderm can itself act as an evocating stimulus. Moreover, it is difficult to see how nonspecific sublethal cytolysis could be the mechanism of induction during normal development. The phenomenon of autoneuralization does mean, however, that all factors necessary for neuralization must be present in ectoderm, and these may be activated by proper treatment. For instance, untreated gastrula ectoderm has no inducing capacity but when heated, it acquires archencephalic-inducing properties and when exposed to phenol, ecto-

derm induces head muscle and even some notochord. These treatments seem to release or activate preexisting but inactive inducers in ectoderm. Hence the question whether the inducing stimulus in normal development is itself instructive, or whether it, like many nonspecific agents, elicits a predetermined response within the reacting ectoderm, is still an open one.

*Question 5:* What is the chemical nature of the inducer?

With the realization that tissue extracts could elicit specific responses, a search was on to isolate and identify the organizer. For more than 30 years, one or another class of substances was touted as the natural organizer. A group in England focused on lipids and sterols since ether extracts of amphibian neurulae exhibited high inductive potential. Besides, many sterols could induce secondary neural structures. The significance of sterols was questioned by Spemann's group, which was reporting that organic acids such as oleic acid or even DNA could induce; according to them, low pH was inductive. Since autoneuralization is easily produced by changes in pH, we would expect many different acids and compounds to produce an inductive response. Sterols and acids were discarded when Barth and Graf showed that nucleoproteins promoted strong inductions, and Brachet reported that nucleic acid, particularly RNA, was the natural inducer.

Accordingly, as scientific fashions shifted toward nucleic acids, two informational macromolecules, protein and RNA, were proposed as the natural inducers. This dichotomy persists today; some investigators believe the natural inducer is RNA while others propose protein. There is evidence for both, most of it, however, favoring protein.

Efforts to identify the chemical nature of the inducer involve isolation and purification of heterogeneous inducers from adult tissues. Work on the embryo itself has been less extensive and more controversial. One difficulty with most of these studies is that the specter of autoneuralization hangs over the experiments since it is unavoidable in most test situations.

Let us first examine some experiments on the embryo itself and then explore the work done on heterogeneous inducers obtained from adult tissues.

Experiments by Niu and Twitty in 1953 pointed to RNA as the natural inducer. When small bits of ectoderm (about 15 to 20 cells) were cultured in mi-

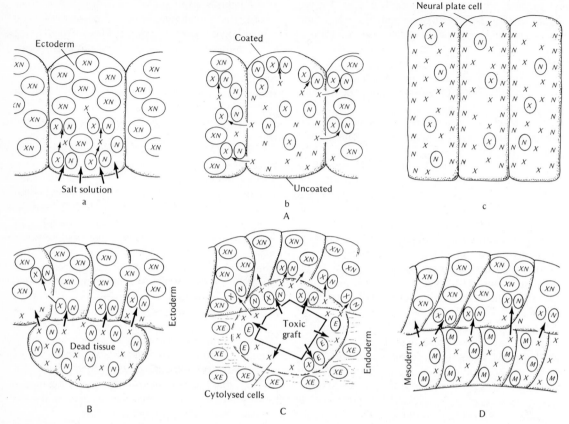

**Figure 12-34**  *Autoneuralization and evocation mechanisms of neural induction according to J. Holtfreter (Symp. Soc. Exp. Biol. 2:17–49, 1948). (A) Explanted ectoderm in a hypertonic salt solution. The ectoderm is assumed to contain its own inducer (evocator), X, in an inactive bound manner (XN). The salt ions activate the complex, release the inducer, X, which in turn by a cascade effect activates other inactive complexes. The N (neural factor) is released and proceeds to convert the low columnar epithelial cell into the elongated neural plate cell shown in (c). (B) Likewise, a dead graft of ectoderm containing an activated inducer X (phenol treatment, for example) will induce the overlying ectoderm as in (A). (C) A toxic graft in endoderm will release an inducer from its inactive complex XE by a cytolytic action. The inducer will diffuse into ectodermal cells and release the evocator X in its own cells. (D) The natural induction which occurs between mesoderm and overlying ectoderm. In this case it is assumed that the XM complex is somehow activated, releasing the X inducer into the ectoderm and again inducing a cascade effect within the ectoderm. (From J. Holtfreter, Symp. Soc. Exp. Biol. 2:17–49, 1948.)*

crodrops along with pieces of chordamesoderm, almost 90 percent differentiated, either as neural cells, pigment cells, or myoblasts. In the absence of chordamesoderm, no differentiation was observed; the ectoderm developed as a typical flattened epithelium suggesting that inducer had diffused into the medium. Later, Niu cultured explants of chordamesoderm for 10 days to "condition" the medium, removed inducer, and replaced it with small bits of ectoderm;

neural differentiation occurred in 24 hours. Longer periods of conditioning (12 to 15 days) induced pigment cells, nerve, and myoblasts. Unconditioned media or media conditioned by noninducer tissues had no effect on ectodermal differentiation. It was concluded that inducing tissues release different substances into the medium during growth; those released earliest have neural-inducing tendencies, whereas those appearing later induce mesoderm.

There are several difficulties with these experiments. First, conditions for maintaining these cells were not the most favorable for growth. Ectodermal cells in these culture conditions were barely managing to keep alive, let alone grow. The degree and quality of differentiation in these small cell populations were also poor. Finally, sublethal cytolysis could have occurred, resulting in autoneuralization.

Later, Niu found conditioned medium rich in RNA and concluded that RNA was the inducer because treatment with ribonuclease reduced inductive capacity. The results were equivocal, however, since activity was also eliminated by a combined treatment with deoxyribonuclease and ribonuclease, while in some cases chymotrypsin alone destroyed the inductive capacity of the medium.

These results are compelling in view of other claims that tissue-specific properties may be transferred to cells by RNA. Since RNA is an informational molecule, it seems reasonable to expect it to be a vehicle for information transfer during induction. But more than 90 percent of the RNA extracted from any tissue is noninformational ribosomal RNA; only a small proportion is messenger RNA, and it is unlikely that penetration of these few molecules into a cell is enough to instruct it to carry out a function it is itself incapable of doing. To date, most claims for information transfer via RNA extracts have not stood up to rigorous analysis.

Attempts to extract natural inducer from amphibian embryonic tissues have yielded mixed results. Although the chemical nature of the natural inducer has not been determined, it seems that two inducing factors may be present in the embryo, a neuralizing factor and a mesodermalizing factor. Results of experiments on heterogeneous inducers from embryonic and adult tissues have been consistent with this hypothesis and may be summarized as follows.

1. Two classes of substances with different inductive effects have been extracted from a variety of tissues. Neuralizing substances, ribonucleoprotein in nature, extracted from such tissues as guinea pig liver, induce a high proportion of archencephalic structures, whereas proteins extracted from kidney and bone marrow show proportionately more hindbrain and spino-caudal or mesodermalizing inductions. These actions resemble, in part, the archencephalic and spino-caudal inductions produced by anterior and posterior fragments of archenteron roof.

2. The inductive activity of most fractions is unaffected by treatment with ribonuclease but is altered by proteolytic enzymes such as trypsin. This suggests the active principle in both classes of inducers is a protein.

3. Activity of fractions varies according to treatment. A mesodermalizing factor exposed to high temperature or prolonged alcohol treatment will become more neuralizing. Proteins may undergo changes in configuration with concomitant changes in inductive properties, or, alternatively, two proteins may be present in each fraction with each treatment favoring the expression of one over the other.

We cannot determine whether heterogeneous tissue inducers are identical to the natural inducer since they may act in test situations by promoting autoneuralization. Two classes of substances with actions like natural inducers may reveal only the limits of an ectodermal response pattern. A class of inducers may favor expression of only a part of a wide range of developmental expression. Besides, most heterogeneous inducers are extracted from adult tissues with alcohol or petroleum ether, both substances able to convert ectoderm into an inducer.

Finally, the phenomenon of autoneuralization suggests that both classes of inductive agents may be present in ectoderm in an inactive form and that each adult factor specifically evokes only one type of inducer.

Though these studies have provided some clues as to the chemical nature of heterogeneous inducers, they have told us nothing about the natural inducer or its mechanism of action.

## Models of induction

A few models of induction will be reviewed to illustrate how some of the facts are used. The two-gradient hypothesis of Toivonen and Saxen postulates two natural inducers, neuralizing and mesodermalizing (Figure 12-35). The combined action of both factors distributed as a dual gradient over chordamesoderm specifies the nervous system. Schematically, the neuralizing factor is distributed equally along the cephalocaudal axis of the entire sheet, highest in concentration in the dorsal midline. The mesodermalizing factor, absent in the anterior region, is distributed as a gradient beginning in the midbrain area, increasing caudally. Where neuralizing and mesodermalizing factors overlap, mid- and

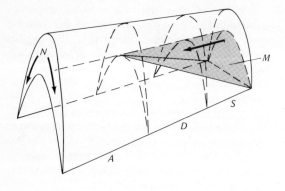

tive proportions of neural and spino-caudal derivatives were then assayed. When neural tissue predominated, no·mid- or hindbrain tissues developed, but these latter structures did appear when neural and mesoderm cells were mixed in equal proportions, a result consistent with the hypothesis.

The Nieuwkoop activation-transformation hypothesis emphasizes the temporal aspects of induction (Figure 12-36). It proposes two successive phases of induction; the first involves a generalized "activation" throughout the presumptive neural ectoderm by nonspecific factors derived from mesoderm. All ectoderm is determined to a level of forebrain differenti-

**Figure 12-35**  *Two-gradient hypothesis of Saxen and Toivonen. The inducer tissue (chordamesoderm) contains two inducer components, a neuralizing factor (N) distributed in gradient fashion dorsal-ventrally as shown, with no gradient along the anterior-posterior axis. A mesodermalizing factor (M), however, is also present as a gradient with a peak at the posterior grading toward the center of the embryonic axis. Interaction of the two gradients leads to deuteroencephalic (D) and spino-caudal (S) inductions while in anterior regions, in the absence of any mesodermalizing factors, only archencephalic inductions (A) result. (From S. Toivonen and L. Saxen,* Ann. Akad. Sci. fenn. Ser. A IV *30:1–29, 1955.)*

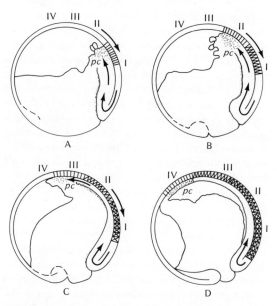

**Figure 12-36**  *Activation-transformation hypothesis of Nieuwkoop. (A–D). Diagrams illustrating oppositely directed movements of invaginating archenteron roof and epibolically extending neural ectoderm during gastrulation. The interaction of both layers first leads to a caudo-craniad spread of an activating inductive principle in the ectoderm when exposed to the prechordal portion of the archenteron roof (pc). This primary action is followed by a caudo-craniad spreading of a superimposed transforming inductive principle emanating with increasing intensity from successively more caudal regions of the archenteron roof. Vertical lines = archencephalic activation; diagonal lines = deuteroencephalic; cross-hatched = spino-caudal. (From P. D. Nieuwkoop,* International Lecture Course on Cell Differentiation and Morphogenesis. *New York: Wiley, 1966.)*

hindbrain structures are induced. Anteriorly, in the head region, in the absence of the mesodermalizing factor, only archencephalic structures are induced. Posteriorly, where the mesodermalizing factor is in highest concentration, trunk and tail structures are induced. The model also explains the sequential nature of induction; the initial interaction induces neural tissue and only after exposure to mesodermalizing factors does head-trunk regionalization take place. Though each inducer alone gives archencephalic or spino-caudal structures respectively, when they are combined, all midbrain and hindbrain structures should develop. To test this prediction, dissociated neural and mesodermal cells from *Triturus* early neural plate stages were mixed in varying proportions and allowed to develop in vitro for 14 days. The rela-

ation as its intrinsic neural factors are activated. In the second phase, the activated ectoderm is regionalized by a secondary inductive action called "transformation." A graded spino-caudal inducing signal from mesoderm now spreads over the already activated ectoderm. The differentiation of a region is a balance between the initial archencephalic tendencies and the caudal transformation wave impinging from the outside.

The model is tested by implanting folds of ventral ectoderm into different regions of an invaginating gastrula. These folds differentiate to varying degrees, depending on their position and the time at which they receive the activation-transformation wave from the host. For instance, the prechordal region, which never loses its archencephalic potencies (a transformation wave never reaches it), always induces anterior brain structures. Though most results are consistent with the sequence of activation-transformation, the two events may be separated experimentally and even reversed. For example, isolated pieces of a neural plate region normally differentiate into hindbrain vesicles. If these pieces are placed in contact with prechordal plate in an early neurula stage, they are transformed into archencephalic brain structures instead. We see that a piece already committed toward a more posterior pattern of differentiation can be shifted to a more anterior forebrain type of induction. This result suggests that the activation and transforming waves may be reversed; one need not be a precondition for the other.

Can the two models be reconciled; that is, can two types of inducers be fitted to the temporal qualities of the process? Conceivably, neuralizing and mesodermalizing factors may be different-sized molecules with different rates of diffusion or times of release from invaginating chordamesoderm. Neuralizing factors may be the first to diffuse into reacting tissue and induce archencephalic structures. Another possibility is that competence of ectoderm to each inducer stimulus also changes. For instance, neural competence may be lost first, whereas competence to mesodermalizing factors (transforming stimuli) could persist for longer periods.

Initially, induction evokes neural tendencies which are secondarily stabilized into regional specificities. Two diffusible signals seem to be involved, either separated in time or operating simultaneously but diffusing at different rates over the tissues. These specify the developmental program of a region of ec-

toderm by changing cell properties such as mitotic activity, adhesiveness, motility, and shape. How all these cellular events are translated into rolling up a two-dimensional sheet into a tube still remains one of the major problems of embryogenesis.

But even if we could identify the chemical nature of the inducer, its mechanism of action would also have to be worked out. Perhaps one of our difficulties is that in the past we focused on the inducing stimulus and its instructive qualities, neglecting the behavior of responding ectoderm. The phenomenon of autoneuralization, though alerting us to the nonspecific nature of the inducing signal, is still unexplained. In recent years, however, as the regulatory role of ions is recognized and the importance of membrane and contact relationships increases, ion effects on primary induction have been examined with renewed interest.

P. Weiss, in the late 1940s, was particularly sensitive to the importance of surface interaction in induction when he formulated the concept of "molecular ecology." He claimed that the spatial arrangement of molecules on extracellular membranes has important implications for cell behavior. According to his model, interfacial forces between partly immiscible molecular populations in cell surfaces concentrate certain selected molecular species in the plasma membrane. By their surface positions, these molecules become ordered into distinct configurations and thereby exercise control over cell metabolism. This is due partly to their common orientation relative to the interface, which increases their structural coherence and chemical combining faculty, and partly to their control over substances exchanged across the cell surface.

This concept may be closer to the truth than was originally believed, particularly in view of the behavior of fluid membranes and their importance in regulation of many cellular processes. (See Chapter 10.)

Weiss proposed that the inducing cell surface organizes a specific array of surface molecules within the responding ectoderm (Figure 12-37). This new surface arrangement, perhaps by changing membrane permeability, sets off a sequence of events within the ectoderm leading to neural determination. Mesoderm cells with slightly different surface configurations could rearrange the ectoderm surface to favor a mesodermalizing pattern of differentiation. Changes in surface configuration (distribution of receptors or recognition sites) may set off a train of

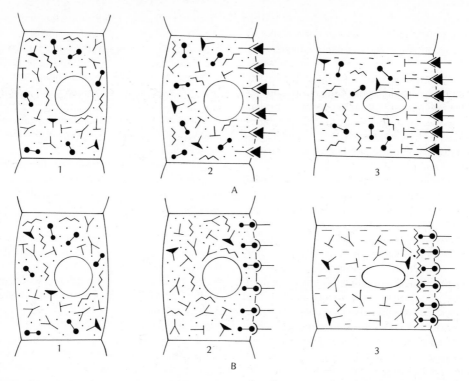

**Figure 12-37** *Role of the surface in induction; the molecular ecology model of P. Weiss (Quart. Rev. Biol. 25:177, 1950). Two ectodermal cells are shown at an early gastrula stage, each containing the same population of macromolecules. (A) The cell comes in contact at its surface with one type of inducing molecule with a distinct structural configuration. As a result the notch-shaped molecules are entrapped at the cell surface, thereby altering surface properties. These changed surfaces may result in a stabilization of the surface change by the accumulation of T-shaped molecules within the surface. This will program the cell and lead to one pattern of cell differentiation (archencephalic?). (B) The same cell shown in (A) but exposed to a different inducing molecule with a different structure. This will entrap another (ball-shaped) molecule in the surface and lead to new surface configurations that program the cells toward another pathway of differentiation (mesodermal?).*

determining events in the ectoderm. Such surface-mediated inductive phenomena remind one of the contact interactions in prespore cell differentiation in *Dictyostelium* (see Chapter 9). In that system interactions between specific membrane sites seem to be essential to differentiation.

### Ions, membranes, and induction

Initial inductive interactions may involve permeability changes in membranes of neural ectoderm; this may facilitate ion flux. In fact, patterns of induction are affected by the ionic environment. Competent ectoderm differentiates into neural tissue when pH is altered or when ammonia or salt ions are added to

the medium. Though these may be cases of sublethal cytolysis, all involve membrane permeability. Certain ions, such as lithium, have variable effects depending on concentration.

Lithium inhibits amphibian morphogenesis; in its presence, chordamesoderm and endoderm do not invaginate but spread outside, forming an exogastrula. The ectoderm, deprived of its normal inducing tissues, does not differentiate a neural tube. In this case, lithium interferes with the spreading tendency of epithelium and inhibits invagination. At lower concentrations, gastrulation is incomplete and prechordal plate does not advance into the anterior head region. As a result, head development is abnormal, eye primordia fail to separate, and microcephalic or cycloptic larvae develop. The inductive action of the dorsal lip is also inhibited by exposure to lithium ions. Pretreatment of a dorsal lip fragment before implantation into a blastocoele prevents induction.

The mode of action of lithium on gastrula cells in vitro differs from its action on the whole embryo. Small aggregates (about 250 cells) of presumptive epidermis isolated from early or mid-gastrulae *(Rana pipiens)* develop into an epithelial sheet. When incubated in low concentrations of lithium chloride, nerve cells differentiate, but at higher concentrations, the cells rapidly spread out as flat sheets and accumulate pigment. Under these conditions lithium seems to be acting as an inducer of nerve and pigment cells. Sublethal cytolysis is not involved because the concentrations used are below those usually causing cytolysis. Lithium seems to have a sequential effect; with increasing concentration, first nerve cells and then pigment cells differentiate.

Other ions are also active. Potassium and sodium induce mucous and nerve cells but not pigment cells. In fact, induction is not possible in this system unless cells are exposed to sodium ion at some point in the procedure; it seems to be essential in all inductions. Sodium at 88 mM will induce but at lower concentrations (44 mM) no inductions occur. If, however, cells are first cultured in sucrose with $Ca^{++}$ ions for several hours, then inductions are obtained even after the addition of only 44 mM sodium. Though high internal concentrations of sodium ion are essential for induction, a prior exposure to calcium ions seems to compensate in some manner.

If sodium is the principal ion, how do we explain the lithium effect? We know that lithium combines with proteins and alters enzymatic function, perhaps by changing protein configuration. Lithium also replaces sodium in nerve membranes during impulse transmission, suggesting that lithium may, during induction, exert its effect on the cell membrane. Apparently, both ions interact as they penetrate the plasma membrane since sodium uptake is enhanced in the presence of lithium. Lithium may make the cell surface more permeable to sodium. In the inductive situation in vitro, lithium acts synergistically with sodium and stimulates neural inductions. Both sodium and lithium induce neural cells, but sodium is less effective when alone.

It should be noted that induction of presumptive epidermis by mesoderm in vitro also depends upon sodium ion concentration. At the usual sodium concentration (88 mM), mesoderm induces functional nerve and pigment cells, in addition to muscle and mesenchyme. At a reduced sodium concentration (44 mM), no nerve or pigment cells differentiate, but muscle and mesenchyme develop normally. It is possible that mesoderm acts in vivo by preparing the presumptive neural plate for induction by sodium ions in the blastocoele fluid. According to this model, anything that promotes sodium uptake will have a neuralizing tendency. Proteins and nucleoprotein inducers from adult tissues may act by changing membrane permeability to sodium ions. Moreover, the autoneuralizing phenomenon in isolated ectoderm could also be interpreted as an increased flux of sodium ions as permeability of ectodermal cells increases.

Extrapolating from these results on explanted cells, Barth proposed a model of induction during normal gastrulation (Figure 12-38). In the blastula, ectodermal cells surrounding the blastocoele cavity release ions into the cavity from their inner surfaces. The ions accumulating in the cavity include $Na^+$, $Ca^{++}$, and $Mg^{++}$. During gastrulation, the invaginating chordamesoderm becomes closely applied to the inner surface of the ectoderm overlying the archenteron roof and establishes an extracellular matrix between the two layers. While ventral ectodermal cells still lining the reduced blastocoele cavity continue to release ions into the cavity, the ions released from the inner surface of the presumptive neural ectoderm accumulate between the two layers instead, and the extracellular concentration of $Na^+$ (and other ions) probably increases sufficiently to insure a higher concentration inside the cells, enough to induce. Ectoderm cells lining the blastocoele cavity, containing a lower concentration of sodium ions, will not develop into neural tissue but become epidermis instead. The ion trap in the extracellular matrix between the two layers may also increase the $Ca^{++}$ ion to a point sufficient to activate the contractile microfilaments which bring about the shape changes associated with neurulation.

As we learn more about cell membranes, many phenomena that were difficult to interpret before are now beginning to make sense. The structure of the cell surface undergoes transient changes. Enzyme systems assigned to the cell surface participate in and regulate intracellular metabolism. An example of such a system is the enzyme adenylcyclase which regulates intracellular levels of cyclic AMP. We know

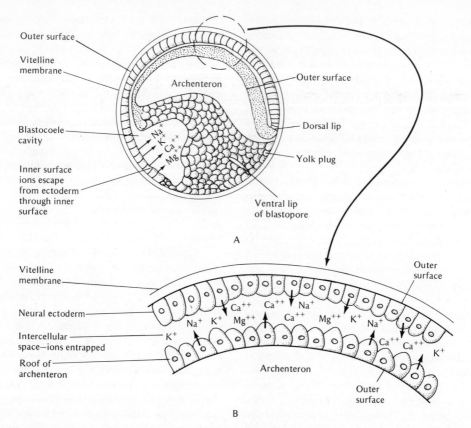

**Figure 12-38**   *Role of ions in primary induction according to model proposed by L. G. Barth and L. J. Barth (Develop. Biol. 39:1–22, 1974). (A) Late gastrula of* Rana pipiens, *when roof of archenteron is inducing neural plate in overlying ectoderm. Ions ($Na^+$, $K^+$, $Ca^{++}$, and $Mg^{++}$) escape from permeable inner surfaces of ectoderm and collect in blastocoele fluid. Since ions are lost from ventral ectoderm no induction occurs and the cells differentiate as epidermis. In the dorsal half of the embryo the roof of the archenteron and overlying ectoderm are in close proximity and share an extracellular matrix in which ions escaping from ectoderm and roof of archenteron are entrapped (by extracellular glycoproteins?). (B) An enlarged view of the region of induction shows these ions collecting within the extracellular matrix with a consequent higher ion concentration within the ectoderm cells themselves. This initiates neural induction. The outer surface of the gastrula is continuous with the surface of the invaginated archenteron roof. This surface is relatively impermeable to ions which cannot diffuse into the archenteron or external medium.*

that cyclic AMP affects cell permeability; cation flow is altered by changes in levels of cyclic AMP inside or outside the cell. Many ions, in turn, regulate adenylcyclase activity. Such mutual interactions could affect cell division and morphogenetic movement. The lithium ion is a good example since it has numerous morphogenetic effects. It inhibits adenylcyclase activity in nervous tissue which, in turn, diminishes nerve conduction, particularly at synaptic terminals. We can see that lithium inhibition of adenylcyclase activity at the surface of ectoderm cells could be

translated into changes in cyclic nucleotide levels which secondarily alter some regulatory systems controlling phenotypic expression. Contact between cells in different tissue layers or the interactions between cell surfaces and surrounding glycoprotein matrices could also modify enzyme activity on the cell surface. This may be translated into intracellular fluctuations in cyclic nucleotides or in the flow of ions among various cellular compartments. Even specificity of responses by cells and sensitivity to different inducers could reside in specific membrane receptors that recognize small regulatory molecules. If we concentrate on the cell surface as a regulatory site, the problem of primary induction will become clearer.

## DEVELOPMENT OF THE BRAIN

Primary induction leads to a series of secondary and tertiary inductions while the neural ectoderm becomes a functional nervous system (Figure 12-39). As the nervous system is specified, portions of the neural tube acquire inductive properties which secondarily stimulate sense organ development. This occurs as the brain vesicles expand and differentiate.

The anterior prosencephalon constricts into two chambers, a *telencephalon* follwed by a *diencepha-*

*lon*. The most anterior portion of the telencephalon develops into the olfactory center of the brain, emerging as a pair of small outpocketings, the olfactory bulbs. The olfactory nerves grow out of these bulbs toward the periphery and contact the *nasal placodes,* thickenings in the head ectoderm, which invaginate to become the organ of smell.

The greater part of the telencephalon divides into two lateral lobes, the cerebral hemispheres, which expand into the most important region of the vertebrate brain. In higher vertebrates, expansion of the lobes is so great that they overgrow the rest of the brain; fish and amphibian cerebral hemispheres do not show this overgrowth and retain their anterior position.

The next chamber, the diencephalon, produces

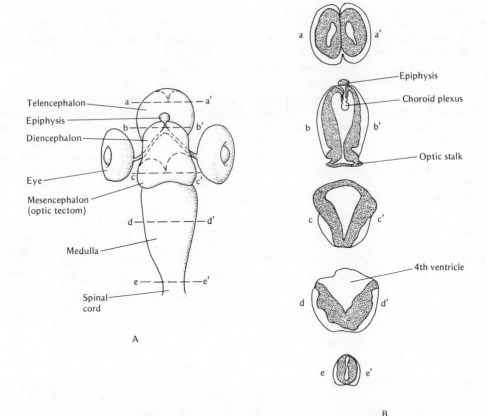

**Figure 12-39**   *Brain of a young tadpole. (A) Whole view of brain from dorsal side showing eyes and optic stalks as well as major brain regions. (B) Cross sections of brain in regions shown, showing arrangement of cell bodies or gray matter (black) and region of fiber tracts (white). (From B. I. Balinsky,* Introduction to Embryology. *Philadelphia: Saunders, 1975.)*

two lateral outpocketings, the optic vesicles, which become the future eyes and optic stalks. The thickened wall of the diencephalon differentiates as the optic lobes. Its thinned-out roof ultimately becomes a nonnervous choroid plexus, the region where oxygen and nutrients are exchanged between brain cavities and blood vessels. More posteriorly, in the diencephalon roof, small outpocketings emerge with short stalks. These become the accessory organs of the brain, the *pineal body*, which in lower vertebrates develops into a pigmented, light-receptive organ, and the *parietal body*. These organs have a hormonal role in higher vertebrates, but their precise relationship to other endocrine glands and to the nervous system is still uncertain.

One of the most important outpocketings of the brain in vertebrates is the pituitary gland, a ventral outgrowth from the floor of the diencephalon. A funnel-like depression called the *infundibulum* detaches from the floor of the brain and fuses with an invagination from the roof of the developing mouth of the embryo. These two tissues with differing origins join to form the pituitary gland, the chief endocrine gland of vertebrates. The adjacent region of the diencephalon becomes the *hypothalamus*, the neurosecretory brain center concerned with autonomic control of most bodily functions such as respiration, circulation, and endocrine regulation. The midbrain or mesencephalon remains small and undifferentiated compared with the forebrain region. The roof of the midbrain

differentiates into the *tectum*, the visual center in fish and amphibians.

The hindbrain or rhombencephalon gives rise to the *medulla oblongata* and *metencephalon*. In this region the ear vesicle develops from an *auditory placode*. Major cranial nerves emerge from the ventral region of the metencephalon, while the dorsal region expands and folds into the cerebellum, the region of the brain controlling equilibrium and balance.

Additional cranial nerves extend from the medulla, which joins with the spinal cord. While these major brain regions are developing, they interact with adjacent mesoderm and ectoderm of the head and secondarily induce sense organs.

### Eye development

Eye development displays the full morphogenetic repertoire of embryonic sheets, including inductive interactions, sheet movement, changes in cell shape, and, finally, differentiation of specific cell phenotypes of lens and retina.

The eyes of the vertebrate embryo are determined at the neural plate stage. Vital dye mapping of the amphibian neural plate shows the eye-forming region as two small primordia in the anterior attached to each other by a short stalk immediately behind the transverse neural fold (Figure 12-40). As neural folds rise, the eye primordia are displaced laterally, retaining their connections by the small

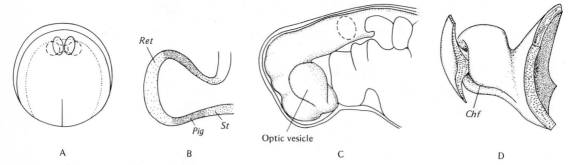

A          B          Optic vesicle          Chf
                          C          D

**Figure 12-40**   *Eye development in urodeles. (A) Presumptive eye region in neural plate stage showing paired anlagen at early stage (solid lines) and anlagen separating (dotted lines) as neural folds and neural tube form. (B) Primary optic vesicle with presumptive retina (Ret), pigment epithelium (Pig), and eye stalk regions (St). (C) Eversion of primary optic vesicle from presumptive diencephalon. (A–C from A. Kühn, Lectures on Developmental Physiology. New York: Springer-Verlag, 1971.) (D) Model of eye cup and lens thickening of a 5.5-mm human embryo showing invagination of eye vesicle to form eye cup and the deepening of the lens pit. Chf = choroid fissure. (From I. Mann, Development of the Human Eye. London: British Medical Association, 1949.)*

median strip of tissue which develops into the *optic chiasma,* the region of crossing optic nerves.

Shortly before neural folds fuse in the midline, the future eyes emerge as two lateral outpocketings of the diencephalon, the optic vesicles. These balloon out toward the head ectoderm, the most distal portion thickening into a columnar layer while the more proximal region thins out. Gradually the optic vesicles contact the head ectoderm, whereupon they transform into cuplike structures as the thickened layers collapse inward and contact the inner surface of the thin layers. The thickened layer develops into the sensory layer of the retina, whereas the proximal thinner layer becomes the pigmented retina.

Contact between optic cup and ectoderm is essential to retina development. If the presumptive retina fails to contact the outer ectoderm, it thins out, forms pigmented epithelium instead, and the eye is abnormal. The cuplike structure enlarges by mitotic activity while in contact with head epidermis. In addition, adhesion between cup and head ectoderm may induce the outer layers to thicken into a retina.

The determination of the eye anlage involves inductive interaction of the most anterior mesodermal substrate, the prechordal plate, with overlying neural ectoderm. If neural ectoderm with attached median mesoderm is grafted onto the flank of a frog embryo, eyes develop 100 percent of the time; but eyes develop only 75 percent of the time if the graft is made without mesoderm. If the same region is combined with lateral mesoderm instead, eyes arise in only 33 percent of the cases, as if lateral mesoderm inhibits induction. Likewise, transplants of posterior neural plate over the prechordal mesoderm develop into eyes, but reciprocal grafts of anterior neural ectoderm over posterior mesoderm do not.

Mesoderm may play two roles, inductive in one case and inhibitory in another, both essential to normal eye development. Transplants of anterior neural plate without mesoderm often develop into a single or cycloptic eye. The prechordal plate must be present for both eyes to form; otherwise, the anlagen fuse into a large single eye. Moreover, if prechordal plate is rotated 180 degrees, the axis of the head is correspondingly shifted. In normal situations the median portion of the prechordal mesoderm first induces a single-eye field, then inhibits its expression in the middle of the brain floor as the neural tube forms. Its presence splits the eye anlage, and two eyes develop.

The anterior neural plate behaves like a morphogenetic field, a region in which certain properties, graded according to position or time, or both, may lead to increased morphological complexity. A "field" is purely an operational definition based on the following procedures in the case of the eye field:

1. Transplantation of different parts of the anterior neural plate into a host embryo shows the greatest eye-forming capabilities in the median portion grading off laterally; the eye field exhibits maximum morphogenetic intensity at its center.
2. Removing half an eye field at the neural plate stage does not prevent eye development. Remaining cells in the anlage regulate into a normal but smaller eye.
3. Even in a late neurula, an eye vesicle grafted adjacent to a host eye vesicle results in fusion of the two into a large single eye. The vesicle retains developmental plasticity well after it is determined.

Each tissue of the eye vesicle is labile in early neurulae; sensory and pigmented layers are interconvertible. A fragment of any layer transplanted into an embryo's head often reorganizes into both tissues, sometimes as a small eye cup. In fact, the pigmented layer of adult urodele eyes still retains the capacity to reorganize into a retina. The retina, however, loses its potency as it differentiates into sensory layers and is unable to differentiate into pigmented epithelium. It still exhibits some lability in the early neurula after it has differentiated inasmuch as an inverted retina develops normally; all retinal layers emerge in proper orientation with rods and cones facing toward the lens. Lateral-medial polarity of the retina seems to be fixed much later in development.

The eye cup deepens and the thickened retina invaginates and fuses with pigmented epithelium. The cavity of the cup remains open ventrally, forming a groove extending to the optic stalk. This develops as the *choroid fissure* and fixes the ventral region of the eye. The hyaloid blood vessels course along the choroid fissure in the mature eye.

It is at this stage that the eye becomes polarized along its anterior-posterior and dorsal-lateral axes. The pattern of determination was studied by rotating eyes in newts *(Triturus).* The eye vesicle may be removed, rotated 180 degrees, and replaced in its new orientation in the orbit. In the newt the visual axes of the retina are determined sequentially; the anterior-posterior axis is determined first, at least before the neural tube forms. Rotation of the eye primordia at the neural plate stage results in normal development and a normal visual field. Rotation of the eye at a neural fold stage results in a reversed visual field; the larva behaves as if the anterior-posterior field is inverted and snaps at a lure in the wrong part of the field. A lure placed in its nasal field (in front) is snapped at behind the animal as if the animal sees the bait in its temporal field. The animal still sees correctly in the dorsal-ventral aspect of the field. The dorsal-ventral axis is specified after eye cup formation and rotation of the eye after this stage results in complete inversion of the eye field in the larval or adult eye. The animal now sees the world completely inverted and reversed; the rotated retina cannot readjust its polar axes after the periods of specification during embryonic life, no matter how long the animal is retrained to see correctly. The mediolateral axis of the eye is established last, at a swimming stage.

Each axis seems to be determined independently of the others at different developmental stages. By the time the retina begins to differentiate into sensory layers (rods, cones, and so on), all three visual axes characterizing functional expression have been established.

## Lens induction

The lens develops after the retinal axes have been polarized. Lens formation is one of the first signs of secondary induction by neural tissue. A lens in many amphibians and birds differentiates after head ectoderm receives an inductive stimulus from the adjacent optic vesicle (Figure 12-41). If the optic cup is removed before it contacts head ectoderm, no lens develops. Virtually any competent ectoderm responds to the inductive stimulus. Lenses are induced when an optic cup is transplanted underneath foreign ectoderm or when pieces of flank ectoderm are placed over the optic cup in the head. Competence of ectoderm to form lens declines at different rates in amphibians. In *Rana sylvatica*, competence to form lens is found in the entire epidermis, whereas in *Rana esculenta* only head ectoderm is competent. And in general, competence is maximal at late gastrula and minimal at closure of neural folds.

Lenses may develop without induction by the optic cup in some amphibians, but in these cases

head mesoderm or, more likely, archenteron roof, plays the inductive role. Mesoderm seems to be important in all cases of lens induction, even when the optic vesicle is involved. For example, in the newt *Triturus alpestris*, head ectoderm isolated from a neurula along with underlying endomesoderm develops into a small lens without being induced by an optic vesicle. Lens induction is a two-stage process; the first "conditions" ectoderm by a mesodermal interaction so that it can respond to a second inductive stimulus from the optic vesicle. In some species, the initial mesodermal conditioning satisfies all requirements for lens induction with the eye cup only localizing the lens-forming region. In other species, the mesodermal stimulus is insufficient; a second intimate interaction with eye cup is a prerequisite for normal formation.

At the beginning of lens induction, cells of the reacting ectoderm elongate and orient in parallel toward the adjacent optic cup, a process that correlates with oriented microtubules. The layer thickens into a placode which invaginates inside the optic cup. As the lens rudiment invaginates and pinches off, cell number increases. The anterior outer face is mitotically active, acting as a germinative layer. Cells produced in this region migrate into the center, stop dividing, and differentiate into lens fiber cells by synthesizing several crystalline proteins. These high-molecular-weight proteins account for the light-transmitting properties of the lens. Cell membranes fuse as lens fiber synthesis continues, nuclei degenerate, and a clear homogeneous lens appears in the eye chamber.

Is the inductive interaction between epidermis and optic vesicle based on cell contact or diffusion? It was believed at first that contact was necessary; a cellophane membrane interposed between interacting tissues in the chick embryo prevented lens formation. But cellophane would also inhibit diffusion of inducer molecules. When porous membranes are used instead, lens formation occurs, suggesting passage of molecules from optic cup to ectoderm. This, of course, does not eliminate the alternative explanation of contact via fine cytoplasmic filaments. On the other hand, tissue extracts that induce archencephalic structures also induce independent lenses in regions removed from other neural tissue. A diffusible inducer seems to be acting, possibly by passing directly from inducing into responding cells through close contact points such as

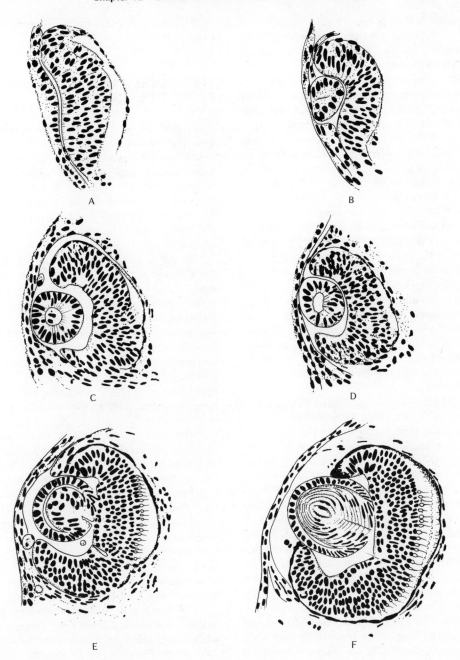

**Figure 12-41**   *Lens Induction in Axolotl eye. (A) Thickened presumptive retina makes contact with head ectoderm and induces epidermal cells to elongate into a thickened lens placode. (B) Lens placode invaginates into a lens vesicle as retina becomes convex, its outer surface making intimate contact with a single layer of pigmented epithelium. (C, D) Lens vesicle enlarges; retina becomes more cuplike. (E) Lens crystalline proteins appear in central lens cells undergoing differentiation. (F) Layering of retina begins and rods and cones differentiate. (From H. Spemann,* Embryonic Development and Induction. *New Haven: Yale University Press. Copyright © 1938 by Yale University Press.)*

gap junctions. A mucopolysaccharide matrix several microns in thickness develops between the tissue layers, which may also affect surface interactions.

If a diffusible inducer exists, we would expect the ectodermal response to vary with concentration. This is suggested by experiments showing that the size of a lens is directly related to the extent of ectoderm in contact with optic cup; large optic cups induce large lenses. Recruitment of ectodermal tissue for lens differentiation seems to depend upon the magnitude of the signal emanating from the inducer.

Summarizing, we see that eye development is a program of *sequential inductions*. The eye field is determined during primary induction, when anterior archenteron roof interacts with overlying neural ectoderm. The eye primordium is restricted to two small anlagen which emerge from the brain as optic vesicles. These grow out toward the head and make contact with the head ectoderm. Having acquired inductive capabilities, the optic cup, acting now as a secondary inducer, directs lens differentiation in the ectoderm. The lens, in turn, acquires inductive capabilities and influences differentiation of accessory eye structures such as the cornea, the transparent epidermal layer that completely covers the eye in front of the lens.

The cornea, like skin, is derived from an outer epidermal layer. In contrast to normal skin, cornea becomes transparent because pigment and gland cells do not develop and mesenchyme below remains unvascularized. In amphibian and chick embryos, the cornea becomes transparent after lens detaches from the epidermis. Presence of the lens is necessary for corneal development. If the lens and eye cup are removed at early stages, the head ectoderm undergoes normal skin differentiation and becomes suffused with glands and pigment. Presumptive skin transplanted from other regions to the eye is induced by lens to differentiate into normal cornea, containing collagen fibrils oriented in an orthogonal latticework.

Cellular interactions during eye development are typical for most sense organ morphogenesis. Sense organ rudiments start as "fields" in the neural plate, which, by inductive interactions, recruit overlying ectoderm to differentiate as sensory placodes resembling the lens placode. In the ear and nasal placodes cells change shape, adhere closely, orient toward the inducing tissue, and round up into vesicles before they establish connections with the brain.

## THE NEURAL CREST: A MIGRATORY CELL POPULATION

Many derivatives of the vertebrate nervous system owe their origin to a population of independent migratory cells that arise during closure of the neural tube. Sensory and sympathetic ganglia lining the spinal column, the cranial nerves of the head, and the supportive cells of nerves such as Schwann cells and glia are among a large group of tissues derived from the neural crest, a unique population of cells exhibiting directed migration.

In the development of the vertebrate embryo, the directed movement of cells to specific organ sites is based on cell recognition. Cells follow prescribed pathways and eventually invade predetermined organs. Here recognition calls for cells picking up cues in the environment along the route and homing in on their destination. Neural crest cells journey great distances in the embryo, homing in on specific tissue sites before they differentiate. The routes taken by these peripatetic cells are highly specific, and as soon as they reach their goal, they stop migrating.

The neural crest arises in the dorsal midline along the neural axis (Figure 12-42). The first sign of neural crest is the emergence of lateral neural folds at the edges of the medullary plate. The folds converge toward one another to form the neural tube. Neural crest cells arise dorsally, detaching from either side of the neural tube, where ectoderm and neural tube wall are joined. The cells migrate away in two streams. The first, a dorsal stream, migrates in tissue spaces between the neural tube and the overlying ectoderm. A second stream migrates ventrally between neural tube and somites, using the basal lamina covering the neural epithelium as a substratum. The earliest crest cells emerge anteriorly while new populations arise posteriorly as the neural tube closes. The cells migrate along basement membranes and within collagen-filled extracellular matrices, interacting with neighboring tissues. They eventually stop migrating in specific sites where they may aggregate into compact masses, or disperse in neighboring tissues.

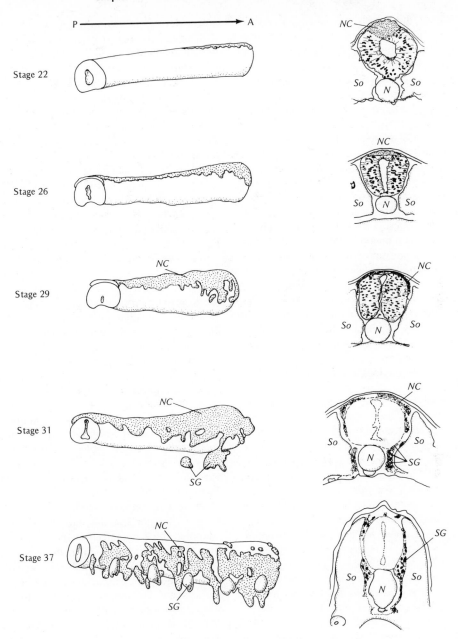

**Figure 12-42**  *Development of neural crest in Ambystoma punctatum. On the left are views of the neural tube and developing spinal cord with neural crest (stippled) at different developmental stages. Anterior-posterior axis shown at top, along segments 2–7. On right are cross sections made at level of third segment in each stage. NC = neural crest; SG = spinal ganglia; N = notochord; So = somite. At stages 29 and 31 the neural crest can be seen migrating in two streams, a dorsal and a ventral, the latter moving between neural tube and somite. (From L. R. Detwiler, Am. J. Anat. 61:63–94, 1937.)*

The fate of these cells has been studied by several techniques such as ablation or transplantation. Crest cells are labeled in order to follow migration patterns and ultimate destinations. Cells are marked by labeling their nuclei with radioactive thymidine or the neural crest of one species, such as the Japanese quail, may be grafted onto chick hosts. Quail cells are easily distinguished from chick cells because their interphase nuclei contain several heterochromatic blobs that are absent in chick nuclei.

With these techniques the fate of many derivatives of the neural crest has been determined. For example, the anterior or head neural crest that migrates from the hindbrain will develop into cranial nerves, and even mesodermal derivatives such as cartilages of the face or loose mesenchyme of the head. Ablation of the anterior neural folds of an amphibian embryo, before neural crest migrates out, usually eliminates most of these derivatives.

The dorsal streams of the mid or trunk neural crest migrate between the neural tube and the overlying ectoderm until they invade the skin and differentiate as pigment cells (melanophores, xanthophores, and iridophores) each containing melanin pigment along with pterines and carotinoid derivatives.

The ventral streams of the trunk invade the somitic environment and become segmentally arranged. The ventral streams may stop in any of three different places depending on their origin in the neural tube. For example, using quail-chick isotypic or isochronic grafting techniques (grafts from host to donor come from the same region or are at the same stage of development), LeDouarin and Teillet demonstrated that crest cells from the region of the first seven somites invade the entire gut and differentiate as enteric ganglia. They become cholinergic neurons of the autonomic nervous system and secrete acetylcholine as a transmitter. On the other hand, crest cells from the 8- to 28-somite area give rise to adrenergic orthosympathetic ganglia of the gut; these neurons synthesize catecholamines such as epinephrine. In addition, a specific population of ventral crest cells from somite regions 18 to 24 migrate out and invade the developing adrenal gland where they become the adrenal medulla, secreting adrenalin. The neural crest from the posterior segments behind somite 28 become sympathetic ganglia of the postumbilical gut.

Some neural crest cells in the ventral streams do not migrate as far ventrally and will aggregate into segmentally arranged sympathetic ganglia running the full length of the dorsal aorta. These ganglia contain adrenergic neurons secreting catecholamines. Finally, the last ventral neural crest to leave remains attached to the neural tube and becomes dorsal sensory ganglia along the spinal cord. Other crest cells in this population differentiate as glial cells and associate with ventral spinal nerves growing out of the spinal cord.

Though all neural crest cells have a common origin in the neural tube (presumably all are presumptive neural tissues), they follow different migratory pathways through the tissues, invade specific regions, and differentiate into a wide range of tissue types, from cartilage to neurons (Table 12-2). This raises several questions concerning the developmental behavior of these cells. (1) Is the directed migratory behavior of neural crest due to some intrinsic genetic property of the migrating cells or do they follow predetermined pathways in the surrounding tissues? (2) Under what conditions do dispersed cells stop migration; that is, are neural crest cells attracted to a specific region such as the skin or adrenal gland? (3) Is the fate of a specific neural crest cell population already determined at the moment it leaves the neural tube or does the cell population express a specific cell phenotype depending upon its destination. In other words, is the neural crest a pluripotent cell population, able to differentiate into any cell type depending upon the nature of its interactions along its migratory pathway or with tissues at its final destination?

Directed migration seems to be modulated by the environment and not intrinsic to the cell. For example, chick neural crest migrates more rapidly through somite tissue than in the spaces between somites, which may explain the segmented arrangement of neural crest; it invades a segmented somite environment. If an inverted neural tube is transplanted into a nonsegmented mesenchymal environment of the lateral plate mesoderm, no segmentally arranged collections of neural crest develop. The segmented patterning of these cells is probably imposed by a structured environment into which cells migrate rather than by any predetermined tendency of neural crest cells to migrate out in a spaced fashion.

Furthermore, the initial orientation of the migrating crest cells is related to the orientation of the

TABLE 12-2   Summary of Normal Neural Crest Fates

| | Nervous System | | Skeletal and Connective Tissue |
|---|---|---|---|
| Pigment Cells | Sensory | Autonomic | |

*Trunk Crest (Including Cervical Crest)*

| | | | |
|---|---|---|---|
| 1. Melanophores<br>2. Xanthophores (erythrophores)<br>3. Iridophores (guanophores)<br>    in dermis, epidermis, and<br>    epidermal derivatives | 1. Spinal ganglia<br>2. Some contribution to<br>    vagal (X) root<br>    ganglia | 3. Sympathetic<br>    Superior cervical<br>    ganglion<br>    Prevertebral ganglia<br>    Paravertebral ganglia<br>    Adrenal medulla<br>4. Parasympathetic<br>    Remak's ganglion<br>    Pelvic plexus<br>    Visceral and enteric<br>    ganglia | Mesenchyme of dorsal fin<br>    in Amphibia |
| | 5. Some supportive cells<br>    Glia (oligodendroglia)<br>    Schwann sheath cells<br>    Some contributions to meninges | | |

*Cranial Crest*

| | | | |
|---|---|---|---|
| Small, belated contribution | 1. Trigeminal (V)[a]<br>2. Facial (VII) root<br>3. Glossopharyngeal<br>    (IX) root (superior<br>    ganglia)<br>4. Vagal (X) root<br>    (jugular ganglia)<br>6. Supportive cells[a] | 5. Parasympathetic ganglia<br>    Ciliary<br>    Ethmoidal<br>    Sphenopalatine<br>    Submandibular<br>    Intrinsic ganglia of viscera | 1. Visceral cartilages (except basibranchial 2)<br>2. Trabeculae craniae (anterior)<br>3. Contributes cells to posterior trabeculae, basal plate, parachordal cartilages<br>4. Odontoblasts<br>5. Head mesenchyme (membrane bones) |

[a] Also receives contribution from placodes in lateral ectoderm. Trunk ganglia of nerves VII, IX, and X derived from ectodermal epibranchial placodes. Some supporting cells of these ganglia presumably also derived from placodal ectoderm.
Source: From J. Weston, *Adv. Morphogen.* 8:41–114, 1970.

neural tube. If neural tube is placed upside down, two streams emerge in proper orientation to the neural tube but not to the surrounding environment. Directionality is conferred on the emerging neural crest cells by the neural tube, but how this is accomplished is obscure.

The direction of crest migration is away from the neural tube, dorsally toward the periphery or ventrally into the somite region. In each case cells interact with extracellular matrices filled with collagen fibrils and glycosaminoglycans such as chondroitin sulfate. These substrates may promote cell migration, perhaps because cells make preferred adhesive contacts with the matrix rather than with one another. Cells may continue to migrate as long as adhesion to the substrate is greater than mutual cohesiveness. Migration may stop and cells aggregate when cell cohesion is greater than cell-substratum adhesion. Since the extracellular matrix undergoes developmental changes and is probably of variable composition in different regions traversed by crest cells, we can see how interactions between a migrating cell population and its surrounding matrix might affect the behavior of the population.

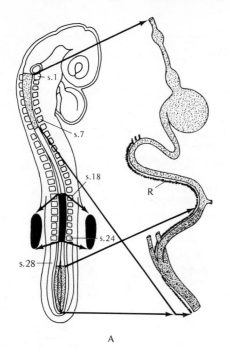

A

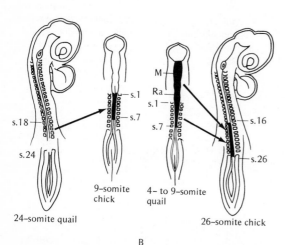

B

**Figure 12-43** *Developmental fate of neural crest and its experimental manipulation. (A) Diagram showing origin of some neural crest derivatives in the chick embryo. Neural crest from region of anterior somites 1 to 7 colonize the enteric ganglia of the whole gut. Those from segments posterior to somite 28 contribute to ganglia of the postumbilical gut. Neural crest from the region of somites 8 to 28 do not contribute to enteric ganglia but give rise instead to adrenergic orthosympathetic neurons. Finally, those from somites 18 to 24 give*

If crest cells localize in different regions of the embryo, are they predetermined as to site of localization at the time they leave the neural tube? To answer this question, LeDouarin and Teillet transplanted quail neural tubes from different regions into chick hosts of different ages (Figure 12-43). When tubes from the 18- to 24-somite region (presumptive adrenal medulla) were placed in the first 7-somite region of a chick host, the neural crest cells invaded the entire gut and differentiated as cholinergic enteric ganglia. These cells followed the available pathways to the gut and differentiated according to the region they occupied rather than their origins. Likewise, a transplant of quail hindbrain into the 16- to 26-somite region of a chick yielded quail crest cells invading the adrenal gland where they differentiated as adrenal medulla along with cartilage and mesenchyme. The cells are not committed to follow a specific route or localize in a specific tissue; they migrate out at random but predetermined pathways guide them to specific sites where they differentiate.

These experiments suggest that the neural crest is a pluripotent cell population that differentiates one of several phenotypes, depending upon the tissue environment with which it interacts, either during migration or at its destination. If cells are in the skin or dermis, they become pigmented; in the gut, they differentiate as enteric ganglia while in the head they become cartilage or cranial nerve. In each case the local matrix and tissue environment are different and may favor one cell phenotype over the others. But these experiments do not eliminate the alternative possibility that many different cell types are determined before cells migrate from the neural tube; the environment in this instance would favor proliferation and survival of only one type.

Attempts to distinguish these alternatives have been unsuccessful so far. Chick neural crest may be grown in vitro to see if phenotypic expression can be experimentally manipulated. When grown in dense cultures, in media that favor aggregation and minimal outgrowth, most cells differentiate as neural and glial

*rise to adrenomedullary cells. R = rhombencephalon. (B) Operation scheme in which quail neural fragments are grafted to chick embryos of different developmental stages (heterochronic) and in different sites (heterotopic). M = mesencephalon; Ra = anterior rhombencephalon. See text for details. (From N. M. LeDouarin and M. M. Teillet, Develop. Biol. 41:162–184, 1974.)*

supporting cells, although some pigment cells are also seen at the periphery of such cultures. The pigment cell phenotype is promoted by growing cells under dispersed conditions, in less dense cultures. Since the cultures are derived from a mixed cell population it is difficult to decide whether the same cell is responding differently to two environmental conditions or whether two different cell populations are involved.

The most effective test is that of cloning; that is, produce a population of cells derived from a single progenitor. In this way we are certain the cells are identical and we can determine whether the population will respond differently to different environmental signals. Such clones have been prepared from Japanese quail neural crest and three different clonal populations are obtained: a pigmented clone of small compact colonies, a mixed population of pigmented and unpigmented cells, and a nonpigmented clone of cells that form diffuse colonies. Each clone seems to maintain its characteristic morphology over many cell generations, which suggests that three different cell types may be included in the initial primary outgrowth of crest cells. No nervous tissues have been obtained from such cultures, although the primary culture from which clones are derived will develop neurons after several weeks. This successful cloning of neural crest offers an opportunity to test its pluripotentiality, but so far the definitive experiments have not been carried out. There seems to be no intrinsic specificity in neural crest cells. They wander out at random, assuming a directed quality only because of environmental factors which modulate migratory and adhesive behavior. Cells are not "instructed" about either pathways to follow or regions to populate. They become oriented within an environment which already has developed a structured, patterned quality. This patterning impinges upon a neural crest cell population and directs its developmental fate.

## Differentiation of neurons

Determination of brain, spinal cord, and sense organs is the first stage in nervous system development. Neuronal centers in brain and cord are wired up to peripheral sense and effector organs (muscle, endocrine glands, and viscera) to establish a functionally integrated system. This is accomplished by the differentiation of nerve cells or neurons with their long axonal processes. Cells of sheets and tubes, now behaving as members of regionally discrete populations, attach and grow out independently to connect with one another into a functional system. Making the right connections is a fundamental requirement of all newly arising neurons. The wiring program, that is, the manner whereby the elaborate network of neuronal connections is constructed, is essential to development of integrated behavioral patterns. Sheet properties now give way to the growth behavior of individual axon fibers searching through the tissue fabric of the embryo.

Vertebrate spinal nerves grow from the neural tube toward peripheral effector organs such as muscles. Instead of migrating away like neural crest, the cell bodies of the axons within these nerves remain in the walls of the neural tube. These neurons arise by mitotic activity of cells in the pseudostratified epithelia lining the lumen of the neural tube and brain (Figure 12-44). The elongated cells are attached at basal surfaces while their nuclei move to and fro within the extended cytoplasm, where they undergo mitosis. Germinal cell nuclei begin a cell cycle at the edge of the lumen after a telophase and migrate to the periphery to replicate DNA. They then return to the border of the lumen to enter mitosis and divide. This migratory behavior of nuclei can be inhibited by colchicine and cytochalasin which suggests microtubules and microfilaments may be involved in this phenomenon of intermitotic nuclear migration. Firm desmosome attachments are seen between cells near the lumen and when nuclei enter mitosis the cytoplasmic processes at the periphery retract and the cells round up to divide. After division a cytoplasmic process is again extended to the periphery with the nucleus migrating behind. Presumably a prescribed number of divisions are completed before neurons begin to differentiate. At the completion of the last mitosis, a cytoplasmic process is again extended to the periphery but this time the desmosome attachment at the luminal surface detaches and the cell is released. It assumes a bipolar morphology as the nucleus migrates toward the periphery. The proximal cytoplasmic process is withdrawn and the cell becomes unipolar, its distal process now extending as an axon with a growth cone at the tip. The growth cone is the active growing point of the neuron and is the first to make contact with the surrounding matrix environment. It is this region that determines the direction of axon growth.

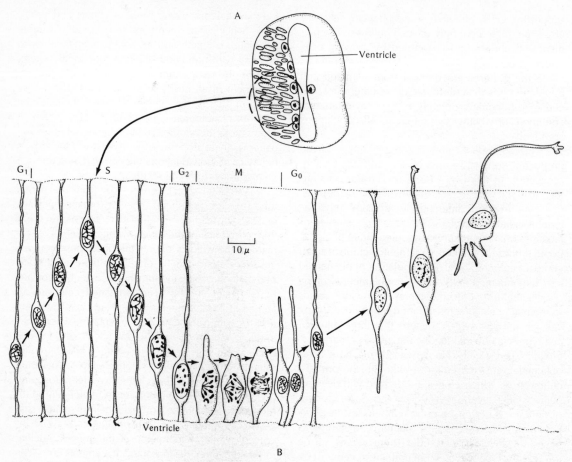

**Figure 12-44**  *Intermitotic migration of neuroepithelial nuclei in wall of developing neural tube of the chick embryo and cycle of neuron differentiation. (A) Cross section of neural tube showing ventricle and location of mitotic figures at ventricle border. (B) An enlarged view of the neural tube wall in which the intermitotic migration is followed at half-hour intervals. Cells enter S as the nuclei migrate away from the lumen after a previous division. After completing DNA replication the nuclei migrate back toward the lumen and enter mitosis. See text for additional details. (Adapted from F. C. Sauer, J. Comp. Neurol. 62:377–405, 1935.)*

Mitotic activity varies in different regions of the neural tube, being most active in regions of maximum nerve outgrowth, that is, where limbs emerge. As young neurons wander out toward the periphery of the wall, the neural tube differentiates into distinct layers; an innermost zone of cells in mitosis, an intermediate zone (mantle) of young neurons, and an outer marginal zone of peripheral cytoplasmic processes (growing axons) extending from cells in the germinal layer.

Young neurons differentiate as a filament of cytoplasm, the incipient axon, grows out of the peripheral end. This was first seen by Ross Harrison in 1907 after he explanted a piece of the neural tube of a frog embryo and watched the axon grow in vitro. He noted the broadened growth zone at the axon tip from which many fine filopodia extended and retracted rapidly in all directions, as if surveying the immediate environment. The axon tip acts like a ruffled membrane gliding over the substratum, presumably making and breaking minute adhesive contacts.

One characteristic feature of axon cytoplasm is the presence of oriented microtubules (neurotubules) and microfilaments (neurofilaments). Within the growth cone, no microtubules are present; instead the microspikes or fine extensions of the cone are filled with a lattice of fine filaments. They may account for the extensive contractile activity of the growing tip since they resemble the actin lattices seen in motile fibroblasts (see Chapter 8). Longitudinal extension of the axon is probably supported by neurotubules acting as a skeletal framework, while neurofilaments may contribute to the flow of axon cytoplasm. An axon grows as new membrane is synthesized behind the growth cone from vesicles assembled in the cell body. These vesicles flow through the axon cylinder from the cell body to the growth cone. Depending on temperature, axons grow rapidly, at an hourly rate of 15 to 56 $\mu$ in amphibians and at roughly the same rate in chick neurons. Axon tips extend, contract, and branch, an indication of the dynamic changes occurring in their surfaces.

How does polarity of the neuron become established; that is, why does an axon grow out of one end? Orientation of the neuroblast in the wall of the brain or spinal cord may determine the site of outgrowth. We would expect cytoplasmic outgrowth to originate from the free peripheral end since the other end is attached to the inner surface of the lumen by desmosomes.

Migration and outgrowth of individual neurons account for the intricate wiring of the nervous system. Growing axons make specific connections with the central nervous system, migrating long distances and by-passing many neighboring neurons to do so. They reach out to muscles and sense organs as if guided by local signals. Neurons acquire this recognition property early in neurogenesis, possibly in the form of surface qualities that assist them in seeking out specific pathways to peripheral organs.

One of the most dramatic cases of directed growth of axons is demonstrated by the giant Mauthner neurons in fish and amphibian brains. A pair of these giant neurons is located in the hindbrain of amphibian larvae (Figure 12-45). They receive many afferent inputs from the so-called lateral-line sensory system and the vestibular nerves coming from the ear. The neuron sends out a large axon that first crosses to the opposite or contralateral side, then grows caudally almost the full length of the body. It sends out collateral efferent axons that innervate the ventral motor neurons in the spinal cord. Hence the Mauthner cell is an interneuron that conveys impulses (water vibrations) from the lateral line and auditory

receptors to the motor output system of the cord. The neuron mediates the "startle response" in fish and larval amphibians, a sudden flexion and extension of the body in response to the stimulus. It is an escape reaction, essential to aquatic forms.

Mauthner cells are determined early in the amphibian neurula. Ablation of a portion of the presumptive hindbrain of the neural tube containing the future Mauthner cells produces an embryo completely lacking in these cells. On the other hand, when a piece of neural tube containing presumptive Mauthner neurons is transplanted into a host neurula, a larva with two additional Mauthner cells develops. If the anterior-posterior orientation of the graft is main-

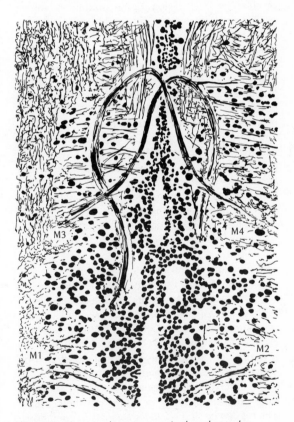

**Figure 12-45** *Mauthner neurons in the salamander* Pleurodeles waltlii. *M1 and M2 are the normal Mauthner's neurons of the host (duplicated on left side) while M3 and M4 are Mauthner's neurons in a medulla graft with anterior-posterior axis reversed. The axons of the grafted neurons decussate, curve back, and grow in the correct direction. (From E. Hibbard, Exp. Neurol. 13:289–301, 1965.)*

tained in the same orientation as the anterior-posterior axis of the host, the grafted Mauthner cells develop a normal axon outgrowth which decussates and extends caudally, just as do the host neurons. If the graft is rotated 180 degrees, its anterior-posterior orientation is inverted with respect to the host body axis. Mauthner cell axons emerge as they do normally, with proper polarity, and grow in an anterior direction toward the head for a short while, but soon they loop around, decussate, and extend caudally in proper orientation with respect to the host body as if growing tips recognize preferred pathways on which to grow.

This does not seem to be the case, at least in the Janus head experiment carried out by Hibbard. If two presumptive head regions of the amphibian embryo are cut from an early neurula and fused together so that each anterior portion faces in an opposite direction, a few of these monsters will develop into double-headed embryos which survive long enough for Mauthner cell development. An examination of the Mauthner cells in these embryos indicates that they develop in their normal position and send their axons posteriorly in the normal fashion after decussation. The fibers penetrate the tissues of the opposing head and move anteriorly, totally contrary to the normal posteriorly directed growth in an intact embryo. It appears as if the growing tip cannot distinguish a normal from an abnormal environment, which suggests that once the intrinsic polarity of the neuron is established and axon growth is initiated, there is no change in the direction of growth under these conditions.

In addition to the intrinsic polarity of the neuron, exogenous factors also influence the direction of axonal growth; these factors are chemical cues in tissue spaces, on basement membranes, and in extracellular matrices. The growth cone surfaces, by virtue of their special motile and contact properties, respond with directed growth. This can be demonstrated by growing embryonic neurons in vitro. Explants of chick embryo dorsal spinal ganglia will develop large numbers of growing axons and their growth patterns can be studied under a variety of conditions. First, it can be shown that the elongation of the axon—or, more exactly, its motile behavior—differs significantly from the movement of neighboring glial cells, fibroblastlike cells that accompany neurons in large numbers. Whereas glial cells are pulled over the surface by their ruffled membrane, only the grow-

ing tip of the axon is motile. The neuron cell body is attached to the substratum while the highly active growth cone attaches and deadheres as it "searches" the environment. It throws out numerous microspikes filled with a microfilament network, and these exhibit active pinocytosis and exocytosis. Meanwhile, new membrane is added to the axon, behind the growth cone, as the axon elongates.

The nature of the surface contacted by the growth cone seems to affect axon growth. Many more axons grow out more rapidly on a substratum of glial cells or even heart fibroblasts than on plastic or glass. In fact, axon growth is stimulated if the dish is coated with a layer of cartilage or gelatin, indicating that the axon "prefers" certain substrata over others. This is further demonstrated by comparing axon growth on dishes coated with palladium, polylysine, or polyornithine. The polyanions are preferred substrata and promote axon elongation. Orthogonal patterns of polylysine may be placed over a plastic or palladium background, and the fibers tend to migrate within the polylysine regions and trace out the patterns. Motility of growth cones is promoted by specific substrata, suggesting that in vivo, specific pathways are sought out during axon growth. This may be similar to migrating neural crest cells which follow directed pathways to the skin and other target tissues. Inasmuch as glial cells are constant companions of growing axons, it is likely that glial cell surfaces (or the extracellular matrix surrounding glial cells) serve as pathways for axon growth in the developing nervous system.

As the neuron develops its characteristic function, it is incorporated into a neuron action system that underlies behavioral networks. Functional patterns emerge, at first generalized, but later these are refined and become more focused and adaptive. The intrinsic response patterns built into the genome of a developing nervous system are modified by the environment as an organism learns. In a short time, a two-dimensional sheet is transformed into an information-transforming system capable, in man, of the most abstract creative expression.

## MESODERMAL TUBES: FORMATION OF THE HEART

Major derivatives of the mesodermal sheet in vertebrates are the heart and its accompanying blood ves-

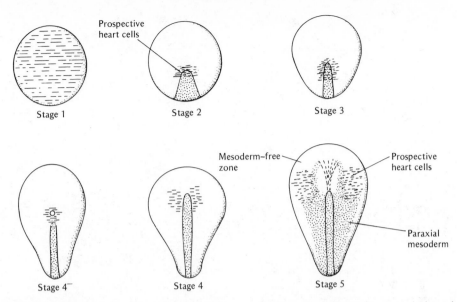

**Figure 12-46** *Progressive localization of heart-forming cells in the chick embryo. Hatched area marks heart-forming cells. (Based on D. Rudnick,* J. Exp. Zool. *79:399–425, 1935; and M. Rawles,* Physiol. Zool. *16:22–42, 1943.)*

sels. Like the ectoderm, parts of this sheet transform into tubes and retain this tubal structure throughout the life of the organism. The morphogenetic factors underlying heart formation differ only slightly from those underlying neural tube formation in ectoderm.

Heart morphogenesis begins early in the chick embryo. Before 24 hours of incubation, in prestreak and early primitive streak stages, potential heart-forming cells are found throughout the blastodisc. Any part of the blastoderm explanted in vitro will give rise to pulsating heartlike tissue (Figure 12-46). With the formation of the primitive streak and mesodermal invagination, however, heart-forming capacity is restricted to the posterior half of the embryo. Later, when the head process lifts off the yolk, heart-forming cells are limited to a pair of oval zones, one on each side of the embryonic axis. Heart formation is another instance of progressive restriction of developmental potency with advancing embryonic age.

At 24 hours of incubation, the thickened portion of the mesodermal sheet near the anterior midline delaminates into two layers, a *splanchnic* and a *somatic mesoderm* separated by the coelomic cavity. The heart primordia develop within the widened portion of the coelomic cavity; these are known as the *amniocardiac vesicles* located on either side of the protruding foregut.

The heart forms from a pair of narrow vessels that fuse into a single tube (Figure 12-47). The vessels develop from loose mesenchymatous cells which detach from the inner splanchnic layer and aggregate into a thin-walled tube. Shortly thereafter, each tube is partially surrounded by a thickened layer of columnar cells coming from the splanchnic mesoderm. Between 27 and 29 hours, the thin double endocardial tubes fuse in the midline below the foregut, and by 30 hours, a single tubular heart appears, made up of a thin endocardial layer surrounded by a thick-walled myocardium.

The paired heart primordia fuse in a cephalocaudal direction; the anterior ventricular region fuses before the auricular and sinus regions. The heart continues to elongate, bulging and twisting into an S-shaped tube beneath the foregut. In the amphibian, this tendency to change shape is intrinsic to the heart and not to any mechanical pressures exerted by the surrounding tissue environment. Isolated pieces of heart primordia in vitro transform into typical S-shaped tubes with bulges in the right places.

Fusion is incomplete anteriorly, and the original pair of tubes forms the ventral aortae. These vessels curve upward over the edge of the foregut and continue posteriorly as a pair of dorsal aortae. At the posterior end of the heart, in the region of the *anterior*

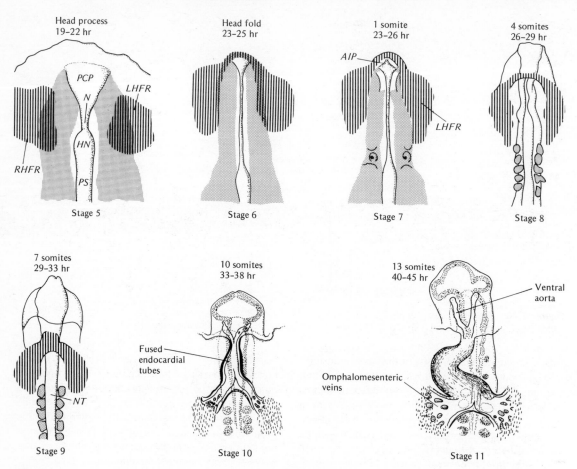

**Figure 12-47** *Early development of chick heart (Hamburger-Hamilton stages 5–11). Paraxial mesoderm and somite derivatives shown in horizontal hatching. Primitive streak (PS), notochord (N), and prechordal plate (PCP) are in midline. Vertical hatching indicates cardiac mesoderm in left (LHFR) and right (RHFR) heart-forming regions. NT = neural tube; AIP = anterior intestinal portal; HN = Hensen's node. (Based on R. DeHaan,* Acta Embryol. Morphol. Exp. 6:26–38, *1963.)*

*intestinal portal*, fusion is also incomplete, and the tubes develop into two large veins, the omphalomesenteric veins, that drain the extraembryonic circulation already developing on the yolk.

Myocardial cells differentiate when oriented myofilbrils appear and the typical striated structure of heart muscle spontaneously assembles. Contractility begins precociously in the chick embryo to facilitate yolk absorption through the extraembryonic circulation. Between 34 and 40 hours of incubation, the first arrhythmic heart beats are noted; within a few more hours, a rhythm is established.

At early streak stages, heart primordia behave as organ-forming fields; that is, double hearts develop if the two lateral primordia are prevented from fusing. Regionalization begins in older embryos; anterior ventricular regions differentiate from posterior auricular regions. At these later stages, isolation of fragments gives only partial hearts. Presumably, this polarity is specified early. In the early streak stages, polarity is labile; a heart primordium rotated 180 degrees develops with normal orientation. If rotation is carried out at a later stage then polarity is already fixed and inverted hearts develop. As in the nervous system, regional specification is a progressive process; potency is lost as different regions are determined.

Heart formation is not an autonomous property of mesoderm. Here, too, inductive influences play an important role. Interactions with the neighboring endoderm are essential to heart formation. During amphibian gastrulation, the prospective heart mesoderm comes into contact with the underlying endoderm as it invaginates into the blastoporal lip. The paired heart-forming regions can be identified in an early gastrula from fate maps, and their morphogenesis depends on continuing inductive influences from endoderm. For example, if the entire endoderm is excised from a neurula, heart formation is prevented. If lateral plate mesoderm from a gastrula is implanted in the heart-forming area of a neurula, it is induced to form heart though originally fated to form other mesodermal derivatives. Finally, fragments of precardial tissue from a gastrula or neurula develop into hearts in vitro only if endoderm is carried along with them.

The nature of the inductive influence is not clear. DeHaan suggests that endoderm in the chick furnishes a substratum on which precardiac mesoderm migrates. If a piece of endoderm is removed from the embryo with precardiac mesoderm and explanted in vitro, mesoderm migrates along the endodermal surface and differentiates into a pulsating heart vesicle in 24 hours. If the endodermal tissue pattern is altered by explanting it on a plastic surface rather than on nutrient agar, then the mesoderm fails to migrate and does not differentiate. This suggests that mesoderm migration is preprogrammed in endodermal surface features. The early embryonic endoderm is a squamous epithelium, but later, in the precardiac area, cells elongate parallel to the embryonic axis, becoming spindle-shaped and forming an anterior crescent. DeHaan proposes that oriented endoderm furnishes the polarity for migration of the mesodermal sheet and thereby insures the accumulation of the minimal number of cells needed to form a heart. The inductive influence of the endoderm may reside entirely in surface and matrix configurations that permit migration of a critical mass of precardiac mesoderm into the heart region.

## Cell proliferation in heart formation

Until recently, cell proliferation was considered important in heart morphogenesis inasmuch as cell division is a constant feature in the developing heart. It was assumed that twisting of the heart was due to differential proliferation of the myocardium and that heart asymmetry, where the left side dominates the right, was also due to more active proliferation in the former region. We find, however, that this is not so. Mitotic activity in right and left sides of the developing heart is similar, which means that heart morphogenesis is entirely the outcome of migration of sheets and changes in cell shape.

Although mitotic gradients appear within different regions of the heart, these do not correlate with morphogenetic changes. Rather, they relate to muscle differentiation and growth. Cells stop dividing before they differentiate into contractile units. Mitotic activity persists until hatching, albeit at a declining rate as the heart continues to grow.

Two cell types may be isolated in vitro from embryonic hearts taken at different stages, a spindle-shaped myoblast undergoing fibrillogenesis and another mitotically active fibroblastlike cell. The former cells begin to contract spontaneously in vitro and behave as pacemakers of heart function. The latter are more mobile and move about freely. We do not know whether two distinct cell populations co-exist in the developing heart or the two cell types denote two different stages of heart cell differentiation.

## Pacemaker cells

Spontaneous beating of the chick heart is detectable at the 10-somite stage, after the preventricular portion of the tube has differentiated. Even in 9- and 8-somite embryos, before contractility begins, rhythmic action potentials can be recorded in the posterior end of the tubular heart. Heartbeats begin in the ventricular region first and spread throughout the heart as each new region differentiates. The beat is spasmodic at first, but a rhythmic rate of about 30 to 40 beats per minute develops in a few hours. By the time the ventricle has looped, heart rate increases to 80 to 90 beats per minute, and after about 60 hours of incubation, the heart rate reaches the normal embryonic rate of 110 to 120 beats per minute.

The heart rate is region-specific; that is, each region has an autonomous beat. If a heart contracting at 120 beats per minute is cut into three fragments, only the sinoauricular region retains the rapid rhythm, while the other two fragments revert back to their slower rates. This implicates the sinoauricular region as the pacemaker. How is the intrinsic rhythm

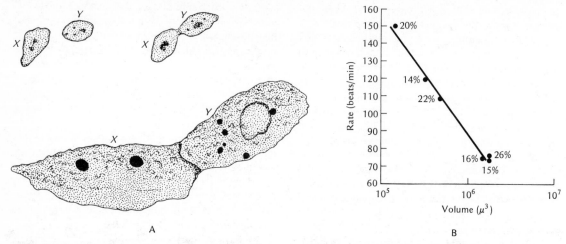

**Figure 12-48**    *Pacemaker cells and the effect of aggregate size on contraction rate. (A) Two M cells are beating, Y at a higher rate than X. When the two make contact, the rate of both cells is intermediate. The contact between cells is intimate, with evidence of membrane specializations between cells (desmosomes?). (B) Pulsation rate as a function of aggregate volume of 7-day heart cell aggregates. The percentage of cells without myofibrils is also given for each of six aggregates. (From H. G. Sachs and R. L. DeHaan, Develop. Biol. 30:233–240, 1973, and R. L. DeHaan and R. Hirakow, Exp. Cell Res. 70:214–220, 1972.)*

of cells that make up the region synchronized? One assumption is that pacemaker cells entrain all adjacent cells. A beat initiated in pacemaker cells is communicated instantly to neighboring cells, possibly across special contact points known as *intercallated discs*.

When embryonic heart cells are isolated in vitro, they exhibit a range of intrinsic rhythms. Some cells beat more frequently than others. The true mark of a pacemaker cell would be the induction of a higher frequency beat to neighboring cells on contact. In most cases, as two beating cells make contact, they establish a beat intermediate between the two instead of both beating at the higher rate (Figure 12-48). These cells may be made to aggregate into spherical masses of several hundred cells. The aggregates beat at rates inversely proportional to aggregate volume and not to the higher frequency of presumed pacemaker cells. Large aggregates have slower rates than small aggregates. The cells in such aggregates are tightly coupled electrically, presumably because of gap junctions. An electrode placed in one region will pick up currents generated in more distant regions without any decline. The system is isopotential, behaving almost like a single cell. Though rate and pattern of impulses generated in beating cells may

be genetically programmed, the properties of a beat are affected by many factors including aggregate size and the ionic content of the medium.

What conditions favor differentiation of intrinsic contractility? To answer this question, dissociated cardiac cells from 7-day embryos were studied. Roughly 50 percent of the cells beat spontaneously at widely different rates from 50 to 290 beats per minute, with a median beat of 116 beats per minute. DeHaan found that potassium ion concentration is the most important variable affecting contractility; the proportion of beating cells is inversely proportional to potassium concentration. By optimizing culture conditions, DeHaan showed that the highest proportion of beating cells (about 50 percent) occurs in 7-day hearts and progressively declines to about 14 percent in 18-day hearts. These studies on the origin of the heartbeat imply that the old view of a pacemaker cell entraining all neighboring cells is too simple. The nature and size of tissue organization may play a far more important role since these factors affect the overall beat, possibly by controlling rate of impulse transmission from cell to cell.

Delamination and reaggregation are the formative factors in heart morphogenesis. Mesenchyme cells detach from a two-dimensional sheet and reaggregate into small clusters of cells which, after migration, fuse into paired endocardial tubes. Sheet migration initiates the process as heart-forming regions condense from lateral mesoderm. Changes in cell shape transform the tube into a looped structure constricted into specific regions. After the tube forms, proliferative activity diminishes, and contractile cells differentiate and beat spontaneously. A rhythmic beat, established early, well before the heart assumes its mature form, continues throughout the life of the organism.

## SOMITE FORMATION: CONDENSATION OF A SHEET INTO SEGMENTALLY ARRANGED BLOCKS

An important morphogenetic event in vertebrate development is the appearance of segmentally arranged somites in the embryonic mesoderm. A sheet of loosely bound cells condenses into a series of compact blocks of tissue on either side of the developing neural tube and notochord. These early events mark the first signs of segmentation, a basic feature of the vertebrate body plan. The process has been examined in the chick embryo by Lipton and Jacobson (Figure 12-49).

Regression of the primitive streak, as we have seen, is an essential feature of chick gastrulation. As the streak regresses, the thickened Hensen's node moves posteriorly, and mesoderm cells ingress freely from the streak into spaces between epiblast and hypoblast. As the node regresses, cells detach from the nodal region anteriorly and condense into the presumptive notochord, segregating the mesoblast into right and left halves. Meanwhile, cells in the epiblast above migrate toward the midline and begin to condense into the presumptive neural plate. These cells elongate into columnar shapes and are distinguished from the more lateral epidermal cells which are shorter and less compact. The migrating mesoblast cells, beneath the epiblast, attach to the basal lamina on the inner surface of condensing neural plate cells and condense into a presumptive somite by forming tight junctions. These somite cells elongate after attaching to the basal lamina while the more flattened, unattached mesoblast cells below continue to migrate. More cells accumulate in the future somite area, possibly by virtue of their greater adhesive tendencies. Nonsomite cells beneath the presumptive epidermis show few signs of condensation or packing, which suggests that the neural plate cells influence the behavior of somite cells below.

As the neural tube begins to roll up, the adherent somite cells remain for awhile but soon detach from the basal lamina and a space, probably filled with matrix materials, appears between the somite and tube. The formation of segmentally arranged clefts in the somite block begins dorsally as mesoderm cells detach from the undersurface of the neural plate at periodic intervals. Clefts also appear from the lateral side and cut inward toward the midline to carve out the tissue block. Presumably in cleft regions, cells become less adhesive with neighbors on one side and join the more compact mass of the somite. Later, the surface of the somite is surrounded by a basal lamina, to give it additional integrity, and long fibers extend from the somite toward the notochord developing between a pair of somites, probably stabilizing somite patterns.

What factors contribute to somite formation? Many experiments have implicated the notochord, the node or a pair of somite-forming centers on either side of the node, or even regression of the primitive streak as responsible for somite formation (Figure 12-50). No somites develop if the node and notochord are removed, if somite-forming areas are removed, or if the entire anterior portion of the streak is ablated. Though each model implicates a different factor, it is likely that no single factor is primarily responsible. Lipton and Jacobson suggest a combination of several factors including node regression, notochord formation, and neural plate induction. To test these models they carried out a series of experiments on somite-forming mesoderm in chick embryos.

To obtain mesoderm tissue unaffected by any inductive influences from neighboring tissues, Lipton and Jacobson divided a head process blastoderm (before any somite formation) into two fragments, fragment I containing the node, primitive streak, head process, and somite-forming centers, and fragment II lacking all (see Figure 12-50). Each fragment was cultured to determine whether somites would form. Fragment I with all tissues present was expected to form somites while fragment II, lacking any influence, was expected to form no somites. Surprisingly, three to seven somites appeared in fragment II in addition to some neural plate, but no notochord developed. These somites appeared within 10 hours after

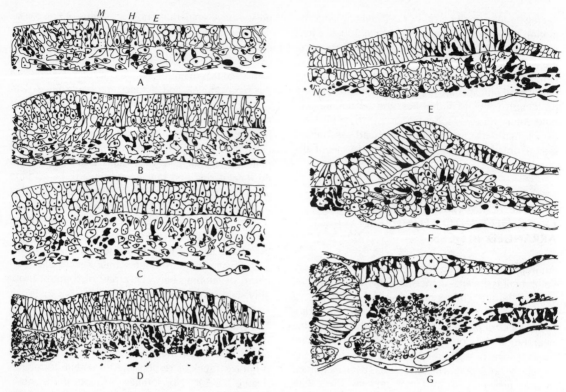

**Figure 12-49**  *Development of somites in the chick embryo. Transverse sections through level of third somite. All sections at same magnification with midline of embryo at left. (A) Primitive streak stage. Mesoblast (M) cells fixed during lateral migration from the streak are bound to epiblast (E) and hypoblast (H). Epiblast is presumptive neural plate, with flat basal lamina surface which differs from irregular basal surface of primitive streak cells on left merging into mesoblast. (B) Head process stage. (C) Head fold stage. Presumptive somite mesoderm (pS) forms a loosely arranged columnar epithelium whose basal surface is bound to basement lamina of neural plate cells above. Flattened mesenchymal cells are probably lateral plate mesoderm. (D) One-somite stage. Neural plate has narrowed toward midline in advance of neural tube closure. The adherent presumptive somite cells below have become more closely packed. (E) Three-somite stage. Cells of third somite packed as anterior and posterior borders are forming. No extracellular space seen between cells at this magnification. Somite cells have made new junctions at apical ends as well as at basal lamina. (F) Four-somite stage. Neural fold forming while somite cells below have formed a rosette configuration. (G) Eight-somite stage. Neural tube closed. Third somite well defined but has not organized into sclerotome, dermatome, and myotome. (From B. H. LIpton and A. G. Jacobson, Develop. Biol. 38:73–90, 1974.)*

the operation, but when cultured longer (an additional 14 to 20 hours), the somites broke down and the entire mesoblast layer seemed to disappear.

In another experiment, the node and head process were removed, including anterior tissue and somite-forming centers. The only tissue remaining was posterior streak, and in this case, the embryo was split into two longitudinal segments with somites appearing at each side. Since neither experiment in-

cluded a node or somite-forming centers, how do we explain somite formation in each?

The one thing in common in both was separation into a right and left longitudinal fragment, a process that seems to occur normally in the embryo as the node and streak regress and separate the mesoblast into right and left halves. According to Lipton and Jacobson, the mesoblast, before regression of the node, is prepatterned into a series of segmentally arranged blocks which cannot express this tendency unless released by splitting the mesoblast. Presumably the shearing action of node regression splits the mesoblast and stimulates the condensation into definitive somites. As a test of this hypothesis, some posterior fragments of the type prepared in the previous experiment were split and others maintained in an unsplit condition. Only the split fragments developed somites, and these lacked a notochord. Experimental splitting of the fragments substituted for the regressing node and notochord in separating the mesoblast into right and left halves. Surprisingly, all somites appeared simultaneously in split fragments, whereas normally, somites develop progressively in

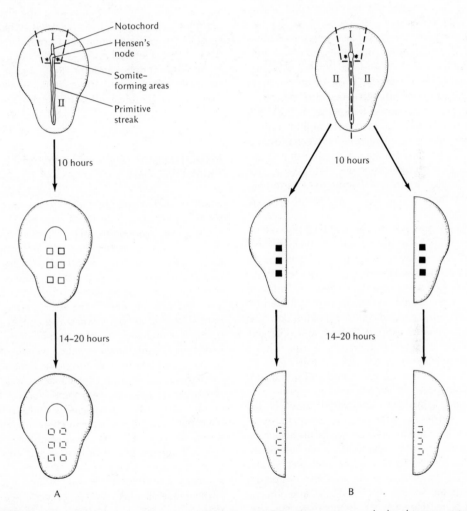

**Figure 12-50**  *Experiments to study factors in somite formation. (A) Operation removes node; head process putative somite centers (\*) and a portion of the streak from a stage-5 embryo (head process). This represents fragment I. The remaining fragment II is cultured for 10 hours and develops somites. With an additional 14 to 20 hours of culture, somites disappear. (B) Operation is similar to that in (A) except that blastoderm is cut down the middle of the primitive streak as well to form two fragment II's. Each develops as a longitudinal strip containing somites, all appearing simultaneously but disappearing after additional culture. (Based on B. H. Lipton and A. G. Jacobson, Develop. Biol. 38:91–103, 1974.)*

an anterior-posterior gradient as the node regresses. This is consistent with the argument that node regression and notochord formation split the mesoblast and activate somite development.

The role of the neural plate is also clear from these experiments. The neural plate is already defined before the axial mesoderm is established. If a piece of epiblast is removed from a head process embryo, cleansed of attached mesoblast cells, and replaced in a 180-degree rotated position over the lateral mesoderm, the mesoblast below is induced to develop into somites. Attachment to neural plate basal lamina seems to induce somite formation. The neural plate may be responsible for setting the prepattern segmental arrangement in the mesoderm, while the shearing action of the regressing node may stimulate the mesoderm to condense into somites. Apparently, inductive interactions between presumptive neural plate and somite mesoderm result in mutual changes in surface properties; neural plate cells elongate while somite cells condense. This interaction is the basis of the repeat pattern of the axial mesoderm.

The subsequent developmental fate of somites is complex. Within the somite mass, which thickens dorsoventrally, a cavity develops. The walls of the cavity differentiate in later development into distinct layers. The outer wall (toward the epidermis) thickens and becomes the *dermatome*. Its cells are the progenitors of all the connective tissues of the skin. The inner wall becomes extremely thick and enlarged, segregating into two discrete masses, an inner loose mass of mesenchymatous cells, the *sclerotome,* which migrate and surround the notochord and spinal cord to develop into the segmented vertebral column (bone and cartilage). A more condensed mass of the wall, the *myotome,* becomes the body musculature. Since the somites are segmentally arranged, the differentiating muscle cells organize into longitudinal parallel bundles in the trunk of the lower aquatic vertebrates. These bundles become the swimming muscles of the tail in fish and larval amphibians. Each set of muscle segments is innervated by segmentally arranged spinal nerves leaving the spinal cord. In these forms, the myotome represents the bulk of the somite, whereas in the higher amniotes, the sclerotome predominates. The myotomes lose their segmental arrangement as amniote embryos adapt for a different life on land with new locomotory patterns.

In the embryo the rodlike notochord is the principal support of the axial system. The notochordal cells, initially condensed into a lengthening rod, begin to enlarge and become highly vacuolated as they synthesize chondroitin mucoproteins. This structure supports the surrounding somites as they develop into longitudinal muscles, and the subsequent swimming behavior of larval fish and amphibians is sustained by the long notochordal rod surrounded by segmentally arranged trunk and tail muscles. Later in development, the cartilage-forming areas of the sclerotome surround the notochord and become the vertebral column, the principal supporting structure of body musculature, replacing the notochord. These cartilage-forming areas are replaced by bone in most fish and higher vertebrates.

## ENDODERMAL TUBULATION: THE PRIMITIVE GUT

Ectoderm and mesoderm start out as two-dimensional sheets and develop into tubes. A major tube is also formed from endoderm, the alimentary tract. It is derived from the archenteron, the endoderm-lined cavity appearing during gastrulation. An invaginating cell sheet becomes a tubular structure as a finger of endoderm tissue penetrates into the embryo (Figure 12-51). The tissue lining the cavity is entirely endodermal, though its outer walls also consist of mesoderm. The archenteron migrates anteriorly into the head region until it makes contact with the ectoderm of the outer body wall. At this point, a mouth breaks through as a stomadeal invagination while at the posterior end of all deuterostome embryos (echinoderms, vertebrates), the anus is derived from the blastopore leading into the archenteron. In protostome embryos (annelids, mollusks), the mouth develops from the blastopore while the anus breaks through as a separate opening.

Soon after it forms, the invaginated archenteron is regionally specified by constrictions into foregut, midgut, an hindgut. Each becomes a different organ system in the vertebrate embryo; foregut develops into pharynx, lung, liver, and pancreas, midgut into stomach, and hindgut into intestine. Cells change shape and local tissues expand into buds which interact with mesenchyme wandering in from the neighboring splanchnic mesoderm to form specific organs such as liver, lungs and thymus. Meanwhile, constrictions appearing along the length of the tube coincide with changes in wall thickness and convert the alimentary tract into chambers with different functional roles.

Early studies of amphibian embryos seemed to indicate an autonomous endodermal differentiation. Isolated pieces of gastrula endoderm differentiated according to their prospective fate without induction. Later, it was realized that these pieces were contaminated by mesoderm. Pure fragments of endoderm isolated from gastrulae and early neurulae did not self-differentiate in vitro even if cultured for more than 50 days. If, on the other hand, a fragment of endoderm from the primitive gut at late gastrula or middle neurula stages was explanted with a piece of chorda-mesoderm, various endodermal tissues developed. The type of differentiation depended on the source and age of the mesoderm. Any endodermal fragment cultured with notochord always differentiated pharyngeal epithelium, whereas liver differentiation was inhibited. Lateral ventral mesoderm favored intestinal epithelium differentiation, and anterior-lateral mesoderm favored liver development. As in neural tube regionalization, specification of endodermal regions depends on inductive influences from neighboring mesoderm.

At the neurula stage, presumptive stomach tissue occupies a narrow band at the anterior region of the archenteron. The embryo lengthens, and as the tube extends in a longitudinal direction, this ring contracts. The band eventually enlarges into a barrel-shaped structure, the future stomach. A short stalk connecting the stomach to the pharyngeal region becomes the primordium of the esophagus.

The most anterior chamber develops into the pharyngeal cavity as ventral epithelium expands laterally into pouches. The pharyngeal pouches become external gills in aquatic vertebrates. Amphibian gills are composed of ectoderm and mesoderm, but morphogenesis is induced by underlying endoderm; if endoderm is removed from this region, gills do not develop. Pharyngeal pouches grow out toward the overlying ectoderm of the head, fusing with the epidermis at secondary constrictions to form gill clefts.

Similar local fusions between endoderm and ectoderm explain the origin of oral and anal openings. The process seems to be identical in both vertebrates and invertebrates. The mouth forms late in vertebrate development, after most primary organ rudiments are established. A mouth breaks through

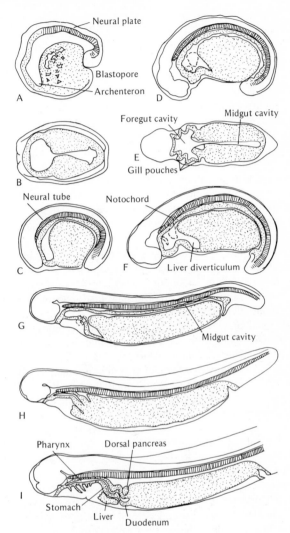

**Figure 12-51**  *Development of endodermal organs in the newt* Triturus taeniatus *from the neurula stage to the swimming larva. (A, C, D, F–I) Saggital sections. (B, E) Frontal sections. (From B. I. Balinsky,* Wilhelm Roux' Arch. Entwicklungsmech. Organ. *143:126–166, 1947.)*

where the endodermal tip of the foregut joins with the invaginating ectoderm in the anterior region of the head. The ectodermal invagination, known as the stomadeum, is a region of thickened columnar epithelium. Two cell layers fuse at the stomadeal invagination into a membranouslike structure which perforates and becomes the mouth. The lining of this newly formed oral cavity is derived from the invaginating ectoderm and endoderm, growing in to fill the surfaces.

The mouth forms because of inductive interactions between the archenteron tip and overlying ectoderm. The inductive stimulus is short, usually expressed at the neurula stage. After induction, removal of endoderm does not affect stomadeal invagination. The induced ectoderm transplanted to another region of the embryo will invaginate as a stomadeum without connecting with underlying gut as long as a small piece of endoderm is also included. Mouth invagination bears certain similarities to lens formation; a stomadeal invagination resembles a thickened lens placode with columnar cells oriented in parallel array.

Lungs, liver, and pancreas arise as diverticula of the endodermal tube. The lung and liver develop as ventral evaginations, whereas the pancreas sprouts as a dorsal bud. The lung first appears as a single diverticulum that bifurcates at the tip into two lateral branches. The unpaired region becomes the trachea and bronchioles leading into two saclike lungs. The important feature of lung development, particularly in lungs of higher vertebrates adapted to terrestrial respiration, is the elaborate branching of the bronchial passages to form numerous small bronchioles leading into air sacs.

The pancreas develops from two rudiments, a ventral and a dorsal bud. Ultimately these two buds fuse and branch into tubules or grapelike clusters of epithelial cells, a process resembling air-sac development in the lung. Both lung and pancreas require an inductive interaction between surrounding mesenchymal cells and endodermal epithelium. Inductive interaction between an endodermal bud and surrounding mesenchyme to form an elaborate branching epithelium is common to many organs derived from the endodermal tube.

## Salivary gland morphogenesis: significance of extracellular matrices

The salivary gland, like many other endodermal derivatives, arises as a small epithelial bud that grows out into a tissue space filled with mesodermal mesenchyme. The mesenchyme surrounds the epithelium and promotes branching growth of grapelike clusters of tubules that make up the salivary gland (Figure 12-52). The interaction resembles an induction between the mesenchymal inducing tissue and the responding endodermal epithelium. Branching results from the continuous appearance of clefts in expanding epithelial lobules. Cleft formation is also the principal mode of morphogenesis in epithelia such as lung.

The entire process can be studied in vitro. A 12- to 13-day mouse salivary bud is isolated at an early stage, before tubule morphogenesis, and the endoderm epithelium is freed from surrounding mesenchyme by mild trypsin treatment. The epithelium may be cultured separately in vitro on a millipore filter in plasma and serum, or recombined with mesenchyme.

When the whole bud is grown in vitro, it continues to develop a characteristic cluster of tubles in 2 to 3 days. Mesenchyme is essential to this pattern of morphogenesis. Epithelium grown without mesenchyme expands on the surface of the filter and fails to form tubules. On the other hand, if salivary mesenchyme is placed on the opposite side of the filter, the epithelium does not expand. The tissue condenses instead, and tubules and terminal clefts appear at about the same time they would develop in vivo.

Only salivary mesoderm has this property of inducing tubule morphogenesis. In the presence of heterogeneous mesenchyme (bronchial mesoderm), the epithelium rounds up into a compact spherical mass but no tubules appear. This indicates a high degree of specificity in induction. Since the average porosity of these filters is less than 0.5 $\mu$, no whole cells can penetrate, but cytoplasmic processes are able to grow to the epithelial cells on the other side, particularly if the pores are straight-sided and evenly arranged. The interior of the filter is filled with large amounts of extracellular materials, including collagen fibrils, presumably produced by both tissues during morphogenesis. Extracellular materials between tissue layers are essential to this inductive process. Surface contact amidst a matrix-filled space seems to mediate the interaction between the two tissues.

The interface between mesenchyme and epithelium is complex, consisting of a basal lamina made up of collagen fibrils and mucopolysaccharides or glycosaminoglycans (GAG) containing sulfate ions (chondroitin sulfates). Superficial to this well-defined lamina is a more diffuse region of extracellular matrix containing other GAG derivatives and

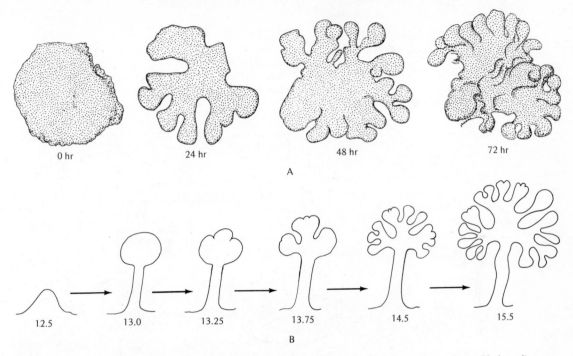

**Figure 12-52** *Salivary morphogenesis. (A) An explant of a whole 13¼-day embryonic mouse submandibular salivary gland grown in vitro for three days. The epithelium continues to undergo typical branching and budding within a mesenchymal mass of tissue. (From M. R. Bernfield, R. H. Cohn, and S. D. Banerjee, Am. Zool. 13:1067–1083, 1973.) (B) A schematic drawing of the branching behavior of salivary epithelium as it develops in vivo in the three days from the first appearance of the salivary bud rudiment to a fully branched epithelium with duct. (From B. Spooner, Am. Zool. 13:1007–1022, 1973.)*

additional collagens. These matrix materials, derived from both epithelium and mesenchyme in a cooperative synthesis, can be specifically stained with dyes such as ruthenium red. In addition, the distribution of newly synthesized GAG can be visualized by radioautography, using such GAG precursors as radioactive glucosamine or sulfate. GAG is detected in the entire basal lamina region, but the most morphogenetically active regions, the buds and lobules that undergo cleft formation, are usually the sites of deposition of new GAG, whereas little or no new GAG is found on cleft surfaces between branches and stalk. Active deposition of GAG in the basal lamina seems to be important in epithelial branching.

Salivary gland rudiments treated with trypsin or collagenase lose the surrounding mesenchyme and the superficial GAG layer (Figure 12-53). When these

gland rudiments are placed in culture with fresh mesenchyme, lobules and clefts disappear and the epithelium becomes smooth and spherical. Later, it develops branches and resumes normal morphogenesis. On the other hand, treatment with low concentrations of collagenase only removes mesenchyme from the rudiments, leaving most of the GAG layer intact. Such epithelia recombined with mesenchyme do not round up; they retain their normal lobulated appearance and continue morphogenesis. It seems that an intact superficial layer of GAG sustains the branching morphology of the epithelium.

With different enzyme treatments, specific GAG components are selectively removed to determine which are most important in morphogenesis. Highly purified enzymes such as hyaluronidase, or chondroitinase, remove GAG components from the basal lamina. Its structure is disorganized and epi-

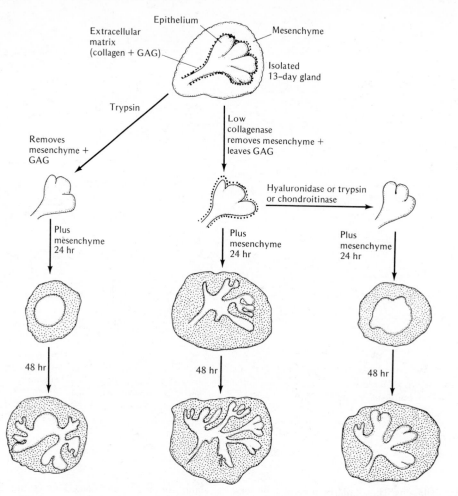

**Figure 12-53**   *Role of extracellular matrix materials in salivary gland morphogenesis. Isolated salivary gland rudiments consist of an epithelial bud surrounded by mesenchyme. The extracellular space between epithelial and mesenchyme cells is separated by a basal lamina (made up of collagen and glycoproteins) plus glycosaminoglycans (GAG) in the matrix. See text for explanation.*

thelia round up when recombined with mesenchyme. No cleft or lobules persist under these conditions suggesting that the integrity of the basal lamina depends, in part, upon its GAG constituents; only after new GAG components are synthesized does morphogenesis resume. Since collagenase also disrupts the basal lamina, we conclude that branching of salivary epithelium depends upon an intact basal lamina. Morphogenesis also depends on cell division since removal of the basal layer also inhibits cell division in the epithelium.

The mesenchyme is essential for continued

morphogenesis; it is required for synthesis of a basal lamina. A low-collagenase treated epithelium with intact basal layer loses lobules and clefts if placed in culture without mesenchyme. The mesenchyme provides some essential components for normal morphogenesis.

How does branching take place, and what is responsible for cleft formation? We have already noted that orientation of epithelial cells is a function of the distribution of microtubules and microfilaments. Microtubules in salivary epithelial cells are oriented parallel to the long axis of the cell. Microfilaments are present in apical and basal regions of the cell, arranged as bands perpendicular to the long axis. Microtubules are not important in controlling shape of these cells. Though epithelia treated with colchicine do not continue morphogenesis, they retain clefts and lobulated morphology; cell shape persists. On the other hand, treatment of epithelia with cytochalasin B, an agent presumed to disrupt microfilaments, does inhibit morphogenesis and all evidence of clefts and contours is eliminated; the epithelium flattens.

The effect is reversible; washing out cytochalasin restores morphogenesis, and tubule formation resumes. Moreover, washing out cytochalasin followed by colchicine treatment does not prevent cleft formation, which can occur in the absence of organized microtubules.

Microfilaments in apical and basal ends disappear when tissues are treated with cytochalasin. Randomly arranged granular components and short lengths of filaments are seen instead, which is consistent with the hypothesis that microfilaments are responsible for cleft formation. During recovery, after cytochalasin is washed out, microfilament bands reappear before morphogenesis resumes. We see then a strong correlation between the integrity of microfilaments and tubule morphogenesis in this epithelium. According to the model, contraction of microfilaments in epithelial cells results in cleft

formation. Continued contraction deepens the clefts, which transform into many epithelial folds and convolutions.

One can envision a stepwise process of tubule morphogenesis (Figure 12-54): (1) a phase of cleft initiation at distinct sites in the epithelium, (2) the elaboration of clefts by contraction of microfilaments in small populations of epithelial cells, and (3) the stabilization of the entire process by extracellular materials including collagens and mucopolysaccharides. The specific inductive role of salivary mesenchyme is still not understood. It does not seem to be essential for microfilament formation, since salivary epithelia exposed to noninducing bronchial mesoderm possess microfilaments but do not undergo cleft formation. Perhaps its initial function is to activate those sites of the epithelium which organize into clefts. We know from colchicine studies that mitotic activity also contributes to tubule morphogenesis. Mesenchyme seems to operate by inhibiting the natural spreading tendencies of epithelium when in isolation and by promoting mitotic activity. We see in salivary gland morphogenesis the cooperative interaction of two tissues which exchange signals that affect cell motility, shape, and mitotic activity while at the same time producing matrix-filled spaces that orient exchanges between cells and stabilize all shape changes that occur.

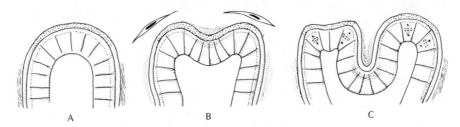

A                          B                          C

**Figure 12-54** *Diagram showing the relationship between extracellular matrix and cleft formation in salivary morphogenesis. (A) Primary lobule. At distal tip (stippled area) newly synthesized GAG accumulates in basal lamina; no new GAG accumulates at lateral edges where collagen fibrils are seen on lamina surface. Microfilaments are seen in basal ends of cells. (B) Early cleft formation. Microfilaments contract within distal cells to produce wedge-shaped cells, and tissue folds in to form secondary lobules. Cytochalasin prevents this contraction. Tropocollagen from mesenchyme assembles into fibrils at surface of incipient cleft. (C) Deepening and stabilization of the cleft and growth of secondary lobules. Cleft deepens because microfilaments continue to contract, and mitotic activity contributes to growth of secondary lobules. Fibrogenesis in primary cleft continues as cleft is stabilized at branch point. This cleft is now insensitive to cytochalasin. Distal ends of secondary lobules accumulate new GAG in basal lamina, and secondary clefts begin to appear as microfilaments contract in basal ends of tip cells. Fibrogenesis in secondary clefts resembles the pattern in primary clefts as each cleft is stabilized. (From M. R. Bernfield, R. H. Cohn, and S. D. Banerjee, Am. Zool. 13:1067–1083, 1973.)*

# LIMB MORPHOGENESIS: A DIALOG BETWEEN AN ECTODERMAL SHEET AND MESODERM

In many inductive interactions, one tissue acts as an inducer and the other responds; the communication seems to be one-way. This is only an operational distinction however, since in all inductive systems, the participating tissues mutually interact. A reciprocal dialog is established with each tissue acting as both inducer and responder, as the fate of both tissues is decided. In eye development, the interaction between optic cup (presumptive retina) and head ectoderm (presumptive lens) involved exchanges in two directions. A similar dialog emerges during vertebrate limb development, but in this instance a two-dimensional sheet, the outer ectoderm, interacts with a solid mass of mesoderm in the center of a limb bud.

## Chick limb development

The chick limb field is determined very early in development, after the neural tube closes and well before mesoderm accumulates as a limb bud. The limb bud arises initially from a condensation of mesenchyme detached from the lateral plate mesoderm (Figure 12-55) in the region of the future wings and legs. Initially the limb field extends widely over the flank of the embryo, covering the area between the future forelimb and hindlimb. An inducing tissue (spinal cord) inserted into any site in the flank may stimulate mesoderm to accumulate as a compact mass under the epidermis, and a limb bud emerges. The determinative event for the future limb bud is usually completed by the 2-somite stage. Rotation and grafting experiments of presumptive limb bud mesoderm show that, as in the eye, the anterior-posterior axis of the future limb is the first to be determined (5-somite stage), whereas the dorsal-ventral axis is determined at 13 somites (see Chapter 16). At the 14-somite stage, the limb bud becomes morphologically visible as a mass of mesenchyme covered by epidermis. Early in development, before the limb bud appears, the future symmetry relations of the adult limb are established.

A dialog between ectoderm and mesoderm begins immediately after a limb bud is stabilized in the direction of limb morphogenesis. The condensed mesenchyme (prospective limb mesoderm) induces

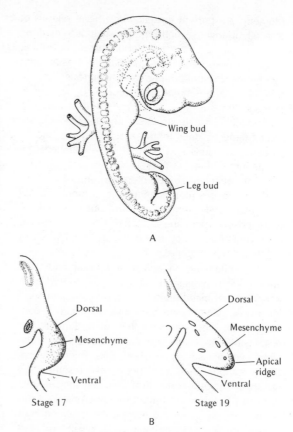

A

Stage 17   Stage 19

B

**Figure 12-55**   *Limb morphogenesis in the chick embryo. (A) Stage 21 with fore- and hindlimb buds extending from body wall. (B) Cross sections of limb buds at stages 17 and 19 showing differences in thickness of ectoderm (apical ridge) surrounding limb mesenchyme at dorsal and ventral faces.*

the overlying epidermis to thicken at the most apical end into a crest of tissue known as the *apical ectoderm ridge* (AER). Epidermal cells divide and migrate distally up the limb bud, piling up at the apex to form the ridge. The basal cells are compressed into a pseudo-stratified layer. The small limb bud elongates from the body wall as a cylindrical structure, a mesodermal core surrounded by an epidermal covering, with the prominent ectodermal ridge restricted to the distal end of the bud. Elongation continues because of mitotic activity in the mesoderm and cell movement and the distal end broadens out into a flattened paddle-shaped structure. The final contours of the limb are sculptured by morphogenetic movements and a pattern of cell destruction that leads to the formation of the digits. The principal tissues of the limb — cartilage, muscle, and bone — begin to differentiate after the mesoderm condenses into discrete masses that delineate the bones and muscles of distal and proximal segments.

### The apical ectodermal ridge

The inductive action of mesoderm is a primary event. Pieces of prospective limb mesoderm, transplanted under flank epidermis, induce the overlying epidermis to thicken, and a supernumerary limb grows out (Figure 12-56). Epidermis from almost any part of the body may cooperate with early presumptive limb mesoderm to form a limb bud. Before the supernumerary limb emerges, the epidermis, in each case, is first transformed into a thickened apical ridge.

Although the dialog is started by mesoderm, it is now picked up by the ectoderm. A naked mesoderm cannot develop into a limb; it must be covered with ectoderm organized as an apical ridge. The apical ectodermal ridge, in its turn, exercises control over the outgrowth of the underlying mesoderm. Saunders found that surgical removal of the ridge prevents outgrowth of the limb and inhibits distal limb differentiation. The earlier the operation, the more drastic the effect; fewer distal structures develop, although most proximal structures (shoulder girdle) are normal and in proper proportion. It seems that mesoderm cannot complete distal morphogenesis unless it is in contact with and receiving signals from an ectodermal ridge.

Zwilling separated sheets of ectoderm surrounding a limb bud from the underlying mesoblast by trypsin treatment. The ectodermal jacket could then be stuffed with mesoderm of different ages or from other sources and reimplanted into the flank of another embryo or grown on the chorioallantoic membrane. He could show, for example, that small blocks of naked limb mesoderm implanted in chorioallantoic membranes do not form limb structures; at best, small nodules of cartilage develop. If similar blocks of mesoderm are first placed in an ectodermal jacket (with an AER) before transplantation to the chorioallantoic site, they develop almost complete distal limb structures (Figure 12-57). Wing mesodermal fragments form wing, and leg mesoderm produces typical hindlimb structures. If, on the other

hand, mesoderm fragments are enveloped in flank ectoderm rather than in limb ectoderm, no limb elements develop. The AER plays an essential inductive role; it promotes morphogenesis and differentiation of distal mesodermal structures. A reciprocal dialog between the two tissues is established.

One advantage of the limb system is the availability of limb mutations that interfere with morphogenesis. Some mutations affect the ectodermal ridge. One such mutation is *wingless*, a homozygous recessive exhibiting no wings and reduced eyes. A limb bud of the wingless mutant emerges at the normal time (three days), but its apical ridge regresses by the fourth day and no limb bud outgrowth occurs. Similarly, in another mutant, *eudiplopodia*, a secondary apical ridge develops which induces duplication of all distal structures.

The exact nature of the AER inductive influence is unknown. Its inductive properties are exerted continuously during limb bud outgrowth but are lost by stage 29. An ectodermal jacket from a stage-29 limb will not induce outgrowth of young mesoblast implanted in it. The signal released by an AER is not instructive; that is, it is nonspecific and brings in no morphogenetic information. It appears to elicit a predetermined program of mesodermal outgrowth and morphogenesis. Its inductive influence may affect the proliferative rate or pattern of the underlying mesoderm, which, in early stages (stage 18-23), is in the form of a proximo-distal mitotic gradient with the high point at the distal end. The gradient persists into later stages, particularly in the central core region, the future chondrogenic cells. This may polarize limb outgrowth, either by promoting directional growth, or by affecting movement of cells in the mesoderm. So far the mechanism of action of the AER is still obscure.

### Role of mesoderm

Survival of the AER is also dependent upon a continued dialog with the mesoderm. If limb ectoderm with AER is combined with nonlimb mesoderm, the ridge flattens and disappears with a concomitant inhibition of mesodermal outgrowth. Moreover, if a mixture of limb and nonlimb mesoderm tissue fragments is stuffed into an ectodermal jacket, only those portions of the ectoderm in contact with limb mesoderm retain an elevated ridge; all other ectodermal regions flatten out. The thickened condition of the

A

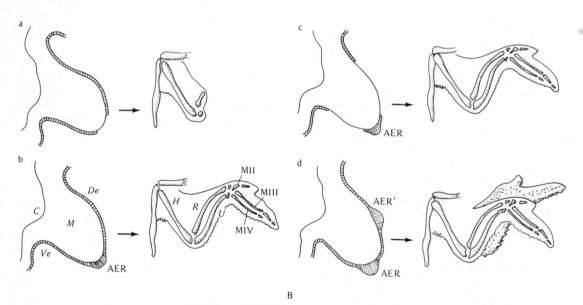

B

**Figure 12-56**   *Role of apical ectodermal ridge in limb morphogenesis. (A) A cross section of early wing bud showing mesodermal mass and overlying ectoderm with thickened apical ridge. (B) Experimental operations on apical ridge. On the left are diagrams showing the nature of the operation performed, and on the right the results obtained.*
*(a) Excision of the ridge results in deficiencies of terminal parts; no outgrowth takes place. (b) In presence of normal ectoderm, wing develops in normal proximal distal sequence and pattern. (c) Removal of dorsal and ventral ectoderm permits outgrowth to occur as long as apical ridge is intact. (d) Grafting an additional apical ridge to the dorsal surface induces outgrowth of another wing. AER = apical ectodermal ridge; AER' = grafted apical ridge; C = coelom; De = ectoderm of dorsal face of wing bud; H = humerus, M = mesenchymal core; R = radius; U = ulna; Ve = ectoderm of ventral face. MII, MIII, and MIV are metacarpals II, III, and IV respectively. (From J. W. Saunders,* Patterns and Principles of Animal Development. *New York: Macmillan. Copyright © 1970 by John W. Saunders.)*

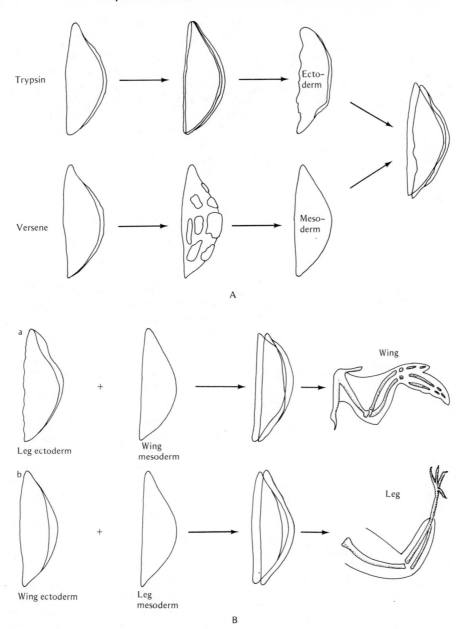

**Figure 12-57**  *Importance of mesoderm in limb morphogenesis. Ectodermal jacket procedure developed by Zwilling. (A) Preparation of test system. A wing or limb bud is treated with trypsin to remove the intact ectoderm jacket or with versene to denude mesodermal core. Then the two components are combined with mesoderm inside ectoderm. The tissues are grafted to the coelom of a host embryo and allowed to develop. (B) Combinations are made between ectoderm and mesoderm from wing and leg buds. (a) Leg ectoderm surrounding a wing mesoderm gives rise to wing structures. (b) Wing ectoderm combined with leg mesoderm produces leg structures, indicating that the specific nature of the limb outgrowth is determined by the mesoderm.*

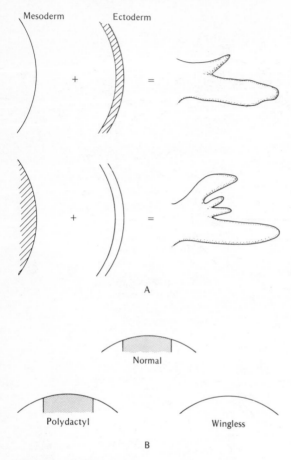

**Figure 12-58** *Use of mutants in analyzing limb morphogenesis. (A) Example of reciprocal interactions between ectoderm and mesoderm of normal and polydactyl (hatched area) limb buds. Polydactyly results when mutant mesoderm is used, indicating that the mutant defect resides in the mesoderm. (B) Probable distribution of apical maintenance factor in normal, polydactyl, and wingless limb buds. (From N. J. Berrill, Developmental Biology. Copyright © 1971 by McGraw-Hill, Inc. Used with permission of McGraw-Hill Book Company.)*

ectodermal ridge depends upon factors coming from the underlying mesoderm. The ability of limb mesoderm to sustain the ridge also diminishes with age. It is strongest in early stages (stages 16 to 19) when it can induce any ectoderm to become an AER. Subsequently, its inductive capacity falls off; by stages 24 to 25 of limb outgrowth, the AER maintenance prop-

erty of the mesoderm disappears. Aging affects both competence and the inductive properties of these interacting tissues, just as it does in primary induction.

Mutations interfere with the dialog by affecting mesoderm or ectoderm independently (Figure 12-58). Polydactyly is a mutation in which a limb exhibits many additional digits. A polydactylous mesoderm with a normal ectodermal jacket induces numerous thickenings and ridges in the ectoderm, exactly as it does in vivo. Many distal limb structures grow out in response to the multiple AERs. Reciprocal combinations of polydactylous ectoderm and normal mesoderm result in a normal single ridge, suggesting that the mutant defect resides in the mesoderm. In the mutant, either more ectodermal maintenance factor is synthesized, or its distribution in the mesoblast is abnormal so as to induce more apical ridges and additional sites of distal morphogenesis.

The *eudiplopodia* mutant, however, affects the ectoderm; combinations of eudiplopodia ectoderm with normal mesoblast result in a secondary ridge in the dorsal region and the production of additional digits. Normal ectoderm with a eudiplopodia mesoblast is completely normal. In this mutant, therefore, the ectoderm is considerably more responsive to the mesodermal inducing signal (apical ectoderm ridge factor), and a secondary ridge develops.

Mesoderm also governs ectodermal differentiation. Normally, chick hindlimb ectoderm differentiates into scales, whereas forelimb ectoderm differentiates into wing feathers. When hindlimb mesoderm is combined with wing ectoderm, the outgrowth is typically that of the hindlimb, and the ectoderm develops scales. The reciprocal combination also gives the expected result; that is, hindlimb ectoderm exposed to wing mesoderm develops into a typical wing with feathers instead of scales. In other words, the ectoderm can display two differentiation programs; the one that is followed depends on an instructive signal furnished by the underlying mesoblast. The mesoblast contains the information for the specificity of limb morphogenesis. This specificity persists late into development and is manifested in interaction between dermis and epidermis (Chapter 17).

Morphogenesis of limb tissue is dependent on a continuing exchange of inductive signals. Three factors seem to be implicated; an AER inducing factor from the mesoderm, an AER maintenance factor arising in mesoderm, and a mesodermal polarizing factor coming from the ectoderm. In addition, there is the mesodermal factor that determines feather or scale ectodermal differentiation. We know little about these factors or their mode of action. The AER inducing and maintenance factor is instructive; young mesoderm can induce foreign nonlimb ectoderm to organize as an apical ridge and promote limb outgrowth. On the other hand, the mesodermal polarizing factor is noninstructive; it elicits a program of morphogenesis intrinsic only to limb mesoderm; no other mesoderm can respond. It is not clear whether the mesodermal instruction that directs epidermis to produce feathers or scales is the same or different from the ectodermal maintenance factor acting earlier in limb morphogenesis.

The dialog takes place in the matrix-filled extracellular space that separates the tissues. The nature of the interface is important in the interaction. At the under surface of ectodermal cells a basement membrane develops, consisting of a matrix of mucopolysaccharides, within which collagen fibrils are organized into a complex lamellar pattern. Layers of fibers run at 90-degree angles, overlapping and alternating, forming a woven fabric. Presumably, in the young limb bud the basement lamella is simple at first but becomes progressively complex as more fibrillar layers are laid down. Both ectoderm and mesoderm contribute to its formation; collagen fibrils from the mesoderm are embedded in a matrix, derived, in part, from ectoderm. Morphogenetic interactions between ectoderm and mesoderm are most effective when the basement membrane is immature. After a mature basement membrane develops, mesoderm comes to depend upon continued interactions. If ectoderm is now removed, without affecting the basement membrane covering the mesoderm, then limb growth and morphogenesis are constrained. Ectoderm added to such a system also fails to induce outgrowth. It is only when ectoderm is applied to a denuded mesoblast, one without a basement membrane, that outgrowth occurs. The ectoderm exerts its inductive effect as a new basal membrane is established. Presumably, only when a joint membrane is established can the ectoderm continue to exist.

The property of "limbness" is confined to the limb mesoderm. Although other mesodermal areas such as somites are equally capable of forming muscle, bone, and cartilage, none forms limb structures even when combined with limb ectodermal ridge and explanted in a host embryo. Cartilaginous nod-

ules appear instead. Zwilling suggested that this difference is due to an early determination of cell surface recognition properties. Cells of limb or somite mesoderm can be labeled with radioactive thymidine and dissociated. Labeled and unlabeled cells from somite and limb are allowed to reassociate into random cell mixtures; then they are packed into compact masses for insertion into a limb ectodermal jacket. Their development is followed after transplantation to the flanks of host embryos. Several days later, the graft is removed and the location of labeled and unlabeled tissue is noted. Limb mesoderm segregates from somite mesoderm, usually at the periphery with the somite in the interior. Moreover, limb mesodermal cells tend to accumulate underneath an ectodermal ridge. When these grafts differentiate, digits appear only within limb mesodermal regions, whereas somite mesoderm, displaced at the base of the grafts, usually forms distorted cartilages segregated from the long bones of the limb. Sorting out occurs with the tissues of the same or different ages and is seen between mesoderms of fore- and hindlimbs. The degree of sorting depends upon the proportion of each in the mixed aggregate, with the minor component displaced away from the major region of differentiation.

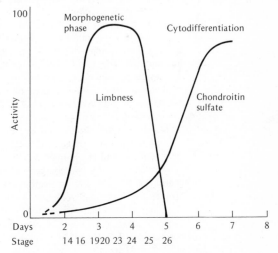

**Figure 12-59** *Time sequence of morphogenesis and differentiation in the avian limb. The morphogenetic phase of limbness precedes cytodifferentiation as assayed by the synthesis of chondroitin sulfate and cartilage formation. (From E. Zwilling,* Emergence of Order in Developing Systems, *ed. M. Locke. New York: Academic Press, 1968, pp. 184–207.)*

By stages 18 and 19 of limb development, this property of "limbness," manifested as specific recognition in sorting out, is restricted to limb mesoderm. Though cells from somite and limb mesoderm are histogenically equivalent, they still segregate on the basis of prospective fate although each synthesizes identical products and differentiates into cartilage and bone. A surface recognition property of the membrane or matrix may account for sorting out and may be the initial step in limb determination.

According to Zwilling, the distinctive surface properties that segregate limb from somite mesoderm do not persist in the late embryo when these tissues have already formed cartilages. If limb and vertebrate cartilages of 8-day embryos are dissociated and recombined in random mixtures, a mosaic of cartilage develops with labeled and unlabeled cells; there is no sorting out.

"Limbness" is a morphogenetic property that peaks early in development and diminishes when cytodifferentiation begins (Figure 12-59). The early morphogenetic phase depends on cell surface and proliferative behavior that determines cell position in the outgrowing limb. Later, it appears as if the specific program of differentiation followed by mesodermal cells within the outgrowth depends upon their position within the limb (see Chapter 17).

**REFERENCES**

Barth, L. G., and Barth, L. J. 1974. Ionic regulation of embryonic induction of cell differential in *Rana pipiens. Develop. Biol.* 39:1–23.

Bernfield, M. R., Cohn, R. H., and Banerjee, S. D. 1973. Glycosaminoglycans and epithelial organ formation. *Am. Zool.* 13:1067–1083.

Burnside, B. 1973. Microtubules and microfilaments in amphibian neurulation. *Am. Zool.* 13:989–1006.

Cohen, A. M., and Konigsberg, I. R. 1975. A clonal approach to the problem of neural crest determination. *Develop. Biol.* 46:262–280.

DeHaan, R. L. 1968. Emergence of form and function in the embryonic heart. *The emergence of order in developing systems,* ed. M. Locke. New York: Academic Press, pp. 208–250.

DeHaan, R. L., and Sachs, H. G. 1972. Cell coupling in developing systems: The heart cell paradigm. *Current topics in developmental biology,* ed. A. A. Moscona and A. Monroy. Vol. 7. New York: Academic Press, pp. 193–228.

Gibbins, I. R., Tilney, L. G., and Porter, K. R. 1968. Microtubules in the formation and development of the primary mesenchyme of *Arbacia punctulata. Develop. Biol.* 18:523–531.

Guidice, G. 1973. *Developmental biology of the sea urchin embryo.* New York: Academic Press.

Gustafson, T., and Wolpert, L. 1967. Cellular movements and contact in sea urchin morphogenesis. *Biol. Rev.* 42:442–498.

Hibbard, E. 1965. Orientation and directed growth of Mauthner's cell axons from duplicated vestibular nerve roots. *Exp. Neurol.* 13:289–301.

Holtfreter, J., and Hamburger, V. 1955. Amphibians. *Analysis of development,* ed. B. Willier, P. Weiss, and V. Hamburger. Philadelphia: Saunders, pp. 230–296.

Karfunkel, P. 1973. The mechanisms of neural tube formation. *Int. Rev. Cytol.* 38:245–272.

LeDouarin, N. M., and Teillet, M. A. 1974. Experimental analysis of the migration and differentiation of neuroblasts of the autonomic nervous system and of neuroectodermal mesenchymal derivatives using a biological cell marking technique. *Develop. Biol.* 41:162–184.

Letourneau, P. C. 1975. Cell to substratum adhesions and guidance of axonal elongation. *Develop. Biol.* 44:92–101.

Lipton, B. H., and Jacobson, A. G. 1974. Experimental analysis of the mechanisms of somite morphogenesis. *Develop. Biol.* 38:91–103.

Lopashov, G. V., and Stroeva, O. G. 1961. Morphogenesis of the vertebrate eye. *Adv. Morphogen.* 1:331–378.

Lovtrup. S. 1965. Morphogenesis in the amphibian embryo. Determination of the site of invagination. *Acta Zool.* 46:119–165.

Manasek, F. J. 1975. The extracellular matrix: A dynamic component of the developing embryo. *Current topics in developmental biology,* ed. A. A. Moscona and A. Monroy. New York: Academic Press, pp. 35–102.

Nicolet, G. 1971. Avian gastrulation. *Adv. Morphogen.* 9:231–262.

Nieuwkoop. P. D. 1973. The "organization center" of the amphibian embryo: Its origin, spatial organization and morphogenetic action. *Adv. Morphogen.* 10:1–39.

Perry, M. M. 1975. Microfilaments in the external surface layer of the early amphibian embryo. *J. Embryol. Exp. Morphol.* 33:127–146.

Roach, F. C. 1945. Differentiation of the central nervous system after axial reversals of the medullary plate of Ambystoma. *J. Exp. Zool.* 99:53–77.

Sanders, E. J., and Zalik, S. E. 1976. Aggregation of cells from early chick blastoderm. *Differentiation* 6:1–11.

Saunders, J. W., Jr., and Gasseling, M. T. 1968. Ectodermal-mesenchymal interactions in the origin of limb symmetry. *Epithelial-mesenchymal interaction,* ed. R. Fleischmajer and R. Billingham. Baltimore, Md.: Williams and Wilkins, pp. 78–97.

Saxen, L., and Toivonen, S. 1962. *Primary embryonic induction.* London: Logos Press.

Spemann, H. 1938. *Embryonic development and induction.* New Haven: Yale University Press.

Spooner, B. 1973. Microfilaments, cell shape change and morphogenesis of salivary epithelium. *Am. Zool.* 13:1007–1022.

Toivonen, S., Tarin, D., and Saxen, L. 1976. The transmission of morphogenetic signals from amphibian mesoderm to ectoderm in primary induction. *Differentiation* 5:19–25.

Trinkaus, J. P. 1969. *Cells into organs.* Englewood Cliffs, N. J.: Prentice-Hall.

Watterson, R. 1965. Structure and mitotic behavior of the early neural tube. *Organogenesis,* ed. R. L. DeHaan and H. Ursprung. New York: Holt, Rinehart and Winston, pp. 129–159.

Weiss, P. 1950. Perspectives in the field of morphogenesis. *Quart. Rev. Biol.* 25:177–198.

Weston, J. A. 1970. The migration and differentiation of neural crest cells. *Adv. Morphogen.* 8:41–114.

Zwilling, E. 1968. Morphogenetic phases in development. *The emergence of order in developing systems,* ed. M. Locke. New York: Academic Press, pp. 184–207.

# 13

# Information Flow in
# Early Embryonic Development

Morphogenesis depends on the flow of information from gene to protein in embryonic cells. Such processes as directed cell movement or selective adhesion and recognition are possible by virtue of cell surface proteins encoded in the genome. The ordering of morphogenetic activities in cell populations may result from the temporal expression of these genes. At critical developmental periods receptor or recognition proteins are inserted into cell surfaces. Later such proteins may be removed and new ones added, thereby changing cell surface properties. This implies that the information encoded in genes is transcribed at different times, and if we could understand how this is regulated we could resolve the causal factors underlying morphogenesis.

How can we get at the problem of information flow during embryogenesis? We may ask whether all or a few genes are transcribing at a certain stage. We may study how maternal information stored in the egg during oogenesis is used at different times during development. Which morphogenetic processes de-

pend upon maternal information and which rely on newly transcribed information from the embryonic genome? We may also ask about the regulation of translation. For instance, are all embryonic messages translated into proteins as soon as they are transcribed or is there evidence of regulation of message utilization? If so, how does this correlate with the specific morphogenetic events in progress?

A biochemical approach assumes that molecular events of information flow are causally related to gross morphological changes in embryos. Embryogenesis is a continuous process and we arbitrarily divide it into distinct phases such as fertilization, cleavage, gastrulation, and neurulation, each expressing a different morphogenetic activity. For example, we would expect division genes to be expressed during cleavage and genes coding for surface proteins and extracellular matrices to be most active during gastrulation and neurulation.

It is possible to study the biochemical events of fertilization since large populations of relatively synchronized unfertilized and fertilized eggs can be compared. Likewise, the events of cleavage may also be studied, particularly during the synchronous phase where most cells are dividing and doing little else. With the onset of mitotic gradients, however, cell populations become heterogeneous; cell cycles vary and cell sheets are regionalized into discrete local populations, each in a different developmental mode. Some may be proliferating, others may be changing shape and becoming motile, while still others are moving into new areas. This is true of gastrulae and neurulae which means it is almost meaningless to analyze whole embryos at these stages and assume the biochemical parameters measured are in any way typical of the cell population as a whole. The problem becomes more acute later when cell populations differentiate into specific functional types, each proceeding independently according to its own developmental program. In such situations, we isolate small fragments of tissues, presumptive organ primordia, to study their biochemical parameters. We cannot do this easily in gastrulation and neurulation stages, which means that most biochemical studies of whole embryos are limited to the events of fertilization, cleavage, and blastulation up to the initiation of gastrulation. In this chapter we will examine some of the biochemical activities characteristic of these phases. Though many processes have been described, in embryos the real problem is to establish causal relationships with morphogenetic events. But we cannot correlate the events of information flow with the specific behavior of cells in division or in tissue displacement, selective adhesion, or sorting out. The connecting statements between the biochemical and morphological levels have not yet been formulated.

## METABOLIC ACTIVATION AT FERTILIZATION

During oogenesis, the metabolically active oocyte synthesizes and accumulates nutrients, yolk, and informational molecules. When fully mature, many eggs becomes less active; levels of macromolecular synthesis decline, and even respiratory activity falls off. Mature eggs are metabolically unstable and tend to "age" or "ripen," surviving in the unfertilized state for hours or, at most, only a few days. If overripened eggs are fertilized, they develop with bizarre chromosomal abnormalities. Metabolically, eggs run

downhill, and unless they are activated by sperm, they die.

Most changes at fertilization reflect a general activation of the egg's metabolic machinery. The repressed egg is "rejuvenated"; its metabolic activity becomes more vigorous. The egg divides rapidly and synchronously and cell size is reduced within a matter of hours. Respiratory metabolism is activated and energy is provided for protein and nucleic acid synthesis as an activation program is initiated at fertilization.

### An activation program

As noted in Chapter 11, the events of fertilization, such as lifting the fertilization membrane, result from a sequence of intracellular changes that initiate development. In the sea urchin embryo, these changes follow an activation program that is completed within a few minutes after insemination (Figure 13-1). The unfertilized sea urchin egg is metabolically repressed, but it is immediately derepressed as several independent metabolic systems are stimulated. The program of sea urchin activation is divided into early events, beginning at 1 to 3 seconds after insemination and lasting for 5 minutes, and late events, beginning at 5 minutes and extending until first cleavage.

Early events begin with an action potential resulting from a sudden influx of sodium ions after the acrosome filament makes contact with the egg surface. The action potential may be preceded by a release of calcium sequestered in the cortex and cytoplasm of the egg, perhaps in the first few seconds. These ionic changes precede the cortical reaction, sperm entry, and the block to polyspermy, which begin at 25 to 30 seconds. Shortly thereafter, there is a burst in activity of the respiratory enzyme, NAD kinase, and within 15 seconds this is followed by a short spurt of oxygen uptake.

The late reactions involve changes in potassium conductance and activation of amino acid and phosphate transport. These are followed by increased protein synthesis along with changes in adenylcyclase, the membrane-bound enzyme regulating intracellular cyclic AMP. Eventually, DNA synthesis is stimulated and the egg begins preparations for first cleavage. Though this sequence of events suggests that late reactions are dependent upon early ones, it is also possible that some reactions occur autonomously.

**Figure 13–1** *Program of activation, or postfertilization changes in eggs of Strongylocentrotus purpuratus at 17°C. Events for which times (minutes) are known are on the right only. (From D. Epel, Am. Zool. 15:507–522, 1975.)*

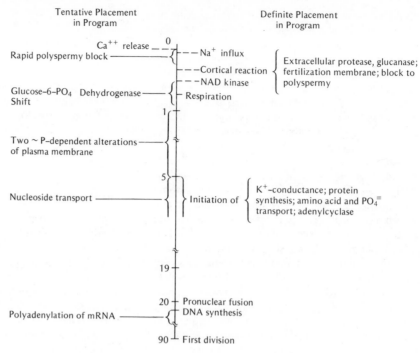

The extent to which late events depend upon early events has been examined in ammonia-activated eggs. Ammonium chloride at pH 8 activates sea urchin eggs, but only late responses are induced. Such eggs show no evidence of Na influx nor do they undergo the cortical reaction or an increase in respiration. They do show such late responses as increased potassium conductivity, DNA synthesis, and increased protein synthesis. Many late events can be activated independently of such early events as respiration: the former are not dependent upon the latter, which means the program of activation is not a single chain as shown in the figure but involves activation of several independent pathways. From all indications it appears that the early events do act as a chain of reactions, perhaps dependent upon Ca++ release.

The intracellular release of bound calcium may be the primary activating signal in many eggs. Parthenogenetic activation of sea urchin and frog eggs occurs when a calcium ionophore is applied in minute concentrations. The ionophore promotes the flux of calcium ions through membranes, and induces membrane elevation, cortical granule breakdown, and an action potential, as well as protein and DNA

synthesis. Just as calcium release activates muscle contraction or a nerve impulse, the initial changes in egg activation depend upon the release of calcium. Free calcium may trigger such early events as cortical granule breakdown and sodium influx, and may account for the transient increase in respiration by activating the NAD kinase.

### Respiratory metabolism

One of the earliest metabolic changes in the sea urchin embryo is activation of respiratory metabolism. At one time it was thought that fertilization activated respiratory metabolism of all eggs. This was based on Warburg's demonstration in 1911 that seconds after fertilization of the sea urchin egg, the rate of oxygen consumption increased 4 to 5 times (Figure 13-2). The initial burst seemed to correlate with changes occurring at the surface, that is, cortical granule breakdown and the cortical reaction. It was thought that fertilization restores an egg's respiratory metabolism to its normal state. This idea, however, proved erroneous; it was found that the pattern of oxygen consumption after fertilization varied in eggs of different species. Some eggs showed absolutely no changes (the clam *Spisula*), whereas oxygen consumption in other eggs (the worm *Chaetopterus*) declined at fertilization. Since oxygen consumption was measured at different times after fertilization in these species, some early activation events were missed and comparisons were meaningless. The only generalization is that rates of oxygen consumption following fertilization are comparable to those of normally growing cells; any changes observed are not necessarily related to activation.

In the sea urchin egg, however, the burst of oxygen consumption may be a key change within the first few seconds after sperm entry. It lasts for 2 to 3 minutes, returns almost to the prefertilization level, and begins to rise again.

We do not understand why oxygen consumption suddenly increases after fertilization, since most oxidative enzymes are unaffected. The activity of some soluble glycolytic enzymes increases slightly and, according to one hypothesis, metabolic activation results from structural rearrangements of inactive enzyme systems. Cofactors or enzymes formerly bound in compartments are released and activated. For instance, glucose-6-phosphate dehydrogenase is released from a particulate fraction to the supernatant in sea urchin embryos. There is also a significant increase in the concentration of NADP and NADPH after fertilization, probably because NAD kinase is activated, converting NAD to NADP.

Enzyme systems may reorganize in a new ionic environment created by increased membrane permeability. The release of protein-bound ions may also change protein conformation and activate enzymes, rearrange their topological relationships, and expose cofactors and activators, thereby stimulating respiratory metabolism.

The initial events of development require energy supplied from stored ATP or activated respiratory metabolism, or both. This energy is needed for the synthesis of macromolecules, particularly those necessary for chromosome replication.

### Nuclear activation: the onset of DNA synthesis

Fertilization activates nuclei and triggers DNA replication. Both sperm and egg nuclei undergo morphological changes prior to first cleavage (see Chapter 11). The former rearranges most dramatically since it enters the egg cytoplasm in a highly compacted, almost crystalline, state. It expands rapidly as DNA unravels and proteins and other cytoplasmic constituents enter a swelling pronucleus. The egg nucleus also enlarges into a pronucleus and as both pronuclei swell, DNA replication begins, as shown by radioactive thymidine incorporation (Figure 13-3).

Other cytoplasmic factors bring the nuclei together, preparing them for fusion, or *syngamy*, prior to first cleavage. Pronuclei migrate toward the central spindle region, a movement that seems to be assisted by microtubules attached to the outer nuclear membranes. DNA replication is usually completed before fusion.

What stimulates DNA synthesis after fertilization? In the sea urchin, thymidine is incorporated periodically into DNA during a 15-minute S phase beginning in telophase of a preceding cleavage division. Initiation of DNA replication may occur as DNA polymerase enters nuclei from the cytoplasm. After activation, the cytoplasmic state seems to be primed for DNA synthesis, but we do not know how (see Chapter 11). Most eggs contain large stores of nucleotide precursors which are utilized immediately for DNA synthesis. In addition, a pool of cytoplasmic tubulin and actin are available for microtubule and microfilament assembly.

Activation commits the egg to DNA replication and division. As a result, the egg's developmental potential is partitioned into cleaving blastomeres. Chromosomes, spindles, and membranes are synthesized and assembled while energy from respiratory metabolism is channeled to pathways for macromo-

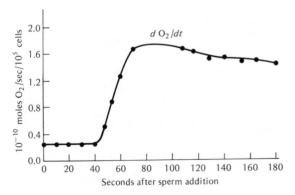

**Figure 13-2** *Rate of respiration after fertilization of eggs of* Strongylocentrotus purpuratus *at 17°C. (From D. Epel,* Biochem. Biophys. Res. Commun. *17:69–73, 1964.)*

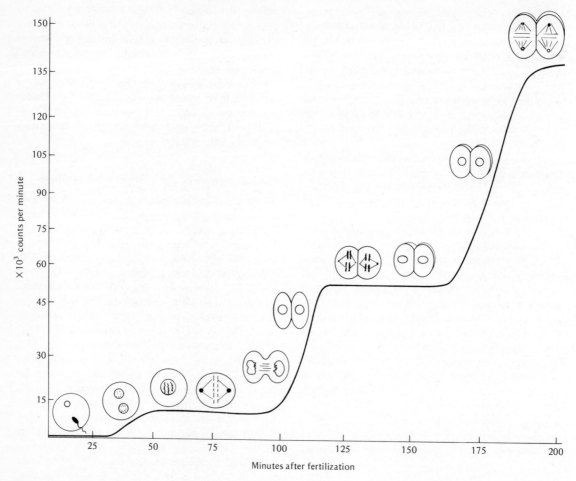

**Figure 13-3**  *Synthesis of DNA after fertilization of sea urchin eggs as measured by radioactive thymidine incorporation. Ordinate is counts per minute incorporated. The first three cleavage divisions are shown to indicate when DNA synthesis occurs. (Based on R. T. Hinegardner, B. Rao, and D. E. Feldman,* Exp. Cell Res. 36:53–61, 1964.)

lecular synthesis. The short generation times characteristic of cleavage provide little time for anything but genome replication.

**Protein synthesis**

Most biochemical studies have been carried out on sea urchin and frog embryos because these are easiest to study at early stages. As more information about other embryos is obtained, we find that no generalizations can be made about macromolecular changes after activation; most biochemical events are unique

for each species. Only a few biochemical patterns are common to all embryos. Even among mammals, where we would expect similarities because of the constancy of embryonic and fetal environments, significant species differences exist.

Since synchronous division is the principal activity of cells immediately after activation, most protein synthesis is concerned with chromosome replication and mitosis. Chromosomal proteins and spindle proteins must be synthesized and assembled to carry out mitosis. Membrane proteins and lipids are also made to construct new membranes during division. It is estimated that the first cleavage division requires a de novo synthesis of about 25 percent new membrane materials.

The amounts of newly synthesized protein are comparatively small after activation. There is no net protein increase; instead, yolk proteins turn over and are transformed into enzyme and structural proteins of the developing embryo.

The sea urchin egg will serve as our principal illustrative material inasmuch as fertilization activates protein synthesis in this system. The unfertilized egg incorporates amino acids into proteins at extremely low levels, but fertilization is followed by a burst of incorporation, as if activation has switched on some mechanism to accelerate synthesis (Figure 13-4). Hours afterwards, the rate reaches levels 6 to 15 times the rate in the unfertilized egg.

The enhanced incorporation in fertilized eggs is not entirely due to an increased permeability to amino acids. Something more fundamental in the regulation of protein synthesis is involved. The burst

of protein synthesis is essential for development. If it is prevented by an inhibitor like puromycin, cell division is blocked and the embryo dies. Though it contains a store of proteins essential to early development, the egg still requires new proteins to initiate the division process.

Although the initiation of protein synthesis follows respiratory activation after insemination, the two may be uncoupled by ammonia, which indicates that energy released by respiration does not trigger protein synthesis.

Some hypotheses regarding the occurrence of activation are the following:

1. New message templates are transcribed before translation can occur; that is, fertilization activates messenger RNA synthesis.
2. Inactive ribosomes of unfertilized eggs may be activated at fertilization.
3. Certain transfer RNAs and activating enzymes may be absent or inactive in the unfertilized egg.
4. Some soluble cofactors, ions or enzymes necessary for protein synthesis, may be absent or inactive in the unfertilized egg.
5. Message templates are present, but inactive, unable to combine with ribosomes into functional polysomes. Templates and ribosomes may be spatially separated (compartmentalized) or are structurally unable to associate. Somehow fertilization changes their structure or spatial distribution, bringing them together into a functional polyribosome.

One or several of these possibilities could account for the low protein-synthesizing capacity of the unfertilized sea urchin egg. Of the above alternatives, number 5 concerning the failure of message and ribosome to join into a functional polyribosome may explain the metabolic block in the unfertilized egg.

At first it was found that protein synthesis was stimulated 10 to 20 times in unfertilized egg ribosomes by the addition of a synthetic messenger, polyuridylic acid; ribosomes from fertilized eggs were only slightly stimulated. This suggested that the defect in the unfertilized egg was the absence of RNA messages and that fertilization stimulated message transcription. This turned out to be incorrect since genome transcription could be inhibited by

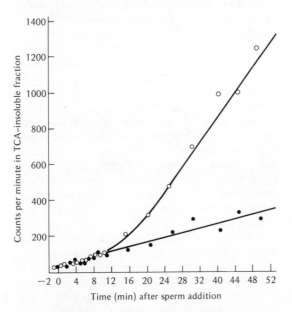

**Figure 13-4**  *Protein synthesis after fertilization. Incorporation of leucine-$^{14}$C into unfertilized (closed circles) and fertilized (open circles) sea urchin eggs. Unfertilized eggs preloaded with labeled amino acid, washed, and one portion was fertilized. Incorporation into acid-precipitable material. (From D. Epel, Proc. Nat. Acad. Sci. U.S. 57:899–906, 1967.)*

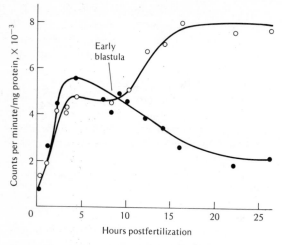

**Figure 13-5**  *Effect of actinomycin D on incorporation of radioactive valine into protein after fertilization of sea urchin eggs. Controls: open circles in 20-minute pulse incorporations; closed circles, embryos pretreated for 3 hours before fertilization with actinomycin D. Incorporation similar until early blastula stage. (From P. R. Gross, L. I. Malkin, and W. A. Moyer, Proc. Nat. Acad. Sci. U.S. 51:407, 1964.)*

actinomycin D without affecting protein synthesis (Figure 13-5). Stimulation of protein synthesis after fertilization does not depend upon simultaneous gene transcription, which implies that RNA templates are already present in the unfertilized egg but are not available for translation, or that ribosomes are inactive. But ribosomes from unfertilized and fertilized eggs were found to be equally efficient in combining with synthetic messenger RNA in vitro and promoting amino acid incorporation into protein. Evidently unfertilized egg ribosomes are normal, able to translate any messages provided.

In unfertilized eggs virtually all ribosomes are present as free ribosomes; few polysomes are present. On the other hand, the fertilized egg contains a large polysome component. Between 1 to 2 hours after fertilization, 20 to 40 percent of the free ribosomes gradually accumulate as polysomes, and protein synthesis increases.

Ribosomes and maternal messages are present in the unfertilized egg, but they associate only after fertilization. Since unfertilized egg ribosomes are functional in vitro, we can conclude that the message component is inactive or masked in some way, unable to organize into polysomes. This alone may explain the low level of protein synthesis in unfertilized eggs.

The masked messenger hypothesis provides an explanation of the low level of protein synthesis in the unfertilized sea urchin egg and its stimulation after fertilization. Messenger molecules are stored in a protected state, sheltered from degrading enzymes such as ribonuclease. Some event at fertilization must "unmask" messenger molecules, either by digesting a protein coat or by changing messenger conformation, making them accessible to the ribosome.

In extracts of sea urchin eggs particles that sediment at 25 to 65S have been found that are rich in protein, with a small amount of RNA resembling messenger RNA. Spirin called these particles *informasomes* because they contain informational RNA in an inactive complex with proteins, and suggests they may be the masked messages that are activated at fertilization. Some believe that they are artifacts of the fractionation procedure, artificial complexes of newly synthesized RNA with protein or perhaps a complex of messenger RNA with a ribosomal subunit. Since we know little about the nature of ribosomal subunit formation and assembly — or about the manner in which messenger molecules are transported out of the nucleus into the cytoplasm — it is difficult to reconcile these differences.

The so-called informasome particle may really be messenger RNA in transport from nucleus to cytoplasm. Furthermore, the informasome may be nuclear rather than cytoplasmic, part of the rapidly synthesized nuclear RNA which turns over without being transported to the cytoplasm. If these suppositions are true, then the informasome is not doing what its proponents claim.

Though we know that maternal messages are stored in the unfertilized egg, we do not know how, nor where. They may be in heavy polyribosomal fractions or in the postribosomal particulates, the informasomes, or even in other, still unidentified, particulate components in cytoplasm or nucleus.

Other changes in sea urchin egg RNA may be equally important in activating protein synthesis. Shortly after fertilization the amount of RNA containing polyadenylic acid doubles. It is generally assumed (see Chapter 3) that the addition of poly A residues to messenger RNA is a prerequisite for its use as an active message. In most cells the addition of poly A occurs in the nucleus, as part of message processing and transport to the cytoplasm. But in eggs, polyadenylation also occurs in the cytoplasm since enucleated merogones exhibit the same doubling of poly A message RNA after fertilization.

Poly A-containing messages are also stored during oogenesis. Shortly after fertilization these messages are transferred from a subribosomal fraction to the mature ribosomes, presumably for translation. It is after this that RNA polyadenylation occurs in the cytoplasm, and these newly adenylated messages also associate with ribosomes. According to Slater, the egg contains a population of maternal adenylated and nonadenylated messages, both associated with subribosomal particles. These are recruited to ribosomes after cytoplasmic adenylation at fertilization, and these changes may be responsible for the activation of protein synthesis.

The factors triggering protein synthesis in sea urchin eggs are still unresolved. Any one of the manifold changes at fertilization could act as a stimulus. The magnesium ion is particularly important because it stabilizes both polyribosomes and ribosomes. In its absence, these particles fall apart. Changes in magnesium ion flux, its compartmentalization or its absorption to proteins, could affect the entire translation capacity of the sea urchin egg at the moment of fertilization. We need more information before we can understand the chain of events from the moment of sperm attachment to the activation of protein synthesis.

## RNA synthesis: transcription of the embryonic genome at fertilization

We showed above that protein synthesis is initiated at fertilization and continues unabated in the absence of any RNA synthesis. Enucleated sea urchin eggs can be activated, will synthesize protein, and will cleave. Chemical enucleation with actinomycin has no effect on cleavage or protein synthesis. Transcription of new messenger RNA molecules is unnecessary for cleavage, which continues at the same rate and rhythm even after complete shutdown of the embryonic genome. Yet activation after fertilization does initiate genome transcription. RNA synthesis, albeit at low levels, is detected in many embryos even before first cleavage. It is studied by following the incorporation of radioactive precursors ($^3$H-uridine) into the different RNA classes. The amount of RNA made is

small compared with the large bulk of RNA already stored in the egg as ribosomal RNA. Consequently, there is no measurable net synthesis of RNA in early stages of most embryos. RNA begins to accumulate when stable ribosomal and transfer RNA are synthesized, usually at late blastula or gastrula stages.

The unfertilized eggs of sea urchins, frogs, and fish show low levels of RNA synthesis (Figure 13-6). Following fertilization in the amphibian embryo RNA synthesis becomes detectable, particularly during cleavage. The RNA synthesized at this time is heterogeneous nuclear RNA (hnRNA), resembling DNA in base composition. Ribosomal and transfer RNA are not formed in detectable amounts in these early stages. More than 90 percent of this hnRNA is broken down in the nucleus, whereas only 5 percent or less is transported to the cytoplasm. Most hnRNA seems to consist of transcripts from reiterated sequences in the genome; it is not a precursor for ribosomal RNA. It has been suggested that most of this RNA is regulatory and only a few transcripts of single-copy genes enter the cytoplasm for translation. The importance of this RNA to the embryo is uncertain since its synthesis can be blocked with actinomycin D without inhibiting cleavage or development to the blastula stage.

In mammals, the pattern is different. Immediately after fertilization, all types of RNA are synthesized. The mouse egg, for example, will not cleave in the presence of actinomycin since the RNA synthesized after fertilization is required for development. Ribosomal RNA is actively synthesized by the four-cell stage when nucleoli appear. The sensitivity of the mouse egg to actinomycin is probably due to its dependence on this early synthesis of ribosomal RNA. Presumably the store of ribosomes in the unfertilized egg is unable to sustain protein synthesis after fertilization.

Finally, the large store of egg mitochondria is an additional source of RNA synthesis after fertilization. Parthenogenetically activated and enucleated sea urchin eggs continue to synthesize RNA, presumably on mitochondrial DNA templates. Transcription begins within 40 minutes after insemination and may contribute 40 to 80 percent of total RNA synthesis in early cleavage. The RNA produced makes up size classes between 11S and 17S, but here too these transcripts are unessential for early development since enucleated sea urchin eggs cleave in the presence of actinomycin D, after all RNA synthesis is shut down.

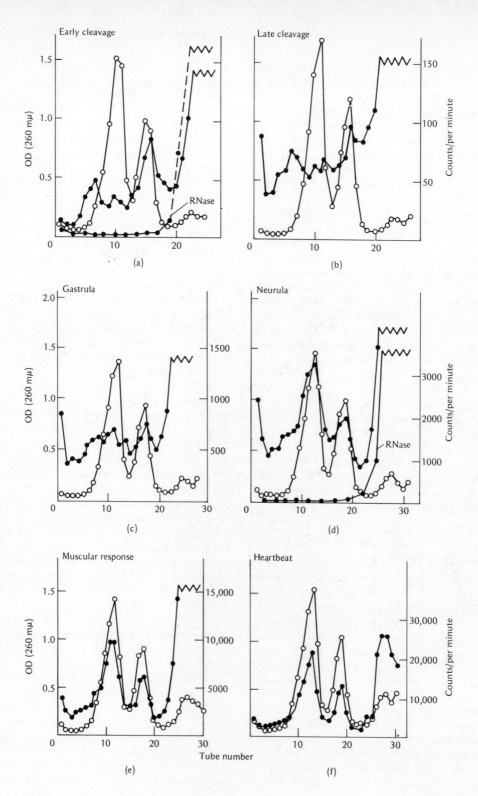

We find no common pattern of RNA synthesis after egg activation, nor can we correlate specific features of RNA synthesis with egg organization. At fertilization the mammalian embryonic genome begins to transcribe products essential for continued development. In other eggs, the embryonic genome may be activated at fertilization but the products, though used during cleavage, are not essential because the egg contains a maternal store of similar RNA messages. In some instances, the newly synthesized RNA may be stored for use later in development, possibly at the end of cleavage. Finally, there are eggs in which the embryonic genome does not transcribe actively after fertilization and is essentially unnecessary until the end of cleavage when morphogenesis begins.

## BIOCHEMICAL ACTIVITIES DURING CLEAVAGE

Immediately after fertilization, eggs enter a period of short cell cycles and synchronous division. Early cleavage cycles vary from 10 minutes in the *Drosophila* egg to 10 hours in the mammalian egg. In virtually all instances the $G_1$ period is either absent or brief. For example, in the frog *Xenopus*, a cell cycle of 25 to 30 minutes is typical at the early cleavage divisions, with 15 minutes devoted to S, approximately 8 minutes to $G_2$, and 5 to 7 minutes to mitosis. At later stages (Figure 13-7) when several hundred cells are present, the cycle is even shorter (about 15 minutes), with an S period of 10 minutes and about 5 minutes or less for mitosis. In this case, there may be no $G_2$. Likewise, in the mouse embryo no $G_1$ phase is detected until the eight-cell stage. Most of the cleavage cycle is devoted to DNA replication, with S phases usually less than an hour.

Since there is no cell growth between divisions, there are no preparations for division. Apparently the egg contains reserves of all the precursors and coenzymes necessary for replication. The synchronous period probably depletes these stores, and asynchronous division begins with lengthening cell cycles including longer $G_1$ and $G_2$ phases. Most changes occur in the $G_1$ phase. The metabolic activities during

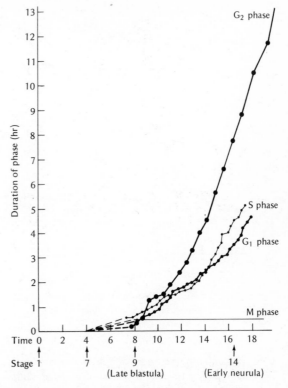

**Figure 13-7**  *Changes in cell cycle during* Xenopus *development. Estimated lengths of each phase of the cell cycle in endoderm cells. (From C. F. Graham and R. W. Morgan, Develop. Biol. 14:439–460, 1966.)*

cleavage to blastulation depend on maternal information stored during oogenesis.

### Respiration during cleavage

In virtually all embryos, oxygen consumption gradually increases over the cleavage period, paralleling the increase in cell number (Figure 13-8). The increase appears smooth because measurements on large numbers of eggs conceal fluctuations within individual embryos. When single sea urchin embryos are measured, an intrinsic respiratory rhythm is detected which matches the cell cycle. The rate of oxy-

**Figure 13-6**  *Patterns of RNA synthesis in* Xenopus *development as assayed by sucrose density gradient centrifugation of RNA after pulse labeling with radioactive phosphate. Open circles, optical density readings; closed circles, radioactivity as counts per minute. RNase curves indicate level of counts after treatment with RNase. (From D. D. Brown and E. Littna, J. Mol. Biol. 8:669–687, 1964.)*

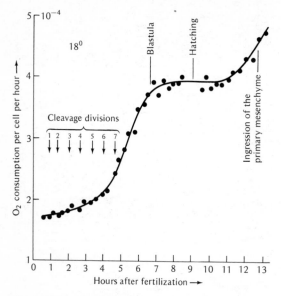

**Figure 13-8**  *Respiratory rate during early development of sea urchin,* Psammechinus miliaris. *(After H. Borei, Biol. Bull.* 95:124, 1948.)

gen consumption increases during interphase and decreases during mitosis, a pattern correlated with fluctuations in inorganic phosphate uptake. This suggests that the rate of oxidative phosphorylation oscillates over the cell cycle, periodically restoring energy (ATP) during interphase for utilization in the next mitosis. We know that division in the sea urchin embryo is dependent upon the amount of ATP; if it is reduced by 50 percent, division is inhibited.

The overall oxygen consumption curve for most embryos exhibits a logarithmic increase in respiratory rate followed by a leveling off at the blastula stage. A new spurt of activity begins at the gastrula stage but usually at a lower rate than before. The slight plateau at the blastula stage coincides with the time at which asynchronous division and mitotic gradients appear. It is at this point that the overall cell cycle lengthens as more time is spent in interphase, and it is also at this point that the rate of increase in cell number and DNA synthesis diminish. Clearly this stage is a turning point in morphogenesis, marking the transition from a replicating to a morphogenetic phase.

Two questions arise concerning respiratory regulation during cleavage. (1) What accounts for the increasing rate of respiratory activity? (2) Why does the rate level off at the blastula stage? The answer to the

first may be trivial; as cell number increases during cleavage the amount of respiratory surface area for gaseous exchange also increases as new membranes are synthesized at each division. The similarity between the oxygen consumption curves and the increase in cell number tends to support this hypothesis.

On the other hand, if intrinsic changes in respiratory metabolism account for increased oxygen consumption, we might expect activation or synthesis of respiratory enzymes or coenzymes, with the obvious candidates being those in the mitochondria. Accordingly some have proposed that increased oxygen consumption during cleavage is due to an increase in the number of mitochondria or an increase in their respiratory activity. But neither of these is correct since measurements of mitochondrial DNA synthesis suggest that no new mitochondrial DNA is made until after hatching in either the sea urchin or the frog embryo. Moreover, egg mitochondria may appear structurally more simple than somatic cell mitochondria but they contain most respiratory enzymes and are equally active. For instance, cytochrome oxidase activity per mitochondrial protein is constant until hatching in frog embryos.

We conclude that mitochondria, like egg ribosomes, are a reserve of energy-generating organelles to be partitioned out to cells of the cleaving embryo. Changes in respiratory metabolism during cleavage are not due to increased mitochondrial numbers nor to their functional maturation.

This does not rule out the possibility that subtle regulatory mechanisms associated with compartmentalization of mitochondrial enzymes, substrates, and cofactors account for the respiratory changes. For example, respiratory enzyme activity in homogenized embryos is far in excess of anything observed in the intact embryo. Obviously factors in the intact cell maintain enzyme activities at levels adjusted precisely to the needs of the embryo. Subtle changes in mitochondrial permeability could change the distribution of substrates such as ADP and coenzymes, and these, rather than mitochondrial maturation, could account for respiratory increase during cleavage.

Probably the simplest explanation for the increased oxygen consumption is the one mentioned earlier; that is, more surface area for gaseous exchange is provided as new cells are produced during cleavage. Oxygen consumption, calculated on a per cell basis, is constant throughout cleavage, which means the overall increase per embryo reflects an increase in cell number.

This may also answer our second question about the leveling off of respiratory metabolism at the blastula stage. Asynchrony begins, mitotic activity diminishes, and the cell cycle lengthens in the blastula. The increase in cell number declines, which means that the rate of synthesis of new membrane surface also decreases. But this is only part of the answer, since the nature of respiratory activity also changes at the blastula stage. In some embryos, a shift occurs from one energy-generating system to another.

Amphibian and fish embryos cleave normally to a blastula stage in the complete absence of oxygen. In fact, frog embryos cleave in the presence of inhibitors of oxidative phosphorylation such as cyanide. The energy generated through anaerobic glycolysis, coupled to whatever energy reserves the egg contains in the form of ATP, is enough to carry the embryo through cleavage. This means mitochondria are bypassed; that is, increased oxygen consumption is unrelated to mitochondrial energy-generating systems. Evidently, in these embryos extramitochondrial glycolysis is sufficient to carry the embryo to the blastula stage. Beyond this point, however, oxidative phosphorylation is required for continued development. The energy demands of tissue morphogenesis are met by a switch to active aerobic respiration.

The mammalian egg shows a similar shift in respiratory activity. After fertilization, as the cleaving embryo migrates down the oviduct, its respiration is anaerobic because oxygen tension in the lumen is low. It reaches the uterus at a blastocyst stage, equivalent to an early blastula, where it implants in the uterine wall. As soon as it makes contact with the high oxygen tension of uterine tissue, it switches to aerobic respiration and oxidative phosphorylation. The changes in respiratory metabolism reflect the particular adaptations of the embryo to its respiratory needs.

Different parts of the embryo respire at different rates as regional properties are established. Mitochondria are segregated randomly into cleaving cells, some cells acquiring more mitochondria than others. As we have already indicated in our discussion of dye reduction gradients in the sea urchin embryo, gradients in respiratory activity could reflect a disproportionate distribution of mitochondria. But this doesn't seem to be the case inasmuch as animal and vegetal halves of the echinoderm embryo consume oxygen at equal rates.

Energy-generating systems in the embryo utilize a store of precursors. Glycogen is the principal car-bohydrate store and is broken down almost immediately after fertilization. Lipids and proteins are also utilized for energy metabolism. Most fish eggs have a rich lipid reserve, usually in the form of large oil droplets. These too are oxidized to provide energy, not necessarily at the same time as carbohydrates.

## Macromolecular synthesis: information transfer during cleavage

Developmental information is stored in eggs during oogenesis as nucleic acids coded by the maternal genome. After fertilization, DNA replication begins, and its rate and pattern are controlled by maternal information. The synthesis of DNA is extremely rapid in early cleavage. In a cleaving blastomere the same amount of DNA may be replicated in 15 minutes as would take 6 to 8 hours in a normally dividing somatic cell in the adult. The reasons for this rapid in early cleavage. In a cleaving blastomere the already seen (see Chapter 3) that more initiation sites may be present in chromosomes of cleaving embryonic cells.

The pattern of DNA synthesis during early amphibian development is typical of many embryos (Figure 13-9). An initial period of constancy, when the total amount of DNA per embryo does not change, is followed by a logarithmic increase to the end of blastulation. The rate of synthesis declines before gastrulation begins and continues at a lower rate thereafter. The surprising result is that no net synthesis of DNA is seen during early stages, until several hundred cells are produced. Yet new nuclei with a full complement of DNA increase logarithmically. How do we explain this paradox?

The answer lies in the fact that most eggs start out with a DNA excess in the cytoplasm, the DNA in mitochondria. The amount of mitochondrial DNA exceeds the total nuclear DNA of the zygote by several orders of magnitude in some eggs. In the *Rana pipiens* egg, for example, the total amount of DNA in the embryo is 0.02 $\mu$g, of which only 0.001 $\mu$g is located in the zygote nucleus; the remainder is found in the mitochondria. Our analytical methods are unable to detect the small amounts of nuclear DNA synthesized against a large background of mitochondrial DNA. Only after nuclear DNA reaches detectable levels do we begin to see evidence of any net increase in total DNA per embryo. The time at which this occurs varies with the size of each egg and its store of mitochondria. In frogs, it is measurable after

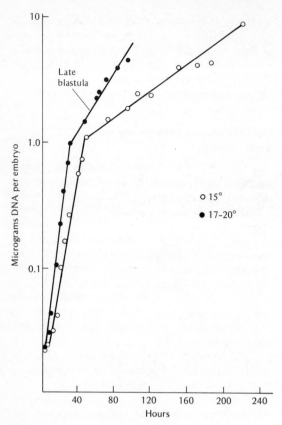

Late
blastula

o 15°
● 17-20°

Hours

**Figure 13-9**  *A semilog plot of DNA synthesis during
early development of* Rana pipiens. *(From P. Grant,
J. Cell. Comp. Physiol. 52:227–248, 1958.)*

500 to 1000 cells are produced; in sea urchins, after
32 cells.

The principal precursors for DNA synthesis in
amphibian cleavage are ribonucleotides that are also
used for RNA synthesis. Both pathways of nucleic
acid synthesis compete for the same substrates. Enzymes, such as the ribonucleotide reductases, convert ribonucleotides to deoxynucleotides for DNA
synthesis, which explains why radioactive ribonucleotides are preferentially incorporated into DNA during early cleavage of many embryos. At the end of
cleavage, DNA synthesis decreases and most nucleotides are diverted to a more active RNA synthesis.
Small amounts of deoxynucleotides and coenzymes
are also present in the fertilized egg, but these are exhausted by a mid-blastula stage, when asynchronous
division begins.

Because of this precursor reserve, the cleaving
embryo is insensitive to most inhibitors of DNA synthesis and cell division. For example, amethopterin,
an inhibitor of folic acid metabolism, does not inhibit frog egg cleavage. The egg contains high levels
of folic acid coenzymes which are involved in purine
and pyrimidine biosynthesis and are important in re-
plenishing the nucleotide pool. Development stops at
the blastula stage when folic coenzymes reach limiting concentrations. Even strong mutagens such as
nitrogen mustard do not alter the cleavage rate in
amphibians, and development stops just before gastrulation.

The stage at which an inhibitor stops development varies with the amount of reserves in the egg.
The sea urchin egg cleaves to the eight-cell stage in
the presence of very high doses of FUDR (5-fluorode-
oxyuridine), a powerful inhibitor of DNA synthesis.
Inhibition can be reversed by the addition of thymidine, which means that the limiting factor in sensitivity is the size of the thymidine pool in the egg.

Summarizing, the transition from synchrony to
asynchrony in many embryos coincides with the exhaustion of precursors and coenzymes required for
DNA synthesis. This turning point also correlates
with the time at which respiratory activity begins to
level off. The similarity of the respiratory and DNA
synthesis patterns suggests that the two metabolic
pathways are coupled. Inhibitors of respiratory metabolism stop cell division, and, conversely, inhibitors of DNA replication also decrease oxygen consumption. One explanation is that both pathways
draw on and compete for a common nucleotide pool.
For instance, the nucleotides ATP and UTP are essential to both pathways, which means that respiration
can regulate rates of DNA synthesis by controlling
precursor availability.

### Utilization of maternal information

In most interspecific hybrids in frogs and sea urchins,
cleavage rates and patterns are determined by the
maternal genome. In some instances, cleavage is
possible, albeit abnormal, in the complete absence of
any nuclei. Enucleated sea urchin eggs, activated
parthenogenetically, will cleave into an irregular collection of blastomeres; the mechanism for cytokinesis
is preprogrammed in the egg.

An example of a maternally inherited gene controlling cleavage is found in the salamander. *Axolotl*

females, homozygous for the gene *cl*, produce eggs exhibiting abnormal cleavage. Cleavage furrows appear only in the egg's animal half while the vegetal half does not cleave. Though the animal half is fully cleaved, partial blastulae result which do not gastrulate. The mutant effect is maternally inherited; only the homozygous *cl/cl* females give rise to such eggs, as if the necessary factors for furrowing (microfilaments?) are not localized in vegetal halves of the egg during oogenesis.

In most embryos, cleavage can occur without information from the embryonic genome. Actinomycin D does not inhibit cleavage in sea urchin and frog embryos even if more than 95 percent of the embryonic genome is shut down. Early development in these embryos depends entirely on information stored in the egg during oogenesis.

The RNA stored during oogenesis may be the maternal messages used in cleavage. In Chapter 11 we saw that in *Xenopus* up to 5 percent of the total repetitive DNA and 1.2 percent of the nonrepetitive fraction were transcribed and stored as message in oocytes at the lampbrush stage. Although the nonrepetitive RNA represents only $10^{-3}$ to $10^{-4}$ of the total RNA present, its heterogeneity suggests that up

to 20,000 diverse structural genes may have been copied. DNA-RNA competition hybridization experiments tell us whether some of this maternal information is used in cleavage.

We can assay the similarity of messages synthesized at the lampbrush stage with those from different embryonic stages and also determine how much of the total repetitive RNA at a cleavage stage is identical to lampbrush messages. For example, nonradioactive or "cold" RNA from mature oocytes and two-cell and early stage-7 blastulae is made to compete with labeled "hot" lampbrush-stage RNA for complementary sites on *Xenopus* DNA under conditions that assay for only the repetitive, nonribosomal sequences (Figure 13-10). If mature-oocyte and cleavage-stage RNA is identical to lampbrush RNA, then it should show the same degree of competition as cold lampbrush RNA. The results with mature oocytes show that 65 percent of its transcripts are homologous to lampbrush RNA; approximately 35 percent of the transcripts are lost during the completion of egg maturation, presumably to support protein synthesis. What is most interesting, however, is that the same 65 percent fraction is found in the two-cell and stage-7 blastulae; early

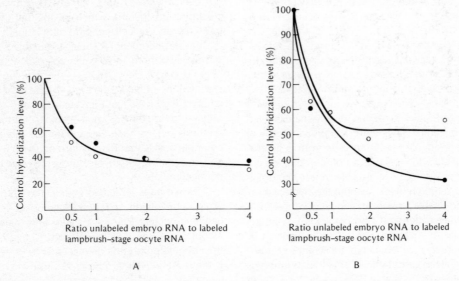

A                    B

**Figure 13-10**   *(A) Competition hybridization experiment comparing sequence homology of cold RNA from stage-2 (filled circles) and stage-7 (open circles) Xenopus embryos with $^{32}$P-RNA and $^{3}$H-DNA from lampbrush-stage oocytes. (B) Competition experiment between increasing amounts of cold stage-7 RNA (closed circles) and stage-9 RNA (open circles) added to $^{32}$P-RNA and $^{3}$H-DNA from lampbrush-stage oocytes. (From M. Crippa, E. H. Davidson, and A. E. Mirsky, Proc. Nat. Acad. Sci. U.S. 57:885, 1967.)*

blastula embryos contain RNA homologous with 65 percent of that synthesized at the lampbrush stage. In other words, from the beginning of fertilization to a blastula embryo of approximately 2000 cells, there is no apparent loss of maternal information. It is not until the mid and late blastula stages (stage 9) that this maternal information is used; some of the original lampbrush RNA declines. Late blastula RNA in competition with hot lampbrush RNA shows that much less homologous RNA is present at the later stage; the degree of competition is reduced. Approximately 18 percent of lampbrush RNA complementary to redundant portions of the genome is lost. These results tell us nothing about the single-copy message fraction, but they point to a significant change in the information flow system at the late blastula that correlates with other nuclear events. Asynchronous divisions begin, the cell cycle lengthens, the embryonic genome is activated, and nucleoli appear as ribosome synthesis begins in earnest.

This brings us to another important maternal contribution to the egg, namely, the rich supply of ribosomes synthesized in oogenesis. What happens to these during cleavage? Studies of the anucleolate mutant (o-nu) of the frog *Xenopus* show that the mutation involves a deletion of the nucleolar organizer region of the chromosome; that portion of the DNA coding for ribosomal RNA is lacking. Homozygous individuals (o-nu/o-nu) cannot synthesize ribosomal RNA and behave as developmental lethals, usually dying after hatching. Heterozygous individuals (+/o-nu), though possessing only one nucleolus, synthesize normal amounts of ribosomal RNA.

A mating of a heterozygous female (+/o-nu) with a heterozygous (+/o-nu) male will give the expected one-quarter homozygous embryos surviving only to the hatched tadpole stage. All eggs from a heterozygous female will stockpile a normal complement of ribosomes during oogenesis. Therefore, homozygous (o-nu/o-nu) embryos, without the nucleolar organizer and unable to make their own ribosomes, are used to study patterns of ribosomal RNA synthesis and the fate of maternal ribosomes. In contrast to normal embryos, which begin to synthesize ribosomal RNA at the gastrula stage, anucleolate embryos do not synthesize ribosomal RNA at any stage. Yet in spite of this, mutant embryos develop normally to the swimming tadpole stage, carrying out all protein synthesis and differentiation on the original reserve of maternal ribosomes. A long period of development is

possible with "old" maternal ribosomes made during oogenesis. New "embryonic" ribosomes are required for larval development only. Maternal ribosomes are stable for a long time. The amount of radioactive precursors incorporated into ribosomal RNA during oogenesis is identical to the amount of radioactivity in ribosomes extracted from eggs shed 6 months later. In other words, maternal ribosomes made in oogenesis do not turn over.

Ribosomal RNA synthesis does not occur until the gastrula stage. This is consistent with the appearance of nucleoli at the same time and suggests that the embryo genome is activated only at the end of cleavage. The embryo genome, however, is not completely silent during cleavage. Cleavage nuclei actively transcribe RNA, but, as we have seen, most of this information never gets into the cytoplasm for translation.

### Transcription of the embryonic genome: RNA synthesis in cleavage

In sea urchins, the embryonic genome transcribes before and after fertilization but at barely detectable rates. No significant increase in the rate of RNA synthesis by the embryo is seen until after the 8- to 16-cell stage when micromeres appear. Cell number, however, increases more rapidly than RNA synthesis during this period, which means that the rate of transcription per cell genome may actually decline during development. Cleavage nuclei may transcribe RNA more rapidly than nuclei of later embryonic stages. Nevertheless, there is no net accumulation of RNA in the embryo; the total RNA per embryo is constant during cleavage and increases only at gastrulation, when stable ribosomal RNA synthesis begins. One explanation for this apparent contradiction is that RNA is turning over; it is made and degraded at a high rate during cleavage and does not accumulate.

As we have indicated earlier, the RNA synthesized in cleaving sea urchin embryos is heterogeneous nuclear RNA, with very little entering the cytoplasm as translatable messages. Though it makes up a minute proportion of total embryo RNA, hnRNA is the most actively synthesized informational molecule during cleavage. Most transcribed, reiterated sequences are degraded, but a posttranscriptional control mechanism selects a few for transport to the cytoplasm where they are translated into functional proteins. Most of this newly synthesized RNA in the

**Figure 13-11** *Transcriptional complexity of RNA molecules complementary to repeated and single-copy DNA sequences at different stages of mouse and rabbit development. A large excess of RNA is reacted with a small amount of labeled single-copy DNA fragments. (From R. B. Church and G. A. Schultz,* Current Topics in Developmental Biology, *ed. A. A. Moscona and A. Monroy, vol. 8. New York: Academic Press, pp. 179–202, 1974.)*

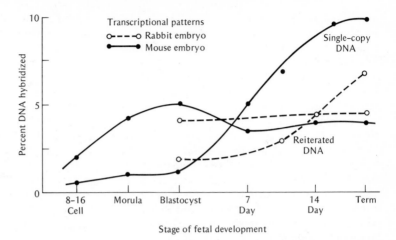

cytoplasm is associated with active polysomes of all sizes and is used for immediate translation.

Heterogeneous nuclear RNA seems to be the only species of RNA synthesized by cleaving embryos in sea urchins, frogs, and fish. In these, ribosomal RNA synthesis is not detected until the late blastula or gastrula stages. This stable species of RNA begins to accumulate at the late blastula when a definitive nucleolus appears, indicating that genes coding for ribosomal RNA are turned on at the end of cleavage. But this issue is still controversial since some claim (Emerson and Humphreys) that ribosomal RNA is, in fact, transcribed in cleavage nuclei at all times. They point out that ribosomal RNA synthesis is not detected in cleavage because too few nuclei are present and its low rate of synthesis is swamped out by the intense synthesis of nuclear RNA. In later stages, when heterogeneous nuclear RNA synthesis diminishes, and many more nuclei are present, synthesis of ribosomal RNA becomes detectable. If these studies are correct, it suggests that ribosomal RNA genes transcribe at all times. It implies that embryos make small amounts of ribosomes in early cleavage but too few to be detected against the large bulk of stored maternal ribosomes.

We know that mammalian embryos synthesize ribosomal RNA from the beginning of development. Ribosomal RNA synthesis is detectable in mouse embryos as early as the 2-cell stage, and nucleoli appear at the 4-cell stage. Ribosomal RNA synthesis increases rapidly from the 8- to 16-cell stage, until a high level is reached at the blastocyst stage, before implanta-

tion. Roughly 60 to 80 percent of the total RNA synthesized during these early cleavage stages in mouse embryos is ribosomal RNA, 6 to 10 percent is transfer RNA, and the remaining 8 to 18 percent is measurable as heterodisperse DNA-like RNA. Development in the mouse is directly dependent upon transcription from the embryonic genome from the beginning of development. In contrast to the frog and sea urchin, the mouse embryo relies on the ribosomes synthesized during the first few cleavage divisions rather than on a store of maternal ribosomes to get through early development.

A study of the transcriptional complexity during early mammalian development indicates that different components of the mouse genome are being transcribed at different stages. In the mouse, three broad classes of the genome with differing degrees of redundancy can be identified. The most redundant satellite DNA represents approximately 6 to 9 percent of the total genome and, in general, this is not transcribed at any time in development. Approximately 20 to 25 percent of the genome is a class of moderately redundant sequences that are distributed throughout the genome. Finally, the remaining 60 to 70 percent of the genome represents the single-copy component. The latter two components are most important in developing systems.

The transcriptional patterns for the mouse embryo were analyzed by DNA-RNA hybridization experiments at different developmental stages (Figure 13-11). First, we should note that no more than 10 percent of the total single-copy component of the genome is transcribed by the end of development, and only 5 percent of the redundant portion appears; most of the genome is not transcribed. Second, the bulk of transcription in early stages, until implantation (late blastula or early gastrula stage), is on the redundant portion. After implantation and during morphogenesis and differentiation, the single-copy component of the genome predominates. Presumably, proteins involved in morphogenesis and tissue differentiation are made on single-copy transcripts. The significance of the early transcription of reiterated sequences is not understood. According to some models of gene action, these are regulatory genes, perhaps reflecting transcription of portions of the maternal genome that had been transcribed earlier. The mammalian embryonic genome is actively transcribing from the beginning of development, but the complexity of the transcripts changes at later stages as more single-copy components of the genome are expressed.

The problem of RNA synthesis in the embryo, particularly during early cleavage, is also complicated by RNA synthesis in cytoplasmic organelles such as mitochondria. Enucleated sea urchin embryos, stimulated to develop parthenogenetically, cleave and synthesize RNA. This RNA is ribosomal and transfer RNA typical of mitochondria. It may also include messenger RNA with base sequences more homologous to mitochondrial than to nuclear DNA. This means that mitochondria are synthesizing RNA in early development, which may be an important informational contribution. We see that the activation signal at fertilization stimulates both a nuclear and a cytoplasmic genome to initiate transcription almost simultaneously.

We have been concentrating on transcription during early cleavage when cell division is synchronous. What about the period of asynchrony at the blastula stage? Does a different pattern of RNA synthesis begin at this time? In the amphibian embryo, the embryonic genome becomes activated between the early and late blastula stages, toward the end of the cleavage period. Radioautographs of early blastulae show only few grains marking the incorporation of labeled precursors into nuclear RNA, but in mid to late blastulae, much more incorporation is detectable. There is also a regional difference in activation; it appears that cells on the dorsal side (future embryonic axis) are more active than those on the ventral side. Regionalization of the embryo is correlated with different patterns of genome activation.

The importance of embryonic information tran-

scribed in the blastula is shown in the actinomycin inhibition studies in sea urchin embryogenesis. Many embryos at different cleavage stages are exposed to actinomycin D for varying lengths of time, removed, and washed. This process shows when messages synthesized during the period of treatment are used in development. If early cleavage stages are treated, development is normal. Messages made in cleavage do not seem to be important for later development. If, however, the embryo is treated at an early blastula stage, development ceases. On the other hand, if the embryo is treated after this stage, then gastrulation proceeds. It appears that informational RNAs transcribed in the blastula stage are necessary for development through gastrulation. Then development can continue for a short time, even in the presence of actinomycin. This informational RNA is transcribed early, 6 to 11 hours after fertilization, but its translation is delayed until gastrulation begins. One recurrent feature of information flow in embryos is the transcription of messages at one stage and their retention in an inactive state for translation at a later stage.

The changing pattern of RNA transcription during early development can be visualized in the electron microscope. Using techniques of gentle spreading of interphase nuclei, DNA strands in the act of transcription can be examined. In Figure 11-9 we have already seen the appearance of ribosomal RNA transcripts in amphibian lampbrush oocytes, and McKnight and Miller have used this technique in *Drosophila* embryos. They find that at 1 to 2 hours after fertilization, at the stage of approximately 64 cleavage nuclei, the spread chromatin appears structurally compact with a beaded configuration. Only a few short ribonucleoprotein (RNP) fibrils, or transcripts, are seen attached to this inactive chromatin. An estimate of the relative amount of transcription indicates that about 1.07 percent of the contour length of DNA examined has RNP-associated fibrils. The structure of these transcripts is different from rRNA transcripts, suggesting that ribosomal RNA synthesis has not begun at this time. Since the replication cycle in these nuclei is about 9.6 minutes, it is estimated that only small RNP transcripts can be synthesized in this short time; most of the cyle, about 6 minutes, is devoted to mitosis.

At 2 hours after fertilization, the syncytial nuclei migrate to the periphery where they complete three additional synchronous divisions. Following the last division, nuclear cycles lengthen to well over an hour, and cell membranes form to establish the cellular blastoderm. Spread chromatin from nuclei before the final syncytial division still contains no ribosomal RNA transcripts. Only after the last syncytial division, just as the first true cell cycle has begun, do rRNA matrices appear. The total contour length of these typical rRNA transcripts suggests that no more than 50 rRNA genes are active per nucleolar organizer region at this time. Though ribosomal genes are turned on at this stage, the number of genes activated is only a small proportion of the total available.

At 2½ to 3 hours the cellular blastoderm stage is clearly established. The spread chromatin is considerably more active in nuclei from this stage. In addition to rRNA matrices and the few short RNP matrices seen in cleavage, a new class of RNP matrices appears with different structure; these are less dense than the early matrices, contain longer fibrils, and occupy a greater proportion of the contour length of DNA, about 5.52 percent. New sets of genes are transcribed at this later stage, after the cell cycle has lengthened. We see how temporal control of DNA replication and the cell cycle also affects the proportion of time available for the synthesis of large transcripts. This pattern of temporal regulation seems to determine which genes are transcribing at particular stages.

Though maternal information dominates macromolecular synthesis in early development, we see that the embryonic genome is not entirely silent. The question now arises, how is this mixture of genetic information used in the synthesis of proteins? Are maternal and embryonic messages translated simultaneously, or are embryonic messages (the few that enter the cytoplasm) stored for use later in development? We will see that both occur during cleavage.

### Translation during cleavage: protein synthesis

We know that protein synthesis is activated after fertilization in the sea urchin egg to levels 5 times that in unfertilized eggs, at least until first cleavage. After this, the rate remains constant until the blastula stage. At one time it was thought that the rate of protein synthesis was activated 30-fold, increasing during cleavage to the blastula stage. When these results are corrected for changes in amino acid permeability and precursor pool sizes, the rate of protein synthesis turns out to be constant during cleavage until the

blastula stage. Later, during gastrulation, rates of protein synthesis do increase.

Although protein synthesis is activated in cleavage, the net synthesis of protein is virtually nil. In fact, the total protein content per embryo is constant throughout most of development. Proteins turn over; that is, yolk proteins are broken down to amino acids which are reutilized for synthesis of other embryonic proteins. Proteins synthesized during cleavage are negligible in amount compared with the bulk proteins stored in the egg.

Most messages made available for translation during cleavage are stored maternal messages which are unmasked and translated on new populations of polyribosomes appearing after fertilization. As development proceeds, more ribosomes are engaged as polysomes in two major size classes, small numbers of light polysomes and a larger group of heavy polysomes. Both become functional at fertilization and continue to increase in numbers in later stages.

The unmasking of maternal message is not the only source of informational RNA for polysome formation. As we saw earlier, newly synthesized embryonic RNA associates with all size classes of polysomes, preferentially with the light-polysome class. From this polysome class a 9S RNA can be isolated. In the presence of actinomycin D, this class of light polysomes decreases, suggesting it is this class of polysomes that is assembled on newly synthesized embryonic message.

What kinds of proteins are synthesized during cleavage? The inhibitor studies tell us that the small amounts of proteins synthesized in early cleavage are essential to development. Since chromosome replication is the major synthetic activity during cleavage, we would expect chromosomal proteins, histones, and acid proteins to be the most important macromolecules to be synthesized, second to DNA. Kedes and Gross, studying incorporation of amino acids into early cleavage stages of the sea urchin, found that most of the labeled precursor concentrated in the cytoplasm at first, but shortly thereafter it localized in nuclei, over the chromosomes. Virtually 40 to 60 percent of the proteins synthesized in these stages are histonelike, acid-soluble nuclear proteins. Their synthesis is tied to the replication of DNA, which is another indication of their histone nature.

These proteins are preferentially synthesized on light polysomes, and their synthesis is most sensitive to actinomycin D. The reduction in this polysome class in the presence of actinomycin means that embryo templates normally coding for these molecules are not being synthesized. This evidence suggests that the embryonic genome is transcribing messages for the synthesis of histones. Five separate 9S RNA message classes can be isolated and each has been shown to code for one of the five histone classes. Each is transcribed at equivalent rates and associates into small polysomes of two to five ribosomes, about the right size for small histone proteins. Apparently histone messages are detectable because they are coded by multiple copies of histone genes. Since complete inhibition of these polysomes by actinomycin does not stop cleavage and nuclear protein synthesis, we must conclude that maternal messages also code for histone synthesis, and both types are utilized simultaneously during cleavage. Approximately 60 percent of protein synthesis in the cleaving embryo is carried out on maternal messages and the remainder on embryonic templates.

Histones are not the only proteins synthesized during cleavage. Whole classes of proteins are synthesized that differ from the profile of bulk proteins already present in the egg. They are produced in minute amounts, and none of these newly synthesized proteins has been identified. Most enzymatic proteins do not change during early development. The embryo completes development with a complement of maternal enzymes stockpiled in oogenesis, and few new ones are made before hatching.

The qualitative nature of the proteins synthesized changes at the end of cleavage, during the blastula stage. New classes of proteins appear, reflecting the heightened activity of the embryonic genome and the production of new species of informational molecules. It is usually at this point that some paternal genes are first expressed. This is a biochemical transition point in anticipation of the major morphogenetic events to follow, when cell sheets transform into a functioning organism.

## NUCLEAR ACTIVATION AT GASTRULATION

Restriction of developmental fate of embryonic cells may correlate with certain nuclear changes that become apparent just before gastrulation. First, we note that rates of nuclear replication decrease sharply in gastrula stages of many embryos. For example, length of the cell cycle increases from 15 to 30 minutes in

*Xenopus* cleavage to 3.5 hours in the gastrula stage (Figure 13-7). More than 60 percent of toad blastula cells are in some mitotic phase, but only 7 percent of gastrula cells are in mitosis. As the cell cycle lengthens, the rate of DNA synthesis diminishes. Its rate decreases fivefold in the frog embryo, a decrease due, in part, to the exhaustion of nucleotide reserves at the end of the blastula stage.

All phases of the cell cycle lengthen, including mitosis. The $G_1$ and $G_2$ periods exhibit the greatest change, so that the proportion of the cycle devoted to S or replication is very much reduced. This implies that more time is available for transcription of DNA, and, as we have seen, the embryonic genome becomes active at this time.

Nucleoli first make their appearance during late blastula or early gastrula stages (Table 13-1). This marks the onset of stable ribosomal RNA accumulation, a visible sign that the embryonic genome is actively transcribing. Still another sign of an activated embryonic genome is the appearance of paternal proteins at gastrulation. For example, in a hybrid cross between *Paracentrotus lividus* females and *Psammechinus microtuberculatus* males, no paternal antigens are found during cleavage, but 24 hours after fertilization, at the mesenchyme blastula stage, paternal antigens are detected in hybrid embryos. In this cross, messages from the paternal genome are probably synthesized and translated after cleavage is over.

Gastrulation also marks a critical metabolic turning point; this stage is most sensitive to metabolic inhibitors. Whereas cleavage of some eggs (for example, amphibian) persists under anaerobiosis, in the presence of actinomycin D or even nitrogen mustard, gastrulation cannot continue. Amphibian cleavage is insensitive to most metabolic inhibitors, except inhibitors of protein synthesis, because cleavage relies on a store of maternal information in egg cytoplasm. It may continue even if chromosomes are damaged or eliminated. On the other hand, gastrulation depends upon the presence of a balanced genome; most chromosomal abnormalities and aneuploidy prevent gastrulation. Sheet morphogenesis demands strict genetic and nucleocytoplasmic compatibility; otherwise, development ceases.

What types of new genetic information are expressed at the gastrula stage? We can try to answer this question with the techniques of DNA-RNA hybridization which measure the proportion of the

**TABLE 13-1    Appearance of the Nucleolus during Development**

| Category | Genus | Stage of Appearance |
|---|---|---|
| Mollusk | *Limnaea* | 24-cell stage |
| | *Helix* | Blastula |
| | *Chiton* | Prototroch, apical tuft |
| | *Cyclops* | 2- to 8-cell stage |
| Tunicate | *Ciona* | Neurula |
| | *Ascidia* | Metamorphosis |
| Echinoderm | Sea urchin | Mesenchyme blastula |
| Insect | *Chironomus* | Blastoderm |
| | *Drosophila* | Blastoderm |
| Amphibian | *Triturus* | Gastrula |
| | *Rana* | Gastrula |
| | *Xenopus* | Gastrula |
| Fish | *Fundulus* | Gastrula |
| | *Coregonus* (whitefish) | Gastrula |
| | *Misgurnus* (loach) | Mid-blastula |
| Bird | Chicken | Late blastula |
| Mammal | Mouse | 2-cell stage |
| | Rat | 8-cell stage |

Source: From D. D. Brown, *Nat. Cancer Inst.,* Monogr. No. 23, 1966, pp. 297–310.

genome transcribed at particular stages of development and show whether genes transcribed at one stage are also transcribed at other stages. We can now analyze nonrepetitive sequences coding for structural genes as well as repetitive sequences described earlier. An evaluation of structural gene expression is more relevant to the problem of gene expression during development.

Galau and his group have recently carried out such an analysis in the sea urchin embryo. They isolated the total messenger RNA from gastrula polysomes and obtained a population of working message transcripts characteristic of the gastrula stage. Most of this message fraction is transcribed on the embryonic genome and probably contains some stored maternal message as well. Since this RNA hybridizes exclusively with the nonrepetitive DNA of the gastrula, it is probably derived from structural genes.

Two subclasses of messages have been identified in this population, a prevalent message class (about 90 percent of the total) made up of many transcripts of each of several hundred genes, and a complex class (the remaining 10 percent) derived from 10,000 to 15,000 different genes. Each mRNA transcript in the complex class is probably represented only 1 to 10 times per gastrula cell, whereas those from the prevalent class are present at least 100 times more frequently.

Sets of single-copy DNA sequences complementary to gastrula mRNA are isolated by hybridization with total gastrula RNA. In this manner, the single-copy sequences are enriched and can be used to assay for the presence of homologous RNA transcripts in other tissues. By hybridizing this DNA fraction with polysomal RNAs from oocytes, blastula and pluteus embryos, and adult tissues we can ask whether message sequences are shared with those

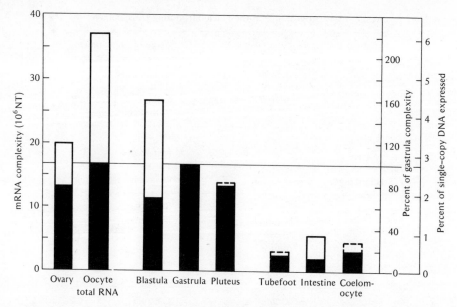

**Figure 13-12**   *Comparison of structural genes active in sea urchin embryos and in adult tissues. The bars show the proportion of single-copy genes expressed in various tissues. With gastrula taken as the standard, the solid bars represent the amount of single-copy sequences shared between gastrula mRNA and RNA from each tissue. Open bars indicate those single-copy sequences found in tissue RNAs but absent in gastrula mRNA. Dashed bars mark the maximum level of single-copy sequences that could have escaped detection in those tissues. Gene expression is shown as a percentage of gastrula mRNA complexity or as a total genome single-copy sequences, or nucleotides of single-copy sequences. (From G. A. Galau et al., Cell 7:487–505, 1976.)*

in the gastrula, how much is shared, and whether new sequences are produced.

Under the conditions of the experiment, only the mRNA of the complex class was examined. The results shown in Figure 13-12 indicate that some transcripts of the gastrula are also present in adult tissues; approximately 15 to 35 percent of transcripts are shared with those in the gastrula, and probably contain identical messages in each tissue. Shared transcripts may represent the so-called "housekeeping" genes that code for those enzyme systems required by all tissues for respiratory metabolism and glycolysis. If the minimal level of shared transcripts is taken as a measure of this gene population, it is calculated that housekeeping genes make up 0.3 percent of the single-copy DNA sequences, or 1000 to 1500 structural genes.

Each tissue is also characterized by a unique set of transcripts; they contain transcripts that are absent in the gastrula. This suggests that each set of mRNAs is a highly restricted sample of the total structural gene information, probably correlating with the pheno-

typic heterogeneity of a tissue (the number of different cell types it contains) as well as its functional state.

The ovary contains up to 80 percent of homologous transcripts while the mature oocyte contains all gastrula transcripts including a large population of transcripts that are not present in the gastrula. It is interesting that the gastrula embryo synthesizes sets of transcripts identical to those already present in the mature oocyte, presumably stored as maternal messages.

The total complexity of transcripts is also tissue-specific. In all cases the proportion of the single-copy genome transcribed is small, varying from less than 1 percent in adult tissues to about 6 percent in the oocyte. If we assume the oocyte starts development with information stored as sets of RNA transcripts, it can be calculated that the complexity of its message population corresponds to the expression of 20,000 to 30,000 diverse structural genes. During development the complexity diminishes; a smaller percentage of the structural gene information is transcribed at later embryonic stages until the adult when each tissue seems to operate with information from about 5000 genes. In fact, the pluteus larva, containing many more cell types than the gastrula, seems to be functioning with a set of transcripts almost identical to that of the gastrula. This implies that the genome is transcribed at the gastrula stage but the messages are translated much later as pluteus tissues differentiate. Here, again, we see that embryonic genes, transcribed at one time, are stored in an inactive state for translation at a later stage.

These data indicate that more genetic information is available for use in early embryonic stages and that tissue differentiation in later stages is accompanied by progressive restriction of utilizable information. The information flow system at the molecular level seems to be causally related to the behavior of tissues at the level of the whole embryo. We have seen how certain tissues such as the neural plate and neural tube show a progressive restriction in developmental potential as they acquire regional properties and develop into specific organs. This question is examined further in Chapter 17.

## SUMMARY

The early developmental events depend upon maternal information stored in the egg during oogenesis. The utilization of this information is initiated at fertilization, when an activation program sets off a train of reactions leading to gene activation and protein synthesis. A metabolically unstable egg is triggered to replicate DNA, transcribe RNA, and synthesize proteins required for cell division.

The maternal information is stored as inactive messenger RNA that becomes active at fertilization. The activated messenger combines with ribosomes to form polysomes and protein synthesis is initiated. This pattern is true for the sea urchin embryo; other embryos are characterized by different patterns of activation.

The embryonic genome is also activated and begins to transcribe RNA. In most embryos, the pattern of RNA synthesis indicates that gene expression is regulated even during early development; some genes are transcribed only at specific stages. Ribosomal RNA is synthesized at high rates at gastrulation in sea urchin, frog, and fish embryos. In the mammalian embryo it is synthesized during cleavage, presumably because the egg contains lesser amounts of stored information and relies on the embryonic genome immediately after fertilization.

In most embryos the proteins translated during early stages are those relevant to cell division, particularly the histones that make up newly replicating chromosomes. This synthesis supplements the utilization of stored precursors such as histone, tubulin, and actin that are assembled into spindle apparatus and microfilaments during mitosis and cytokinesis. In the sea urchin embryo, and perhaps in others, histone synthesis is dependent upon two message populations, inactive maternal messages stored during oogenesis and the newly transcribed messages that come from the embryonic genome. Finally, a key feature of embryonic message production is the transcription of message at one stage for translation at later stages in embryogenesis. This seems to be characteristic of all developmental systems.

**REFERENCES**   Brown, D. D., and Gurdon, J. B. 1964. Absence of ribosomal RNA synthesis in the anucleolate mutant of *Xenopus laevis. Proc. Nat. Acad. Sci. U.S.* 51:139.

Church, R. B., and Schultz, G. A. 1974. Differential gene activity in the pre- and postimplantation mammalian embryo. *Current topics in developmental biology,* ed. A. A. Moscona and A. Monroy. Vol. 8. New York: Academic Press, pp. 179–202.

Davidson, E. 1977. *Gene activity in early development.* 2d ed. New York: Academic Press.

Emerson, C. P., and Humphreys, T. H. 1970. Regulation of DNA like RNA, and the apparent activation of ribosomal RNA synthesis in sea urchin embryos: Quantitative measurement of newly synthesized RNA. *Develop. Biol.* 23:86–94.

Epel, D. 1975. The program of and mechanisms of fertilization in the echinoderm egg. *Am. Zool.* 15:507–522.

Galau, G. A., Klein, W. H., Davis, M. M., Wald, B. J. Britten, R. J., and Davidson, E. H. 1976. Structural gene sets active in embryos and adult tissues of the sea urchin. *Cell* 7:487–505.

Graham, C. E., and Morgan, R. W. 1966. Changes in the cell cycle during early amphibian development. *Develop. Biol.* 14:439–460.

Grant, P. 1967. Informational molecules and embryonic development. *The biochemistry of animal development,* ed. R. Weber. New York: Academic Press, pp. 483–593.

Gross, P. R. 1967. The control of protein synthesis in embryonic development and differentiation. *Current topics in developmental biology,* ed. A. A. Moscona and A. Monroy. Vol. 2. New York: Academic Press, pp. 1–46.

Hinegardner, R. T., Rao, B., and Feldman, D. E. 1964. The DNA synthetic period during early development of the sea urchin egg. *Exp. Cell Res.* 36:53–61.

Humphreys, T. 1969. Efficiency of translation of messenger RNA before and after fertilization in sea urchins. *Develop. Biol.* 20:435–458.

Kedes, L. H. 1976. Histone messenger and histone genes. *Cell* 8:321–331.

Kojima, S., and Wilt, F. H. 1969. Rate of nuclear ribonucleic acid turnover in sea urchin embryos. *J. Mol. Biol.* 40:235–246.

McKnight, S. L., and Miller, O. L., Jr. 1976. Ultrastructural patterns of RNA synthesis during early embryogenesis of *Drosophila melanogaster. Cell* 8:305–319.

Nemer, M. 1974. Molecular basis of embryogenesis. *Concepts in development,* ed. J. Lash and J. R. Whittaker. Stamford, Conn.: Sinauer Associates, pp. 119–127.

Nemer, M., and Surrey, S. 1976. mRNAs containing and lacking Poly (A) function as separate and distinct classes during embryonic development. *Prog. Nucleic Acid Res. Mol. Biol.* 19:119–122.

Ruderman, J., and Gross, P. R. 1974. Histones and histone synthesis in sea urchin development. *Develop. Biol.* 36:286–298.

Spirin, A. S. 1966. Masked forms of messenger RNA in early embryogenesis and other differentiating systems. *Current topics in developmental biology,* ed. A. A. Moscona and A. Monroy. Vol. 1. New York: Academic Press, pp. 2–38.

Whiteley, H. R., and Whiteley, A. H. 1975. Changing populations of reiterated DNA transcripts during early echinoderm development. *Current topics in developmental biology,* ed. A. A. Moscona and A. Monroy. Vol. 9. New York: Academic Press, pp. 39–88.

Wilt, F. H. 1973. Polyadenylation of maternal RNA of sea urchin eggs after fertilization. *Proc. Nat. Acad. Sci. U.S.* 70:2345–2349.

# 14

# Asexual Buds

In our fascination with eggs and embryos we often tend to ignore other developmental systems. Asexual reproduction also poses important developmental questions inasmuch as small groups of somatic cells in the adult retain developmental potential and proliferate into buds that develop into adult organisms. Buds exhibit simpler developmental programs than eggs. Many eggs pass through a sequence of intermediary larval stages before becoming adults, while a bud develops directly into an adult. Asexual buds are also more successful reproductively; both males and females are able to produce offspring. Large numbers of genetically identical organisms are generated within a parental body.

The egg is primarily a gamete, adapted for sexual recombination, whereas the asexual bud is both a developmental and a reproductive adaptation producing many individuals from a limited number of cells.

An asexual bud is initiated after a small group of cells (10 to 50) escapes parental control. Physiologically released, the cells acquire independent mor-

phogenetic potentialities and develop into a new individual that separates from the parental body. Sometimes new individuals remain attached to the parent and organize as a colony, a mode of asexual development typical of many coelenterates and tunicates. Colonies are made up of genetically identical individuals but, in some, a phenotypic polymorphism has evolved; that is, morphologically distinct individuals display different specialized functions such as feeding or reproduction.

Eggs have no more than two developmental programs (larva and adult); the developmental repertoire of the asexual bud is richer. For example, a coelenterate egg produces a simple ciliated larva (planula) with a few tissue types. An asexual bud of the same species can develop directly into one of many different functional adults, specialized for feeding, reproduction, or protection. In this case, asexual buds pose a developmental problem that is absent in the egg: what determines the specific developmental program taken by a bud in polymorphic colonies?

Three patterns of asexual development or *blas-*

**445**

*togenesis* have appeared: budding, fission, and *gemmula* or *statoblast* formation. The latter two are budding modifications, and all three are widely distributed among most phyla but absent in insects and vertebrates. Budding is found principally in sponges, coelenterates, flatworms, and tunicates. Fission, or the division of a mature organism into two or more fragments, is common in flatworms and annelids. Smaller groups like the sponges and bryozoans display all asexual modes of development, including gemmula or statoblasts, populations of undifferentiated somatic cells derived from adult tissues and surrounded by a protective shell. These insure survival of the species during adverse conditions of drought or extreme temperatures that kill the parental body. After conditions become favorable, they resume development into new adults.

Many developmental questions arise in budding systems. For instance, what cells give rise to the bud? Do they come from a few somatic cells of the adult which are reprogrammed to develop into all somatic cells in a new organism? Or do buds develop from undifferentiated embryonic cells retained in adult tissues for purposes of asexual development? Genetic control would be different for each since somatic cell genomes would have to be reprogrammed to acquire pluripotency, whereas undifferentiated embryonic cells would develop directly into all somatic cell phenotypes

Other questions concern initiation and siting of buds. How are cells released from parental domination to acquire morphogenetic autonomy? Why do buds always appear in certain regions, spaced in definite patterns with respect to dominant growth centers in the parental body? A description of some budding patterns will provide some answers to these questions.

## BUDDING PATTERNS

In general, buds arise from adult epithelial sheets functioning as membranes or walls of vesicles and cavities. Buds also emerge as collections of unspecialized cells accumulating by migration or entrapment. A small number of cells, less than 10, is enough to initiate a bud. Once it arises, a bud usually expands by proliferation or tissue movements into a small, two-layered epithelial disc that later evaginates from the parental body wall into a vesicle resembling an echinoderm gastrula. By a series of folds, con-

strictions, evaginations, and expansions, the epithelial layers develop into an adult organism.

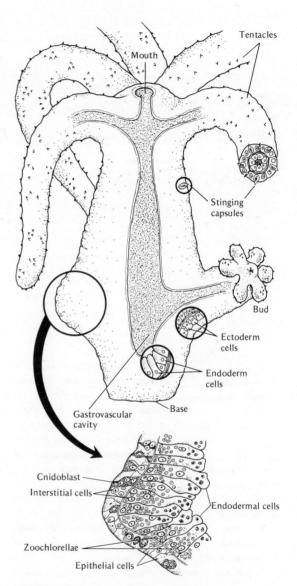

**Figure 14-1**   *Cross-sectional view of a hydra to show its internal structure and budding pattern. Bud evagination on left shows appearance at initiation of bud with interstitial cells on both sides of the mesoglea. Symbiotic zoochlorellae proliferate, indicating localized activity. Bud on right has already developed hypostome and tentacles.*

## Budding in Hydra

To illustrate how a bud develops, we begin with a simple coelenterate, the fresh water *Hydra*. A mature *Hydra* consists of two layers of cells, an outer ectoderm and an inner endoderm made up of gastroepithelial cells lining an enteric cavity (Figure 14-1). These two layers are separated by a mesoglea, an extracellular matrix containing mucopolysaccharides and collagen fibrils. It resembles the basement membrane found in vertebrate tissues and may have a similar morphogenetic role. The cylindrical body is topped by a circle of tentacles surrounding a raised portion, or hypostome, containing the mouth. At the aboral end is a foot or pedal disc which attaches to the substratum, a twig or the underside of a leaf. Besides gastric cells, four other basic cell types make up the hydra body: (1) epitheliomuscular cells of the ectoderm, (2) nerve cells in a diffuse net throughout the body, (3) cnidoblasts or stinging cells in clusters on all tentacles, and (4) interstitial cells, undifferentiated cells with large nuclei that are the progenitors of nerve cells and stinging cells. Interstitial cells are distributed in both cell layers, either as single large cells or as nests of small cells in multiples of two. These nests are different stages in nerve cell or cnidoblast formation.

A bud always appears in the budding zone, a little more than halfway between hypostome and foot. Blastogenesis begins as a local thickening of ectoderm and endoderm, possibly as a result of local changes in cell shape and adhesiveness. The small bud elongates by migration of parental sheets into the bud zone rather than by cell division. Mitotic activity in the bud is not significantly higher than in adjacent regions. Budding is only an extension of the normal growth pattern in *Hydra*, in which proliferating cells are found distributed throughout the organism, with no apparent concentration of mitotic activity in any region (Figure 14-2).

Both ectoderm and endoderm expand along the linear axis from hypostome to foot in a growing hydra. Two factors contribute to this movement, passive displacement due to growth and a more active tissue migration, with each layer moving at different rates. The siting of a bud in the budding region involves diversion of these expanding tissues; both cell layers change polarity 90 degrees at the junction between parent and bud as the latter grows with its own axial symmetry.

The gastrovascular space in the bud endoderm is continuous with the parental cavity. The bud assumes a cone shape with interstitial cells at the tip, under the thinning epithelial muscular layer. This apical region becomes the future hypostome of the new individual as cells migrate distally and proximally from it. Some interstitial cells in the prospective hypostome region differentiate into nerve cells which gradually anastomose into an irregular net, sweeping downward from the hypostome into the lengthening column. As we will see, nerve cells have a trophic role in growth and nerve net differentiation may correlate with the progressive outgrowth of the bud. Tentacles appear at the base of the oral cone while a mouth breaks through at its center, and the new individual, now able to feed and live an independent existence,

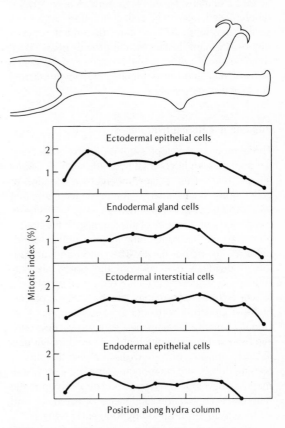

**Figure 14-2** *Distribution of cell proliferation along the body column of Hydra littoralis in four cell types. No mitotic activity occurs in the tentacles. Note the absence of any gradient. (From R. D. Campbell, Am. Zool. 14:523–535, 1974.)*

detaches at its base and separates from the parental body. The entire budding process takes 50 hours at a temperature of 26°C, and a population of budding hydras will double its number every 3 1/3 days if well fed.

In some species of *Hydra*, budding and sexual reproduction are alternative morphogenetic pathways. At low temperatures in *Hydra oligatis*, gonads rather than asexual buds appear in the budding zone. Interstitial cells proliferate into a diffuse plate of cells at the base of the ectoderm and differentiate into a testis or ovary with a single large egg. When sexuality is induced, budding is inhibited; if, however, the temperature is raised early enough, a developing gonad may be shifted into asexual budding. Such systems of alternative developmental pathways are characteristic of many asexual organisms. A relatively undifferentiated interstitial cell population carries out several developmental programs depending on conditions. In many instances, the growth phase (budding) competes with a sexual program and the switch from one to another is initiated by environmental cues such as temperature.

### Budding in colonial hydroids

The alternative developmental programs of an asexual bud are seen in colonial hydroids such as *Obelia* (Figure 14-3). This species undergoes alternation of generations, an asexual hydranth stage alternating with a free-living sexual medusa stage. The basic morphogenetic structure of this organism is the stolon, a two-layered horizontal tube (coenosarc) surrounding a gastrovascular cavity. The stolon and all budding individuals derived from the stolon are surrounded by a rigid extracellular secretion, or perisarc, to which the coenosarc may attach at several places. Stolonic growth is horizontal on a substratum, occasionally sending out branches. As it grows, new perisarc is secreted. Two different functional individuals extend vertically from bud discs along the stolon, feeding polyps and reproductive gonangia on which medusae develop. The type of bud depends upon the differential contraction tendencies of ectoderm and endoderm in response to environmental signals.

The principal growth zone in a stolon is the terminal end where ectoderm and endoderm cells elongate (Figure 14-4). As in *Hydra*, proliferation occurs throughout the stolon in both layers and is no higher at the tip. Although some believe the terminal

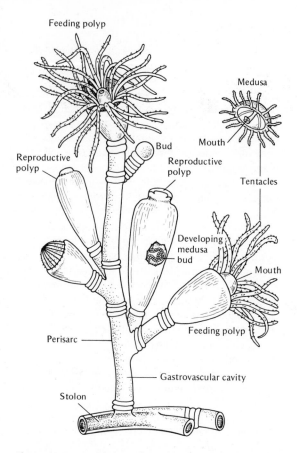

**Figure 14-3**  *An* Obelia *colony showing reproductive and feeding polyps and medusa stage.*

region acts as a mitotic center, most evidence suggests that cell movement rather than proliferation is directly responsible for growth and morphogenesis. For instance, stolonic growth is unaffected by X-ray irradiation, a treatment known to block mitosis.

The zone behind the tip pulsates rhythmically to maintain fluid (hydroplasm) flow in the gastrovascular cavity. The contractile zone is several millimeters long and its activity insures nutrient circulation to all tissues. Periodic contractions at the "growing tip" result in tissue migration and are recorded as annular constrictions in the surrounding perisarc. Growth is marked by a series of pulsations, accompanied by changes in shape of ectoderm cells. Time-lapse recordings of tip growth show it advancing in short pulsations, extending forward for a few minutes, followed by a short retraction and an advance again after a short delay to a more distant position. This leads to extension of the stolon. This pulsatile ''growth'' is independent of contractile zone activity since the two are completely out of phase.

In a related organism, *Campanularia,* the growth pulses establish a cyclic growth pattern with the duration of each cycle being constant (6.5 minutes on the average). The distance advanced each time is approximately 20 to 30 $\mu$, that is, the difference between the distance extended and the distance retracted at each pulse. Average cycle times decrease with increasing temperature whereas average growth per cycle is uniform over a wide temperature range. Temperature seems to affect the rate at which the growth cycle is completed but does not alter the amount of growth achieved.

An isolated tip (300 to 500 $\mu$ in length) will exhibit the same growth cycle for several hours with only slight modification, which means that stolon growth is an intrinsic property of the tissues in the tip and is independent of hydroplasm pressure changes. In normal growth, pressure changes within the hydroplasm do affect the retractile phase of the cycle, presumably by creating low pressure.

Accompanying these movements are extensive cyclic changes in thickness of the epidermis, varying

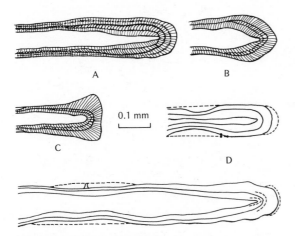

**Figure 14-4** *Stolonic motility in* Obelia. *(A) End of fast-growing free stolon. (B) End of slow-growing free stolon. (C) An attached slow-growing stolon showing laterodistal spread of epidermis. (D) Pulsatile activity at ends of free-growing stolon showing contractile zone and pulsations at tip. Note change of cell shape in epidermis in each case. (Based on N. J. Berrill,* Quart J. Microscop. Sci. *90:235–264, 1949.)*

from 5 to 30 $\mu$ at the thinnest phase of the cycle to 35 to 60 $\mu$ at the thickest, the average difference being 25 to 30 $\mu$. Such changes are probably produced by contraction of microfilaments within cells accompanied by changes in lateral adhesion. The growth-cycle pulse of extension and retraction is produced by contraction and relaxation of myofibrils in epithelial muscular cells, whereas shape changes in cells are produced by microfilaments. Growth of a stolon is dependent on extension and relaxation of epithelial sheets accompanied by changes in cell shape. Similar expansions of epithelial sheets have already been noted in epibolic movements during gastrulation.

Such patterns of horizontal growth may continue indefinitely, but buds usually originate behind the contractile zone and emerge vertically as an upright feeding hydrant. The bud rudiment in *Obelia* arises vertically because the ectoderm thickens in the region. Compared with a stolon tip, the bud is wider and grows at a slower rate. Growth (extension by pulsatile growth) continues at the widening distal end with both cell layers expanding and contracting just as they do at the stolon tip. The tip widens further, and a broadened hydranth develops on top of an elongated stalk (Figure 14-5).

The center of tissue expansion is disclike, radiating in all directions. The disc expands and becomes polygonal with a tentacle bud appearing at each corner. A raised conical region in the center becomes the oral cone or manubrium through which the mouth breaks.

Budding patterns are controlled in part by temperature, because ectoderm and endoderm respond differentially to temperature changes. At high temperatures, both layers in *Obelia* grow rapidly and synchronously, typical of stolonic growth. At lower temperatures, however, each layer grows at a different rate, producing wide, slow-growing terminal growths from which tentacle rings and hydranths develop. At still lower temperatures, linear growth and expansion of the endoderm are suppressed in relation to the ectoderm, and a gonangial individual appears.

A gonangium or sexually reproducing individual develops at the terminus of the stolon. From the beginning, however, its growth differs from that of a hydranth. In a gonangial bud, the epidermis always maintains a more rapid rate of expansion than does the underlying endoderm cap. The gonangial rudiment, the last of a series of growth surges, expands

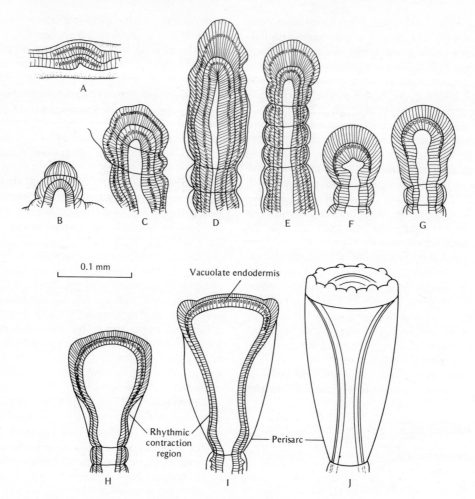

**Figure 14-5** *Pulsatile growth in vertical stem and development of hydranth in Obelia. (A) Bud initiation; elongation of cells in ectoderm and endoderm. (B) Early outgrowth showing second annulus. Epidermal blob (elongated cells) in advance of endoderm. (C) Endoderm expands into blob at two-annulus stage. (D) Later stage after first phase of annular growth. Note elongation of epidermis into blob initiating next phase of growth. (E) Formation of hydranth rudiment at end of second annular phase. (F, G) Enlargement and elongation of rudiment after endoderm expands into epidermal blob. (H, I) Progressive growth and expansion with proximal formation of rhythmic contraction region; coenosarc detaches from perisarc. (J) Slightly later stage. Complete separation of coenosarc from perisarc and appearance of tentacle rudiments. (Based on N. J. Berrill, Quart. J. Microscop. Sci. 90:235–264, 1949.)*

transversely more rapidly than it elongates but does not develop tentacles or a manubrium. Instead, the lateral walls detach from the perisarc and evaginate into two-layered folds of tissue which become medusa buds.

Medusae are morphologically different individuals though derived from cells with identical genomes. The budding pattern changes to form a new arrange-

ment of cell layers which develop into flattened discs with a modified hypostome, tentacles, and a manubrium. These sexually reproducing medusae detach from the gonangium and take up an independent free-living existence. Gonads develop along the radial canals of the gastrointestinal cavity. Eggs are fertilized internally, cleave, and gastrulate into a ciliated planula larva. Larvae escape into the sea, swim about for a short period, attach to a substratum, and transform into a stolon that resumes growth as a colony with feeding hydranths and gonangia.

## Strobilation in Aurelia

Among the Scyphozoa (true jelly fish), the sexual phase, or medusa, is most important whereas the polyp generation is very much reduced. Here, too, elaborate adaptations for asexual budding have evolved. The life cycle of *Aurelia* illustrates some of the principal features of reproduction.

The medusa is a pelagic jellyfish found widely distributed in temperate waters. Sexes are separate with gonads developing on radial canals. In the breeding season, sperm are shed into the sea and find their way into the gastral cavities of females, where they fertilize the eggs. These develop into ciliated planula larvae, escape, and swim about for 4 to 5 days while ectoderm and endoderm undergo histogenesis. The larva loses its cilia, settles to the bottom, and attaches by its broad end to form a pedal disc. It gradually transforms into a polyp or scyphistoma with a mouth and 20 to 24 tentacles with nematocysts and hypostome. It feeds actively, increasing in size, budding asexually into additional scyphistoma, and even sending out hollow stolons from its base.

As polyps mature, another striking change becomes evident along their slender tubular length. A series of transverse constrictions appears at regular intervals along the cylinder. These constrictions deepen, giving the polyp the appearance of a stack of inverted plates. These gradually transform into radially symmetrical buds with marginal lappets, a manubrium, gastral filaments, and rudimentary sense organs. The uppermost bud is most advanced, with others following behind in decreasing order of maturity until those near the base are least differentiated. The polyp is now called a *strobila*, and the process of plate formation is known as *strobilation*. Strobilation is a modified form of budding in which a series of buds is simultaneously carved out of parental tissues in a progressive order of development (Figure 14-6).

Each strobila bud becomes mature at the apical end and eventually detaches as an ephyra larva. As soon as one escapes, the next one in line matures and is released in its turn. The ephyra is a microscopic animal that swims freely in the sea. It grows and

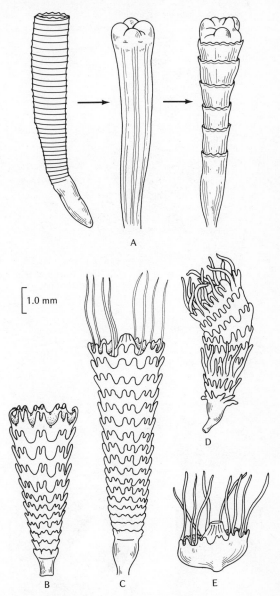

**Figure 14-6**  *Strobilation in* Scyphomedusa. *(A) Annulated theca (outer perisarc), prestrobila, and strobila of the jellyfish* Nausithoe. *(B–D) Strobilation in* Aurelia. *(E) An ephyra stage released from a strobila. (Based on* Growth, Development, and Pattern *by N. J. Berrill. W. H. Freeman and Company. Copyright © 1961.)*

gradually transforms into the medusoid stage, which it attains when only 1 cm in diameter. The young medusa feeds and grows into the mature adult.

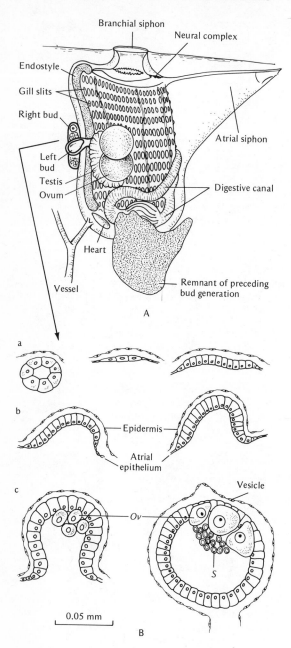

**Figure 14-7** *Budding in the colonial ascidian* Botryllus. *(A) One mature zooid in the colony with mature testis and ovum and two buds, left and right, of the next generation and a remnant of the parental zooid of a previous bud generation. (B) Within the atrial epithelium of the developing left bud another bud disc is appearing to develop into the next generation. (a) Surface and side views of initial disc stage and growth to maximum disc size. (b) Evagination of disc into a hemisphere. Cells change shape to do so. (c) Conversion of hemisphere into closed vesicle surrounded by protective epidermis. Segregation of oocytes first and then spermatogonia from vesicle wall. Ov = ova; S = mass of spermatogonia. (From N. J. Berrill,* Biol. Bull. *80:169–184, 1941.)*

What is notable about these life histories is the developmental versatility of asexual budding. The same genome is able to express several different developmental programs, each determined by previous history, nutrition, and temperature. No intrinsic genetic clocks seem to trigger each event.

In summary, asexual budding in coelenterates is an example of morphogenesis largely divorced from growth. In fact, tissue motility rather than cell division is responsible for stolonic growth and the emergence of hydranths and gonangia. Tissues at "growing points" exhibit pulsatile activity accompanied by tissue extension. The motive forces for these changes are still unclear, but they resemble morphogenetic movements in epithelial sheets that are mediated by contractile microfilaments. The unique feature of coelenterate budding is the two-layered condition surrounding a vascular cavity in which a dynamic hydroplasm flow occurs. The two layers, separated by an extracellular matrix, the mesoglea, move together or independently, depending on conditions. Such movements lead to form changes, emerging as tentacles and hypostome. The mesoglea, containing glycoproteins and collagen, resembles the basement membrane of most vertebrate tissues, and, like the basement membrane, it may be responsible for the morphogenetic form and motile activity of adjacent tissues. Morphogenetic mechanisms of these simple epithelia are similar to those responsible for movements of epithelial sheets in vertebrate embryos.

## Budding in ascidians

Other types of budding patterns are displayed by ascidians. As we have seen, ascidian eggs develop into swimming tadpoles which, in turn, metamorphose into a single sessile adult or a colony of individuals. Only adult ascidians bud, a process distinguished by high regulative capacity; a bud adjusts its developmental pattern to the number of cells originating in the initial bud disc. The fewer the number of cells in the disc, the simpler is its final organization, and only a few essential organ systems develop. Development proceeds irrespective of the number of cells in the disc, and size regulation insures that functional zooids will emerge.

The ascidian bud consists of two layers. The outer layer is from the adult epidermis. The underlying layer may be derived from any one of three different epithelia: (1) the epithelium lining the atrial cavity on either side of the body, (2) the lining of the epicardium or perivisceral cavity adjacent to the heart, or (3) the mesenchymal septum separating afferent blood flow in a stolonlike growth. The outer epidermis never participates in bud morphogenesis but acts as a protective envelope surrounding the totipotent cell layer below. Its role is limited to gross changes as opposed to the detailed sculpturing of a new individual. Bud morphogenesis is a property of the inner cell sheet, transforming into all adult organs by direct development.

Budding patterns are almost as diverse as the number of species. We will describe a simple pattern in *Botryllus,* a colonial form consisting of mature zooids connected via common cloacal cavities into a star-shaped disc pattern (Figure 14-7). A bud first appears as a local enlargement of 5 to 8 cells of atrial epithelium in the body wall of a growing zooid. These cells proliferate rapidly into a small population of columnar cells surrounded by a squamous epidermis. As soon as the disc of cells achieves a maximum size of 20 to 150 cells, it begins morphogenesis by evaginating into a hemisphere that eventually becomes a hollow sphere. The overlying epidermis,

never in contact, conforms to produce a two-layered vesicle, the initial stage for all morphogenesis. Subsequent development is rapid and direct. The adult body plan is blocked out by a series of simple folds and constrictions. Two lateral folds invaginate to divide the cavity into three chambers. Then, by a series of smaller local invaginations in the central compartment, the presumptive heart, intestine, and neural complex appear almost at once, establishing all primary organ systems. At this stage, the entire bud may be only 0.1 mm in diameter and consist of a few hundred cells. This develops into an adult by further elaborations of the primary rudiments. Rows of gill slits break through in the central compartment, joining with atrial chambers, and the new zooid is now capable of feeding (Figure 14-8).

A bud appears at an early stage in the life of a zooid, usually after the heart rudiment becomes distinct. The new budding generation is initiated even before the parental zooid is completely functional or sexually mature. Budding individuals remain connected to form the typical colony morphology. A parent undergoes dissolution before the next generation of buds appears.

The initial size of a disc, or its cell number, is determined at random before it folds into a vesicle. The size of a vesicle is directly related to the size of a disc because the division program is fixed in these cells. This is the essence of size regulation. About one and a half divisions intervene between disc and vesicle. A 20-cell disc produces a 33-cell vesicle, a 50-cell disc becomes a 75-cell vesicle, and a 150-cell disc pro-

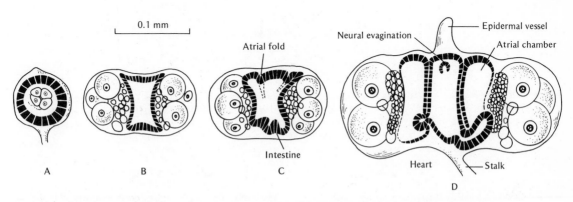

**Figure 14-8**   *Continued development of a* Botryllus *bud disc of maximal size. (A, B) Stages in gonad development on each side. (C) By a series of evaginations, foldings, and constrictions, inner vesicle divides into central and lateral compartments. (D) Further evaginations from central chamber form intestine, heart, and neural complex. (From N. J. Berrill,* Biol. Bull. *80:169–184, 1941.)*

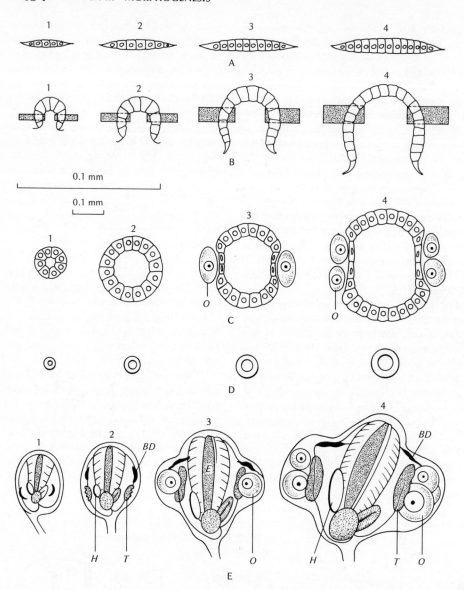

**Figure 14-9** *Size regulation in budding: disc size determines gonad segregation in Botryllus. (A) Discs of various sizes. (B) Vesicles developing from each disc showing presumptive gonad region in black. (C) Closed vesicles developing which show no ova segregating in (1) and (2), single ova in (3), and double ova in (4), corresponding to number of cells included in gonad primordium above. (D) Sphere stages of (C) reduced to one-fifth scale. (E) Continuation of development on reduced scale to show gonad development in maturing bud. BD = bud disc; H = heart; E = endostyle; O = ovum; T = testis. (From N. J. Berrill, Growth Symp. 3:89, 1941.)*

duces about 210 cells (Figure 14-9). The timing of division and morphogenesis is precisely controlled and constant, irrespective of cell number or vesicle size. Certain folds and constrictions always appear at precisely the correct time; consequently, the size of the organ formed depends upon the number of cells available at that time. Segregation of gonads will illustrate this phenomenon of size regulation.

Most ascidians are hermaphroditic; male and female gonads develop within each growing bud. Gonads appear early in bud morphogenesis, even before the vesicle stage. The presumptive ova are the first cells to segregate from the lateral wall of the hemisphere stage, before it closes into a vesicle. The presumptive testes appear shortly thereafter in the vesicle stage. Gonads must segregate at this time or not at all, and the number of disc cells available determines whether any gonads develop. In large discs of 150 cells, large oocytes segregate first, followed by testicular cells. In discs of less than 60 cells, reproductive cells never appear. Among discs of intermediate size, testicular components may appear either alone or with a few nongrowing oocytes. Only large discs produce functional oocytes; the limiting factor in ovum segregation is disc cell number at maximum size.

Since a minimal number of disc cells is required for morphogenesis of such vital organs as heart and intestine, only large discs can segregate cells as ova. A disc two-thirds maximum size can produce only one ovum, but no cells are available for ova in smaller discs. Later, in the vesicle stage, when total cell number has doubled, testes may be segregated. Failure of ova to appear does not exclude differentiation of testes because ova segregate at the earliest stage, before the full complement of disc cells is produced. The moment for female sexuality is transitory and is limited entirely by cell number in the disc.

Size and growth of presumptive bud regions seem to be regulated independently of growth of the entire zooid: for instance, we might expect competition between parent and bud to result in dissolution of the parent zooid when the bud attains a critical size. If this were true, then removal of late stage buds should prolong the life of the parent zooid, but this does not occur; no zooid persists more than 24 hours beyond the untreated controls. Death of the parental bud is intrinsic to its tissues and does not result from competition between aging and "youthful" bud tissues.

Dissolution of the parent promotes a premature functional maturity of the bud. In this sense, buds do seem to compete with one another. If buds at one side are extirpated, the buds surviving become larger since more nutrients are available without competition. Removal of buds on one side may induce an arrested bud on the opposite side to begin rapid growth. If bud removal is carried out at an early disc stage, it may even induce the remaining bud to divide into two bud discs, each developing into a zooid because of increased opportunity for growth.

## ANALYSIS OF ASEXUAL BUDDING

In general, asexual budding is an extension of the normal growth and reproductive cycle of the individual. Most asexually reproducing organisms are characterized by high growth and regenerative potential. Their somatic tissues are constantly undergoing a process of cell renewal; in a sense, the whole organism behaves as a "renewal organism." Periodic cycles of growth, budding, and dissolution at the organismic level are indicative of a system of massive cell replacement. In most cases, organisms contain mitotically active cells distributed randomly throughout the body. As new daughter cells appear, they tend to migrate to more mature regions where they replace tissues depleted by continuous cell death. Most somatic cells are pluripotent, able to develop into more than one cell type, whereas some cells (for example, those in segmentation zones) are totipotent, able to develop into a complete organism. An analysis of normal growth patterns in *Hydra* illustrates some of these points.

### Dominance and polarity in Hydra

The *Hydra* body plan, as we have seen, consists of five regions arranged along the linear axis: tentacles, hypostome, digestive cavity, budding zone, and pedal disc. Although it appears as if a hydra grows from the subhypostomal region, no mitotically active growing zones are demonstrable in any region; dividing cells are seen throughout the body. Tissue migration rather than localized mitotic activity accounts for the overall growth pattern. If cells of the subhypostome are stained with a vital dye and grafted back in place, some stained cells move proximally along the body until they accumulate at the basal disc and eventually disappear (Figure 14-10). Others migrate distally out to the tentacles. Cells do not move independently, but each layer expands as a coherent sheet at different rates. The jellylike mesoglea between layers may serve as a substratum on which sheets of ectoderm and endoderm migrate, proximally toward the basal disc and distally into the tentacles. Both are regions of cell loss since cells age and die in

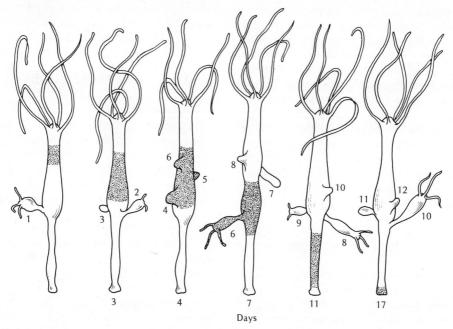

Days

**Figure 14-10**   *Tissue movements in normal growth of* Hydra. *A stained segment is grafted into the column beneath the hypostome and is progressively displaced toward the pedal disc, incorporated, in part, into buds as it enters the budding zone. It is sloughed off at the pedal disc. A succession of 10 buds developed during the 17 days of the experiment. (After P. Brien and M. Reniers-Decoen, Bull. Biol. Fr. et Belg. 83:293–386, 1949.)*

the basal disc, and batteries of nematocysts in the tentacles are constantly used up in feeding. These sites act as "cell sinks," continuously replaced by a flow of cells from other parts of the body.

The patterns of *Hydra* growth reflect a balance between cell loss at pedal disc and tentacles, and tissue migration from the subhypostomal region. Cells are replenished uniformly throughout the body as a result of random mitotic activity. If mitotic activity is inhibited by X-rays, the balance is upset and no new cells are produced. The remaining cells grow old and die within a few weeks, and the animal gradually shrinks and degenerates. *Hydra* is a system of rejuvenating tissues that undergo complete cell turnover.

Polarized migration of tissue layers in *Hydra* is not understood. In fact, "polarity" is an operational term; it describes the results of experiments on regeneration of *Hydra* fragments (see Chapter 15). For example, if the hypostome region is removed, a new hypostome with tentacles regenerates at the cut surface. If, however, the hypostome is only partially cut and remains attached to the body, no new hypostome

regenerates at the cut surface. The presence of a hypostome seems to inhibit regeneration of a new one. Polarity extends through the *Hydra* body; fragments, excluding the pedal disc, always regenerate a hypostome at distal ends. Each fragment retains the original distal-proximal polarity of the intact organism. The hypostome is the high point of a regenerative field; fragments behind the hypostome regenerate more rapidly than fragments nearer the base. The hypostome dominates the growth pattern in this organism.

Dominance seems to depend on cell-cell contact, since unattached hypostomes inserted into the enteric cavity do not prevent hypostome regeneration in decapitated hosts. Likewise, disruption of cell continuity of a host hypostome by chemical disaggregation destroys its ability to inhibit hypostome formation in other regions. These results suggest that hypostomal dominance is not exerted by an inhibitor diffusing through the gastric cavity but is mediated by an inhibitory signal passing from cell to cell. Furthermore, it has been shown that the rates of signal transmission can be accounted for entirely by rates of chemical diffusion. This suggests that a gradient of an inhibitor substance, from a high point at the hypostome decreasing gradually along the body to a low point at the basal disc, could explain both hypostome dominance and polarity of morphogenesis in *Hydra*.

The nature of the inhibitory signal is unknown. In fact, two substances must be postulated to account for morphogenesis, one inhibitory and another stimulatory, since a hypostome fragment also induces new outgrowths when grafted (see in Chapter 15). A clue as to the source of such factors comes from studies of the distribution of nerve cells within the *Hydra* body. Some of the earliest demonstrations of nerve cells in *Hydra* showed a diffuse nerve net with the greatest concentration of neurons in the hypostome region, decreasing toward the basal disc (Figure 14-11). This in situ staining pattern correlates with direct counts of nerve cells obtained from macerated tissues from different regions of the body wall. The spatial distribution of nerve cells corresponds with the gradient predicted for the inhibitor substance.

Most nerve cells in *Hydra* are multipolar, with many long branching processes penetrating the tissues. Secretory granules are seen within the nerve

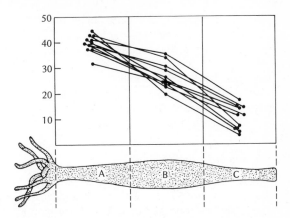

**Figure 14-12**  *Gradient distribution of interstitial cells (I cells) in the ectodermal layer of* Hydra viridissima *in 11 individuals. Ordinate: average number of I cells per 6400 $\mu^2$ in each third of the body. (Based on P. Tardent, Wilhelm Roux' Arch. Entwicklungsmech. Organ. 146:593–649, 1954.)*

cell body and during morphogenesis granules seem to leave nerve cell processes. Growth stimulation by nerve cells is suggested by the fact that extracts of hypostomes (highest nerve cell density) stimulate supernumerary growth of tentacles at distal cut surfaces.

Nerve cells may be responsible for both stimulatory and inhibitory control. Dominance by hypostomal regions may be due to neural control of tissue contractility. In most hydroids, the epitheliomuscular cells of the ectoderm are organized as a coherent sheet and conduct slow electrical impulses to generate rhythmic contractions. Stimulation of the nerve net inhibits these slow contractions. Inasmuch as growth of *Hydra* depends upon tissue movements, inhibition of contractility by nerve may account for failure of most body regions to initiate a bud; they do not possess the intrinsic motile behavior to organize a new axis of outgrowth.

The interstitial cells (I cells) also play a role in *Hydra* morphogenesis. They are the progenitors of both nematocyst and nerve cells and are distributed in a gradient along the body wall with a high point at the hypostome (Figure 14-12). From this region, new nematocyst-containing cells are produced to replace those lost from the tentacles during feeding, while interstitial cells transform into nerve cells in a more proximal direction. The distribution of I-cell progenitors determines the distribution of nerve cells,

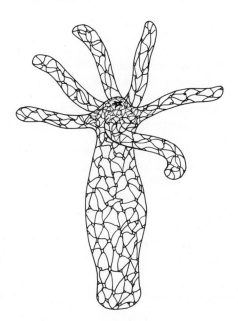

**Figure 14-11**  *Nerve net in* Hydra *showing greatest concentration in hypostome around the mouth.*

and since the former are concentrated in the hypostome, we find nerve cells most numerous in that region. We see that a two-cell system, interstitial cells and nerve cells, contributes to the polarized growth patterns within the *Hydra* body.

How does this pattern of hypostome dominance explain the initiation and siting of buds, and where do bud cells originate?

## Initiation and siting of buds

Buds always emerge from the *Hydra* body wall in a localized budding zone. The morphology of the *Hydra* budding zone, however, is no different from adjacent regions prior to the initiation of a bud. The same distribution of cell types is seen, and there is no evidence of local accumulation of interstitial or nerve cells, nor is mitotic activity higher in the bud region.

We assume that bud initiation is based on factors controlling cell shape and tissue migration. A bud first appears as a thickening of the two layers, either as a change in cell shape or as a result of increased cell adhesiveness in the budding zone. Because of enhanced adhesive contacts, the region becomes a "sink" into which parental tissue layers preferentially migrate. Cell division is not involved; bud growth seems to be entirely due to a polarized migration of tissues from the parental body into a secondary axial outgrowth, now oriented perpendicular to the main axis.

We do not understand why cells in the budding zone acquire new surface properties in association with a new polarity. The fact that it occurs always at a certain distance from the parental hypostome suggests that hypostome dominance is important.

If a hypostome tip is grafted into more proximal regions of the *Hydra* body closer to the budding region, a secondary axis is induced even in the presence of the host hypostome. At this distance from the hypostome, the graft escapes the inhibitory influence of the host. Increasing the distance by insertion of a fragment of digestive zone from another animal between host hypostome and graft results in a large secondary outgrowth. It appears, therefore, that a bud normally emerges at that distance from the hypostome where it just escapes hypostomal dominance. Since the nerve net is most diffuse in this region, an inhibitory influence from nerve cells is minimal. Contractility is stimulated locally and tissues begin to migrate. As more cells enter an incipient bud, the

population of interstitial cells increases and new nerve cells develop. Under these conditions, trophic factors promote hypostome differentiation and bud outgrowth is induced even further. Once a dominant hypostome center is established, a new bud is stabilized. Interstitial cells proceed to differentiate stinging cells as bud tentacles emerge. When a region is released from inhibitory control, it is free to express its innate morphogenetic potential.

## Control of strobilation in Amaroucium

Most asexual budding is under exogenous control, initiated in response to environmental signals such as temperature. The interaction of these signals with the endogenous control system to initiate strobilation in an ascidian was examined by Freeman.

The colonial ascidian *Amaroucium* is organized into three body regions: a thorax with branchial basket, an abdomen containing stomach-intestine, and an elongate postabdominal region (Figure 14-13A). These organisms strobilate by a series of ectodermal constrictions which cut the body abdominal regions into small fragments or *blastozooids* that develop into new zooids.

Strobilation in nature is initiated under stress conditions, usually during the winter when water temperature drops. The parental bodies regress as they strobilate, while each new fragment resumes the morphogenetic cycle. Although environmental stress triggers strobilation in nature, in the laboratory it occurs spontaneously under conditions that are still not understood.

Brien suggested that growth is the factor initiating strobilation; zooids must reach a critical size to trigger the process. Freeman tested Brien's hypothesis by inhibiting zooid growth with X-rays and determined the time at which strobilation occurred. He found that although growth is depressed in irradiated animals, all strobilate at the same time irrespective of X-ray dose. The experiment indicates that neither growth nor size of the animal is the trigger for strobilation. Strobilation probably reflects a rhythmic contractile process, very much like the pattern of rhythmic growth in coelenterate stolons.

By removing parts of the animal's body, Freeman determined the region that controls the timing of strobilation (Figure 14-13B). Removal of the thorax triggered strobilation within 24 hours in animals that normally would strobilate 6 to 8 days later. This indicated that strobilation itself was under some

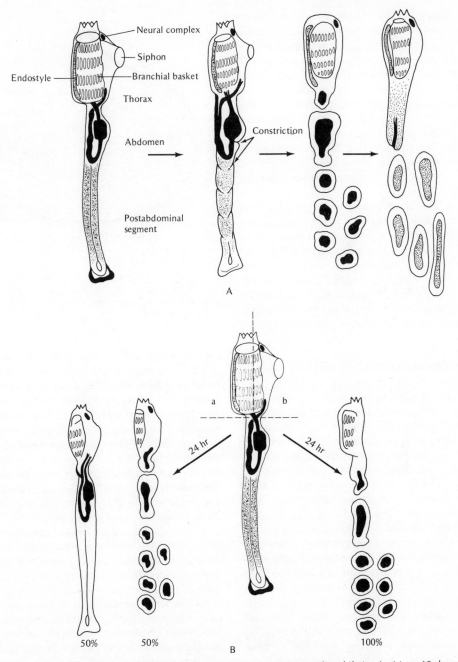

**Figure 14-13** *Control of strobilation in the ascidian Amaroucium. (A) Pattern of strobilation in 14- to 15-day-old (days after metamorphosis) individual. First constrictions appear posterior to stomach mass in abdomen; then within a few hours the entire region behind the thorax is constricted into contracted fragments, each of which soon extends and regenerates into a new individual. (B) Precocious strobilation induced by removing neural complex. Nine-day-old individual is operated as shown, operation b removing neural complex and dorsal cord. Within 24 hr all animals strobilate while only 50 percent of those with the neural complex intact strobilate. The remainder regenerate and strobilate at the same time as controls. (Based on G. Freeman, J. Exp. Zool. 178:433–456, 1971.)*

kind of inhibitory control from factors emanating from the dorsal thorax. The region contains the dorsal nerve cord and neural complex, and Freeman suggested that the dorsal cord region contains cells that control strobilation. According to his hypothesis, all ectoderm has the tendency to constrict but is prevented from doing so in growing zooids by cells in the dorsal cord. The mechanism of inhibition is not known, but when animals reach a certain size, they acquire the ability to strobilate. Presumably an inducing environmental signal received by these cells releases inhibition, and the ectoderm immediately constricts.

The control of strobilation in *Amaroucium* resembles regulation of budding in *Hydra;* both are regulated by an inhibitory nerve center elsewhere in the parental body. These results suggest that release from domination by the parental body permits the morphogenetic potential of a region to be expressed. This is consistent with the observation that budding regions always establish autonomous morphogenetic axes once they escape inhibitory influences.

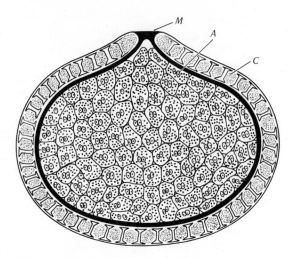

**Figure 14-14**    *A gemmule of the sponge* Ephydatia fluviatilis. *M = micropyle; C = outer protective coat; A = archeocytes filled with yolklike material, the undifferentiated progenitors of a sponge when the gemmule germinates.*

## SOURCE OF BLASTOGENIC CELLS

We have seen that in *Hydra* and *Obelia* cells of the bud come from parental somatic tissues migrating into a new morphogenetic zone with an independent axial gradient system. All differentiated ectoderm and endoderm cells, including interstitial cells, participate in bud outgrowth. Interstitial cells furnish the nematocysts and nerve cells of the emerging bud.

Each tissue layer in coelenterates retains morphogenetic plasticity. The two layers may be separated, and each can regenerate cell types of the other layer, even to the point of reconstituting a new animal. Cells dedifferentiate before they become morphogenetically active as a bud and acquire large nuclei and nucleoli with RNA-rich cytoplasm, a characteristic of actively growing cells.

The situation in ascidians is similar. The adult epithelia supply all bud disc cells including germ cells that differentiate into a mature zooid. In these groups of organisms, asexual budding does not depend on a reserve of undifferentiated embryonic cells; rather, most somatic cells preserve their pluripotent nature in spite of their specialized function and transform into a bud when they escape inhibition of the parental body.

In other groups of organisms, however, such as sponges, undifferentiated reserves of totipotent cells do participate in asexual budding. These cells do not exhibit any specialized function in the adult but appear to be retained for budding and regeneration. An example of these cells is the *archeocyte* in the sponge. Archeocytes, undifferentiated cells found throughout the sponge body (Figure 14-14), can differentiate into all adult cell types during gemmule production. A gemmule is an isolated aggregation of archeocytes and nutritive cells that develops in response to adverse conditions (changes in temperature or moisture). Protected by a thick layer of spongin, it survives complete drying while the remainder of the sponge dies. When released from the degenerating parental body, it reconstitutes a completely new sponge as soon as conditions become favorable.

In general, somatic cells of lower invertebrates have high reproductive potential. Except for highly specialized cells such as nerve cells and cnidoblasts, most somatic cells are labile and pluripotent. This property must be viewed within the framework of life cycles alternating between sexually and asexually reproducing generations. One major difference between these organisms, and insects and vertebrates which exhibit no asexual development, is reflected in the origin of totipotent germ cells. In the latter groups, the distinction between germ and soma is sharp. Primordial germ cells are segregated very early in embryonic development and serve as the sole source of all gametes in the adult. In asexually reproducing invertebrates, germ and soma are not distinct entities. Germ cells can arise from interstitial cells in ectoderm and endoderm in coelenterates and can also arise from epithelial somatic cells in the developing ascidian bud. It is not surprising that somatic cells can participate in asexual reproduction; many retain the full developmental potential of germ cells. Since they are retained within the adult body and fully nourished by it, they do not require the elaborate cytoplasmic organization so typical of eggs which are usually released as independent cells into the environment.

## BLASTOGENESIS AND EMBRYOGENESIS COMPARED

Most asexual organisms exhibit an alternation of generations — an asexually reproducing stage that buds or strobilates into sexually mature individuals. Embryogenesis and blastogenesis are found at different stages in the life cycle. We have also seen that some ascidian embryos undergo asexual development even before they complete embryogenesis. The egg and the asexual bud are developmental systems that employ similar morphogenetic processes. They both produce two-dimensional cell sheets that fold into a double-layered vesicle. This double-layered vesicle is the starting point for the construction of a multicellular organism in both, with its basic tube-within-a-tube body plan. From this point the similarity vanishes.

The two sheets in the bud differ morphogenetically from comparable sheets in the gastrula. The ectoderm of an ascidian bud is protective and makes only a modest contribution to the adult body. The bud endoderm, however, folds into most of the organ systems of the adult, including the nervous system. By contrast, only gastrula ectoderm gives rise to the nervous system. Most eggs contain the information for two developmental programs, larval and adult. Embryogenesis, in general, produces an efficient larval organization because this stage serves an important adaptive function, such as dispersal or site selection, in the life cycle of the species (see Chapter 18). Where the egg is organized to make a larva, potential adult structures remain undifferentiated, as in insects. The final adult organization is achieved after an elaborate tissue reconstruction during metamorphosis. Most larval tissues are lost, never incorporated into the adult body, while other tissues organize into the adult body plan.

The bud, with a single purpose, develops directly into the adult. It seems to be the more primitive developmental system since it does not include a larval program.

## SUMMARY

Most asexually reproducing invertebrates are renewal organisms; their tissues undergo continuous cycles of growth, morphogenesis, and death. Asexual reproduction relies on morphogenetic movement rather than on cell proliferation. Prior to bud formation, cells change shape, acquire greater contractile and adhesive properties, and emerge as centers of morphogenesis. When asexual buds appear, the normal growth dynamics of an organism are diverted into new axes of tissue expansion and outgrowth.

All adult somatic cells participate in budding. They exhibit a high degree of plasticity and develop into a variety of cell types. Though populations of undifferentiated or interstitial cells exist, these are not the primary source of cells for a bud; rather, they are the progenitors of mitotically inactive specialized cells such as nerve cells and nematocysts in coelenterates. They also give rise to gametes during the sexual phase of the life cycle.

Buds emerge when they escape from dominance of more active centers in the organism. Dominance seems to depend on gradients of inhibitor substances that block the inherent contractile tendencies of tissues, the basis of morphogenesis in these forms. Inhibitory substances may be derived from nerve cells which also produce growth-promoting substances that induce bud outgrowth.

Blastogenesis differs from embryogenesis in that it is simple and more direct. Embryogenesis is often interrupted by the insertion of a larval stage in the life cycle, but the final outcome is the same; identical adult body forms are produced. The morphogenetic potential of a bud seems to exceed that of the egg in that all bud tissues retain developmental potencies and will resume morphogenesis when isolated from the body. They exhibit vigorous regenerative capabilities, as we will see in the next chapter.

**REFERENCES**   Beloussov, L. V., and Dorfman, J. G. 1974. On the mechanics of growth and morphogenesis in hydroid polyps. *Am. Zool.* 14:719–734.

Berrill, N. J. 1941. Spatial and temporal growth patterns in colonial organisms. *Growth* 5(Suppl.):89–111.

———. 1951. Regeneration and budding in tunicates. *Biol. Rev.* 26:456–475.

———. 1961. *Growth, development and pattern.* San Francisco: Freeman.

Bode, H., Berking, S., David, C. N., Gierer, A., Schaller, H., and Trenkner, E. 1973. Quantitative analysis of cell types during growth and morphogenesis in *Hydra. Wilhelm Roux' Arch. Entwicklungsmech. Organ.* 171:269–285.

Brien, P. 1968. Blastogenesis and morphogenesis. *Adv. Morphogen.* 7:151–204.

Campbell, R. D. 1974. Cell movements in *Hydra. Am. Zool.* 14:523–535.

Freeman, G. 1971. A study of the intrinsic factors which control the initiation of asexual reproduction in the tunicate *Amaroucium constellatum. J. Exp. Zool.* 178:433–456.

Gierer, A. 1974. *Hydra* as a model for the development of biological form. *Sci. Am.* 231 (6):44–67.

Webster, G. 1971. Morphogenesis and pattern formation in hydroids. *Biol. Rev.* 46:1–46.

Wyttenbach, C. R. 1974. Cell movements associated with terminal growth in colonial hydroids. *Am. Zool.* 14:699–717.

# Regeneration: A Developmental Reprise

# 15

We have already seen that complete developmental programs can be elicited from adult tissues in asexually reproducing species. Asexual budding, however, is only one feature of adult morphogenesis, since many organisms also regenerate body parts after injury or removal. Regeneration is an aspect of asexual reproduction. Wounding stimulates growth and morphogenesis; a site of injury may become a center for replay of developmental programs. In general, the magnitude of a regeneration response varies directly with a species' potentiality for asexual reproduction. Adult coelenterates and ascidians, which bud continuously, also regenerate readily; whole organisms can be reconstituted from small fragments. Among adult insects and vertebrates, however, with no asexual reproduction, recovery of lost parts does not occur or is restricted to a few organs.

In restoring lost parts, a regenerative process usually recapitulates the developmental program normal for that part in the embryo, but with some differences. For instance, a regenerating adult tissue is responsive to hormonal signals, which are usually ab-

sent in the embryo. Recruitment of cells to replace injured parts may also differ from normal ontogeny and may follow one of two paths. In simple metazoa such as coelenterates, the injured part is reorganized by migration of cells to the wound site from adjacent regions, a pattern known as *morpholaxis*. This, in essence, is a recapitulation of normal growth processes. In *epimorphic* regeneration, new structures are derived from local cells in the region which proliferate and accumulate at the wound site as a regeneration *blastema*. Wound healing, the resealing of open or injured surfaces by migrating epidermis, is, of course, a simple form of regeneration in all organisms.

The adaptive significance of regeneration is obvious. An animal restores functional competence by regenerating the lost part. For lower invertebrates, this means survival. Regeneration is an essential process in the life cycle of renewal animals, inasmuch as they continuously slough off aging body parts and replace them with new tissues. Regeneration has limited adaptive significance in higher forms where survival does not hinge on the retention of every body part. Complex organisms have evolved efficient homeostatic mechanisms and survival is not usually threatened by injury or loss of parts. Regeneration potential is progressively lost in higher vertebrates. Only fish and amphibians can replace appendages, while birds and mammals can regenerate only a few organs, such as liver.

## PATTERNS OF REGENERATION

Single cells regenerate; an oral apparatus of *Stentor* is regenerated after amputation, just as it is reconstituted during cell division. In these cells, regeneration depends on communication between nucleus and cortex. As the cell grows, an equilibrium is reached between the organelle pattern in the cortex and rates of precursor synthesis controlled by the nucleus. When portions of the cortex are removed, the equilibrium is upset; the injury is communicated to the nuclear genome and gene expression is triggered. The stimulus to cell regeneration seems to involve the plasma membrane which, through injury, undergoes a temporary change in ion permeability. This could stimulate genome expression and new protein synthesis. Regeneration continues until the injured surface is replaced by a new membrane and cortex with its complement of ciliary organelles. Once this state is reestablished, equilibrium is restored, nuclear function is diminished, and regeneration stops.

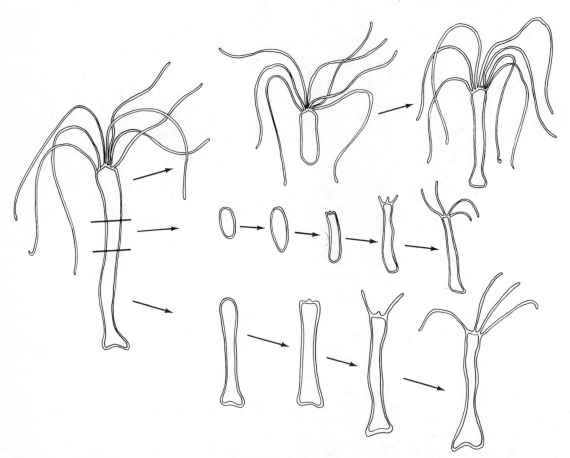

**Figure 15-1**  *Regeneration in Hydra showing polarity. Distal fragments with hypostome regenerate a pedal disc. Proximal fragments regenerate a hypostome with tentacles. Intermediate body fragments regenerate a hypostome at the distal end and a pedal disc at the proximal end.*

This is only a model of homeostatic mechanisms that may govern organelle regeneration in single-celled protists. The model implicates injury-induced membrane changes as the trigger to regeneration at the cell level. Such changes in membrane permeability are a common feature of all regenerative phenomena, including those at the multicellular level.

## Morpholaxis

As we have already seen, the life of a hydra is one of constant cell renewal. While the organism grows, both ectoderm and endoderm migrate down the body and newly differentiated cells appear. All cells, except nematocytes and nerve cells, retain mitotic potential, while those reaching the basal disc die and slough off. As long as cells are capable of continued mitotic activity and migration, a hydra is virtually immortal. Cells lost by budding or by sloughing at extremities are replenished by mitotic

activity throughout the body column, which explains why such renewal organisms have high regenerative capacity.

The first experiments on regeneration in *Hydra* were done in 1744 by Abraham Trembley. His observations remain among the most incisive studies of regeneration. He named the organism "hydra" because of its ability to regenerate new tentacles. Except for the tentacles, most fragments of a hydra's body can regenerate a whole animal. Distal fragments can regenerate proximal portions, and proximal parts produce distal structures (Figure 15-1). If a hydra is cut into very small pieces, these reaggregate and reconstitute an intact animal, sorting out as if they "know" exactly where to locate in the reconstituting mass.

Regenerative events at a cut surface in *Tubularia,* a marine hydroid, are typical of most renewal organisms where tissue migration is the principal source of replacement cells. The *Tubularia* hydranth is equipped with two rings of tentacles encircling a

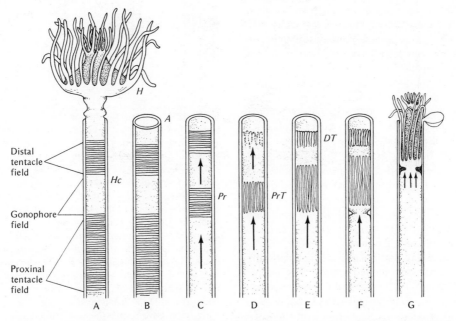

**Figure 15-2** *Regeneration in* Tubularia *showing distal migration of tissues. (A) An intact polyp showing distal and proximal tentacle fields in horizontal shading. Gonophore field between them. (B) Hydranth removed. (C) Migration of tissues displaces both tentacle fields toward distal end. Primordia become visible as accumulations of pigment granules and greater density of tissue. (D) Proximal tentacle primordia begin to emerge. (E) Distal tentacles appear. (F) Constriction of base of hydranth. (G) Emergence of fully regenerated hydranth. H = hydranth; Hc = hydrocaulus or perisarc; A = level of amputation; DT = distal tentacles; Pr, PrT = proximal tentacles.*

hypostome. The hydranth stalk is surrounded by a perisarc whose thickness varies along the length of the stem. The thinnest portion surrounds the young tissue of the animal beneath the hypostome, whereas the thickest portion surrounds the older, more proximal regions.

After hypostome and tentacles are removed, cells adjacent to the cut surface rapidly expand over the exposed area and cover the wound (Figure 15-2). Mitotic activity is not stimulated, which means that cell proliferation is unessential in wound healing. Several hours later, primordia of the proximal tentacles appear as a pigment accumulation at the distal end. In time, distal tentacle primordia become visible as proximal tentacles grow out. Eventually a complete head with a dual ring of tentacles emerges from the perisarc. Regeneration is completed in 24 hours at 20°C. Again, mitotic activity is unimportant; regenerating cells are furnished by migration of both tissue layers (mostly endoderm) from the more proximal regions in the stem. By experimentally inducing repeated regeneration, such migrating sheets may be exhausted. If tentacles are cut off as soon as they appear after each regeneration, the body of the animal eventually thins out as more and more of its cells migrate distally to regenerate. Eventually the few remaining cells are unable to reconstitute a hydranth.

Regenerating tissue fragments retain the animal's original polarity; distal cut surfaces always regenerate hydranths and hypostomes, whereas proximal surfaces always regenerate basal discs. Polarity is also manifested by differential rates of regeneration along the body axis; distal pieces regenerate more rapidly than proximal fragments (Figure 15-3). The most posterior region rarely regenerates; most of its cells are terminal, and not enough cells can be recruited for regeneration. Again, gradients of regeneration do not reflect a mitotic gradient along the stem.

But it turns out that polarity is a cellular rather than a tissue property. In *Hydra*, polarity is retained even if the tissues are mechanically dissociated into individual cells. Cells reaggregate into spherical masses with ectoderm on the outside and endoderm inside. Within two days these regenerate into a multiheaded structure with several hypostomes and tentacles. Using this dissociation procedure, Gierer compared the regenerative behavior of fragments from different gastric regions. Distal fragments close to the head and neighboring proximal pieces were

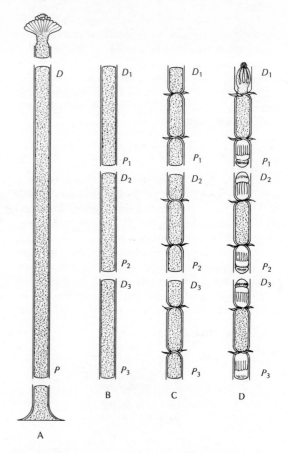

**Figure 15-3** *Regeneration gradient in* Tubularia *as shown by differential rates of distal and proximal regeneration. (A) A long fragment of stem is isolated, DP. (B) It is cut into three pieces and each is numbered in distal-proximal order. (C) A few millimeters of stem at each end are isolated by ligature. This eliminates dominance interactions of distal over proximal ends in each piece and allows the intrinsic regenerative potential to be expressed. (D) A distal-to-proximal gradient in regenerative rate is seen. $D_1$ has a completed hydranth with distal and proximal tentacles. $D_2$ has only proximal tentacles with distal ridges and is slightly behind; $D_3$ is behind with distal primordia just beginning to appear. $P_1$ is most advanced of proximal ends but is only equivalent to $D_3$. $P_2$ is intermediate, and $P_3$ is in a very early stage showing only the proximal tentacle primordium. (From L. G. Barth, Biol. Bull. 74:155, 1938.)*

dissociated and each cell suspension was allowed to reaggregate. The aggregates were then fused in combinations, with a distal aggregate placed on either side of a proximal aggregate, or placed between two proximal aggregates. In each case tentacles always emerged from the ends of distal fragments irrespective of their spatial relation to neighboring pieces. Cells retained their original polarity after dissociation and reaggregation; those from distal regions tended to regenerate hypostomes more readily in such combinations.

One explanation is that individual cells are polarized and when reconstituting, they rearrange themselves with "head" and "feet" aligned. Another possibility is that cells produce a growth-promoting morphogen at different rates, with distal cells most active. This would result in a distal-proximal gradient of a growth-promoting substance. A similar gradient could arise from a graded distribution of a cell type (nerve cell?) synthesizing morphogen at the same rate, with the greatest number of these cells in distal regions. Distal fragments enriched with these cells would regenerate more rapidly than proximal aggregates, and as soon as hypostomes appeared in fragment combination experiments, they would inhibit additional hypostome formation in neighboring regions. As we will see, the last of these possibilities can account for polarity in *Hydra*. Nerve cells are indeed distributed in a graded fashion with a high point at the hypostome, and these cells do contain growth-promoting substances.

Polarity is reversible in *Tubularia*. If a distal end of a *Tubularia* fragment is ligated with a fine thread, a hydranth appears at the proximal end instead. Cells now migrate from the ligated distal end to the proximal end to regenerate. Any agent that inhibits distal-end regeneration converts the proximal end into a hypostome-regenerating site. Since polarity is also reversed by lowered oxygen concentration at distal surfaces, it seems that tissues tend to migrate toward regions of high oxygen tension.

Polarity may be altered by increasing the length of the cut fragment (Figure 15-4). *Tubularia* pieces greater than 10 mm, for example, regenerate at both ends, albeit at differing rates, with distal ends regenerating more rapidly than proximal. It is possible that distal portions that normally inhibit regeneration of proximal ends cannot do so at distances exceeding a threshold value, or longer pieces have enough cells for regeneration at both ends. On the other hand, very short pieces of stem, about a millimeter or two in length, also regenerate hydranths from both ends simultaneously. This eliminates tissue mass as a factor and suggests that inhibitory influences from

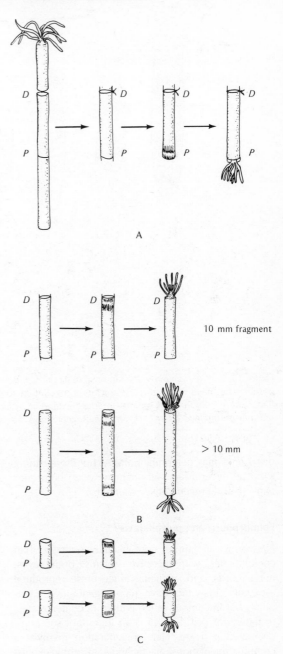

**Figure 15-4** *Reversal of polarity in* Tubularia *regeneration. (A) Tight ligature placed at distal (D) end of fragment allows proximal (P) end to regenerate. (B) Increasing the length of a fragment beyond 10 mm removes proximal end from distal dominance and both ends regenerate. (C) Very short fragments will regenerate partially at one end or may regenerate at both ends. In each case only one ring of tentacles develops.*

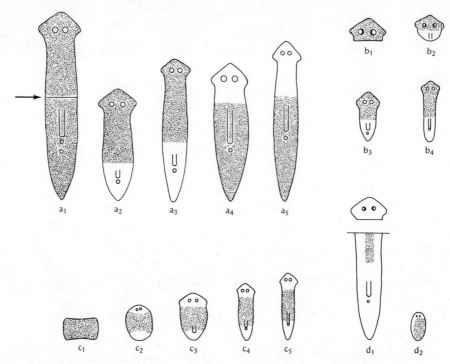

**Figure 15-5** *Regeneration in* Planaria maculata *(a–c) and in* Curtisia simplicissima *(d). (a) Regeneration of anterior and posterior fragments after transverse section in the middle. (b) Regeneration of an excised head. (c) Regeneration of a transverse section from the middle. (d) Regeneration of a piece removed from the center of the body. Regeneration portions unstippled. (From T. H. Morgan,* Regeneration, *2d ed. New York: Macmillan, 1907.)*

distal ends may determine polarity, but these cannot exert an effect at a small-fragment level; both ends express distal regenerative potential.

### Epimorphosis and morpholaxis

Flatworms are more complicated metazoa than coelenterates, with digestive systems, brain, eyes, nerve cords, and genitalia. Since these reproduce asexually, they too have high regenerative capability; but in contrast to the pattern in coelenterates, in flatworms cell division is an important source of regenerating tissue. Here, regeneration involves a combination of morpholactic and epimorphic events.

If a planarian is cut in half perpendicular to the anterior-posterior axis, the anterior fragment regenerates a tail at the cut surface and the posterior piece regenerates a head (Figure 15-5). Regeneration occurs at both wound surfaces, not only by migration of coherent sheets as in coelenterates but also by the proliferation of undifferentiated cells at the cut surface into a blastema that differentiates new head or tail structures.

As in coelenterates, polarity is an important feature of flatworm regeneration. Anterior fragments regenerate heads more rapidly than posterior pieces, although this property varies from species exhibiting no head regeneration to others exhibiting no gradient where all fragments regenerate heads at the same rate.

Polarity expresses itself in many ways. If, for example, the wound surface of a posterior fragment is longitudinally cut into two components which are not allowed to fuse (Figure 15-6), each portion regenerates a complete head. A multiheaded monster may be produced by making several parallel cuts at the surface. On the other hand, two heads may be fused to form one. The wound surface behaves as a morphogenetic field; that is, when disturbed, it regulates and restores a normal developmental pattern. Polarity seems to be intrinsic to the piece; changing its position by implanting it at different

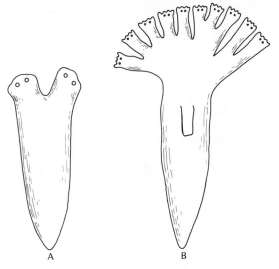

**Figure 15-6**  *Multiple head regeneration. (A) Twin heads after decapitation and a medial incision in* Curtisia simplicissima. *(B) Numerous heads after decapitation and multiple incisions in* Dendrocoelum lacteum.

rather than tails on the cut surface, very much like short pieces of *Tubularia* hydranth.

After a blastema appears, it is committed either to head or to tail development. When 2-day-old blastemas are isolated, they develop into heads or tails respectively in vitro. At no time does either blastema revert to the other type, nor do they develop into intact animals. If, however, a head and tail blastema are fused, a small intact worm often develops; the combined blastemal masses reorganize a complete organism. Within the first day after blastema formation, cells have received their morphogenetic instructions, presumably from neighboring adult tissues. Anterior fragments signal tail blastema to differentiate parenchymal and epidermal cells, while posterior fragments signal brain, eyes, and head differentiation. This suggests that an identical stimulus has qualitatively different effects in the two regions because each occupies a different position relative to the anterior-posterior gradient of each fragment. One is at the high point of the gradient while the other is at the low point.

sites along the longitudinal axis does not alter its rate of regeneration.

Anterior-posterior polarity is reversible; in *Dugesia*, each thin strip from the central region produces a double-headed monster, or Janus head. Even anterior head pieces of this species, cut just in front of the brain and eyes, often regenerate heads

### Regeneration of amphibian limbs

No other regenerating system has attracted as much attention as the regeneration of limbs in amphibians. Spallanzani in 1768 had already published many experiments on this phenomenon, and by 1901 Thomas Hunt Morgan collected enough literature to publish a book on the subject.

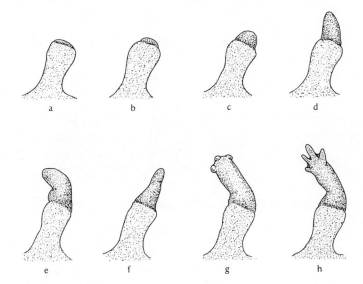

**Figure 15-7**  *Stages in regeneration of the forelimb of the adult newt* Triturus. *(From M. Singer,* Quart. Rev. Biol. *27:169–200, 1952.)*

Larval and adult urodeles and anuran tadpoles, before metamorphosis, are able to regenerate limbs and tails. Limb regenerative potential disappears at metamorphosis in anurans, for reasons not well understood.

Regeneration of adult urodele limbs takes about a month (less in larvae) and has been arbitrarily divided into four phases: (1) wound healing, (2) demolition or tissue histolysis, (3) blastemal cell recruitment and proliferation, and (4) morphogenesis and differentiation (Figure 15-7). Amputation is followed immediately by overgrowth of the wound by epidermal sheet expansion and some proliferative activity. This

takes 1 to 2 days. Soon tissues in the stump region undergo massive breakdown or histolysis; first cartilage, then muscle and bone degenerate. The rate of tissue histolysis is dictated by temperature, age of the animal, and size of the limb. Larvae regenerate more rapidly than adults because the period of histolysis is shorter.

During histolysis, stump tissues dedifferentiate. Muscle bundles break down and slough off their myofibrillar architecture to become single mononucleate cells (Figure 15-8). Meanwhile, cartilage matrix dissolves, releasing individual cartilage cells into tissue spaces, but these do not die; instead they begin to

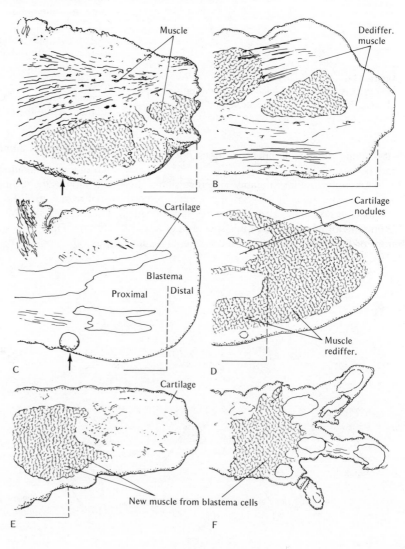

**Figure 15-8** *Cross section of forelimb regeneration in the newt* Notophthalamus viridescens. *Dotted line indicates amputation site. The black line is fixed to the same position on the limb throughout. Figures stress muscle dedifferentiation and redifferentiation. (A) A large amount of muscle remains in stump; extent of dedifferentiation will reach level at arrow. (B) Eighteen days; distal tip contains some blastemal cells dedifferentiated from muscle. (C) Twenty-one days; level of dedifferentiation at arrow. Proximal tissues of blastema from various dedifferentiated stump tissues. Distal blastema has begun to proliferate. Cartilage has already begun to differentiate. (D) Muscle has begun to differentiate from blastemal tissues anterior to cartilage. (E, F) More cartilage elements differentiate within blastemal mass. Muscle continues to differentiate as the digits are carved out. (From E. Hay,* Concepts of Development, *ed. J. Lash and J. R. Whittaker. Stamford, Conn.: Sinauer Associates, 1974, pp. 404–428.)*

proliferate. After discarding their specialized architecture, cells enter a growth-duplication cycle, with large nuclei and basophilic cytoplasm. Such cells are released into the presumptive blastemal area from connective tissue, cartilage, and muscle. Nerve and epidermal cells also grow out but remain attached to their sites of origin.

After the dedifferentiation phase of 6 to 7 days, small blastema cells collect beneath wound epidermis and proliferate rapidly. A conical-shaped blastema grows into a cylinder of tissue, with maximum growth reached by 16 days after amputation. Growth slows down as the distal portion of the rudiment flattens into the so-called "paddle" stage. Cells in the interior segregate into muscle, connective tissue, and cartilage-forming areas, and the process of histological differentiation begins. The first digits are cut out at 20 to 28 days and elongate while bone deposition begins and continues until the final limb pattern is attained.

Some have proposed that a reserve of embryonic cells, located elsewhere in the body, migrates to the amputation site to form the blastema. This turned out to be incorrect after it was shown that blastema cells are derived from stump tissues. X-irradiation of a limb prior to amputation prevents regeneration even if the remainder of the body is unirradiated. Moreover, if a normal limb is transplanted to a totally irradiated animal and then amputated, it regenerates completely in spite of the inability of host cells to proliferate and participate in regeneration. Regeneration is entirely a local phenomenon.

### Problems of regeneration

These few examples raise several questions that are common to all regenerating systems. We will outline them first and then proceed to a causal analysis of regeneration.

1. What initiates regeneration? Is it the wound stimulus, physiological isolation from the rest of the body, or a combination of both?
2. What is the source of cells for replacement of lost parts? We have already seen that morpholaxis and epimorphosis differ; in the former, cells migrate from adjacent tissues, whereas in the latter, cell proliferation is the key. But the real question in morpholaxis is, which cells migrate in, adjacent somatic cells or totipotent embryonic cells (neoblasts) able to reconstruct complete organisms. As for epimorphosis, do proliferating cells of a blastema come from embryonic reserve cells, or do they come exclusively from dedifferentiated somatic cells in the region?
3. How is control exercised at the site of regeneration? To what extent are local and systemic control factors involved in regulating the sequence of morphogenetic events? How does regeneration of a part compare to its normal embryogenesis?
4. What is the basis for the polarity of regeneration? Why do parts usually regenerate more distal structures? What is the basis for pattern formation in a regenerate?
5. Why do some species exhibit regenerative capacity while other, closely related species do not? What is the reason for the loss of regenerative capacity when a larval anuran metamorphoses into an adult? Why have mammals lost the ability to regenerate appendages?

We will see that many questions are still unanswered.

## INITIATION OF REGENERATION

Amputation of a body part produces three fundamental changes, each of which can act as a trigger to regeneration: (1) injury and tissue destruction, (2) isolation from tissues with which parts normally interact, and (3) exposure of injured cell surfaces to new environmental stimuli. These three factors—injury, physiological isolation, and membrane changes—usually act in concert to initiate a regenerative process.

### Injury

Injury is a prerequisite to regeneration. A regenerative stimulus is elicited by tissue destruction and release of cell contents from injured or dead cells. Regeneration can be mimicked by injecting foreign tissue or tissue extracts into a limb site or by local tissue trauma. In all cases, tissue extracts and degenerating cells at cut surfaces or in traumatized regions may release growth-promoting substances. For example, a supernumerary limb can be induced to grow out of a nonamputated urodele limb by implanting a fragment of foreign tissue such as kidney into an epidermal pocket. Extensive trauma at the site of the implanted tissue stimulates growth and partial limb morphogenesis as digits appear with bone and muscle. It is not certain, however, if the implanted tissues themselves contain a growth-promoting substance or if host systemic responses are implicated. Fragments of frog kidney, for example, implanted into a urodele limb, induce supernumerary limbs only after the host experiences an immunologic response that promotes histolysis of limb tissues. If frog kidney is heated, its antigenicity is eliminated and its ability to induce supernumerary limbs is also destroyed. Evidently local immunologic reactions associated with trauma may act as a regeneration stimulus.

Tissue trauma alone will promote outgrowth even in structures that never regenerate. Incipient regeneration of an adult frog forelimb can be provoked by continuous pricking and scoring of the amputation surface with a needle or by exposing it to hypertonic salt solutions. A small outgrowth with one digit may appear in such adult limbs.

Growth-promoting factors at the wound site may initiate regeneration. Products of dying or injured cells, released into local tissue spaces, can stimulate proliferation in surviving cells. Embryo extracts, for example, are known to be active sources of growth-promoting factors in cell and tissue culture. Other local factors are also important in initiation. The wound epidermis growing over the exposed site plays a role. The epithelium covering the wound must come from the wound site. If wound epithelium is replaced by head skin, limb regeneration is inhibited because head skin does not form undifferentiated epidermal cells, and cells in the underlying stump fail to aggregate into a blastema.

Local wounding is complicated by other systemic effects that are important in regeneration. For example, trauma elicits hormonal responses in vertebrate organisms; stress hormones (cortisone) may cooperate with local factors to promote a regenerative response.

Tissue breakdown is not essential for regeneration in some organisms. For example, certain arthropod crustaceans normally autotomize limbs throughout their growing cycle; that is, they spontaneously amputate an appendage. No tissue breakdown occurs at the site of amputation, but a protective epithelium covers the wound site and a limb regenerates.

Where tissue histolysis is important, factors from dying cells may act as attractants for other cells migrating to the wound area, a factor in recruiting cells for the blastema. For the present, however, we can only speculate about such factors since none has been isolated. The behavior of arthropod limbs, however, indicates that more than one trigger must be involved in initiating a regeneration response.

### Physiological isolation

Physiological isolation may be as important as injury as a stimulus to regeneration. Tissues released from inhibitory influences emanating from more dominant growth regions often regenerate. This is most evident in plants. Vascular plants regenerate readily because of their high potentiality for asexual growth. Cuttings of leaves, stems, and roots are able to regenerate whole plants. One difficulty in comparing regeneration in plants and animals is that meristematic regions or blastemas preexist in plants as buds or cambium layers in various parts of the plant body. Many of these regions are cryptic; they are dormant and unexpressed in the mature plant but become morphogenetically active only after they are cut away from physiologically more dominant regions.

The most dominant regions are the apices of shoots. A shoot contains many morphogenetically active regions called lateral buds whose growth is inhibited as long as an apical bud is present. If the apical bud is amputated, lateral buds grow out, with more distal buds emerging first (Figure 15-9). Once released from apical dominance, lateral buds proliferate and complete morphogenesis. Cells in lateral buds are morphogenetically inactive because of a hormone produced in apical regions. Auxin or indole acetic acid, synthesized at high rates in the apex, is translocated to lateral bud tissues where it inhibits growth. If a block of agar soaked in auxin is placed over the apical cut surface of a stem, lateral bud growth is inhibited. In plants, morphogenesis occurs in many plant tissues as soon as they are isolated from inhibiting hormones of more dominant regions. Even isolated cells can express an intrinsic morphogenetic potential if removed from the parent body. Cells isolated from different parts of the plant body will develop into a complete plant behaving in every way like a zygote (see Chapter 17).

Physiological isolation is also important in animal regeneration, as illustrated by lens regeneration in certain urodeles *(Triturus)* (Figure 15-10). In this case, initiation of development is triggered by removal of a lens. The epithelial tissue bordering the dorsal rim of the iris responds and regenerates a new lens in several weeks. The ability to form a lens is an intrinsic property of iris epithelium, yet it is never realized unless lens is removed. Only dorsal iris has lens-regenerating potential; fragments of dorsal iris transplanted into a lensectomized eye will regenerate a lens at the same time that one appears at the host dorsal iris (Figure 15-11). Moreover, a dorsal iris from *Triturus,* the species that can regenerate lens, placed in lensectomized eyes of *Ambystoma,* a species that does not, still yields a lens. The reciprocal exchange is not successful; an *Ambystoma* iris in the *Triturus* eye will not form a lens, even in an environment that normally supports lens regeneration. Some unique property of *Triturus* dorsal iris epithelium permits it to regenerate lenses in the eye chamber of other species, as long as a lens is absent.

If removal of the lens is the only factor that stimulated dorsal iris to regenerate, then dorsal iris would regenerate in vitro, but it does not. Iris differentiates a lens in vitro only if explanted after lens formation has already begun. At an earlier prolifera-

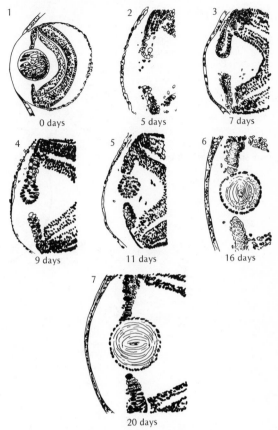

**Figure 15-10** *Stages in lens regeneration after removal of a lens in the newt* Triturus. *(1) Normal eye. (2) Loss of pigment from cells in dorsal iris. (3) Proliferation of dorsal iris cells. (4) Outgrowth of epithelium into a small vesicle; proliferation continues at edge of iris. (5) Vesicle enlarges. Cells in center begin to differentiate into lens cells. (6) Further growth of lens; lens cells synthesize crystalline proteins. (7) Almost fully formed lens.*

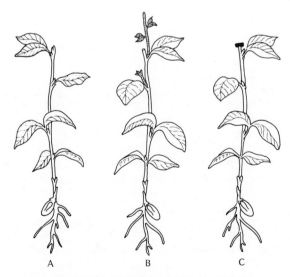

**Figure 15-9** *Apical dominance in plants. The terminal bud inhibits growth of axial buds on leaves. (A) Growing young pea plant. (B) Growth of axial buds after removing terminal bud. (C) Inhibition by an agar block containing auxin.*

tive phase, it is unable to complete differentiation in vitro. Evidently the eye chamber itself contains factors essential to regeneration, probably coming from the retina.

Dorsal iris implanted subcutaneously does not regenerate a lens, but if combined with a piece of pigmented retina or exposed to neural retina, a lentoidlike structure or incipient lens develops. Lens regeneration is suppressed by removing neural retina, pigmented retina, or choroid, together with the lens. Most eye tissues provide factors essential to lens regeneration.

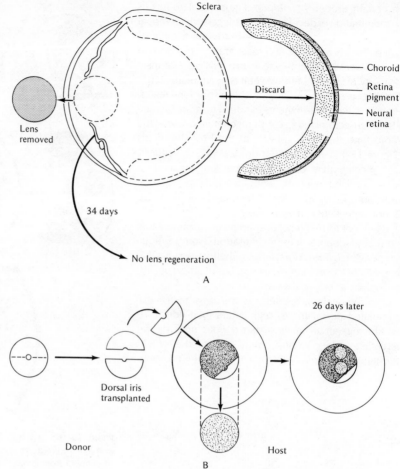

**Figure 15-11** *Role of dorsal iris and neural retina in lens regeneration. (A) Operation in which lens, sclera, neural and pigmented retina were removed. No lens regeneration from the remaining dorsal iris after 34 days. (B) Dorsal iris then was implanted into a host eye from which the lens had been removed. A lens regenerated from host and implanted dorsal iris. (From L. S. Stone, Anat. Rec. 131:151–172, 1958.)*

The mechanism of inhibition by intact lens on surrounding dorsal iris is not understood. Lens fibers may produce an inhibitor that prevents iris epithelium from expressing any morphogenetic properties other than pigment synthesis. In fact, separation of dorsal iris from lens by an impermeable film will stimulate development of a small lens; an inhibitor seems to be present in the aqueous humor (Figure 15-12). No systemic factor is involved; presence of a lens in one eye does not prevent lens regeneration in the other; the inhibition is local. On the other hand, these same results can be explained by a positive control mechanism. The eye chamber might normally contain factors that stimulate lens regeneration (retinal factors), but the lens itself metabolizes these materials, making them unavailable for iris epithelium. As soon as lens is removed, the factors accumulate and lens formation is stimulated. We do not know whether physiological isolation involves escape from an inhibitor-producing dominant region or whether it involves accumulation of trophic factors normally removed by such regions.

## Membrane changes

In coelenterates we have seen that a cut surface becomes a site of higher oxygen tension, and neighboring tissues migrate toward this metabolically more active region and regenerate. In these systems, a regenerating signal stimulates metabolic activity, perhaps by changing membrane permeability. The fact that the polarity of regeneration in coelenterates is reversed by an electric field suggests that changes in ion permeability may indeed play an important role in regeneration. Isolated fragments of *Obelia* stems are electrically polarized; the outer surface is positive in relation to the inner gastrovascular cavity (Figure 15-13). In addition, charge differences have been detected between distal and proximal ends of a fragment; the distal end is more positive. The polarity of fragments can be reversed by an electric field, if the polarity of the imposed field is different from that of the intrinsic bioelectric field within the tissues. A proximal end facing a positive pole in the field will regenerate instead of the distal end. Rates of regeneration depend upon charge density, the magnitude of the field, and polarity of the test fragment, which suggests that changes in the direction of ion flow across membranes may govern regeneration.

Even nonregenerating limbs of adult frogs and mammals may be stimulated to partial regeneration by direct application of electric current to the stump. Rengeration is most often incomplete; only small outgrowths are induced with some distal structures.

A regenerativelike process in tissue culture also depends on changes in the cell surface. Many cells proliferate into a confluent monolayer in vitro before cell division stops (see Chapter 5). If a region is cut

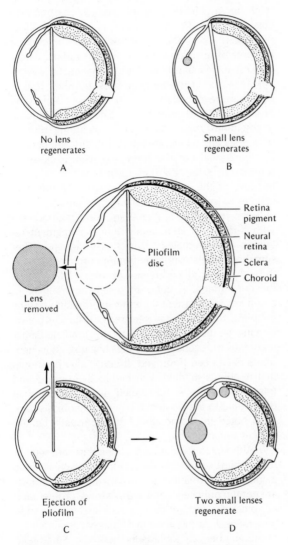

Figure 15-12 *Importance of interaction with retinal tissue for lens regeneration. In center, a diagram of the operation. Lens is removed and a pliofilm membrane is implanted to isolate iris from retina. (A) Complete isolation, no lens regenerates. (B) Incomplete isolation, a small lens regenerates. (C, D) If pliofilm was partially or completely extruded from eye, a lens regenerated from dorsal iris and often smaller lenses regenerated from the edge of the wound. (From L. S. Stone, J. Exp. Zool. 139:69–81, 1958.)*

Labels in figure 15-12:
- No lens regenerates — A
- Small lens regenerates — B
- Lens removed
- Pliofilm disc
- Retina pigment
- Neural retina
- Sclera
- Choroid
- Ejection of pliofilm — C
- Two small lenses regenerate — D

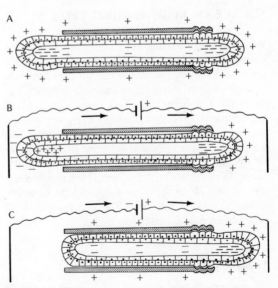

Figure 15-13 *Effect of an electric field on polarity of regeneration in Obelia. (A) Normal bioelectric field pattern; original apical end to the left. (B) Reversal of field by applied current at apical end. Proximal end exposed to positive pole. (C) Final result; proximal end grows out and regenerates while distal end is inhibited. (From E. J. Lund, J. Exp. Zool. 41:155–190, 1925.)*

out of the monolayer, the cells near the cut surface are stimulated to proliferate and refill the gap. This is a regenerative response similar to epithelial wound healing. Release of cells from contact inhibition stimulates cell division. Regeneration in this system may be mediated by adenylcyclase in the cell surface and its control over the levels of cyclic AMP.

Is contact inhibition a factor in epimorphic regenerating systems? Contact relations of cells are severely altered after amputation and demolition. Cells are released from intact tissues during histolysis and assume an independent existence. These events precede the proliferative phase and may, in fact, trigger a cell's reentry into the growth-duplication cycle. In morpholaxis, on the other hand, tissue migration is the principal response to amputation and this, too, may be initiated by a release from contact-inhibition of cell movement. Both cell properties depend on changes in the cell surface, as we have seen, which suggests that altered surfaces may trigger regenerative phenomena in multicellular organisms as well as in single cells.

## SOURCE OF REGENERATING CELLS

Where do regenerating cells come from? In some cases, as in coelenterate morpholaxis, tissues migrate from adjacent regions to the cut surface. Most somatic cells in these organisms retain phenotypic plasticity and can differentiate into a new range of cell types when conditions warrant. Even an isolated piece of ectoderm or gastrodermis reconstitutes into a complete hydra without stimulation of cell division or migration. The tissue transforms into a double-layered vesicle, while gastrodermal cells dedifferentiate, slough off some of their mucoid or secretory granules, become interstitial-like, and regenerate all other cell types of a complete hydra.

### Embryonic reserve: neoblasts

At one time, it was assumed that an embryonic reserve of pluripotent interstitial cells was responsible for such a pattern of regeneration. But this assumption is not entirely true, since all cells of a tissue layer participate. Interstitial cells included in each tissue layer behave in regeneration as they do in budding, as progenitors of nematocysts and neurons of the regenerating hydranth and tentacles.

An embryonic reserve of neoblasts has been proposed as a source of regenerating cells for planarian flatworms. Neoblasts are spherical basophilic cells with large nuclei, sometimes resembling nerve cells because of their cytoplasmic processes (Figure 15-14). Dubois suggests that these pluripotent cells, distributed throughout the body of the organism, migrate to the wound surface where they organize a blastema. It is true that neoblast number increases in regenerating sites, but their role in planarian regeneration is still controversial. One problem is that identification of a neoblast is difficult. It resembles many other cells, including mucus-secreting cells. More often than not, they are defined by operational criteria. Neoblasts seem to be more sensitive to X-rays than are most somatic cells, and because of this they are implicated as a principal source of regenerating cells. For example, if an anterior region of a planarian body is irradiated with X-rays and amputated, only the fragment retaining nonirradiated tissues regenerates. Presumably, neoblasts migrate from nonirradiated tissue through the irradiated region to form a blastema at the cut surface. A delay in regeneration of larger fragments is consistent with this interpretation, since neoblasts must migrate over greater distances before arriving at the wound. The totally irradiated fragment fails to regenerate and dies because all its neoblasts have been destroyed. It could also be shown that neoblasts marked with radioactive uridine and transplanted into a nonlabeled decapitated animal do migrate to the blastema. Dividing neoblasts have been noted in regenerating regions, usually behind the cut surface.

Although large numbers of neoblasts are seen in regenerating regions, evidence for their exclusive role in blastema formation is not convincing. Neoblasts are not the only cells sensitive to X-rays; proliferation of epidermal and gut cells is also affected. In addition, rates of regeneration in a fragment are unrelated to the number of neoblasts they contain. Small fragments with fewer neoblasts regenerate as rapidly as large fragments, suggesting that migrating neoblasts need not be the sole source of regenerating cells. Moreover, presence of neoblasts in tissue does not correlate with regeneration capacity. Posterior tissues of *Dendrocoelum*, which do not regenerate, contain as many neoblasts as anterior tissues which do. Dubois claims that posterior neoblasts migrate to the cut surface as in anterior fragments but die in the wound area rather than construct a blastema.

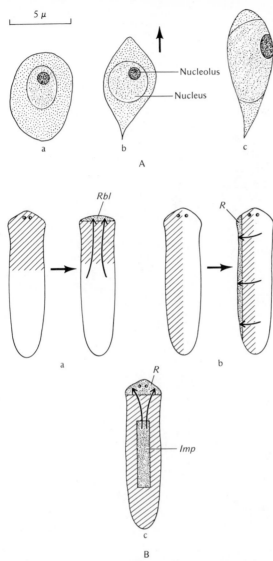

**Figure 15-14** *Role of neoblasts in* Planaria *regeneration according to Dubois. (A) Appearance of neoblasts. (a) Resting. (b, c) Migrating in direction shown by arrow. (B) Radiation experiment to show importance of neoblast migration to regeneration in* Euplanaria lugubris. *(a) Anterior region (hatched) irradiated with X-rays followed by removal of head. Regeneration takes a much longer time than in unirradiated controls, but a regeneration blastema appears, after neoblasts migrate from non-irradiated posterior regions (arrows). (b) Irradiation of lateral edge and formation of regeneration blastema from migration of neoblasts. (c) Total body irradiation, decapitation, and implantation of normal fragment in center leads to regeneration. R = regenerant; Imp = implant; Rbl = regeneration blastema.*

He suggests that local tissue properties necessary for neoblast survival and growth are restricted to prepharyngeal fragments.

Finally, the presence of neoblasts at the cut surface does not prove they migrated there. Local dedifferentiation of somatic cells into smaller basophilic cells resembling neoblasts could also explain an apparent accumulation of neoblasts. A histological study of the wound surface in planaria indicates that wound epithelium comes entirely from old epidermis, and tissues underneath the epidermis such as muscle and gland epithelium may also dedifferentiate into blastemal cells. This means that neoblasts are by no means the sole source of cells of the regenerating blastema, and dedifferentiation of local tissues into morphogenetically active cells could serve as an equally important reservoir of blastema cells.

## Source of blastemal cells in amphibian limb regeneration

Dedifferentiation is also a prerequisite for cell proliferation in amphibian limb regeneration. Most functional tissues in the stump shed their specialized organelles before they proliferate. Cells of a limb blastema are of heterogeneous origin, arising from dedifferentiating muscle, bone, cartilage, and connective tissues of the stump.

What is the nature of blastemal cells? Do dedifferentiated cells revert to a totipotent embryonic state, or do they return to a specific state of determination, expressing only the limited development repertoire characteristic of their tissue of origin? Blastemal cells from muscle could retain a muscle cell bias; those from cartilage continue to differentiate as cartilage, and so on. On the other hand, cells might undergo a metaplastic transformation. A dedifferentiated muscle cell could become cartilage in the developing blastema.

The question of blastemal cell transformation (metaplasia) is most important in the analysis of vertebrate limb regeneration. We know that X-irradiation of a limb before amputation prevents regeneration; that is, irradiated stump tissues are unable to proliferate into a regeneration blastema. We can insert a piece of unirradiated tissue such as muscle or bone into an irradiated amputated limb site to see whether regeneration occurs. If it does, we may conclude that all cell types in the regenerate are derived from the inserted tissue and that metaplastic transformation has occurred; that is, one cell phenotype is replaced

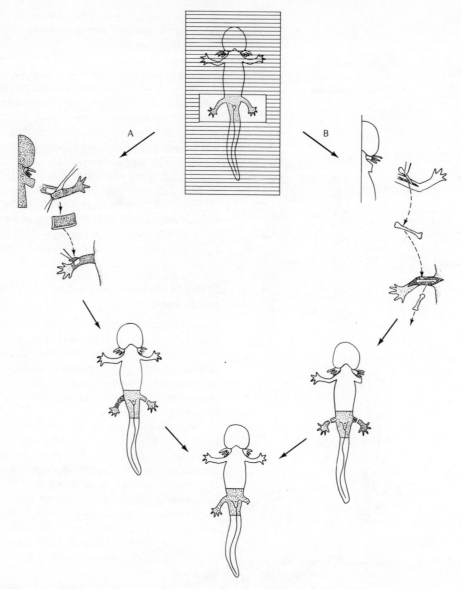

**Figure 15-15** *Source of blastemal cells in limb regeneration. The hindlimbs of a newt are irradiated while the rest of the body is shielded. (A) A fragment of skin from the forelimb is grafted onto the left hindlimb, after which both hindlimbs are amputated at the same level. Only the hindlimb with skin graft regenerates. (B) In this instance, a humerus from the forelimb is implanted in the hindlimb after irradiation. Both limbs are amputated through the graft region and only the limb with the nonirradiated bone graft regenerates.*

by another. When a fragment of muscle, bone, or connective tissue is implanted into an irradiated limb, regeneration occurs in each case, and a small limb outgrowth is obtained (see Figure 15-15). Does this

mean that all stump tissues contributing to a blastema are metaplastic? Probably not, since irradiated tissues can recover proliferative capacity after interacting with normal tissues. In fact, dead tissues or tissue homogenates, when implanted into X-irradiated *Axolotl* limbs, also promote regeneration if the limbs are repeatedly amputated. Moreover, repeated amputations of irradiated control limbs, without implanted tissue, also induce regeneration!

These results suggest that continuous trauma (that is, repeated amputations) in a region of irradiated limb tissues releases factors that repair irradiation damage. In other words, the regenerated limbs obtained in the above experiments could have come from host rather than from graft tissues. Another possibility is that repeated amputations of irradiated limbs stimulate normal fibroblasts in distant unirradiated regions (for example, the shoulder) to migrate into the irradiated limb site and thereby supply the cells of the blastema. If the shoulder is irradiated along with the limb, repeated amputations do not elicit regeneration. We must conclude, therefore, that local irradiation of the limb does not prevent migration of host cells from unirradiated regions or may have no effect on repair and recruitment of locally irradiated cells by factors diffusing from normal tissues.

One way to distinguish host from graft tissues is to label graft cell nuclei with radioactive thymidine. The labeled graft tissues (muscle or cartilage) are then inserted into a limb before amputation to determine whether cells of a regenerate come from graft or host. Using the double-label procedure, graft tissues, obtained from the limbs of triploid animals that have regenerated in the presence of radioactive thymidine, are implanted into limbs of diploid animals and can be distinguished from host tissues because graft cell nuclei are larger, contain three nucleoli rather than two, and are radioactive. All graft cells must be both triploid and radioactive; any thymidine-labeled diploid cells in a regenerate must have come from host cells incorporating radioactive thymidine released from degenerating graft cells.

The metaplastic properties of three tissues, cartilage, muscle, and connective tissue, were tested. Cartilage was least metaplastic. A graft of double-labeled cartilage tissue implanted in place of the humerus in a stump of an amputated limb produced regenerates in which most graft cells gave rise to cartilage. A few cells of the perichondrium and some fibroblasts beneath the skin were also labeled, suggesting a modest amount of metaplasia. Most dedifferentiated cartilage cells become part of the blastema and differentiate into cartilage again, even after

several divisions; that is, their cartilage bias is stable and heritable.

The stability of cartilage has been questioned in studies of cartilage implants in irradiated limbs of *Axolotl*. In this case, a cartilage graft from a pigmented (or nonpigmented) donor was implanted in the irradiated limb of a host of different pigment genotype. Small single outgrowths were obtained, some with digits, and it appeared that in this instance cartilage did transform into other cell types, including muscle. Since it is more difficult to distinguish graft from host tissues in this experiment (the pigmented tissues became diluted during growth), the argument for cell transformation from cartilage is not proven.

Muscle is more labile. If labeled muscle tissue is implanted into a normal limb before amputation, it dedifferentiates into single myoblasts, which make up the bulk of the blastema and differentiate into most cell types in the regenerate including cartilage. Blastemal mesenchymal cells seem to retain the potentialities of some earlier embryonic state when limb mesenchyme could form cartilage or muscle. But muscle tissue is usually contaminated with connective tissue elements, the most metaplastic of all limb tissues, which means we cannot conclude that myoblasts have transformed into other cell types.

A graft of labeled connective tissue from the tail fin of a salamander inserted into an irradiated amputated limb develops into a limb in which most tissues are labeled. This means that connective tissue can be a progenitor of cartilage, perichondrium, connective tissue, and muscle, a true metaplasia of one specialized phenotype differentiating into a spectrum of mesodermal derivatives when challenged in a regenerating field. If labeled whole skin is grafted into an X-irradiated limb, only its connective-tissue derivatives contribute to the tissues of the regenerate. Since most tissues are normally surrounded by connective tissue elements, we assume that most cases of metaplasia reported in limb regeneration probably result from contamination by connective-tissue cells.

The relative stability of cartilage cells reflects the peculiarities of limb regeneration. Cartilage is the first tissue to differentiate, long before the appearance of muscle or connective tissue. Consequently, blastemal cells from cartilage do not have the opportunity to display other patterns of differentiation since they are committed early to cartilage matrix synthesis. Once matrix is formed, it is unlikely that cells can escape to transform into other cell phenotypes.

In summary, cells of a regenerating limb blastema come from dedifferentiated cells derived from stump tissues. The blastemal cells, however, are not all pluripotent; some, such as cartilage, are stable and redifferentiate into cartilage. Others (those derived from connective tissue) are more labile and differentiate into a limited range of cell phenotypes of mesodermal origin, such as fibroblasts, muscle, cartilage, and perichondrium. As in other regenerating systems, cell metaplasia may be an important feature of limb regeneration. Dedifferentiation, the shedding of the specialized architecture of a cell, triggers proliferation which, in turn, may be a prerequisite for reprogramming genomes of some cell types (see Chapter 17).

### Stability of blastemal type

Once a blastema forms, its cells are committed to a specific developmental program. Blastemal cells only replace lost distal structures. Though all blastemas look alike, they do possess a regional specificity. A tail blastema produces tail tissues; if transplanted to a limb stump, it develops into a mixture of limb and tail structures (Figure 15-16). If the limb stump is irradiated, the tail blastema develops independently into a tail in spite of adjacent limb stump tissue patterns. Though limb tissues are histogenically similar to tail (they contain cartilage, muscle, and so on), they differ genetically from tail tissues and are so recognized.

Regional determination of a blastema occurs early in regeneration. Before it is determined, a young amphibian tail blastema, if grafted to foreign sites, may be induced to form such tissues as lens, a sign of remarkable tissue plasticity. A young urodele forelimb blastema transplanted to the stump of a hindlimb develops according to its new site rather than according to its previous location and yields hindlimb morphology. Young blastemas behave as morphogenetic fields; they are labile and exhibit developmental plasticity. For example, two limb blastemas, when fused, form a single giant regenerate. Young blastemal tissues regulate and adjust to loss or gain of tissue by reorganizing into a single morphogenetic field. In time, however, the blastema is determined; its lability changes, and it behaves as a mosaic. Fusion of two determined blastemas results in two limbs, or cutting a blastema in half results in only half a limb. When transplanted to other sites, these blas-

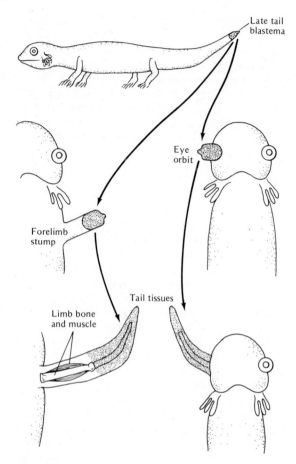

**Figure 15-16** *Stable determination of a tail blastema. A late tail blastema is implanted on a forelimb stump in one case and in the orbit of the eye in another. In each, the blastema develops autonomously into tail structures in spite of local tissue influences.*

temas develop according to their origin; a tail blastema in a limb site develops only as tail.

If a young limb blastema is placed in a neutral site such as the dorsal tail fin, it will self-differentiate into limblike structures. Tissues in the stump need not be present to promote normal limb patterning since both proximal and distal limb elements develop from these grafts. We will see later (Chapter 16) that stump tissue can affect axial patterns of regenerating blastemas.

A blastema undergoes progressive changes in its state of determination very much like embryonic tissues. Initially, it can regulate and will develop according to its surroundings. Soon its developmental fate is stabilized along a specific pathway, and it behaves as a mosaic able to differentiate a limited array of tissue types such as muscle, cartilage, and bone. A few tissues such as fibroblasts may respond to reprogramming cues; they dedifferentiate into blastemal cells that transform into other cell phenotypes. At least for some limb tissues in amphibians, the differentiated state is reversible.

## LOCAL AND SYSTEMIC REGULATION OF REGENERATION

Most regeneration systems recapitulate ontogenetic events. A limb blastema repeats most morphogenetic sequences of limb embryogenesis. As in normal limb ontogeny, populations of pluripotent cells are determined, establish a morphogenetic field, and differentiate into an organized assembly of muscle and bone.

We now ask if there is a similarity between the regulation of limb regeneration and limb formation in the embryo.

### Apical cap: a local factor

The blastemal cone in a limb regenerate resembles the embryonic limb bud even to the extent of forming an apical ridge (see Chapter 12). The ridge directs limb bud outgrowth, as does its counterpart in a regenerating blastema. If the apical cap is removed from a blastema, regeneration is inhibited. Since an apical cap regenerates rapidly, it has to be removed repeatedly. This prevents blastemal formation and regeneration, while control limbs, from which adjacent ectoderm is removed, do regenerate.

The position of the apical cap determines the direction of blastemal outgrowth. A secondary wound on the skin, off to the side of an amputated stump, stimulates growth of new wound epithelium which carries with it the preexisting apical cap. Once displaced, the resulting blastema organizes directly under the apical cap and in no other region. An apical cap stimulates blastemal cell division and accumulation of cells in its immediate vicinity, and an asymmetrical regenerate develops from the acentric blastema.

Apical caps, transplanted to the base of a limb blastema, induce secondary outgrowths and accessory limbs, while other epidermal grafts have no effect. When apical caps are grafted to new locations, the mitotic index in the area increases while the mitotic index decreases in the original site. At least with respect to the role of the apical cap, the embryonic program is replayed during regeneration; apical ectoderm promotes mesodermal outgrowth.

Although the blastema behaves like an embryonic limb bud, there are differences between the two. First, only missing parts are regenerated. At no time does a completely new limb develop; no matter where the amputation site is, only those parts distal to the stump tissues appear. Some control is exercised by stump tissues over blastemal morphogenesis, but its nature is not understood. Blastemal morphogenesis shows a distal-proximal gradient; that is, distal parts or digits regenerate first, progressively forming proximal structures until they meet up with stump tissue. Distal regions seem to behave as organizing centers in blastemal outgrowth. In the embryonic limb, however, morphogenesis occurs in a proximal-distal direction; proximal tissues (shoulder, upper limb) are determined first and progressively the more distal tissues become determined as the limb elongates.

Under certain conditions, the potentialities of isolated blastemas exceed their prospective fate. More proximal structures may develop from distal tissues. For example, if a single aggregate of mesenchyme from the hand plate of a paddle stage regenerate is transplanted to a neutral site (the eye orbit), it differentiates only hand elements. If many such distal mesenchymal masses are collected as one large aggregate and transplanted, then more proximal structures develop, including forearm and humerus. An increase in tissue mass results in the formation of both proximal and distal structures from tissues destined to form only distal elements. Sometimes only proximal structures will develop from whole isolated blastemas developing in the dorsal fin. Proximal structures may develop in the absence of digits, which suggests that distal dominance is lost during regeneration of blastemas separated from stump tissues.

Another important distinction is that regeneration is dependent upon hormones and trophic influences from nerve, whereas the embryonic limb bud is never exposed to hormones and can develop without any apparent neural influences.

## The role of the nervous system in regeneration: a local factor

The nervous system is essential in practically all regenerating systems. Its action is independent of impulse transmission but is more directly related to neurosecretory behavior, that is, the secretion of growth-promoting factors.

The diffuse nerve net in coelenterates, as we have seen, coordinates motor responses of the whole organism and controls growth and morphogenesis (Chapter 14). A nerve cell gradient correlates with the apical-basal gradient of regenerative capacity. In addition, low-molecular-weight substances have been isolated from neurosecretory granules which stimulate tentacle development.

Though a graded distribution of such growth-promoting substances may account for polarity of regeneration in *Hydra*, the nervous system is also implicated in hypostome dominance. This seems to raise the paradox of the same cell acting both as inhibitor and stimulator. But we know that nervous systems behave in just this way; they contain stimulatory and inhibitory neurons secreting different transmitter substances. Sometimes the same transmitter can stimulate or inhibit, depending on the type of cell with which it interacts. The nature of a response to a neural transmitter is an intrinsic property of the responding cell and not the triggering neuron. Accordingly, we can imagine in the primitive nerve net of *Hydra* the presence of at least two cell types, one synthesizing a growth-promoting morphogen

while another controls cell motile behavior by its action as an inhibitory neuron. Such a dual control system could account for growth and regeneration in *Hydra*.

Nervous tissue has an important inductive influence on flatworm regeneration. In the species *Polycelis nigra*, many eyes are distributed in a row along the anterior and lateral edges of the head (Figure 15-17). If eyes are removed, they regenerate in their normal location. If, however, the brain is also removed and prevented from growing back by repeatedly removing its blastema, then eyes fail to regenerate; the margin of the head shows no re-

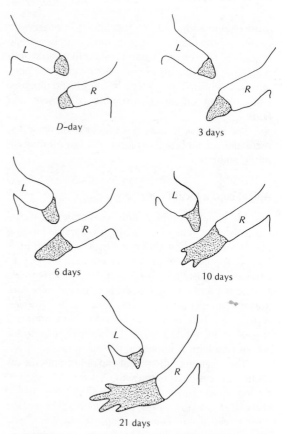

Figure 15-18   *Role of nerve in limb regeneration in* Ambystoma *larva. Limb is denervated 7 days after amputation. The denervated regenerate (left forelimb, L, compared with right limb, R) grew slowly, stopped, and regressed. Drawn 3, 6, 10, and 21 days after denervation, D-day. (From O. E. Schotté and E. G. Butler,* J. Exp. Zool. *97:95–121, 1944.)*

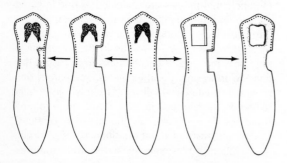

Figure 15-17   *Role of the brain in eye regeneration in* Polycelis nigra. *Eyes regenerate in the margin of the head only if the brain is present. Removal of the brain prevents any regeneration of the lateral edge. (From R. Goss,* Principles of Regeneration. *New York: Academic Press, 1968.)*

growth. Brain exerts a trophic effect since regeneration occurs even if axons between brain and eye are cut, or if the brain is destroyed by X-rays. Necrotic brains or brains from other species of flatworms exert this same inductive effect, which means that a substance released from the nervous system induces eye formation. Extracts of heads will induce eye regeneration in brainless animals, whereas extracts from tails exert no such effect. This same extract inhibits brain regeneration in a normally regenerating form, yet it acts as a specific eye inducer in a brainless animal. The active material probably comes from neurosecretory cells in the brain as their number increases rapidly within the first 3 days of regeneration.

The ventral nerve cord along the length of the annelid worm also seems to be essential to regeneration. For instance, polarity of the regenerate is a function of nerve cord orientation. An anterior piece of nerve cord which has been inverted induces a tail instead of a head when in contact with the body wall. Conversely, a head instead of a tail regenerates when a posterior cord fragment is inverted. In both cases, no amputation is necessary; a wounded surface is the signal to regenerate as long as nerve is implanted in or near the wound. Though any region of the body can regenerate, its polarity seems to depend upon the orientation of its subjacent neural cord and the direction of release of a neural factor.

The regenerating amphibian limb is also dependent upon trophic influences from nerves; limbs without nerves fail to regenerate (Figure 15-18). A normal limb bud acquires neural dependence during embryonic development and its tissues retain a persistent need for neural stimulation throughout life. Growth and contractility of muscle require continuous low-level trophic stimulation from innervating nerves.

How do nerves participate in limb regeneration? First, we know that neurons do not exert their effects by impulse transmission. The neuronal effect is local and trophic. If only the ventral roots of nerves are cut, isolating all limb neurons from central motor connections (but retaining sensory connections), limbs still regenerate in the absence of stimulation. Evidently sensory neurons are as effective as motor neurons at an amputation site. In fact any nerves diverted from other sites will supply the necessary trophic factors for limb regeneration.

A minimum number of nerve endings must penetrate the limb area to furnish threshold concentrations of a trophic factor for blastemal outgrowth (Figure 15-19). Limbs that regenerate best have most nerve fibers, as in *Triturus*, whereas adult anuran limbs, which fail to regenerate, contain too few fibers to meet even the threshold required for regeneration in *Triturus*. In general, the greater the number of nerve fibers, the more trophic factors are secreted. Exceptions are found in *Xenopus* limbs which possess few nerves but exhibit some regeneration, possibly because large-diameter nerves supply the limb. Even a denervated regenerating stump forms an outgrowth. No regeneration blastema appears, the out-

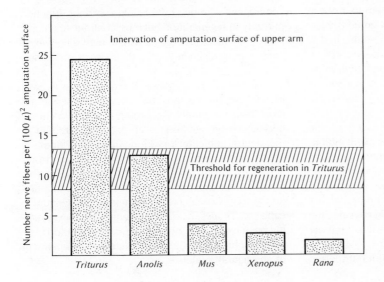

Figure 15-19  *Number of fibers per unit area in upper arm of various invertebrates correlates with regeneration potential. Hatched bar represents threshold level of fibers in* Triturus, *above which regeneration occurs and below which no regeneration occurs. Threshold for regrowth is probably different for each species. (Based on K. Rezhak and M. Singer,* J. Exp. Zool. *162:15–22, 1966.)*

growth emerges directly from cartilage in the stump, which suggests that regeneration in this species is less dependent on nerve.

Sensitivity to the nerve influence varies with the age of the organism. For example, newt larvae have lower thresholds; fewer nerve fibers are necessary for normal regeneration. On the other hand, anuran tadpoles which regenerate hindlimbs have as many fibers as adult frog limbs which do not regenerate. Apparently, at metamorphosis, anuran regenerative capacity is lost in spite of a rich nerve supply, possibly as a result of an increased threshold requirement for neuronal factors. Modest limb regeneration may be evoked in young adult frogs if the total number of nerve fibers in the limb is increased by diverting the sciatic nerve to the forelimb stump.

Not every phase of limb regeneration is equally dependent upon nerve. Both wound healing and demolition phases do not require any innervation. During these early phases, the nerve stump itself undergoes some degree of degeneration. In fact, limbs deprived of nerves show more tissue histolysis than normal; the presence of nerve may limit the amount of dedifferentiation.

Formation of the blastema is most sensitive to nerve influences. Blastema formation is a two-step process: migration of dedifferentiated cells to the local site followed by proliferation, and both seem to be under neural control. No blastema forms in denervated limbs because few cells accumulate. Those that do may synthesize DNA at the same rate as innervated blastemas during the first few days but soon stop and do not divide. They seem to arrest in $G_2$. Denervation before proliferation inhibits blastemal growth, but once a blastema is established, it becomes independent of nerve. Limbs denervated after blastema formation complete regeneration, forming a small limb. Evidently recruitment and proliferation of blastemal cells are both sensitive to neurotrophic influences.

At first acetylcholine was implicated as the trophic substance because atropine, an agent inhibiting acetylcholine activity, also prevented regeneration. Further analysis showed that this was not the case. Direct infusion of acetylcholine or related compounds into denervated, amputated limbs did not induce regeneration, and to date we still have no clue about the nature of the nerve factors that promote regeneration.

Many different macromolecules are transported down the nerve axon for release at the synaptic terminal. Amino acids incorporated into neuron proteins can be traced into the axon synaptic terminal and even the postsynaptic cell. This process of macromolecular transport and release is probably mediated by contractile neurofilaments which facilitate the flow of transmitter substances as well as any trophic agents. Between nerve and muscle a continuous trophic interaction takes place.

Some experiments on aneurogenic limbs show that regeneration is possible in the absence of nerves. Limbs are made aneurogenic in *Ambystoma* embryos by excising those portions of the neural tube which normally send out nerves to a limb region. Such an animal is so severely affected that when it develops, it cannot feed and must be parabiotically joined with another embryo to insure its survival. Limbs develop and regenerate in the absence of nerves. Never having been in contact with nerves, these limbs behave as if they are independent of trophic influences from nerve and can regenerate without it.

Not all limb tissues acquire neurotrophic dependence during development. For example, if skin of an aneurogenic limb is replaced by normally innervated skin followed by amputation, no regeneration occurs. In fact, the limb stump regresses. On the other hand, if mesodermal tissues of aneurogenic limbs are replaced by comparable tissues from normally innervated ones, regeneration proceeds at the expense of the normal tissues. It seems that skin becomes "addicted" to the presence of nerves in normally developing limbs, but muscle does not. When placed in aneurogenic limbs, normal skin cannot function properly in regeneration but must continue to receive nerve influences. This suggests that nerves may exert control over regeneration by their effects on epidermis.

Marcus Singer proposed that all limb tissues synthesize trophic factors, but nerves synthesize far more than the others. According to this hypothesis, nervous tissue synthesizes an excessive amount of the neurotrophic factor during limb development, which spills over into neighboring tissues and prevents them from synthesizing this same factor. This creates a dependence by these tissues on continued neural supply of the factor for their own regenerative capacity. Tissues in an aneurogenic limb, never having been exposed to nerve, continue to develop their own trophic factors and are therefore independent of any neural influence.

An addiction to neural factors may be acquired by more mature tissues in the adult. Aneurogenic limbs transplanted into limb sites in normally innervated animals fill up with nerves growing in from stump tissue and gradually acquire a dependence upon nerve trophic factors for later regeneration. If these limbs are denervated and amputated at different times after nerves have penetrated, regenerative capacity declines until no regenerates occur. The presence of neurotrophic factors seems to inhibit synthesis of similar factors in other tissues, and the tissues become dependent on nerve.

There may be some relationship between nerve trophic factors and the injury stimulus. Recall that supernumerary or accessory limb structures are induced by injury, implantation of tissue extracts, or even inert substances such as bits of cellophane. Wounding may release trophic factors from injured cells which insures that the trigger to regeneration is tissue destruction rather than factors supplied continuously by nerve. Otherwise, limb tissues would be under continuous pressure to regenerate or to produce supernumerary limbs.

Hormones also play an important role in regeneration. In contrast to limb development in the embryo where hormones are not involved, hormones are present throughout regeneration in tadpoles and adults and exercise a controlling influence on the course of regeneration.

## The role of hormones in regeneration: systemic factors

Most neurosecretory effects in regeneration are integrated into a neuroendocrine feedback system. It is often difficult to distinguish hormonal from neural influences during regeneration, particularly in invertebrates, but the two systems are more easily dissociated in vertebrates.

Several hormones are important to amphibian limb regeneration, each controlling a different phase of the process. Furthermore, the hormonal response varies with the age of the animal. For example, larval urodeles regenerate limbs in total absence of pituitary hormones, whereas the adult is completely dependent upon pituitary for limb regeneration, a dependence acquired during metamorphosis.

Regulation of regeneration by the pituitary gland is most difficult to unravel because of its feedback interactions with other endocrine glands. Surgical re-

moval of the pituitary (hypophysectomy) at the time of limb amputation inhibits regeneration. If, however, hypophysectomy is delayed until several days after amputation, some regeneration occurs, its extent increasing the longer the delay between amputation and the operation. After a delay of 3 days, some limbs regenerate; but if amputation precedes hypophysectomy by 13 days or more, virtually all limbs regenerate. The best interpretation of these results is that pituitary hormones act only during the very early stages of regeneration, namely, wound healing and dedifferentiation. Blastemal growth and differentiation do not require a continuous supply of pituitary hormones.

Several hormones, principally ACTH, growth hormone, and even prolactin, stimulate limb regeneration of hypophysectomized animals. Since pituitary hormones control other endocrine glands, it is difficult to decide whether their effect on regeneration is direct or results secondarily from failure to stimulate another gland whose hormones are necessary in regeneration. This is true for ACTH, the hormone regulating adrenal gland functions.

Experiments on amphibian adrenal glands are difficult since the adrenal is a diffuse organ distributed in kidney tissues and cannot be removed. A drug amphenon, which inhibits coritsone secretion, chemically "excises" all adrenal tissue. When administered, it inhibits limb regeneration, but its effects are reversed by cortisone and not by ACTH. The pituitary effect on regeneration and its alleviation by ACTH is mediated entirely through the adrenal gland's production of cortisone. Cortisone's action in regeneration is consistent with the well-known fact that cortisone promotes wound healing. In its absence, wound healing fails, and a thick dermal pad forms at the stump tip, preventing further regeneration.

Cortisone effects are complex because replacement therapy of hypophysectomized animals with cortisone does not always improve regeneration. Cortisone regulates ACTH production by the pituitary and also controls sodium and potassium balance. Moreover, both ACTH and cortisone increase after local tissue stress, which may explain why stress stimulates regeneration and supernumerary limb formation in amphibians.

Thyroxin, the hormone controlling metamorphosis, also affects regeneration, particularly in anurans where the larval-adult transformation is most dramatic. But the thyroxin effect is poorly understood

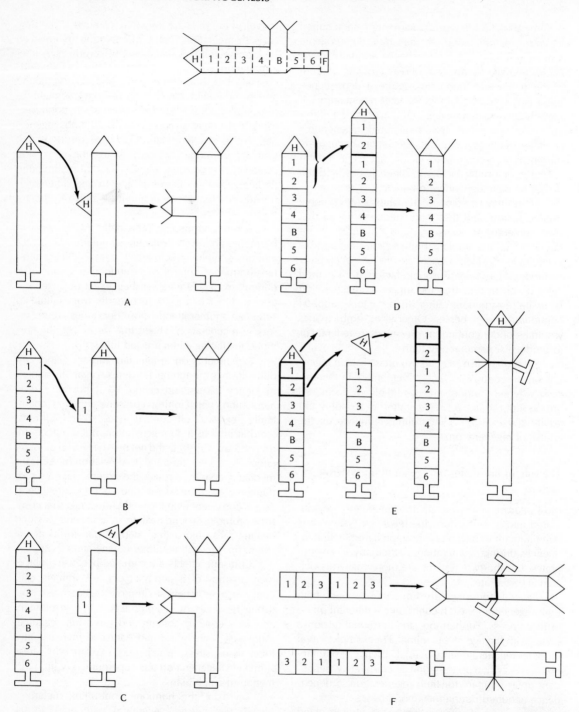

**Figure 15-21** *Experiments to demonstrate hypostome dominance and polarity interactions during regeneration of Hydra. Top figure shows regions of Hydra body. H = hypostome; B = budding site; F = foot. (A) Induction of new outgrowth by*

since it inhibits tadpole limb regeneration if administered before amputation but accelerates morphogenesis if given at blastemal stages. Thyroxin inhibition is consistent with loss of regenerative capacity after a larva transforms into an adult. Thyroxin stimulates limb morphogenesis in the anuran tadpole, but once formed, a limb loses thyroxin dependency and its regeneration is inhibited by thyroxin instead.

Hormones join with local factors in controlling regeneration. One factor is the extracellular matrix filling tissue spaces. In morphogenesis the extracellular matrix plays an important part, as we have seen, either as a substratum on which motile cells or tissues move or as the glue that keeps cells together or stabilized into distinct shapes. Changes in the matrix of regenerating limbs do occur in the demolition phase as cartilage and bone dedifferentiate and a new matrix is established within the interstices of the developing blastema. The appearance of the new matrix is accompanied by synthesis of hyaluronate (Figure 15-20), reaching peak levels at the most distal portion of the tip and grading off into more proximal regions of the stump. Afterward, the level of hyaluronate decreases rapidly, to be replaced by another matrix material, chondroitin sulfate, the principal matrix of differentiating cartilage and bone. The early accumulation of hyaluronate may be essential to blastemal cell proliferation and cell migration, including the recruitment of cells from dedifferentiating stump tissues. Until we learn how hormones affect matrix synthesis and blastema formation, we are left with the feeling that hormones exercise some control over regeneration, but we do not know how.

## POLARITY IN REGENERATION

Regeneration always restores tissues in normal axial relationships with stump tissues and adjusts their size and symmetry to the rest of the body. An intrinsic polarity of the system seems to direct the overall repatterning of the amputation site. We cannot account for tissue polarity in most cases nor do we understand

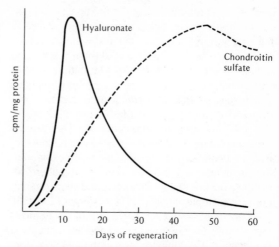

**Figure 15-20**   *Role of the extracellular matrix in limb regeneration. Changes in the synthesis of hyaluronate and chondroitin sulfate in a regenerating limb as measured by labeled precursor incorporation. The decline in hyaluronate results from an increase in the enzyme hyaluronidase. (Based on B. P. Toole and J. Gross, Develop. Biol. 25:57–77, 1971.)*

how it controls the reestablishment of a normal pattern. But models of polarity determination have been proposed based on studies of *Hydra* regeneration.

To describe these experiments on polarity, the *Hydra* body is arbitrarily divided into a series of regions, starting at the distal end, designated in linear order as H1234B56F, with hypostome and tentacles as (H), gastric region (1234), budding zone (B), peduncle (56), and foot (F) (Figure 15-21). In Chapter 14 we saw that the H fragment, the most apical region, is dominant over all others; it inhibits regeneration of more proximal regions and it alone has inductive ability. A graft of an H region to any part of the gastric zone induces a new hypostome outgrowth, even in the presence of the original host hypostome. The new growth consists of graft and host tissue, which means the graft organizes a new axial system.

Other regions can induce outgrowths if the host

*grafted hypostome. (B) Failure of gastric fragment to induce because of hypostome inhibition. (C) Removal of hypostome permits induction by gastric fragment. (D) Fusion of hypostome and adjacent regions 1 and 2 to another host; only a single hypostome develops. (E) Removal of dominant hypostome from graft and fusion as in (D) above leads to head and tail regeneration at junction as well as hypostome growth at distal end. (F) Regeneration at junction when gradients are opposed. (Based on L. Wolpert, J. Hicklin, and A. Hornbruch, Symp. Soc. Exp. Biol. 25:391–416, 1971.)*

hypostome is removed. A graft of region 1 made on such a host induces a secondary hypostome axis as the new host hypstome regenerates. A graft of region 1 into region 6 of an intact animal (with hypostome) induces a secondary outgrowth. Inhibition by host hypostome falls off at more distant body regions, which suggests that an inhibitor produced in hypostomes decreases in gradient fashion to the foot.

The inhibitor gradient polarizes the *Hydra* body. Fragments of the *Hydra* body may be grafted together in any orientation by impaling them on a fine hair. Regeneration occurs after tissue fusion, its pattern depending on the regions involved and their orientation. For example, a graft of H12 to a larger 1234B56F fragment (represented as H12/1234B56F) reorganizes into a single hydrant and no regeneration occurs at the junction. The inhibitor level at the junction (from the H12 piece) prevents any outgrowth. Removal of the head (H) region (a 12/1234B56F union) eliminates distal dominance. Though a head usually develops distally in about 50 percent of the combinations, heads and feet also appear at the junction. In the absence of inhibition from an H region, two high points emerge in the gradient (at 1 in each case), and each regenerates according to its original polarity.

If the grafts are held in opposite polarity, such as 123/321, each fragment acts independently; feet form at the junction, and heads regenerate at the ends. A 321/123 combination leads to heads at the junction with feet at the ends. These results show that if polarities of two body parts of the same axial system are opposed (oriented in opposite directions), then no dominance is exerted by one part over the other. Inhibitor seems to diffuse in only one direction, from distal to proximal, and cannot pass a boundary that is not oriented in the same direction.

Many other combinations indicate that polarity can be reversed by grafting hypostomes; new axes of outgrowth are initiated wherever these are placed. To explain these results, Wolpert has proposed a dual gradient model which, he believes, may also account for axial patterning in other morphogenetic systems (Figure 15-22). One gradient is based on an inhibitor substance flowing from distal to proximal regions,

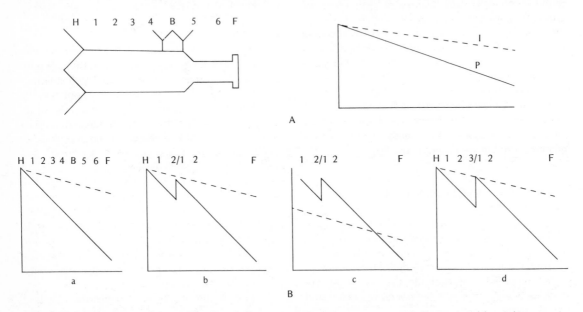

**Figure 15-22** *Wolpert model of positional gradients to explain grafting experiments. (A) Regions used for grafting experiments. Positional value represented by a single gradient, P, and the inhibitor gradient is shown as I. They interact as explained in text. H = hypostome; B = budding site; F = foot. (B) Interaction of gradients. (a) The gradient as seen in the normal animal. (b) Changes in gradient system to explain H12/12 combination. No regeneration at junction. (c) In 12/12 combination, two high points of positional value exceed inhibitor gradient and regeneration occurs at junction. (d) No regeneration at junction in this combination. (From L. Wolpert, J. Hicklin, and A. Hornbruch, Symp. Soc. Exp. Biol. 25:391–416, 1971.)*

and a second gradient defines the position of each cell. The positional gradient permits cells in two-dimensional epithelia to assess their relative positions to one another and to the boundaries of the system. A model of a positional gradient based on communication through gap junctions has been described in Chapter 9. Both inhibitor (I) and positional (P) gradients fall off from head to foot, the latter at a steeper slope. The inhibitor is synthesized in the head, diffuses rapidly in one direction down the gradient, and is labile. In the absence of the head, the inhibitor gradient decreases sharply, and if at any point the I value is greater than the P value, no additional head-end boundary forms, and a head fails to regenerate. If I is less than P, then a head-end boundary forms at the high point of P and cells at this level regenerate a head while feet form at the low point of the positional gradient.

Grafting experiments can be explained according to the model. For example, in the H12/12 interaction, P is never higher than I, and no heads or tails form at the junction. In the absence of a head, however, the inhibitor gradient falls, and two new positional values are established for head and tail regeneration at the junction.

A difficulty with this model is that the origin and maintenance of a gradient cannot easily be explained in terms of known cell properties, nor is it possible to design experiments to prove the existence of gradients. But real gradients are found in *Hydra* in the form of nerve and interstitial cell distribution along the body axis (Figure 15-23). Both cell types are most abundant in the hypostome and decline progressively along the gastric region. Moreover, in buds and during regeneration, nerve cells increase in number in morphogenetically active regions. Dominance and polarity both correlate with these cell gradients.

Direct cellular continuity is necessary in establishing gradients in regenerating systems. In most instances, inhibition by dominant regions requires communication between adjacent cells.

For example, in the cockroach *Leucophaea*, amputation of a limb results in regeneration, but a cut or simple wound fails to induce any outgrowth because the distal region still exerts its inhibitory effect on the wound site. If, however, a wedge of tissue is cut out of the tibia (Figure 15-24), then a regenerate appears at the proximal cut surface in spite of the presence of distal limb structures. If diffusion per se was involved, the gap could have

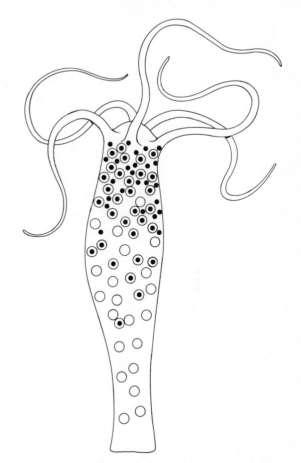

**Figure 15-23** *Distribution of growth-regulating substances in* Hydra *correlate with distribution of nerve and interstitial cells. Stimulators (open circles) and inhibitors (black dots) are both in high concentration in hypostome region where inhibitors inactivate stimulators (dots in circles). Inhibitors disappear halfway down the stalk, leaving stimulators free to initiate bud outgrowth. Toward base stimulator concentration declines. Budding occurs because stimulators exceed inhibitors. (From R. Goss,* Principles of Regeneration. *New York: Academic Press, 1968.)*

been by-passed, but the fact that regeneration was not inhibited means that information from distal structures cannot by-pass gaps by making lateral detours. The information that determines polarity in this system is transmitted in a polarized direction, from cell to cell.

Tissue contacts in *Leucophaea* must retain their original polarity for inhibitory information to pass

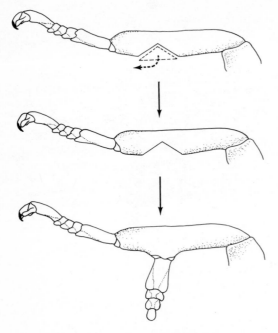

**Figure 15-24**   *Effect of cutting a V-shaped wedge from the tibia of* Leucophaea. *The next instar develops a lateral regenerate. Release from physical continuity with a dominant region allows regeneration. (From H. Bohn, Wilhelm Roux' Arch. Entwicklungsmech. Organ. 156: 449–503, 1965.)*

from distal regions. Amputation of a limb followed by regrafting of the distal portion back on the stump results in a normal limb outgrowth but no regeneration at the junction. Both graft and host structures are arranged in comparable polarity. If, however, a leg from the other side is rotated and grafted on, the information from distal structures cannot pass from the stump, presumably because of topographical incompatibilities, and two supernumerary regenerates emerge from the junction. This resembles the *Hydra* grafting experiments and suggests that information inhibiting regeneration is transmitted only between cells sharing equivalent polarities.

Polarity in vertebrate limb regeneration differs in that no all-over dominance relationships are exhibited. There is no distal dominance as in the embryonic limb bud (see Chapter 12). Polarity is a local phenomenon; the regenerate restores distal structures in proper proportion to the underlying stump.

There is a certain degree of intrinsic polarity in

an amphibian limb. After amputation, polarity is imposed by the orientation of peripheral nerves leading to the limb. Nerves grow out toward the peripheral wound epidermis and penetrate the blastema from one direction. This fixed polarity influences the pattern of the entire limb, since, as we have seen, innervation is important during early phases of regeneration.

Besides nerve, the remaining stump tissues may have an organizing influence on the subsequent growth of the regenerate. Bone, muscle, connective tissue, and cartilage maintain a pattern in proximal regions of the stump in spite of excessive demolition of tissues beneath wound epithelium. "Organizing information" could be transmitted from the stump to the blastema to promote morphogenesis along correct axes of symmetry and insure proper replacement of lost parts. But presence of a specific bone or muscle seems unessential for regeneration of that structure. If, for example, the ulna from the forelimb of an adult newt is removed before the limb is amputated, the regeneration of the forelimb does not reflect the deficiency of the ulna in the arm. Although the ulna is not completely replaced, the distal portion of the regenerate does include a portion of the ulna in its proper location (Figure 15-25). We have also seen that isolated blastemas in neutral sites do develop most limb tissues. Can we conclude that stump tissue does not participate in its own replacement?

The answer to this question is not simple. If, for example, an extra ulna is introduced into the forearm, followed by amputation, the regenerate that grows out includes three bones, two ulnas, and one radius, suggesting that the implanted bone did indeed direct the outgrowth of its own replacement. In this case, stump skeleton exerts a directing influence on bone formation in the morphogenetically active blastema. Although the limb compensates for skeletal deficiencies, it seems to have no mechanisms to correct for any superabundance of structural subunits. Other tissues may take over in the absence of bone, but in its presence it determines its own replacement. Only live bone exerts an inductive effect because dead bones inserted in the limb do not. An additional viable bone surface in the stump may behave as a nucleating site favoring aggregation of presumptive bone cells and elaboration of a supernumerary part.

Stump tissues may also influence the polarity of

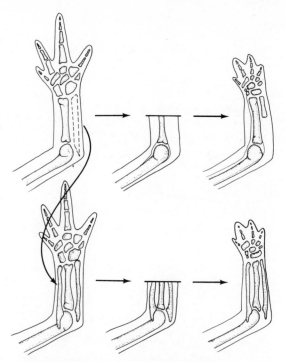

**Figure 15-25** *Role of bone in limb morphogenesis. (Upper) Removal of ulna before amputation does not prevent regeneration of a cartilaginous ulna in operated region. (Lower) Insertion of an extra ulna into the forearm before amputation leads to production of corresponding extra skeletal element in regenerate. (Based on R. Goss,* Principles of Regeneration. *New York: Academic Press, 1968.)*

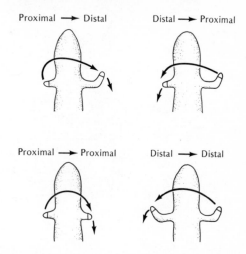

**Figure 15-26** *Diagram of four different types of transplant operations. In each case the dorsal-ventral axis of the graft coincided with that of the host limb. Only the anterior-posterior axis of the graft was reversed compared to the host stump. (From L. E. Iten and S. V. Bryant,* Develop. Biol. *44:119–147, 1975.)*

the regenerate. The time at which anterior-posterior and proximal-distal axes of the regenerating limb are established is still uncertain. Some believe polarity is fixed as soon as the blastema appears while others claim that axes are determined later. Interactions between stump and blastema seem to determine axial polarity.

This is shown by grafting blastemas of different ages onto stump tissues in various orientations (Figure 15-26). Proximal blastemas, obtained after amputation of the upper arms, are grafted onto distal stumps (amputation site in the forearm or wrist) and distal blastemas are grafted onto proximal stumps. In all cases, anterior-posterior axes of graft and stump are reversed while dorsal-ventral axes are aligned normally, and at the end of regeneration the type and polarity of skeletal elements are examined. If early, middle, and late blastemal buds are used, we can determine when blastemal polarity is established and is no longer responsive to polarizing cues from stump tissues.

All transplants of proximal blastemas at the medium bud to early digit stages to distal stumps retain their original anterior-posterior polarity. Only early bud stages respond to the polarity of the stump, which means that by the beginning of the medium bud stage, the anterior-posterior axis is determined. The stable behavior of proximal buds contrasts with the lability of distal buds grafted onto proximal stumps. Virtually all grafts in this instance develop the anterior-posterior polarity of the stump; irrespective of stage, distal polarity is always reversed. Since all distal grafts undergo radical dedifferentiation after transplantation, whereas proximal grafts show no disorganization and retain their original pattern, tissue disorganization seems to make cells responsive to new polarity cues emanating from stump tissues.

In many proximal to distal transplants, supernumerary limbs developed as reduplications of regenerating limbs with the same polarity. The best model to explain such reduplications is shown in Figure 15-27. Presumably a blastema acquires an-

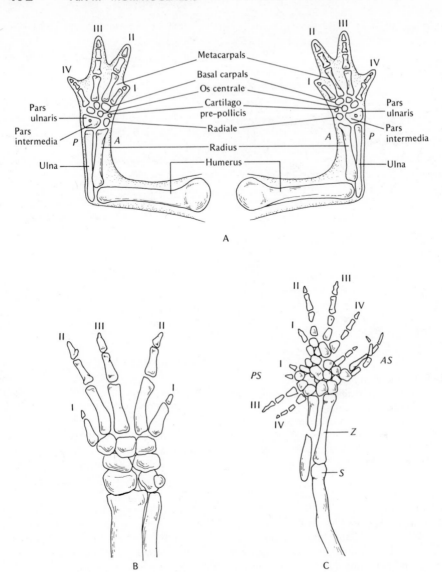

**Figure 15-27**  *Examples of types of regenerates developing after operations. (A) Diagram of right and left forelimbs showing skeletal elements after normal regeneration. A = anterior; P = posterior. (B) Skeletal preparation of a double anterior limb formed after transplantation of an early bud blastema from left to right (proximal to distal). Both pre- and postaxial margins of the limb consist of anterior digits and carpals. Posterior digits and carpals are located in the center of the hand. (C) Skeletal preparation of a limb showing anterior and posterior supernumerary regenerates. A blastema at the stage of early digits was transplanted from right to left (distal to proximal). A single humerus as well as a radius and ulna have regenerated with left (stump-type) asymmetry. The transplant retained its right-handedness. Arrow indicates level to which transplant was made. Z = regenerated zeugopodium; S = regenerated head of the stylopodium. I, II, III, and IV identify digits. PS = posterior supernumerary regenerate; AS = anterior supernumerary regenerate. (From L. E. Iten and S. V. Bryant,* Develop. Biol. *44:119–147, 1975.)*

terior-posterior polarity gradually. In early proximal buds or in any distal bud, the polarity is unstable and a spectrum of limbs is produced with polarity patterns intermediate between graft and stump. In order to establish these patterns, either new cells come from the stump with their own polarity or the blastema reorganizes to eliminate the discrepancy between levels of tissue organization at the graft site. At later stages, the proximal blastemal axis is stabilized. When these are transplanted, no intermediates develop. Instead, supernumerary limbs develop with distinct polarities.

Determination of proximal-distal polarity is more complicated. Proximal blastemas grafted to distal sites (at least by medium bud stages) resulted in serial duplications of skeletal elements in the proximal-distal axis. This indicates that the proximal boundary of the blastema is specified by that stage and that it is not modified by contact with stump at a more distal level.

In the reciprocal combination, all distal to proximal grafts reorganized and developed in accordance with proximal-distal orientation of the stump. No duplications or supernumerary structures developed. Since this is true for late bud grafts as well, it indicates the grafts become respecified according to the new proximal-distal level of the stump.

A model to explain these results invokes a gradient of developmental capacity, polarized proximal-distal (Figure 15-28). Note that the model resembles the positional information model proposed to explain *Hydra* polarity during regeneration. If a distal part is removed, the remaining tissue, at the high point of the gradient, can easily regenerate all distal structures lower in the gradient. On the other hand, a distal bud grafted to a proximal stump retains the same gradient orientation and will reorganize or regulate to reestablish the gradient between high and low points, and a normal proximal-distal sequence of structures will regenerate. Finally, a proximal graft to a distal stump places a barrier at the site with two equal high points. Since the proximal border is fixed at his time, tandem duplications of elements will develop, and no blastemas form at the site.

Polarity is an elusive phenomenon usually explained by diffusion gradients or active transport of inhibitors from dominant regions. The nature of inhibitors and the origin and maintenance of gradient

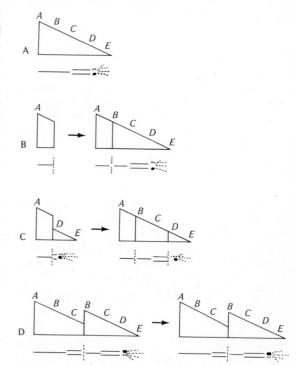

**Figure 15-28** *Diagram of proximal-distal developmental gradient in regenerating newt limb. Values A, B, C, and so on correspond to particular portions of the limb, as shown beneath the gradients. (A) A complete gradient, representing the entire appendage. (B) Normal regeneration from a proximal level. (C) Distal to proximal transplant leading to intercalary regeneration. (D) Proximal to distal transplant leading to tandem duplication of limb elements. (From L. E. Iten and S. V. Bryant,* Develop. Biol. *44:119–147, 1975.)*

systems are poorly understood, but, as we will see, such models are now popular in explaining pattern formation in most morphogenetic systems (Chapter 16). Ultimately polarity may be explained in terms of polarized flow of ions between cells, possibly through gap junctions. Or it may be based on a graded distribution of specific cell types such as neurons which can exert trophic influences on morphogenesis. The intrinsic anatomical polarity of nerve cells coupled to their importance in regeneration suggests they may organize polarity of restored tissues. But each regenerating system is unique, and other factors are undoubtedly involved in polarizing a morphogenetic field.

## SUMMARY

Regeneration of lost parts is common in most asexually reproducing organisms, a manifestation of high morphogenetic potential retained by most somatic cells of the adult. It is found in higher vertebrates but is usually restricted to a few organs such as limbs, tails, and eyes. The trigger to regeneration is a release from inhibitory control by more dominant regions of the body. Once isolated from the parent body or cut off from dominant centers, the intrinsic morphogenetic potential of a region is expressed.

In most cases regeneration recapitulates the ontogenetic history of the organ. Except for a few differences in hormonal and neural control, limb regeneration in amphibians seems to follow the same sequence of events and abides by rules similar to those governing normal embryonic events. Cells for regeneration are derived from local tissues which dedifferentiate but retain their phenotypic bias and restore the original tissue. Some tissue may undergo metaplasia and differentiate into other cell types. In some systems undifferentiated neoblasts also contribute to regeneration, providing some essential cell types such as nerve cells.

The nervous system is essential to regeneration, furnishing trophic factors that support growth and morphogenesis. Factors derived from nerve exert stimulatory or inhibitory control over regenerating tissues and in some cases may account for the dominance and polarity of regeneration.

Polarity controls the restoration of lost parts, which are always replaced in a definite pattern in relation to underlying stump tissues. The establishment of axes of symmetry in regenerating systems may depend upon diffusion gradients which assist cells in ascertaining their position in space. Although gradient models can account for some of the patterning of regeneration, the nature of such gradients and their mode of origin are still not understood.

**REFERENCES**

Barth, L. G. 1934. The effect of constant electric current on the regeneration of certain hydroids. *Physiol. Zool.* 7:340–364.

Brønstedt, H. V. 1970. *Planarian regeneration.* Elmsford, N.Y.: Pergamon.

Chandebois, R. 1976. Histogenesis and morphogenesis in planarian regeneration. Ed. A. Wolsky. Basel: Karger.

Gierer, A., Berking, S., Bode, H., David, C. N., Flick, K., Hansmann, G., Schaller, H., and Trenker, E. 1972. Regeneration of *Hydra* from reaggregated cells. *Nature, New Biology* 239:98.

Goss, R. G. 1968. *Principles of regeneration.* New York: Academic Press.

Hay, E. 1966. *Regeneration.* New York: Holt, Rinehart and Winston.

Iten, L. E., and Bryant, S. V. 1975. The interaction between the blastema and stump in the establishment of the anterior-posterior and proximo-distal organization of the limb regenerate. *Develop. Biol.* 14:119–147.

Jabaily, J. A., and Singer, M. 1977. Neurotrophic stimulation of DNA synthesis in the regenerating forelimb of the newt, *Triturus. J. Exp. Zool.* 199:251–256.

Kiortsis, K., and Trampusch, H. A. L., eds. 1965. *Regeneration in animals and related problems.* Amsterdam: North-Holland.

Schaller, H., and Gierer, A. 1973. Distribution of the head-activating substance in *Hydra* and its localization in membranous particles in nerve cells. *J. Embryol. Exp. Morphol.* 29:39–52.

Singer, M. 1974. Neurotrophic control of limb regeneration in the newt. *Ann. N.Y. Acad. Sci.* 228:308–321.

Steen, T. P. 1970. Origin and differentiative capacities of cells in the blastema of the regenerating salamander limb. *Am. Zool.* 10:119–133.

Stocum, D. L. 1968. The urodele limb regeneration blastema: A self-organizing system. I. Differentiation in vitro. *Develop. Biol.* 18:441–456.

Thornton, C. S. 1968. Amphibian limb regeneration. *Adv. Morphogen.* 7:205–249.

————. 1970. Amphibian limb regeneration and its relation to nerves. *Am. Zool.* 10:113–118.

Toole, B. P., and Gross, J. 1971. The extracellular matrix of the regenerating newt limb: Synthesis and removal of hyaluronate prior to differentiation. *Develop. Biol.* 25:57–77.

Wolpert, L., Hornbruch, A., and Clarke, M. R. B. 1974. Positional information and positional signalling in *Hydra. Am. Zool.* 14:647–663.

# 16

# Morphogenesis of Patterns

Whole-organism patterns are the final outcome of developmental processes. Morphogenesis of cell sheets leads to an orderly arrangement of body parts: for example, the paired appendages of mammals, the repeated segments in a worm, the symmetry of a flower. The ordered arrays of hairs, bristles, and feathers in animal integuments are faithfully reproduced each generation. As patterns develop, cellular boundaries seem to be transcended by "global factors"; cell groups, often separated by great distances, cooperate to assemble a whole organism.

Pattern formation calls for the action of many genes and is easily perturbed by single gene mutations. It is unlikely that the pattern itself is encoded in the genome; rather, the genome furnishes instructions for pattern building. The patterning process, however, is not rigidly programmed; cells are responsive to environmental cues and make periodic adjustments as they carry out genetic instructions. In other words, pattern formation has two attributes: mosaicism as a consequence of heritable gene programs, and regulation. As cells monitor their positions in space, they make ordered rearrangements in response to exogenous influences.

We may define pattern as ordered space. In the genesis of pattern, cells behave as if they "know" their position in space; they recognize their neighbors, make specific associations with other cells, migrate to what appear to be predestined positions in the organism, and differentiate into a distinctive cell type. Besides, spatial location seems to define the phenotypic expression of a cell. "Inside" cells develop differently from cells on the surface. In plants where cell position is rigidly fixed by cell walls, functional activity will vary in a graded fashion along an axis of symmetry. In these cases, metabolic changes, or patterns in time, are translated into morphological changes, or patterns in space. How this comes about is the essential problem of pattern formation. In this chapter, we will first examine gene regulation of patterns, then see how cell properties such as mitosis and cell movement are translated into morphological patterns, and finally discuss some models of pattern regulation.

## PATTERNING GENES

Some genes may be called patterning genes; they specify all-over body patterns such as hair and bristle distribution in the epidermis or the arrangement of scales in an insect wing. They do so by determining cell position in two-dimensional sheets.

Most body patterns in insects appear in the cuticular integument secreted by a layer of epidermal cells. Interspersed with bristles, hairs, and sense organs, the cuticle records the developmental history of epidermal cells below. Each bristle and hair is a derivative of an epidermal stem cell and is precisely located in the integument; that is, each bristle position is genetically determined and invariant in a species. Mutant genes modify bristle patterns by preventing certain hairs and bristles from appearing or by inducing extra structures.

For example, in *Drosophila* a set of alleles at the *scute* locus in wild type flies determines the location of certain bristles in the head and thorax (Figure 16-1). Mutations modify the pattern by inhibiting bristle formation at these sites but not at others. In

the mutant $sc^4$ virtually all bristles in the head and eight pairs of bristles in the thorax are absent, whereas $sc^2$-mutant flies lack a posterior pair of bristles in the head, two pairs in the prothorax, and two pairs in the scutellum. We see that genes change the total pattern by controlling the development of specific bristles.

Two hypotheses have been proposed to explain how these pattern mutants function. One states that every mutant gene determines a different "prepattern," that is, an underlying physiological "blueprint" specifying the location of each bristle in the epidermis. In this model, the pattern gene generates "metabolic singularities" in specific sites in the epidermis where cells are induced to differentiate as a bristle rather than as an epidermal cell. Each mutant gene codes for a different prepattern which precedes development of the realized pattern. An alternative hypothesis places mutant gene action in the responding cell; that is, there is no variant prepattern for each mutant. Overall conditions for pattern formation are the same in both mutants and wild type. Differences between mutants reside in epidermal

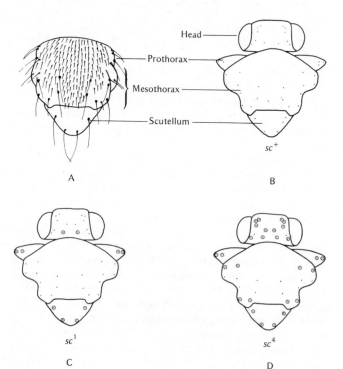

**Figure 16-1** *Bristle patterns in Drosophila. (A) Dorsal view of thorax showing arrangement of bristles on each segment in wild type fly (female). (B) Diagram of bristle distribution in head and thorax of wild type. Each dot is the position of a single large bristle shown at left. (C)* sc¹ *mutant; circled dots are sites where bristles do not usually differentiate. (D)* sc⁴ *mutant. (Modified from C. Stern,* Cold Spring Harbor Symp. Quant. Biol. *21:375, 1956.)*

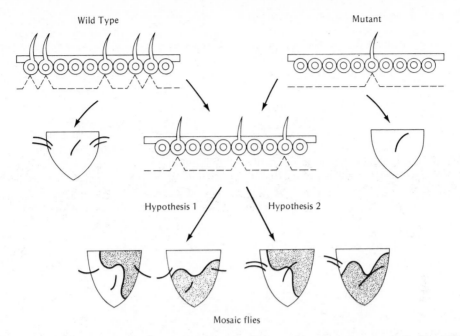

**Figure 16-2**  *Comparison of two models to account for the development of bristle patterns in mosaic flies. Wild type and mutant epidermis are shown with respective bristle patterns. A prepattern for each is indicated by the dashed line below. The hypothetical prepattern is shown in the center and its predictions according to each hypothesis are diagrammed in the mosaic patterns. See text for details. Shaded area = mutant tissue; unshaded area = wild type tissue.*

cells, with each responding autonomously according to its own genotype. In this model, only wild type epidermal cells "recognize" their position in a common pattern and differentiate a bristle. Mutant cells in the same position either cannot recognize the position as a bristle-forming site or are unable to generate a bristle if they do.

Each hypothesis makes a prediction that is testable (Figure 16-2). The first, or prepattern model, states that combinations of two prepatterning mutant genes within the same organism should generate novel bristle patterns. Each genotype would contribute its own prepattern that would, in turn, interact to create a new prepattern with a different distribution of metabolic singularities. In the figure a wild type prepattern is shown specifying five bristles in the thorax. A *scute* prepattern lacks two bristle pairs but specifies the bristle in the center. The prepattern in a mosaic fly carrying both genotypes is novel. As a result of the interaction between both prepatterns, three bristles are specified, one central bristle and only one at each edge. The model predicts

that in mosaics of wild type and mutant tissue, all three bristles would differentiate irrespective of the genotype of the tissue covering the bristle region. Both genotypes should respond equally to the new prepattern. Any combination of mutant and wild type epidermis should give rise to three bristles arranged as shown.

The second model predicts that a genetic mosaic would not generate novel bristle patterns; each tissue would respond to a common wild type pattern according to its genotype. Three bristles would arise in the mosaic patterns in a different arrangement; the central one would be present as before, but a pair would develop at one edge with no bristle emerging at the opposite margin. Wild type tissue would always produce bristles at the bristle sites according to the wild type pattern. Mutant tissue would not respond to any site except the central one and would not differentiate marginal bristles. The prepattern model is an induction model with each allele specifying a different pattern of bristle induction sites, whereas the second model is a responding model with each allele determining a response to a common pattern of inducing signals.

To test the models, sex gynandromorphs are made in *Drosophila* as genetic mosaics. These individuals contain a mixture of male and female tissues of different genotype, each expressing a different sex-specific bristle pattern. Sex in *Drosophila* is determined by the X chromosome; two X's make a female and one X, either in the normal XY combination or in the abnormal X/O, is male. Cells always express their sex genotype autonomously, irrespective of background. Given an appropriate genetic marker on the X chromosome, we can easily distinguish male from female tissues. For example, male-female

gynanders appear with high frequency in a strain of female flies carrying a ring X chromosome aberration; that is, one X chromosome is in the form of a ring rather than a normal rod (Figure 16-3). A small percentage (1 to 2 percent) of female zygotes with this chromosomal aberration develops as genetic mosaics because one ring X chromosome segregates abnormally in mitosis and is randomly lost from nuclei during cleavage. Consequently, some cells acquire a complement of two X chromosomes (one rod and one ring X) and develop as female tissues, whereas others receive only the rod X and develop as male tissues. The time at which this event occurs determines the amount of mosaic tissues in the adult.

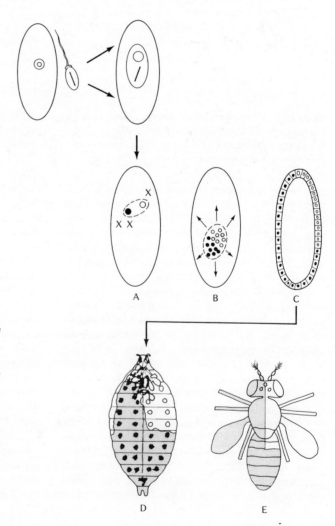

**Figure 16-3** *Development of a gynandromorph mosaic fly from a zygote bearing a ring X chromosome. Ring X (circle) is in egg nucleus and sperm carries normal rod X. (A) In the first division of the zygote, one ring chromosome is lost, resulting in a cleavage nucleus with only one X chromosome (open circle) and a nucleus with both X's (filled circle). (B, C) Subsequent cleavage and blastoderm formation showing distribution of male (X) and female (XX) nuclei. (D) A larva split open dorsally and folded back to reveal imaginal discs and the mosaic character of its own tissues. The discs develop into adult structures after metamorphosis. (E) The adult, which is a composite formed from some male and some female discs. See text for additional details.*

If it occurs at first cleavage, then half the tissue is male and half female. If segregation occurs later, only small patches of male tissue arise.

Mutant tissue is marked by linking marker genes such as yellow body color (y) to the bristle pattern gene (scute) in a heterozygous combination. Both mutant alleles are in the rod X chromosome, while the ring X carries the dominant wild type allele for both genes. Therefore, each female cell with two X chromosomes will express the wild type bristle pattern and dark body color, while each male cell without the ring X chromosome will express the mutant phenotype, scute bristle pattern and yellow body.

Various combinations of scute bristle pattern mosaics were analyzed and all gave the same result; no interactions were observed, nor did any new bristle patterns appear. Instead, presence or absence of a bristle in a specific site depended on the genotype of the tissue covering it. If the species bristle pattern designates a bristle at a specific site, then in mosaics a bristle develops if the site is covered by wild type tissue (female), irrespective of the amount of surrounding mutant male tissue. Even the smallest patch of wild type tissue over the site results in a bristle. Conversely, male yellow tissue, which forms no bristles, does not do so in a mosaic regardless of the size of the mutant patch or the amount of surrounding normal tissue. Each tissue behaved autonomously in the pattern according to its own genotype; it did not interact with neighboring tissues. Cells were obviously unaffected by prepatterns, but behaved according to the prediction of the second model. Scute genes do not determine a prepattern but alter a cell's genotype so that it cannot produce a bristle even if given a signal to do so. The genes have no global effect but act within each cell.

There is, however, one case which seems to support the prepattern model, the eyeless dominant (ey$^D$) pleiotropic mutant. This allele is on the X chromosome and in hemizygous males, the gene induces development of supernumerary sex combs on the foreleg. In the wild type the last two rows of bristles in the male basitarsus differentiate into 10 to 13 sex-comb teeth (Figure 16-4). In the ey$^D$ mutant, additional sex combs appear with 40 to 50 teeth; that is, a much larger area of the basitarsus is committed to sex-comb differentiation. The gene affects pattern by expanding the epidermal field available for sex-comb differentiation.

If, indeed, this is a prepattern gene, then a mosaic composed of non-ey$^D$ female tissue with black bristles and ey$^D$ male tissue with yellow bristles should exhibit nonautonomous expression of the phenotypic pattern. A small patch of non-ey$^D$ tissue surrounded by a larger area of ey$^D$ tissue should respond to the mutant prepattern and form sex combs irrespective of its own genotype. The results bear out the prediction; in 33 of a total of 47 critical mosaics, a nonautonomous expression of sex combs in non-ey$^D$ tissue was observed. Mosaic sex combs were found containing both black and yellow bristles. Besides, more than one row of sex-comb teeth developed from wild type female tissue, which meant that female tissue responded to the expanded prepattern of the eyeless dominant gene and formed supernumerary sex combs. New prepatterns established by the eyeless dominant gene induced a sex-comb response in tissue that, under normal conditions, was unable to do so.

Though these observations are consistent with the prepattern model, other hypotheses might also explain the results. For example, in the ey$^D$ mutant described, a two-dimensional gradient of sex-comb substance in the epidermis may be expanded by the mutant gene. Or the mutant gene may alter the boundaries of the sex-comb field by increasing the number of stem cells in the epidermis committed to bristle differentiation. In this case, the gene may regulate rates of cell division, or the orientation of division planes in imaginal disc cells that develop into adult forelimb epidermis.

Additional experiments are needed to pin down the prepattern nature of the eyeless dominant mutant. If it is found that most patterning genes behave autonomously, then the problem of genetic determination of pattern becomes a problem of cell differentiation (see Chapter 17). Since all cells in wild type epidermis share an identical genome, how do some cells "know" when and where to form a bristle? Presumably, a predetermined physiological state specifies sites of bristle formation in a two-dimensional epithelium. Only those cells genetically equipped to respond to the signal synthesize a bristle. Mutants lacking the specific alleles do not respond. The factors that establish the initial discontinuities in embryonic epithelia may be similar to those responsible for cell differentiation.

A genetic prepattern is the constellation of regionally specific local and systemic factors within

**Figure 16-4**  *Eyeless dominant mutant showing nonautonomous expression of sex combs in male mosaics. (A) Basitarsus of normal wild type leg showing single row of sex-comb teeth. (B) Eyeless dominant male with several rows of many sex-comb teeth displaced on the basitarsus. (C) Two examples of mosaics. The stippled teeth are derived from non-ey$^D$ female tissue and the open teeth are yellow, from male mutant tissue. In both cases, wild type female tissue develops sex combs in abundance.*

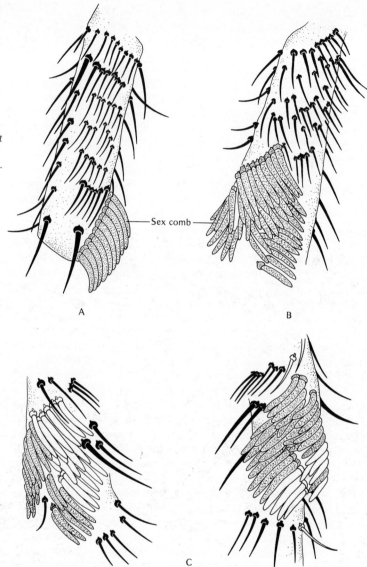

the organism. The mosaic studies tell us that a cell's response is controlled by its genome at a particular time in the ontogeny of these physiological states. In response to the same signal, some cells synthesize hairs, others scales, and still others bristles, while a large proportion of epidermal cells do not respond at all. Cells respond to specific environmental cues which define a cell's position along the axes of symmetry in the organism.

## POLARITY OF TWO-DIMENSIONAL SHEETS

Cells in a two-dimensional epidermis are polarized; they have basal and apical surfaces. Basal surfaces usually abut on a fibrillar basement membrane separating epidermis from connective tissue below. Apical surfaces are exposed and, in insects, are the sites of cuticle deposition. The cuticle itself has polarity. Cuticular fibers run parallel to the anterior-posterior axis of major body segments, or the principal proximal-distal axis in appendages. Thus the fibrillar organization of the cuticle correlates with the overall polarity of a two-dimensional sheet. Embryonic axes of symmetry (anterior-posterior, dorsal-ventral) may preexist in egg organization or may emerge at fertilization or cleavage (see Chapter 11). Subsequently, overall body polarity is somehow imprinted on the polarity of individual cells in epithelial sheets. For most structures (eye, limb), determination of polarity is a progressive process, with the anterior-posterior axis established first, followed by dorsal-ventral symmetry. These axes are the major prepatterns to which cells respond. An example of axial patterning is seen in the behavior of insect imaginal discs.

## Polarity in insect imaginal discs

The adult body plan of the holometabolous insect is constructed from imaginal discs, collections of undifferentiated cells, each determined for a specific organ such as leg, genitalia, antenna-eye, or wing. Discs are sequestered in the larval body and differentiate at metamorphosis into adult organs with axial symmetries. The factors regulating mediolateral polarity of such discs have been studied.

Mediolateral polarity is an intrinsic property of cells making up the leg disc (Figure 16-5). If a normal leg disc of a third-instar larva is bisected into medial and lateral halves and transplanted to the abdomen of a third-instar larva just before metamorphosis, each fragment develops into a medial or lateral half-leg. Though the entire disc is determined as a leg, the disc is a mosaic of regionally specified clones, each committed to develop into restricted portions of the leg. Cells of the leg disc have acquired a positional property, distinguishing lateral from medial. The intrinsic polarity of these cells is not irreversibly fixed. Cells can acquire new polarities under different conditions. For example, the mutant gene *reduplicated (rdp)* causes reduplication of the leg disc when homozygous. Two legs develop, or parts of two legs appear that are mirror images of each other, with each duplicate including both medial and lateral limb parts. In such mirror-image duplications, presumptive medial structures also develop into lateral structures. In the mutant, a genetic change disturbs the rules governing pattern formation, perhaps by affecting cell proliferation.

Positional properties of normal disc cells are

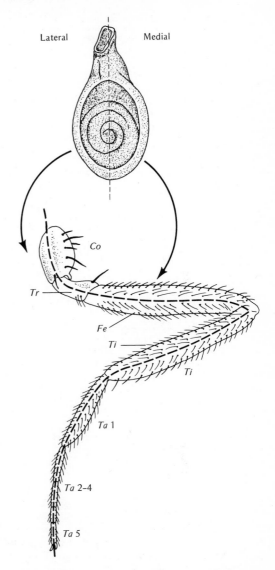

**Figure 16-5** *Mosaic development of leg imaginal disc in Drosophila. Above is a diagram of a leg imaginal disc dissected from a larva. The lateral medial axis is shown. Below is a leg showing lateral medial axis (dotted line through middle of leg). When medial and lateral halves of the disc are isolated and placed in the abdomen of a third-instar larva after metamorphosis they differentiate into medial or lateral half structures as shown. Each half disc displays its characteristic bristle pattern along with the typical structures found in each. At this stage the disc behaves as a mosaic. Co = coxa; Tr = trochanter; Fe = femur; Ti = tibia; and Ta 1–5 = first to fifth tarsal segments.*

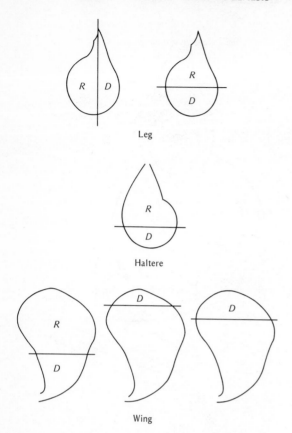

**Figure 16-6** *Developmental fate of fragments of imaginal discs cultured for a long term in adult abdomens before metamorphosing in third-instar larvae. R = regeneration; D = duplication.*

labile after they are reprogrammed by long-term proliferation. Since adults lack molting hormone, fragments of disc cells may be grown in adult abdomens for much longer periods than in third-instar larvae which undergo metamorphosis. Imaginal tissues may be grown indefinitely by serial transfer in adult abdomens, then placed in third-instar larvae to induce metamorphosis to see what pattern differentiates. If half-fragments of leg discs are cultured for many generations in adult abdomens and then transplanted to a third-instar abdomen to metamorphose, they exhibit greater morphogenetic potential than before. Now the medial half regenerates a completely intact leg with normal symmetries (it behaves like medial tissue in a reduplicated mutant) while the lateral half reduplicates a mirror

image of itself (see Figure 16-6). After many divisions medial cell progeny acquire new polarities and can replace missing lateral parts. Lateral cells, on the other hand, produce daughters that retain their original polarities; consequently, they reduplicate lateral parts with mirror-image symmetry. Similar dual potentialities are seen in other experimental situations in which disc fragments are maintained for long periods of proliferation before inducing metamorphosis in third-instar larvae. For example, upper halves of the leg disc regenerate while lower halves reduplicate. Three-quarter fragments of halter or male genital discs always regenerate, whereas one-quarter fragments reduplicate. Wing discs are somewhat anomalous; apical halves regenerate but small apical fragments duplicate instead. Discs may contain two distinct populations of cells: medial cells which possess high regulative capacities and regenerate, and more stable lateral cells which seem to retain their polarities and reduplicate lateral structures. Total tissue mass also seems important in some cases in that large fragments tend to regenerate whole structures while small fragments duplicate.

A gradient model has been proposed to account for differences in polarity within the disc. Each disc expresses a gradient of developmental capacity; the position of each cell in the gradient determines its capacity to develop into certain structures (Figure 16-7). Cells at the high point of the gradient possess the greatest ability to recover lost parts and only replace parts lower in the gradient. In the leg disc, the high point of the gradient is the apical, medial portion while lateral regions occupy lower gradient levels. If the disc is bisected into medial and lateral halves, the medial half, at the high point, can regenerate all lower parts of the gradient and replace all missing parts. The ABC region simply replaces regions D and E. The lateral half, only able to generate parts lower in the gradient, develops a mirror-image duplication. Levels CDE reconstitute regions D and E in the mirror-image plane. We see that regeneration and reduplication are identical phenomena, differing only in the topographical relationships of the cut surfaces with respect to an all-over gradient system. When new daughter cells arise, they are polarized according to their position in the remaining gradient system.

The gradient model is not different from the prepattern model mentioned earlier, the prepattern being a gradient (chemical, metabolic, or proliferative) to which cells respond. Presumably the gradient system supplies the same positional cues to all tissues, and they respond by expressing graded developmental potential. For example, in a mutant "antennapedia," the antennae of homozygous individuals develop into legs. The polarity and positional values of leg structures correspond to equivalent positions in the antenna. Distal leg structures always emerge where distal antennal structures would appear. Genetically different cells exhibit autonomous responses to identical signals in a prepattern.

According to the gradient model, the mutant gene *rdp* functions by (1) inducing cell death, thereby removing part of the gradient in the imaginal disc while remaining tissue regenerates mirror-image duplications to replace dead cells, or (2) altering the factors responsible for the gradient. Though we cannot distinguish between these alternatives, we do know that certain mutations which cause cell death, such as *eyeless*, do often result in duplica-

tions. The eye tissue in the eye-antennal discs of such mutants degenerates, and this is usually accompanied by reduplication of neighboring antennal parts.

Gradients of developmental capacity have been invoked to explain pattern formation in many biological systems. Alternatively, asymmetries in development might reflect differences in mitotic potential within the disc; for example, cells in medial halves may replicate at greater rates than those in lateral halves. We find that medial halves of a leg disc, grown for long periods in adult abdomens, do transform into other disc structures, such as wing or thorax, while lateral fragments never show any signs of change. This type of transformation depends on cell division activity (Chapter 17) and suggests that differential rates or patterns of cell division may account for differences in regenerative versus reduplicative growth. Cells may retain their regional specificities over the first few divisions but reprogram specificities after many more divisions. Tissues with only limited mitotic potential (lateral halves) would be incapable of regeneration or transformation. Though this does not answer the question as to how growth gradients are established within a developing disc, it does show that asymmetries in cell division potential can generate pattern.

### Diffusion gradients as a prepattern

In only a few cases have morphogenetic gradients been related to biological or chemical properties of cells. Except for *Hydra* where regeneration potential and polarity correlate with the distribution of nerve and interstitial cells, most other gradient systems remain theoretical models of patterning behavior of cells. Some gradient systems are based on diffusion of chemical substances from a more metabolically active "source" at one tissue boundary to a lower point, or "sink," at the opposite boundary. Presumably cells arrayed in such concentration gradients can determine their relative position within the boundary and respond phenotypically. A system that seems to develop its pattern according to such a model is found in the insect cuticle of the blood-sucking bug *Rhodnius*.

As we have seen, the insect cuticle is a lattice of fibrillar elements organized into layers. Like cellulose wall synthesis in plants, cuticle precursors organize into parallel fibers on the epidermal sur-

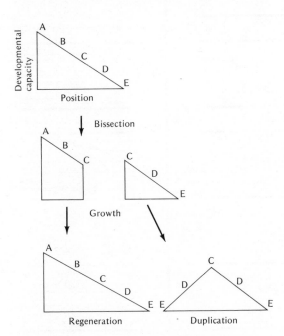

**Figure 16-7** *Positional gradient model to explain developmental capacity of disc fragments. Portions at high points in the gradient regenerate lower points. Fragments at lower points can only duplicate since they can only restore low points in the gradient. (From P. J. Bryant, Develop. Biol. 26:637–651, 1971.)*

face. The outermost cuticle layer, or cuticulin, forms surface patterns in the abdomen of larval and adult insects. These result from differential rates of cuticle secretion by epidermal cells in different regions.

In each segment of the larval abdomen of *Rhodnius*, cuticulin is thrown into a stellate pattern (Figure 16-8), but in the last molt (fifth) it is transformed into adult cuticle with transverse ripples. The

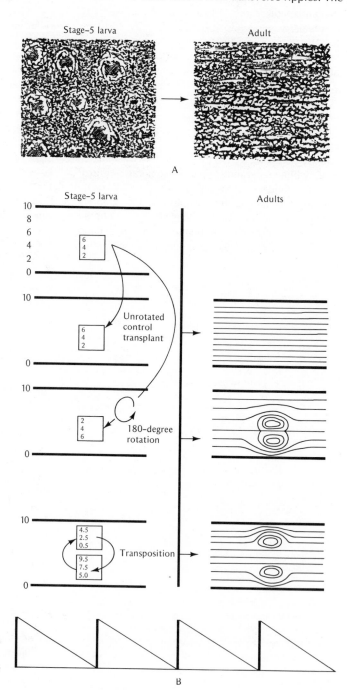

**Figure 16-8** *Cuticle regeneration in* Rhodnius. *A gradient model explains formation of adult cuticle patterns. (A) At the final molt, the stellate larval cuticle pattern is replaced by transverse ripples of the adult pattern. (B) A series of experiments showing the operation of the gradient. Numbers on left indicate height of the segmental gradient, anterior at the top. Lateral displacement of a graft or an intersegmental graft placed at the same level in the gradient produces a normal ripple pattern. Rotated grafts or grafts transposed to different gradient levels result in distorted contour patterns. The gradient pattern in a series of adjacent segments is shown below. Dark lines are intersegmental membranes. (From P. Lawrence,* Adv. Insect Physiol. *7:197–266, 1970; based on M. Locke,* J. Exp. Biol. *36:459–477, 1959.)*

ripple pattern is interrupted by an intersegmental membrane, a thickened structure that borders each segment. The rippled surface results from differential rates of cuticulin synthesis and deposition.

The origin of ripple patterns in *Rhodnius* has been analyzed by grafting pieces of last-stage larval epidermis in different orientations within an abdominal segment before the final molt. We ask, what new adult ripple pattern is synthesized by displaced epidermal cells in the larval-adult molt?

If a graft is displaced laterally within a segment without changing its relative anterior-posterior position, it regenerates a normal ripple pattern, as though it has not been moved at all. If, however, the graft is shifted along the anterior-posterior axis, the resulting ripple patterns are distorted at the boundaries between graft and host. Ripple patterns in the graft connect with the host via a series of curved instead of transverse lines. Likewise, if a graft is retained in position but rotated 90 or 180 degrees, the ripple pattern is distorted into concentric contour lines.

Locke proposed that each abdominal segment contains an identical diffusible gradient system with the same high point, low point, and slope, decreasing linearly along an anterior-posterior axis. Epidermal cells at the same position in each segment share a common level of the gradient and exhibit identical rhythms of cuticle secretion. Anterior cells occupy higher points than posterior cells, which means ripples at different levels in the gradient may be assigned a positional value corresponding to their place along the axis. Epidermal cells respond to the gradient with a rate of cuticle synthesis that corresponds to the level occupied. This model can explain all graft cuticular patterns.

For example, a laterally displaced graft regenerates a normal transverse pattern because its position is undisturbed within the segmental gradient. Likewise, grafts between adjacent segments produce normal ripple patterns only if the graft is placed in its corresponding position in the gradient. On the other hand, a rotated graft or a graft displaced along the anterior-posterior axis develops a distorted or contour pattern because a stable gradient in graft tissues interacts at its edges with different levels of the host gradient. Graft and host cells physiologically reunite in cuticle synthesis only at corresponding levels in each gradient. Cells at gradient level "six" in the graft join with "six-level" cells in the host gradient and synthesize a common ripple pattern.

To explain physiological integration of cells at the same gradient level, Locke suggested that epidermal cells are joined to one another by septate desmosomes at lateral edges but not at anterior or posterior edges. This means that chemical regulators can diffuse laterally between cells but not anteriorly or posteriorly. Lateral continuity means that all cells occupying a specific level of the gradient act coordinately.

To establish the gradient, a "source and sink" model has been used, the "source" at the high point and a "sink" at the low point with a linear decline between the two. An active factor from the source diffuses in a polarized manner and is either lost or used up at the sink. Lawrence tested this simple diffusion model with a rotated graft in an abdominal segment. He reasoned that since simple diffusion is time-dependent, a longer period between making the graft and formation of adult cuticle should result in a gradual rather than an abrupt gradient, and the contour line pattern should be smooth between graft and host tissues. He lengthened the larval to adult molt by starving the animals and compared the ripple pattern with a control pattern in unstarved animals but found no differences. The gradient was not more relaxed as a result of extending the time of cuticle formation. These results suggested that simple diffusion alone could not account for the observed ripple patterns.

He then applied a homeostatic gradient model proposed by Crick, where each cell may behave as either a source or a sink, depending on its position and the concentration of diffusible substances. A cell becomes a source and begins to synthesize the diffusible substance if its concentration is reduced; it becomes a sink and destroys the substance if its concentration increases. Cells, functioning as homeostatic systems, try to maintain original preset concentrations of the substance. Lawrence found that the experimentally produced contour patterns were much closer fits to this homeostatic model than to other models tested. Some believe that not only does this model account for results of transplantation experiments in the *Rhodnius* cuticle; it may also be applicable to all situations in which cells acquire positional information.

Diffusible chemical gradients may also initiate other types of graded behavior by cells. In response to graded ionic signals or such regulatory molecules as cyclic AMP, cells may exhibit differential rates of division. In this manner, a chemical gradient acting as a prepattern is translated into a cell division gradient, which emerges as patterned growth of a tissue.

## CELL LINEAGE AND PATTERNING IN TWO-DIMENSIONAL SHEETS

Patterns based on cell division asymmetries are found in plants and insects. The rigid cell walls in plants fix the position of every cell after division. The complete division history or cell lineage is revealed as a longitudinal file of cells all derived from a common point. Such cell lineage patterns play an equally important role in patterning of animal systems, and again the insect integument furnishes the best examples.

The location of stem cell populations, the orientation of division planes, and the number of cell divisions define the spatial distribution of bristles, hairs, and scales in the insect integument. As imaginal disc epidermis differentiates into adult epidermis, it leaves a record of its lineage in the cuticle. The position of each cell in the sheet is fixed by its last division before it secretes cuticle; no tissue migration disturbs the spatial arrangement of cells.

### Number of cell divisions determine scale patterns

The development of wing scales in Lepidoptera suggests that the number of cell replications of a clone is responsible for a pattern; each scale type follows a specific division program. The wings are composed of two chitinous lamellae and their underlying epidermal layers. The scales, developing from wing epidermis, contribute to the diverse patterns of color and texture typical of lepidopteran wings (Figure 16-9).

Three types of scales protect the wings of the moth *Ephestia:* the uppermost *cover* scales, *middle* scales directly below, and a layer of *ground* scales underneath all. Where all three types are present in the wing, scales overlap like shingles, in recurrent identical groups. Cover scales are largest, ground scales are smallest. There is a correlation between scale size and chromosome ploidy in the nucleus. Cover scales are 32N, middle scale ploidy is 16N, ground scales are either 4N or 8N, and all surrounding epidermal cells are diploid (2N). Is there a causal relationship between chromosome replication and scale development?

The overall scale pattern depends on the division history of each scale cell in a wing disc. A scale develops just like a bristle. The progenitor of each scale is an undifferentiated epithelial scale cell (Figure 16-10) which divides unequally into a scale and an epithelial stem cell. The scale progenitor divides again into a large superficial cell above and a smaller cell below, which degenerates. The superficial cell then divides into a small collar cell in the epidermis and a larger, lower scale cell. The collar cell differentiates into a doughnut-shaped structure through which the expanding scale grows out.

As a scale cell differentiates, its nucleus replicates endomitotically (chromosome replication without cell division). The large cover scale cell com-

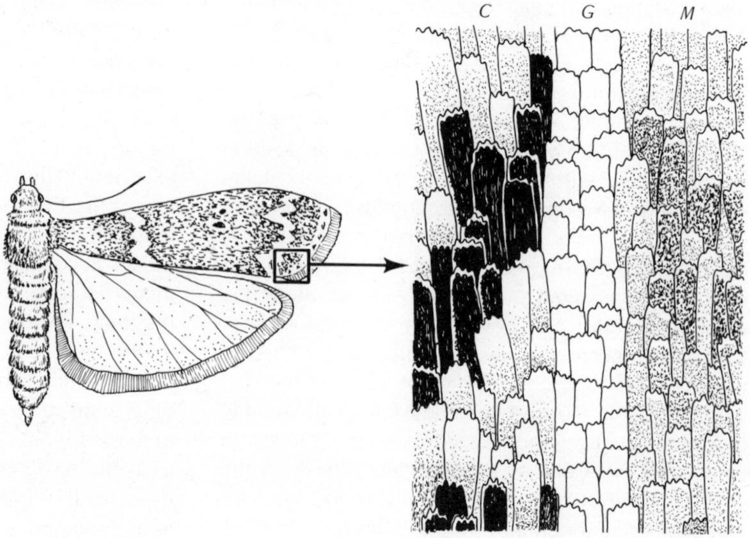

Figure 16-9  *Scale patterns in the wing of the moth* Ephestia kuhniella. *Adult showing wing color pattern due to distribution of overlapping scales. Part of forewing upper surface enlarged. C = cover scales; M = middle scales after cover scales have been removed; G = ground scales after removal of cover and middle scales.*

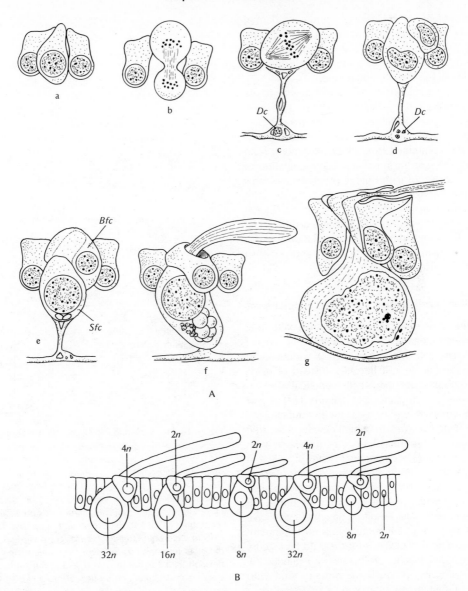

**Figure 16-10** *Development of wing scales in* Ephestia *and resulting ploidy patterns in various scale types. (A) Scale formation in the epidermis. (a) Primary scale stem cell. (b) First differential mitosis. (c) Second mitosis of large superficial cell; smaller cell below degenerates (Dc). (d–g) Scale-forming cell (Sfc) and socket-forming cell (Bfc). (B) Epidermis with polyploid nuclear classes. Cover scales = 32n; middle scales = 16n; ground scales = 8n; scale sockets = 2n and 4n; and epidermal cells = 2n. (From A. Kühn,* Lectures on Developmental Physiology. *New York: Springer-Verlag, 1971; based on M. Stossberg, Z. Morph. und Okol. Tiere 34:113, 1938.)*

pletes four endomitotic replications and becomes 32N, a middle scale 3(16N), and a ground scale only 1(4N). Meanwhile, the neighboring epidermal stem cell divides several times, the number of its divisions matching endomitotic replications in accompanying scale cells. Epithelial stem cells adjacent to ground scales complete four divisions, middle epidermal cells complete two divisions, and in cover scale regions, each cell usually divides once. For each scale cell lineage, the number of epidermal cell divisions plus scale cell replications equals five.

Mitosis of epidermal stem cells seems to be linked to endomitotic replications in scale stem cells. Each daughter cell acquires a different division program, one mitotic and epidermal, the other endomitotic and committed to scale development. Both stem cells may be in junctional communication to explain coupling of their mitotic programs. It seems that the number of divisions in a clone correlates with the pattern of scales in the moth wing. At several points in development of this lineage, a differential division defines two different cell populations, whose division programs affect cell position and establish an all-over scale pattern.

### Orientation of cell division determines wing patterns

Orientation of division planes also controls patterning in the insect integument. To study the ontogeny of such patterns, cells must be marked in some way to produce identifiable cell lineages. Instead of using gynandromorphs, we mark cells genetically by inducing somatic crossing over (Figure 16-11). For example, embryos heterozygous for the mutant gene *multiple wing hairs (mwh),* which develop into wild type adults with normal hairs, are irradiated with X-rays at different developmental stages to induce a low frequency of random somatic crossing over. In cells so affected, homologous chromosomes pair as in meiosis and chromatids break and rejoin, often exchanging segments. Resulting daughter cells have different genotypes, one homozygous for the mutant gene, the other homozygous for the wild type gene. Each cell continues to proliferate into genetically identical progeny, producing a mosaic adult with patches of mutant multiple wing hair tissue on a background of wild type.

The size of mutant patches depends on the developmental stage at which somatic crossing over occurs; if it occurs early in development, large mutant patches appear because more cells have had an opportunity to develop from the progenitor cell undergoing somatic crossover. This precise method of genetic marking makes it possible to trace the lineage of a cell during morphogenesis.

A cell lineage model for pattern generation in

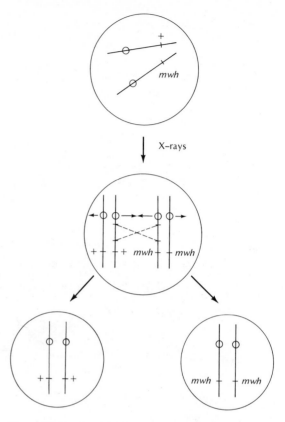

**Figure 16-11**   *Somatic crossing over in* Drosophila *induced by X-rays. Cell in heterozygote for multiple wing hair (mwh) shows wild type and* mwh *genes in homologous chromosomes. X-irradiation at different larval stages randomly induces some cells in wing imaginal discs to undergo somatic crossing over. Homologous chromosomes pair at mitosis; chromatids break and recombine as shown, then separate to opposite daughter cells to yield one daughter homozygous wild type and the other homozygous for mwh. A mosaic wing will develop with patches of wild type and mutant tissue since each daughter cell replicates into a clone expressing the mutant or wild type phenotype.*

the wing predicts that mosaic patches of mutant tissue should correlate with the major axes of growth. On the other hand, if cell division is unrelated to integumentary patterns, then mutant patches should be distributed at random with respect to the wing's normal growth pattern.

The results show that the spatial distribution of

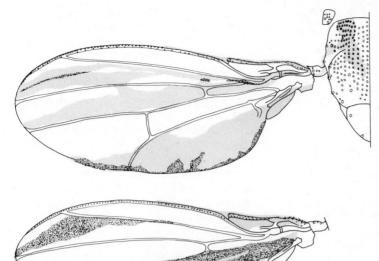

**Figure 16-12**   *Outline of mutant (shaded) and wild type clones on wing and thorax. Dorsal surface of wing in upper diagram, ventral below. Shaded areas include results from yellow and multiple-wing-hair phenotypes. X-irradiation at 2 hrs following oviposition. Some mutant tissue overlaps dorsal and ventral surfaces and is shown at the margins. (From P. Bryant, Develop. Biol. 22:390, 1970.)*

mutant clone tissue in mosaic animals is not randomly arranged but appears as longitudinal stripes in the wing parallel to the principal axis of growth (Figure 16-12). This implies that mitoses are oriented parallel to the principal longitudinal axis of growth. Clones are nonrandom in shape; they are many cells long and only a few cells wide. The progenitors in the disc are radially aligned to produce these stripe patterns of oriented growth.

Since size of mutant patches is directly related to the time of irradiation, it was estimated that the wing disc is determined at 2 hours of embryonic development, when it consists of only 11 cells in the cellulated blastoderm. Simultaneously, a predetermined division program is established in the wing disc which accounts for the final pattern of the wing. Morphogenesis of the insect integument resembles plant morphogenesis in its dependence on cell division rather than on tissue migration. In both, the position of every cell is fixed and its division history is recorded. Plant cells secrete a rigid cellulose wall, while the insect epidermis synthesizes an extracellular cuticle that marks the position of all cells below. In fact, variegated stripe patterns in leaves resemble the artificially produced somatic mosaics in Drosophila.

## MIGRATION PATTERNS

Tissue migration is probably more important in determining tissue patterns in systems other than insects. Simple systems such as the slime mold *Dictyostelium* suggest how directed movements of large numbers of aggregating cells can generate tissue patterns.

Patterns arise during aggregation of slime mold amoebae. In response to graded chemotactic signals, streams of aggregating cells converge toward "centers" and form pulsating concentric ring patterns (Figure 16-13).

A rhythmic production of chemotactic stimuli creates a pattern-forming system. Chemical stimuli produced at a center in periodic pulses are conducted as waves toward the periphery. As a result, a large area, or aggregation field, is controlled by a single center. Wavelike patterns, or surges of motile behavior, spread from the center toward the periphery.

Chemotactic activity is rhythmic during aggregation because orientation and speed of movement of responding cells fluctuate periodically. Using mutants, it was demonstrated that two chemotactic factors are responsible; one, secreted periodically, orients cell movement, while the other, secreted

**Figure 16-13**  *Concentric ring migration patterns in dense cell layers of* Dictyostelium discoideum. *(Based on photo in G. Gerisch,* Current Topics in Developmental Biology, *ed. A. A. Moscona and A. Monroy, vol. 3. New York: Academic Press, 1968, pp. 157–197.)*

continuously, controls migration rate. A combination of these two seems to determine the wave patterns.

The chemotactic factor controlling motility is probably cyclic AMP, and an oscillating release and enzymatic destruction of this substance within centers and in migrating cells generates the pattern. The pattern is sustained by a relay system; the signal generated at the center is conducted without decrement across a cell layer in which all members also secrete cyclic AMP. Cells respond by rhythmic movements toward their neighbors, in the direction of the center, and a concentric ring pattern appears. Regulation of this pattern is complicated inasmuch as inhibitors may also be involved. For example, aggregation centers are equally spaced in a disc. In fact, the boundary lines seen in some patterns are points where waves from two adjacent centers meet. Spacing may be

controlled by a diffusible inhibitor acting over an aggregation field, preventing additional center formation within a certain radius. Though no inhibitor has been found in *Dictyostelium*, the cyclic AMP oscillation-conduction system may, in itself, space a center. As centers are randomly established within a dense population, oscillating wave patterns are generated from each. Competition between waves emanating from these different foci ultimately leads to the extinction of most centers, leaving only a few widely spaced active centers, each with concentric ring patterns. The conduction-oscillation system demands that each cell experience a refractory period of unresponsiveness to additional stimuli, a property all slime mold cells display.

## COAT COLOR PATTERNS IN MAMMALS

Certain patterns, particularly the color patterns in the integuments of birds and mammals, arise because of independent cell migrations of the neural crest (see Chapter 12). These cells emerge from the neural tube and wander in segmentally arranged streams to the dorsal and ventral body surfaces where they differentiate into melanoblasts that invade hair follicles or feathers. Patterns arise because of interactions between the migrating neural crest and surrounding tissues.

Coat color patterns in mammals are controlled by many different genes acting at various stages of neural crest development. Some of these genes control the time of origin of neural crest from the neural tube. Others may affect the rate of proliferation as they migrate in streams toward the skin. This would affect the number of neural crest cells that finally invade skin tissues. Still other genes may alter cell migration rate or pattern as they make their way through the tissues. Once cells invade a tissue, other genes regulate the rate and rhythm of pigment synthesis, the type of pigment produced, and its distribution within the pigment granules. Any one or a combination of these many factors will affect the final color pattern.

Some of these genes act autonomously within the neural crest cell itself. The albino mutant gene, for example, results in the loss of the enzyme tyrosinase, which is essential to melanin synthesis. Such nonpigmented melanocytes invade hair follicles and produce white hairs all over the body. These genes contribute little to the development of pattern.

Patterning genes, those responsible for spotting or stripes, probably do not act autonomously within the melanocyte; they act as prepatterning genes within the tissue environments through which neural crest cells migrate, affecting direction and rate of migration, proliferation, and survival. Such genes, by altering the tissue environment invaded by neural crest, generate the exogenous cues to initiate pigment synthesis. Patterns are established when these genes act in some regions and not in others. Melanocytes express a pigmented phenotype in one area but not in the other, and a stripe or a spot develops. A description of some coat color patterns will show how some genes control the development of color patterns by a migrating cell population.

There are 13 different gene loci associated with white spotting in the mouse, one of which, the *piebald* gene, causes irregular white spotting in ventral regions and on extremities of the body. No melanocytes are found in white areas, suggesting that a limited number of neural crest cells have been produced, and these populate the dorsal regions (the first region they enter), leaving none to migrate into ventral regions and extremities, sites furthest from the neural tube. The wild type allele of the *piebald* gene probably controls neural crest cell proliferation. It is also possible that enough melanocytes are produced, but ventral skin is less able than dorsal skin to support melanocyte survival or pigment synthesis, or both.

Developmental changes in a region may prevent migration of late arriving cells. In the hooded Norway rat, migration of melanoblasts to the epidermis is delayed 1 to 2 days. Though head and shoulder regions receive melanoblasts a day late, enough cells arrive to populate hair follicles and produce pigment since neural crest arises first in anterior regions. In the remaining skin of the body the density of connective tissue increases as skin differentiates. Late appearing melanoblasts cannot penetrate the skin in these areas and populate hair follicles, leaving trunk and limbs white. Though the gene affects the timing of neural crest emergence, temporal changes in the tissue surround make it unfavorable for migration, and cells cannot reach their destination. Delayed migration may be an intrinsic defect of melanoblast motility or may result from deficiencies in the substratum on which they migrate.

Some genes affecting pigment synthesis act in the surrounding skin or hair follicles invaded by melanocytes (Figure 16-14). In these cases, gene products synthesized by hair follicle cells are utilized by melanocytes for pigment synthesis; pattern in this case is a function of the tissue background. The best example is the wild type agouti locus in the mouse, AA.

In this mouse, the dorsal side is darker than the gray ventral side. An agouti dorsal hair is black at the tip with an intermediate band of yellow followed by more black in the remainder of the hair shaft. The ventral hair is lighter, with a black tip, a larger band of yellow, and a smaller dark region. Two types of melanocytes occupy the agouti hair follicle, those secreting yellow pigment and those secreting black. Although each agouti melanocyte contains the wild type allele, it synthesizes only one pigment at a time, black or yellow, depending on local conditions. The agouti hair pattern depends, therefore, on the timing of secretion of each type of pigment granule. Black pigment is secreted first, followed by yellow, ending with black, reflecting the changing state of the hair follicle. Because of physiological differences in ventral epidermis, timing of pigment synthesis varies. Yellow pigment is secreted for a longer time as ventral hairs grow out; these hairs have more yellow pigment and are lighter.

Several mutant alleles at the agouti locus affect this pattern. For example, the nonagouti gene inhibits the agouti pattern; yellow pigment is not synthesized in dorsal or ventral hairs; a nonagouti black genotype gives a uniform pattern of dark color in all hairs. Inability to synthesize yellow pigment could be due to (1) a genetic defect in melanocyte biosynthetic pathways or (2) deficiencies in surrounding hair follicles; that is, conditions in the follicle are not favorable for yellow pigment synthesis.

Skin-grafting experiments distinguish between these alternatives. For example, one of the agouti gene alleles, a dominant yellow mutant, $A^Y$, produces a uniform yellow coat. Yellow fetal skin can be transplanted onto a black, nonagouti genotype (another agouti allele) well before follicles are invaded by melanocytes. Recall that the black nonagouti cells do not synthesize yellow pigment. Nevertheless, black melanocytes that invade the yellow skin follicles at the graft border synthesize yellow pigment rather than black. Synthesis of yellow pigment depends upon factors produced in the follicle

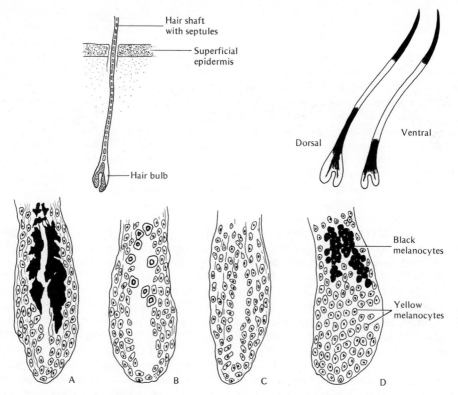

**Figure 16-14**  *Action of some coat-color genes in mammals. Hair-bulb regions of different genotypes showing different melanocytes. (A) In wild type, melanocytes are so numerous they fill matrix of the bulb with black pigment. (B) A hair follicle has been invaded by melanocytes in an albino. The melanocytes are present but cannot synthesize any pigment. (C) The follicle contains no melanocytes, a condition that is found in many spotting mutants such as piebald. Melanocytes fail to invade follicles because of deficiencies in proliferation, migration, or even survival. (D) Agouti hair follicle. Dorsal and ventral hairs compared. The agouti follicle contains yellow and black pigment synthesizing melanocytes that are active at different times during hair growth.*

cells and not in the melanocyte, and these are absent in nonagouti hair follicles.

The nature of such factors is not known. Tissue culture studies suggest that certain substances favor one or the other pathway. Yellow melanocytes in vitro change their color to black but revert to yellow pigment synthesis if sulfhydryl compounds are added to the medium. Sulfhydryl compounds such as gluta-thione may accumulate in hair follicles, inhibit eumelanin synthesis, and stimulate yellow pigment synthesis instead.

Do similar interactions between melanocytes and environment explain the more complicated stripe patterns of the tabby cat or the zebra? Is there a

striped prepattern within the tissues to which geno-typically identical melanocytes respond? A genetic mosaic procedure, similar to that used in analyzing *Drosophila* patterns, has provided some clues. *Allophenic* mice are genetic mosaics, a mixture of cells of two different genotypes. These are prepared experi-mentally by fusing, in vitro, early cleavage embryos (morulae) derived from genotypically different strains and reimplanting the resulting giant blastocyst into a foster mother (Figure 16-15). Approximately 30 percent of such reimplanted blastocysts develop into normal offspring. Since the pups are of normal size, regulation must have occurred early in development, probably by elimination of cells from the early embryo. Cell loss from mosaics is not random; cells of some genotypes seem to be preferentially eliminated in mixtures.

Allophenic mice obtained after fusion of a colored embryo with an albino exhibit striped coat color patterns. About one-third of the offspring are colored, one-third albino, and one-third allophenic, showing both colors. Single-colored mice are explained by size regulation in early development (cell loss), after the giant blastocyst is implanted in the foster mother. An unknown selection mechanism preferentially eliminates melanoblast progenitors of one or another genotype, early in development. Since other tissues in these mice are mosaics (for example, blood), we conclude that complete elimination of cells of one genotype does not occur. Cell loss is selective in only certain tissues.

The two-colored allophenic mice showed a full spectrum of color patterns, the most distinctive patterns being those with alternating light and dark transverse stripes on either side of the dorsal midline. All seem to be derivatives of one standard pattern which consists of 3 bands in the head, 6 in the body, and 8 in the tail, a total of 34 bands in all, 17 on each side. Bands on either side of the midline need not be continuous in color, suggesting that right and left patterns develop autonomously. To explain the regular banding patterns in these mosaics, Mintz suggested that each band is a clone of cells descended from a single melanoblast precursor of a specific genotype. According to this hypothesis, there are initially 34 primordial neural crest cells, 17 on each side, each giving rise to a clone. Stripes develop as a result of dorsal to midventral migration and proliferation of these neural crest clones. The hypothesis, however, raises some fundamental questions about melanoblast origin and behavior in allophenic mice.

What accounts for alternate stripes of clonal outgrowth? Do two genotypes of clonal-initiator cells take up alternate positions along the longitudinal axis of the embryo after they are determined as neural crest cells, or do progenitors migrate out at random, to be selected by a preexisting pattern in the substratum? The hypothesis favors the second of these alternatives; that is, cells of different genotype are randomly arranged, but after they acquire phenotypic differences (for example, surface specificities or recognition sites), they are preferentially selected according to a preexisting pattern in surrounding tissues. Some tissue environments favor the albino

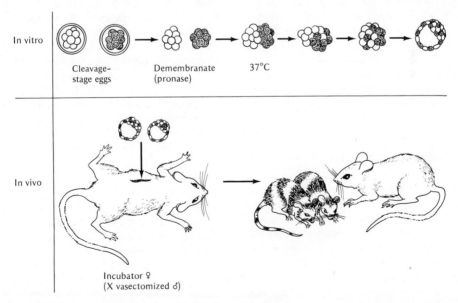

**Figure 16-15** *Production of allophenic mice and the progeny that develop when embryos of black and albino genotypes are fused in vitro. Embryos isolated in vitro are fused at the morula stage and reimplanted into the uterus of a foster mother (incubator female) until pups are born. Thirty percent of the offspring are mosaic with a striped coat. (Based on B. Mintz, Proc. Nat. Acad. Sci. U.S. 58:334–351, 1962.)*

phenotype while others select the colored phenotype. The fact that stripes are not sharp but blurred at the edges suggests that selection occurs at different times during the proliferation and migration phases. Since the cells are not synthesizing pigment when they migrate out, the cell trait selected cannot be pigment synthesis but must be some other property such as cell migration, adhesion, or proliferation. To explain the banded patterns, the trait that is selected must be linked to color genes.

Cell viability may be another melanoblast trait that is selected during clonal outgrowth. We know that cell death is a normal concomitant of embryogenesis (see below). Cells are genetically programmed to die in definite locations at specific times. Mintz was able to show that white areas in certain single-genotype allophenic mice result from death of melanoblasts colonizing these regions. Melanoblasts differentiate in the region but soon die, not as a result of an unfavorable skin environment but because of some intrinsic cellular change. The cells are programmed either to die or perhaps to divide less frequently, thereby depleting certain regions of melanocytes of a specific phenotype.

Viability of a phenotype may also depend on the environment. Genotypically mixed populations of melanoblast cells may make up segmentally arranged migrating streams in the normal fashion but the surrounding tissue environment may favor survival of one genotype which colonizes hair follicles. The tissue-surround may be prepatterned into regions differing in ability to maintain specific melanoblast phenotypes and thereby produces alternating band patterns.

## PROGRAMMED CELL DEATH

A prepattern accounts for body patterns by controlling such cell properties as mitotic activity, orientation of division planes, migration, and even specific protein synthesis. Cells respond to singularities in a prepattern (or to gradients) according to their genotype, and a pattern results. A prepattern may also define conditions in the extracellular environment (matrix materials?) that are unfavorable for survival of certain cell phenotypes. In this way the spatial distribution of dying cell populations becomes a patterning device.

Conditions favoring cell death need not be lo-calized within the prepattern. Cells may be genetically programmed to die at a certain time, as if they carry a death clock, preset for self-destruction.

There are many instances of cell death occurring normally in developing systems. The nervous system, the eye, and even the developing digestive system contain many dead and dying cells, characterized by pycnotic nuclei. In the nervous system, many more neurons are produced than are needed and those that fail to become included in functional networks eventually die. The destruction of many larval tissues during metamorphosis (for example, the tadpole tail) illustrates how tissue destruction can be an important adult patterning mechanism. Besides such gross tissue losses, there are more subtle programs where death of local cell populations is exploited for fine sculpturing of an organ. Carving out the form of the vertebrate limb is an example of this process.

The ontogeny of the vertebrate limb illustrates how many factors interact to achieve its final shape. Bone, muscle, and cartilage are arranged along a proximal-distal axis from shoulder to digits with dorsal-ventral as well as anterior-posterior axes of symmetry. As we have seen in Chapter 12, the form of the limb depends upon a sequence of morphogenetic interactions between two tissues, the internal mesodermal core and the overlying ectodermal jacket. As the limb bud grows out, certain areas of collagen condensation mark the sites of the future bones. In addition, form of the limb results, in part, from localized death of cell populations in the regions between the digits as well as in portions of the forelimb.

A comparison of morphogenesis of wing and leg in the chick and the duck illustrates how patterns of dying cells contribute to the ultimate shape of the limb (Figure 16-16). The pattern of necrosis is stage-specific and characteristic of each appendage. At stage 24 in the chick, a zone of necrosis appears at the posterior margin of the wing bud while at the same time a larger zone of necrotic cells appears along the preaxial margin in the leg primordium. Cell death in these regions contributes to contour modeling of the upper arm and thigh. Note that no necrosis appears in the duck limb until stage 25. In later wing stages, the spatial distribution of necrotic zones is similar in both species, but it is in the modeling of the leg that significant differences arise in carving out the toes. The massive necrosis of interdigital zones in the chick foot separates the individual digits, whereas in the duck, necrosis is limited to peripheral regions between digits 1 and 2 and most epidermis remains as webbing between the toes. Interdigital necrosis is found in mouse, rat, and human embryos, where separate digits are fashioned.

Saunders believes that cells in necrotic zones are genetically programmed early in development to carry out a degenerative change leading to death. His conclusions are based on transplantation experiments in which presumptive necrotic regions are placed in different sites. Between 96 and 106 hours of development, up to 2000 cells die in the posterior necrotic zone of the wing bud and are phagocytized by macrophages. If this region, as early as stage 22 (before any sign of necrosis), is excised and grafted to the dorsal side of the embryo, its cells become necrotic on schedule. They show degenerative changes at exactly the same time, as if they had remained in situ on the developing limb. Control tissues from the dorsal side of the wing bud grafted in identical locations do not degenerate. Saunders concludes that posterior necrotic zone cells express an internal death clock. The setting of the clock must occur at or before stage 17, because cells that overgrow the extirpated region on the wing bud do not themselves undergo necrosis, nor does necrosis occur in any tissues grafted to this region. The expression of the death clock is reversible between stages 17 and 22. The tissues are still quite labile, and transplantation within these stages will permit development without showing any degenerative changes. Between stages 22 and 23, however, an alteration occurs. At this time the death program is fixed and the cells are irreversibly committed to degenerate. No external signals such as hormones promote death; rather, cells seem to respond to an intrinsic death signal. Even if the posterior necrotic zone is explanted in vitro it dies on schedule, degenerating at the same time as comparable regions in the developing donor embryo in situ.

The patterning of cell death involves interactions between limb mesodermal core and the overlying

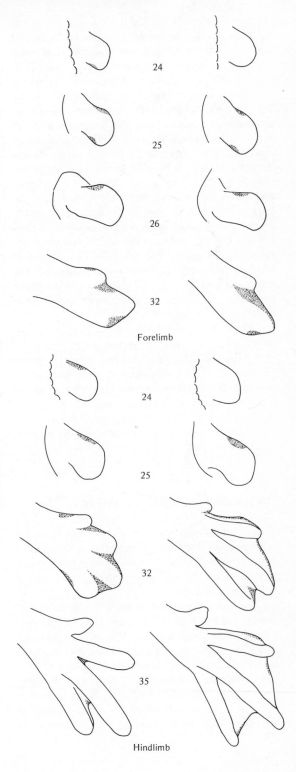

**Figure 16-16** *Patterns of cell death during development of chick (left) and duck (right) limbs. Superficial mesoderm undergoes programmed cell death as detected by vital staining with Nile blue. Shaded areas are necrotic. Numbers indicate Hamburger-Hamilton stages. (From J. W. Saunders, Jr., and J. F. Fallon,* Major Problems in Developmental Biology, *ed. M. Locke. New York: Academic Press, 1966, pp. 289–316.)*

epidermis. Saunders isolated the leg mesoderm from leg buds of chick and duck at stages 18 to 21 and covered these in chimeric combinations with ectodermal hulls stripped with trypsin from the wing buds of chick and duck. A duck leg mesoderm is covered with chick wing ectoderm and compared to chick leg mesoderm covered with duck wing ectoderm after both have grown into limb structures as flank grafts in host chick embryos. Limbs composed of duck mesoderm covered by chick wing epidermis display a pattern of necrosis resembling that of the normal duck leg and produce the typical webbing pattern. In the other chimera, chick mesoderm induces interdigital necrosis in duck wing ectoderm, but the extent of cell death is much less than in control chick limbs. Although the overall pattern of necrotic regions seems to be established by the underlying mesoderm, the outcome also depends upon the responsiveness of epidermal cells of a specific genotype; chick epidermis responds more readily than duck.

These results suggest that the underlying mesoderm acts as an inductive prepattern to which the epidermis responds. The spatial distribution of singularities arises within mesoderm to elicit local death responses in ectoderm, but the properties of mesoderm responsible for this induction are not understood. Later in this chapter we will see that mesoderm inductive actions determine patterns in other systems including the development of scales and feathers.

## ONTOGENY OF LIMB SYMMETRY

The polarity of regenerating limbs has been discussed in Chapter 15. Local stump factors seem to organize the polarity of tissues in a limb blastema. What factors determine the axes of symmetry during normal ontogeny of the limb?

Experiments on salamander limbs have demonstrated that the axes of limb symmetry are specified in sequence (Figure 16-17). The anterior-posterior axis of the limb is specified early, at the end of neurulation before the limb bud appears. Inversion of a limb-forming area at this stage produces a limb transplant facing posteriorly but with proper dorsal-ventral orientation; the dorsal ventral axis is not specified at this stage. Later at the tail-bud stage, inversion of

the limb primordium results in a limb with the plantar surface of the hand facing upwards, indicating that the dorsal-ventral axis has already been specified and fixed. The limb primordium at the tail-bud stage is no longer responsive to dorsal-ventral positional cues from the surrounding environment. Finally, after the limb bud appears, the proximal-distal pattern of the limb is determined. In carrying out these latter operations, the limb bud has to be transplanted inside-out, the apical portion of the limb growing into the body. Though many transplants do not develop, those that do are abnormal. Often the limb field splits in these procedures and duplicated limbs emerge, with normal axial orientation corresponding to the host axes of symmetry. As in the *Drosophila* imaginal discs, the occurrence of mirror image duplications indicates that cells specified according to one set of polarizing axes are still labile. After a period of growth their progeny can respond to new polarizing influences. Axial specification is not a cellular property that is transmitted to progeny; it seems to be a global property of the entire region.

In the chick limb, axial specification involves inductive signals, although their mode of operation is obscure. As early as 2 days before a visible bud is detected, anterior-posterior and dorsal-ventral axes of both fore- and hindlimbs have already been determined. Because of this precocious determination, it is difficult to study axial specification, but in some recent experiments, Saunders was able to control axial specification in supernumerary limbs (Figure 16-18).

Though the posterior necrotic zone is a site of programmed cell death, it seems to have properties totally unrelated to necrosis. If a fragment of mesoderm from this region is grafted to a host limb bud subjacent to apical ectoderm, it will induce a supernumerary limb outgrowth from the host preaxial mesoderm. No other limb tissue has this inductive capability. Because of this property, the region has been called "zone of polarizing activity" (ZPA) by Saunders. The anterior-posterior symmetry of the supernumerary limb seems to depend upon the position occupied by the graft in the host limb bud. If placed under the apical ectoderm of a right limb bud, a supernumerary limb outgrowth with right-hand symmetry emerges, just like the host. In this instance, the posterior border of the wing outgrowth faces the graft site. If the graft is inserted into the anterior preaxial border of the wing bud, a supernumerary limb with left-wing anterior-posterior symmetry is produced, a mirror image of the right wing of the host. In this case, the polarity of the supernumerary limb is unrelated to the orientation of the grafted tissue.

Though ZPA exerts an orienting role on neighboring mesoderm, its removal from a normal limb has no effect on development. We do not know how to interpret this other than to suggest that necrosis of the mesoderm may act as a signal inducing supernumerary limb outgrowth in graft situations. Its orienting influence is difficult to understand unless we assume that mirror-image duplications result when large numbers of cells die. We have here an

instance of a small tissue region that can influence the polarity of an induced structure. Is this region one of the singularities in a prepattern? Could prepatterns consist of differential distribution of inductive influences which direct growth in an overlying responsive tissue? Though there is no answer to these questions, we do know that inductive interactions play an important part in determining the axial properties of the vertebrate nervous system (Chapter 12). The organization of brain and spinal cord depends upon signals coming from the underlying mesoderm, and we will see that the axial polarity of the vertebrate epidermis arises in response to graded properties in an underlying inducing tissue.

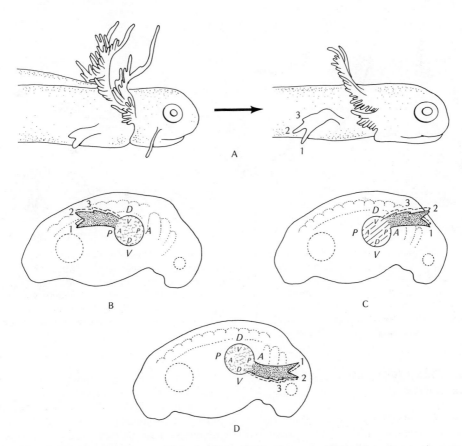

**Figure 16-17** *Axial specification of the amphibian forelimb. (A) Development of limb at late stage to show normal orientation and arrangement of digits. (B) Operation at end of neurulation before limb primordium has been specified; 180-degree rotation of primordium region at an early stage has no effect on limb symmetry. (C) Rotation at a later stage after AP axis is specified but DV axis still labile. (D) Rotation after limb is fully specified. AP and DV axes are inverted.*

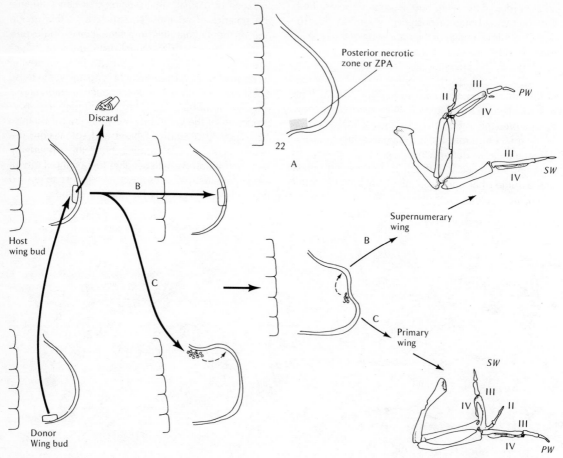

**Figure 16-18**  *Influence of ZPA region of chick embryonic limb on axial properties of induced supernumerary limbs. (A) Stage-22 wing bud showing site of posterior necrotic zone, the source of the inducing tissue (ZPA). (B, C) Implantation of ZPA region beneath apical cap of host wing bud to induce supernumerary outgrowth. In (B), the ZPA is placed directly beneath the apical cap and the supernumerary wing has the same right asymmetry as the primary wing. If in (C) the ZPA is implanted on the preaxial border of the right wing bud, the resulting supernumerary wing has mirror-image left asymmetry. In this case both wings share digit II. SW = supernumerary wing; PW = primary wing. (Based on J. W. Saunders, Jr., Ann. N.Y. Acad. Sci. 193:29–42, 1972.)*

## THE CHORDAMESODERM: A PREPATTERN FOR THE EPIDERMIS?

During development of the amphibian embryo, the outer epidermis becomes ciliated and the embryo rotates slowly within the perivitelline space. Cilia appear during neurulation and their polarized movements can be studied by following the flow of carmine particles placed along the surface (Figure 16-19). In the neurula, cilia beat forward to back, thus creating an anterior-posterior flow.

Polarity of ciliary beat was examined by removing a rectangular piece of vitally stained presumptive epidermis from an early gastrula, rotating it 180 degrees, and allowing it to heal in place in a host embryo. At the neurula stage the direction of ciliary beat of the graft was identical to its surrounding tissues. Direction of ciliary beat is undetermined

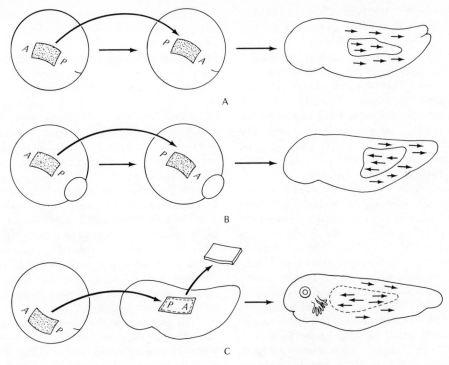

**Figure 16-19** *Determination of the direction of ciliary beat in the amphibian epidermis. (A) Transplant of ectoderm from an early gastrula into a host of the same age. Ectoderm rotated 180 degrees. Resulting ciliary beat is normal, the same as host epidermis. (B) Transplant of rotated ectoderm from a late yolk plug gastrula. Ciliary beat of graft epidermis is now reversed. (C) Transplant of ectoderm in normal AP orientation from early gastrula into a pocket on the flank from which host ectoderm has been removed. Underlying mesoderm rotated 180 degrees. Ciliary beat of graft ectoderm is reversed in most portions compared to host epidermis. (Based on T. C. Tung and Y. F. Y. Tung, Arch. Biol. 51:203–218, 1940.)*

at the early gastrula stage. If, however, a graft from a late gastrula is rotated, the direction of ciliary movement at the neurula is opposite to that of surrounding host epidermis. The reversed ciliary beat is unstable since the direction of beat gradually shifts within a few days until it corresponds to the surrounding tissue. We see that anterior-posterior orientation of ciliary beat is established during gastrulation. But how?

In another series of experiments, a piece of presumptive epidermis from an early gastrula was placed over the flank in a neurula from which an epidermal flap had been removed and underlying mesoderm rotated 180 degrees. The resulting ciliary beat of the fragment was reversed and corresponded to the direction of the rotated mesoderm. Anterior-

posterior axis of the embryo seems to be set by the invagination of chordamesoderm during gastrulation. The direction of ciliary beat or cell polarity is likewise determined by the position of the cell within an epidermal sheet on which axial information has been imposed by underlying mesoderm. We can say that the intrinsic polarity of the underlying mesoderm is the "prepattern" to which epidermal cells conform during differentiation of their individual polarities.

Invaginated chordamesoderm is also the prepattern for the nervous system of vertebrate embryos. The axial symmetries established during these early inductive interactions polarize the neural tube and probably furnish the primary cues for the development of connectivity patterns in the nervous system.

## PATTERNING IN THE NERVOUS SYSTEM: CONNECTIONS BETWEEN EYE AND BRAIN

The cell patterns in the nervous system are probably the most complicated patterns found anywhere. The billions of neurons that make up the brain and spinal cord are wired into intricate networks with each synaptic contact playing an essential part in the behaviorial circuitry. The overwhelming complexity and almost infinite variety of cell types, ganglia, bundles, tracts, networks, and synapses underlie those exquisite aspects of behavior known as memory, learning, and thought. How does such complex anatomical patterning arise from the simple neural tube? What rules of connectivity govern the establishment of neuronal circuits as the nervous system develops?

Connectivity, we will see, depends upon several factors: (1) the time and place of neuron origin, (2) the intrinsic polarity of neurons as they differentiate, (3) the direction of axonal and dendritic outgrowth, and (4) the ability of growing tips to recognize and make proper synaptic connections. We have already seen that neurons acquire polarity as they differentiate after a terminal division in the brain or spinal cord. The distal end of the cell becomes the growing axon while the opposite end becomes the dendrite. This initial polarity defines axonal growth patterns and contact relationships with other neurons at some distance. For instance, sensory cells in the eye receive spatial information from the environment and establish point-to-point connections with distant neurons in brain centers. The connections made by retinal neurons permit them to accurately receive, process, and transmit ordered information about space to the brain.

### Retinotectal fiber connections

The amphibian eye develops as an outgrowth of the brain, the optic vesicle, which transforms into the retina (see Chapter 12). As the eye develops, ganglion cells in the innermost layer of the retina send axons through the optic stalk toward the brain. These cross at the optic chiasm beneath the diencephalon and connect with the cells in the contralateral optic tectum (Figure 16-20). Fibers from the right eye enter the superficial fiber layer of the left tectum and the left optic nerve crosses the chiasma and connects with the right tectum in an identical manner.

The retina is a polarized two-dimensional sheet organized as a hemisphere. The temporal or posterior portion of the retina looks out into the front of the visual field, while the anterior or nasal retina looks out in back. Different parts of the visual field impinge on different regions of the retina. The visual field of the frog or tadpole is almost 360 degrees around.

Ganglion cells are topographically polarized, and their fibers establish ordered connections with tectal cells in the brain. Precise maps of the fiber projections of the retina on the tectum are made by making small lesions in the retina. These result in specific patterns of fiber degeneration which show that certain retinal quadrants connect with specific tectal regions. Fibers in the anterior or nasal portion of the eye connect with caudal regions of the tectum, while those from the temporal (or posterior) retina are found in the rostral (anterior) half of the tectum. The fibers from the dorsal quadrant connect with the ventral-lateral tectum and those from the ventral retina to the dorsal tectum.

The precision of fiber ordering can be demonstrated by placing recording electrodes in the tectum while stimulating points in the visual field of the contralateral eye with a spot of light. Each tectal site

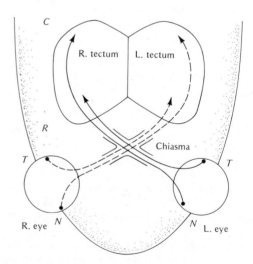

**Figure 16-20**  *Pattern of connections between eye and brain in the adult frog. Left eye sends fibers through the optic chiasma to the right tectum. Fibers from temporal retina (T) connect at the front or rostral half (R) of the tectum, while fibers from the nasal retina (N) connect with the caudal half (C). The right eye is connected to the left tectum in a similar manner.*

responds to stimulation from a specific point in the field. By moving the electrode in a coordinate pattern over the tectal surface, we obtain a map of the visual projection. The electrode detects presynaptic impulses coming only from small populations of stimulated ganglion cells. Such visuotectal maps show points in visual space projecting on specific cells of the retina, which connect in a highly ordered array with cells in the tectum. The physiological map corresponds to the anatomical map; that is, points arrayed from front to back in the visual field project in front-to-back order on the tectum. Likewise, points arrayed up to down in visual space project medio-laterally on the tectum (Figure 16-21).

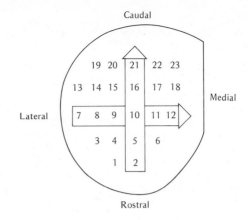

## Optic nerve regeneration: neuronal specificity

How do such ordered patterns arise during development? Is specific connectivity an intrinsic property of each neuron, or does the environment through which axons grow impart specificity?

The optic nerve in adult amphibians is able to regenerate. After the nerve is cut, the distal fibers degenerate. The proximal fibers (those still attached to the ganglion cell bodies) regenerate and eventually grow back to the brain. After regeneration the animal exhibits normal visually controlled behavior; it successfully locates lures in its visual field. This implies that ganglion cell fibers in different parts of the retina reestablish normal neural circuitry in brain centers after they regenerate. There are several ways to interpret this result.

1. Fibers grow back to their original positions in the tectum and restore their normal retinotectal connectivity.
2. Fibers grow back at random, thoroughly mixed, and make the wrong connections in the tectum. But visual relearning restores normal behavior. As the retina receives new stimulation, readjustments are made in the tectum to excite the proper motor response as the eye relearns the visual field.
3. It makes no difference where the fibers end up on the tectum since visual information is encoded in a specific pattern of impulses from each retinal region which is decoded in the tectum and transmitted to the appropriate motor system. This hypothesis could be called impulse pattern specificity.

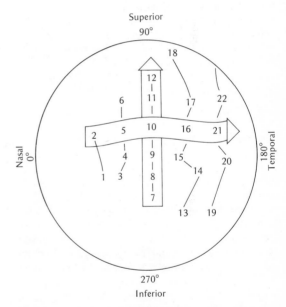

**Figure 16-21** *A normal retinotectal map in* Xenopus *as determined by electrophysiological recording. The left visual field (lower) is projecting on the right tectum (upper). Electrode positions on the tectum are numbered in order of recording. Positions in the visual field that elicit a response at the numbered tectal positions are given a corresponding number in the field. Because the frog sits on a platform during recording, most of its inferior visual field is blocked, which accounts for the absence of points in this region. Note that the spatial order in the visual field corresponds to a similar ordering of positions on the tectum as indicated by the arrows. (Based on R. M. Gaze et al., J. Physiol. 165:484–499, 1963.)*

**Figure 16-22** *Misguided striking behavior of a frog following rotation of the eye and regeneration of the optic nerve. (A) After 180-degree rotation with lure in temporal field, frog strikes toward imagined prey in the nasal field in front and down. Both axes are inverted. (B) After dorsoventral inversion of the eye, animal strikes toward lure in nasal field (the anterior-posterior axis of the eye is normal), but in the inferior rather than the superior field. (C) After nasotemporal inversion of the eye, only AP axis is reversed, and frog strikes correctly in the dorsal or superior field but opposite the actual position of the prey. (From R. W. Sperry,* Handbook of Experimental Physiology, *ed. S. S. Stevens. New York: Wiley, 1951, pp. 236–280.)*

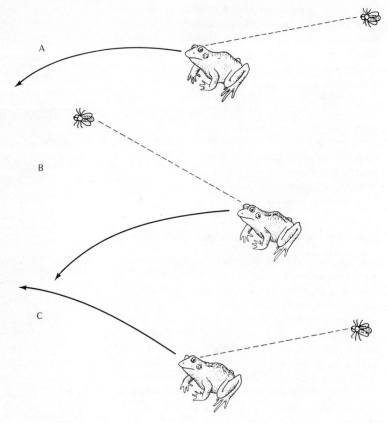

Eye rotation experiments eliminate hypothesis 2. If the eye is rotated 180 degrees at the same time the optic nerve is cut, regeneration still occurs and vision is restored after several weeks, but this time the animal exhibits abnormal responses to lures (Figure 16-22). A lure placed in back of the animal (in its temporal field) is mistakenly struck at in front. Similarly, if the lure is placed in the dorsal field, the animal strikes downwards; the animal consistently strikes away from the prey, in a direction reversed or inverted from the position of the lure. It persists in this abnormal behavior indefinitely, never relearning proper behavior. No readjustments are made in the brain by visual function.

Hypothesis 3 is eliminated by experiments involving direct electrical stimulation of different quadrants of the retina in the eye of the teleost fish while recording from different regions of the tectum. Here again a specific pattern is observed and persists, irrespective of the nature of the stimulus pattern. This means that impulse coding could not be specific for each region of the eye and suggests that the point-to-point correspondence of visual behavior and visuotectal mapping relies on anatomical connections rather than on impulse coding.

The first hypothesis provides the best explanation for these results; that is, in spite of the rotated eye, fibers from each retinal quadrant grow back to their original positions on the tectum. Because central connections between tectum and motor output remain normal, an inverted visual field evokes misdirected strike behavior. With a rotated eye, points in the anterior visual field are now seen by the nasal retina which connects to the caudal tectum, a region that always directs motor output toward the opposite, posterior visual field.

This interpretation is confirmed by visuotectal mapping (Figure 16-23). Here we see that the reversed visual field corresponds to the same ordering of positions in the tectum as existed before eye rotation. This confirms that ganglion cell fibers from different regions of the rotated eye have grown back to their original positions on the tectum. Ability to distinguish correct from incorrect positions on the tectum suggests that regenerating fibers possess a form of recognition specificity (see Chapter 10).

Ordered connectivity can be due to (1) directed growth of fibers back to the brain following environmental cues that lead them to correct sites for synaptic connections (see the discussion of axon growth in Chapter 10), (2) random growth of fibers to the tectum, while axon tips "seek out" correct cells with which to connect, or (3) random growth of an excess of fibers to the tectum, making both correct and incorrect synapses. Only correct connections survive while all others degenerate. In each case, some form of specific cell recognition is involved, either recognition of pathways for axon growth or identification of correct tectal cells with which to organize functional synapses, or both.

Sperry has proposed a hypothesis of neuronal specificity to explain these results. The regenerating fibers behave as if they had acquired a specific recognition property during development. Each cell in the retinal ganglion layer is distinguished from all other cells by its unique position in the sheet. Cells differ with respect to this spatial property in a graded fashion; cells farther apart in the retina are more different from each other than are cells close together. These positional properties are acquired in a stepwise manner during embryonic development, just as the axial qualities of the vertebrate limb are specified. First, cells are polarized along the organism's anterior-posterior axis and then according to its dorsal-ventral symmetry. In this way each ganglion cell is specified in an x,y grid and acquires a specific positional label. According to the hypothesis each tectal cell also acquires a unique positional specificity, presumably matching those in the retina. Each neuron develops a chemospecific label on its surface, depending on its position at the time of retinal or tectal polarization. Such a model insures that fibers from ganglion cells can hook up only with tectal cells with matching chemospecificities and thereby establish an ordered pattern of connections between eye and brain.

If optic fibers are specified, as proposed, they should exhibit directed growth, homing in on the correct tectal site. Attardi and Sperry showed that fibers do display directed growth in the goldfish. They sectioned the optic nerve and simultaneously removed a portion of the retina (Figure 16-24). After the temporal half of the retina was removed, the fibers from the nasal half regenerated through both

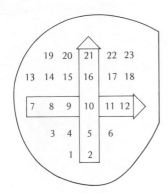

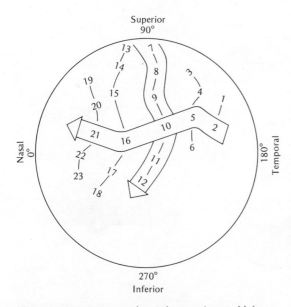

**Figure 16-23** *Retinotectal map from an inverted left eye projecting to the contralateral right tectum. The entire visual field is inverted. The retinal cells in the inverted eye now receive visual input from opposite parts of the visual field than originally, but continue to connect with the same tectal region. For example, position 1–2 in the temporal visual field in this inverted eye is seen by the posterior retina, which normally receives input from the nasal field (see position 1–2 in the normal projection in Figure 16-21). (Based on M. Jacobson,* Develop. Biol. *17:202–218, 1968.)*

**Figure 16-24** *Attardi and Sperry experiment on the goldfish eye. (A) Anterior (A) and posterior (P) halves of the retina ablated and the optic nerves cut. Fibers grow back to the eye according to their original specification. Anterior fibers grow to the caudal tectum and posterior fibers to the rostral tectum. (B) A similar experiment in which dorsal (D) and ventral (V) retinas were ablated. M = medial; L = lateral. (From D. G. Attardi and R. W. Sperry,* Exp. Neurol. *7:46–64, 1963.)*

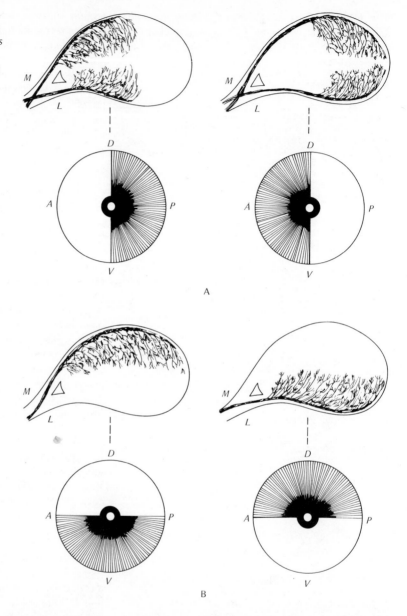

medial and lateral branches of the regenerated optic nerve and extended over the tectum, penetrating the optic layer only in caudal regions. Likewise, temporal ganglion cells returned only to the rostral region of the tectum. These experiments, though limited to fibers visible in the light microscope, showed that ganglion cell fibers seek out the most direct pathways to connect with matching tectal cells. Such routes may be provided by degenerating optic nerve sheaths, filled with glial cells that persist as preferred substrata on which regenerating fibers migrate. Since smaller fibers could not be followed, there would be no way to determine whether they grew back in a directed or random fashion. Only an electron microscopic study of regenerating fibers could settle this question.

## Axial polarization of the retina

Like the limb, axial specification of the embryonic retina, when retinal cells acquire the positional information determining their connections to the tectum, takes place in steps. Retino-tectal maps of *Xenopus* eyes were recorded weeks after eye primordia were rotated 180 degrees at different embryonic stages, before any optic fibers reached the tectum. Eyes rotated at stage 28 gave rise to normal visual fields, but eyes rotated at stages 29/30 (only 2 hours later) showed an inverted anterior-posterior projection and a normal dorsoventral projection. At this stage, the anterior-posterior axis of the retina is determined but the dorsoventral axis is still labile. A few hours later, at stage 32, the dorsoventral axis becomes fixed.

Once the retina is polarized at stage 32, it seems to retain its axial bias irrespective of any other positional cues. Stage-32 eyes transplanted onto the flank of younger embryos, or maintained in vitro for several days, when reimplanted into a host eye orbit show no change in polarity. Even unspecified stage-22 or -25 eyes placed in the ventral abdominal region of another embryo for several days and transplanted to later stage host orbits remained unchanged. Specification seems to be intrinsic to the eye and independent of any axial cues from the host body. The axial information acquired by ganglion cells seems to be a stable property.

Retinal polarization depends on conditions in the retinal sheet at the critical stages. It occurs after cells in the center of the retina complete a last division before maturation and fiber outgrowth. At this time retinal cells are coupled to one another by gap junctions and are in ionic communication. Later these gap junctions disappear, which suggests that the cells of the sheet are "communicating" only when they acquire their axial cues.

This observation is consistent with the model proposed by Loewenstein to explain how cells establish their position in a two-dimensional sheet (see Chapter 9). A retinal sheet organized into two chemical gradient systems, one anterior-posterior and the other dorsal-ventral, could furnish the positional cues to ganglion cells exchanging information through gap junctions, a model consistent with the Sperry hypothesis. Each ganglion cell would be specified according to its position in the orthogonal gradients.

## Ontogeny of the retinotectal projection: sliding connections

Different rules seem to govern the ontogeny of connections between eye and brain in the embryo as contrasted with regeneration of optic nerve fibers. The specificity of retinal cells is not as irreversibly fixed when their fibers are growing on the embryonic tectum making synaptic connections.

The development of visuotectal maps in *Xenopus* was studied by Gaze, recording from tadpoles at different developmental stages (Figure 16-25). If ganglion cells are rigidly specified, they should make only one functional connection in the tectum. Yet this does not seem to be the case. The earliest maps obtained are not ordered and come from all portions of the visual field. In other words, the first fibers that reach the tectum do not necessarily "home in" on their correct position. One reason is that the tectum matures more slowly than the retina, with the front or rostral end maturing first. Because the optic tract first arrives at the rostral tectum, all optic fibers make initial contacts with rostral tectal cells. As development proceeds, more tectal surface is covered with newly arriving opitc fibers and the ordered retinotectal map extends progressively over the tectum from front to back. To explain this pattern it was suggested that fibers continuously make and break connections as they grow over the optic layer. The original fibers that connect in the young tectum are displaced by new fibers coming in from the growing retina and make new connections in more posterior regions of the tectum. This pattern of "sliding connections" implies that the initial specification of a ganglion cell (and/or tectal cell) is sufficiently labile to permit a sequence of connections to be made as the eye and brain grow. At metamorphosis, the mature map is established and the entire tectum is covered by fibers.

This suggests that cells are not uniquely labeled by chemospecific markers but are able, at each stage, to determine their relative position and make connections as long as there are cues as to anterior-posterior and mediolateral axes. No additional information is required. As the system grows, new cells are added and these, in turn, ascertain their position in the gradient to connect in proper order on an enlarging tectum. At each stage, the relative position of fibers in the map is always retained; that is, nasal fibers are always at the most posterior region of the

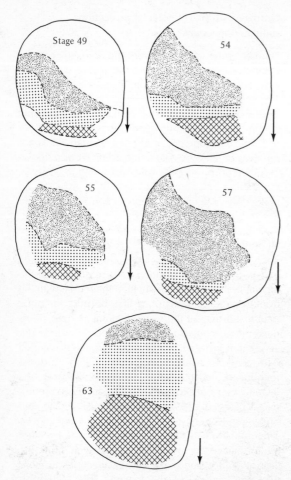

**Figure 16-25** *Development of the visuotectal projection in* Xenopus. *Diagrams show the average field projections on the tectum at various larval stages. Visual responses are first detected at stages 44 to 45, and metamorphosis is complete at stage 66. Cross-hatching indicates nasal field; horizontal dots show the central part of the temporal field; and stippling indicates the most temporal sixth of the field. Arrow points to anterior. (From R. M. Gaze, M. J. Keating, and S. H. Chung,* Proc. Roy. Soc. London Ser. B. *185:301–330, 1974.)*

tectum while temporal fibers are most rostral.

A lability of connections has been observed in other experimental situations such as double-nasal or double-temporal compound eyes. Such eyes are made by cutting a specified stage-32 eye in half, removing the nasal or temporal portion, and replacing it with a nasal or temporal half from the opposite eye.

The cut edges fuse and normal-appearing eyes develop, which can be mapped after metamorphosis. If retinal ganglion cells retain their positional specification, each nasal half (of a double-nasal eye) will project only on the caudal half of the tectum, while each temporal half (of a double-temporal eye) will spread only over the rostral half of the tectum. The maps obtained do not follow the prediction, however; each nasal (or temporal) half spreads over the entire tectum, forming two mirror-image visual fields, and all points on the tectum receive input from at least two different points in the visual field. Though nasal and temporal poles of these compound eyes project to their proper regions in the tectum, fibers from the center of each half-eye (in the region of fusion) spread over the tectum, occupying completely new positions (Figure 16-26). The cells in the center of the eye, the oldest and first ganglion cells to be specified, are not committed to specific positions on the tectum; depending upon the situation, they may make synapses at different tectal sites. Neuronal specification is not rigid nor irreversible; cells display new specificities when challenged in different contextual situations. A cell's positional properties depend upon the total state of the system. Chemical gradients may be important for setting positional properties at one time, but other asymmetries might arise in a tissue sheet, such as gradients of cell division, differential rates of axon fiber outgrowth, or differential numbers of axon fibers, all of which together or individually may induce ganglion fibers to form ordered visuotectal projections.

### Nature and origin of tectal specificities

For the Sperry hypothesis to be correct, all tectal cells should acquire positional specificities corresponding to those in the retina. They may exhibit similar stepwise polarization and should retain these specificities in most experimental situations.

An attempt has been made to determine when tectal cells are specified in normal development of *Xenopus*. At neurula and late tail-bud stages, a portion of the presumptive midbrain was rotated 180 degrees. The tadpoles that developed displayed visuotectal projections that corresponded with the orientation of the tectum with respect to the neighboring diencephalon. If as a result of the operation the diencephalon was behind the tectum, the retinotectal projection was inverted, but if the dien-

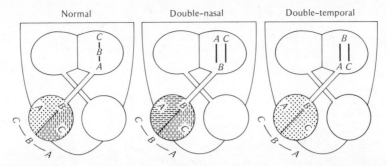

**Figure 16-26**  *Relationship between visual field, retina, and tectum in* Xenopus *with normal, double, and double-temporal compound eyes. Normally points A, B, C in the visual field project in a front to back order on the tectum. In the compound eye, projection from each half-retina has spread on the tectum. The order of projection for the original half-eye is normal, but that for the grafted half-retina is reversed. The center of the eye, B, projects rostrally in the double-nasal eye (it is now the temporal pole for both half-eyes). (From R. M. Gaze, J. Physiol. 165:484–499, 1963.)*

cephalon was still anterior to the tectum, the projection was normal. There was no evidence, at the stages examined, of any fixed determination of tectal polarity as was obtained in the retina. This suggests that if the tectum is polarized, it must take place at a later stage, perhaps at the time optic fibers are growing into the superficial layers in the tectum. The diencephalon, in these experiments, seemed to behave as an organizer, directing the fiber ordering on the tectum.

The mature amphibian tectum is more rigidly specified than the retina. If parts of the amphibian tectum are rotated in place 90 or 180 degrees, then fibers regenerate into the graft and reestablish connections. The restored visuotectal projection is distorted 90 or 180 degrees in conformity with the rotated graft. There is no lability and no regulation occurs; the original specificities of tectal cells seem to be retained and correct connections are made with appropriate retinal quadrants.

**Spatio-temporal ordering of connectivity: development of the compound eye in Daphnia**

Though Sperry's hypothesis of neuronal specificity is the most attractive explanation for eye-brain connectivity, other aspects of synaptic connectivity, such as relative rates of fiber growth or the time of fiber emergence, may play a part in establishing neuronal patterns. Spatial patterns may emerge because developmental events are ordered in time. The best example of spatio-temporal ordering of connections is the development of neuronal patterns between

the compound eye and brain in the water flea *Daphnia*. The single compound eye in this crustacean sits in the midline of the head and consists of only 22 ommatidia, 11 on each side. Because of *Daphnia*'s small size and limited cell number, it is possible to make serial sections of the head region for study in the electron microscope, to reconstruct the fiber patterns growing from the eye to the brain. Using computer methods to analyze the thousands of photographs required, growth of individual fibers could be followed during development and the patterns of connections could be traced. Since *Daphnia* reproduces parthenogenetically, isogenic clones of animals could be obtained that developed synchronously, with minimal variation from one animal to another.

Each ommatidium is a complex structure consisting of lens and cone cells surrounded by eight receptor or retinula cells (Figure 16-27). Cells of one ommatidium may be derived from a common stem cell, and each cluster of cells develops sequentially from a mitotically active growing region in the eye anlage. As a result, the developing eye contains ommatidia arranged in a sequence of maturation, the most mature ommatidia being close to the midline of the eye while the least mature are emerging from the mitotic center. On each side of the midline 11 ommatidia are lined up in decreasing order of maturation.

As an ommatidium matures, each retinula cell sends an axon toward the brain, entering the optic lamina at the midline. Eight retinula fibers grow as a cluster from a single ommatidium with one fiber,

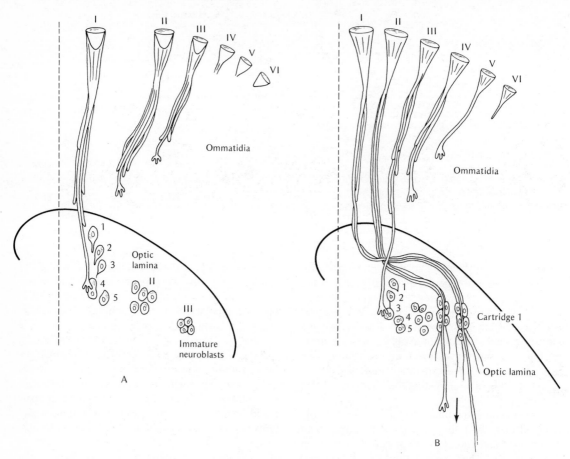

**Figure 16-27** *Spatio-temporal ordering of connections between compound eye and brain in the water flea Daphnia. (A) An early stage in the ingrowth of retinula fibers into the optic lamina from the developing eye. The first ommatidium, the most mature, has already reached the optic lamina in the midline. Its lead fiber has contacted optic lamina neurons 1, 2, 3 and is contacting neuron 4. Other ommatidia are beginning to send their fibers to the lamina. (B) A later stage; ommatidia I and II have already formed a visual cartridge with a set of five lamina neurons and have shifted laterally. Their respective lead fibers have continued to grow into the medulla of the brain. Meanwhile the lead fiber of ommatidium III is contacting its set of optic lamina neurons in sequence as the following fibers grow behind. (Based on V. Lopresti, E. R. Macagno, and C. Levinthal. Proc. Nat. Acad. Sci. U.S. 70:433–437, 1973.)*

the lead fiber, preceding all others. The lead fiber, the only one with an active growth cone, acts as the "pioneer" while the others follow along its surface. As the lead fiber enters the lamina, a cluster of five immature lamina neurons appears near the midline.

The ingrowth of retinula fibers is ordered spatio-temporally so that they will contact a cluster of five lamina neurons in sequential order. The first ommatidium to mature sends its eight retinula fibers into the

brain to meet the first group of lamina cells. As the eight fibers make contact with lamina neurons, the entire collection, called a visual cartridge, is displaced laterally, making room for another cluster of five lamina neurons to appear at the midline, ready to receive the second set of retinula fibers from the next ommatidium. Eleven ommatidia on each side of the midline grow in sequential order into the brain in this manner. Because of precise timing of fiber growth and lamina neuron production, an ordered array of visual cartridges is established within the brain.

The encounter between the lead fiber and each lamina neuron is the most interesting feature of the process. As the tip of the lead fiber enters the lamina, it first contacts the uppermost (most anterior) lamina neuron (see Figure 16-27). During this brief encounter, the lamina neuron wraps itself about the fiber and forms transient gap junctions. The lead fiber contacts the remaining four lamina neurons in sequence, each enveloping the fiber in its turn and forming transient gap junctions. After interacting with the lead fiber, each lamina neuron, in order, sprouts an axon, as if contact with the lead fiber signaled fiber maturation. Functional synapses are ultimately made between these neurons and following fibers, while the lead fiber grows into the medulla of the brain to make functional synapses.

Specific information may be transferred during the brief interval of junctional communication between lead fiber and neuron. Recognition specificity may arise at this time as contact with the lead fiber organizes all lamina neurons into a specific visual cartridge.

We see that elaborate wiring patterns arise, not because of any prior chemospecific cell recognition, but because cells are in the right place at the right time. Spatio-temporal ordering of cell division patterns, cell maturation, fiber growth, and synapse formation generate anatomically precise neuronal connections. Because of the timing of developmental events, cells come to occupy positions in the embryo which insure that they will make contacts with other cells at the correct stage of maturation. Similar spatio-temporal ordering of connections have been observed in the more complex networks of vertebrate brains, and we assume that insight into neuronal patterning will come when these ordering mechanisms are fully understood.

## MODELS OF PATTERN FORMATION

According to Wolpert, patterns develop because cells "recognize" their neighbors and "know" their positions within a tissue field. Though several models for the development of patterns have been proposed, none satisfactorily explains all patterning systems. In fact, we may find as many patterning mechanisms as there are patterns. The simplest assumption, however, is that basic patterning rules are followed by all morphogenetic systems. One feature that seems to be common to most systems is the presence of some type of gradient.

### Wolpert's gradient model

Wolpert formulates his model in terms of a positional gradient. We have already seen how this model explains regeneration patterns in *Hydra* (see Chapter 15). Related gradient models have already been described for ripple patterns in *Rhodnius* cuticle.

Wolpert's model for pattern generation involves acquisition of positional information by a cell within a positional gradient. Its position relative to the boundaries of the field has a spatial value that is transduced by a cell into phenotypic expression. The positional signal may be a diffusible substance or a more complex cellular property.

According to Wolpert, this is a universal patterning model for all morphogenetic systems, but the specific response to a positional value in a gradient is a genetically determined characteristic of each species. For example, frog embryo tissue placed in a segmental gradient system of the *Rhodnius* abdomen would respond to its anterior-posterior polarity by differentiating corresponding anterior-posterior frog structures.

An essential problem in the model is how and when the positional gradient is established. Using simple diffusion kinetics, it was calculated that a gradient steep enough to furnish positional information could be set up within 10 hours over a distance of 50 to 100 average-sized cells. This is consistent with the size of many cell populations when they are determined in embryonic development before they carry out specific morphogenetic programs.

Although this model explains many polarization phenomena in regenerating *Hydra*, retinal specification, and imaginal disc reduplication, it fails to show how positional information gradients are generated. As a result it is difficult to design experiments to test it.

### A phase-shift model for spatial organization

Another model for acquiring positional information has been proposed by Goodwin and Cohen based on oscillating physiological behavior exhibited by cells and tissues. Biological rhythms in growth and respiration are common phenomena. Such rhythms may have diurnal or seasonal cycles. A common periodic property of cells is mitosis, and it may be the principal "clock" setting many cellular activities. Pattern generation then becomes a problem of translating temporal ordering of cell division into the spatial ordering of morphological patterns.

This model is sometimes called the "lightning and thunder model" because positional information is acquired by cells in a manner analogous to our calculation of distance from a lightning flash by comparing the temporal difference between the arrival of two signals, light and sound. The model postulates that cells in a tissue respond to two time-dependent signals, S and P. The time between detection of these signals, or the *phase shift,* furnishes a cue for positional information for any cell. One event, the S or signaling event, is initiated by a pacemaker cell in the tissue. This biochemical event acts as a brief signal, which, by functional coupling with neighboring cells (gap junctions?), is transmitted in a radial pattern. Similar S events are induced in neighboring cells which in turn pass on the signal, thereby propagating an S wave outward from the pacemaker cell. The action of such a pacemaker is seen in heart cells; the one with the most rapid frequency can entrain neighbors to beat at a higher rate when contact is made. Polarity arises in this S wave since a refractory period occurs immediately after the S event, preventing the initiation of another S event in that cell until it recovers. The wave can pass in only one direction.

The second event, the positional determining event, or P, provides the fine details of positional information. In the pacemaker cell, the P event is driven by the S event. The P event propagates more slowly from the pacemaker than the S wave, and the phase of occurrence of P relative to S in each cell will increase monotonically with the distance of the cell from the pacemaker. A cell occupies a site over which two different metabolic waves pass, and it acquires its positional information from the shift in phase between one wave and another.

This model lends itself to experimental manipulation and points to the kinds of biochemical events that may be responsible for establishing gradients or coupled oscillations. Inasmuch as cells communicate through gap junctions, we have here a structural basis for the establishment of gradient systems or propagating waves. The extent of such fields depends on the distribution of intercellular connections, with boundaries of a field defined by the absence of such junctional communication. This implies that spatial patterns of junctional contacts will affect the transmission of oscillating states in a field. Genetic control of pattern acts at two sites—the determination of a prepattern in the form of gradient fields or oscillating physiological states, and the setting of a response pattern by the cell's own genome.

## The Turing hypothesis

Finally, there are models which generate patterns entirely at random from rates of chemical reactions and simple laws of diffusion. Such a model was developed by Turing, a mathematician interested in topological problems. In 1952 he proposed a stochastic model for pattern generation in biological systems. According to this purely chemical model, periodic discontinuities are produced in homogeneous systems and become organized into stable patterns. It is a "condition-generated" patterning device, depending entirely on reaction and diffusion rates of chemical substances. Concentrations of these substances vary randomly in a homogeneous population of cells, but fluctuations in reaction and diffusion rates can lead to differential cell responses that become fixed into repeated morphological patterns. The model may explain generation of so-called "singularities" in the prepattern of insect epidermis.

The Turing model is only a first approximation, but it highlights the fact that patterning can be generated at random from an initially homogeneous system. Problems arise with the model since it does not direct itself to the nature and mode of action of the morphogenetic substances. It treats them abstractly, as would a mathematician rather than a biologist. The usefulness of the Turing model lies in its demonstration that stochastic processes in homogeneous systems can generate periodicities and patterns.

Turing's equations create positional information through random fluctuations in metabolic reaction systems. The ordering process arises accidentally and is independent of special mechanisms for producing spatial information. It is entirely possible that chemical periodicities do arise randomly in biological systems and are translated into morphological patterns.

## Polar coordinate model for epimorphic fields

The patterning behavior of imaginal discs is similar to that of regenerating systems described in Chapter 15. Both behave as epimorphic fields; that is, removal of parts results in epimorphic regeneration

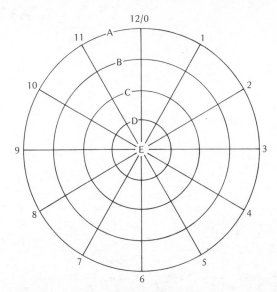

**Figure 16-28** *Polar coordinate system for positional information. Each cell acquires information with respect to its position on a radius (A through E) and its position around the circle (0 through 12). Positions 12 and 0 are identical, which means that the sequence is continuous. The number of values between positions is arbitrary. (From S. V. Bryant and L. E. Iten,* Develop. Biol. 50:212, 1976.)

with cells proliferating to replace lost structures in proper symmetry. Such fields tend to regulate in one of two ways, either by regeneration of lost parts with the original polarity, or duplication of existing parts, often with mirror-image symmetries. This, as we have seen, is typical of imaginal disc regulation.

Rules governing such pattern restoration have been proposed in an all-inclusive model by Vernon French and co-workers. The model treats each system as a two-dimensional sheet, an assumption that is more appropriate for imaginal discs than for amphibian limbs. Nevertheless, the model manages to explain patterns of amphibian limb regeneration in spite of their three-dimensional nature.

The positional information system in this model differs from those described for other models; it is based on a system of polar coordinates in which cells acquire two positional values, one defined by a position on a circle and the other by a radius (an angular dimension) (Figure 16-28). The numbering values and their spacing are arbitrary. In amphibian and insect limbs the outermost circle corresponds to the

proximal boundary of the limb field while the center represents the distal tip of the limb field. In imaginal discs, the outer circle represents the disc boundary while the center becomes the most distal tip of those discs producing limbs.

In addition to the system cueing positional information, the model invokes two simple rules of cell growth behavior in response to surgical perturbation: (1) *rule of shortest intercalation* and (2) *complete circle rule for distal transformation*. According to the first, removal of tissue from a field results in wound healing which brings nonadjacent positional values into contact. As a result, growth at the junction occurs between the regions to intercalate cells with all intermediate positional values. Once this is completed, growth ceases. Because number values are continuous around a circle at the 12/0 boundary, intercalary growth can occur in either one of two directions. For example, removal of a wedge of tissue with positional values 4,5 brings regions 3 and 6 into juxtaposition after wound healing. Growth can fill in the intermediate sequences in either direction, 3(4,5)6 or 3(2,1,12/0, 11,10,9, 8,7)6. The intercalary rule states that the shorter of these patterns of growth is followed in all instances.

The rule for distal transformation states that a circular sequence of positional values at any radius may undergo distal transformation and produce all positional values central to it. This can occur only when a complete circle of values is exposed at an amputation site or emerges because of intercalary growth during regeneration.

Application of these rules to limb and imaginal disc regeneration predicts the pattern of structures to emerge during growth. Only a few of these will be described to show how the model works.

Proximal-distal intercalation explains the regeneration of cockroach legs (Figure 16-29). For example, fusion of a distal portion of the tibia with a more proximal tibial stump so that nonadjacent positional values are juxtaposed results in intercalary growth and restoration of intermediate values to produce a normal limb. A graft of a more proximal tibia region onto a distal stump leads to reversed proximo-distal polarity of the regenerate because of the intercalary rule. These regenerating patterns are similar to those obtained after grafting distal and proximal limb blastemas in amphibians (see Chapter 15). Even grafts between femur and tibia promote intercalary growth as long as the wound junction brings incongruent positional values into contact.

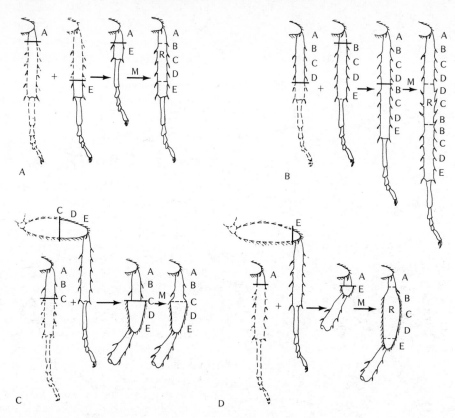

**Figure 16-29**   *Intercalary regeneration and distal transformation in proximal-distal grafts between fragments of tibia and femur of cockroach legs. Letters A to E show physical levels of leg segment which correspond to positional values in the radial sequence shown in Figure 16-28. The graft combination is shown along with the result after two molts (M). Intercalation is shown as R. (A, B) Regeneration between two nonadjacent positional values in tibia-tibia graft combinations. (C) No regeneration occurs in a tibia-femur graft combination if graft and host are joined at homologous positional values. (D) Intercalary regeneration occurs, however, between tibia and femur fragments if nonadjacent positional values are juxtaposed. (From V. French, P. J. Bryant, and S. V. Bryant,* Science *193:969–981, 1976. Copyright © 1976 by the American Association for the Advancement of Science.)*

Regeneration of intermediate femur structures results in an abnormally long limb. On the other hand, if femur-tibia grafts are joined at identical positional values, than no regeneration occurs: no intercalary growth is needed to restore any positional values.

The rules also explain supernumerary limb formation in amphibians (Figure 16-30). For example, if a distal left-limb regeneration blastema is grafted to a right proximal stump blastema, anterior and posterior axes of the graft are reversed with respect to those of the host. Under these conditions both the rules for shortest intercalary growth and complete circle distal transformation account for the appear-

ance of two supernumeraries at anterior and posterior borders of the host limb. Intercalary growth occurs between graft and host borders along the shortest pathways between each point of incongruity. At the regions of maximum incongruity, between values 9 and 3 on graft and host respectively, the short routes are the same in both directions. As a result, a complete circle of values is produced instead, and a boundary for a new limb site is established. Since a free circle of values is created, the distal transformation rule fills in more distal structures in the center of the circle. A supernumerary limb grows out with polarities shown in the figure.

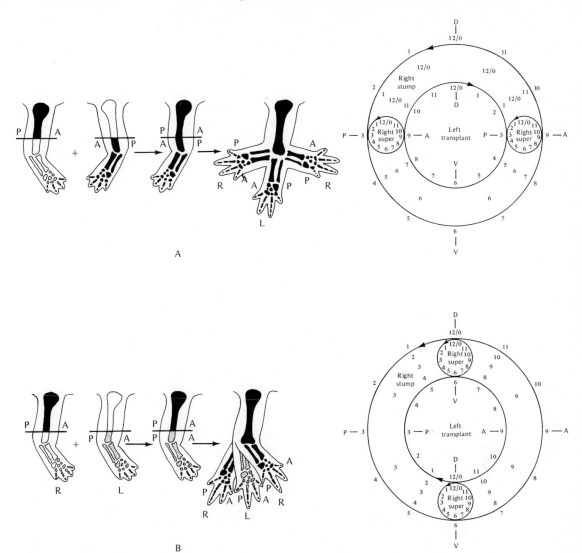

**Figure 16-30**   *Model explains production of supernumerary limbs in newt regeneration following contralateral transplantations of regeneration blastemas with either (A) anterior-posterior (A, P) or (B) dorsal-ventral (D, V) axes opposed (see Chapter 15). Supernumerary (super) limbs form at points of maximum incongruity between positional values. On right in each is a schematic cross section at the graft-host junction to show the relationship of positional values. Outer circle is host circumference; inner circle is graft circumference with diameters differing in size for clarity only. Circular sequence is marked around the circle by numbers 0 to 12 while numbers between circles show values generated by intercalary regeneration via the shorter route between graft and host sets of confronted values. The shorter route is different on the two sides of each point of maximum incongruity, and a complete circle of positional values is generated. Subsequent distal transformation leads to supernumerary limb regeneration in these regions. Arrangement of positional values around supernumerary limbs is determined by intercalary regeneration by the shortest route rule. Direction of intercalation adjacent to supernumerary limbs gives their handedness and orientation. In each case, both supernumerary limbs are of stump-handedness, oriented in same way as the limb stump and in mirror-image symmetry with the transplant. (From V. French, P. J. Bryant, and S. V. Bryant, Science 193:969–981, 1976. Copyright © 1976 by the American Association for the Advancement of Science.)*

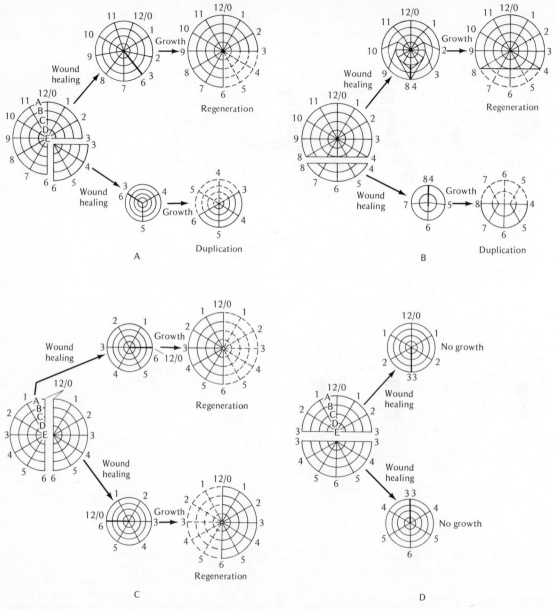

**Figure 16-31**  *Model explains regeneration and duplication in the growth of Drosophila imaginal disc fragments. (A) Result of culturing 90° and 270° sectors of a wing disc. In both cases, cut edges at wound surfaces bring positional values 3 and 6 together. Intercalary growth of values 4 and 5 by the shorter route results in regeneration of the larger fragment and duplication of the shorter fragment. (B) Large and small fragments of a wing disc, following the same rule, regenerate and duplicate, respectively. In both cases, positional values 4 and 8 are brought together at the point of healing. (C) Cutting a genital disc parallel to the axis of bilateral symmetry separates it into two equivalent medial halves which regenerate identically. (D) Cuts perpendicular to the axis of symmetry lead to juxtaposition of homologous positional value at the wound site, and neither regeneration nor duplication occurs. (From V. French, P. J. Bryant, and S. V. Bryant, Science 193:969–981, 1976. Copyright © 1976 by the American Association for the Advancement of Science.)*

Regeneration of imaginal disc patterns is also predicted by the model. Imaginal discs may be cultured indefinitely in adult abdomens. This permits their fullest growth before they are induced to differentiate into adult structures by transferring them to the abdomens of metamorphosing larvae (see Chapter 17). Fragments of discs grown in abdomens either regenerate or duplicate, as we have seen, and these patterns also emerge in the model (Figure 16-31). For instance, removal of a 90-degree wedge from a wing disc results, after wound healing, in juxtaposition of tissues with incongruent positional values in both large and small fragments. Following intercalary growth, the large fragment regenerates and fills in lost parts while the small fragment duplicates. Likewise, unequal section of the wing disc leads to a large fragment which regenerates and a small fragment which duplicates. Section of a bilaterally symmetrical genital disc along the axis of symmetry results in two medial fragments which regenerate. There is no difference in their mode of regulation since healing brings the same positional values together in each. If, however, a lateral fragment is removed, the result is the same as with large and small fragments; lateral fragments duplicate and medial fragments regenerate. Finally, a genital disc cut through the center, perpendicular to the axis of symmetry, leads to two halves that heal to identical positional values, a situation that promotes no intercalary growth since incongruities of positional value do not exist.

How do we account for formation of a polar coordinate system of positional values at the cellular or molecular level? Though gradient models seem to produce positional information along linear axes, they cannot generate a circle of positional values unless more complicated double gradient systems are invoked. The phase-shift model described above, however, could generate a circular system of values. Cells at a given radial level undergo metabolic oscillations with similar amplitudes and frequencies while those at different positions on the circle differ progressively in phase with respect to these oscillations. But, here, too, molecular mechanisms are difficult to come by.

The polar coordinate model could apply to the specification of positional values in other developing systems such as the vertebrate retina. Ganglion cell fibers growing in the optic fiber layer of the *Xenopus* retina are arranged in bundles radiating from the periphery toward the optic nerve head at the center, a pattern identical to the polar system described here. Specification of polar positional values rather than orthogonal grid values may be a more accurate representation of the axes of symmetry in the embryonic retina. Since the model seems to have such general application, it may lead to some understanding of the molecular mechanisms underlying pattern formation in development.

## SUMMARY

Morphogenesis generates patterns. Organisms exhibit symmetrical arrangement of parts as if some whole-body blueprint is guiding the movements and shape changes of epithelial sheets. The position occupied by a cell within a sheet seems to determine its behavior in the overall body plan of the organism. To account for patterning, the concept of positional information has been proposed, a cellular property that determines the spatial relationships of cells and their future progeny.

Pattern formation is genetically determined; there are patterning genes that seem to define tissue shape and the position of specific cell populations within a tissue. Bristles, hairs, scales, and other integumentary structures are assigned invariant places, and their development follows specific patterning rules. A most common rule is that cells exhibit gradient properties of movement, replication, or synthesis, and these, because of intrinsic polarity, with high and low points, generate patterns. Activity of high points dominates the behavior of low points in a gradient, a principal factor in establishing morphogenetic axes in whole organisms or in body parts.

The origin and nature of morphogenetic gradients are not understood, although diffusion of regulatory molecules such as cyclic AMP between cells coupled in junctional communication serves as a working hypothesis for such gradient systems. Cells respond to gradients by adjusting their metabolic activities accordingly, and morphogenetic patterns develop.

The regulation of rates of cell division and the orientation of division planes in proliferating cell sheets also generate patterns, and these need not be defined by underlying gradient prepatterns. The derivatives of a single cell, or a cell lineage, oriented along a single axis, all exhibiting specific phenotypic expression (pigment production), evoke patterned tissues.

The migration of cell populations such as vertebrate neural crest also generates patterns. Here, prepatterning genes seem to be implicated, acting in the tissue or matrix-filled environment through which a migratory population passes. These genes define a patterned substratum that evokes differential migratory, proliferative, or synthetic behavior that is translated into coat color patterns in mammals.

The most complex patterns are those established in the nervous system as neurons connect into behaviorial networks. The strict anatomical specificity of these connections communicates information about the ordered nature of the environment. These intricate patterns may be based on chemospecificity regulating point-to-point interactions between neurons. It makes possible the ontogeny of ordered networks by matching specificities of those neurons essential to an impulse chain.

All patterning rules that regulate the behavior of cells in sheets may be part of the larger spatio-temporal ordering of developmental events. Order in time can generate order in space, and it is possible that the clock that defines the sequence of developmental events is the primary architectural principle governing the morphogenesis of patterns.

**REFERENCES**

Bryant, P. J. 1970. Cell lineage relationships in the imaginal wing disc of *Drosophila melanogaster*. *Develop. Biol.* 22:389–411.

————. 1974. Determination and pattern formation in the imaginal discs of *Drosophila*. *Current topics in developmental biology*, ed. A. A. Moscona and A. Monroy. Vol. 8. New York: Academic Press, pp. 41–80.

Caveney, S. 1974. Intercellular communication in a positional field: Movement of small ions between insect epidermal cells. *Develop. Biol.* 40:311–322.

Crick, F. 1970. Diffusion in embryogenesis. *Nature* 225:420–422.

French, V., Bryant, P. J., and Bryant, S. V. 1976. Pattern regulation in epimorphic fields. *Science* 193:969–981.

Gaze, R. M. 1970. *The formation of nerve connections.* New York: Academic Press.

Gaze, R. M., Keating, M. J., and Chung, S. H. 1974. The evolution of the retinotectal projection during development in *Xenopus laevis*. *Proc. Roy. Soc. London Ser. B* 185:301–330.

Goodwin, B. C., and Cohen, M. H. 1969. A phase shift model for the spatial and temporal organization of developing systems. *J. Theoret. Biol.* 25:49–107.

Harrison, R. 1921. On relations of symmetry in transplanted limbs. *J. Exp. Zool.* 32:1–136.

Henke, K., and Pohley, H. J. 1952. Differentielle Zellteilungen und Polyploidie bei der Schuppenbildung der Mehlmotte *Ephestia kuhniella*. *Zeit. Naturforsch.* 7b:65–79.

Hunt, R. K., and Jacobson, M. 1974. Neuronal specificity revisited. *Current topics in developmental biology*, ed. A. A. Moscona and A. Monroy. Vol. 8. New York: Academic Press, pp. 202–259.

Jacobson, M. 1968. Development of neuronal specificity in retinal ganglion cells. *Develop. Biol.* 17:202–218.

Lawrence, P. A. 1973. The development of spatial patterns in the integument of insects. *Developmental systems: Insects,* ed. S. J. Counce and C. H. Waddington. New York: Academic Press, pp. 157–209.

Locke, M. 1967. The development of patterns in the integument of insects. *Adv. Morphogen.* 6:33–88.

Lopresti, V., Macagno, E. R., and Levinthal, C. 1973. Structure and development of the optic lamina of *Daphnia*. *Proc. Nat. Acad. Sci. U.S.* 70:433–437.

Mintz, B. 1971. Clonal basis of mammalian differentiation. *Symp. Soc. Exp. Biol.* 25:345–369.

Poole, T. W. 1974. Dermal-epidermal interactions and the site of action of the yellow (Ay) and non-agouti (a) coat color genes in the mouse. *Develop. Biol.* 36:208–211.

Saunders, J. W., Jr. 1972. Developmental control of three-dimensional polarity in the avian limb. *Ann. N.Y. Acad. Sci.* 193:29–42.

Saunders, J. W., Jr., and Fallon, J. J. 1966. Cell death in morphogenesis. *Major problems in developmental biology,* ed. M. Locke. New York: Academic Press, pp. 289–316.

Silvers, W. K., and Russell, E. S. 1955. An experimental approach to action of genes at the agouti locus in the mouse. *J. Exp. Zool.* 130:199–220.

Sperry, R. W. 1965. Embryogenesis of behavioral nerve nets. *Organogenesis,* ed. R. L. DeHaan and H. Ursprung. New York: Holt, Rinehart and Winston, pp. 161–186.

Stern, C. 1968. *Genetic mosaics and other essays.* Cambridge, Mass.: Harvard University Press.

Stern, C., and Tokunaga, C. 1967. Non-autonomy in differentiation of pattern-determining genes in *Drosophila.* I. The sex-comb of eyeless-dominant. *Proc. Nat. Acad. Sci. U.S.* 57:658–664.

Tokunaga, C. 1972. Autonomy or non-autonomy of gene effects in mosaics. *Proc. Nat. Acad. Sci. U.S.* 69:3283–3286.

Tung, T. C., and Tung, Y. F. Y. 1940. Experimental studies on the determination of polarity of ciliary action of anuran embryos. *Arch. Biol.* 51:203–218.

Turing, A. M. 1954. The chemical basis of morphogenesis. *Phil. Trans. Roy. Soc. London Ser. B* 237:37–72.

Ursprung, H. 1966. The formation of patterns in development. *Major problems in developmental biology,* ed. M. Locke. New York: Academic Press, pp. 177–216.

Wolpert, L., Hicklin, J., and Hornbruch, A. 1971. Positional information and pattern regulation in regeneration of *Hydra. Symp. Soc. Exp. Biol.* 25:391–415.

# DIFFERENTIATION

Differentiation involves the appearance of functionally specialized cell populations from more homogeneous beginnings. Growth, through cell division, produces millions of cells which organize into discrete populations with different developmental fates. Each cell population is a clone derived from small groups of progenitor cells and is defined by its morphology, its specific proteins, and its functional behavior. Since all cells in an organism share identical information, how do the many different cellular phenotypes arise from a zygote? We will see that each specialized tissue expresses only part of the total genetic information available to it.

When do cells become committed to expressing only restricted portions of the genome? Embryonic cells pass through a sequence of genetic states in which the availability of the genome for expression undergoes progressive restriction, from a state of complete accessibility (totipotency), through periods of partial availability (determination), to terminal states when cell specific protein synthesis leads to functional specialization (differentiation).

Most knowledge about gene regulation comes from prokaryote systems which exhibit limited developmental repertoires. We assume information flow in all biological systems is basically the same, and that eukaryote modes of gene action are similar but modified to account for their greater developmental potential. Now we must ask whether basic rules of gene regulation are common to all developmental systems.

**539**

# 17

# Totipotency, Determination, and Differentiation

The emergence of an organism from a zygote requires the production of hundreds of different clonal populations, each specialized for a single function such as contractility, secretion, or impulse conduction. During its history each clone has passed through various genetic states, from totipotency to the fully differentiated state when only 5 to 10 percent of its genetic information is actually expressed.

A cell is defined as totipotent if it can give rise to all cell phenotypes of an organism when challenged in different environments. In some instances such cells may produce an entire organism; they behave as zygotes. A pluripotent cell shows some restrictions and develops into a limited range of cell phenotypes. In time, pluripotent cell progeny become determined and express the same phenotype in different environments. The determined state cannot be distinguished by any morphological or biochemical criteria; the cells posses no specialized organelles or biochemical functions. Though they all look alike they display different developmental potentialities.

Differentiation is the ultimate fate of these populations. The progeny of determined cells develop into specific phenotypes; they synthesize cell-specific proteins and assemble functional organelles. Mitotic potential is usually lost or diminished as permissive conditions evoke the differentiated state. Inductive signals from neighboring tissues or specific hormones may trigger differentiation. Since most experimental tests of differentiation are carried out in cell populations, we are really measuring a population property rather than the behavior of single cells. We will see that cell and tissue "states" differ significantly under identical conditions.

First, we will describe each state and examine the genetic mechanisms that may account for its stability. In addition, we will try to identify the factors that switch cells from one state to another. Finally, we want to know how large portions of a cell's genome are progressively shut down and how the remaining information is utilized when a cell phenotype differentiates.

**541**

# THE TOTIPOTENT STATE

The most important totipotent cell is the zygote; from it are derived all cell phenotypes of an organism. Totipotency is transmitted to the first few cleaving blastomeres in many embryos; that is, isolated blastomeres can develop into an intact organism. Some blastula tissues are also totipotent since a complete organism develops from half-embryos cut along the animal-vegetal axis. A blastula cut across the equator, however, produces halves unable to develop into a normal embryo. Totipotency, here, is a population property, dependent upon a balance of animal and vegetal factors and is not intrinsic to individual cells. Isolated blastula cells do not develop into an embryo yet their nuclei, when transplanted into egg cytoplasm, are able to support a complete developmental program (see Chapter 11).

A totipotent cell usually possesses high mitotic potential, though many mitotically active cells are not totipotent. Proliferative potential is a necessary but not a sufficient condition for totipotency. In general most totipotent cells proliferate into morphogenetically active cell sheets which transform into an organism.

Totipotent cells are unspecialized; morphologically they resemble actively growing cells with large nuclei. Except for genes controlling cell division and growth, most genes are inactive inasmuch as no primary or secondary gene products can be detected. The totipotent genome is fully responsive to cues that elicit specific developmental programs. It is this complete accessibility that distinguishes the genomes of totipotent cells from those of determined or differentiated cells.

## Embryoid culture and totipotence in plants

Root and shoot apices in vascular plants behave as centers of growth and morphogenesis from which all specialized tissues of the plant arise. Tissues in these regions, known as meristems, retain division potential throughout the life of the plant. Cells in shoot apices become leaves while those in roots become vascular elements or conducting tubes. Under appropriate conditions, shoot apices behave as totipotent systems; they can be induced to form roots which develop into an intact plant.

Even nonmeristematic plant tissues can be induced to undergo morphogenesis. Virtually all mature somatic tissues, roots, stems, leaves, and even parts of flowers may be cultured in vitro. Small explants of tissue placed on a solid agar medium will proliferate into calluslike growths that can be perpetuated indefinitely. These tissues usually grow as disorganized masses for many transplant generations, but if the nutritional and hormonal conditions are changed, their innate morphogenetic potential is expressed and well-differentiated shoots and roots emerge. Pieces of pith from the stem of a tobacco plant, for example, grow as a callus in a synthetic medium of salts, sucrose, and vitamins. The addition of plant hormones induces these disorganized tissues to form roots, stems, and even whole plants. A precise hormone balance governs the morphogenetic program. High ratios of auxin to kinins favor root development while the converse favors shoot development. High concentrations of both hormones upset the balance and callus growth resumes. Most callus cells are not dividing. Centers of disorganized cell division are spaced at random within the mass; exposure to a balanced hormone environment organizes them into apical meristems capable of root or shoot morphogenesis.

Plant somatic cells can be induced to become meristematic and totipotent. Either the plant genome is not permanently modified during development and is labile and responsive to morphogenetic cues or random foci of totipotent cells may preexist in all plant tissues which may proliferate when released from systemic controls by wounding or excision.

Sometimes somatic cells become totipotent; they behave like eggs and exhibit normal embryogenesis. For example, carrot phloem in roots is a conducting tissue that transports nutrients from leaves. When explanted in a basic glucose-salt medium, it grows slowly as a callus (Figure 17-1). Growth is accelerated by the addition of coconut milk, a rich source of growth-promoting factors, but no differentiation occurs. These masses may be dissociated and cultured as free cell suspensions in a coconut milk medium. Under these conditions, many cells, presumably the progeny of differentiated phloem, undergo division patterns typical of early embryos. As isolated cells, they are not subject to constraints from their neighbors and are free to respond to all morphogenetic signals. Apparently, they revert to a totipotent state and behave exactly like a zygote, giving rise to minute embryolike structures. Thousands of these embryoids develop roots and shoots like a normal embryo and may be nurtured into completely intact plants with flowers and seeds.

Embryoids have been obtained from a variety of mature plant tissues such as stems, leaves, floral buds, and excised seed embryos, but it is still uncertain whether all embryoids arise from differentiated cells that have acquired totipotence. Many somatic plant tissues do contain some meristematic cells (lateral

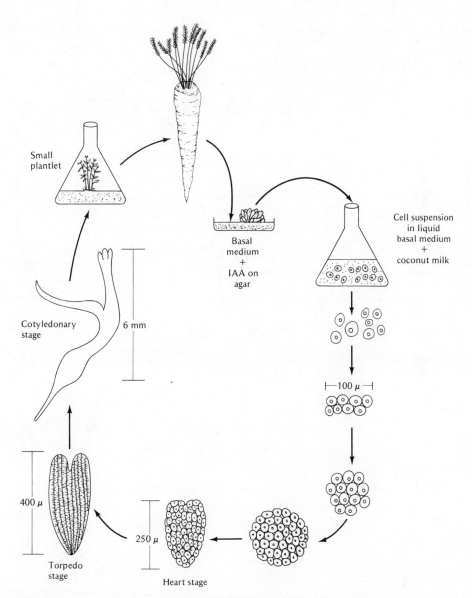

**Figure 17-1** *Totipotent behavior of carrot root cells. Fragments of carrot root phloem grow as a callus on agar in basal medium supplemented with indoleacetic acid (auxin). Samples of callus dissociated in a liquid medium supplemented with coconut milk grow as a free cell suspension. Thousands of these are stimulated to develop into embryoids which pass through typical "heart" and "torpedo" stages of embryogenesis to become small plantlets. These mature into normal plants with flowers.*

**Figure 17-2** *Alternative developmental pathways of microspores in vitro. Microspores are haploid products of meiosis and in vitro they grow as a cell suspension. If the first division is unequal, tube and generative nuclei are formed and a pollen developmental pathway follows. Equal division in a different plane leads to embryoid development and growth of a mature haploid flowering plant that produces no seed.*

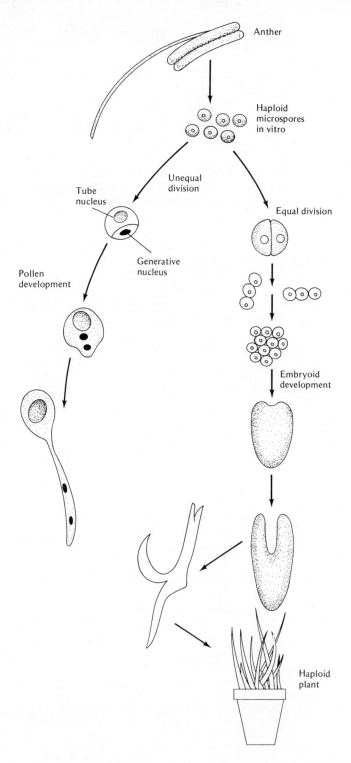

buds, lateral root initiation sites) which develop as embryos when stimulated to grow in vitro.

Embryoid formation depends on a specific order of nutrient addition. Reversion of specialized cells to the totipotent state requires a series of proliferative events in a correct sequence. Each physiological state triggers a new state which, in turn, requires a new set of conditions before morphogenesis can be resumed.

The specific orientations of cell division may also be important in attaining totipotency in plants, a phenomenon seen in the behavior of pollen grains grown in vitro. A pollen grain is a meiotic product of a microspore mother cell in the anther of a flower (Figure 17-2). Derived from a haploid microspore, it divides unequally into a large vegetative or tube nucleus and a smaller condensed generative nucleus. When the pollen grain germinates, the tube nucleus regulates growth of the pollen tube while the smaller generative nucleus divides once again to form two haploid sperm nuclei that eventually fuse with nuclei in the female embryo sac in the ovule of the flower. In normal ontogeny, pollen grain nuclei are destined to differentiate as haploid sperm.

Microspores grown in vitro exhibit totipotence and develop into small plantlets that flower. They do not form seed since meiosis is completely abnormal in haploid plants. The fact that all cell phenotypes of the intact plant emerge, even when haploid, indicates that only one complete set of instructions is necessary for normal development.

During in vitro development of the microspore the first division is probably most critical since it determines the future developmental pathway. If the division is unequal, a normal pollen tube grows out with sperm nuclei. If the first division is oriented in a different plane and is equal, both equal-sized cells continue to develop as embryos with chlorophyll and emerge as small plantlets. The frequency of equal divisions is increased by lowering the temperature to 5°C so that more than 80 percent of the microspores can be induced to develop into plantlets. Sometimes divisions are reversed; the generative nucleus comes to occupy the larger cell and the tube nucleus is smaller. Under these conditions, no chlorophyll develops and plantlets emerge that exhibit an abnormal pattern of morphogenesis.

These results suggest that initial division patterns determine the developmental expression of a haploid microspore. An unequal division leads to normal morphogenesis of a pollen tube, a restricted developmental program where only a small portion of the genome is expressed. An equal division, on the other hand, tends to unmask the entire genome for expression and the cell behaves as a totipotent zygote, undergoing a typical embryogenic sequence. One interpretation is that a microspore nucleus is totipotent, responsive to cytoplasmic signals. Some factors localized in the cytoplasm are segregated by an unequal division. High concentrations of these factors cause the generative nucleus to condense; most genes are silenced and a pollen tube grows out. On the other hand, shifting the division plane 90 degrees results in an equal division, and each cell receives a smaller but equal amount of factor, not enough to change nuclear behavior. Under these conditions, nuclei do not condense and express a full developmental repertoire.

Factors governing plant development do not produce irreversible changes in the genome, which is more labile than the animal genome. The complete repertoire of development may be expressed by most plant cells when released from their repressed state in the plant as long as they are provided with a proper balance of nutrients and hormones. Two events seem to be prerequisite for the acquisition of totipotence; cells must be isolated from growth regulation in the intact plant and they must be stimulated to rapid proliferation to reprogram their genomes.

### Totipotent animal systems

In Chapter 14 we saw that some animal systems, the asexual buds, are totipotent. Resembling plant meristems, they undergo morphogenesis and produce all cell phenotypes. Each cell of an asexual bud should be able to develop into an organism in isolation since there are no genetic reasons that preclude development of somatic animal cells into embryos just as carrot phloem cells develop. But to date no animal somatic cells have been stimulated to undergo embryogenesis. Cells of most asexual organisms are difficult to culture in vitro. Besides, animal cell nutrition is more exacting than the autotrophic nutrition of plant cells grown on simple media supplemented with hormones. Finally, most animal culture media favor unrestricted proliferation, a condition often unsuitable for the expression of specialized cell phenotypes.

One animal system that comes close to being

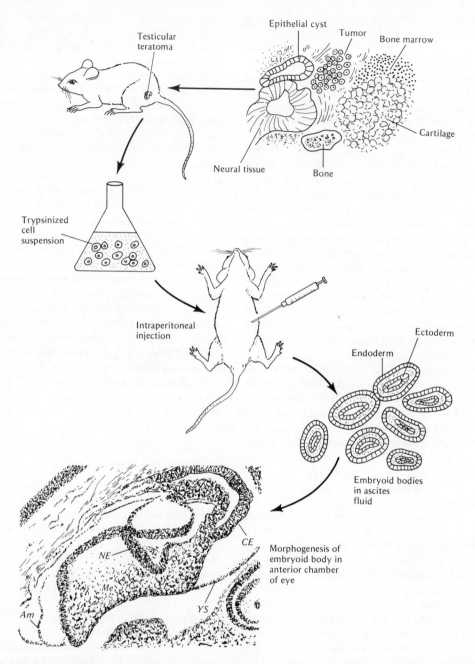

**Figure 17-3**    *Totipotent properties of a mouse testicular teratoma. Teratoma consists of many disorganized but differentiated tissues mixed with tumor cells. A trypsinized cell suspension injected into the peritoneal fluid of the mouse will develop into large numbers of embryoid bodies that resemble early embryonic stages. An outer endoderm surrounds an inner ectoderm and mesenchymatous tissue. If these are grown in the anterior chamber of the eye they undergo extensive morphogenesis and sometimes resemble more advanced embryonic stages with neural epithelium (NE), mesenchyme in somite region (Me), yolk sac epithelium (YS), resembling coelomic epithelium (CE), and amnion (Am).*

totipotent is a tumor of the mammalian gonad known as a *teratoma* or embryo carcinoma. Most spontaneous teratomas consist of a jumble of somatic tissues in various stages of histogenesis, intermingled with malignant tumor cells (Figure 17-3). The multipotential nature of the tumor cell is demonstrated by clonal isolation experiments with a mouse testicular teratocarcinoma. This tumor contains 12 different somatic tissues such as bone, muscle, and pigment along with tumor cells. Single carcinoma cells may be isolated and injected into peritoneal cavities of mice to establish individual clones. About 10 percent of these single cell grafts produce teratocarcinomas with various somatic tissues. Each tumor clone develops a specific spectrum of somatic tissues, suggesting that each pluripotent progenitor cell has different developmental potential.

Some carcinoma cells are pluripotent rather than totipotent since they do not differentiate into all tissue types. For example, some teratomas may form neural tissue primarily as if they are committed to only one of many phenotypic programs. The conditions under which such tumors are grown may limit expression of their totipotency. Illmansee and Mintz have improved conditions by injecting single teratoma cells from mice of one genotype into early blastocysts of mice obtained from another genotype and re-implanting these blastocysts into foster mothers. Under these conditions, the injected teratoma cell develops as a clone of cells which can be distinguished phenotypically (by enzymatic patterns) from host cells. The transplanted blastocysts develop into normal young, many of which contain mosaic tissues, with cells derived from the original teratoma cell. A wide range of tissues contained progeny from the tumor cell, indicating it is indeed totipotent.

Such teratomas may be descendents of totipotent germ cells. First, they arise initially in testicular tubules of fetal mice. Second, mice bred for the congenital absence of germ cells do not develop teratomas. And, finally, genital ridges from 12-day fetuses transplanted into the testes of adult mice will induce carcinomas, whereas genital ridges from earlier stages, before invasion by primordial germ cells, fail to induce carcinomas when transplanted.

Normal embryonic cells develop into teratomas when transplanted into testes of adult mice. Disaggregated 6-day-old mouse embryos transplanted into testes develop into disorganized teratomatous growths which may be transplanted indefinitely, consistently giving rise to the same pattern of disorganized growth. Embryos older than 6 days do not produce a teratoma when transplanted. At these later stages, most embryonic cells are determined and have lost totipotency, a property essential for the establishment of the teratoma.

Sometimes teratomas display organized growth, almost resembling the early development of the mammalian embryo. When teratomas are disaggregated and injected intraperitoneally into mice, they develop into thousands of free-floating embryoids, small vesicles consisting of ectodermal and endodermal tissues organized just like the early mouse embryo of 5 to 6 days gestation. Mesenchyme cells are found in some, and each tissue layer gives rise to its typical derivatives, but again in a disorganized manner. Though cells are neoplastic, they may recapitulate the early phases of embryogenesis just as somatic cells of a carrot root recapitulate embryogenesis in culture.

This dual origin of teratomas from primordial germ cells and undetermined embryonic cells demonstrates that the totipotent state is itself transmissible through many generations without alteration of genetic potential. Abnormal neoplastic cells with high mitotic potential perpetuate the totipotent state. They serve as a stem cell population, continuously producing clones of cells that specialize, albeit in a disorganized manner.

We will see that totipotency may be a cryptic potentiality of all cells, animal as well as plant. Since every cell in an organism has a complete set of genetic instructions, there are some cues that may evoke expression of more genetic information than is usually expressed in a single phenotype and new cell phenotypes arise. Such cues are present within egg cytoplasm and can be studied by transplanting somatic cell nuclei into eggs.

### Nuclear transplantation and totipotent nuclei

Nuclear transplantation tests the stability of nuclear changes accompanying cell differentiation in frog embryos (see Chapter 11). Nuclei from cells at different embryonic stages can be transplanted into enucleated amphibian eggs to test their developmental potential. We have already seen that blastula nuclei are totipotent; that is, up to 80 percent of transplanted blastula nuclei initiate normal development through hatching.

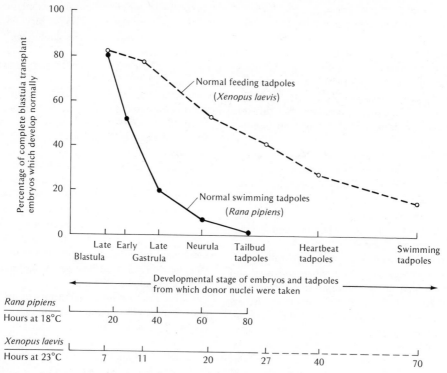

**Figure 17-4** *Nuclei from progressively older embryonic cells exhibit a declining ability to promote normal tadpole development after transplantation into egg cytoplasm. Vegetal hemisphere and endoderm nuclei were transplanted from embryos at stages shown. Data are expressed in terms of percentage of complete blastula transplant embryos. (From J. B. Gurdon, Quart. Rev. Biol. 38:54, 1963. Reproduced with permission of The Quarterly Review of Biology.)*

As a frog embryo develops, nuclei seem to lose developmental capacity. Only 50 percent of early gastrula endoderm nuclei give normal development after transplantation, and by the neurula stage less than 10 percent of transplanted endoderm nuclei yield tadpoles (Figure 17-4). Endoderm nuclei from tail-bud embryos cannot support normal development after transplantation. This pattern of decreasing developmental expression of nuclei from older embryonic cells seems to correlate with a tissue's loss of developmental capacity as it is determined.

One hypothesis to explain the decline in developmental potential is that stable genomic changes cause restriction in potency; that is, when cells are determined or differentiated, their nuclei experience an irreversible genetic change that silences large blocks of genetic information.

A test of this hypothesis is not simple because many factors contribute to restrictions in developmental capacity. For example, nuclei from older embryonic cells replicate more slowly than cleavage or blastula nuclei. Generation time of a cleaving blastomere is 25 minutes to 1 hour. Generation times of 2 hours or less characterize blastula cells and cell cycle times lengthen thereafter. Slowly replicating older nuclei transplanted into eggs, unable to keep pace with the rapid division cycle of egg cytoplasm, replicate abnormally. Chromosomal aberrations are often produced in the first few divisions after nuclear transplantation, and aneuploidy is typical of most abnormal embryos. This accounts for the failure of some nuclei from late embryonic tissues to support normal development. The apparent loss of developmental potentiality may be an artifact of the transplantation procedure and not a genetic change associated with differentiation.

Though most abnormal transplant embryos do contain visible karyotypic abnormalities, some aberrant embryos develop with visibly normal karyotypes; that is, chromosome number and gross chromosome structure are unchanged through several serial transplant generations. For example, first-generation transfers of late endoderm nuclei produce a spectrum of abnormalities, from failure to cleave to defective tadpoles (Figure 17-5). If some of the normal blastulae are used as donors for second-generation transfers, their nuclei give rise to clones, all developing identically, expressing the full potency of the original donor endoderm nucleus. Some clones develop only to the gastrula stage; other clones emerge as abnormal tadpoles with microcephalic syndromes. By serial transplantation, such clones may be carried through several transplant generations and in most cases normal karyotypes are seen. No gross changes in chromosome number are noted, in spite of the abnormal developmental pattern. Loss of developmental potentiality in the original endoderm nucleus may involve a stable genetic change, but this is only a tentative conclusion since most karyotype analyses cannot detect small chromosomal deletions or translocations. Small nonvisible chromosomal aberrations could account for the loss in developmental capacity of nuclei from these cells.

Some support for the hypothesis that stable genetic changes accompany determination and differentiation of embryonic nuclei has come indirectly from transplants of primordial germ cell nuclei. Primordial germ cells of the frog are large, yolk-filled cells migrating along the developing gut in the early tadpole (see Chapter 11). If primordial germ cell nuclei are transplanted into enucleated eggs, more than 40 percent of the embryos develop normally; at this stage germ cell nuclei are totipotent. By contrast, endoderm cells taken from adjacent gut tissues, though comparable in size and developmental age, do not give any normal embryos after transplantation (Table 17-1). One possible difference between endoderm and primordial germ cells is that the latter are mitotically active at this stage, which may explain their success after transfer. Their nuclei easily enter the rapid cell cycle of egg cytoplasm and fewer chromosomal abnormalities result.

Contrary results were obtained by Gurdon who

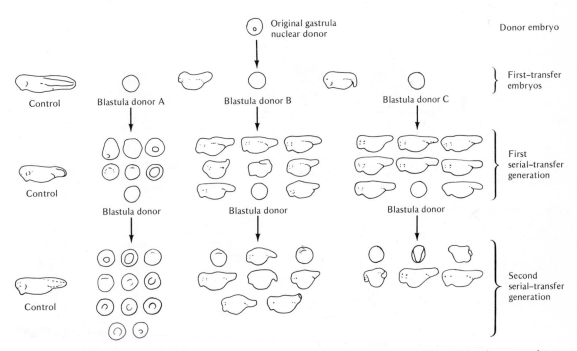

**Figure 17-5**  *Serial transplantation of endoderm nuclei in* Rana pipiens *to demonstrate stability of clonal patterns of development. All serial-transfer embryos from one donor resemble one another as do those in the second serial-transfer generation. In a few of these serial transfer clones, chromosomal karyotypes were normal. (From T. J. King and R. Briggs, Cold Spring Harbor Symp. Quant. Biol. 21:271–290, 1956.)*

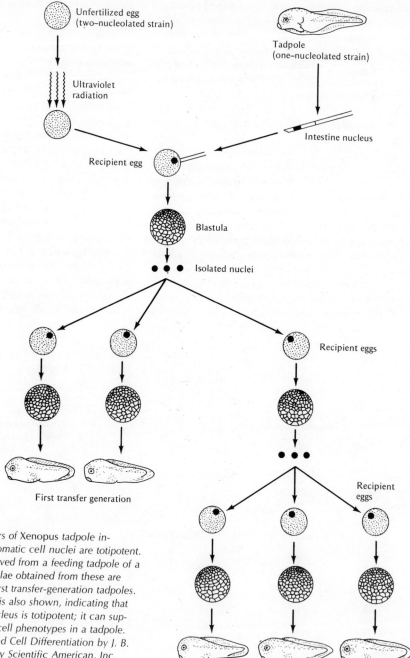

**Figure 17-6** *Nuclear transfers of Xenopus tadpole intestinal nuclei indicate that somatic cell nuclei are totipotent. The primary transfers are derived from a feeding tadpole of a one-nucleolated strain. Blastulae obtained from these are used to produce a clone of first transfer-generation tadpoles. A second transfer generation is also shown, indicating that the original intestinal cell nucleus is totipotent; it can support the differentiation of all cell phenotypes in a tadpole. (From Transplanted Nuclei and Cell Differentiation by J. B. Gordon. Copyright © 1968 by Scientific American, Inc. All rights reserved.)*

**TABLE 17-1  Developmental Capacity of Primordial Germ Cell Nuclei as Tested by Nuclear Transplantation (Rana pipiens)**

| Nuclear Donor | Number of Nuclear Transplants | Percentage of Normal Blastulae | Percentage of Normal Blastulae That Develop into Normal Stage-25 Tadpoles |
|---|---|---|---|
| Blastula, stage 8[a] | 193 | 74 | 74 |
| Germ cell, stage 25[a] | 410 | 19 | 40 |
| Endoderm, stage 23[a] | 382 | 7 | 0 |

[a] Stage 8, mid-blastula; stage 25, feeding tadpole; stage 23, swimming tadpole.
Source: Modified from L. D. Smith, *Proc. Nat. Acad. Sci. U.S.* 54:101–107, 1965.

showed that somatic cell nuclei are totipotent. He transplanted nuclei from visibly differentiated tissues of *Xenopus* tadpoles, that is, cells of the striated border of the intestinal epithelium of feeding tadpoles (Figure 17-6). These differentiated cells possess a row of cilia at their distal ends. Using the anucleolate mutant to mark donor nuclei, he obtained a low frequency of normal development after nuclear transfer.

Ten embryos out of 700 transfers (approximately 1.5 percent) developed normally into swimming tadpoles and some were reared to sexually mature adults (Table 17-2). These nuclei were totipotent and had not changed irreversibly during differentiation. Gurdon claims that the low percentage of totipotent nuclei is a technical artifact, since most donor cells are small and their chromosomes are easily damaged during transplantation. Moreover, most donor cells are probably postmitotic with long cell cycles, ill-prepared to replicate normally in egg cytoplasm.

To eliminate this problem, Gurdon entrained donor intestinal nuclei to a rapid replicative cycle by transfer to an egg. Some of these developed into normal blastulae with cell cycles more typical of early stages. These blastulae were then used as donors for serial transplantation. All nuclei from such blastulae made up a nuclear clone derived from one original intestinal cell nucleus. Nuclear transfers from these blastulae gave many more normal embryos, and Gurdon calculated that approximately 70 percent of all intestinal epithelial nuclei were able to support development to a tadpole stage. According to his interpretation, most nuclei of this tissue are totipotent; that is, in egg cytoplasm they give rise to all cell phenotypes. Others point out, however, that intestinal epithelium at this stage still contains

wandering primordial germ cells. The small percentage of successful development after first-generation transfer could have come from totipotent germ cells mistakenly sampled as striated border cells.

Gurdon and Laskey countered this objection by transplanting nuclei from tail-fin epithelial cells cultured in vitro. In this instance, epithelial cells grow out uncontaminated by germ cells. Using serial transplantation to entrain donor nuclei to the cell division rate of a cleaving egg, they could show that here, too, a small percentage of donor nuclei is totipotent. Again, making allowance for artifacts, they calculated that most epithelial cells growing in vitro contain totipotent nuclei.

**TABLE 17-2  Development of Embryos Prepared by Transplanting Nuclei from Specialized Cells into Enucleated Unfertilized Eggs of Xenopus laevis**

| Donor Cells | Percentage of Total Transfers Reaching Tadpoles | | |
|---|---|---|---|
| | Total Transfers | First Transfers | First and Serial Transfers |
| Intestinal epithelial cells of feeding tadpoles | 726 | 1.5 | 7 |
| Cells grown from adult frog skin | 3546 | 0.03 | 8 |
| Blastula or gastrula | 279 | 36 | 57 |

Source: From J. B. Gurdon, *The Control of Gene Expression in Animal Development.* Oxford: Clarendon Press, 1974. Data from J. B. Gurdon, *J. Embryol. Exp. Morphol.* 10:622–640, 1962, and J. B. Gurdon and R. Laskey, *J. Embryol. Exp. Morphol.* 24:227–248, 1970.

The experiments suggest that if stable nuclear changes are associated with cell differentiation, they are not irreversible. Nuclei from specialized cells may, after many replications, be reprogrammed by exposure to egg cytoplasm. As in plants, it would appear that the genome is unaltered in animal development; the amount and kind of genetic information is constant in all cells. This supports the argument that animal cells are potentially totipotent.

If stable genetic changes do, in fact, accompany the determined and differentiated states, they may be reversed when nuclei are exposed to egg cytoplasm. Earlier (Chapter 11), we showed that egg cytoplasm can reprogram stable nuclear genomes; red blood cell and brain cell nuclei are induced to replicate DNA, indicating that stringent genetic controls can be reversed by replicative signals from egg cytoplasm. Embryonic nuclei, even those taken from late tadpoles, do revert and enter cleavage mitosis. Adult brain nuclei, however, though stimulated to synthesize DNA in egg cytoplasm, do not engage in mitosis. These adult nuclei seem less able to revert to an embryonic state, suggesting that their genomes are more stable.

Summarizing, nuclear transplantation demonstrates that somatic cell nuclei do not suffer irreversible genetic changes during development. Neither the amount of genetic information nor its organization is permanently altered as cells differentiate. Given proper conditions, genetic information may be reprogrammed to evoke a totipotent pattern of development. Some somatic cell nuclei are more sensitive to reprogramming cues than others, possibly because they are in a more responsive part of the cell cycle when transplanted. Totipotency is never lost; it is simply repressed, as cells enter the determined state.

## THE DETERMINED STATE

Early in spirally cleaving embryos, individual blastomeres lose developmental potency and become committed to a specific phenotype (see Chapter 11). Likewise, in vertebrate development, tissues become determined during gastrulation. Determined tissues (or cells) differentiate according to their prospective fate irrespective of the environment in which they are placed; they are committed to carrying out limited developmental programs. They resemble totipotent cells; they are not functionally specialized

(neither morphologically nor biochemically) but behave as progenitors of cell populations that do visibly differentiate.

The determined state is stable and transmissible over many cell generations without change. Transitions from totipotent to determined states must, therefore, involve stable genetic changes. By most operational tests, changes in state occur in quantal jumps; cells seem to become "more" determined with each successive division. After the first step, most cells continue to express a wide range of phenotypes; they are pluripotent. Presumptive neural tissue, for instance, at an early neural plate stage can produce any tissues of the brain or spinal cord. After regionalization, anterior tissue is determined as brain and posterior tissue is committed to spinal cord; they are not interchangeable. In time, cells are further diverted into a single developmental track toward one cell phenotype. Brain is segregated into regionally specific domains and sense organs. At each step, the total genome available for expression is progressively restricted. These changes are probably most critical during development of a cell clone. How do these changes come about?

### Insect imaginal discs: a determined cell population

Insect imaginal discs are probably the best examples of determined cell populations. They arise at the end of cleavage after several thousand nuclei migrate to the periphery to form a cellulated blastoderm. Most blastoderm cells develop into larval tissues but some are determined to become adult tissue after metamorphosis. These (usually small groups of 2 to 30 cells) are sequestered as primordia for organ-specific imaginal discs such as leg, wing, eye-antenna, or genitalia, and remain relatively quiescent during embryogenesis. After hatching, they proliferate actively into visible imaginal discs, single or paired anlages distributed throughout the larval body (Figure 17-7). Disc cells proliferate during each molting cycle (three larval instars in *Drosophila*) until metamorphosis when they differentiate into specific adult structures.

Again, determination is defined operationally. A disc of one genetic strain transplanted into the abdomen of a larva from another strain metamorphoses into a strain-specific organ and no other; a wing disc always becomes a wing, a leg disc an adult leg. Neither the genetic constitution of the host nor the site of implantation has any effect on autono-

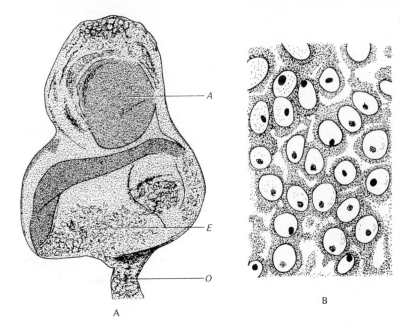

**Figure 17-7**  *(A) Eye antennal disc of* Drosophila melanogaster *in late third-instar larva (before pupation). Living preparation. A = antennal disc; E = eye disc; O = optic stalk (mag. 160X). (B) Example of imaginal disc cells with large nuclei, typical of all discs (mag. 1000X).*

mous differentiation of a disc. Its cells are committed to a specific phenotypic program and they express it irrespective of influences from the host. By such criteria, tissues are defined as determined.

All imaginal disc cells look alike; they are small, with large nuclei, and morphologically undifferentiated. Only at metamorphosis is their intrinsic phenotypic bias revealed. A male genital disc, for example, transplanted into a third-instar host, develops into adult genitalia consisting of a pair of vasa deferentia that hook up with a single testis, a pair of accessory glands called paragonia that empty into a single common structure, the contractile ductus. This transports sperm into a small gland, the sperm pump, that propels sperm into the chitinous external genitalia, consisting of a single penis, paired claspers, and anal plates (Figure 17-8). The disc differentiates into paired structures (anal plates, paragonia, vasa, and claspers) and unpaired structures (ductus, penis, single testis).

The state of determination of parts of the third instar disc is evaluated by cutting a genital disc into fragments and implanting each into a third instar host where they differentiate into adult structures when the larva metamorphoses. Each part is autonomous, differentiating into a specific structure of the genitalia according to a pattern. Paired structures come from the two lateral enlargements while single structures such as sperm pump and testis come from medial regions. Each region of the mature disc is committed to develop into a specific organ of the adult genitalia; the disc is a mosaic of organ-forming regions (see Chapter 16).

Determination is a state assigned to each disc cell and is not a property of a population or region. Mixtures of genetically marked wing and genital disc cells, when reimplanted into a larval abdomen and allowed to metamorphose, will sort out and develop according to their respective fates. No wing cells participate in formation of a genital structure; the cells aggregate according to disc-specific affinities. Even within the same imaginal disc, region-specific sorting out occurs; cells from the distal part of a wing disc do not cooperate in tissue formation with cells from more proximal regions.

The developmental expression of a third-instar imaginal disc tissue is dependent upon the mass of cells available at the time of metamorphosis (Figure 17-9). A half-fragment of a genital disc implanted into a third instar develops as a mosaic; several structures are absent. If the fragment is implanted into a young larva (second instar) or repeatedly transplanted into old host larvae and allowed to grow, then complete regeneration occurs. Regenerative growth of medial portions probably restores the lost parts, as described in Chapter 16. This disc behaves as other discs;

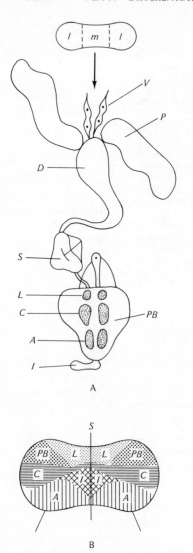

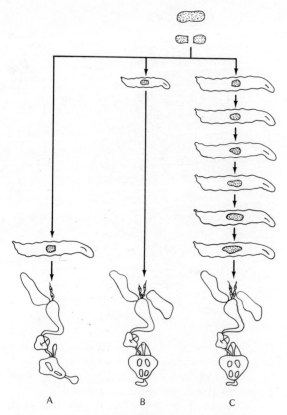

**Figure 17-9**   *Regenerative potential of genital disc tissues. See text for details. (Based on H. Ursprung, Develop. Biol. 4:22–39, 1962.)*

**Figure 17-8**   *Mosaic development of a late third-instar genital disc of Drosophila as tested by transplanting into abdomens of third-instar larvae. (A) Diagram of the disc showing lateral (l) and medial (m) regions and the genital structures derived from it. A = anal plate; C = clasper; D = ejaculatory duct; I = intestine; L = lateral plate; P = paragonium; PB = peripheral bristles; S = sperm pump; V = vas deferens. (B) A mosaic fate map of some soft and chitinous parts derived from the disc.*

medial parts regenerate while lateral parts duplicate.

When do these cells become determined?

A test of determination for early blastoderm tissues was developed by Chan and Gehring, who

took advantage of the fact that embryonic and larval tissues can be cultured indefinitely in adult *Drosophila* abdomens (Figure 17-10) without differentiating. Fragments of imaginal discs are serially transplanted from one abdomen into another where they grow vigorously. Disc tissues can be made to differentiate only in response to the metamorphic hormone ecdysone, which is absent in adults but present in larvae. If cultured tissues are transplanted into third-instar larval abdomen, just before metamorphosis, all undifferentiated imaginal disc cells will then express their adult fate.

Chan and Gehring isolated anterior and posterior halves of genetically marked blastoderm stages, dissociated the cells, and mixed them with cells derived from whole blastoderms of a different genetically marked strain (Figure 17-11). The cells were pelleted by centrifugation and fragments were implanted into adult abdomens for 10 to 14 days

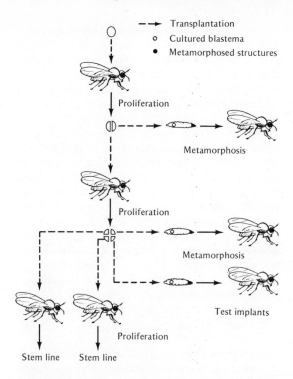

Transplantation

Cultured blastema

Metamorphosed structures

Proliferation

Metamorphosis

Proliferation

Metamorphosis

Test implants

Proliferation

Stem line    Stem line

**Figure 17-10** *Method for long-term culturing of imaginal disc fragments in vivo in adult abdomens where metamorphosis does not occur. Differentiation of test implants obtained by growing in abdomen of third-instar larvae which undergo metamorphosis under the influence of the molting hormone ecdysome. (From W. Gehring,* The Stability of the Differentiated State, *ed. H. Ursprung. New York: Springer-Verlag, 1968, pp. 136–154.)*

growth, then removed (now a mixture of larval and imaginal disc tissue) and reimplanted into third-instar larva for metamorphosis and differentiation. If anterior and posterior cells of the blastoderm are indeed totipotent, then the resulting tissues should be whole body mosaics with cells derived from the anterior region able to form whole tissue chimeras of both head and abdominal structures. If, however, anterior and posterior cells are each determined for head and abdomen respectively, then anterior cells will form only head chimeras while posterior cells develop chimeras for posterior organs. The results are consistent with the second prediction; anterior cells are already committed to form adult head structures while posterior cells are committed to form abdominal organs. By the blastoderm stage, determination of imaginal disc cells has already taken place along the anterior-posterior axis of the embryo, with head structures at the anterior pole, thorax in the middle, and abdominal structures at the posterior end.

How can we reconcile this result with the results of nuclear transplant studies which reveal that blastoderm nuclei are totipotent? One possibility is that nuclei are reprogrammed when exposed to egg cytoplasm. Stable nuclear changes accompanying determination might persist as long as nuclei continue to function in their own cytoplasm. Only intact cells or blastoderm tissues, therefore, would behave as if they are determined, as, in fact, they do. On the other hand, nuclei removed from a specific cytoplasm and transplanted into an egg would be

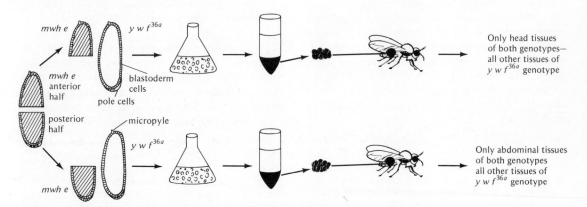

$mwh\ e$    $y\ w\ f^{36a}$

$mwh\ e$
anterior
half

blastoderm
cells

posterior
half

pole cells

Only head tissues
of both genotypes—
all other tissues of
$y\ w\ f^{36a}$ genotype

micropyle

$y\ w\ f^{36a}$

$mwh\ e$

Only abdominal tissues
of both genotypes
all other tissues of
$y\ w\ f^{36a}$ genotype

**Figure 17-11** *Early determination of imaginal disc tissues. Fragments of anterior and posterior halves of the multiple-wing-hair genotype (mwh e) were mixed with whole blastoderms of another genotype containing yellow body genes (y w f³⁶ª). Cells were pelleted and cultured in adult abdomens before metamorphosis in third-instar larvae. (Based on L. N. Chan and W. Gehring,* Proc. Nat. Acad. Sci. U.S. *68:2217–2221, 1971.)*

subjected to reprogramming signals that might erase phenotypic biases and restore totipotency. Phenotypic stability seems to depend upon cytoplasmic conditions (states?) favoring expression of specific gene sets. Such states may be stabilized as cells interact in tissue populations. Isolation of a cell from a tissue may make it more labile and responsive to environmental cues.

## Transdetermination

If the determined state is stable, then imaginal disc cells cultured for long periods of time should give progeny with the original phenotypic bias. We will see that this is usually the case but some lability is evoked after long-term culture.

Hadorn maintained *Drosophila* imaginal disc tissues in adult abdomens as vigorously growing undifferentiated cells for many transfer generations. Their capacity to differentiate into adult structures was tested by reimplantation of tissue fragments into third-instar larval bodies before metamorphosis. By carrying out long-term culture in this way, Hadorn could test the stability of the determined state. Persistence of a specific disc phenotype over many cell generations would indicate that the determined state is based on stable genetic changes. A typical transfer pattern is shown in Figure 17-12, in which part of a genital disc was passed through many transfer generations in adult abdomens. Up until the sixth transfer generation, disc tissues always developed into anal plate structures. Evidently the original fragment isolated from the disc was made up largely of anal plate progenitor cells. Developmental expression seemed to depend on fragment size; small fragments differentiated into fewer structures than large fragments. Anal plate determination behaves as a stable genetic state, which Hadorn called *autotypic determination*. Other discs showed similar patterns of phenotypic stability in the first 6 to 10 transfer generations.

Not all imaginal disc sublines were conservative. In many clones, the state of determination changed after a few generations of autotypic expression. Some fragments metamorphosed into structures foreign to the original disc. For instance, between the sixth and tenth transfer generation of anal plate lines, legs, and palps appeared about the same time and with similar frequency among the various fragments. Later, wings appeared and then, finally, in some lines, thorax structures developed. Each new phenotype that appeared was itself stable for several transplant generations before another phenotype appeared. Hadorn called these newly emerging determined states *allotypic determinations*. These were perpetuated over many cell generations without change until they, too, underwent a transformation or, in Hadorn's words, a *transdetermination*. Evidently, after many replications a determined state becomes labile and acquires new properties.

Transdetermination was not entirely random; certain patterns emerged:

1. A shift to a new state was abrupt; there were no intermediate stages containing mixtures of cells of two phenotypes.
2. Transdetermination involved a shift to broader phenotypic expression — for example, from "anal plate" to "leg" or "wing" and not to a leg part such as trochanter.
3. Some transdeterminations followed a predictable sequence of phenotypes before a terminal state was stabilized. Some were reversible while others were not; some occurred with great frequency, others were rare.

Transdetermination correlates with rapid proliferation. Genetically marked leg and wing discs mixed before implantation in an adult host exhibit a high proliferation rate and 53 percent of the test implants are transdetermined. The same discs grown separately proliferate slowly and only 3 percent

**Figure 17-12**   *A transdetermination lineage for fragments of genital discs determined as anal plate structures. During first six transfer generations in adult abdomens, anal plate differentiation is stable. Subsequently, between the sixth and tenth transplant generations, some fragments give rise to leg and head structures. These transdetermined cell lines are stable for several generations, but some of these begin to transdetermine into wing and thorax structures. Other fragments retain their original biases (anal plate) for many more generations, while some lines may switch to other tissue types several times. In the inset is a summary of transdetermination sequences. Short arrows indicate rare determinations. Dotted arrow indicates that transdetermination to genital cells was observed but the cell type of origin is unknown. (Based on E. Hadorn, Develop. Biol. 14:424–509, 1966.)*

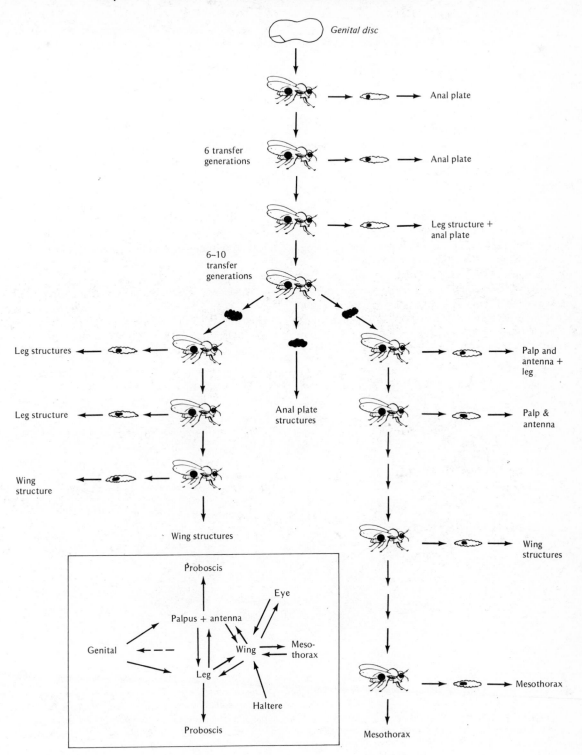

show any transdetermination. A linear correlation relates frequency of transdetermination with the number of cell generations. How, then, does cell division increase the rate of transdetermination?

Hadorn proposed that relatively stable cytoplasmic systems maintain the determined state over many cell generations but these are diluted out in rapidly proliferating populations. A new cytoplasmic state replaces the old at random and triggers different sets of genes to produce a new phenotype. The fact that no intermediate states are observed during transdetermination suggests that a switch coordinately activates sets of "leg genes" as distinguished from sets of "wing genes." Cell division may activate these switches by stripping DNA of its regulatory proteins during replication and allowing new gene sets to be regulated at random.

Several naturally occurring mutant genes in *Drosophila* spontaneously transdetermine during development. One such homeotic mutant is *aristapedia,* in which an antennal disc develops leg structures in place of a portion of the antenna. The mutation turns off antennal differentiation and turns on a leg program instead. Hadorn suggested that the aristapedia gene accelerates cell division of antennal disc cells, which favors leg gene expression. Some homeotic mutants do, in fact, respond to cell division inhibitors. Aristapedia discs treated with colchicine differentiate as antennae rather than legs, as if inhibition of cell proliferation favors expression of antennal genes.

Does a single cell or a small cell group actually transdetermine, or are we dealing with a selective process? An antenna disc, for example, may contain small numbers of leg and wing cells which, if allowed to proliferate, would produce populations large enough to differentiate. In other words, does a cell actually undergo genetic reprogramming, or is selection of preexisting cell types taking place when discs are cultured in adult abdomens for many generations?

To answer this question, clones of genetically marked cells from one disc type must be traced during transdetermination. Such marked clones are made by inducing somatic crossing over in early imaginal disc cells with X-irradiation (Figure 17-13).

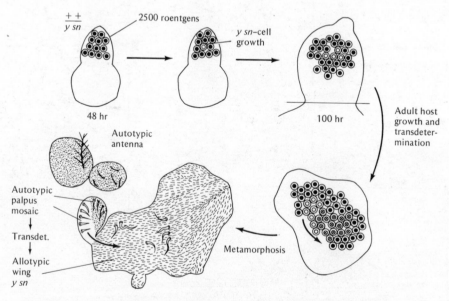

**Figure 17-13**  *Transdetermination (eye-antenna disc) occurs in a single cell developing as a clone. Black nuclei are heterozygous wild type for yellow and singed (y sn). X-irradiation induces somatic crossing over and a homozygous yellow-singed cell (light nuclei) is produced. Disc fragments proliferate in adult abdomens and undergo transdetermination. Allotypic wing tissue with yellow-singed genotype and autotypic antennal structures of two phenotypes appear, which means that progeny of a single clone have transdetermined. (From E. Hadorn,* Major Problems in Developmental Biology, *ed. M. Locke. New York: Academic Press, 1966, pp. 85–104.)*

As a result, a cell clone, bearing mutant marker genes, yellow body with singed bristles, in a background of wild type body color and bristle shape was obtained. An irradiated antennal disc cultured in adult abdomens transdetermines, and both normal autotypic tissue (antennal structure) and transdetermined allotypic tissue (wing structure) are marked by the genetic markers. Since crossing over is an infrequent event that occurs at random, both tissues could be derived only from the same marked cell. This eliminates selection of contaminating wing cells in antennal discs and means that one heritable state has transformed into another.

In summary, progressive determination means a cell clone loses the ability to retrieve the full range of phenotypic programming available to it in its genome. A genetic event accompanying determination seems to repress most genetic information; only a small portion is available for expression. The change is stable and heritable, but it is not irreversible since reprogramming can take place after rapid proliferation. How, then, are large portions of the genome silenced during embryogenesis?

## Cytoplasmic factors in determination

We do not understand the genetic mechanisms underlying determination; the state, itself, is defined only operationally. Boveri, in 1902, proposed that a cell is determined after its nucleus interacts with different cytoplasmic components of the egg. According to his model, the partitioning of cytoplasm during cleavage segregates genetically identical nuclei into different cytoplasms which stimulate nuclear changes. These, in turn, modify the cytoplasm so that subsequent progeny nuclei will undergo additional changes. This model of nucleocytoplasmic interactions is generally accepted today, although the evidence for it is incomplete.

In determinately cleaving eggs special cytoplasms may be qualitatively segregated into blastomeres and induce a precocious determination of cell fate (Chapter 11). Insect cells are determined after cleavage nuclei interact with peripheral egg cytoplasm during blastoderm formation. Destruction of certain peripheral cytoplasmic regions before they interact with cleavage nuclei inhibits development of specific structures in the larva or adult. Primordial germ cell nuclei are determined after interacting with germinal cytoplasms, a model of nucleocyto-

plasmic interactions that is applicable to many determining events. Since all egg cytoplasms are encoded in the maternal genome, it suggests that early determination patterns may be maternally inherited.

One classic case of maternal inheritance is the genetics of coiling in the shells of gastropod mollusks. Among these snails the direction of coiling of the shell is either clockwise (dextral) as viewed from the apex, or counterclockwise (sinistral) (Figure 17-14).

In the fresh-water snail *Limnaea*, all individuals are dextral, but infrequently sinistral types arise as recessive mutations. A cross between eggs of a dextral individual with sperm from a sinistral are genotypically *Dd* and all exhibit the dextral coiling pattern. On the other hand, eggs from a sinistral individual fertilized by sperm from a dextral individual, though of the same *Dd* genotype, are all phenotypically sinistral. Nonidentity in reciprocal crosses always suggests maternal inheritance. In both crosses above, the $F_1$ are phenotypically identical to the maternal parent, evidently controlled by the maternal genotypes, *DD* and *dd*, respectively.

An $F_2$ obtained by breeding dextral *Dd* individuals from the above cross does not exhibit the typical 1:2:1 segregation. Instead, all individuals are phenotypically dextral. Though genotypic segregation does take place, a ratio of 1 *DD* to 2 *Dd* to 1 *dd* individuals in the population, all display the dextral phenotype because all come from maternal parents with the genotype *Dd*, with the dominant wild type gene *D* responsible for dextral coiling in the $F_2$.

In the $F_3$ segregation finally occurs, this time according to broods, reflecting the genotype of the maternal parent. All *DD* individuals above will produce dextral offspring, all *Dd* heterozygotes will also produce dextral $F_3$, while the one-quarter of the $F_2$ that are homozygous *dd* will yield all sinistral offspring in the $F_3$. The $F_3$ ratio will be three dextral to one sinistral, an $F_3$ phenotypic ratio that reveals the true genotypic segregation of the $F_2$.

Most significantly, the coiling pattern of the adult shell correlates with the pattern of spiral cleavage exhibited early in development. Eggs from dextral females exhibit dextral spiral cleavage at the second division while eggs from sinistral females display a counterclockwise or sinistral cleavage pattern. The position of the D blastomere is different in each case. Since this blastomere gives rise to the major internal organs of the adult, such as heart, kidney, and gonads, its position at the end of cleavage will also determine shell asymmetry. The primary time of gene action is during cleavage. Individuals carrying the wild type allele produce eggs with a *D* factor that favors dextral cleavage. Individuals lack-

ing the gene either fail to introduce the *D* factor or produce a "*d* factor" instead which favors sinistral cleavage, and eventually sinistral coiling. Such factors may control the orientation of centrioles in

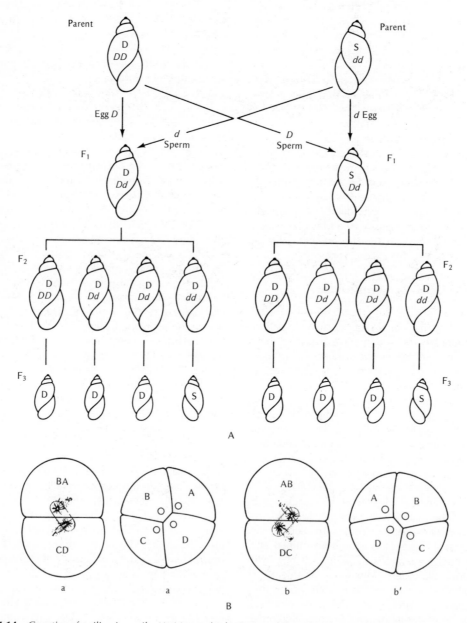

**Figure 17-14**   *Genetics of coiling in snails. (A) Maternal inheritance of dextral and sinistral coiling in* Limnaea peregra. *D = dextral phenotype; S = sinistral phenotype; D = gene for dextrality; d = gene for sinistrality. (B) Beginning and end of second cleavage in eggs of sinistral (a, a') and dextral (b, b') snails.*

cleavage divisions. In this instance, early interactions between cytoplasmic components and the genome establish future planes of symmetry in the adult.

A more recent example of maternal control of early developmental events has been analyzed in *Axolotl,* which has a developmental mutation known as the *o* gene.

The *o* gene is a maternally inherited, single Mendelian gene. This means the mutant phenotype is only expressed in offspring derived from homozygous mothers, *o/o.* All crosses between heterozygous individuals (+/o) yield normal offspring, but the *o/o* progeny show some slight abnormalities at the larval stage and these can be selected for breeding. All *o/o* males are sterile; they do not differentiate spermatocytes (Figure 17-15).

A cross between an *o/o* female and normal +/+ male produces +/o zygotes. All cleave and develop into normal-appearing blastulae that are, however, unable to gastrulate. They remain viable for awhile and eventually die. The reciprocal cross of an *o/+* male with a heterozygote *o/+* female always results in normal development even among the quarter of the population that are homozygous *o/o.* It is the maternal genotype and not that of the zygote that determines the phenotype. If the maternal parent possesses a wild type allele, all her eggs develop normally, irrespective of their genotype after fertilization. This suggests that homozygous *o/o* females fail to produce a normal egg during oogenesis; that is, these eggs probably lack a cytoplasmic factor found in normal eggs.

To test this hypothesis cytoplasm from normal eggs was injected into eggs obtained from *o/o* females. Most of these eggs gastrulated and some developed to abnormal neurulae (Table 17-3). These

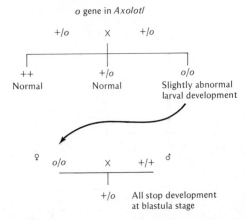

Figure 17-15 *Maternal inheritance of o gene in Axolotl. Almost all offspring from cross of heterozygotes are normal irrespective of genotype and develop into fertile adults. The o/o genotype can be distinguished because its developmental pattern is slightly delayed and abnormal. All offspring of female homozygotes o/o fail to develop beyond the blastula stage irrespective of heterozygous genotype.*

results showed that a component in normal egg cytoplasm could partially correct the defect of an *o/o* egg. It was further shown that large amounts of *o* factor were localized in the germinal vesicles of wild type oocytes since germinal vesicle contents injected into *o/o* eggs induced many to develop to tadpole stages. Evidently, during oogenesis, *o/o* oocytes do not accumulate a specific morphogen that is necessary for gastrulation. Wild type eggs obtain this factor when the germinal vesicle breaks down at the beginning of polar body formation.

**TABLE 17-3   Effect of Cytoplasm from Normal Mature Eggs on the Development of Eggs of o/o Females**

| Material Injected | Number of Eggs Injected | Number of Complete Blastulae | Development of Blastulae | | | |
|---|---|---|---|---|---|---|
| | | | Arrested Blastulae and Gastrulae | Gastrulae | Neurulae | Postneurulae |
| Control (nothing injected) | 0 | 1208 | 1208 | | | |
| Cytoplasm from eggs of normal females | 267 | 186 | 4 | 6 | 29 | 147 |

Source: From R. Briggs and G. Cassens, *Proc. Nat. Acad. Sci. U.S.* 55:1103, 1966.

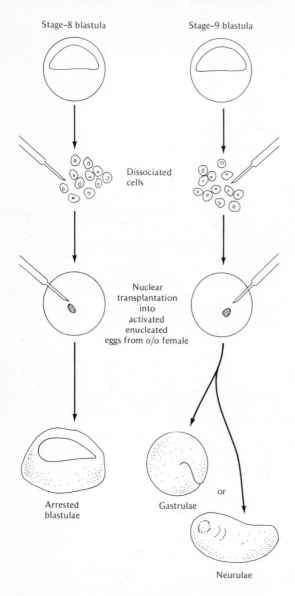

Stage-8 blastula

Stage-9 blastula

Dissociated cells

Nuclear transplantation into activated enucleated eggs from o/o female

Arrested blastulae

Gastrulae

or

Neurulae

**Figure 17-16** *Nucleocytoplasmic interactions in axolotls in which the o factor seems to induce stable changes in donor nuclei. Normal blastula nuclei from stage 8 and stage 9 are transplanted into enucleated o/o eggs. Only those from the late stage blastula overcome the block at the blastula stage and develop into abnormal gastrulae or neurulae. A stable change has taken place in nuclei of a late blastula. (Based on A. J. Brothers,* Developmental Biology, *ed. D. McMahon and C. F. Fox, ICN-UCLA Symp. Mol. Cell. Biol., vol. 2. Menlo Park, Calif.: Benjamin, 1975, pp. 338–343.)*

So far, attempts to characterize this factor have been only partially successful; it seems to be a protein. It is not species-specific, however; germinal vesicles from other amphibian species also rescue o/o eggs after injection. The factor acts before gastrulation since nuclear RNA synthesis is not activated on time in o/o blastulae. This means the "o" factor may be necessary for transcription in embryonic nuclei at the blastula stage.

We do not know how "o" factor works, but some recent experiments suggest that the factor might be involved in nucleocytoplasmic interactions (Figure 17-16). Wild type nuclei have been transplanted from normal donors into enucleated o/o eggs. Donor nuclei taken from stage-8 blastula cells have no effect on the development of o/o eggs; they fail to gastrulate. On the other hand, nuclei transplanted from stage-9 blastulae, only several hours older, do support development of o/o eggs beyond gastrulation and may even result in abnormal tadpoles. Evidently a normal blastula nucleus is stabilized between stages 8 and 9 and is capable of correcting development of o/o eggs; they gastrulate even in the absence of o factor. One possibility is that blastula nuclei interact with o factor and are programmed into a stable functional state that is independent of further interactions with o factor. The results do indicate that a normal late blastula nucleus is determined and can support development of defective o/o eggs through gastrulation. Possibly after interacting with cytoplasmic o material during cleavage nuclei are stabilized into a new developmental state and are not reprogrammed by replication in defective o/o cytoplasm. A stable nuclear change seems to be induced by a maternally inherited cytoplasmic factor.

## Extrinsic controls of determination: epigenetic signals

Determinative signals may be exchanged between cells or between cells and the surrounding environment. Neural induction is only the first of a series of tissue dialogs that determine cell fate. In this instance, determinative signals come from the chordamesoderm. In contrast to imaginal discs, these determinative states are transitory; neurally determined tissues shift into more stable states when brain, spinal cord, and sense organs differentiate. A few cell populations in vertebrate embryos remain determined and pluripotent for some time, displaying a repertoire of phenotypes in response to environmental cues. One of these, as we have seen, is neural crest (see Chapter 12).

Though the pluripotent nature of neural crest is still unproven, its behavior in vitro suggests it can differentiate as pigment cells or sensory ganglia. Chick embryo sensory ganglia produce a mixture of neurons and glial and pigment cells in vitro. The compact central region favors neuron differentiation whereas the less dense periphery, where cells make fewer contacts, promotes melanocyte differentiation. This effect of cell density suggests that some surface

change may signal a shift in the phenotypic expression of a pluripotent cell population.

Most neural crest phenotypes share a common metabolic pathway. Melanin pigments in melanocytes, the catecholamines in sympathetic ganglia, and noradrenalin of the adrenal medulla are all derivatives of tyrosine and phenylalanine metabolism (Figure 17-17). The pathways diverge at different points to give rise to specific end products characterizing each phenotype. Commitment to one or another phenotypic pathway may result from single switches in a common pathway. For example, a pigment cell utilizes the substrate dopa, oxidizing it to dopa quinone with the enzyme dopa oxidase. The

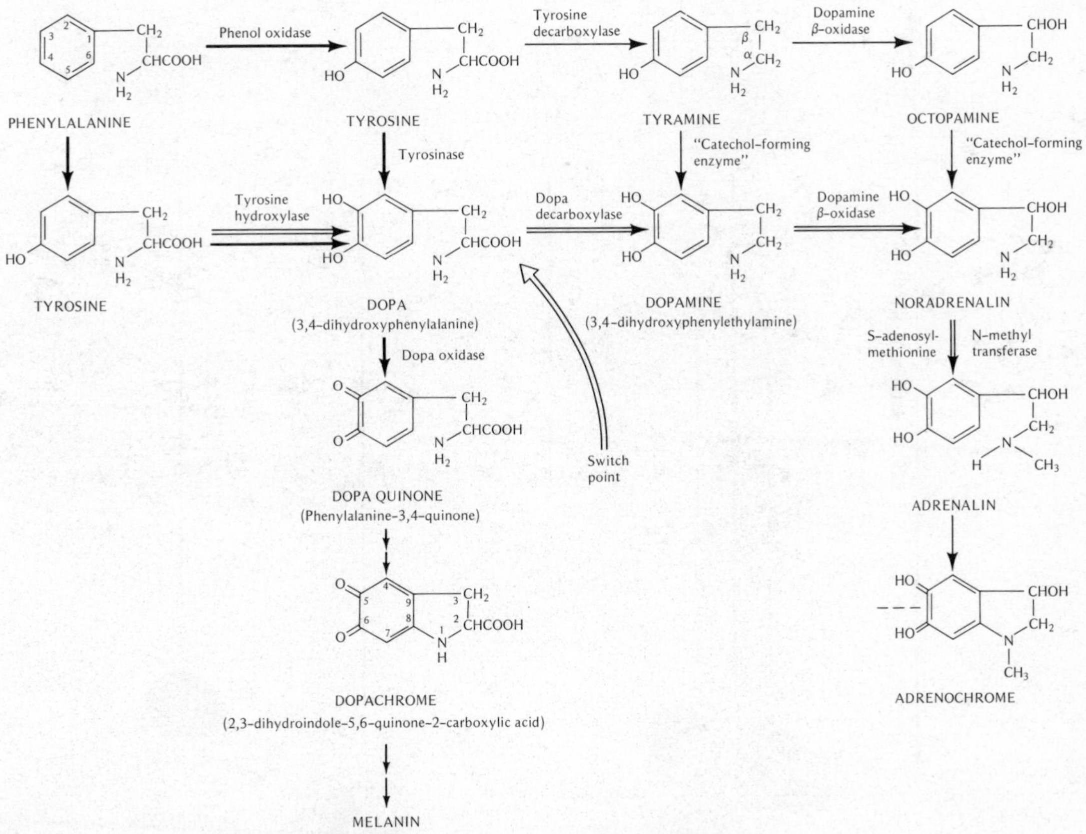

**Figure 17-17**  *Alternative metabolic pathways of tyrosine and phenylalanine metabolism that may operate in neural crest cells. Thick black arrows represent the melanin pathway in melanocytes. Open arrows show a pathway in catecholamine synthesis which may exist in sensory or sympathetic ganglion cells. Thin arrows are possible alternative pathways. A major switch point involves the metabolism of dopa; it may be oxidized to dopa quinone and thence to melanin or decarboxylated to dopamine and thence to noradrenalin.*

oxidase is either absent or present in low amounts in sympathetic ganglion cells. Ganglion cells decarboxylate the substrate dopa into dopamine instead and then to the catecholamines, noradrenalin, and adrenalin. Changes at the cell surface may trigger enzyme changes in a common metabolic pathway.

Pluripotent cell populations with a limited range of phenotypic commitments persist into the adult animal. In vertebrate bone marrow and spleen, undifferentiated but multipotent stem cells produce a variety of blood cell phenotypes. These progenitor cells are determined in a broad sense as "blood-forming cells." A stem cell may play out any one of five

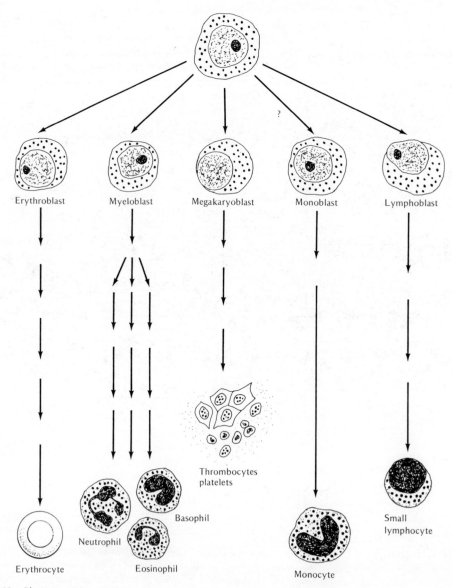

**Figure 17-18**   *Pluripotentiality of stem cell in bone marrow or spleen. Five different phenotypic pathways are shown, each with several intermediate stages indicated by a number of arrows. Question mark means pathway is uncertain.*

different blood cell programs, including the erythrocyte and myelocyte, a principal line of white blood cells (Figure 17-18). When committed to a phenotypic pathway, a stem cell completes a programmed number of divisions before accumulating large amounts of specialized cell products.

The multipotentiality of stem cells has been demonstrated by injecting bone marrow cells into mice whose blood-forming cells have been destroyed by X-irradiation. The injected cells populate the host spleen and establish foci of isolated colonies, each presumably derived from a single stem cell (or colony-forming unit, CFU). Each colony contains three different blood cell lines, erythrocytic, myelocytic, and megakaryocytic. If injected bone marrow cells are irradiated with low doses of X-rays to induce chromosomal aberrations as markers, the three different blood cell types all carry the same chromosome marker. This implies that all three cell lines come from the same stem cell, and supports the concept of stem cell pluripotency.

According to most models, a hormone or combination of hormones triggers each phenotypic pathway. The stem cells make up a renewal population and replenish their differentiated progeny in response to feedback signals in the circulation (see Chapter 6). Most undifferentiated stem cells are arrested in the $G_0$ phase but are still responsive to signals that shift them into $G_1$. The shift is associated with a determinative event controlled by a specific hormone; a stem cell becomes a committed progenitor.

Erythropoiesis, the pathway leading to erythrocyte differentiation, is controlled by a glycoprotein hormone erythropoietin, whose level in the blood is affected by oxygen tension. Under conditions of hypoxia, or in anemic blood, erythropoietin is in high concentration, and as blood percolates through bone marrow, stem cells respond to erythropoietin by acquiring an erythrocyte "bias." Reticulocytes are produced in large numbers after a period of intensive proliferation. It is equally possible that committed progenitor cells, already present in the population, respond to erythropoietin and proceed to proliferate. Within a few days spleen and plasma are populated by large numbers of red blood cells in different stages of maturation, and a new erythrocyte population is differentiated.

A specific hormone impinging on a multipotent cell population "selects" one of five possible phenotypic pathways. The determinative event may occur as cells shift into $G_1$. Erythropoietin may act directly on the genome, perhaps by stimulating transcription, inasmuch as the first detectable event after injection of erythropoietin is an increased synthesis of RNA. On the other hand, erythropoietin may first affect the cell surface, a change that is subsequently translated into secondary genome signals.

## THE DIFFERENTIATED STATE

The differentiated state is the terminal state of a determined cell clone and is characterized by specific morphological and biochemical properties. Whereas all determined cells resemble those in growth-duplication cycles, differentiated cells are postmitotic, synthesize specific proteins, assemble distinctive cytoarchitectures, and carry out specialized functions. Operationally, we define differentiation as that moment when tissue-specific proteins begin to accumulate and are first detected.

All cells of an organism share the same genotype and express certain gene sets in common, such as those coding for enzymes of basic respiratory metabolism, or protein synthesis. These are the "housekeeping" proteins required by all cells for survival and growth. Certain gene sets are unique to each cell phenotype; that is, they are usually expressed in one clone of cells and no other. This means that specific protein is actively synthesized in some clones while it is absent or barely detectable in most others. Genes coding for specialized proteins, the so-called luxury molecules such as hemoglobin, myosin, and amylase, are responsible for the visible phenotypic differences among cells.

Only a small proportion of the total genome is expressed in a differentiated cell. Besides housekeeping molecules, a specialized cell synthesizes only a small sample of its repertoire of luxury molecules. Information for all other luxury molecules encoded in the genome is usually repressed. If we assume that a vertebrate organism contains 100 different cell phenotypes, each with 10 cell-specific luxury proteins, the genome will therefore encode 1000 cell-specific proteins. Each cell phenotype, however, will synthesize only 10 of these or 1 percent of the total repertoire. This is an inevitable outcome of a process of determination that has silenced most of the genome.

A determined cell population, as we have seen, is committed to express one of many phenotypic pathways but has not done so. A differentiated population, on the other hand, has gone one step further and has transformed into an identifiable phenotype. It is the visible manifestation of the determined state. At the genetic level, a determined cell population has a limited number of gene sets available for expression. When exposed to permissive conditions cells use this information, synthesize specific proteins, and differentiate.

### Erythrocyte development: a paradigm of differentiation?

In studying differentiation, we first want a determined cell population that can be induced to differentiate at an appropriate signal. Second, we want cells with an easily characterized set of luxury molecules such as hemoglobin, or muscle proteins, to see how the relevant genes are regulated. Is the shift from the determined to the differentiated state an "all-or-none" phenomenon, or does the determined cell pass through a sequence of maturation stages, each progressively more stable and differentiated? If the former is true, then a differentiating tissue should consist of only two cell types, cells that synthesize luxury molecules at maximal rates and cells that are nonsynthesizing but determined. The second alternative implies cell heterogeneity with cells at various stages of differentiation, synthesizing luxury molecules at different rates.

We will find that most differentiating systems pass through a sequence of well-defined stages, or a cell lineage. Determined cells proliferate into daughter cells that progressively acquire functional properties of the specialized state; they occupy intermediate states of differentiation. These intermediate states differ from the terminal, mature phenotype in that they lack the full machinery of a mature cell.

Since our end point for differentiation is synthesis of specific luxury molecules, enough differentiated cells must accumulate before we can detect the protein product. We are, therefore, limited by the sensitivity of our assay. Synthesis may begin in a small number of partially differentiated cells, at rates too low to be detected. Only later, after many more cells differentiate and accumulate product, can we detect the first signs of differentiation. It is therefore difficult to decide when progeny of determined cells switch into a differentiated program.

Ideally, we would like to detect the first message molecules as they are transcribed, but this is difficult. Usually we can assay for the protein product, which is primarily a measure of translation and not when genes are acting. And in many systems there is often a delay between the moment of gene transcription and the time of message translation, and this in itself may be a principal mode of regulation in differentiating cells. A most sensitive assay for the detection of proteins is an immunological one. Specific antibodies are prepared by injecting purified preparations of luxury molecules into rabbits. The purified antibody is used as a highly sensitive reagent that can detect small numbers of antigenic protein molecules as they appear within a tissue. Tagging antibodies with a fluorescent dye makes them sensitive cytological reagents capable of localizing differentiated cells and their specialized products.

In most vertebrate embryos, erythrocyte differentiation is coupled to changing physiological requirements for oxygen transport. Embryo, fetus (or larva), and adult have different respiratory needs and have evolved special hemoglobins adapted to each developmental stage. As development proceeds, physiological demands change and embryonic or larval types of hemoglobin disappear as adult hemoglobins are synthesized. For example, in mammals, the earliest embryonic hemoglobins are replaced by a fetal variety (in humans) and ultimately at birth adult hemoglobin appears. The fetus, depending upon gaseous diffusion across placental membranes, must have a hemoglobin with a higher affinity for oxygen than maternal hemoglobin to expedite exchange. Such shifts in hemoglobin synthesis indicate precise regulation of gene expression and raise the following question: Is each hemoglobin species synthesized by a completely different cell lineage, or are embryonic hemotocytoblasts progenitors of all erythrocyte cell types? If the latter is true, then several sets of hemoglobin genes—embryonic, fetal, and adult—should switch on and off during growth of one erythroid cell clone. On the other hand, if different clones develop, then only one set of hemoglobin genes would be regulated in each clone. An answer emerges from studies of erythropoiesis in mammalian embryos.

Adult mammalian hemoglobin, as we have seen, is a tetramer, two alpha chains associated with two beta chains, each coded by unlinked genes, A and B. Early in mouse development, several different embryonic hemoglobins appear, $E_I$ ($xy$), $E_{II}$ ($\alpha y$), and $E_{III}$ ($\alpha z$), only two of which have alpha chains. Only the A gene is active in the embryo while other genes coding for three different polypeptide chains—x, y, and z—are transcribed instead. Only late in fetal development does adult hemoglobin appear as the B gene is activated Since the A gene is active throughout development, changes in hemoglobin result from the sequential regulation of genes coding for only one subunit.

The first signs of erythropoiesis in the mouse are detected in the blood islands of the yolk sac at the eighth day of gestation when erythroid cell precursors proliferate and hemoglobin appears (Figure 17-19). Nucleated erythroblasts enter the embryonic circulation on the ninth day, synthesizing all three embryonic hemoglobins. These yolk sac erythroid cells (YSEC) complete four cycles of division before they

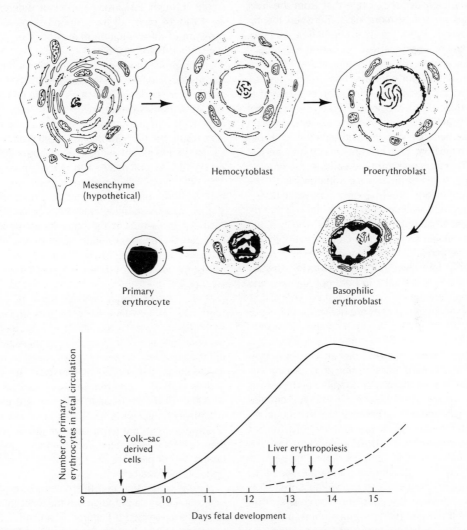

**Figure 17-19**  *Differentiation of yolk-sac-derived hemocytoblasts in fetal mouse. The morphological changes are shown in the diagram, with some hypothetical stages included. Only one of the intermediate stages between the basophilic erythroblast and erythrocyte is shown. Below is a curve showing the increase of YSEC during fetal development (solid line). Dashed line shows beginning of liver-derived erythrocytes. (Based on A. Fantoni, M. Lunadei, and E. Ullo,* Current Topics in Developmental Biology, *ed. A. A. Moscona and A. Monroy, vol. 9. New York: Academic Press, 1975, pp. 15–38.)*

mature into erythrocytes with condensed nuclei. During the twelfth day of gestation, erythropoiesis begins in the fetal liver where a second stem cell population differentiates into nonnucleated reticulocytes synthesizing adult hemoglobin A. Erythroid cells from the yolk sac are produced through the fourteenth day, and for awhile, fetal circulation contains a mixture of cells with different hemoglobins. The YSEC population soon declines while that from the liver proliferates for a few more days. Between the fifteenth day and birth, the liver population declines, and spleen and bone marrow become the definitive centers for adult erythrocyte manufacture.

In answer to our question, two distinct populations of erythroid cells arise in different sites and at different stages. Yolk sac progenitors probably do not seed the fetal liver to initiate erythropoiesis, since no embryonic hemoglobins are synthesized in cells derived from the liver. Both cell lines may have emerged from a common ancestral type early in the embryo and exhibit different programs of differentiation. The yolk sac population produces only embryonic hemoglobins and differentiates into a terminal cell that retains a condensed but inactive nucleus. The liver and spleen populations synthesize adult hemoglobin and differentiate into anucleated reticulocytes which mature rapidly into the typical erythrocyte. The sequence of events is similar in both populations, and we will concentrate on that found in the yolk sac.

Yolk sac erythroid cells are derived from hemocytoblasts in blood islands which behave like a stem cell population. They are committed to erythroid differentiation before the eighth day and begin to synthesize hemoglobin, transforming into proerythroblasts. On the ninth day, basophilic erythroblasts leave the blood islands and enter the fetal circulation. Though synthesizing hemoglobin, they continue to proliferate and complete four cycles of division between day 10 and day 14. The doubling time for each division increases from 12 hours during the tenth to eleventh day to 24-plus hours during the twelfth to thirteenth day. Most cells stop dividing on day 13. With each division the population of cells synthesizing all three embryonic hemoglobins increases rapidly. Meanwhile the nature of transcription and translation undergoes progressive changes.

The erythroblast starts out with a rich supply of ribosomes, which accounts for its basophilia. The rate of ribosome gene transcription decreases twentyfold between 10 and 14 days and the total ribosome population declines to about one-sixth its original amount. But more than enough ribosomes survive for protein synthesis.

The principal transcriptive event in this program is synthesis of globin messenger RNA prior to day 10. From day 11 to 15, synthesis of globin chains is insensitive to actinomycin D, which indicates that the message transcripts made before day 10 are stable through the 5-day period of differentiation. By 13 to 14 days, all transcription stops; the entire program of differentiation may depend on messages transcribed before 10 days.

The process of message translation matures also during this period. All housekeeping proteins are synthesized on labile messages produced up to day 13. These transcripts turn over rapidly since synthesis of their protein products is sensitive to actinomycin D. As a result, the proportion of total cell protein synthesis devoted to globin increases from about 20 percent on day 10 to practically 100 percent from day 13.

What stabilizes globin messages? Some suggest that addition of poly A sequences to newly transcribed mRNA stabilizes messages before transport to the cytoplasm. On the other hand, stabilization may be due to transcription of an excess of globin messages which are slowly degraded, with enough persisting to support a 5-day program of protein synthesis.

The rate of globin message translation per cell seems to be constant early in differentiation but soon this declines. From about day 12, polysomes become less efficient in translating messages. Their number declines as fewer polysomes are recruited to globin synthesis. The mature cells are terminal and as cells begin to die, many polysomes are inactivated in an aggregated state and are unable to complete polypeptide synthesis and release free ribosomes. With the termination of globin synthesis, these cells are eliminated from the fetus, presumably by the end of the fifteenth day, leaving only the liver population of red cells to carry on.

Can we assume that this pattern of erythrocyte differentiation is a paradigm for the differentiation of all cell phenotypes? It features (1) an early transcriptive event in which genes for a specific luxury protein are copied into a large population of stable messages; (2) a program of limited cell divisions during which the population of committed cells is rapidly increased while the cells continue to synthesize differentiated products; (3) progressive changes in message utilization and the efficiency of cytoplasmic machinery devoted to synthesis. The proportion of luxury protein increases, largely because its messages are more stable relative to all other messages. Posttranscriptional controls seem to be as important in determining the nature of the phenotype as do events at the transcriptional level.

Since the erythrocyte is a terminal cell, many of these features may be uniquely related to its limited life. Heterochromatic inactivation of an entire nucleus is not typical of most differentiating cells. Nevertheless, similar regulatory mechanisms are probably utilized in other differentiating systems where, as in the erythrocyte, less than 10 percent of the total genetic information is expressed. Let us examine other systems and compare them to erythropoiesis.

## SIGNALS PROGRAMMING GENE EXPRESSION

We know that differentiation is a social phenomenon; cells are members of populations, interacting with one another and with neighboring cells in other clones. Signals that activate a genomic program of differentiation may arise within the cell population itself (homotypic) or may arrive from exogenous sources, the extracellular environment or other cell types (heterotypic). Heterotypic signals may come from adjacent, unrelated clones, or arrive as humoral factors produced by distant cell populations. We have seen that erythropoiesis, at least in adult bone marrow, is triggered by a hormone, erythropoietin.

### Contact, cell density, and extracellular matrices

The differentiation of cartilage (chondrogenesis) is an example of homotypic differentiation. A clone of cells, determined in the early embryo, produces progeny which synthesize cartilage, an extracellular matrix composed of collagen fibers and chondroitin sulfate. Substances like chondroitin sulfate are not specific to chondroblasts since other cell phenotypes synthesize similar matrix materials. Nevertheless it is the marker molecule used for cartilage differentiation because it is produced in large amounts. Operationally, cartilage differentiation is often defined by synthesis of chondroitin sulfate as measured by radioactive sulfate incorporation into total cell proteins.

Somites differentiate into vertebral cartilages, with anterior somites maturing first. Early experiments in chick embryos showed that somite cartilage differentiation resulted from an inductive interaction with spinal cord or notochord. Isolated somites from stage-17 chick embryos are unable to differentiate into cartilage but do so if a piece of embryonic notochord or spinal cord is added (Figure 17-20).

Somites can differentiate into cartilage in the absence of notochord if conditions are changed, particularly if total cell mass is increased. Though a few somites cultured alone do not synthesize cartilage, many somites cultured together will self-differentiate under the same conditions. An increased somite mass seems to supply the same signals as notochord. Even a small number of somites differentiate into cartilage if the medium is enriched with fetal calf serum instead of horse serum. Progeny of determined somite cells differentiate into cartilage only under permissive conditions, with notochord furnishing the optimal environment for differentiation. The notochord signal is not instructive; like fetal calf serum or increased somite mass, it supplies permissive conditions for cartilage differentiation. What are these conditions?

Cartilage develops in vivo after chondrogenic cells condense into a compact mass. Dissociated chondrogenic blastemas of embryonic chick limbs do not differentiate, but if the cells are first allowed to aggregate, then cartilage develops. Intimate contact between chondrogenic cells, or a high cell density, seems a prerequisite for cartilage differentiation; only then does matrix accumulate in intercellular spaces. Dispersed chondrocytes may synthesize matrix precursors at all times, but these may diffuse into the culture medium, unable to form cartilage. In an aggregate, however, precursors accumulate and are stabilized into a visible matrix. For example, self-differentiation of chick somites in vitro occurs only after stage 16. If stage-13 somites are wrapped in a vitelline membrane and cultured in the same medium, then self-differentiation of cartilage takes place. Permissive conditions arise within the confining membrane, possibly by concentrating matrix precursors escaping from cartilage cells.

The behavior of mature chondrocytes from older embryos, that is, cells already synthesizing cartilage, is consistent with a model of differentiation based on cell contact. When such chondrocytes are grown under highly dense conditions, they retain a polyg-

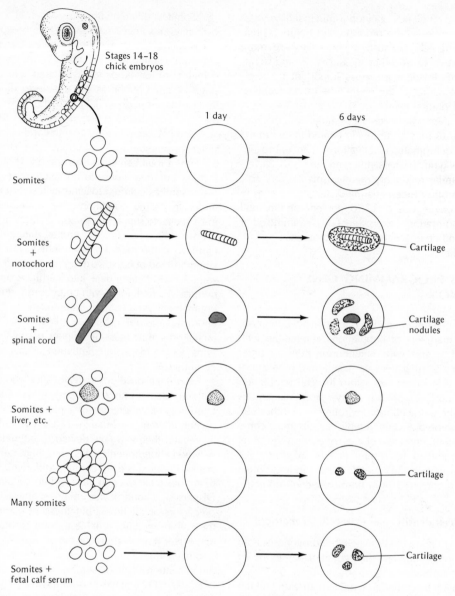

**Figure 17-20**   *Cartilage differentiation from somites in vitro. Somites isolated from stage 14–18 chick embryos and cultured in vitro under conditions shown. In all except the last, horse serum was used in the medium. (Based in part on H. Holtzer,* Synthesis of Molecular and Cellular Structures, *ed. D. Rudnick. New York: Ronald Press, 1961, pp. 35–88.)*

onal shape in the center of the culture, deposit large amounts of matrix, and synthesize chrondroitin sulfate. On the other hand, isolated, dispersed cells at the periphery are stellate or fibroblastic and do not secrete matrix; they synthesize DNA and divide instead. The shape of the cell and the amount of exposed surface seems to correlate with cartilage differentiation.

The role of the cell surface in cartilage differentiation is further seen in experiments with 5-bromo-deoxyuridine (BUDR). This agent is incorporated into replicating DNA in place of thymidine. Though it does not inhibit DNA replication, it affects a cell's phenotypic expression when incorporated into DNA. After BUDR treatment chondrocytes change shape. Though viable and actively proliferating, they do not synthesize matrix. The cells become more adhesive to the substratum than to one another, attach, and flatten out into proliferating stellate fibroblasts. No spherical or polygonal cells appear. After BUDR is washed out and cells are restored to the normal medium, their polygonal shape reappears and they pile up and accumulate cartilage.

Although chondrocytes display surface changes in BUDR, we cannot assume that BUDR inhibits differentiation by altering the cell surface. Correlations between chondrocyte shape, contact relationships, and cartilage synthesis are consistent with the hypothesis that surface interactions between identical cells provide permissive conditions for the expression of a specific tissue phenotype.

## Cell fusion and muscle differentiation

As in the erythrocyte, differentiation of skeletal muscle is a gradual process (Figure 17-21). Muscle develops from the somite myotome and from lateral plate mesoderm. At 2 days of incubation in the chick embryo, most myotome cells are spherical and mitotically active. Within a few hours a small population of elongate, spindle-shaped myoblasts appears containing fine myosin filaments, and by the fourth day most myoblasts contain cross-striated myofibrils. The population is heterogeneous with all stages present from proliferating myoblasts to cells with myofibrils. Later on these cells fuse into multinucleated myotubes with cross striations.

In these early somites, definitive synthesis and assembly of a myofibrillar apparatus take place in mononucleated myoblasts. Gradually the more complex fibrillar elements with A and Z bands are organized as filaments assemble into sarcomeres within the myotubes.

Developing muscle in limbs of older embryos shows a different pattern; myofibrils appear only in multinucleated myotubes and are rarely seen in single myoblasts. Here, fusion of myoblasts into myotubes seems to be a prerequisite for differentiation of contractile organelles. The reason for this difference is not understood.

Most studies relating fusion to differentiation have been carried out with embryonic chick limb musculature. By the fourth day of limb outgrowth mesoderm acquires its definitive commitment to myogenesis. Progenitor cells in the peripheral regions of the limb mesoderm proliferate into myoblasts. Mitosis declines between the seventh to eleventh day, and as myoblasts leave the cell cycle they fuse into small myotubes. Fusion is a gradual process in vivo and by the eighteenth day most myoblasts have fused into differentiated muscle, leaving a small population of mitotically active mononucleated myoblasts. At any stage limb muscle is a mixture of dividing fibroblasts, myoblasts, postmitotic myoblasts, and multinucleated myotubes in various stages of fusion and myofibril differentiation.

A cell suspension from trypsin-dissociated limb muscle (11 days old) supplies a mixed population of fibroblasts and myoblasts that can be used to study muscle differentiation in vitro (Figure 17-22). Myoblasts are spindle-shaped cells, smaller and less motile than fibroblasts. Each cell type proliferates on agar into individual colonies; myoblasts fuse into irregular networks of myotubes while fibroblasts develop into smooth, spherical colonies. Each muscle colony is a clone derived from a single myoblast.

In vitro myoblast fusion seems to be a necessary condition for muscle differentiation. No free myoblasts contain myosin nor do they possess any fibrils. Fusion in vitro is rapid and occurs after only 24 hours, whereas it takes days in vivo. In vitro differentiation depends on high cell density, which promotes cell fusion; myotubes appear after only 1 day in dense cultures but are delayed 5 to 6 days in less dense populations. Dilution of a myoblast population at constant cell density delays fusion, but this dilution effect may be overcome by the addition of media from cultures of fusing myoblasts. A diffusible factor produced by fusing myoblasts accumulates in media and promotes cell fusion.

Cell fusion and myoblast proliferation are also correlated. Fusion occurs only after cells have left the mitotic cycle. Myoblasts synthesizing DNA do not fuse, nor do nuclei in multinucleated myotubes synthesize DNA. Completion of a terminal cell division is a prerequisite for fusion; after division, myoblast daughter cells enter a $G_1$ phase of ap-

**Figure 17-21**  *Muscle differentiation from myotome in the early chick embryo. Somite segregates into three tissue masses, dermatome (dermis of skin), myotome, and sclerotome. Myotome cells become spindle-shaped myoblasts, some of which may contain myosin as detected by immune fluorescence. Myoblasts fuse and fibrillogenesis is stimulated, particularly as many myoblasts fuse into long myotubes. Cross striations are seen in these multinucleated structures which mature into typical skeletal muscle.*

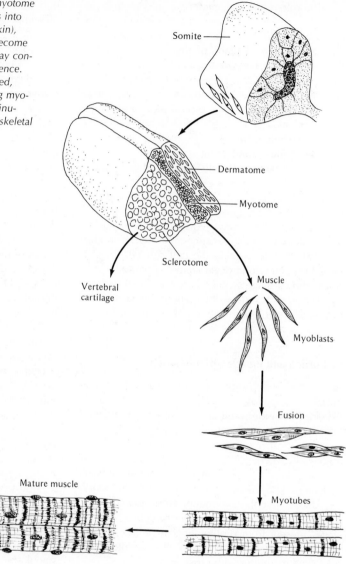

proximately 6 to 8 hours while they prepare for fusion. Presumably during $G_1$, the cell surface acquires properties that make fusion possible. In contrast to erythrocyte differentiation, division precedes differentiation.

Is fusion an absolute requirement for muscle differentiation? In some instances, differentiation does not occur if fusion is prevented. The inhibitor 5-bromodeoxyuridine (BUDR) reversibly inhibits myoblast fusion without affecting cell division or DNA synthesis. Cells that incorporate BUDR into DNA during replication are unable to fuse and do not synthesize myosin or assemble fibrils. The BUDR effect is reversible by growing cells in thymidine which replaces BUDR after several rounds of DNA replication. After a few replications cells fuse and differentiate myofibrils, which suggests that fusion is essential for differentiation. On the other hand,

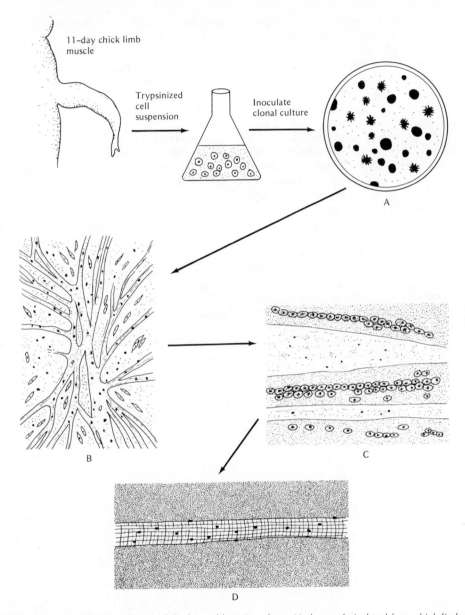

**Figure 17-22**   *Differentiation of 11-day chick limb myoblasts in culture. Limb muscle isolated from chick limb is dissociated in trypsin and plated out on agar. (A) Two types of colonies are produced, smooth round fibroblast colonies and irregular muscle colonies, both derived from single cells. (B) Enlarged view of colony showing branching myotubes surrounded by many single and fusing myoblasts. (C) Enlarged view of myotube showing many myoblast nuclei crowding the same cytoplasm. (D) A later stage; cross striations are seen in myofibrils within myotubes, with fluorescent antibody prepared against myosin.*

fusion and muscle protein synthesis are uncoupled by chelating agents that bind divalent cations. Under these conditions, individual myoblasts synthesize myosin and assemble myofibrils, a situation seen in early chick somites. This means that fusion is not an absolute condition for muscle differentiation, even in vitro. When it occurs, however, it seems to promote fibrillogenesis.

Potentialities for muscle differentiation are not lost even after long-term proliferation over many cell generations. Yaffe established several cell lines of mammalian myoblasts which do not synthesize myosin while proliferating but differentiate muscle when grown under permissive conditions. Myoblasts remain in a proliferative state indefinitely if cultured in an enriched medium, but if this medium is replaced by a less rich medium, then proliferation slows down, and within 18 hours cell fusion and fibrillogenesis begin (Figure 17-23). Myosin synthesis begins and contractile organelles are self-assembled. The activity of muscle enzymes such as creatine kinase also increases, paralleling the kinetics of fusion.

We do not understand how fusion regulates muscle differentiation. All muscle proteins increase after fusion, indicating that sets of structural genes are coordinately turned on by a single switch. On the other hand, fusion may stabilize or amplify a pro-

gram already switched on but operating at a minimal level. Muscle enzymes such as creatine kinase are indeed detectable at low levels before fusion, which suggests that some genes have transcribed RNA messages. Fusion seems to enhance these levels and activate other genes in the program.

It is during the 18-hour lag before fusion that the principal transcriptive events take place. During this cryptic phase, genes coding for muscle proteins are probably transcribed since actinomycin D inhibits synthesis of muscle proteins if applied in the first 12 hours of this cryptic phase. After this point, the inhibitor has little effect on fusion and muscle protein synthesis (Figure 17-24). As in erythrocytes, stable messages for luxury molecule synthesis are transcribed within the first 12 hours and are translated after myoblasts have fused.

Myoblast fusion may stabilize a cytoplasmic state partially committed to muscle protein synthesis. Within a lengthening myotube, fibrillogenesis can occur, with myosin and actin subunits self-assembling into hexagonal arrays of the contractile apparatus. The elaboration of cross-striated organization may, itself, accelerate protein synthesis and information flow (see Figure 8-10).

Homotypic differentiation of cartilage and muscle is dependent in part on surface factors synthesized by cells of a clone. These membrane com-

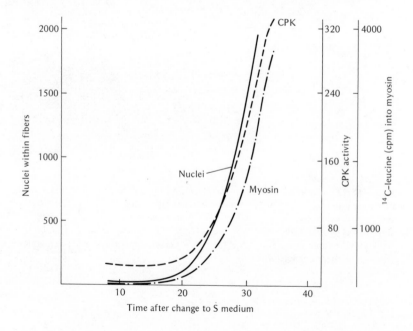

**Figure 17-23** *Synthesis of myosin and creatinine phosphokinase (CPK) after fusion of myoblasts into myotubes in vitro. Fusion measured as the increase in number of nuclei in multinucleated fibers. Myosin synthesis measured as incorporation of $^{14}$C-leucine into myosin isolated by gel electrophoresis. CPK activity on a per culture plate basis. S medium is fusion-inducing medium. (Based on D. Yaffe and H. Dym,* Cold Spring Harbor Symp. Quant. Biol. *37:543–548, 1972.)*

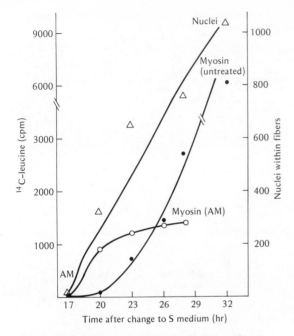

**Figure 17-24** *Effect of actinomycin D (AM) on fusion and myosin synthesis in vitro. After addition of actino-mycin at 17 to 18 hr, fusion is completely unaffected and myosin synthesis is initiated and maintained at control rates for several hours before it levels off. (From D. Yaffe and H. Dym,* Cold Spring Harbor Symp. Quant. Biol. *37:543–548, 1972.)*

ponents or extracellular matrices may affect cell proliferative activity, which usually declines before cells synthesize specific proteins. Surface changes need not trigger gene action but may stabilize or amplify a flow of genetic information already underway, possibly through posttranscriptional regulation of message utilization. Large amounts of luxury proteins accumulate and are assembled into a mature phenotype.

## Genes and surface proteins

As we have seen, cell morphogenetic behavior seems to depend upon surface properties such as motility, recognition, and cellular interactions (Chapter 12). Since the nature of the cell surface is defined by genes coding for protein constituents, control of gene expression will affect surface behavior. A system that has proven useful in studying

genetic control of surface properties is the T-locus in the mouse.

The T-locus is a site on the IXth chromosome with many alleles affecting development of the chordamesoderm, neural tube, and embryonic axis. The *T* allele is a semidominant mutant which is lethal when homozygous (*T/T*). In such embryos the posterior half does not develop and the embryo dies at about 11 days, halfway through development. Its expression in heterozygotes (*T/t*) is less severe; short-tailed offspring result.

Recessive *t* alleles have also been obtained at this locus, arising from recombination abnormalities that occur frequently in the region. The mutation rate is about $2 \times 10^{-3}$ compared with a rate for most other genes of $10^{-5}$. The recessive *t* alleles enhance the expression of the *T* gene when heterozygous (*T/t^n*); in almost all cases tailless offspring result.

Homozygosity of recessive genes (*t^n/t^n*) is invariably lethal for almost all alleles. As a result, a cross between a heterozygous *T/t^n* with another *T/t^n* is a balanced lethal system; it produces only viable heterozygotes (*T/t^n*) because all homozygotes derived from such a cross (*T/T* and *t^n/t^n*) die.

Each recessive allele expresses its lethality at different developmental stages (Figure 17-25). The first allele to act, the *t^{12}* allele (*t^{12}/t^{12}*), disrupts the early morula stage, and a loose arrangement of blastomeres fails to form a coherent blastocyst. Ribosome metabolism is also disturbed inasmuch as abnormal nucleoli develop and the amount of cytoplasmic ribosomal RNA is markedly reduced.

The *t^0/t^0* homozygote acts after implantation; it prevents separation of the inner cell mass into embryonic and extraembryonic ectoderm and the egg cylinder stage aborts at 6 to 7 days. Still other alleles, such as *t^{w18}*, cause overgrowth of the neural tube and an excessive amount of neural tissue in relation to notochord develops. Such embryos abort at 8 to 10 days of development.

Electron microscopic studies of the tissues in these homozygous embryos suggest that cell surface relationships are disturbed. The cohesiveness and stability of tissue is usually affected, as are the spatial relationships between notochord and neural tube. Apparently it is these disturbances that lead to the overall effects on the axial system.

The mutant genes have pleiotropic effects, particularly on male fertility. Heterozygotes, with the wild type allele (*t/t^n*), show an unusual distortion

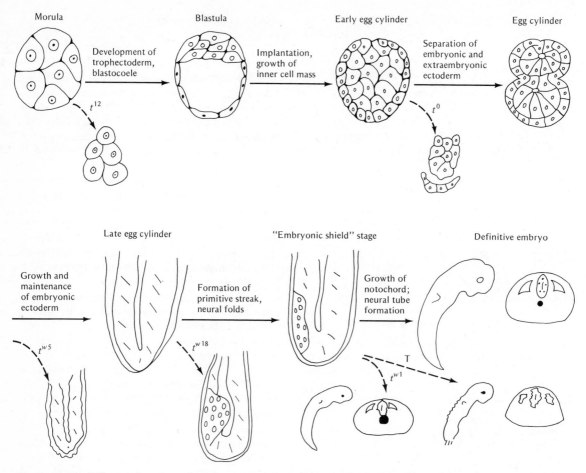

**Figure 17-25**   *Differentiation of ectodermal and nervous structures of mouse embryos showing the points at which the recessive* t *alleles interfere with normal development. (From L. C. Dunn and D. Bennett,* Science *144:260–267, 1964. Copyright © 1964 by the American Association for the Advancement of Science.)*

of gamete segregation ratios in males. In such combinations, virtually all sperm carry the recessive $t^n$ allele; those with the wild type gene do not appear in the offspring. Moreover, when two t alleles are combined as in a $t^x/t^y$ combination, no functional gametes are produced in males. The females are unaffected in all instances by these combinations.

Apparently, the T-locus is a region that codes for a sperm surface antigen. The antigen is not specific for sperm inasmuch as the same antigen has been detected on cleavage cells and those of early morulae. No other adult tissue carries this specific antigen, which differs from the histocompatibility antigens found on all somatic cells. Moreover, the

same antigen is found on cell lines that have been derived from a testicular teratocarcinoma. These cells, as we have seen, are neoplastic, having come from totipotent primordial germ cells, and can differentiate into a variety of tissue types. It is interesting that these cells produce antibodies when injected into mice that specifically affect the primitive teratocarcinoma cells (PTC) and also interact with sperm and cleavage cells. We have here a common surface antigen coded by a single gene locus that defines properties of normal embryonic cells, teratomas (derived from germ cells), and sperm (also germ cell derivatives). Because the antigen is not detected on any mature, differentiated cells of the adult, its presence in the surface of these cells seems to correlate with their totipotent or pluripotent properties. In fact, the antigen disappears from embryonic mouse cells at day 9 or 10, shortly after most cells have been determined.

The mechanism of action of the gene product in sperm function, in early embryos, or in primary teratocarcinoma cells is not understood. The important point is that a gene coding for a surface protein can exercise such diverse effects on the early differentiation of neural tube and notochordal structures.

### Hormone signals

In heterotypic differentiation, signals that program genes often come from unrelated tissues across intercellular spaces as in inductive interactions, or at a distance, from tissues secreting hormones into the systemic circulation. The cell surface is first to receive the signal which is then transduced through the cytoplasm to impinge on the flow of information from gene to protein. Here, too, the cell surface may determine permissive conditions for differentiation by interacting with exogenous signals.

Hormone stimulation of protein synthesis in adult cells is often used as a model of gene regulation in differentiation. Since these are fully differentiated cells, the hormone is only modulating levels of protein synthesis in response to physiological conditions and is not inducing a determined cell population to differentiate. There are, however, a few cases of hormonal programming of information flow which

may tell us something about gene regulation during embryonic differentiation.

In hormone-induced cell modulations, some genes may be switched on to transcribe, but in most cases low levels of gene expression already underway are only stimulated by hormones; once hormone is removed, activity reverts to its original low state.

Because many proteins in eukaryote cells undergo turnover, the effects of hormones are often complicated by the balance that exists between rates of enzyme synthesis and rates of degradation. For example, adult rat liver cells respond to hydrocortisone by synthesizing four new enzymes (Figure 17-26). In contrast to total cell protein synthesis, which remains constant, synthesis of these four enzymes is selectively stimulated by the hormone at what appear to be different rates. The curves are misleading; on careful analysis, the enzymes are not being synthesized at different rates. Rather, the curves reflect the fact that each enzyme starts out with a different basal rate of synthesis and widely different stability in normal unstimulated liver. Each is undergoing turnover and is being degraded at different rates. Though the hormone seems to induce synthesis of all four enzymes at different rates, the observed pattern for each enzyme represents a summation of hormone-stimulated basal rates of synthesis minus rates of degradation (Table 17-4). The level of stimulation may, in fact, be the same for all enzymes, an example of coordinated regulation of multigene expression by a single hormone, but the rates of degradation differ. Though transcription is stimulated (synthesis of all major classes of RNA is elevated), this does not mean

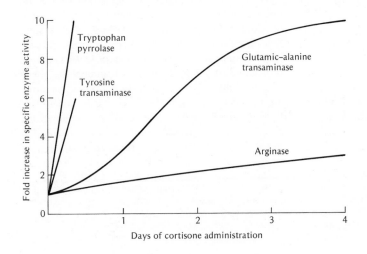

**Figure 17-26** *Cortisone stimulation of basal levels of four rat liver enzymes. (From R. T. Schimke*, Nat. Cancer Inst., *Monogr. No. 27:301–314, 1967.)*

**TABLE 17-4   Enzyme Turnover and Response to Cortisone of Four Rat Liver Enzymes**

| Enzyme | Enzyme Synthesized | | |
| --- | --- | --- | --- |
| | Basal, units/hr | Cortisone, units/hr | Ratio Cortisone: Normal |
| Tryptophan pyrrolase | 0.84 | 3.4 | 4.0 |
| Tyrosine-glutamic transaminase | 27 | 114 | 4.2 |
| Glutamic-alanine transaminase | 2.0 | 14.4 | 7.2 |
| Arginase | 138 | 534 | 3.9 |

Source: From R. T. Schimke, *Nat. Cancer Inst.*, Monogr. No. 27, 1967, pp. 301–314.

that the hormone has switched genes on. Low levels of enzyme activity are always detected in unstimulated liver; that is, genes are transcribing but at very low levels. Hydrocortisone may enhance the rate of transcription, or it may, on the other hand, stabilize messenger RNAs with normally short half-lives. The result in each case is the same, the apparent induction of enzyme synthesis.

Though this pattern of coordinated induction of enzyme synthesis does not involve gene switching, it could serve as a model for differentiation of determined tissues already synthesizing low basal levels of cell-specific proteins. During the stepwise process of differentiation, rates of protein synthesis and accumulation may increase at each stage until mature levels are reached and the phenotype is expressed.

## Multiple hormone signals and a single cell phenotype

Sometimes acquisition of a specific cell phenotype calls for several hormone signals, acting sequentially or synergistically on a determined cell population.

Each hormone plays a specific role, with the primary event being the acquisition of specific hormone receptor sites on cell surfaces or within the cytoplasm. Cells become targets for specific hormone stimulation. In the following example, a stem cell, dividing in the presence of two hormones, produces daughter cells responsive to a third hormone that induces protein synthesis and differentiation.

Differentiation of the alveolar cells of the mouse mammary gland is marked by the synthesis of milk proteins, principally casein. Before differentiation, the mammary gland is a simple ductile tissue arising in the fetus that grows during adolescence and the first half of pregnancy (Figure 17-27). In the second half of pregnancy, alveolar cells differentiate in response to multiple hormonal stimuli, a synergistic action of three hormones, insulin, hydrocortisone, and prolactin.

The action of these hormones on mammary gland epithelium can be studied in vitro. Mammary tissue from preadolescent mice cultured in the presence of insulin, hydrocortisone, and prolactin differentiate into alveolar structures containing secretory material. All three hormones are required inasmuch as removal of any one inhibits morphological differentiation of tubules and casein synthesis.

When all hormones are present, a period of cell proliferation precedes milk protein synthesis and alveolar differentiation. Since undifferentiated mammary glands secrete low levels of casein in the absence of hormones, how does each hormone work in this system?

First, incubation in insulin stimulates DNA synthesis but has no effect on differentiation. Of the three hormones, only insulin stimulates cell division of mammary epithelium. Furthermore, unless insulin first stimulates cell division, hydrocortisone and prolactin cannot act. Since the peak of DNA synthesis is reached at 20 to 24 hours of incubation, while casein synthesis is maximal at 48 hours of incubation, the rise in casein synthesis follows new cell formation. If cell division is inhibited by colchicine, the

**Figure 17-27**   *Effect of hormones on the differentiation of mouse mammary epithelium in vitro. (A) Histological appearance of explant after various treatments for 48 hr. (B) Synthesis of casein as measured by incorporation of $^{32}P$. Each point represents incorporation in preceding 4-hr pulse-labeling period. I = insulin; HC = hydrocortisone; P = prolactin; NH = no hormone. (C) Outline of scheme of hormone interactions in differentiation of mammary epithelium. Tissue contains stem cells and secretory cells synthesizing at basal levels only. Insulin alone stimulates cell division but in the presence of hydrocortisone, daughter cells become receptive to prolactin. (From R. Turkington,* Current Topics in Developmental Biology, *ed. A. A. Moscona and A. Monroy, vol. 3. New York: Academic Press, 1968, p. 199.)*

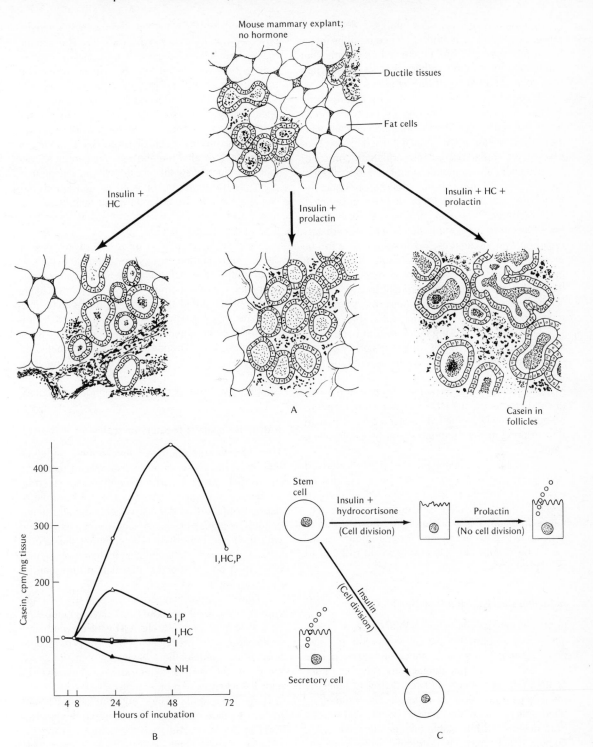

Mouse mammary explant; no hormone

Ductile tissues

Fat cells

Insulin + HC

Insulin + prolactin

Insulin + HC + prolactin

Casein in follicles

A

I,HC,P

I,P

I,HC

I

NH

Casein, cpm/mg tissue

400

300

200

100

4 8    24    48    72

Hours of incubation

B

Stem cell

Insulin + hydrocortisone (Cell division)

Prolactin (No cell division)

Insulin (Cell division)

Secretory cell

C

rise in casein synthesis is also inhibited, even in the presence of all three hormones. Since baseline levels of casein synthesis are unaffected, it seems that stimulation of casein synthesis occurs only when new cells are produced from stem cells.

It would seem that old casein-secreting cells are not stimulated by the hormones. Progenitor cells are stimulated by insulin to proliferate into cells with zero levels of casein synthesis. If these divisions take place in the absence of prolactin and hydrocortisone, the daughter cells cannot be stimulated to synthesize casein, even if hormones are added postmitotically. If, however, progenitors proliferate in the presence of hydrocortisone, then their progeny do react to the hormone prolactin postmitotically and immediately begin to synthesize milk. Hydrocortisone must be present at some time during the insulin-induced cell cycle for the daughter cells to be receptive to the terminal synthesizing signal brought in by prolactin hormone. Prolactin probably acts by stimulating transcription of genes coding for milk proteins.

Daughter cells arising in the insulin-hydrocortisone medium are covertly differentiated; they become determined for casein synthesis, but only prolactin elicits this synthetic activity. The action of hydrocortisone in this system is not clear. It may induce synthesis of prolactin-receptor sites in membranes of newly formed daughter cells, which in turn respond to the hormone signal by synthesizing casein.

Evidently the phenotype of daughters derived from stem cell epithelium is set by the hormonal environment in which division occurs. Cell division is a prerequisite for the expression of new differentiative function; the genome is programmed during division to express new genes. Programming may include the acquisition of competence to respond to regulatory signals.

## Inductive signals: cell interactions

Inductive interactions between cell sheets of different origin account for the differentiation of most organ systems in the vertebrate embryo (see Chapter 11). Some heterotypic interactions rely on direct contact between sheets, but the dialog in others is mediated by diffusion of molecules across matrix-filled spaces. Such interactions switch cell clones from determined into stable differentiated states. Inducing signals may be instructive, specifying the precise phenotypic program to be expressed by re-

sponding tissues. Or the signal may elicit a genetically preprogrammed response in the reacting tissue. In the former case, the inducing signal is specific; that is, only one tissue acts as inducer, whereas in the latter, many tissues may evoke a response.

We do not understand how inductive signals affect responding tissues. A signal may alter the surface of responding cells, affect the flow of regulatory molecules, stimulate cell division, furnish an essential nutrient, or alter metabolism. We cannot assume that all inductive signals act directly on a gene template, its transcription, or even its translation. Nor can we exclude the possibility that heterotypic signals furnish permissive conditions that secondarily lead to new stable states.

Specificity is also a property of responding tissues; only certain tissues are competent to react to an inductive signal. Immature tissues can respond after they become competent. Newly formed mammary epithelial cells respond to prolactin if they originate in an environment of insulin and hydrocortisone. Again, the nature of this competence is not understood, but hormone models suggest that competence may involve acquisition by cells of receptor proteins either on the cell surface or within the cytoplasm that interact readily with inducing signals.

## Instructive signals: feathers or scales

The epithelio-mesenchymal interactions of chick limb morphogenesis offer an example of induction by instructive signals. The phenotype expressed by overlying epidermis is fixed by the inducing mesoderm. Wing-bud epidermis, which normally differentiates feathers, produces scales when combined with limb bud mesoderm. The epidermis has a dual potentiality and differentiates feathers or scales, depending on the specific information furnished by the inducing dermis.

Feathers are organized in tracts in the skin and appear on the back at about 6 to 7 days of incubation with all feather tracts established by the eleventh day. Scales appear on the leg at about 11 days of incubation. Scales and feathers have a similar ontogeny, arising in the dermis (a mesodermal derivative) as a local thickening of proliferating cells (Figure 17-28). This dermal papilla grows out, interacts with the overlying epidermis, and induces it to differentiate a scale or a feather. Although the dermis is a specific inducer, experimental combinations of embryonic dermis and epidermis from different regions and at different ages show that competence of epidermis is also an important factor in differentiation.

For example, back epidermis normally differentiates feathers, but when back epidermis of 5- to 8½-day embryos is combined with 13-day leg dermis, it forms scales instead (Figure 17-29). The inductive ability of leg dermis is also age-dependent since dermis from young legs does not induce scales in 5- to 8½-day back epidermis. The ability to induce scales is gradually acquired by leg dermis. Other dermal regions also acquire specific inductive capacities; 9-day-old dermis from the site of spur formation on the leg induces spurs in 5- to 8½-day back epidermis, whereas dermis from the beak of 8- to 8½-day embryos induces beak outgrowth instead. In all cases, young back epidermis is phenotypically labile; its response as feather, spur, beak, or scales is determined by an instructive signal from the underlying dermis. Epidermis will even respond to foreign mesoderms from heart and gizzard and develop into heart and gizzard epithelia, respectively.

The competence of back epidermis changes after 8 to 8½ days; it produces only feathers, even when combined with 13- to 15-day leg dermis. The inductive influence from its own dermis alters epidermal responsiveness so that it no longer responds to signals from leg dermis. Though losing scale competence, it will respond to a strong signal from beak dermis and differentiate beak instead of feathers.

How does the inducing signal work? Without dermis skin epithelium in vitro does not proliferate nor do its cells assume a characteristic columnar shape and orientation. Where epidermis is in direct contact with dermis, however, mitotic activity in the basal layer is intiated and the epidermis assumes its typical histodifferentiation. The dermis seems to furnish a factor that changes the shape of epidermal cells and preserves their proliferative activity. Though other factors such as chick embryo extract stimulate epidermal proliferation, proliferation per se is not a prerequisite for differentiation as it is in mammary gland epithelium. Rather, differentiation seems to depend on patterns of local condensation, a property of cell-surface adhesiveness. For example, whole skin of embryos younger than 6 to 6½ days does not differentiate feathers in vitro, but skin cultured after this stage will develop normal feathers. At 6 days in vivo, dermal condensations appear and it is these papillae that actively induce. Dermis undergoes a surface change essential to differentiation.

Cell condensation also precedes epidermal differentiation. If condensation does not occur, as in a scaleless mutant in chickens, no epidermal structures

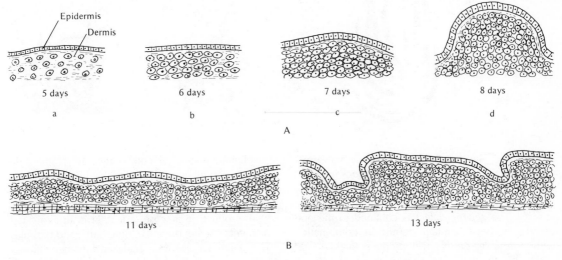

**Figure 17-28** *Early development of feathers and scales in the chick embryo. (A) Initial stages in the development of feathers in back epidermis (transverse sections). Note condensation of dermal cells and changes in thickness of epidermis leading to feather primordium as a domelike elevation on the skin surface. (B) Longitudinal sections through skin of tarsometatarsus (distal) of normal embryos showing early stages of scale development. Scale primordia appear at 11 days at low elevations. (From photographs in M. E. Rawles, J. Exp. Embryol. Morphol. 11:765–789, 1963.)*

**Figure 17-29** *Interactions between dermis and epidermis of different ages and the differentiation of scales and feathers. As epidermis matures, its competence to respond to scale-inducing stimuli from leg dermis diminishes. The inductive potency of dermis also matures since 10-day dermis is unable to induce scales. (Based on M. E. Rawles,* J. Exp. Embryol. Morphol. *11:765–789, 1963.)*

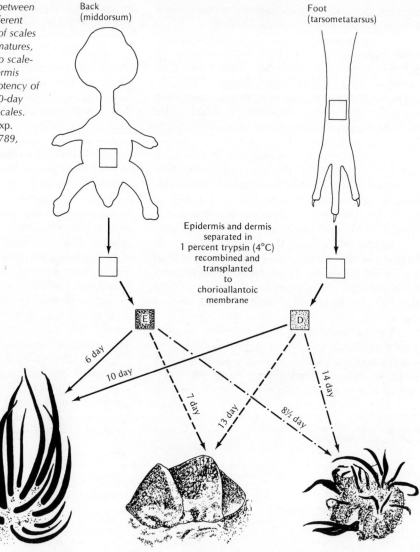

Back
(middorsum)

Foot
(tarsometatarsus)

Epidermis and dermis
separated in
1 percent trypsin (4°C)
recombined and
transplanted
to
chorioallantoic
membrane

E

D

6 day

10 day

7 day

13 day

14 day

8½ day

differentiate. The mutant epidermis does not form a close association with the dermis nor does it exhibit local thickenings. Likewise, the dermis fails to show its usual pattern of condensation and local mitotic activity. The genetic defect is confined to the epidermis, however; mutant dermis in combination with normal epidermis will produce feathers and scales, but the reciprocal combination will not. The basal layer of the epidermis cannot respond to the inductive signal nor change into a columnar shape and proliferate.

Differentiation of both scales and feathers depends on synthesis of keratin, the luxury molecule of skin. Differences between scale and feather reside in the organization and patterning of this phenotypic product. In a sense, a feather is an elongated, pigmented scale, organized into barbs and barbules. The difference between scale and feather differentiation may not reside in structural genes coding for keratin formation, but in the expression of regulatory genes that determine the patterns of cell division, the rhythm of keratin synthesis, and the manner in which the keratin molecules are assembled into organized arrays. The situation again resembles the stem cell population of blood-forming organs in which the direction of phenotypic expression by a pluripotent cell population is determined by exogenous signals. Instructive signals from dermis differ with respect to their information about these processes. Moreover, the signal need not be a specific chemical or hormone. Each mesoderm may induce different proliferative or condensation responses or may secrete qualitatively different kinds of extracellular matrices. Each set of matrix conditions may stimulate different regulatory genes within responding epidermis.

## Evocative signals: pancreas differentiation

In contrast to the multipotentiality of skin epidermis, most responding tissues are programmed to express only a single phenotype. Specificity of induction is not high; a variety of nonspecific conditions elicit the same program of differentiation. The inductive signal is not instructive; it is permissive, triggering a specific program already poised for expression. Epithelio-mesenchymal interaction in pancreas development is an example of such an inductive system.

The mouse pancreatic rudiment appears as an evagination of the duodenal region of the gut on the ninth to tenth day of development. As the bud grows out, it merges with mesenchymal cells overlying the gut region. A basement membrane develops between the mesenchyme and the enlarging endodermal rudiment, with collagen fibers accumulating in the extracellular matrix.

The phenotypes that differentiate from pancreas epithelium are exocrine and endocrine cells. The former organize as tubules composed of cells that synthesize and secrete such enzymes as trypsin, amylase, and chymotrypsin. The latter appear as loose aggregates of insulin-secreting cells or islets in the spaces between the tubules.

The commitment to become pancreas occurs in the dorsal endoderm cells of the gut at 8½ days of development, before the diverticulum is seen (Figure 17-30). The competence of the endoderm to respond to mesenchymal-inducing signals also arises well before the diverticulum emerges, usually at the 14-somite stage (about 9 days), and this requirement for mesoderm persists for some time.

If insulin is used as a marker for endocrine cell differentiation, the first detectable molecules of insulin are noted at the 20-somite stage (9 to 10 days), as the rudiment first appears. At the same time, a few β cells (endocrine) can be seen in the endoderm. Between 10 and 11 days, the rudiment grows profusely as most cells are mitotically active. As the number of β cells increases, insulin content also rises, slowly reaching a plateau. In a few days, a burst of insulin synthesis occurs as many more endocrine cells differentiate, and it is at this time that exocrine cell differentiation also begins.

As we have noted, the need for mesodermal inductive factors is acquired early. An 11-day pancreas rudiment may be separated from its mesenchyme and cultured on a porous filter in vitro (Figure 17-31). The epithelium expands readily into a flattened sheet and does not proliferate nor differentiate. If, however, epithelium is cultured with its own mesenchyme placed on the opposite surface of the filter (a transfilter induction system), it does not spread but differentiates instead. Tubules with zymogen granules appear within 5 days, the same length of time required for pancreas differentiation in vivo. The inductive signal is not specific, since mesenchyme from a variety of tissues also induces. In fact, 13-day salivary gland mesenchyme is a more potent inducer than pancreatic mesenchyme. Even a crude extract of chick embryos promotes differentiation in the absence of mesenchyme. Some protein factor in embryo extract replaces living mesenchymal cells and supplies permissive conditions for differentiation. In this inductive system, specificity resides only in the reacting tissue.

The interaction across the filter may be mediated by (1) release of a diffusible factor from the mesenchyme or (2) direct contact between the interacting tissues by fine cytoplasmic extensions penetrating the pores of the filter. Induction by chick embryo extract supports the diffusion mechanism but does not exclude cell contact. The compact arrangement of epithelium directly over inducing mesenchyme suggests that the latter exercises an orienting influence, perhaps by directed outgrowth of cellular microvilli through the pores of the filter. In the electron microscope the filter is filled with a matrix containing cytoplasmic processes extending from cells on opposite sides of the filter. The glycoprotein matrix is filled with collagen fibrils, which suggests that here, too, as in morphogenesis of salivary epithelium, the extracellular matrix is stabilizing tubule morphogenesis.

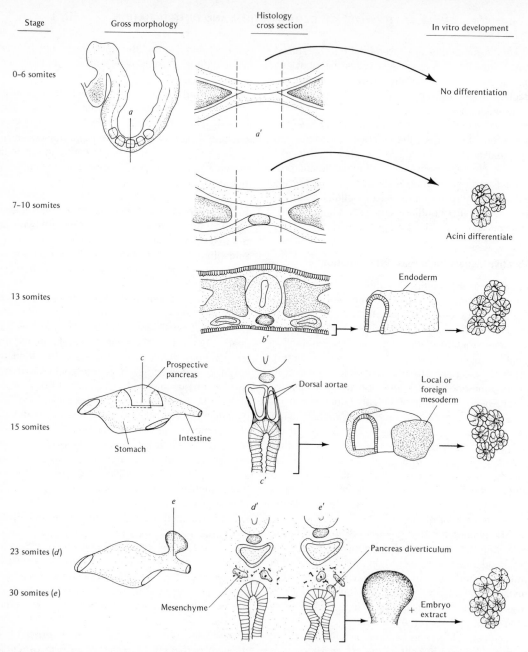

**Stage**     **Gross morphology**     **Histology cross section**     **In vitro development**

0–6 somites     No differentiation

7–10 somites     Acini differentiale

13 somites     Endoderm

15 somites     Prospective pancreas     Dorsal aortae     Local or foreign mesoderm     Stomach     Intestine

23 somites (d)     30 somites (e)     Mesenchyme     Pancreas diverticulum     + Embryo extract

**Figure 17-30** *Development of the mouse exocrine pancreas in vivo and in vitro. Pancreas is determined between somite stages 7 to 10; an explanted central fragment (morphology about the same as in 0–6 somites) differentiates acini. At 13 somites, endoderm separates from notochord and differentiates when explanted with just adherent cells. At 15 somites (about 8 days), prospective pancreas develops below c in diagram. Endoderm plus foreign mesoderm will differentiate. At 23 somites (9 days), mesenchyme accumulates between dorsal aortae, and prospective pancreas emerges as a small bud. Foreign mesoderm can support differentiation at this stage. At 30 somites (about 10 days), diverticulum (e) emerges and differentiation can now occur in the presence of embryo extract, without mesoderm. (Based on N. K. Wessels, Epithelial-Mesenchymal Interactions, ed. R. Fleischmajer and R. E. Billington. Baltimore, Md.: Williams & Wilkens, 1968, pp. 132–151.)*

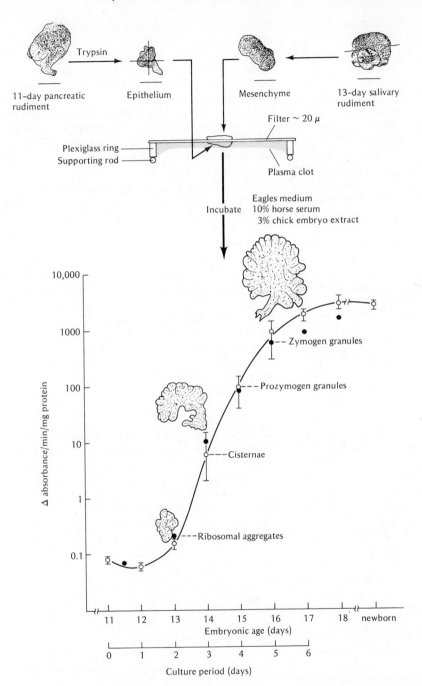

**Figure 17-31** *Amylase activity in differentiating pancreas in vivo and in vitro. In vitro, isolated epithelium and mesenchyme are cultured on opposite sides of a millipore filter. The morphological development of the tissue is shown in the curve below. Open circles represent amylase activity of pancreas rudiments in vivo. Closed circles show activity of cultured epithelia. (Based on W. J. Rutter, N. K. Wessels, and C. Grobstein, Nat. Cancer Inst., Monogr. No. 13:51, 1964.)*

If the inductive signal supplied by mesenchyme is not instructive, what is its function? In the presence of mesenchyme, epithelium first retains its compact appearance, after which cell division begins at a rapid rate. Epithelial surfaces seem to be affected. The mesenchymal signal may inhibit epithelial spreading by altering contact relationships of epithelial cells. It may do so directly by affecting epithelial cell surfaces, or indirectly by changing the nature of the extracellular matrices surrounding epithelial cells. As we have seen, such surface changes may secondarily stimulate cell division (see Chapter 5). As in salivary epithelium, the primary effect of mesoderm is on tubule morphogenesis and not on differentiation.

The interaction between epithelium and mesoderm in vitro is brief. Although differentiation requires 4 to 5 days in vitro, an epithelium can differentiate if exposed to mesenchyme for only 30 hours and cultured separately thereafter. During early induction, epithelium is organized into a condensed mass and cell division is stimulated. A mitotic gradient is established with maximum activity at the periphery, declining toward a central core of postmitotic cells. At 48 hours ribosomes aggregate within some central cells and amylase activity increases. From this point on amylase synthesis increases logarithmically as more cells differentiate and the characteristic zymogen granules of the pancreas collect in the cytoplasm. The epithelium becomes independent of mesenchyme about 18 hours before biochemical differentiation.

Cell division is distributed as a gradient with a high point at the periphery and few or no divisions in the center where the first cells differentiate. Division is necessary for differentiation since continuous exposure to an inhibitor of DNA synthesis (5-fluorodeoxyuridine, or FUDR) from 48 hours of culture invariably inhibits differentiation without affecting RNA or protein synthesis. If treatment is begun at 72 hours, then some postmitotic central cells continue to differentiate and synthesize zymogen granules. After cells complete a division program between 48 and 72 hours they switch to differentiation.

Division and differentiation may be uncoupled by 5-bromodeoxyuridine (BUDR). This analog replaces thymidine in DNA of dividing cells without affecting the rate of DNA synthesis or cell division. Treatment with BUDR up to 48 hours of develop-

ment completely inhibits amylase synthesis but does not stop mitosis or growth of the rudiment. If treatment is begun after 72 hours, then differentiation is noted in some central cells. These, having ceased DNA synthesis before BUDR treatment, do not incorporate the analog and continue to differentiate. Only those dividing cells at the periphery which have incorporated BUDR fail to differentiate. The inhibitor-sensitive period (48 to 72 hours) correlates with the time when cells leave the mitotic cycle and differentiate.

The effect of BUDR on cell differentiation is not understood. The abnormal DNA synthesized at these critical divisions may not transcribe properly, or the physical nature of the template may change so that the transcribing enzymes cannot recognize initiator sites.

The BUDR effect correlates with the effect of actinomycin D on pancreas differentiation. Treatment of pancreas rudiments between 48 and 72 hours inhibits amylase synthesis and differentiation but exposure after 72 hours has only a slight effect on amylase synthesis, suggesting that stable RNA messages are transcribed between 48 and 72 hours. As in erythropoiesis, a critical period of stable message transcription programs differentiation; transcription of luxury protein genes is coupled to a division program.

Pancreas differentiation involves stabilization of whole sets of properties. Besides amylase, approximately ten other enzymes such as trypsin and lipase are synthesized in exocrine cells while endocrine cells synthesize insulin. This means that a minimum of 10 to 12 different structural genes must be turned on. Though the time at which these syntheses are turned on in vivo varies somewhat, the overall pattern of enzyme accumulation suggests that all proteins are synthesized in a coordinate fashion, as if a whole block of genes is turned on simultaneously.

It is noteworthy that low basal levels of enzymes are present well before differentiation has begun. In the earliest pancreatic rudiment that can be isolated from the embryo, virtually all luxury molecules characteristic of the pancreas phenotype are detected, at very low levels, perhaps of the order of $10^4$ molecules per cell. For several days this level is constant, but as soon as differentiation begins, the amount of protein increases rapidly to $10^8$ to $10^9$ molecules per cell, and it is during this phase that the visible phenotype emerges.

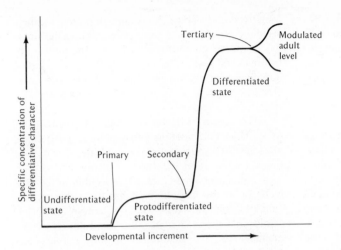

**Figure 17-32**  *Model of the protodifferentiated state as proposed by Rutter. The primary and secondary inductive stimuli are indicated. The undifferentiated state is equivalent to the determined state, presumed to have begun between zero time and the primary inductive signal. (From W. Rutter,* Epithelial-Mesenchymal Interactions, *ed. R. Fleischmajer and R. E. Billington.* © 1968 by The Williams & Wilkens Co., Baltimore.)

According to Rutter, this early period is a *protodifferentiated* state when cytological differentiation is not visible but low levels of luxury molecules are detectable (Figure 17-32). Either specific genes are activated in all presumptive pancreas cells but are transcribing at low rates or a small number of cells have fully differentiated, with all pancreatic genes active and functioning at maximal rates. In the latter case differentiation is an all-or-none process, without intermediate states.

As in erythrocyte differentiation, the first hypothesis assumes that determined progenitor cells are switched on to pancreas differentiation, possibly by an early mesodermal signal. Cells are committed to low levels of luxury protein synthesis (insulin?) since the relevant genes are transcribed. In this protodifferentiated state, cells are still dependent on mesoderm and are mitotically active. A second inductive event occurs, triggering a final division program essential to the fully differentiated state. This occurs in the critical period between 48 and 72 hours in culture. Proliferation slows down and all cells differentiate progressively as the translation machinery is stabilized.

The second hypothesis is based on the model of mammary epithelium differentiation. The progenitor cells of the pancreatic rudiment consist of undifferentiated stem cells and a small population of differentiated cells already synthesizing luxury molecules. Only low levels are detected since there are so few mature cells. The inductive signals from mesoderm stimulate proliferation of already determined progenitors, and all newly formed daughter cells are programmed to synthesize enzymes and hormones, just

as mammary gland cells acquire the ability to synthesize casein after dividing in a balanced hormone medium. The net result is an increase in the level of enzyme activity in the total tissue.

So far, we cannot distinguish between these alternatives. The experiments with BUDR tend to support the second model since lesions in the DNA of newly formed daughter cells prevent their differentiation. If genes were already transcribing as predicted in the first model, we would expect BUDR to have no effect at the posttranscriptional level since it has no effect on the expression of fully differentiated cells. This conclusion, however, is only tentative, inasmuch as we still do not know how BUDR affects the genome.

Let us summarize some of these signaling factors in heterotypic pancreas differentiation and see how they compare to hormones and homotypic interactions.

1. An inductive signal seems to affect the surface of responding cells and promotes close contact relationships, changes in cell shape, and altered permeability. It may do so by changing membrane constituents or by creating an extracellular matrix that facilitates these changes, or both.

2. The signal secondarily stimulates mitotic activity, perhaps by favoring tissue condensation. Cell division always seems to precede cell differentiation, but the relationship between mitosis and differentiation is obscure. It seems that cells that have completed one or several divisions may be the ones to begin differentiation.

3. Differentiation is a population phenomenon; not all cells differentiate at one time. A few pioneer

cells begin and more are progressively recruited. Before any visible morphological changes are noted in cells, some of the cell-specific proteins can be detected at low levels, as if genes have been switched on or are "leaking," that is, transcribing at low rates. Differentiation seems to occur when gene expression and protein synthesis are elevated to much higher levels and stabilized, usually by the assembly of cell-specific organelles. We have to distinguish between signals that switch genes on before cells differentiate and those conditions that mobilize systems of information flow already turned on (modulation). In both cases, products accumulate and assemble into specialized cell organelles, but the former calls for transcriptional control of differentiation and the latter calls for posttranscriptional regulation as well. We still are unable to distinguish these two alternatives in most differentiating systems. Many mechanisms of control are exercised in any one channel of phenotypic expression. The problem is to determine when each regulatory mechanism is operating, and which genes are being regulated—structural genes, regulatory genes, or both.

## Differentiation, stable messenger RNAs, and actinomycin D

In the differentiating systems described, one feature seems common to all. There is a variable lag period while cell machinery is organized for the synthesis of specialized proteins. Early in this cryptic phase, gene transcription occurs and specific message molecules appear. If transcription is inhibited by actinomycin D, the tissue does not differentiate; if the inhibitor is applied after transcription, then differentiation occurs and luxury molecules are synthesized. Genetic information for differentiation is transcribed early but it is translated into phenotype-specific proteins much later. This means that certain populations of messenger RNA molecules are long-lived and are translated for several days as cells differentiate. In contrast to messages coding for housekeeping proteins which are labile and turn over rapidly, those for luxury molecules may be stabilized for long-term use. The existence of such stable messages is demonstrated by inhibition studies with actinomycin D which inhibits all gene transcription. If specific protein synthesis is also inhibited, then the message is short-lived and unstable. If protein synthesis continues in the absence of RNA transcription, we conclude the message is

long-lived and stable. The decline of protein synthesis in the presence of actinomycin measures the kinetics of protein stability and turnover. The technique has revealed long-lived messenger RNAs during differentiation of erythrocytes, lens, muscle, and pancreas, and has shown the presence of masked messages in eggs and seeds.

The nature of message stability is unknown. It may involve combination with proteins which protect messages from degradation by nucleases. Or it may involve association with poly A at one end of the message, a process that takes place either in the nucleus or in the cytoplasm. It has been shown, for example, that messages bound to poly A are more stable than those without poly A in cultured mammalian cells. The attachment of poly A may alter message configuration and make it less accessible to nucleases.

On the other hand, stabilization may simply result from transcription of an excess of message molecules, more than enough to balance the normal rate of turnover. In spite of message decay, enough messages are always present to sustain high levels of protein synthesis. We have seen that this is indeed a possible explanation of message stability in terminal differentiation of the erythrocyte, and similar mechanisms may be used in terminal differentiation of such cells as lens and skin, both of which are also destined to die.

Actinomycin D does not always have predictable effects in some differentiating systems since it inhibits transcription of regulatory as well as structural genes and many of its effects turn out to be posttranscriptional as a result. For example, the enzyme glutamine synthetase (GS) is actively synthesized during differentiation of the chick embryonic retina. A precocious induction of enzyme activity can be obtained by culturing 12-day neural retinas in a medium containing hydrocortisone (HC) (Figure 17-33). The hormone seems to stimulate de novo synthesis of enzyme. Treatment with actinomycin during the first 4 hours prevents enzyme induction but exposure after 4 hours has no effect and almost normal activity levels are reached. Here again long-lived messenger RNA is implicated.

Though this seems to behave like other systems, actinomycin has surprising posttranscriptional effects. Utilization of the message in translation seems to depend upon presence of the hormone. If it is removed after 4 or 6 hours of induction, no increase in

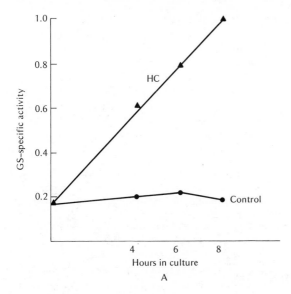

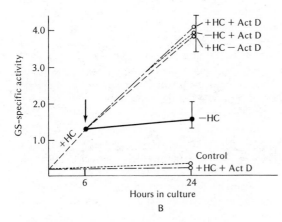

**Figure 17-33**  *Induction of glutamine synthetase (GS) activity by hydrocortisone (HC) in chick embryonic retina in vitro and its response to actinomycin D. (A) Increase in GS-specific activity after 6-hr induction by HC. (B) After an initial induction by HC, GS activity increases even in presence of actinomycin D which replaces continuous exposure to HC. Exposure to actinomycin from the beginning of culture inhibits induction in the presence of HC. Both HC and actinomycin are acting posttranscriptionally. Arrow indicates time of HC withdrawal and/or actinomycin addition. Vertical lines indicate range of values in various experiments. (From M. Moscona, N. Frenkel, and A. A. Moscona, Develop. Biol. 28:229–241, 1972.)*

enzyme beyond the level at 4 hours is noted. Although all messages have been transcribed, they are not translated unless hormone is present. What is unusual is that actinomycin D added to such 4-hour-induced cultures stimulates almost normal levels of enzyme synthesis even in the absence of inducer! Though inducer is required for continued synthesis of the enzyme, it can be replaced by actinomycin D after a short inductive period, as if actinomycin D, which normally inhibits, is itself acting as inducer! In this case, total shutdown of transcription leads to increased protein synthesis. How do we explain this anomalous behavior of actinomycin D?

According to a model proposed by Moscona, actinomycin is acting at posttranscriptional sites of regulation rather than on transcription of genes coding for the enzyme. After stable RNA messages for GS synthesis are transcribed in the first 4 hours, their translation is governed by interactions between the products of the two other regulatory genes. One gene product (an enzyme) destroys GS RNA templates while the other gene product (a desuppressor) prevents this by neutralizing the enzyme. It is assumed that both products are more labile than GS message transcripts. Their stability depends on continued transcription of their own RNA messages, both of which are also less stable than RNA messages for GS.

In the model the hormone inducer activates transcription of GS and desuppressor genes, and stable GS RNA messages are produced. The desuppressor gene product neutralizes any suppressor enzyme, and all new GS templates are translated into enzyme. If HC is withdrawn, transcription of both GS and desuppressor genes stops, suppressor enzyme accumulates, destroys all stable GS templates, and enzyme activity declines.

After 4 hours of induction, some stable GS templates are synthesized. If a high dose of actinomycin is now added, transcription of all genes stops. The more labile suppressor and desuppressor messages are destroyed, and all posttranscriptional control ceases. Meanwhile, the few stable GS templates that have been synthesized are translated at a slow rate and GS enzyme accumulates to maximal levels, completely independent of the inducer. According to this hypothesis, actinomycin acts by inhibiting transcription of those genes regulating message stability and translation.

We do not know whether this type of multigene regulation is responsible for retinal differentiation; it is only one of several models that explain the data. It shows, however, that regulation of eukaryote genomes is complex and often difficult to analyze. Differentiation need not be regulated exclusively at the site of the gene template. Once the initial genetic events have occurred at the template, stability and amplification of the differentiative state seem to be controlled within the cytoplasm.

## GENETIC MECHANISMS OF DETERMINATION AND DIFFERENTIATION

As cell clones progress toward differentiation from the totipotent state, the amount of genetic information available for expression seems to decrease. But this is not really so. The information is there all the time but it is not expressed. We can imagine a genome as a book of instructions for assembly of an organism. During development, each cell receives a complete set of instructions, but at certain stages large portions of the book become inaccessible to each cell clone, leaving only those few instructions necessary for creation of one specific phenotype such as muscle or nerve.

The molecular hybridization studies described in Chapter 3 indicate that each tissue transcribes only a small proportion of the total genetic information in the cell and that each tissue, though sharing many messages with other tissues, also contains a unique population of messages, presumably those programming the specific cell phenotype. The bulk of the somatic cell genome is silent, which means that any mechanisms to account for regulation of the genome during determination and differentiation must explain the wholesale repression of most of the genome while leaving small portions accessible to each cell phenotype for expression.

It would seem that genes are regulated at two different levels, gross and fine, each possibly controlling determination or differentiation. At the gross level, large segments of genetic information are progressively silenced during determination. A small portion of information, that specific for each cell genotype, is protected from this process and it is this set of genes (including all housekeeping genes which should be the same in all phenotypes) that is made accessible for the fine control by activators, repressors, and inducers during differentiation. Models for this type of regulation have been described in Chap-

ter 3. The dual model implies that regulation of the genome during determination is qualitatively different from its regulation during differentiation. In the former case we are probably dealing with physical changes in DNA templates and chromosome structure, either by specific association with histones and nonhistone proteins, supercoiling, or changes in the conformation of DNA-associated proteins by phosphorylation or acetylation of histones. This invokes some type of heterochromatization as a means of gross control during determination.

### Heterochromatin and determination (gross control)

The determined state is difficult to study because there are no morphological or biochemical properties that can be experimentally manipulated. Accordingly, no satisfactory genetic model for determination has been proposed. Tyler suggested that a determinative event occurs when a specific set of genes is transcribed into stable but inactive messenger RNA. Later in development, exogenous factors such as hormones may stimulate translation of these messages to initiate differentiation. This would account for the delay between gene transcription, when a cell is determined, and translation of activated messenger RNA when specialized phenotypes appear. The model, however, does not explain the progressive repression of large portions of the genome.

There are instances of gross regulation at the chromosomal level which silence large blocks of genes. Whole chromosomes, and even whole chromosome sets, are genetically inactivated by heterochromatization (see Chapter 2).

Perhaps the best example of heterochromatic regulation of the expression of several genes coding for a phenotypic program is the Y chromosome in *Drosophila heydei*. The Y chromosome is heterochromatic and inactive in all cells of the male including spermatogonia. Later, in spermatocytes, the Y chromosome displays dramatic changes in its morphology and becomes active. The information in the Y chromosome is utilized when sperm cells differentiate.

The small Y chromosome is completely heterochromatic during development of the germ line, including proliferating spermatogonia in the testis. As meiosis begins in primary spermatocytes, the Y chromosome undergoes an explosive change in morphology and opens out into loops (Figure 17-34), "ribbons,'' and ''nooses'' that resemble the loops on oocyte lampbrush chromosomes. Six loop pairs can be distinguished, each probably representing one fertility gene carried by the Y chromosome. The total length of the loops corresponds to only $\frac{1}{12}$ of the total Y-chromosome DNA, which suggests that most of the DNA is genetically empty.

As loops emerge, RNA synthesis begins, a sign of gene activation resembling the lampbrush stage of oogenesis. RNA synthesis persists for a short period into secondary spermatocytes but shuts down during spermatid differentiation. The RNAs extracted from testes at this time are derived almost exclusively from the extended loops. Most RNA messages are synthesized premeiotically and are probably translated postmeiotically when spermatids differentiate.

All loop-forming regions are essential for differentiation of mature sperm. If portions of the Y chromosome are deleted, certain loops are lost and sperm cells differentiate abnormally. For example, when one of the giant loops is absent, the male fly is sterile. Deficiencies of the long arm of the Y result in spermatids that elongate but do not form a midpiece or flagella. Smaller deletions eliminate some fine threads, and in these cases sperm appear normal but exhibit no motility. The major effects caused by deficiencies of certain loops, however, seem to be regulatory rather than structural. Y chromosome genes may not code for specific structural components in sperm but may control growth and assembly, behaving like assembly genes in bacteriophage T4 (see Chapter 7).

Reversible shutdown of chromosomes in the germ line of D. heydei is probably not applicable to the determination of all somatic cell phenotypes since this phenomenon is found in varying degrees in only a few Drosophila species; it may be a unique adaptation for spermatid differentiation. If heterochromatization is a mechanism that silences large

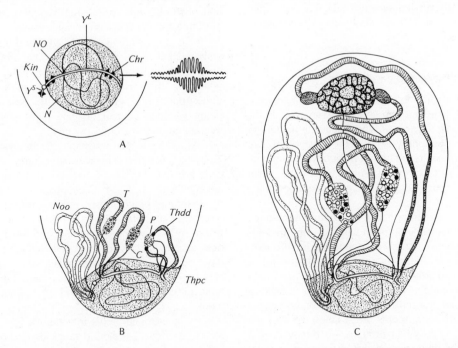

**Figure 17-34** Y chromosome of Drosophila heydei transforms from a condensed heterochromatic state to a univalent lampbrush chromosome during spermatogenesis. (A) Very young spermatocyte showing chromomeric structure of Y chromosome in detail. (B) Early growth stage of primary spermatocyte nucleus with loops unfolding. (C) Full-grown spermatocyte nucleus with fully developed loops. Chr = chromomere; C = clubs; Kin = kinetochore; N = nucleus; NO = nucleolus organizer; Noo = nooses; P = pseudonucleolus; T = tubular ribbons; Thdd, Thpc = diffuse distal and proximal compact sections of the threads; $Y^L$, $Y^S$ = long and short Y arms. (From O. Hess and G. F. Meyer, Adv. Genet. 14:171–223, 1968.)

blocks of genetic information in determination, it should be stable and transmissible through many cell generations. The physical changes in the genome must be replicated at every division to insure that progeny retain a phenotypic bias. Is there any evidence that gross chromosomal changes are transmitted to cell progeny?

We know, for example, that the heterochromatic X chromosome in mammals is replicated at each division during development. Compaction of the X chromosome occurs very early in mammals, perhaps during the first few cleavage divisions. Compaction is a random event in each tissue so that one or the other X is compacted. Once compacted, however, the heterochromatic X chromosome transmits its physical state to its progeny at each replication. The factors responsible for compaction of the X chromosome are intrinsic to the chromosome and are replicated at all subsequent divisions. These factors are not localized in the cytoplasm, since other X chromosomes in the same cell always remain uncompacted; the supercoiled state of the DNA is stabilized in all subsequent replications.

## Chromatin changes in development

Most determinative events occur early during gastrulation and neurulation in vertebrate embryos. In the insect embryo they occur when cleavage nuclei migrate to the periphery to form the blastoderm. Determination may result from a nucleocytoplasmic interaction; that is, cytoplasmic components enter nuclei and change chromosome structure or behavior, or both.

There is evidence that DNA structure is modified in somatic tissues; chromatin isolated from nuclei is in a repressed or "masked" condition. In vitro template properties of chromatin DNA suggest that histones and acid proteins affect transcription (see Chapter 3). Chromatin from different tissues exhibits tissue-specific transcribing capability; unique populations of RNA transcripts are synthesized by each chromatin. Only when proteins are removed from DNA do the differences between tissues disappear. This suggests that large-scale protein-DNA interactions silence blocks of genes in adult tissues. Is there any evidence of progressive repression of the embryonic genome by chromosomal proteins during development?

New histone proteins do appear at critical stages

of sea urchin or amphibian development. For example, a lysine-rich histone predominates during early sea urchin cleavage with new lysine-rich fractions appearing later in development. Acidic, nonhistone proteins also change in sea urchin development. Approximately 24 different proteins have been resolved from nuclei and some of these change at different developmental stages. The changes in these proteins are, at least, consistent with the hypothesis that they regulate genetic expression in development.

On the whole, however, there are few developmental changes that correlate with changes in chromosomal proteins. The template behavior of chromatin extracted from different developmental stages of the sea urchin does show some changes in activity. Chromatins isolated from blastula and pluteus stages differ slightly in composition, particularly as to histone and nonhistone proteins; pluteus chromatin contains twice as much nonhistone protein. Moreover, its template activity in vitro is also greater than that of the blastula, suggesting that more genes are available for transcription in later stages.

It is in differentiating tissues and not early embryos that we see how chromosomal proteins act. As a result of experiments by B. O'Malley, we are beginning to learn how hormones may regulate transcription of specific genes, the best example being estrogen control of gene expression during differentiation of the chick oviduct.

As we have seen in Chapter 5, the oviduct in a 7-day-old hatched chick is immature; epithelial cells are undifferentiated and no luxury molecules are synthesized. Daily injections of estrogen and progesterone precociously induce three new cell types after an initial episode of proliferation, Tubular gland cells secreting ovalbumin and lysozyme appear at 2 to 4 days, ciliated cells differentiate at 6 days, and goblet cells secreting avidin develop later, at 7 to 9 days of treatment. This program of phenotypic differentiation is triggered by hormone treatment.

As in estrogen-uterus interactions (see Chapter 6), the oviduct is a steriod target tissue because it contains estrogen-specific receptors, cytoplasmic proteins that bind to estrogen with high affinity. The cytoplasmic-receptor hormone complex is converted into a nuclear receptor-horome complex by a conformational or metabolic change and enters nuclei where it regulates genome transcription. We know this is true for several reasons: (1) RNA synthesis is stimulated within 2 minutes after hormone addition. All species of RNA are produced and their synthesis may be blocked by actinomycin D. (2) The number of initiation sites for in vitro RNA synthesis by oviduct chromatin is markedly increased after estrogen treatment but their number declines if estrogen is withdrawn.

The DNA template itself may be the direct target of the hormone-receptor complex since the complex binds to DNA with high affinity. But it does so with all DNAs; it lacks specificity under these conditions. This suggests that naked DNA strands are not the principal targets. Rather, it is oviduct chromatin that shows maximum specific binding. The presence of chromosomal proteins enhances hormone-receptor binding and seems to account for its specificity.

Which proteins are involved? If histones are removed from oviduct chromatin, more hormone-receptor complex binds, as if additional binding sites are exposed. If, however, the nonhistone proteins are removed from chromatin, most binding is reduced. If chromatin is reconstituted by adding nonhistone proteins from oviduct chromatin to oviduct DNA, then the binding levels of native chromatin are restored. It appears that binding of the hormone complex in nuclei is dependent upon the nonhistone proteins associated with DNA.

Specificity of binding is also a property of nonhistone protein. For example, if erythrocyte DNA is recombined with erythrocyte histone plus oviduct nonhistone proteins, the high binding affinity characteristic of native oviduct chromatin is obtained. However, oviduct-reconstituted chromatin containing erythrocyte nonhistone proteins will bind to the hormone receptor complex at the low levels typical of native erythrocyte chromatin. In other words, tissue specificity of binding to the hormone receptor complex seems to reside in the nonhistone proteins of nuclei; they behave as the primary target for hormone action.

Now that we know where the hormone acts, the question arises as to whether this binding initiates specific gene transcription. It is possible to extract relatively pure ovalbumin message RNA from estrogen-stimulated oviduct tissue. The purified message is then used in vitro with a reverse transcriptase (RNA-dependent RNA polymerase) which copies RNA templates to make complementary DNA strands (cDNA). These DNA copies, in

effect, represent relatively pure genes coding for the structural protein ovalbumin (cDNA$^{oval}$) and can be used in hybridization experiments to determine how much newly synthesized RNA is complementary to ovalbumin genes. With this cDNA it is possible to compare the transcriptive behavior of chromatin isolated from estrogen-treated chick oviducts with chromatin from unstimulated or withdrawn oviducts.

In the experiment, chromatin is isolated from chick oviducts treated for 2 weeks with estrogen and from oviducts treated with estrogen for 2 weeks but subsequently withdrawn from estrogen for 12 days. The number of initiation sites for in vitro RNA synthesis in withdrawn chromatin is approximately half that of estrogen-stimulated chromatin. Each of these chromatins is allowed to transcribe radioactive RNA in vitro and the proportion of all newly synthesized RNA complementary to ovalbumin DNA is determined by DNA-RNA hybridization. The results show that 0.011 percent of the mRNA from stimulated chromatin is ovalbumin message compared with only 0.0015 percent from withdrawn chromatin. In other words, stimulated chromatin is eight times more active in synthesizing specific ovalbumin messages.

If the chromatins are separated into DNA, histones, and nonhistone proteins and then reconstituted, they behave exactly as native chromatin in their ability to transcribe ovalbumin messages in vitro. On the other hand, reconstitution of withdrawn chromatin with the nonhistone fraction of stimulated chromatin gives a transcriptive activity similar to native stimulated chromatin. Reciprocally, a reconstituted stimulated chromatin with the nonhistone fraction from withdrawn chromatin behaves like native withdrawn chromatin in its transcription of ovalbumin message RNA. Reciprocal exchanges of histones in reconstitution experiments have no effect on chromatin transcription. These results show that specificity of gene function is a property of the nonhistone protein fraction. Combination of a hormone-receptor complex with nonhistone proteins determines the level of transcription of specific ovalbumin genes.

The model of gene regulation that emerges here is one of hormonally induced changes in the configuration of nonhistone proteins associated with specific genes. We do not know how DNA, histones, and nonhistone proteins interact at the level of the gene. What seems clear, however, is that the non-histone proteins of the eukaryote nucleus may be the principal sites for regulation of gene expression in differentiation.

Other examples of gross chromosomal changes show how different genes are activated at different times during development. The giant chromosomes in dipteran tissues undergo visible changes in structure that are tissue-specific and correlate with transcription and translation of cell-specific proteins.

## Chromosome puffs: differential action of genes

Differential gene expression can be visualized in the giant polytene chromosomes of dipterans (Chapter 2). The large size of these chromosomes is due to their polytene nature; they contain 1064 strands of DNA which tends to exaggerate any physical changes that

may occur within small regions. We assume that gross visible changes in giant chromosomes reflect more subtle transformations within single-strand regions.

The banding pattern of these chromosomes corresponds to genetic maps; each band is the site of a gene (see Chapter 3). Giant chromosomes in different larval tissues have identical banding patterns, which means that all tissues share the same complement of genes.

During larval development, individual chromosome bands display a developmental sequence of structural alterations (Figure 17-35). Some bands expand as strands of DNA uncoil and open out into enlarged structures known as "puffs" (Chapter 3). Shortly thereafter the puff regresses while other bands expand into a puff. Puffing patterns are tissue-specific; each tissue displays a different temporal pattern

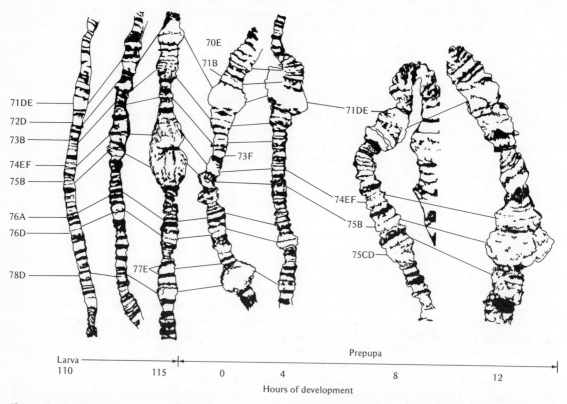

**Figure 17-35** *Puffing sequence in* Drosophila *chromosomes. Numbers refer to specific bands in chromosomes. Puffing sequences in proximal region of long arm of chromosome 3 in* Drosophila *during development from late larva to prepupa. Note that only a small proportion of the bands puff in this interval. Some regions such as 75B puff twice, others only once (77E). The timing of these puffs also varies. (From M. Ashburner,* Chromosoma *21:398–428, 1967.)*

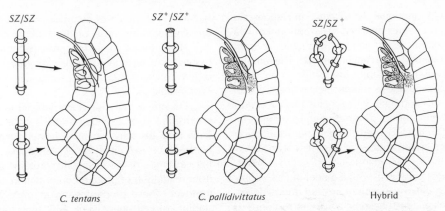

**Figure 17-36** *A puff is a visible manifestation of a gene (SZ) in action. Salivary glands and puffing patterns in two species,* Chironomus tentans *and* C. pallidivittatus. *Distal cells of the latter secrete distinctive large granules that are absent in cells of the former species. Likewise, a Balbiani ring (shaded at end of chromosome) is lacking in* C. tentans. *The hybrid produces fewer granules and displays the Balbiani ring. Because of incomplete pairing due to inversion, the arms of the chromosomes are separated in the hybrid and puffs in the heterozygous arms can be seen. (From A. Kühn,* Lectures on Developmental Physiology. *New York: Springer-Verlag, 1971.)*

of puffing. Peaks of puffing activity usually precede molting episodes by several hours, but some puffs continue during molting. As many as 70 to 100 different puffs may appear within a tissue during larval development, most arising in the late third-instar larva shortly before metamorphosis. Most puffing sites (70 percent or more) are common to all tissues, but the timing of a puff and its regression differs in each. For example, a comparison of puffs in three tissues in *Drosophila heydei* shows that 33 of 116 puffs are tissue-specific; that is, 14 are specific to a single tissue and 19 are specific to two or three tissues.

The fact that a puff involves a band region, a presumed gene locus, suggests that a puff is a gene in action. Most studies have borne out this hypothesis. Puffs contain large amounts of RNA and protein. Moreover, studies with radioactive RNA precursors indicate that puffs are sites of intense RNA synthesis with the rate of $^3$H-uridine incorporation correlating with puff size; large puffs are most active in RNA transcription while puff regression is accompanied by decreased RNA synthesis.

One puff has been isolated from *Drosophila* chromosomes, and its RNA transcript turns out to be a giant molecule, about equivalent in size to the total DNA in the puff. This molecule is transported intact from nucleus to cytoplasm and contains many repeated sequences. It may be a messenger RNA for

some of the salivary gland polypeptides that are actively synthesized at these stages.

An experiment relating puffing to the activation of a known gene was carried out by Beerman. He studied puff patterns in two closely related species of chironomids, *Chironomous pallidivittatus* and *Chironomous tentans*. The salivary glands of these two species differ; large secretory granules are found in a few distal cells in *C. pallidivittatus* but are absent in *C. tentans* (Figure 17-36). The two species, when crossed, form viable hybrids with the phenotypic trait of granule formation inherited in a Mendelian fashion. This gene, mapped within the salivary chromosomes, occupies a band in the fourth chromosome that is shared by both species, but the band puffs only in the distal cells of *C. pallidivittatus* and not in *C. tentans*. In a homozygous *C. pallidivittatus*, two puffs are seen in the homologous chromosomes and none are seen in *C. tentans*. In a first-generation hybrid between the two, only one puff appears, corresponding to the chromosome derived from the *C. pallidivittatus* parent. Moreover, fewer secretory granules are produced in the hybrid, because it has only one functional gene. Here is an example of a correlation between a gene, a puff, synthesis of RNA, and a final phenotypic product in a specific cell.

Puffing is, in effect, the physiological expression of genes in fully differentiated cells. Salivary gland cells undergo functional modulations in response to hormones. The genes involved may represent only a small proportion of the genome available for transcription, roughly 10 percent of the total. The remaining genome is inaccessible, insensitive to any hormone signals. Synthesis of specific polypeptides in salivary glands correlates with specific puffs in *Chironomous* chromosomes. These proteins are included in the "silk" threads used for puparium formation and are synthesized at rates far in excess of any other protein in the salivary gland.

Puffing is consistent with a model of DNA compaction as a means of gene repression during development. When genes become active (in response to a hormone), they transform from a compacted state to a despiralized diffuse condition. Many gene loci in these chromosomes seem to be inactive by virtue of their compacted state as chromomeres. Since everything about the polytene chromosome is exaggerated, we assume the puff itself magnifies what may normally occur in all genes reversibly silenced by compaction. In cases where puffs are induced by raising the temperature, the condensed chromomeres that make up a band do despiralize within 30 minutes into a network of fine threads before the accumulation of ribonucleoprotein granules. It is only on fine fibers that small granules accumulate and eventually coalesce into larger granules of protein and RNA. In addition, nonhistone proteins increase within a puff region, often to more than 200 percent of the original amount. Nonhistone proteins preexist in the nucleus and accumulate at the puff during its expansion. Though the relationship between nonhistone proteins and puffing has not been worked out, we do see in puffing visual evidence of gene activation associated with differential synthesis of specific proteins.

## EPIGENETIC CONTROL OF TISSUE-SPECIFIC PROTEIN PATTERNS

In general, cell phenotypes are distinguished by their repertoire of luxury proteins, which implies that different genes are acting in each tissue. Such proteins are believed to be unique to a tissue, but this is not entirely true. Actin and myosin, though diagnostic of muscle tissue, are ubiquitous and have been found in a wide range of cells. Many different collagens are synthesized and stored in a variety of noncartilage tissues. Certain proteins may become tissue-specific

because their rates of synthesis and accumulation are greater than those of the bulk proteins in the cell. Differences in luxury protein content among tissues may be quantitative rather than qualitative, based on differential rates of message transcription or translation.

Enzyme activity is not a reliable criterion for distinguishing gene expression in tissues because proteins may be activated by epigenetic mechanisms. For instance, even if all genes were active in all cells, the existence of tissue-specific proteins could arise from regulation of protein conformation, stability, or turnover. These properties are as important to the stability of cell phenotypes as is gene transcription.

Most enzyme studies tell us that a specific protein is present in a tissue extract but say nothing about its in situ activity. Extraction changes enzyme function; in vitro, activity is often in excess of that measured in vivo, suggesting that enzymes are under strict regulation in the intact cell. Destruction of cell organization abolishes regulation. Inhibitors, activators, availability of substrates, and intracellular compartmentalization all affect in vivo activity of a protein. These represent epigenetic control of a phenotype after a protein has been synthesized.

We must distinguish between protein synthesis, activation, stabilization, and degradation, since all affect measured levels of activity. Each is regulated differently. Synthesis of new enzyme molecules may result from transcription and translation of new message molecules, or translation of preexisting inactive messages. Enzymes may become active after removal of an inhibitor, or by configurational changes. Changes in enzyme activity also reflect changes in protein turnover or stability. The life of a protein molecule is controlled by proteases that split peptides into amino acids for reuse in protein synthesis. Enzyme activity, as measured in most eukaryotic cells, is a balance between the number of enzyme molecules synthesized, those stabilized into functional configurations (for example, by attachment to membranes), and those degraded. It represents an average state of regulation of virtually all sites in the flow of information from gene to functional protein.

### Isozymes

A special case of epigenetic control relates to factors regulating isozyme content of different tissues. Isozymes are collections of enzyme molecules with similar catalytic function but different physical-chemical properties. Most enzymatic proteins are composed of several subunit polypeptides which may be coded by different genes. A subunit is usually not functional unless it self-assembles with other subunits into the complete protein. Since the assembly process is random, isozymes are produced with similar substrate specificities but different subunit organization.

The enzyme lactic dehydrogenase (LDH), a ubiquitous enzyme, catalyzes the conversion of pyruvic to lactic acid in the absence of oxygen in most cells. The enzyme is a tetramer, composed of two alpha and two beta polypeptide chains (coded, respectively, by an A and B gene) which assemble into five different isozymes; that is, $\beta\beta\beta\beta$, $\beta\beta\beta\alpha$, $\beta\beta\alpha\alpha$, $\beta\alpha\alpha\alpha$, $\alpha\alpha\alpha\alpha$ (LDH1 to LDH5, respectively) (Figure 17-37). Though each isozyme catalyzes the same reaction, they differ in molecular configuration, charge density, and pH optima. Consequently, they

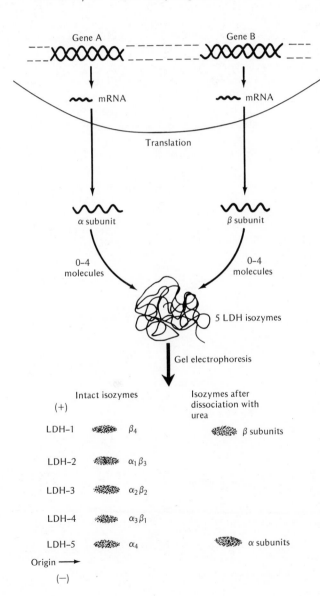

**Figure 17-37** *Epigenetic control of phenotypic expression of isozymes of lactic dehydrogenase (LDH). Subunits $\alpha$ and $\beta$ are each coded by a gene. Differential rates of transcription and translation of their messages lead to different proportions of each subunit in the cytoplasm. This plus conditions in cytoplasm favor random assembly of subunits into isozymes that are separable by gel electrophoresis, as shown to the left.*

have different mobilities in an electric field and can be separated into five distinct enzyme species by gel electrophoresis, with LDH1, the most negatively charged, migrating to the anode. If LDH1 and LDH5 are isolated and subjected to urea denaturation, the enzyme dissociates into its respective subunits, $4\beta$ subunits from LDH1 and $4\alpha$ subunits from LDH5. A mixture of equal amounts of $\alpha$ and $\beta$ subunits automatically reassembles into all five isozymes in a ratio of 1:4:6:4:1. Different ratios of LDH isozymes are reconstituted by varying the proportion of $\alpha$ and $\beta$ chains.

LDH extracted from somatic tissues of an adult rat yields tissue-specific electrophoretic patterns (Figure 17-38). Skeletal muscle shows high amounts of LDH5 whereas heart muscle has an excess of LDH1, 2, and 3. Though A and B genes are active in all tissues, they are not equally active; different proportions of $\alpha$ and $\beta$ chains must be synthesized in each tissue to account for different isozyme ratios. Information flow from each gene is regulated differently at the level of either transcription, message translation, or polypeptide stabilization and assembly. As we have seen in Chapter 7, self-assembly processes are affected by such cytoplasmic factors as pH, ionic content, and even intracellular location. Cytoplasmic states may vary in each tissue or during development of the same tissue, favoring one

or another isozyme pattern. As yet we cannot choose between the alternatives.

The adult isozyme pattern is achieved gradually during differentiation. Different isozyme patterns characterize each stage of tissue development. For example, isozyme patterns for developing mouse heart muscle (Figure 17-39) show that LDH4 and 5 predominate in 9-day fetal mice; the A gene may be more active than the B gene at this stage. As development proceeds, the isozyme pattern shifts; the B gene assumes the dominant role in the adult heart and LDH1, 2, and 3 predominate. Evidently expression of genes for each subunit is regulated independently during the ontogeny of each tissue. Different ontogenetic patterns characterize each tissue, with young embryonic stages showing fewer isozymes than more mature tissues. Most other housekeeping enzymes, such as malic acid and isocitric dehydrogenase, are also isozymic and exhibit tissue-specific patterns at various stages of development.

We see then that differentiation is also modulated by cytoplasmic factors that control rates of subunit translation, accumulation, and assembly. Physiological changes in the state of the cytoplasm may affect isozyme assembly, producing phenotypic changes that are not easily distinguished from those at the genetic level. Control at the template concerns mechanisms that promote expression of ac-

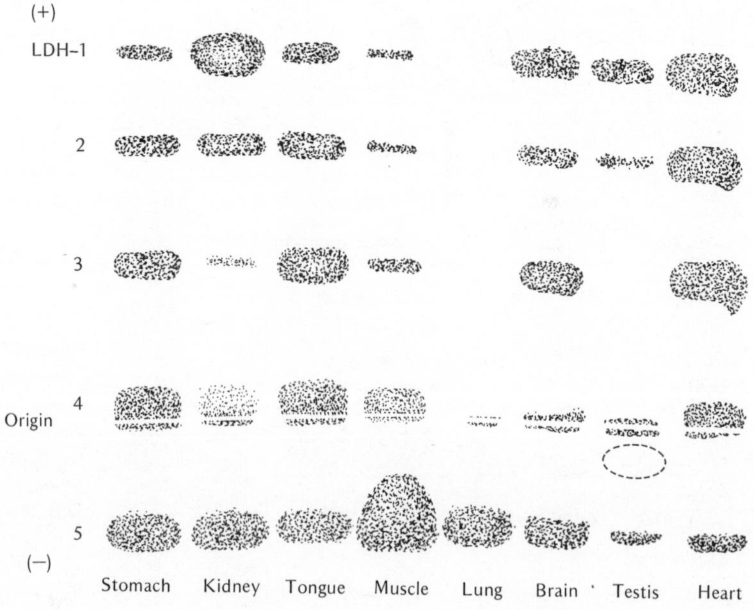

**Figure 17-38** *LDH isozyme patterns in eight tissues of the adult rat. Compare muscle and heart, the former rich in LDH-5 while the latter is richest in LDH-1. Dotted circle in testis column represents position of another LDH isozyme that may be specific for the testis. (From C. L. Markert and H. Ursprung, Developmental Genetics, © 1971, p. 43. Reprinted by permission of Prentice-Hall, Inc., Englewood Cliffs, New Jersey.)*

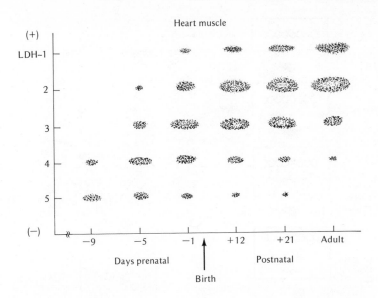

**Figure 17-39**  *Developmental changes in isozyme patterns in heart tissues of the mouse. The pattern shifts gradually from a prominent LDH-5 toward an enrichment at the LDH-1 end of the spectrum. (From C. L. Markert and H. Ursprung, Developmental Genetics, © 1971, p. 44. Reprinted by permission of Prentice-Hall, Inc., Englewood Cliffs, New Jersey.)*

cessible genes. Here we focus on fine tuning of the genome as alternative phenotypic states are stabilized in a clone.

## A VIRAL MODEL OF FINE-TUNING REGULATION OF DIFFERENTIATION

Any genetic model of differentiation must explain how cells sharing identical genomes switch into alternative developmental pathways. In other words, it should show how two cell lines with the same genome can exhibit two different programs of protein synthesis in identical environments. Again, we find that some aspects of genomic regulation in viruses closely resemble the kind of gene switching necessary to choose between alternative genetic states.

Probably the best understood control system in development is that of lambda phage infection of lysogenic strains of *E. coli* (Figure 17-40). Even this simple system, as we will see, involves a network of temporal interaction between several regulatory and structural genes.

When the temperate phage infects *E. coli* it has to "choose" between two alternative pathways of development, the *lytic* or *lysogenic* program. The lytic program involves rapid DNA replication followed by synthesis and assembly of viral head, tail proteins, lysis of the cell, and release of hundreds of new phage particles. If lysogeny is chosen, the bacterial cell survives with phage DNA integrated into and replicating with the host genome. Phage does not appear because it produces a repressor that turns off its lytic genes. Whenever host DNA synthesis is inhibited by an exogenous agent such as UV irradiation, the repressor is inhibited and all genes coding for the lytic pathway are now expressed. New viral particles are made and the cell is lysed.

In these alternative life styles, there are switch points between genetic states and mechanisms for stabilization of a state once acquired. We have here a pattern of gene regulation resembling differentiating eukaryotes—a choice between different phenotypic states and stabilization of that chosen state in all subsequent progeny.

The genes coding for both these pathways have been mapped on the circular phage chromosome (Figure 17-41). Phage DNA is a linear molecule, but after entering the host cell, it forms a circle by base pairing between complementary single-stranded ends. The circular molecule is the informational tape for each phenotypic program and the key regulatory genes are clustered in one segment of the viral chromosome. The N gene is the principal regulatory gene controlling the lytic pathway, but other genes, Cro, CI, CII, and CIII, play important parts in controlling transcription of several key promotor sites, to establish the switch point in the viral life cycle.

The lytic pathway is simpler to describe; it in-

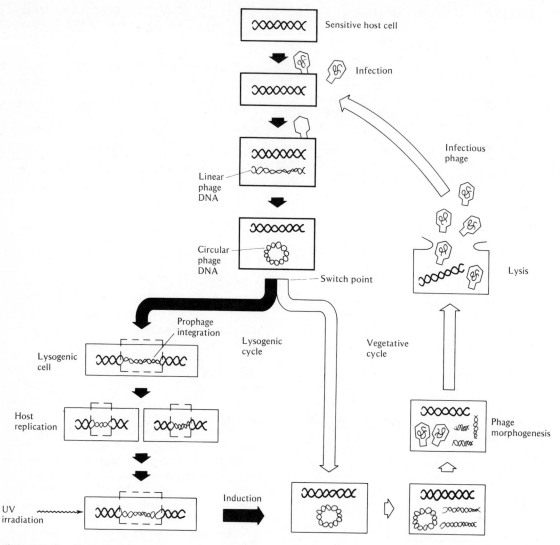

**Figure 17-40**   *Alternative phenotypic pathways of a temperate phage. Lysogenic cycle shown with black arrows, the vegetative cycle with white arrows. See text for details.*

volves fewer gene switches. Three periods of gene transcription have been defined for this short life cycle, the *immediate-early* phases of RNA synthesis (stage 1), the *delayed-early* phase (stage 2), and the *late* phase (stage 3). The first gene to be transcribed (stage 1) is the N gene. Since the template is a circle, transcription is initiated in two directions from the promotors PR and PL, with one DNA strand read in one direction and the opposite strand in the other.

As N protein appears, it activates transcription of stage-2 genes: the recombination genes, DNA replication genes O and P, and gene Q. When the Q gene is activated, its protein, in turn, initiates transcription of genes for tail and head, transcribing sequentially around the template circle. Once all products are synthesized, an intact virus is assembled and released as the cell is lysed.

The temporal features of viral development are

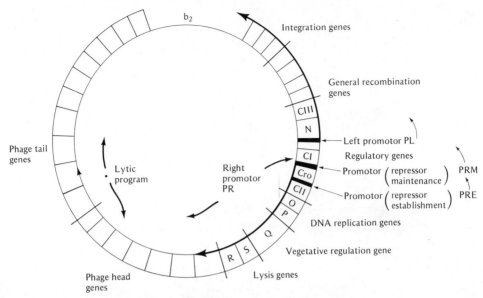

**Figure 17-41** *Simplified genetic map of the phage lambda DNA. Major gene clusters are shown and only some of the regulatory genes are designated by letters—CIII, N, CI, Cro, CII, O, P, Q, and the lysis genes R and S. Sets of genes control recombination and insertion of phage DNA into host DNA. Left promotor site (PL) is the region where transcription of the left DNA strand begins (arrow indicates direction). Right promotor site (PR) is the beginning of right-strand transcription. PRE (promotor, repressor establishment) is the site of transcription initiation for Cro gene to PL. PRM is the promotor repressor maintenance site. The b₂ segment is transcribed in both directions and its function is unknown.*

built into the order of appearance of regulatory proteins, which determines the time other genes are read by RNA polymerase. Timing is also a function of the transcription rate as well as the rate of appearance of specific controlling gene products (N protein or Q protein). This is a type of *cascading regulation,* where the regulatory product of one gene controls other structural and regulatory genes which, in turn, regulate still a third gene set, and so on. A similar cascading model of gene regulation is applicable to gene expression in eukaryotes which display a sequence of stage-specific protein changes.

Control of the lysogenic pathway is another case of cascading regulation (Figure 17-42). To achieve the lysogenic state, the entire lytic program must be repressed and maintained in this repressed condition indefinitely. Of all regulatory aspects in phage development, this is probably best understood. Repression of the lytic pathway results if the CI gene is activated. The CI protein is a repressor protein that combines with promotor sites on either side of the

CI gene and inhibits transcription of all stage-1 genes, including the N gene. If no N protein is synthesized, the entire lytic cycle is thereby stopped; no stage-2 genes are transcribed.

But repression cannot be imposed immediately as soon as phage DNA enters the cell, since activation of the CI gene depends upon products from some stage-2 genes, namely, CII and CIII. In addition, stage-2 recombination genes must be activated before phage DNA can be integrated into the host genome. For a short period after phage DNA enters the host cell, after stage-1 and stage-2 genes have been activated, a precarious balance exists between alternative phenotypic pathways. The factors that influence the choice are still uncertain.

Let us follow the program toward lysogeny from the decision point. Once CII and CIII proteins appear, they form a complex activator protein which combines with the CI gene site to initiate transcription. CI messenger RNAs are translated and as CI protein accumulates, it attaches to the N gene promotor

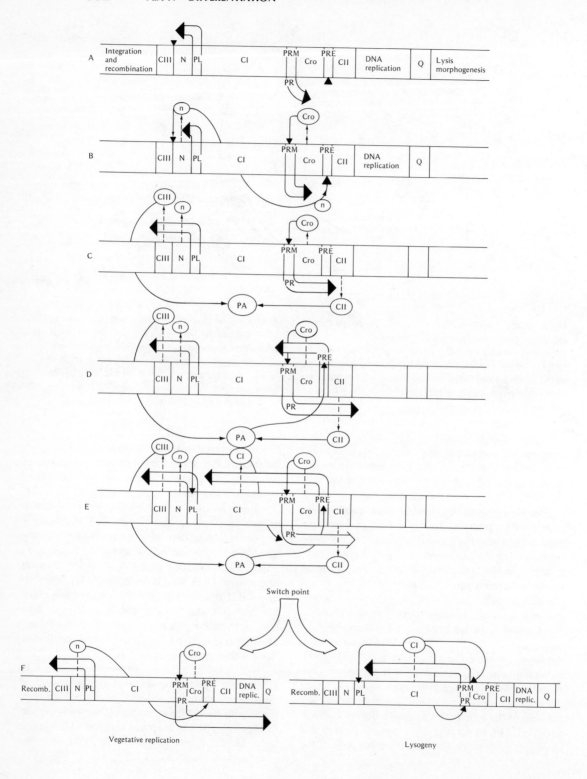

A    Integration and recombination | CIII | N | PL | CI | PRM | Cro | PRE | CII | DNA replication | Q | Lysis morphogenesis

B    | CIII | N | PL | CI | PRM | Cro | PRE | CII | DNA replication | Q

C    | CIII | N | PL | CI | PRM | Cro | PRE | CII

D    | CIII | N | PL | CI | PRM | Cro | PRE | CII

E    | CIII | N | PL | CI | PRM | Cro | PRE | CII

Switch point

F    Recomb. | CIII | N | PL | CI | PRM | Cro | PRE | CII | DNA replic. | Q

Vegetative replication

Recomb. | CIII | N | PL | CI | PRM | Cro | PRE | CII | DNA replic. | Q

Lysogeny

(PL) site and shuts off its transcription, thereby repressing all immediate early genes and, in turn, all other genes in the lytic program.

Meanwhile, since stage-2 genes have been expressed, enough recombination gene products are present to integrate phage DNA into a specific site on the host chromosome by a process of strand recombination between phage and host strands. Integration and lytic repression must be coordinated for the lysogenic state to be successful. A successful commitment to lysogeny depends upon relative rates of transcription of CII and CIII genes and rates of accumulation of CI product compared with rates of accumulation of N protein and activation of the O, P, and Q genes. The environmental factors that influence these decisions and the molecular mechanism involved in differential regulation of rates of transcription and translation of these proteins are not well understood. We do know, however, that host physiological state (age) and the number of phage particles infecting the cell all influence the degree and frequency of lysogeny.

The lysogenic state is stable as long as CI protein is synthesized. Transcription of the CI gene during the stable lysogenic state must be regulated by other mechanisms since the amount of CII and CIII protein declines after the N gene is repressed. Most evidence indicates that the CI protein itself stimulates transcription of its own gene and sustains a so-called "maintenance mode" of synthesis. A cytoplasmic product, by positive feedback regulation, sustains its own synthesis and epigenetic state. It also acts by negative feedback; when the repressor level builds up, it transiently inhibits further transcription of repressor. This is a useful model for stabilizing a eukaryote gene system in a determined state.

A signal to switch to the lytic pathway, as we have indicated, is cessation of host DNA replication. Within 30 minutes after inhibition, some unknown mechanism inactivates the CI protein. This releases N gene repression and the entire lytic cycle is reactivated. The viral DNA must first excise itself from the host DNA, a process initiated by recombination genes, and the late genes of the lytic program are activated as soon as Q protein is synthesized.

A phenotypic program is regulated by transcriptive control of a few regulatory genes. The model is useful for eukaryote systems since it shows how genes may be regulated as determined cell populations differentiate. Cascading regulation can explain the temporal features of a developmental program. The products of one or two regulatory genes control an entire sequence of developmental steps. A single regulatory gene is a switch that distinguishes between two alternative developmental programs. It is tempting to speculate that after each differential division in eukaryote development (quantal mitoses), such switches are thrown so that each daughter cell is channeled into a different developmental pathway, one perhaps remaining in the mitotic state while the other differentiates.

Needless to say, these prokaryote mechanisms are only models from which the more extensive regulatory mechanisms of eukaryote genomes have evolved (see Chapter 2). There is no a priori reason for eliminating any model of gene regulation in eukaryote differentiation. In fact, more than one mechanism may be involved, with some phenotypes more dependent on transcriptional regulation, whereas others might rely on posttranscriptional modes.

A feature of the viral system is that a stable

---

**Figure 17-42**  *Genetic regulation of alternative phenotypic pathways in lambda phage development. The regulatory segment of the chromosome is shown. (A) Transcription is initiated (large open arrows) on both strands when promotors PR and PL are activated. (B) N gene transcribes and n product is produced. This product by-passes terminators (black triangles) and transcription continues. Cro gene also transcribes and its product immediately represses PRM site (promotor repressor maintenance). (C) As transcription continues, CIII and CII genes transcribe their products, combining into an activator, PA. (D) PA activates PRE site (promotor repressor establishment) which initiates CI gene transcription. (E) As CI product accumulates, the critical switch point is reached (about 20 minutes after infection). Control is exercised as a competition between CI and Cro product on the PRM promotor. Cro protein represses the CI protein and activates PRM while repressing PL and PR promotors. If the action of Cro product wins out, the PRM promotor is repressed and the vegetative program is initiated as transcription continues and the gene is activated. On the other hand, if the CI protein wins out, PRM is activated, more CI is synthesized, and PR and PL are repressed. Lysogeny is established and no further transcription is possible except for CI gene, controlled by its own product by positive feedback. (Open arrowheads = activation; black arrowheads = repression; protein products are encircled; Q = vegetative regulation gene.)*

genetic state, lysogeny, is maintained indefinitely by feedback controls of specific gene transcription by gene products. A similar feedback regulation may also stabilize the phenotypic state in eukaryote differentiation over many cell generations.

## STABILITY OF THE DIFFERENTIATED STATE

Adult differentiated cells exhibit a range of phenotypic variation in response to functional demand. Cell function changes as enzyme levels fluctuate in response to hormone signals and inhibitors. These changes in functional states are cell modulations; synthesis of gene products fluctuates.

The degree of modulation is a function of the phenotype itself. An erythrocyte is a terminal cell, destined to die; its phenotype is highly specialized and shows little modulation. At the opposite extreme are the pluripotent stem cells in bone marrow that express a range of phenotypes depending on the signal. Between these extremes, phenotypes of varying stabilities are present.

Stability of a cell phenotype seems to be directly related to its cytoplasmic complexity. Nerve and muscle cells possess highly organized cytoplasms and show the least amount of modulation; their morphology remains relatively constant. Liver and pancreas, with "simpler" cytoplasms, respond readily to changing conditions and exhibit wide phenotypic variations. Stability correlates indirectly with a cell's ability to proliferate; nerve cells, the most stable, do not divide.

This apparent dichotomy between the stability of the differentiated state and proliferation was first noted in the early days of tissue culture. Somatic tissues adapted for growth in vitro after many generations lost all specialized organelles and function; they dedifferentiated. Pigment cells stopped synthesizing pigment, cartilage cells did not accumulate matrix. Irrespective of their origin or state of specialization, cells assumed a simplified morphology in vitro characteristic of cells in a growth-duplication cycle. The more rapidly cells proliferated in vitro the sooner they arrived at a dedifferentiated state.

Before resuming a cell cycle, most specialized cells acquire the morphology of a proliferating cell, with large nucleus and basophilic cytoplasm. They discard tissue-specific organelles such as pigment or secretory granules and transform into a mitotic cell.

Cell division and differentiation seem incompatible; mitotic cells cannot synthesize specialized luxury molecules because their resources are channeled to support cell division instead. Does this mean the differentiated state is unstable under conditions that favor intense proliferation?

Most permanent cell lines are adapted to long-term proliferation in vitro and are functionally unspecialized. But the majority of these cell lines possess abnormal chromosome numbers; they are aneuploid or heteroploid. Such unbalanced genomes usually disturb the regulatory machinery governing phenotypic expression and it is not surprising that most of these cell lines are undifferentiated.

Some permanent cell lines retain phenotypic traits characteristic of their tissue of origin. Several pigment cell tumors (melanomas) continue to synthesize melanin, although they proliferate vigorously in vitro. Tumors of the adrenal gland secrete large amounts of adrenocortical hormones into the culture medium. Sometimes when culture conditions are permissive an undifferentiated cell line begins to synthesize specific proteins. Cells of a mouse fibroblast tumor line do not synthesize collagen in the early stages of proliferation but do so when they become confluent as contact-inhibited monolayers. Often the undifferentiated state of a cell line is an artifact of culture conditions. Genes coding for luxury proteins are expressed in such cells when culture conditions permit stabilization and assembly of gene products.

When specialized somatic tissues are dissociated and placed in primary culture, they lose proteins and appear to dedifferentiate (Figure 17-43). Trypsin dissociation, as we have seen, damages cell membranes and many molecules leak out. Cells repair their surfaces, adjust to the strange media, proliferate, and seem to dedifferentiate. As population density increases, cells make contact in monolayers. New regulatory circuits are activated to partially restore phenotypic expression and tissue-specific proteins may be detected at this time. In other words, cells that appear dedifferentiated when proliferating rapidly need not have lost their phenotypic bias. They may be undergoing modulation in response to artificial culture conditions. By manipulating culture media, we can switch a cell clone from proliferation to differentiation and back again. The dichotomy between differentiation and proliferation is probably an artifact of in vitro growth.

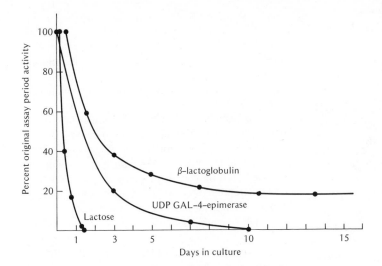

**Figure 17-43** *Loss of specialized functions in primary culture of bovine mammary gland. Three phenotypic traits are shown: lactose synthesis, activity of uridine diphosphogalactose-4-epimerase, and formation of β-lactoglobulin. Each trait declines along a separate time course. (From K. E. Ebner, E. C. Hageman, and B. L. Larson,* Exp. Cell Res. *25:555–570, 1961.)*

For many determined embryonic cells such as myoblasts and chondrocytes commitment to a specific phenotypic program is unaffected by long-term proliferation in vitro; the determined state is indeed stable. Under conditions favoring proliferation, there is little evidence of specific protein synthesis, but when conditions are changed to limit cell division, then specialized proteins are synthesized and accumulate.

When pigmented epithelium of the embryonic chick retina is cultured in vitro, cells grow as monolayers and become totally unpigmented after 2 weeks. Melanin synthesis stops within the first 12 hours in culture; there is no increase in melanin comparable to the increase observed in retinal tissue growing in situ for the same period. In vitro, cells replicate rapidly, and as total cell protein increases, the relative amount of melanin decreases, as if melanin is being diluted out. Loss of melanotic activity in vitro results from loss of the enzyme tyrosinase, an essential first enzyme in melanogenesis. Any tyrosinase that had been present is slowly degraded and no new enzyme molecules are synthesized. Culture conditions favor cell replication at the expense of synthesis of cell-specific proteins.

Low basal levels of enzyme activity persist in the unpigmented monolayers; the relevant genes are not completely turned off. After cells are contact-inhibited or crowded into high cell densities, tyrosinase activity increases and melanin is deposited once again. The cells have not lost their ability to synthesize pigment; they retain their pigment cell bias, even after many generations in culture. We do not know what regulatory switches operate in the transition between states, but the changes in tyrosinase activity suggest control at posttranscriptional sites in vitro.

In some instances, division may be a prerequisite for differentiation. We have already seen that mitosis precedes differentiation in several cell systems. Moreover, in some cases (for example, mammary gland epithelium) only those daughter cells "born" in the presence of a hormone acquire the ability to synthesize cell-specific products. This would suggest that certain divisions are critical in a cell lineage; during or after the division, the genome experiences a genetic change and daughter cells are programmed to express their phenotypic bias. Such divisions have been called *quantal divisions* by Holtzer (Figure 17-44). He distinguishes cells in a proliferative division cycle from cells in a program of quantal divisions. After a quantal division, one or both daughter cells are diverted from a proliferating state to one now switched to phenotypic expression. These cells, though still capable of mitosis, exhibit progressive decline in mitotic activity at each division (generation time usually lengthens) while assembling the machinery for functional specialization and luxury molecule synthesis. The differentiation of the erythrocyte is an example.

In support of this model of quantal mitosis, Holtzer finds that a population of embryonic myo-

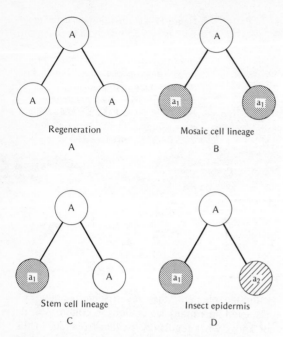

Regeneration

A

Mosaic cell lineage

B

Stem cell lineage

C

Insect epidermis

D

**Figure 17-44**  *Relationship between division and cell differentiation. The possible relationships between mother and daughter cells after a division. (A) Original phenotype perpetuated. It is typical of many determined cell lines in culture, such as myoblasts or chondroblasts, and in regeneration where dedifferentiated cells in a blastema perpetuate the original phenotype. (B–D) Daughters differentiated from mother cell and/or from each other: (B) is found in determinately cleaving eggs; (C) is typical of bone marrow and the germinative layers of skin and intestine; and (D) is a division pattern in insect epidermis during larval, pupal, and adult molts.*

blasts, grown under high-density conditions (contact-inhibited), stop dividing and neither fuse nor differentiate. When this same population is grown under low-density conditions, myoblasts are able to complete at least one more division, and resulting daughter cells fuse and differentiate as myotubes.

It seems that only one critical round of DNA synthesis is essential to prepare myoblasts for fusion and differentiation. Myoblast proliferation is inhibited in the presence of the inhibitor FUDR without affecting viability. If such cells are simultaneously exposed to BUDR they will complete one round of DNA synthesis because BUDR overcomes the inhibition. Such myoblasts do not fuse or differentiate because BUDR has entered the DNA. If thymidine is used instead of BUDR, a round of replication is completed and this time fusion and myotube differentiation occur. Some newly replicated DNA strands are able to initiate differentiation. Evidently, during a critical mitosis, the genome is programmed to express its intrinsic bias to fuse and synthesize muscle proteins.

We do not know why a division is necessary, nor do we know how genes are programmed when the genome switches from replication to nonreplication and differentiation. Nevertheless, we will see that cell division usually precedes all cases in which one differentiated phenotype transforms into another, a phenomenon of *metaplasia*. As in transdetermination, DNA synthesis and cell division are required before cells display a new phenotype, as if one phenotypic program must be erased by cell division before another is expressed.

## CELL TRANSFORMATION: REPROGRAMMING THE GENOME

Cell behavior in vitro indicates that the differentiated state is stable over long-term culture. For a long time, it was assumed that differentiation is irreversible, that once a specific bias is fixed in a clone, it is rarely altered. Cells could undergo short-term regulatory changes or modulations, but over the long term, the genome seemed to be irreversibly committed to a specific phenotype. For some cells (for example, terminal cells destined to die) irreversibility is the case. But irreversibility is only an operational term. It says only that under most conditions to which cells have been exposed, they do not change their phenotypic bias. Nevertheless, even such irreversibly committed genomes as that of a terminal erythrocyte can be reprogrammed by exposure to new cytoplasmic signals in somatic cell heterokaryons (see Chapter 3). Other phenotypes are even more labile; they may exhibit stable genetic changes that switch them from one phenotypic bias to another. For these cells, differentiation is reversible.

Genetic reprogramming or cell transformation has been observed in asexual budding, in regeneration, and in the behavior of totipotent carrot phloem cells (see Chapters 14 and 15). In almost all cases, reprogramming is preceded by nuclear changes—increased nuclear volume, an activated synthesis of total nuclear and cytoplasmic RNA, followed usually by DNA replication and cell division. Meanwhile, cytoplasm dedifferentiates by shedding its specialized structures as the morphology of a proliferating cell is acquired. Apparently cytoplasmic biases residing in one set of organelles (pigment granules, fibrils, or secretory granules) are eliminated before cells replicate. Later, during or after replication, the genome responds to new sets of regulatory signals and is reprogrammed to express a new phenotypic bias.

An apparent cell transformation is illustrated by the effect of vitamin A on differentiation of embryonic skin in vitro. Organ cultures of embryonic chick skin carry out a normal course of phenotypic expression. The germinative epithelium proliferates, producing cells that differentiate progressively toward keratin synthesis. Keratinizing epidermal cells are terminal cells; after keratin is synthesized, cells die and are sloughed off. Stem cells, located in the germinative epithelium, continuously regenerate this population; they divide, and one or both daughter cells proceed to differentiate.

Vitamin A changes the course of differentiation in these cultures. Cell proliferation continues in the germinative epithelium, but in this new environment daughter cells differentiate, instead, as mucus-secreting epithelium. The skin now resembles a respiratory epithelium. Although the terminal muco-protein products differ from keratin, they share a common biosynthetic pathway involving galactose precursors. The metabolic pathway diverges to keratin or mucus, and the switch seems to be controlled by the amount of vitamin A in the immediate environment. At very high concentrations of vitamin A, phenotypic expression is shifted toward mucus secretion, whereas in its absence the keratin-synthesizing phenotype is favored.

The exact role of vitamin A in this process is not known. We do know that excess vitamin A stimulates digestion of mucoprotein matrix in embryonic bone, but the reverse occurs in skin. Instead of matrix breakdown, mucopolysaccharide and matrix synthesis are stimulated by vitamin A.

This example, however, is not a true case of phenotypic transformation because basal cells of the germinative epithelium are pluripotent; they can express several different phenotypes. It resembles the differentiation of blood cell progenitors in which the specific phenotypic bias depends upon the immediate microenvironment in which mitosis occurs.

It suggests that replication and programming of the genome are coordinated.

## Wolffian lens regeneration

A case of true cell transformation—that is, conversion of one fully differentiated phenotype to another—is seen in Wolffian lens regeneration in amphibians (see Chapter 15). This case of cellular metaplasia was first demonstrated by Wolff in 1894. Removal of the lens of a larval or adult urodele initiates profound changes in the cells of the surrounding pigmented iris epithelium. They transform into a new cell phenotype and synthesize lens proteins instead of pigment, to regenerate a new lens.

Lens proteins are terminal products that are normally synthesized only by lens-forming cells. A transformation from melanin synthesis to lens protein synthesis means that some gene sets must be activated while others are repressed.

Before pigmented cells are reprogrammed, they first dedifferentiate (Figure 17-45). Within the first 6 days after lens removal, epithelial cells of the dorsal pupillary margin begin to slough off both cytoplasm and pigment granules; all evidence of the melanotic phenotype is discarded. Meanwhile, RNA and protein synthesis are stimulated. Free ribosomes and rough surface endoplasmic reticulum accumulate in the cytoplasm, and between 6 and 10 days, a dedifferentiated mitotic cell phenotype appears. These cells proliferate rapidly, and all daughter cells move out and transform into a small lens vesicle. The vesicle enlarges as more cells migrate from the proliferating region of the epithelium. Subsequently, cells in the lens vesicle elongate and differentiate into lens fiber-forming cells. After lens proteins are synthesized, nuclei disappear, replaced by translucent fibers of lens matrix. Before cells differentiate as lens, they complete several mitotic cycles, and it is only after the last mitosis that alpha and beta crystalline lens proteins appear.

The increased RNA synthesis noted during the premitotic phases reflects transcription of lens protein messages along with much ribosomal RNA. Use of these stable messages for lens protein synthesis is regulated by posttranscriptional and translational controls.

Summarizing, (1) before a phenotypic transformation can occur in the eye, differentiated cells discard all vestiges of their former specialized state

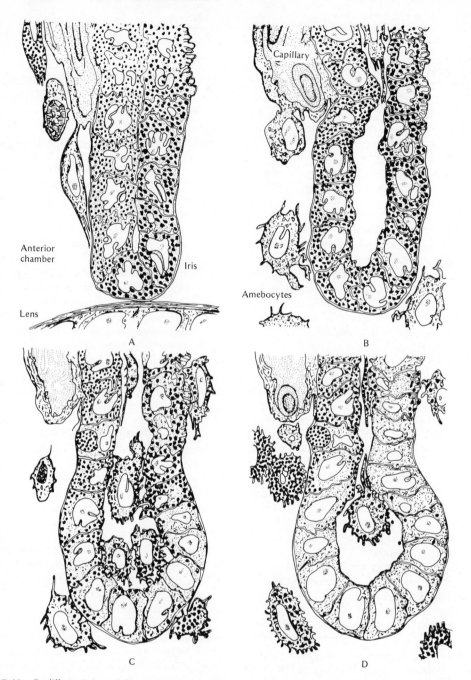

**Figure 17-45**   *Dedifferentiation of dorsal iris in the urodele eye prior to Wolffian lens regeneration. (A) Ultrastructural appearance of double layer of pigmented dorsal iris with lens still present. (B) Three days after lens removal, phagocytes appear. (C) After 8 days phagocytes have engulfed many pigment granules expelled by epithelial cells. (D) On twelfth day most pigment granules are in phagocytes while dedifferentiated epithelial cells become mitotically active and regenerate a new lens.*

and become proliferative, and (2) after lens removal, and in the presence of eye chamber factors, iris epithelium generates lens fiber rather than a pigment-producing phenotype. Proliferation in a specific microenvironment precedes genome reprogramming of newly formed daughter cells.

The crucial changes occur as DNA replicates, when the differentiated state becomes "labile," perhaps in the course of (or after) a critical mitosis. It becomes responsive again to exogenous signals.

The stability of a differentiated state, that is, the ability of cells to transmit phenotype biases to their progeny, depends on the stability of the regulatory circuitry. Most of this regulatory circuitry seems to be confined to the cytoplasm since genome reprogramming is preceded by the shedding of cytoplasmic organelles before proliferation into a new phenotype can take place. Such circuits are stabilized as part of a cell's physiological state and are transmitted to progeny. Somatic cell hybrids permit us to study such circuits just as they help us to understand nucleocytoplasmic interactions.

## SOMATIC CELL HYBRIDS AND GENE REGULATION

By fusing unrelated cells with the Sendai virus procedure, a hybrid is created in which two different genomes share a common cytoplasm (see Chapter 3). The technique is particularly useful for studying regulation of gene expression in differentiated cells. If phenotypic states are stable, fusion of two different cell phenotypes may result in a hybrid in which two genomes are made to interact with different sets of stable cytoplasmic cues. The phenotype expressed in the hybrid clone may reveal how such cytoplasmic states control gene expression.

The technique makes it possible to distinguish among several alternative hypotheses for the stabilization of the differentiated state (Figure 17-46). The first focuses on the autonomous expression of structural genes coding for a specific phenotypic character. A gene template is physically altered by an exogenous trigger and acquires an autonomous steady state of expression. It is active without any cues from the cytoplasm. By removal or addition of chromosomal proteins, template conformation is stabilized into a functional state and remains so during many replications.

The second hypothesis invokes an activator produced continuously in the cell. A gene for an activator molecule is triggered into stable expression by some inductive event, and the activator molecule secondarily regulates expression of a structural gene. The differentiative event in this case is one step removed, at the activator gene, but the model has one gene activating many different structural genes to explain coordinate synthesis of many enzymes in differentiation. In this model, the continuous presence of an activator molecule is essential for the expression of the gene.

The third hypothesis is the opposite of the second and calls for removal or discontinued production of a specific repressor. In this model a repressor gene that is continuously active in development is shut down during differentiation, thereby releasing the structural gene from its inactive state. The repressor gene is assumed to have undergone some conformational change (perhaps according to the first hypothesis) and cannot be reactivated again. In this model, expression of the structural gene depends upon the absence of a diffusible repressor molecule.

Theoretically, hybrids between cells of A and B phenotypes can yield one of the following results:

1. Both A and B phenotypic traits are expressed in the hybrid; there is no change in gene regulation.
2. Both phenotypic traits disappear in the hybrid.
3. Phenotypic traits of one parent are not expressed while the other remains unchanged.
4. New phenotypic traits are expressed in the hybrid by one or both genomes that were not expressed by either parent cell; new genes are turned on.

If result 1 occurs, we assume that the first hypothesis applies since gene expression is unaffected by any diffusible cytoplasmic factor. On the other hand, if result 2 is obtained, then we may assume that the third hypothesis applies since repressor genes from each parental phenotype are acting to shut down respective phenotypic properties in the opposite genome. Result 3 is a modification of 2 and indicates that the second hypothesis is correct. Finally, if result 4 is obtained, we conclude that a diffusible activator exists. Activator genes in one parental genome switch on the expression of the same structural genes in each parental genome. Let us look at some examples.

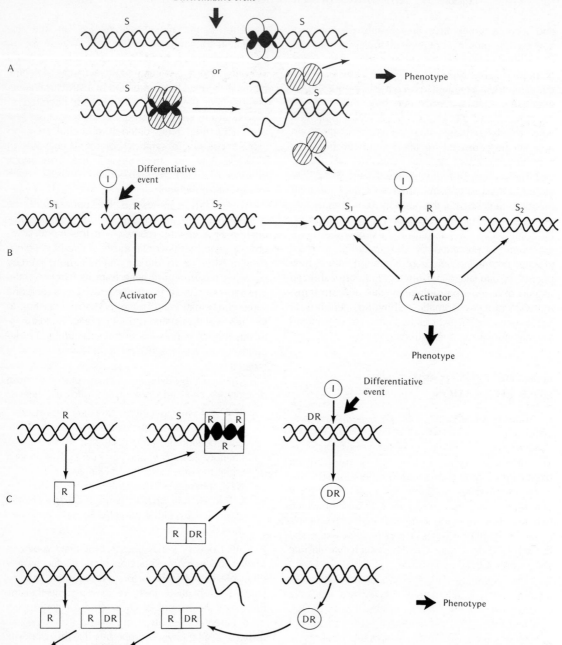

**Figure 17-46** *Models for genetic control of structural genes coding for phenotypic traits during differentiation. (A) Autonomous change of template structure by loss or addition of proteins may trigger a stable gene change leading to phenotypic expression. (B) The differentiative event is production of an activator coded by a regulatory gene that stimulates expression of several structural genes. (C) Critical event in differentiation is expression of a gene that derepresses a structural gene by eliminating an active repressor. Stability of the differentiated state depends upon continued production of derepressor. S, $S_1$, $S_2$ = structural genes; R = regulatory gene; I = inducer; R in box = repressor protein; DR = derepressor gene; DR in box and circle = derepressor gene product.*

## Phenotypic traits remain unchanged in hybrid cells

Most respiratory enzymes are shared by all cells and their relevant structural genes are regulated by common mechanisms. Hamster and mouse cells possess different MDH (malic dehydrogenase) isozymes, and both parental isozymes plus new hybrid varieties are found in mouse-hamster somatic cell hybrids (Figure 17-47). The fact that hybrid isozymes emerge in the expected proportions implies that all parental LDH genes are expressed unchanged in the common hybrid cytoplasm; that is, regulation of these genes is unaffected. Since respiratory enzymes are essential to cell survival, it is not surprising that these house-keeping genes are stabilized for expression at all times.

Some cell-specific traits seem to be regulated in a similar manner. Some tumors of neurons, known as neuroblastomas, retain a specific neuronal phenotype; that is, electrical stimulation elicits characteristic action potentials. A mouse cell line (L cell) lacking these membrane properties was fused with neuroblastoma cells, and the hybrid clones were maintained 10 to 40 generations before their phenotypes were assayed. During growth, chromosomes were lost at random from both parental genotypes but in spite of this, each hybrid clone retained the electrical properties of the neuroblastoma and the enzymes of both parental cells. A membrane property persists in

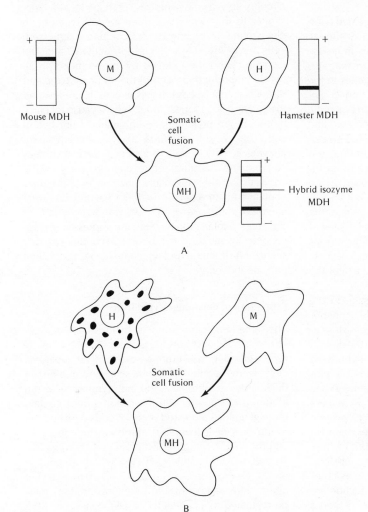

Mouse MDH

Somatic cell fusion

Hamster MDH

MH

Hybrid isozyme MDH

A

Somatic cell fusion

MH

B

**Figure 17-47** *Somatic cell hybrids and the expression of phenotypic traits. (A) Mouse (M) and hamster (H) fibroblast cell lines each contain a species-specific malic dehydrogenase (MDH). The resulting hybrid contains both parental isozymes and new hybrid isozymes formed from combination of subunits from each parent. (B) A pigmented hamster line fused to a mouse fibroblast loses its ability to express the pigmented phenotype in the hybrid.*

the hybrid for many generations in the presence of a foreign genome and cytoplasm; the epigenetic mechanisms that control membrane function are stable in this combination. Here, the relevant genes seem to be expressed autonomously and may be stabilized by some template mechanism.

## Phenotypic traits are lost in hybrid cells

Loss of a phenotypic trait of one parental type is most common in cell hybrids; that is, one parental genome seems to dominate the expression of the other. Regulatory circuits that repress certain structural genes in one phenotype shut down these same genes in the other when the genomes share a common cytoplasm.

Somatic cell hybrids between a pigmented hamster cell line and a mouse fibroblast do not synthesize pigment. The genes for melanin synthesis in mouse fibroblasts are repressed by a mechanism that also inhibits expression of hamster melanin genes in the common cytoplasm. This means repression (1) involves cytoplasmic factors, (2) is not species-specific — both mouse and hamster genes are turned off, and (3) is stable over many cell generations in hybrid cell cytoplasm. One of the key enzymes of melanin synthesis, dopa oxidase is lost in the hybrid, an indication that the synthesis or degradation of the enzyme, or both, has been affected. It is not known whether repression occurs at the template site of the dopa oxidase gene or at posttranscriptional sites of enzyme synthesis and assembly. The third hypothesis best explains the data; that is, the product of a repressor gene in the fibroblast genome inactivates the gene for dopa oxidase in the melanocyte.

The number of genes or gene dosage in the melanoma cell may play an important role in phenotypic expression. Certain melanoma lines are close to being tetraploid; they have almost double the diploid number of chromosomes. Tetraploid cells hybridized with nonpigmented (diploid) mouse fibroblasts produce hybrids containing two hamster genomes to one mouse genome. Under these conditions, 50 percent of the hybrid clones are pigmented and 50 percent are unpigmented. The unpigmented clones maintain their phenotype over many generations, but the pigmented hybrids are unstable, segregating into pigmented and unpigmented clones, perhaps as chromosomes bearing genes for melanin production are lost during proliferation. A double dose of genes

responsible for melanin synthesis can overcome the repressed conditions generated by a fibroblast genome. The enzyme dopa oxidase is as active in these hybrids as it is in the pigmented parent.

In certain somatic cell hybrids, the entire regulatory circuitry for a phenotypic trait, both structural and regulatory genes, may be repressed by a foreign genome. Somatic cell hybrids have been obtained by fusing rat hepatoma cells with mouse fibroblasts. A liver-specific trait retained by the hepatoma is the enzyme tyrosine aminotransferase (TAT). The hepatoma exhibits high basal levels of this enzyme; its structural genes are transcribing but higher levels of activity can be induced by the addition of glucocorticoid hormones. The hormone acts in the regulatory circuitry in enzyme synthesis, perhaps in a manner similar to the model described for GS synthesis in chick retina. TAT in the mouse fibroblast is repressed; cells have low basal levels of activity, and the enzyme cannot be induced by the hormone. In all hybrid clones both aspects of the rat phenotypic trait disappear; that is, basal levels of activity are low and the enzyme is no longer inducible. Because rat chromosomes are preferentially eliminated from hybrid cells, one might conclude that disappearance of enzyme activity is due to loss of rat chromosomes carrying the structural or regulatory genes for the enzyme. But this is unlikely since all hybrid clones contain a mixture of cells with normal and abnormal karyotypes. In spite of this, enzyme activity is low. The best explanation is that the repressed mouse genome generates cytoplasmic conditions that silence rat structural and regulatory genes controlling enzyme induction.

## Expression of new phenotypic traits

The same hepatoma-fibroblast hybrids have demonstrated activation of specific genes that were normally repressed. One liver trait possessed by rat hepatoma cells is the ability to synthesize serum albumin (RSA). Mouse fibroblasts do not synthesize serum albumin (MSA), although the genome must contain these genes in a repressed state. Hybrids derived from fusion of a near tetraploid fibroblast line (about 75 chromosomes) with a near diploid rat hepatoma line synthesize RSA in spite of the loss of 7 to 16 percent of rat chromosomes (Figure 17-48). Since serum albumin genes continue to be expressed, they must be regulated independently of TAT genes. No MSA is synthesized by the hybrid, but the same factors that repress MSA genes are unable to repress RSA genes in the hybrid.

If, however, another tetraploid (99 chromosomes) rat hepatoma line is fused with the same fibroblasts, then three different classes of hybrid clones are produced, each with a different number of chromosomes. The first, with the largest number of hybrid chromosomes, synthesizes both rat and mouse serum albumins. Two hybrid lines with fewer chromosomes synthesize, only MSA but not RSA, while the last line synthesizes no albumin. In these hybrids a large dose of rat genes can activate repressed mouse serum albumin genes. The mouse genome recognizes the rat regulatory signals, but these must be present in at least a double dose to elicit a response, an example of gene dosage involving regulatory rather than structural genes. Those

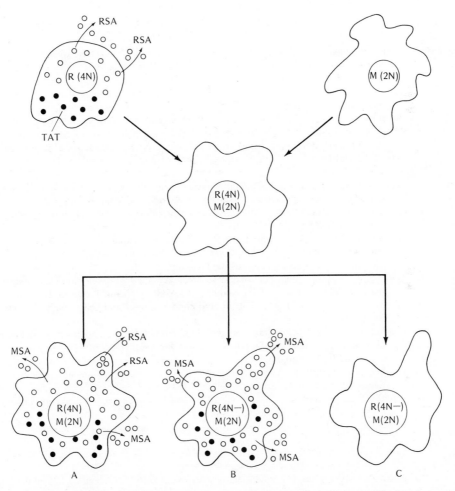

**Figure 17-48** *Somatic cell hybrids and the expression of a new phenotypic trait. A near tetraploid rat hepatoma [R(4N)] is fused to a near diploid mouse fibroblast [M(2N)]. The hepatoma line secretes rat serum albumin (RSA) and also contains basal levels of an inducible enzyme tyrosine amino transferase (TAT). The fusion hybrid gives rise to three different hybrid cell lines with varying numbers of chromosomes. (A) The line with the largest number of chromosomes expresses all traits including synthesis of mouse serum albumin (MSA), a newly expressed trait. (B) The line with fewer chromosomes synthesizes MSA and retains its TAT phenotypic expression. (C) The line having the fewest chromosomes exhibits loss of all traits, indicating that some traits are regulated independently of each other.*

hybrid clones unable to synthesize RSA had probably lost the chromosomes carrying the relevant structural genes.

In summary, it is evident that gene regulation of somatic cell hybrids relies on all three types of mechanisms. Besides structural and regulatory genes, repressors and activators, we see that gene dosage also has important effects on phenotypic expression. To complicate matters, many cell lines used in the experiments are derived from tumors with abnormal chromosome numbers and phenotypic function; the regulatory mechanisms themselves may be abnormal. The experiments do not indicate how repressor and derepressor signals initiate differentiation. They do say, however, that signals diffuse between nucleus and cytoplasm as stable regulatory circuits persisting through many cell generations.

**REFERENCES**

Artzt, K., Bennett, D., and Jacob, F. 1974. Primitive teratocarcinoma cells express a differentiation antigen specified by a gene at the T-locus in the mouse. *Proc. Nat. Acad. Sci. U.S.* 71:811–814.

Bennett, D. 1964. Abnormalities associated with a chromosome region in the mouse. I. The embryological effect of lethal alleles at the t-region. *Science* 144:263–267.

Berendes, H. D. 1973. Synthetic activity of polytene chromosomes. *Int. Rev. Cytol.* 35:61–116.

Briggs, R., and Cassens, G. 1966. Accumulation in the oocyte nucleus of a gene product essential for embryonic development beyond gastrulation. *Proc. Nat. Acad. Sci. U.S.* 55:1103–1109.

Cox, R. P., and King. J. C. 1975. Gene expression in cultured mammalian cells. *Int. Rev. Cytol.* 43:229–281.

Davidson, R. L. 1973. Control of the differentiated state in somatic cell hybrids. *Genetic mechanisms of development,* ed. F. H. Ruddle. New York: Academic Press, pp. 295–328.

Davis, F. M., and Adelberg, E. A. 1973. Use of somatic cell hybrids for analysis of the differentiated state. *Bacteriol. Rev.* 37:197–214.

Fantoni, A., Lunadei, M., and Ullu, E. 1975. Control of gene expression during the terminal differentiation of erythroid cells. *Current topics in developmental biology,* ed. A. A. Moscona and A. Monroy. Vol. 9. New York: Academic Press, pp. 15–38.

Gehring, W. J. 1973. Genetic control of determination in the *Drosophila* embryo. *Genetic mechanisms of development,* ed. F. H. Ruddle. New York: Academic Press, pp. 103–128.

Gordon, J. S., and Lash, J. W. 1974. In vitro chondrogenesis and cell viability. *Develop. Biol.* 36:88–104.

Gurdon, J. B. 1963. Nuclear transplantation in amphibia and the importance of stable nuclear changes in promoting cellular differentiation. *Quart. Rev. Biol.* 38:54.

————. 1974. *The control of gene expression in animal development.* Oxford: Clarendon Press.

Gurdon, J. B., Laskey, R. A., and Reeves, O. R. 1975. The developmental capacity of nuclei transplanted from keratinized skin cells of adult frogs. *J. Embryol. Exp. Morphol.* 34:93–112.

Hadorn, E. 1966. Dynamics of determination. *Major problems in developmental biology,* ed. M. Locke. New York: Academic Press, pp. 85–104.

Herskowitz, I. 1973. Control of gene expression in bacteriophage lambda. *Ann. Rev. Genet.* 7:289–324.

Hess, O., and Meyer, G. F. 1968. Genetic activities of the Y chromosome in *Drosophila* during spermatogenesis. *Adv. Genet.* 14:171–223.

Holtzer, H., Weintraub, H., Mayne, R., and Mochan, B. 1972. The cell cycle, cell lineages and cell differentiation. *Current topics in developmental biology,* ed. A. A. Moscona and A. Monroy. Vol. 7. New York: Academic Press, pp. 229–256.

Illmansee, K., and Mintz, B. 1976. Totipotency and normal differentiation of single teratocarcinoma cells cloned by injection into blastocysts. *Proc. Nat. Acad. Sci. U.S.* 76:549–533.

King, T. J., and Briggs, R. 1956. Serial transplantation of embryonic nuclei. *Cold Spring Harbor Symp. Quant. Biol.* 21:271–290.

Konigsberg, I. R. 1963. Clonal analysis of myogenesis. *Science* 140:1273–1284.

Levitt, D., and Dorfman, A. 1974. Concepts and mechanisms of cartilage differentiation. *Current topics in developmental biology,* ed. A. A. Moscona and A. Monroy. Vol. 8. New York: Academic Press, pp. 103–149.

Markert, C. L. 1963. Epigenetic control of specific protein synthesis in differentiating cells. *Cytodifferentiation and macromolecular synthesis,* ed. M. Locke. New York: Academic Press, pp. 65–84.

Mintz, B. 1974. Genes control mammalian differentiation. *Ann. Rev. Genet.* 8:411–470.

O'Malley, B. W., and Means, A. R. 1976. The mechanism of steroid hormone regulation of transcription of specific eukaryotic genes. *Prog. Nucleic Acid Res. Mol. Biol.* 19:403–419.

Peterson, J., and Weiss, M. 1972. Expression of differentiated functions in hepatoma cell hybrids: Induction of mouse albumin production in rat hepatoma-mouse fibroblast hybrids. *Proc. Nat. Acad. Sci. U.S.* 69:571–578.

Postlethwait, J. H., and Schneiderman, H. A. 1974. Developmental genetics of *Drosophila* imaginal discs. *Ann. Rev. Genet.* 7:381–433.

Rawles, M. E. 1963. Tissue interaction in scale and feather development as studied in dermal-epidermal recombinations. *J. Embryol. Exp. Morphol.* 11:765–789.

Reinert, J., and Holtzer, H., eds. 1975. *Cell cycle and cell differentiation.* New York: Springer-Verlag.

Rutter, W. J., Wessells, N. K., and Grobstein, C. 1964. Control of specific synthesis in the developing pancreas. *Nat. Cancer Inst.,* Monogr. No. 13, p. 51.

Stevens, L. C. 1967. The biology of teratomas. *Adv. Morphogen.* 6:1–31.

Steward, F. C., Kent, A. E., and Mapes, M. O. 1966. The culture of free plant cells and its significance for embryology and morphogenesis. *Current topics in developmental biology,* ed. A. A. Moscona and A. Monroy. Vol. 1. New York: Academic Press, pp. 113–154.

Tsai, S. Y., Harris, S. E., Tsai, M. J., and O'Malley, B. W. 1976. Effects of estrogen on gene expression in chick oviduct. *J. Biol. Chem.* 251:4713–4721.

Turkington, R. W. 1968. Hormone-dependent differentiation of mammary gland in vitro. *Current topics in developmental biology,* ed. A. A. Moscona and A. Monroy. Vol. 3. New York: Academic Press, pp. 199–218.

Wessells, N. K. 1964. DNA synthesis, mitosis and differentiation of pancreatic acinar cells in vitro. *J. Cell Biol.* 20:415.

Yaffe, D., and Dym, H. 1972. Gene expression during differentiation of contractile muscle fibers. *Cold Spring Harbor Symp. Quant. Biol.* 37:543–548.

Yamada, T. 1967. Cellular and sub-cellular events in Wolffian lens regeneration. *Current topics in developmental biology,* ed. A. A. Moscona and A. Monroy. Vol. 2. New York: Academic Press, pp. 247–283.

# 18

# The Immune System:
# A Model of Differentiation

The immune system in vertebrates is an excellent example of a differentiating system in which all steps in the process, from determination to synthesis of functional proteins, can be studied in clonal populations. Within any one clone we find inductive cell interactions, surface signals that program cell division, and, finally, genetic models for the origin of cell diversity. Subtle mechanisms of phenotypic differentiation are amplified in the immune system by virtue of its special function in recognizing and neutralizing foreign macromolecules and cells.

Though simple immune mechanisms are found in invertebrate animals, the true immune response first appears in primitive vertebrates, with the evolution of the lymphocyte, a cell specialized for antibody production. The system develops fully in higher vertebrates, as spleen, bone marrow, and thymus arise as principal organs of lymphocyte production, maturation, and differentiation.

As a differentiating clone, the immune competent cell population exhibits the following features:

1. A determined state in which a receptor protein — an immunoglobulin — is inserted into the plasma membrane of a lymphocyte. As a result it becomes responsive to exogenous signals triggering luxury protein synthesis (immunoglobulins) and differentiation. Such membrane changes may be common to all determined cells.
2. Differentiation is preceded by a period of proliferation stimulated by exogenous signals. Similar proliferation patterns precede most programs of clonal differentiation.
3. Proliferation yields a population of cells committed to synthesis and secretion of specific immunoglobulin antibodies.
4. Interactions among different types of immune competent cells determine the extent and nature of the differentiated state.

## NATURE OF THE IMMUNE RESPONSE

Two different immune response systems have evolved in vertebrates: (1) a humoral mechanism or antibody response system based on circulating antibody molecules which combine specifically with foreign macromolecules, or antigens, and (2) a cell-mediated graft rejection mechanism, in which immune competent cells recognize and eliminate foreign cells, usually without direct involvement of antibody. When a microorganism invades the blood, it invariably produces and secretes macromolecules, some of which act as toxins. In addition, some cells die and degenerate, releasing soluble proteins which act as antigens, the principal target of the antibody response system. Circulating antibodies recognize and combine covalently with antigens and neutralize their action. The graft rejection system is primarily responsive to macromolecules on the surface of intact cells and tissues. Responding immunocytes either engulf the invading cells by phagocytosis or destroy them directly by secreting cytotoxic substances. It is graft rejection that highlights the concept of "self" or the "uniqueness of the individual."

Certain genes, called *histocompatibility genes* (probably found in all organisms), code for specific proteins that are inserted into the plasma membrane of all cells in an organism. Individuality at the molecular level probably resides in these antigens at the cell surface since each individual produces a unique constellation of histocompatibility antigens. This explains why grafts of tissues between individuals are unsuccessful; the immune system of the host recognizes the foreign antigens on the graft and rejects it. Only identical twins can exchange tissue grafts without rejection because they possess identical histocompatibility genes. Most, if not all, histocompatibility antigens are cell surface glycoproteins.

Since every individual synthesizes many histocompatibility proteins, the immune mechanism is capable of recognizing and responding to millions of different antigenic stimuli. Even artificial molecules created in the laboratory elicit specific immune responses when introduced into an organism, which means an almost infinite variety of antibody molecules can be synthesized by an organism.

The cells specialized for the immune response are lymphoid cells derived from proliferating stem cells in the bone marrow (Figure 18-1). Their imma-

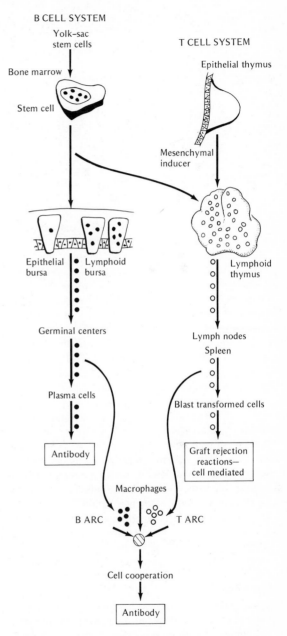

**Figure 18-1**   *Ontogeny of the immune system. Separate development of B cell and T cell systems in the chicken. ARC = antigen recognition cells from the T and B cell systems. See text for details.*

TABLE 18-1  Properties of Human Immunoglobulins

|  | IgG | IgA | IgM | IgD | IgE |
|---|---|---|---|---|---|
| **Whole molecule** |  |  |  |  |  |
| Molecular weight | 150,000 | 180,000 | 900,000 |  | 200,000 |
| Carbohydrate, % | 2.9 | 7.5 | 11.8 |  | 10.7 |
| Heavy chain | $\gamma$ | $\alpha$ | $\mu$ | $\delta$ | $\epsilon$ |
| Molecular weight | 53,000 | 64,000 | 70,000 |  | 75,000 |
| Light chain | $\kappa$ or $\lambda$ | $\kappa$ or $\lambda$ | $\kappa$ or $\lambda$ | $\kappa$ or $\lambda$ | $\kappa$ or $\lambda$ |
| Molecular weight | 22,000 | 22,000 | 22,000 |  | 22,000 |
| Molecular formula | $\gamma_2\kappa_2$ | $\alpha_2\kappa_2$ | $(\mu_2\kappa_2)_5$ | $\delta_2\kappa_2$ | $\epsilon_2\kappa_2$ |
|  | $\gamma_2\lambda_2$ | $\alpha_2\lambda_2$ | $(\mu_2\lambda_2)_5$ | $\delta_2\lambda_2$ | $\epsilon_2\lambda_2$ |

ture progeny are released into the circulation where they "home in" on second-level lymphoid organs (thymus) and differentiate into lymphocytes. Two classes of lymphocytes have been characterized, T and B lymphocytes, based on their site of maturation. T lymphocytes become immunologically competent in the thymus gland in most vertebrates while B lymphocytes, at least in birds, mature in a special organ, the bursa of Fabricius. Though this organ is absent in mammals, it is assumed that others serve as sites for B cell maturation. It is not certain whether the immature cells emerging from the bone marrow are pluripotent, able to differentiate into T or B cells, or whether they are already determined and home in on the specific organ as a result.

After maturation, lymphocytes percolate through the circulation and invade lymphoid tissues such as lymph nodes and spleen. It is within these peripheral organs that cells are exposed to antigens and complete their program of differentiation. In a sense, a mature lymphocyte is determined rather than differentiated; it is competent to respond to a specific signal that evokes a program of proliferation and protein synthesis. The signal in most cases is exposure to the foreign antigen. The B cell system is primarily responsible for synthesizing circulating antibodies in the humoral response mechanism, while T cells are principally involved in cell-mediated graft rejection.

In addition to lymphocytes, there are scavenger cells, the macrophages and reticular cells which capture particulate antigens such as bacterial cells invading lymph nodes through the lymphatic circulation. Reticular cells entrap antigens on their surfaces while lymph streaming into the sinuses carries antigen to macrophages to be engulfed into cytoplasmic vesicles. Scavenger cells also process antigen in preparation for the immune response and interact with lymphocytes.

Cells of the immune system are produced in the lymphoid organs, thymus, spleen, bone marrow, and lymph nodes. The thymus, the first lymphoid organ to develop in the embryo, appears as an epithelial outpocketing from the pharynx. In humans it appears in the third month, in mice on the twelfth day, and in chicks on the ninth day. The epithelium develops into a lobate organ after an inductive interaction with surrounding mesenchyme (similar to pancreas and salivary gland differentiation; see Chapter 17). The mouse thymus begins to produce lymphocytes by day 14 and is fully lymphoid by day 16, about 5 days before birth.

Spleen and lymph node development is completed shortly before birth and depends upon interactions with the thymus. Isolated mouse embryonic spleen does not become lymphoid in vitro but when cultured with thymus, both organs mature more readily and become lymphoid. Bone marrow also fails to develop in isolation and degenerates, but in the presence of thymus it survives for a long period and undergoes extensive lymphoid development. Evidently maturation of immune organs is dependent upon factors produced in the thymus. In fact, removal of the thymus gland from a neonatal animal (mouse or rat) results in the animal's failure to develop immune competence. Differentiation of spleen, bone marrow, and node is thymus-dependent.

The relatively few cell phenotypes produced in these organs recognize all foreign antigens, produce the diversity of antibody molecules, and initiate complex graft rejection reactions. This raises an important question: Does each lymphocyte synthesize antibodies in response to each antigenic stimulus? This

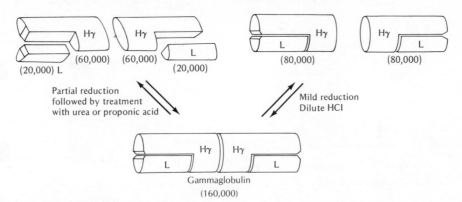

**Figure 18-2**   *Structure of immunoglobulin molecule, gammaglobulin. Two identical heavy chains and two light chains associate into a molecule of 160,000 mol wt. Dissociation of the molecule occurs in urea after disulfide bonds holding chains together are reduced. Mild acid forms two larger fragments. Hγ = heavy chain; L = light chain. Numbers in parentheses are approximate molecular weights.*

implies that all lymphocytes are pluripotent, capable of synthesizing many different antibodies, each specific for an antigenic stimulus. On the other hand, the lymphocyte population may be heterogeneous, with small clones of cells, each capable of responding to one or only a few antigenic stimuli. To resolve this question, let us see how the B cell antibody system operates.

## HUMORAL RESPONSE SYSTEM: THE B CELL

The humoral response system involves specific antibody synthesis by B lymphocytes. Most antigens are proteins, glycoproteins, or polysaccharides, folded into specific tertiary configurations. For a molecule to be antigenic it must be large (no smaller than 4000 mol wt), of rigid structure so that its antigenic sites are fixed, generally soluble, and foreign to the organism. Lymphocytes recognize foreign molecules by virtue of their folded configuration and synthesize antibodies with complementary configurations. Antibody combines covalently with antigen to form a stable complex which neutralizes antigenic activity.

Five classes of antibody molecules or immunoglobulins have been isolated and defined from human sera (Table 18-1). These specific glycoproteins are unique in many ways because they exhibit tremendous structural heterogeneity. An immunoglobulin molecule such as IgG or IgM consists of four poly-peptide chains, two heavy chains joined to two light chains by disulfide bonds (Figure 18-2). The individual chains, obtained after urea denaturation, behave as subunits in a self-assembly system. They reconstitute an intact globulin molecule spontaneously, suggesting that heavy and light chains are made separately in vivo and self-assemble into the complete molecule.

The antigen combining sites are located at the ends of the molecule as crevices in which the proper antigen fits (Figure 18-3). The combining site is complementary in shape to the three-dimensional structure of the antigen, which means that both light and heavy chains contribute to the specificity of the active site. Antibody recognizes antigen by fitting its active site to the shape of the antigenic site. Antibodies, like enzymes, are flexible molecules that bend and twist until the most stable configuration is assumed. In the presence of a recognizable antigen, the antibody acquires a new stable shape.

Amino acid sequence analysis of light and heavy chains tells us how antibody molecules acquire diversity. Different antibodies possess unique amino acid sequences but only in specific portions of the light and heavy chains, that is, those regions which make up the antigen combining site (Figure 18-4). The remaining portion of the molecule is identical in all antibodies. For example, we find that all light chains contain approximately 214 amino acids. The first 105 amino acids make up the variable or V re-

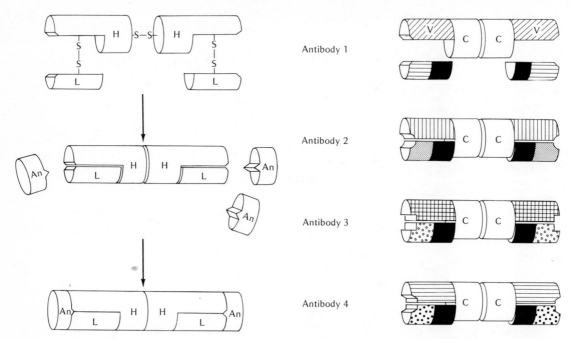

**Figure 18-3** *Diagram of interaction between antibody and antigen. Antigen-combining sites are indicated as notches at the ends of the molecule, shared by light and heavy chains. Antigen (An) has a complementary structure that fits into antibody active site. H = heavy chain; L = light chain; S—S = disulfide bond.*

**Figure 18-4** *Diagram of four different antibody molecules, each with a different antigenic specificity derived from differences in amino acid sequences in the variable portion (V) of the light and heavy chains. Constant portions (C) are white or black and are almost identical in every antibody. Differences arise in amino acid sequences in the variable portion of each light and heavy chain. Variable portions are indicated with lines, crosshatches, or circles.*

gion where antibody differences lie. The remainder of the molecule, from position 106 onwards, is identical in all light chains except for position 189 where amino acid substitutions have been made. This half of the molecule is the constant or C region. This structural duality of antibody molecules, a variable plus a constant region, is unique; no other class of proteins shows this property.

It has been proposed that each polypeptide chain is specified by two different types of genes, a variable class of V genes which have mutated extensively in evolution and another structural gene class of C genes selected for constancy. Only one or a few genes specify the constant region, but the variable portion is coded by many V genes, at least one for each type of antibody molecule made. Since light and heavy chains are made as one piece, a V gene and a C gene together must specify their respective sequences. This, of course, is contrary to the dogma of one gene coding for a single polypeptide chain.

What kind of genetic model can account for V gene diversity and for the synthesis of polypeptide molecules coded by two different genes?

As a first approximation, we assume there cannot be a V gene for each antibody specificity; there are just not enough genes in the entire genome. Since active sites are made up of light and heavy chain regions, the permutations and combinations of light and heavy chains could generate diversity with a smaller number of genes. For example, $10^3$ V genes for light chains and an equivalent number for heavy chains could give rise to $10^5$ different antigen combining sites, if we assume that any given antibody specificity is produced by 10 different combinations of light and heavy chains. Approximately 2000 genes could account for 100,000 different antibodies, a number that is more reasonable than the millions of genes that would be necessary if only one light or heavy chain were involved. The large redundant component of the genome, populations of many genes in tandem, differing by a few nucleotide sequences, can serve as a model for the organization of antibody-forming genes.

An even smaller number of V genes is postulated in the recombination model suggested by Edelman and Gally. Extensive heterogeneity of 20 to 50 tandem duplicated V genes could be achieved if recombination could take place between genes on the DNA template. Each recombined V gene would lead to an entirely new gene sequence to code for the variable segment of the immunoglobulin chain. To date, however, no one model for heterogeneity has been generally accepted.

### Instruction versus clonal selection

How is an antibody response initiated when a foreign antigen invades the organism? We have to explain how a population of antibodies is produced that combines specifically with only one antigen. This brings us back to our initial question concerning the manner in which lymphocytes respond to and synthesize antibody.

Two hypotheses have been proposed to explain antibody synthesis, an *instructive mechanism* and *clonal selection*. The former implies that antigen, acting like a template, brings information about its configuration directly into the immune competent cell. This configuration is somehow stamped on the antibody molecule as it is being synthesized and folded into its tertiary structure. It assumes that the immune cell cannot anticipate antigenic configuration and must receive specific structural information brought in by the antigen. Any immune competent cell synthesizes a complementary antibody as soon as it receives instructions from the incorporated antigen.

Burnet's clonal selection hypothesis, on the other hand, focuses on a recognition mechanism between antigen and lymphoid cell (Figure 18-5). According to this view, maturing lymphocytes are precommitted to the production of a specific antibody structure before exposure to antigen. Specific V and C genes are expressed which code for small amounts of antibody synthesis. These molecules are incorporated into the lymphocyte membrane as a recognition site. This surface configuration is later recognized by an antigen, which means that a population of lympho-

cytes is heterogeneous, made up of small clones predetermined to synthesize a unique antibody. According to the model, any foreign antigen entering the circulation reacts only with those cells carrying a complementary antibody configuration on their

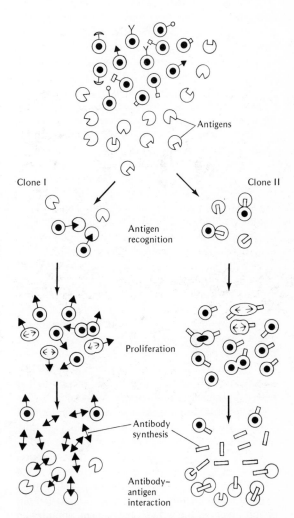

**Figure 18-5** *Clonal selection hypothesis for antibody formation. Seven different lymphocyte populations are shown (with black nuclei), each carrying a specific surface receptor. Two types of antigens are indicated, each combining with lymphocytes with complementary receptors. Following antigen recognition and combination, lymphocytes in each clone are stimulated to proliferate. Large populations of the specific clonal type result, and these begin to synthesize antibody at high rates. Such circulating antibodies combine and neutralize antigen.*

surfaces. After small lymphocytes interact with antigen, they are stimulated to proliferate into a clone of cells able to synthesize the specific antibody. In this way a large population of cells is produced to amplify antibody synthesis.

It is doubtful that the instructive hypothesis has any validity since antigen could not act as a template to instruct the antibody-synthesizing machinery. First, there is nothing unusual about protein synthesis in lymphoid cells. Polypeptide amino acid sequences are specified by the genome in the usual manner, and it is difficult to see how an antigen can modify this mechanism. Second, antigen has never been detected within or on antibody-producing cells, when they are actively synthesizing protein. This leaves the clonal selection hypothesis.

A prediction of the clonal selection hypothesis is that an antibody-producing cell makes one and only one antibody. This has been directly tested by placing single lymphocytes in small microdrops and analyzing the antibody produced. Before plating, the cells are initially challenged in the organism with two related but not cross-reacting antigens. If cells produce more than one antibody, then each microdrop should contain antibodies reacting with both antigens. In thousands of cells examined, only one of the two antibodies was produced by a single cell. This implies that as B lymphocytes mature, thousands of different clones are produced, each expressing a specific combination of V and C genes for light and heavy chains. These molecules are inserted into plasma membranes to act as antigen receptors. The cells are immature, not actively synthesizing or secreting antibody. In a sense, they are in a protodifferentiated state, awaiting an inducing signal from antigen before their synthetic machinery is stimulated.

The selection hypothesis also predicts a period of rapid cell proliferation of the specific cell clone singled out by the antigen for antibody synthesis. As in other differentiating systems, a period of cell replication precedes the active phase of specific protein synthesis accompanying differentiation. Each new daughter cell is recruited to make antibody after division. Accordingly, specific antibody formation should follow exponential kinetics. This, too, has been demonstrated by following the increase in number of mouse spleen cells sensitized to sheep red blood cells (SRD) in an in vitro assay system. During proliferation, antigen seems to be continuously se-

lecting those daughter cell populations which synthesize the most effective antibody. The antibody synthesized on these plasma cells shows increasing "avidity" with time; that is, late appearing antibodies react more vigorously with antigen in vitro. Avidity is a measure of the tightness of fit between antibody and antigen. The mature plasma cell population synthesizes specific antibodies at a tremendous rate, approximately 2000 antibody molecules per second per cell.

We have already seen that lymphocyte proliferation may be stimulated by such agents as concanavalin A which combine with glycoproteins on the cell surface (Chapter 5). Con A seems to act by inhibiting "patch" and "cap" formation in the fluid membrane. The relevance of these surface events to proliferation and differentiation of the clone is not clear, but it is assumed that antigen interaction with receptor at the surface evokes changes that are transduced from the surface into the cytoplasm. Such events may trigger the metabolic changes required for proliferation.

## Development of the B cell system

We can now follow the pathway of B lymphocyte differentiation and antibody synthesis. B cell development in birds begins with the discrete organ, the bursa, which emerges as a saclike evagination on day 5 of development. On days 12 to 13, stem cells migrate from the yolk sac to the bursa and differentiate into B lymphocytes. Inductive signals within the bursa initiate the first phase of lymphocyte development, and B cells appear with immunoglobulin receptors on their surfaces at day 14. These immature cells do not actively synthesize antibody; in a sense they are determined, committed to synthesizing a specific antibody configuration.

The bursa acts as a seeding organ since shortly before or around hatching, B lymphocytes are released into the peripheral circulation where they populate lymph nodes, spleen, and bone marrow. Removal of the bursa before this stage results in an animal unable to develop an immune response. Removal of the bursa after this stage has no effect on the B cell population; B cells in peripheral circulation are self-sustaining and continue to proliferate, particularly in bone marrow, their principal source. Presumably heterogeneous clones of B cells are released from the bursa, each specified with a different antibody receptor on its surface and responsive to only one antigenic configuration.

Differentiation is completed within peripheral organs, after encounter with an antigenic stimulus. The attachment of antigen to the surface receptor triggers several responses. The fluid membrane displaces the antigen-antibody complex into a cap at one pole, followed by pinocytotic engulfment. Redistribution of receptor on the surface may also stimulate replication and a high rate of antibody synthesis. The final phases of differentiation are completed, and mature, antibody-secreting cells are produced.

Response to antigen may take one of several pathways. (1) A B cell may be induced to divide and differentiate into an antibody-secreting cell; (2) it may be paralyzed by the antigen and become immunologically tolerant, no longer competent to respond to a second antigenic challenge; or (3) it may exhibit no response at all. A positive or negative response by a lymphocyte seems to depend upon several factors, among which are antigen concentration, the degree of fit between antigen and antibody receptor sites (avidity), the manner in which antigen is presented (whether antigen is on molecules or attached to cell surfaces), and, finally, whether other lymphocytes are present to help or to suppress a lymphocyte response. Usually concentration of antigen is the most important factor. Very high concentrations usually suppress an antibody response, but low antigen levels, present continuously, below the threshold concentration required for stimulation, can also paralyze the immune system.

## CELL-MEDIATED RESPONSE SYSTEM: THE T CELL

T cells, like B cells, arise from stem cells that migrate from the yolk sac or liver and invade the developing thymus gland around fetal day 11 in the mouse. The first lymphocytes are detected on day 15, as stem cells in the thymus undergo maturation and become competent. They acquire surface antigens specific for T cells (theta factor) and antigenic receptor sites, possibly similar to immunoglobulins, although their exact nature is still controversial. Mature cells are released into the circulation at about birth as a seed population which colonize the peripheral lymph

organs. A thymectomy carried out before birth results in a marked reduction in T cell population, but a similar operation carried out after birth has little effect on the T population, indicating that these, too, are self-sustaining in peripheral organs.

Diverse clones of T cells are probably produced, each committed to respond to a different cellular antigen. The antigen, acting as an inducer, stimulates T cells to secrete nonantigenic-specific factors such as chemotactic agents, cytotoxins, and mitogens, which act directly on the foreign cell. These responses lead to graft rejection, or delayed hypersensitivity, and immunity to some microorganisms.

### T cells as helper cells: T-B interactions

The T and B cell systems are not independent; they interact in many immune situations. T cells actively help B cells in mounting an antibody response. Here we have a classic instance of a bicellular interaction as the basis for differentiated expression of a committed cell. For example, X-irradiated mice which cannot produce antibodies will do so only if repopulated with a mixture of T and B cells. Neither cell type alone is sufficient to produce high levels of specific antibodies, which indicates that a synergistic action between the two cell types is essential.

In vitro studies with spleen organ cultures indicate how this interaction may take place. Spleen organ cultures will produce antibodies in vitro when challenged with immunogens, agents capable of eliciting an immune response. Moreover, synthetic immunogens which elicit specific antibodies to a chemically defined group within the immunogen may also be used in such in vitro systems. Small molecules such as dinitrophenol (DNP), known as haptens, can be conjugated to a protein ($P_1$), and the complex DNP-$P_1$ serves as an immunogen. Spleen organ cultures respond to such molecules by producing antibodies specific to the DNP group.

For instance, does an immune competent cell have to recognize both the hapten (DNP) and the carrier protein before it produces antibodies specific to DNP? If the spleen cells are treated with either free DNP or the carrier protein $P_1$ before the addition of DNP-$P_1$, much less antibody is produced, as if the free components combine with and block specific receptors that prevent the conjugate from attaching and inducing a response. This means the cell must

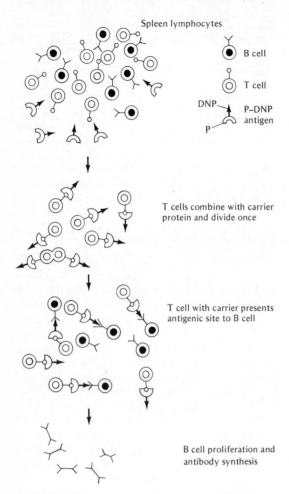

Spleen lymphocytes

● B cell

◎ T cell

DNP → P–DNP antigen
P

T cells combine with carrier protein and divide once

T cell with carrier presents antigenic site to B cell

B cell proliferation and antibody synthesis

**Figure 18-6**   *T cell and B cell helper interactions. A population of spleen lymphocytes contains both T and B cells. Complex antigen consists of carrier protein and DNP hapten, P-DNP. Antigen combines with T cell which recognizes carrier protein, and T cell is stimulated to replicate once. T cell then presents DNP hapten to B cell, which is stimulated to proliferate and synthesize antibodies specific for DNP.*

recognize the DNP-carrier as a unit before it can produce antibody. Experiments showed that two receptors are involved, one able to recognize the hapten and the other the carrier molecule.

Does this mean the same cell has two different receptors, one for the hapten and the other for the carrier, both cooperating to produce the antigenic signal? Or are two cells involved, one with receptors for the hapten and the other for the carrier? The latter

requires bicellular cooperation for the production of an antibody response. The evidence of B and T cell cooperation in X-irradiated mice immediately suggested the second alternative, and experiments were designed to test this in vitro.

According to the hypothesis, T cells would be sensitive to the carrier protein while B cells would be responsive to the DNP hapten and synthesize the DNP antibodies. Three groups of irradiated mice were prepared: (1) repopulated with both T and B cells from normal mice by injecting cells from bone marrow and thymus, (2) repopulated with T cells alone by injections from thymus, and (3) repopulated with B cells by injecting bone marrow. All mice were then injected with RSA (rat serum albumin) to induce a population of cells able to respond to RSA, the protein carrier. The spleens of these animals were isolated in culture 10 days later, and each was challenged with the conjugate DNP-RSA. It could be seen that the greatest amount of antibody was produced only in spleens from animals containing both T and B cells; spleens recolonized with either T or B cells alone did not produce any antibody when challenged to DNP-RSA in vitro.

How does the T cell help the B cell produce antibody (Figure 18-6)? It could be shown that prior treatment of mice with carrier RSA was necessary to enhance the in vitro production of antibody by spleen cells. Spleen had to contain a population of T cells sensitive to carrier (RSA) before B cells in the spleen could mount an antibody response to the DNP-carrier conjugate. The RSA-T cells are derived from the thymus and these must replicate within the spleen before they become competent to assist the B cells in antibody production. The experiments showed that only one critical replication of T cells was essential to their final maturation (a quantal division?), suggesting that after this division, T cells carried sufficient surface receptors to RSA to cooperate with B cells in antibody production.

The interactions are probably more complex than is implied by the simple bicellular model. Several steps may be involved, in which B cells interact with one another and with carrier before T and B cells interact. Furthermore, it is now clear that macrophages also participate in the reaction. The interaction involves three cells—the macrophages, which first receive the antigenic stimulus, process it, and present it to the T cells, which in turn interact with B cells.

These complex cellular interactions are still poorly understood, but they do reveal the extent to which cell communication and recognition are essential in a differentiating system. True, the immune competent cell may be a special case since the system is specifically adapted for recognition of molecular configurations. Similar mechanisms of cell recognition may operate in all inductive interactions with the immune system only a highly specialized version of cell communication via matching surface configurations.

### Graft rejection

The graft rejection mechanism is probably a visible demonstration of the biological phenomenon of "self." Tissues grafted between two individuals, even within the same strain, are rejected if they differ with respect to one or a few histocompatibility genes. Highly inbred isogenic strains of mice will accept skin grafts from one another without rejection because they share these histocompatibility genes, but a skin graft from strain A mice made in strain CBA animals is usually rejected within 10 days. The graft skin may become established within the first few days and appear healthy, but the area is soon invaded by lymphocytes and macrophages which destroy cells, and the degenerating graft is sloughed off (Figure 18-7). Animals exposed to a foreign graft possess an immunological memory; they reject second grafts in 6 days. The primary graft induces a population of memory cells able to recognize the tissues of the donor and mount a more effective rejection mechanism.

The rejection is mediated by T cells and not by antibodies. Immunity to a graft tissue is passively transferred from a stimulated animal (one that has already rejected a graft) to a nonimmunized animal by injection of spleen or lymph cells and not by injection of immune sera (Figure 18-8). Moreover, X-irradiated mice repopulated with spleen cells that have already been exposed to a graft rejection will mount a second-set reponse (an earlier rejection) when challenged by its first graft, again indicating the importance of memory cells in mounting the rejection response.

The importance of the genetic background is demonstrated by crossing A and CBA strain mice to

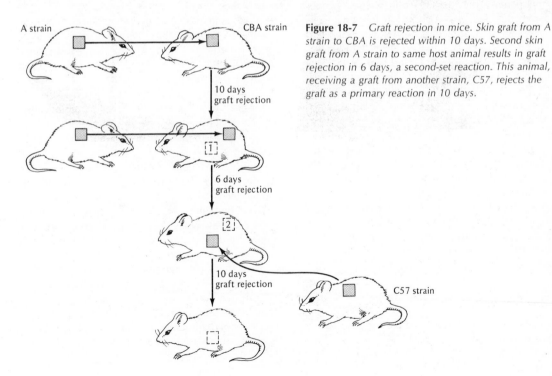

**Figure 18-7**   *Graft rejection in mice. Skin graft from A strain to CBA is rejected within 10 days. Second skin graft from A strain to same host animal results in graft rejection in 6 days, a second-set reaction. This animal, receiving a graft from another strain, C57, rejects the graft as a primary reaction in 10 days.*

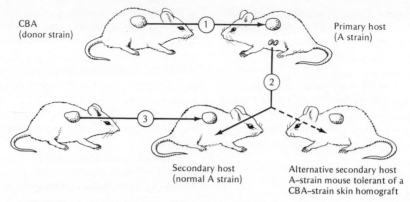

CBA
(donor strain)

Primary host
(A strain)

Secondary host
(normal A strain)

Alternative secondary host
A–strain mouse tolerant of a
CBA–strain skin homograft

**Figure 18-8**  *Transfer of cell-based immunity by means of sensitized lymph-node cells. (1) A normal A-strain mouse is sensitized by a skin graft from a CBA donor. After rejection of the graft, regional lymph nodes are removed from the host and prepared as a cell suspension. (2) These cells are injected into another A-strain host (secondary host) which may be challenged with a skin graft from another CBA donor (3) or an A-strain mouse made tolerant of a CBA skin graft. In the first host the graft rejection is accelerated (a secondary response), while in the second host the donor CBA graft is now rejected and tolerance is abolished. (From R. E. Billingham and W. K. Silvers,* Transplantation of Tissues and Cells. *Philadelphia: Wistar Institute Press, 1961.)*

produce an $F_1$ hybrid. Such hybrid animals accept grafts from either parent since they share the same histocompatibility genes, but a graft of hybrid tissue to either parent is rejected since only one set of genes is shared; the other is recognized as foreign and its tissue products are therefore treated as foreign and rejected. Since there are large numbers of histocompatibility genes in mammals, including man, the possibility that all genes are shared even within the same family is remote. For this reason, all tissue and organ grafts between humans (except for identical twins) are rejected. Even when tests reveal that individuals share many histocompatibility genes, organ transplants are carried out only in situations where the immune response system of the recipient is drastically inhibited, either by X-irradiation or by immune suppressive drugs. Successful kidney transplants in humans are based largely on an initial close match of donor and host histocompatibility antigens and a sustained and continuous monitoring of the immune response pattern of the recipient to see that no signs of specific rejection build up.

## Immune tolerance

The phenomenon of immune tolerance emerges from studies of graft rejection in mammals. Mice of strain A can be made tolerant of grafts from strain CBA

animals if they are exposed to CBA cells prenatally, or just around birth, presumably when the immune system is still immature. Skin grafts made at these early stages persist for long periods. More significantly, animals can be made tolerant by injection of adult spleen cells into a fetus. For example, in a classic experiment, Medawar injected adult spleen cells from strain A mice into a fetus of strain CBA. The mice at 6 months were challenged with a skin graft from strain A which persisted for the life of the organisms. The CBA animals were tolerant because they had been exposed to the foreign A antigens when their immune system was immature; presumably the high dose of A antigens at this stage was enough to suppress any immune response by the developing CBA lymphocytes. It should be emphasized that immune cells can be so overwhelmed by high doses of antigen, or by continuous exposure to low doses, as to be paralyzed, unable to respond to the same antigens when subsequently challenged. The phenomenon of immune tolerance is an example of immune suppression that can normally be induced even in adult animals.

The mechanism of immune suppression is not understood. Inasmuch as antibody production involves cooperation among three cell types—B cells, T cells, and macrophages—antigenic excess at the time of lymphocyte maturation could affect differentiation of any one of these cell types and their ability to respond to antigen.

Some aspects of tolerance may be studied in the in vitro spleen system by culturing spleens from adult animals made tolerant to the carrier protein RSA. As we have seen, it is the T cell population that is responsive to a carrier protein. Such spleens do not make DNP antibody when challenged by DNP-RSA immunogen in vitro but will make antibodies to DNP if challenged by another DNP-carrier system. The spleen contains B cells responsive to DNP but its T cells are lacking because they are tolerant to the carrier RSA. This suggests that tolerance involves the elimination of specific T cell clones by excess RSA. Cooperative interaction between specific B and T cells cannot occur and antibody is not synthesized.

B cells from bone marrow of mice made tolerant to RSA also seem to have been made tolerant; in other words, tolerance may involve more than one cell type. Considering the complexity of the interactions among B and T cells, macrophages, and antigen, it is not surprising that our understanding of tolerance is still incomplete.

## Recognition of self

An organism's immune system must recognize its own wide assortment of native antigens as "self" and display no immune response to its own tissues; otherwise it would destroy its own tissues. Self-recognition develops during the ontogeny of the immune system which becomes tolerant to its own tissue antigens because it is exposed to a large excess of these antigens during its maturation. Clones of lymphocytes specifically responsive to histocompatibility antigens on an organism's own tissues are eliminated or fail to differentiate. An animal's immune system does not attack its own tissue because it lacks those clonal populations carrying receptors to its histocompatibility antigens.

Not all animals within the same isogenic strain accept grafts of tissue readily from one another. Although males in the strain accept grafts from both males and females, females reject grafts from males. Males possess some genes lacking in females, namely, those in the Y chromosomes, and these produce antigenic products that are foreign to the females' immune system. Consequently a female is not tolerant to these antigens (she contains lymphocytes able to recognize male antigenic determinants)

and will reject male grafts as foreign and not self, in spite of the fact that all other antigens are identical.

Even males can mount an immune response to their own tissues. Sperm cells appear late in life, long after the immune system has matured. If one testis of a guinea pig is mascerated, emulsified in oil, and mixed with an adjuvant that enhances the immune response, then injected into the same animal, antibodies will be produced to the sperm cells and the remaining testis will atrophy because all sperm have been eliminated. The animal can make antibodies against those tissues which have been protected from the immune system at its maturation; it can develop autoimmunity against such physiologically isolated tissues as brain and lens. Apparently this is the basis for many human diseases, such as rheumatoid arthritis, where some defect arising genetically or as a result of viral infection or disease leads to the release of tissue antigens (perhaps modified in some way so as to appear foreign to the immune system) which elicit an immune response and subsequent destruction of healthy tissue.

Response to antigenic changes in normal tissues may also serve the organism as a defense against tumor development. A hypothesis for an immune surveillance system has been postulated as an adaptation to detect the incipient origin of cancer cells. Most cancer cells either possess new antigenic groups on their surfaces or have lost antigens carried by normal cells. As a result, new surface configurations are presented by such cells and the immune system recognizes these as foreign and rejects them. This may serve to eliminate incipient cancer foci and protect the organism.

## SUMMARY

The ontogeny of the immune system in vertebrates serves as a model of differentiation for most other systems. A population of stem cells in bone marrow gives rise to progeny which completes maturation in two separate sites, the thymus and in the bursa of Fabricius in birds or its equivalent in mammals. Cells maturing in the thymus (T cells) acquire the ability to respond to antigen signals on the surface of foreign cells and mediate graft rejection and other cellular immunity. B cells, completing their maturation in the bursa, are committed to recognize soluble antigens and synthesize antibodies. The B cell system consists of a heterogeneous population of lymphocyte clones, each responsive to a specific antigenic configuration. The antigen "selects" a population of B cells and triggers a final differentiation.

The mature B cells seem to be in a protodifferentiated state since they require the antigenic stimulus to become fully differentiated. Attachment of antigen to the cell surface results in a surface-modulated change that stimulates a program of proliferation and increases the size of the clonal population. As a final step in differentiation the cells are stabilized to synthesize large amounts of antibody specific for the antigen.

The intiation of this program of differentiation also depends upon cell interactions. T cells and macrophages respond to antigen and present it to B cells to initiate the antibody response. The mechanism of the interaction is not understood but configurational changes of receptor systems in the surfaces of cells are probably involved. We see in this cell system specialized for recognition of molecules a model of surface-modulated changes that program specific protein synthesis.

**REFERENCES**    Abramoff, P., and La Via, M. F. 1970. *Biology of the immune response.* New York: McGraw-Hill.

Auerbach, R. 1971. Towards a developmental theory of immunity: Cell interactions. *Cell interactions and receptor antibodies in immune response,* ed. O. Mahela, A. Cross, and T. E. Kosuhen. New York: Academic Press, pp. 393–398.

Billingham, R. E., Brent, L., and Medawar, P. B. 1953. Actively acquired tolerance of foreign cells. *Nature* 172:603–606.

Billingham, R. E., and Silvers, W. K. 1971. *The immunobiology of transplantation.* Englewood Cliffs, N. J.: Prentice-Hall.

Burnet, M. 1959. *The clonal selection theory of acquired immunity.* London: Cambridge University Press.

Dresser, D. W., and Mitchison, N. A. 1968. The mechanism of immunological paralysis. *Adv. Immunol.* 8:129–181.

Feldman, M., and Globerson, A. 1974. Reception of immunogenic signals by lymphocytes. *Current topics in developmental biology,* ed. A. A. Moscona and A. Monroy. Vol. 8. New York: Academic Press, pp. 1–40.

Feldman, M., and Nossal, G. J. V. 1972. Tolerance, enhancement of the regulation of interactions between T cell, B cell and macrophages. *Transplant. Rev.* 13:3–34.

Greaves, M. F., Owen, J. T., and Raff, M. C. 1975. *T and B lymphocytes: Origins, properties and roles in immune responses.* Amsterdam: Exerpta Medica.

Hood, L., and Prahl, J. 1971. The immune system: A model for differentiation in higher organisms. *Adv. Immunol.* 14:291–351.

Hood, L., and Talmage, D. W. 1970. Mechanism of antibody diversity: Germ line basis for variability. *Science* 168:325–333.

Katz, D. H., and Benacerraf, B. 1972. The regulatory influence of activated T cells on B cell response to antigen *Adv. Immunol.* 15:1–94.

Medawar, P. B. 1958. The immunology of transplantation. *The Harvey lectures,* Series 52. New York: Academic Press.

Mitchison, N. A. 1967. Antigen recognition responsible for the induction in vitro of the secondary response. *Cold Spring Harbor Symp. Quant. Biol.* 32:431–439.

Mitchison, N. A., Rajewsky, K., and Taylor, R. B. 1970. Cooperation of antigenic determinants and of cells in the induction of antibodies. *Developmental aspects of antibody formation and structure,* ed. J. Sterzl and H. Riha. New York: Academic Press. pp. 547–561.

Nossal, G. J. V., and Ada, G. L. 1971. *Antigens, lymphoid cells and the immune response.* New York: Academic Press.

Raff, M. C. 1974. Development and differentiation of lymphocytes. *Developmental aspects of carcinogenesis and immunity,* ed. T. J. King. New York: Academic Press, pp. 161–172.

Singer, S. J. 1974. Molecular biology of cellular membranes with application to immunology. *Adv. Immunol.* 19:1–66.

# 19

# The Larva and Metamorphosis

A larva is an intermediate, often free-living, stage in the life cycle of many organisms, usually the terminal product of embryogenesis. It emerges when the embryo hatches and may be morphologically different from the adult. Following a different life style, it occupies its own ecological niche and has often been mistaken for a distinct species.

For many species with small eggs, a larva is adapted as a motile feeding stage. A limited store of nutrients in a small egg cannot support adult development but is adequate for a simple feeding larva. In such eggs, development is rapid and the hatched larvae are equipped to feed and sustain growth of adult primordia. On the other hand, closely related species with large yolky eggs may exhibit a longer period of embryogenesis and by-pass the larval stage entirely, developing directly into miniature adults. Intermediate situations are found in insects where eggs carry out two developmental programs side by side, larval and adult. The larval program is expressed first, whereas the adult pro-

gram, sequestered in the larval body as imaginal discs, develops slowly as the larva feeds and grows.

If a larval stage evolves in the life cycle of an organism, it may, through natural selection, acquire additional adaptations for species dispersal and site selection. In fact, larvae may have first evolved as adaptations for these latter functions, while their developmental role may have been acquired only secondarily.

Many aquatic organisms are sedentary, burrowing, or sessile, confined to habitats of small size and limited diversity. Once having settled in an area (sand, mud bottom, or rock piling, for example), they do not migrate but tend to proliferate and compete for the limited space and food resources. Larvae have evolved as dispersal mechanisms adapted for invasion and colonization of new habitats. Dispersal prevents overcrowding of stationary populations and minimizes competition for limited food resources. Such larvae are usually pelagic, floating or swimming

**629**

about in the surface waters. Dispersal is brought about by water currents; cilia keep larvae at the surface, preventing them from settling to the bottom too soon.

Larvae have also evolved unique food preferences, exploiting ecological niches that are not in competition with adult populations. Pelagic larvae rely on planktonic food, while sessile adults on the bottom feed on entirely different food organisms. As a result, larva and adult have responded to different selective forces during the course of evolution.

Many free-swimming larvae are also adapted to locate an attachment site, a substratum favorable to metamorphosis into a sessile adult. Accordingly, larvae are equipped with light-sensitive organs and gravity detectors to distinguish favorable from unfavorable substrata, a mud flat from a rocky bottom or a vertical surface (for example, pilings) from a horizontal one. Even the simple larvae of coelenterates possess sensory cells to detect a preferred substratum for attachment. Site selection is the primary function of many ascidian larvae, which barely feed during their brief existence.

Most larval organization is concerned with these functional roles — that is, motility, specialized feeding apparatus, sense organs for site selection, and adhesive structures for attachment to a substratum. Since these structures are not relevant to adult life, they are usually shed during a metamorphic crisis when a larva transforms into an adult.

The metamorphic event is usually traumatic; one entirely different life style is substituted for another. All the appurtenances of larval existence are discarded or transformed as the adult organization emerges. In most cases metamorphic events are hormonally controlled.

The phenomenon of metamorphosis poses several problems of genetic regulation. What initiates metamorphosis and how is the sequential expression of two developmental programs controlled? Where larvae and adult cell populations are segregated early, metamorphosis favors adult primordia differentiation while most larval structures degenerate. In some cases, many larval tissues survive and transform into adult cells; a larval-specific phenotype is shed by a cell (or its progeny) to acquire an adult phenotype. In this case, the same clone plays out two genetic programs. How, then, is each developmental program regulated?

## THE LARVA AS A DEVELOPMENTAL STRATEGY

For bottom-dwelling marine invertebrates, the free-swimming larval stage is one of several strategies for adult development, species dispersal, and site selection (Figure 19-1). As we have indicated, development of an adult body requires enormous energy resources and stored nutrients. One nonlarval strategy involves production of a few large eggs, each containing enough food and energy reserves to develop directly into an adult. Each egg is important to species survival; accordingly, adults copulate, fertilization and embryogenesis are internal, and viviparity is the rule. Immature adults emerge from the female at the end of development. Eggs are often maintained in brooding organs, and adults exhibit brood behavior to protect developing young. These adaptations insure that each egg develops into a functional adult. Many echinoderms, polychaete worms, and mollusks exhibit this pattern of development.

Another evolutionary strategy is to produce a modest number of small eggs surrounded by nurse cells, yolk cells, or a nutrient medium in a cocoon-like capsule. One or many embryos are nourished by nurse cells until late larval or young bottom-dwelling stages develop. The more nurse cells, the better developed are the larvae at hatching. Females may lay between 100 and 5000 eggs per season and exhibit a primitive brooding behavior. This mode of development is again widely distributed among free-living flatworms, echinoderms (crinoids in particular), mollusks, and nemerteans.

The true larval strategy is found in species producing large numbers of small eggs, approximately 1000 to $5 \times 10^8$ eggs per female per season. The risks are greater, however, and egg and larval wastage is high. Much is left to chance, but large numbers insure success. Fertilization is external since sperm and eggs are shed into the sea. Males and females swarm or shed simultaneously, which increases the chances for successful fertilization. In many instances, shedding of sperm is the signal to induce spawning. Embryogenesis is rapid (hours or days), and the hatched larva immediately begins a long pelagic life (1 to 4 weeks in summer or 1 to 3 months in winter) as it is carried about over wide areas by surface currents, feeding on phytoplankton.

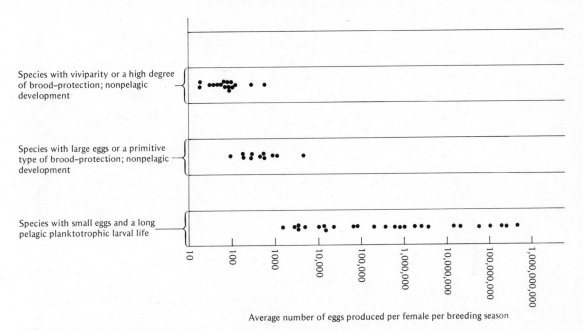

**Figure 19-1**  *Number of eggs produced per female per breeding season of marine bottom invertebrates (echinoderms, nemerteans, polychaetes, decapods, and so on) with different types of development. (From G. Thorson,* Biol. Rev. *25:1–45, 1950.)*

Population size of bottom-dwelling invertebrates with pelagic larvae fluctuates seasonally because larvae are dependent upon such variable environmental factors as phytoplankton availability, sea water temperature, and predators. Since many marine animals are plankton feeders, many larvae in plankton are consumed. It has been estimated that a predator such as the mussel *Mytilus* filters approximately 1.4 liters of water per hour and may consume 100,000 pelagic lamellibranch larvae in 24 hours. Consequently, the longer the pelagic life of a larva, the greater the chances for population decimation by predators and the greater the fluctuation in adult population size.

Most bottom-dwelling invertebrates live on or near the continental shelf, in relatively shallow waters. Since larvae are carried over great distances by surface currents, one problem is to return to shallow water, invade new sites, and metamorphose into adults. Those larvae unable to find a suitable site fail to complete metamorphosis, which is another reason for the excessive wastage of this mode of development.

Pelagic larvae may be divided into two groups, *lecithotrophic* and *planktotrophic:* the former are large floating larvae that feed exclusively on their own yolk reserves; the latter are direct phytoplankton feeders. Roughly 10 percent of the pelagic larvae are of the lecithotrophic type, their primary role being dispersal. They develop from large yolky eggs and are clumsy in shape and unfit for locomotion (Figure 19-2). They possess no effective motile apparatus but depend on flotation devices to keep them in surface currents, where they are carried about passively. They remain pelagic for a long time, nourished entirely by their rich yolk supplies. The amphiblastulae larvae of most sponges are typical of this pattern (Figure 19-3).

These larvae resemble gemmules in their morphology, except that the outer ectodermal surface is ciliated. The central mass of endodermlike cells includes large numbers of yolk cells on which the embryo feeds throughout its pelagic existence. The larvae are dispersed far from their origin and settle onto new sites where they metamorphose into sessile adults. Metamorphosis is simple; larval cells

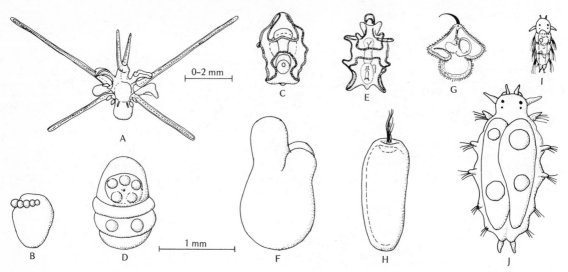

**Figure 19-2**    *Planktotrophic (above) and lecithotrophic (below) larval types of each of the groups of Echinoidea, Holothurioidea, Asteroidea, Nemertini, and Polychaeta. (A) Eucidaris metularia. (B) Heliocidaris erythrogramma. (C) Holothuria arenicola. (D) Psolus phantapus. (E) Asterope carinifera. (F) Henricia sanguinolenta. (G) Pilidium magnum. (H) Oerstedia dorsalis. (I) Nereis dumerili (small larval type). (J) Nereis dumerili (large larval type). Scale below refers to A–G; scale above to I and J. (From G. Thorson, Biol. Rev. 25:1–45, 1950.)*

simply assume adult functions and reorganize themselves into the adult body plan after attachment to a substratum.

The lecithotrophic larva has certain advantages over the planktotrophic type; it is easily dispersed and, being independent of phytoplankton as a source

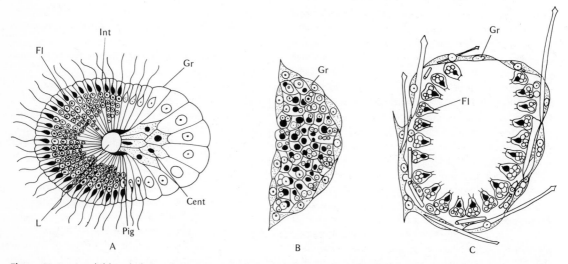

**Figure 19-3**    *Amphiblastula larva of sponges. (A) Free-swimming larva showing flagellated cells (Fl), central cells (Cent), granular cells (Gr), intermediate cells (Int), lens cells (L), and pigment cells (Pig). (B) Larva just attached to a substratum. (C) Young sponge of Leucosolenia variabilis. (From E. W. MacBride, Textbook of Embryology, Invertebrata, vol. I. London: Macmillan, 1914.)*

of food, has a better chance of reaching metamorphosis. Its disadvantage is its large size. This means that smaller numbers are produced, which places more of a metabolic demand on maternal function. Usually species population size remains relatively constant since the limited numbers of larvae are unable to take advantage of any sudden improvement in phytoplankton availability.

More than 70 percent of marine invertebrate pelagic larvae are of the planktotrophic type and include most of the familiar larval types, such as trochophores, veligers, pilidia, and plutei, characteristic of the various phyla. These larvae, derived from small eggs, are equipped with swimming and feeding organs and spend a long time feeding on phytoplankton. Most larvae already possess some adult structures, and metamorphosis is gradual; as adult organs grow, larval structures disappear. On the other hand, some ascidian larvae and the pilidium larvae of nemertean worms undergo a dramatic metamorphosis, shedding virtually all larval structures and replacing them by the metamorphosed adult within a few minutes.

Planktotrophic larvae have many advantages; they are easily produced since they pose no metabolic drain on the female, and their long pelagic life makes for wide dispersal. When food conditions are optimal, large numbers of larvae survive, increasing adult population size. There are disadvantages also; high requirements for food mean absolute dependence on the vagaries of phytoplankton population dynamics. Their longer life also exposes them to predators for longer periods. A description of a few pelagic larvae of this group will show how long-lived feeding behavior makes continued adult development possible while dispersing the species.

### Some larval strategies

Most Protostomia are characterized by a common spiral cleavage pattern and an almost identical larval type, the *trochophore*. The free-living trochophore plan is found only in primitive annelids and mollusks; in most cases, the trochophore becomes highly modified, assuming many adult characteristics before hatching.

Among primitive annelids and mollusks, the trochophore metamorphoses directly into the adult. In more advanced species, however, the trochophore is not free-living and adult structures are formed within embryonic or egg membranes. Metamorphosis continues after hatching and the larva spends a brief period in the pelagic state. In many mollusks, the trochophore is compressed or even eliminated, replaced by another larval type, the *veliger*.

The young trochophore of the primitive annelid worm *Polygordius* consists of two hemispheres: an upper aboral hemisphere with an apical plate and tuft of sensory hairs and a lower hemisphere with mouth, gullet-intestine, anus, and nephridia (Figure 19-4). At the equator, separating the halves, is a continuous band of large cilia, the *prototroch*, the larva's principal locomotory organ. A nerve network encircling the trochophore connects with a brain beneath the apical plate. These larvae feed actively on diatoms and grow over a period of weeks.

Larvae are almost complete organisms in every way, with a functional digestive system and a nervous system to regulate motor and sensory activity. While growing, the larva exhibits complex behavior, responding to light and gravity. It possesses an excre-

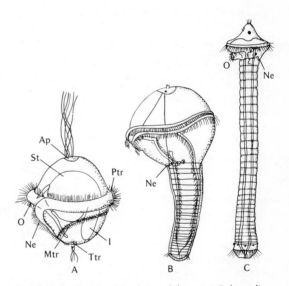

**Figure 19-4** *Trochophore larva of the worm* Polygordius *and its metamorphosis. (A) Free-swimming trochophore larva showing apical plate (Ap), intestine (I), metatroch (Mtr), protonephridium (Ne), mouth (O), prototroch (Ptr), stomach (St), and telotroch (Ttr). (B) Metamorphosis well underway as trunk segments have developed. (C) Metamorphosis almost complete; larval structures at anterior shrinking as adult head forms. (From E. W. MacBride, Textbook of Embryology, Invertebrata, vol. I. London: Macmillan, 1914.)*

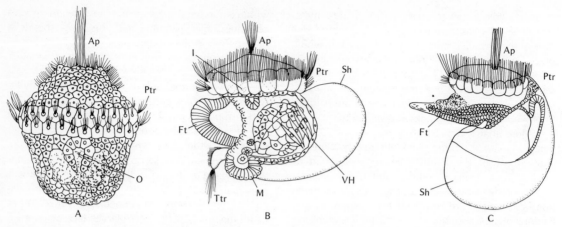

**Figure 19-5** *Larva and metamorphosis in the prosobranch mollusk* Patella. *(A) Trochophore larva showing apical tuft and plate (Ap), prototroch (Ptr), and mouth (O). (B) Larva before metamorphosis; it has assumed molluscan characters with foot (Ft), visceral hump (VH), shell gland and shell (Sh), intestine (I), mantle (M), and telotroch (Ttr). (C) The same as (B) after flexure of the trunk. (From E. W. MacBride,* Textbook of Embryology, Invertebrata, *vol. I. London: Macmillan, 1914.)*

tory system, smaller but functionally similar to that of the adult, and except for the absence of sex organs can be considered a separate species. Meanwhile, within the larval body, adult primordia also grow in preparation for metamorphosis. Metamorphosis is a gradual process of reconstruction of trochophore body parts. The lower half, with mouth and anus, thickens into a ventral plate which breaks up into a series of segments. This half (largely a derivative of D-quadrant cells) lengthens in a plane perpendicular to the primary axis of the trochophore, its segments becoming more distinct. The animal continues to swim about by its prototrochal cilia during metamorphosis. Meanwhile, the upper half is transformed into the preoral lobe of the adult worm (its front end with brain) while the lower half becomes the segmented trunk. Most larval organ systems are reconstructed and retained by the adult although the prototrophic larval swimming organ is lost.

The primitive gastropod body plan is derived directly from a free-living trochophore. Metamorphosis of the *Patella* trochophore follows molluskan architectural principles (Figure 19-5). Mouth and anus first move apart while derivatives of the 2d cell begin forming the shell gland. A thickening develops ventrally in front of the mouth, the primordium of the foot. Meanwhile, endodermal cells of the D quadrant grow out, forming a hump on the dorsal side of the

larva with a loop of intestine growing into it. This becomes the future visceral hump of the mollusk, bearing the shell gland on its upper surface. Because of differential growth, the larval body folds, bringing anus and mouth closer together. The upper hemisphere of the trochophore, as in annelids, forms the head with brain. A mantle develops over the visceral hump and an adult shell is secreted.

The basic plan is modified in most mollusks, where a trochophore develops into a second larval type, or veliger, which more closely resembles a mollusk body plan. In other species, the veliger, develops directly from the embryo as in the gastropod snail *Ilyanassa*. A free-swimming veliger possesses paired velar lobes (a ciliated motile and feeding organ), eyes, foot, larval shell, beating heart, and digestive tract (see Chapter 11). At metamorphosis, the transformation from veliger to adult calls for distortions of the body form. The body undergoes tortion, a 180-degree turn of the visceral hump, so that mouth and anus are brought very close together at the ventral surface, one above the other. The digestive tract also bends markedly toward the ventral side, and growth is asymmetrical with the left side growing more markedly than the right. The visceral mass enlarges, more shell is secreted, and the young adult is fully formed. Metamorphosis is slow, and most larval structures are retained by the adult.

In contrast to the progressive metamorphic changes in the Protostomia larva, metamorphosis of Deuterostomia echinoderm larvae substitutes a new adult body plan and symmetry for a discarded larval organization. Each echinoderm class is characterized by its own larval type, differing in shape and distribution of ciliated bands and lobes. The sea urchins (echinoids) have a pluteus larva, as we have seen (Chapter 11), while starfish (asteroids) and sea cucumbers (holothurians) produce *bipinnaria* and

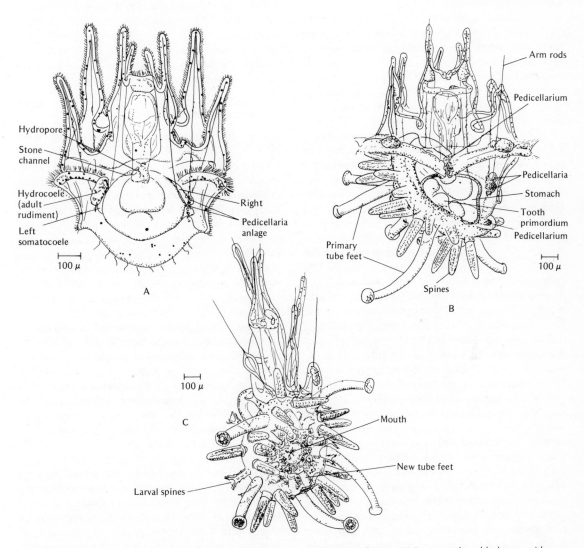

**Figure 19-6** *Metamorphosis of the pluteus larva of the sea urchin* Psammechinus. *(A) Fourteen-day-old pluteus with eight arms and apex rods completely withdrawn. Hydrocoele has developed five protrusions growing into tentacles; this is the echinus rudiment, the anlage of the adult imaginal disc. Opposite is the anlage of the pedicellaria. (B) First stage of metamorphosis in a 33-day-old pluteus. The echinus rudiment has begun to unroll while the ectoderm of the larva is shrinking, withdrawing from the arm rods. Spines and tube feet of rudiment are well formed. (C) One day later. Shrinking of pluteus ectoderm continues. Echinus rudiment seen from ventral side. Between each primary tube foot and the mouth, two additional tube feet have developed in each sector. (From G. Czihak,* Experimental Embryology of Marine and Fresh Water Invertebrates, *ed. G. Reverberi. Amsterdam: North-Holland, 1971, pp. 363–506.)*

*auricularia* larvae, respectively. Inasmuch as most echinoderms are bottom-dwellers, their larvae are of the typical planktotrophic variety, complete organisms unto themselves, living for weeks in the surface waters as their presumptive adult tissues prepare for metamorphosis.

The pluteus larva of the sea urchin transforms a larval body plan with bilateral symmetry to an adult body with radial symmetry adapted to a sessile or sedentary existence (Figure 19-6). Paired coelomic pouches bud from the gut and become the primordia of the echinoderm hydrocoele. This water vascular system controls the tube feet (ambulacra), the principal diagnostic structures of the phylum.

The adult is fashioned entirely from an epithelial bud emerging on the left side of the larva; the right side and most other larval structures (ciliated arms) are discarded. Metamorphosis begins a month after hatching when a shallow depression, the amniotic invagination, appears in the ectoderm on the left side. The invagination expands inwardly and contacts the hydrocoele, arranged as a series of endodermal folds. This combined structure becomes the *echinus rudiment* of the future adult. Five fingerlike projections grow from a ring of the hydrocoele, extending into the tissue of the amnion to become rudiments of the tube feet. This transformation marks the beginning of radial symmetry, and the rudiment assumes the adult form. Radially arranged spines of the future skeleton emerge on the side wall of the amniotic cavity. The subsequent metamorphic crisis is uncomplicated as the amnion over the echinus rudiment is ruptured while the arms of the pluteus are resorbed. The echinus body axis rotates 90 degrees to bring the left side of the animal down to the substratum, and it then walks away on its five tubed feet. Once the miniature adult is formed, it falls to the bottom to take up a functional existence. Most echinoderm metamorphosis occurs at some distance from the shore, which means that young adults must work their way back to the intertidal zone to establish themselves. Miniature adults feed and grow while migrating to a favored ecological niche near the shoreline.

When young, most pelagic larvae are positively phototrophic and remain in the surface where light is most intense. This behavior has a selective advantage since it keeps them in the surface where phytoplankton is in rich supply. As larvae grow older they become negatively phototrophic and tend to settle into deeper waters, presumably in preparation for metamorphosis and descent to the bottom to take up an adult life. Many larvae do not have preferred sites for attachment; they settle at random and, if fortunate, they occupy a favorable site. Otherwise they settle into inappropriate sites, fail to metamorphose or attach, and eventually die. Here, too, larval wastage is high. When successful, however, a random descent to the bottom may colonize large areas with metamorphosing larvae.

### Larva as a site selector

Many pelagic larvae are very selective when it comes to invading a new site. The site must be of the right consistency (mud, sand, rock); otherwise they cannot attach and undergo metamorphosis. For many coelenterate and ascidian larvae, vertical surfaces such as pilings are favored sites. These larvae are adapted to select a preferred site, a role often superseding all other roles. Such larvae are usually locomotory but do not feed. They are dispersed by currents, but their pelagic lives are too short for extensive dissemination. Because it cannot feed, a larva must find a site and metamorphose within a few hours or days; otherwise it dies.

The ciliated planula larva found in most classes of coelenterates seems to behave as a site selector. Its organization is relatively simple as it develops rapidly from the egg (Figure 19-7). The irregular cleavage patterns of many hydrozoa result in a coeloblastula which transforms into a gastrula by unipolar or multipolar ingression of endoderm cells into the blastocoele cavity. The solid endoderm fills the cavity, the ectoderm develops cilia, and the planula larva swims off. It usually lacks feeding structures but is equipped with several sensory structures (light and gravity) and nerve cells at the anterior end. After a free-living existence of a few hours to a few days, the planula affixes itself to some object at its anterior end, presumably having characterized the site by virtue of its light-sensitive cells (Figure 19-7). It may then transform into a hydranth by constriction of its body, growth of tentacle buds, and outgrowth of a stolon.

The ascidian larva has evolved as a most efficient site selector. This tadpolelike larva with neural tube, notochord, myotomes, and pharyngeal gills

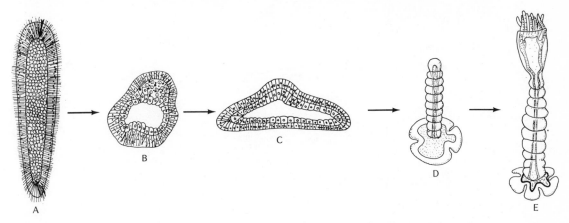

**Figure 19-7** *Metamorphosis of a planula larva of* Gonothyraea, *a hydroid. (A) A young planula with solid endodermal mass and ciliated ectoderm. (B) Planula newly attached to a substratum. (C) Planula at later stage prior to outgrowth. (D) Hydranth growing from attached mass showing annular growth. (E) Hydranth complete. (Based on* The Invertebrates— Protozoa through Ctenophora *by L. H. Hyman. Copyright © 1940 by McGraw-Hill Book Company. Used with permission of McGraw-Hill Book Company.)*

resembles a chordate body plan and is probably the ancestral type for all vertebrates. The swimming life of these forms is short, from a few minutes to 2 days at most. Success of the species depends on the facility with which these larvae find and colonize a new site.

The nature of an ascidian larva correlates with the ecological adaptations of the adult. Most ascidians are sessile, attached to vertical surfaces or under the shaded sides of rocks, growing as large flat colonies. Their larvae are equipped with an *ocellus* or simple eye with lens cells and pigmented retina to distinguish light from dark surfaces. Together with the *otolith*, a gravity receptor, these organs make up the sensory ganglion that controls larval behavior. The larvae move to the water surface early in development, remain for a short time, and subsequently descend. The tail permits sudden horizontal forward movements as the larva descends from surface to bottom. It is these movements that bring the larva in contact with vertical surfaces to which it attaches with its adhesive suckers.

Metamorphosis begins almost explosively in some species. A sudden contraction of longitudinally oriented microfilaments in the dermal cells of the tail results in complete resorption of all tail structures including musculature, notochord, and neural tube (Figure 19-8). In addition to the tail, the ocellus

is lost as the brain vesicle is transformed into an adult cerebral ganglion. Structures adapted for larval existence are discarded as the trunk region transforms into the adult. A rudimentary pharynx or branchial chamber, the principal adult feeding structure, appears in the anterior part of the endodermal organs in the larva. The body rotates, new mouth openings are established, and the general form of the adult sessile organism is assumed.

Many other ascidians (the moleuloids) colonize sand or mud bottoms. These exposed surfaces are so widely distributed that the organisms do not need to be precise site selectors. Their larvae are structurally simple, without eyes and tails, and descend to the bottom in response to gravity, attach indiscriminately, and metamorphose. The existence of tailless larvae among ascidians that do not require preferred sites is additional confirmation that site selection is the principal adaptation for the typical tadpole larva.

Only the larvae possess chordate features; the adults, adapted to a sessile existence, lose all larval locomotory organs and thereby give up all traces of chordate relationships. The ascidian tadpole may have evolved to the vertebrate body plan by repressing metamorphosis, permitting larvae to become sexually mature, a condition known as *neotony*. Many species, such as salamanders, are neotonous adults; they are sexually mature larval forms.

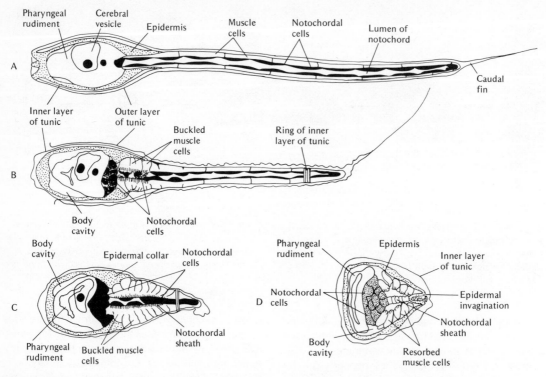

**Figure 19-8** *Dorsal views of the tadpole larva of the ascidian* Boltenia villosa *and its metamorphosis. (A) Mature larva. (B) Two to three minutes after the beginning of metamorphosis. Note contraction of tail due to contraction of micro-filaments in epidermal cells in tail causing compression and buckling of muscle cells. (C) Ten to twelve minutes after beginning of metamorphosis. Tail almost fully contracted. (D) Twenty-nine to thirty minutes after beginning of meta-morphosis. Almost complete; muscle being resorbed as are notochordal cells. (From R. Cloney,* Am. Zool. *1:67–87, 1961.)*

### Larva and symbiosis

Virtually all organisms are themselves hosts for other organisms living on or within their bodies. Living together in this way is called *symbiosis*. It is often mutually beneficial *(commensalism)*, but in many instances it leads to *parasitism*, where the host is damaged by the relationship. A symbiont, in order to survive, must invade new hosts; to do so, it must solve the problems of reaching and penetrating a new host to establish itself. Most of these problems have been solved by larval stages rather than by adults; the egg, embryo, and larva are adapted to host reinfection. Symbiont life cycles often involve two hosts, a *definitive* host occupied by the sexually mature adult stage and one or several *intermediate* hosts, in which larval stages complete development before they reinfect the definitive hosts. Intermediate hosts frequently

serve as vectors, carrying a stage from one host to another. The routes of entry are through the host's mouth by food or drink, or by penetration through the host's skin. Again, developing embryos and larvae have acquired special adaptations to facilitate ingestion by a host or attachment and penetration of host skin.

A simple nematode life cycle involving only one host, the human, illustrates larval specializations for host invasion. Hookworm is a common nematode infection in the southern United States. The sexually mature adults live in the small intestine of the host, sucking blood from the mucosal lining. Fertilized eggs leave the host in the feces where they continue to develop, hatching as immature larvae. These larvae resemble the adult morphologically but lack sex organs. They are free-living for awhile, feeding on feces or other decaying organic material in the soil, molting twice to a third larval stage. Only this stage is infective for a new host and survives in the soil or on a blade of grass for many weeks without any further development. When this larva comes into contact with skin of the definitive host, it sheds its larval skin, attaches, and bores into the subcutaneous tissues by a digestive process. In the tissues, the minute worms enter blood vessels and are carried to the heart, where they are pumped into the lungs. Larvae are entrapped in the capillary bed of the lung and break into alveoli to be carried with mucus to the pharynx, where they are swallowed. The young worms feed and grow in the intestine, molting until they reach sexual maturity and the cycle is repeated.

A free-living larval stage is necessary in this cycle, in that it facilitates invasion of new hosts. In a sense, the hookworm life cycle recapitulates evolution of parasitism in this group. Free-living nematodes living in feces are subjected to frequent crises as feces dry up, and they must find new sources of food. During their evolution a selective advantage was acquired, as adaptations for attachment and penetration of host tissues arose. Once they were established in a favorable environment (host tissues), growth under optimal conditions of food, temperature, and moisture was possible. To complete the cycle, the larvae had to enter the digestive system with easy access to the exterior by discharging eggs in the feces to reinfect new hosts.

More complicated life cycles of trematode parasites, involving two hosts, illustrate how larvae became efficient reproductive individuals by asexual methods. Larval stages with reproductive capabilities vastly increase the number of larvae that can infect new hosts. This is made possible by the retention in larvae of germinal cells that are primordia for development of new individuals.

Adult liver flukes (*Fasciola hepatica*) live in the bile passages of the liver of grazing mammals such as sheep (Figure 19-9). The adults are hermaphroditic; as eggs pass down the oviduct, they are fertilized by sperm stored in a receptaculum, and are subsequently encased in a tough protective shell. Development begins as eggs pass out of the host in the feces. A simple *miracidium* larva is produced, consisting of a ciliated epithelium surrounding a solid mass of endoderm. Embedded in the endoderm are germinal cells that do not differentiate; these are the progenitors of the next developmental stages. The larva does not feed; its intestine is only slightly de-

veloped, but it has an eye spot and ganglion plus a protuberance at the anterior end to penetrate a snail secondary host. This larva is adapted for site selection and must encounter the foot or mantle of a specific snail if it is to survive and continue the developmental cycle. There is considerable larval wastage at this stage, but this is compensated for by the few that successfully invade secondary hosts.

The larvae seem to be attracted to mucous secretions of the snail and attach to the surface epithelium. Penetration is a process of "boring" by the larva, assisted by enzymatic digestion of host epithelium. The larva sheds its ciliated epithelium as the central endodermal mass plus germinal cells enter the softened tissues beneath the skin.

The larva transforms into a resting *sporocyst* that may coil and branch irregularly. Each germinal cell undergoes a typical embryogenesis, forming a morula, blastula, and so on, developing into another larval type, the *redia*, a process known as *polyembryony*. Rediae resemble the adult body plan with mouth, short gut, and disc at the anterior end. Presumably they feed, but this is not their primary role. They too have sequestered large numbers of germinal cells within the central tissue mass and each of these develop (polyembryony again) into second-generation rediae, after the first rediae leave the sporocyst. Usually two, and sometimes three, generations of rediae are produced in this way. The principal function of rediae is to increase the population of larvae within snail tissue so that even a single invasive miracidium larva can generate hundreds of additional redia larvae.

Germinal cells in second-generation rediae shift their developmental program into another channel and produce, by this same process of polyembryony, a third larval type, the *cercaria*. This larva has many more adult features such as suckers, mouth, and intestine, but it retains a larval motile tail. Cercariae leave the snail and become free-living, the tail serving as a locomotory organ, again useful in site selection. Cercariae must locate swamp grass, attach, and metamorphose into an encysted *metacercaria* by shedding their tails and secreting a tough protective cyst. The metacercaria is ingested by a feeding herbivore, and the young worms leave the cyst in the intestine, penetrate the intestinal wall, and pass to the peritoneal surface of the liver. They enter the liver and migrate to the bile ducts, where they grow into sexual maturity.

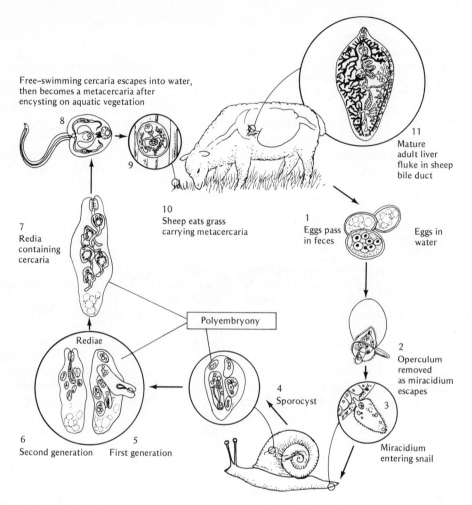

Free-swimming cercaria escapes into water, then becomes a metacercaria after encysting on aquatic vegetation

8

9

11
Mature
adult liver
fluke in sheep
bile duct

7
Redia
containing
cercaria

10
Sheep eats grass
carrying metacercaria

1
Eggs pass
in feces

Eggs in
water

Polyembryony

Rediae

2
Operculum
removed
as miracidium
escapes

3

4
Sporocyst

6
Second generation

5
First generation

Miracidium
entering snail

**Figure 19-9**   *The life cycle of the liver fluke* Fasciola hepatica. *Miracidium larva is adapted for site selection and penetration of secondary host, the snail. Polyembryony within successive generations in snail tissue results in large numbers of secondary larval individuals gradually assuming adult characters: sporocyst, rediae, and cercaria. The latter (with many adult characters) escapes from the snail and seeks a blade of grass to encyst. (From C. Read,* Animal Parasitism, *© 1972, p. 59. Reprinted by permission of Prentice-Hall, Inc., Englewood Cliffs, New Jersey.)*

We see that the challenges of a parasitic existence are solved by the larvae. They have been adapted for dispersal, habitat selection (secondary hosts), and reproductive amplification, insuring that large numbers of infective larvae emerge after a single successful invasion of an intermediate host. Retention of germinal cells by larvae and polyembryony are not unique to parasites, however; polyembryony is found in free-living flatworms, coelenterates, and even ascidians, phyla exhibiting

asexual reproduction. Totipotent embryonic cells separate, each acting as a potential germinal cell able to complete a developmental program.

Other peculiar adaptations also appear among trematode larvae. In a trematode symbiont of birds (*Leucochloridium*), cercariae develop within terrestrial snails but remain within the sporocyst. The sporocyst grows out of the snail surface, decorated with brightly colored bands that attract birds, and thereby insures ingestion by the definitive host.

Here the larvae have developed attractive devices to lure a host to infection!

Parasites acquire infectivity only at specific stages in the cycle. Cercariae dissected from snails cannot invade a secondary or definitive host; they must complete maturation beforehand. The larvae respond to signals produced only by a host. Since many nematode parasites lack the hormone cues essential for development, they must rely on host hormones. In this sense, part of the developmental program of the parasite is written within the genome of the host.

## Larva in evolution

The extended lives of most larvae subject them to selection pressures that lead to evolutionary novelties. Moreover, all adaptations selected in the embryo or larva may also lead to morphological changes in the adult. Acquisition of novelties in ontogenetic patterns accounts for evolutionary change. For instance, neotonous transformation of ascidian larvae, as we have seen, could account for the origin of the vertebrates, a phylogenetic leap of enormous magnitude. Contrary to Haeckel's law of recapitulation, that ontogeny recapitulates phylogeny, no vertebrate ontogeny displays any stage resembling an adult ascidian. Instead, resemblances to the tadpole larva are common. We cannot assume that a characteristic that is embryonic in the ontogeny of an organism reflects some primitive ancestral type. Possession of embryonic characteristics may result from secondary retention of features that are themselves of selective advantage during development yet unrelated to any ancestral type. Accordingly embryos or larvae may differ far more than the corresponding adults (Figure 19-10). We find, therefore, that Haeckel's law of recapitulation, that ontogeny of an organism recapitulates its phylogenetic history, is generally incorrect.

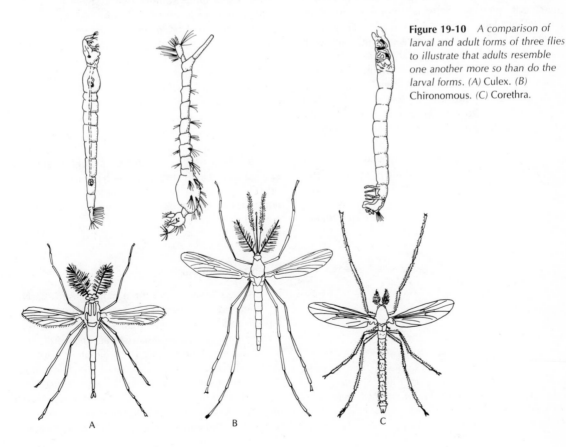

**Figure 19-10** *A comparison of larval and adult forms of three flies to illustrate that adults resemble one another more so than do the larval forms. (A)* Culex. *(B)* Chironomous. *(C)* Corethra.

A                    B                    C

Genetic programming of development involves control of rates of morphogenetic processes. Acceleration or deceleration of these processes by mutation modifies morphological patterns, and we assume that such mutations have accumulated in the genomes of all species to adapt to specific life styles. If such changes result in disproportionate delays in development of somatic tissues while gonads develop at normal rates, then neotony results; larval characteristics are retained as the organism becomes sexually mature.

Other examples of neotony are not as dramatic but are of equivalent validity in evolution of smaller taxonomic groups. The origin of insects from myriapods (millipedes) is an example. Adult millipedes have a head and trunk made up of many body segments, each with two pair of identical jointed walking legs. Insects, however, have a reduced number of body segments organized as head, thorax, and abdomen, with the thorax containing all walking legs and wings. In fact, the larval form of certain myriapods, such as *Glomeris*, resembles an insect at hatching (Figure 19-11); it has a head with six segments, a "thorax" of three segments, each with a pair of walking legs, and nine abdominal segments, each with a pair of reduced nonfunctional appendages. These larvae differ from insects only in the possession of appendages on the abdominal segments and undergo a series of metamorphic molts before they develop the many-segmented body of the adult millipede.

If retardation of abdominal appendages was to persist in the larvae and the addition of body segments was inhibited by mutations, then the body plan would resemble that of some living, wingless insects with vestigial appendages on abdominal

segments. The leap to insects would occur after neotonous transitions of such larvae.

Such phylogenetic changes are contrary to the now-discarded Haeckel concept of recapitulation. The fact that an adult descendant morphologically resembles the young stage of an ancestor rather than the adult is inconsistent with recapitulation. Neotony can be achieved by inhibition of adult metamorphosis, while genital development continues. Factors regulating larval-adult transformations probably play a critical role in these evolutionary adaptations.

## METAMORPHOSIS

Most information about metamorphosis comes from insects and amphibians. Insect larvae nourish small groups of adult tissue primordia, the imaginal discs. In response to a common hormonal signal, most larval cells degenerate as presumptive adult cells differentiate. Only some larval systems are transformed into adult structures. Two separate genetic programs are involved. In amphibian metamorphosis, most larval tissues are reprogrammed and express adult genes. Although certain larval tissues (the larval tail) degenerate, the bulk of the adult organism is derived from preexisting larval structures.

There are many similarities of hormone control of metamorphosis in invertebrates and vertebrates. Both are triggered by environmental signals such as day length or a change in temperature. The signal is detected by sensory cells and transmitted to neurosecretory cells in the brain, a switch point connecting neural controls with the endocrine system. The hormones of the endocrine system are discharged and regulate whole sets of genes that initiate metamorphosis.

### Insect metamorphosis

Insects are classified into three major groups based on metamorphic patterns. *Ametabolous* insects, represented by orders of wingless insects (silverfish), undergo no metamorphosis. Larvae and adults are identical; a hatched larva is a miniature adult in every way except for sex organs, which develop later. *Hemimetabolous* insects, represented by dragonflies, cockroaches, locusts, and the like, undergo incomplete metamorphosis in which larvae hatch into miniature, wingless adults lacking geni-

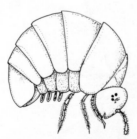

**Figure 19-11** *Larval form of the myriapod* Glomeris *soon after hatching, showing its great resemblance to an insect.*

talia and sex organs. The larvae, known as nymphs, grow through a series of nymphal molts, gradually acquiring wings. Genitalia and sex organs develop at the final so-called adult molt. Finally, the highest orders of insects are *holometabolous,* displaying complete metamorphosis; they include beetles (Coleoptera), moths, and butterflies (Lepidoptera), bees, and flies (Diptera). Larvae hatch without wings; wings develop internally rather than externally in this group. The larva in no way resembles the adult, exhibits a different life style, and is adapted to different ecological niches. Complete metamorphosis calls for three sets of genetic instructions: larval, pupal, and adult. Larvae grow through a series of molting periods and transform into quiescent but morphologically distinct pupae. During the pupal stage, larval gene sets are turned off and pupal genes are expressed. Later, all larval and pupal tissues degenerate while adult primordia read the final set of genetic instructions and differentiate into the adult body plan. The fully formed adult insect emerges after the last pupal-adult molt.

The larval developmental program is precocious, insuring rapid emergence of a functional feeding organism at the end of embryogenesis. Pupal and adult programs depend upon nutrients furnished by the larva. All programs, however, are determined early in embryogenesis, usually when cleavage nuclei migrate to the periphery and form a cellular blastoderm. Though larval and adult tissues are determined at the same time, they develop at different rates. Larval cells differentiate first, but adult primordia persist as determined imaginal discs throughout larval life until they receive the metamorphic signal to differentiate.

How much developmental programming is intrinsic and genetically determined, and how much reflects the hormonal and physiological state of the developing larva?

## Neuroendocrine control of insect metamorphosis

Insect growth and morphogenesis are coupled to the rhythm of the molting cycle; that is, neuroendocrine control of molting underlies all metamorphic events. Molting is an essential feature of insect growth because they have an exoskeleton of limited distensibility; hence growth and morphogenesis occur during intermolt periods.

The cuticle synthesized at each molt reveals all stage-specific external features of the organism. Larva, pupa, and adult cuticles are composed of different layers, each with a visibly different pattern of bristles and hairs. Inasmuch as the epidermis synthesizes the cuticle, it is the principal target tissue of hormone action during molting and reveals its phenotypic response in an ordered array of integumentary derivatives.

An epidermal response to the hormone may include cell hypertrophy coupled to endomitosis (Diptera), cell proliferation, changes in cell shape, and synthesis of a new cuticle. All changes involve synthesis of many new proteins, which means that molting hormones affect the expression of many genes.

Larval molting is controlled by neuroendocrine hormones derived from three major organs: neurosecretory cells of the brain which control two endocrine glands, the *corpora allata* and the *ecdysal* or *prothoracic* gland (Figure 19-12). Brain hormones stimulate corpora allata to synthesize and secrete *juvenile hormone* (JH), an essential larval growth hormone, while prothoracic glands are stimulated to secrete *ecdysone,* the molting hormone.

How do we know the brain is involved in metamorphosis? C. Williams removed the entire brain and other neural elements from *Cecropia* silkworm pupae and found that such aneural preparations survived for many months but did not develop or metamorphose. Implantation of a brain into the thorax or abdomen of this pupa induced metamorphosis 2 to 3 weeks later. The pupa undergoes almost complete development; wings, antennae, limbs, and a typical adult integument are produced. In addition, the abdomen, though devoid of muscles, contains the fully developed gonads and accessory sex organs. A brain, unconnected with the rest of the nervous system, can induce complete metamorphosis except for muscle differentiation. The latter obviously depends upon direct (trophic) influences from the nervous system, perhaps from those motor nerves that normally enervate the muscles.

Though the brain hormone has not been isolated, it is probably synthesized in neurosecretory cells receiving visual signals of photoperiodic changes. Periodic release of brain hormone sets off a sequence of metamorphic events mediated by JH and ecdysone. Juvenile hormone is present in high concentrations throughout larval life, while ecdysone concentration fluctuates, reaching its high point just before the molt. Secretion of ecdysone from prothoracic glands is dependent upon stimulation by brain hormone. In a background of high JH,

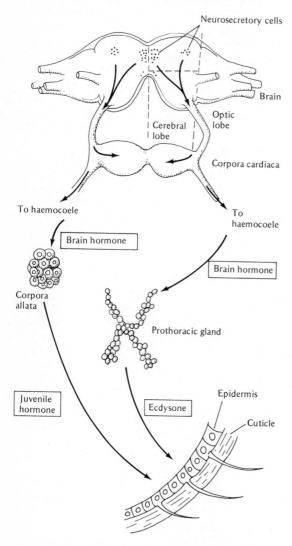

**Figure 19-12** *Neuroendocrine control of molting in insects. Neurosecretory cells in the brain secrete brain hormone which collects in a neurohaemal organ, the corpora cardiaca. The hormone is released into the haemocoele where it stimulates the prothoracic gland in the thorax to secrete ecdysone. In the larvae, juvenile hormone is continuously produced by another set of glands, the corpora allata, which act together on the larval cuticle to stimulate epidermal cell growth and secretion of a new larval cuticle.*

a larva molts into the next larval stage. Ecdysone stimulates epidermal cells to secrete a larval cuticulin pattern. For reasons still not understood (possibly neurohumoral inhibition), corpora allata secretion of JH late in larval life diminishes. As a result, circulating JH concentration declines and at the next molt, epidermis responds to ecdysone by secreting a pupal instead of a larval cuticle. The hormone background now favors expression of pupal and not larval genes.

Complete absence of JH results in a pupal to adult molt. Ecdysone alone induces imaginal disc differentiation, and the epidermis of each disc secretes adult cuticle. By removing corpora allata and JH from the first molting stage, premature metamorphosis in a larva is evoked and a miniature pupa results. We can also interfere with the pupal-to-adult molt by injecting JH into a pupa just before molting; the pupa molts into another pupa and will continue to secrete pupal cuticle as long as JH is present. Juvenile hormone acts as an inhibitor of metamorphosis; only in its absence does a larva transform into an adult.

Because of its role in secreting cuticle, the larval epidermis seems to dominate the metamorphic process. All metamorphic changes are recorded in the cuticle, with its specific patterns of hairs, scales, and bristles. A piece of larval epidermis can be made to express three different programs of cuticle secretion by changing the hormonal background. For example, pieces of larval epidermis from *Ephestia* moths may be transplanted into the abdomen of another larva (Figure 19-13). These fragments develop into inverted epithelial vesicles with epidermis on the outside secreting cuticle into the cavity. The vesicle epidermis proliferates and synthesizes cuticle in response to the changing hormone background of the host. If the host undergoes a larval molt, the vesicle secretes larval cuticle; if the host is metamorphosing into an adult, the vesicle secretes adult cuticle. A vesicle that has completed several larval molts and is prepared to metamorphose into pupa can be induced to molt as a larva again by transplanting it to a young larva. Moreover, the molting sequence can be reversed. A piece of pupal integument placed in a young larva will secrete larval cuticle beneath the original pupal cuticle when the larva molts. A pupal integument already forming adult cuticle with scales can be reversed to pupal integument if implanted into a mature larva undergoing

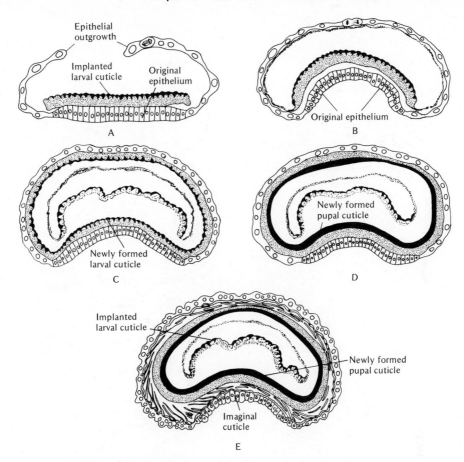

**Figure 19-13**  *Development of epidermal transplant in the moth* Ephestia. *(A, B) Larval integument in abdomen of a final instar transformed into a hollow epithelial vesicle. (C) At the molt, the epidermis secretes a new cuticle and the old larval cuticle lifts off. (D) A piece of integument implanted in larval abdomen undergoing a larval to pupal molt synthesizes pupal cuticle because the ratio of juvenile hormone to ecdysone has decreased. (E) Same type of implant after complete metamorphosis, showing imaginal disc (adult) cuticle inside pupal cuticle and displaced larval cuticle. (From A. Kühn,* Lectures on Developmental Physiology. *New York: Springer-Verlag, 1971.)*

a pupal molt. Even while epidermis is synthesizing adult cuticle, it is labile, competent to respond to hormonal signals and differentiate new cuticle patterns. Its regional nature, however, is determined; it produces a cuticular pattern corresponding to the larval or pupal region from which it was derived.

Metamorphosis in hemimetabolous insects is not as sudden. At each nymphal molt (five in the blood-sucking bug *Rhodnius*), the larva acquires more adult traits such as wings (Figure 19-14). Gradual maturation at each nymphal stage is accompanied by a decline in JH titer, until the JH titer

disappears and the last nymph becomes adult.

An additional larval molt can be induced in a fifth-stage larva by implanting corpora allata. Because the titer of JH is high, the last molt is a larval molt and a sixth-stage larva is produced. When this larva metamorphoses it transforms into a giant adult.

We see that a balance between JH and ecdysone determines the specific pattern of cuticle secretion by epidermal cells. The pupal-to-adult molt takes place in the absence of JH and the emergent adult stops growing. Its prothoracic glands degenerate and ecdysone is no longer synthesized. An adult can

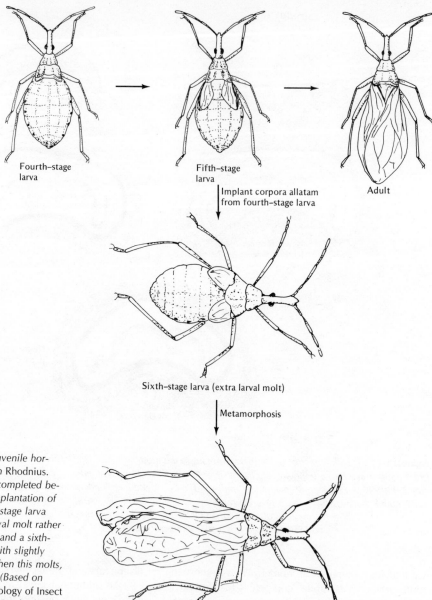

**Figure 19-14**   *Effect of juvenile hormone on larval molting in* Rhodnius. *Five nymphal stages are completed before the adult appears. Implantation of corpora allata into a fifth-stage larva induces an additional larval molt rather than the final adult molt, and a sixth-stage larva is produced with slightly larger wing primordia. When this molts, it produces a giant adult. (Based on L. B. Wigglesworth,* Physiology of Insect Metamorphosis. *London: Cambridge University Press, 1954.)*

be made to molt, however, if ecdysone is injected; the old cuticle is shed and a new adult cuticle is synthesized.

Both JH and ecdysone have been isolated and synthesized. Juvenile hormone is a turpenoid related to many plant turpenoids, some of which exhibit

properties similar to those in insect molting (Figure 19-15). For example, a bug nymph maintained in the laboratory on a bed of filter paper did not molt into an adult at its last stage; instead the nymph went through additional nymphal molts, increasing in size at each stage. It turned out that metamorphosis was prevented by the turpenoids in the paper, acting in place of JH. Any newspaper, or wood chips from certain trees, acted in the same way. This sharing of metabolites by plants and insects is not surprising inasmuch as they evolved together when both invaded land.

Ecdysone turns out to be a steroid, related to most vertebrate steroid hormones. Even urine from pregnant humans exhibits some ecdysone activity. Synthetic analogs of ecdysone may promote or inhibit molting and metamorphosis. Synthesis of insect hormones controlling growth and morphogenesis has opened up new opportunities for biological control of insect populations. Large-scale application of juvenile hormone analogs could prevent metamorphosis of insects that are pests when adult. Conversely, larval pests could be precociously induced to metamorphosis into innocuous adults by application of JH antimetabolites, along with ecdysone. Such agents could be synthesized with high specificity so that only certain pest species would be affected. Moreover, these agents would not have the disturbing side-effects on mammals and birds that chemical pesticides such as DDT have.

## Metamorphic responses of larval tissues

Each tissue of the larva responds differently to metamorphic signals. Many larval tissues degenerate; some survive intact in the adult while imaginal disc primordia differentiate adult tissues. Larval genes are inactivated as adult genes are expressed.

Certain organ systems, such as the digestive system, have to adapt to new feeding habits after metamorphosis. A larva with chewing mouth parts, feeding on leaves, becomes an adult with sucking mouth parts, feeding on blood or nectar. The adult organs develop from imaginal discs in the head as larval mouth parts degenerate. The alimentary tract also adjusts to the new demands; larval crop and salivary glands degenerate and a set of unrelated adult organs differentiate in their place.

Larval muscles also degenerate; the development of wings in the thorax is accompanied by major reshuffling of thorax muscles to accommodate flight muscles. In some insects, however, all larval abdominal muscles are carried over into the adult. In the honey bee, for example, only 15 percent of the larval muscles degenerate; the remaining 85 percent are altered slightly to fit into the adult body plan.

The mechanism of muscle breakdown is not understood because some muscles degenerate while their neighbors may remain intact and pass on into the adult. This process of differential cell death is coordinated by neurohumoral mechanisms. The initial signal is hormonal; that is, intersegmental muscles exposed to ecdysone break down in the emergent adult. It appears that the hormone sensi-

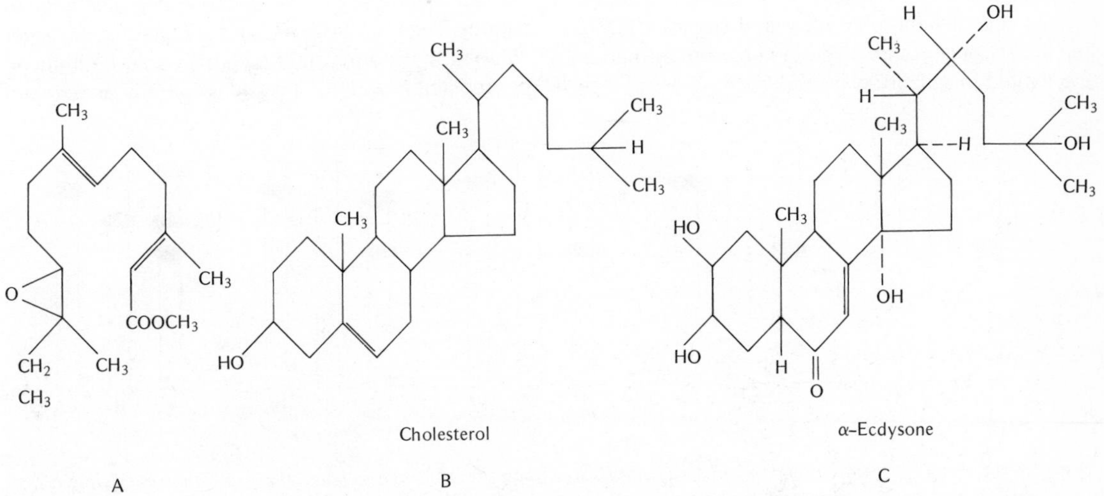

Cholesterol

α–Ecdysone

A                    B                    C

**Figure 19-15** (A) Structural formula for juvenile hormone extracted from male Cecropia moths. (B) Structure of cholesterol, the parent compound of ecdysone. (C) Ecdysone.

tizes the muscle to respond to a neural trigger several days later. Lysosomes break down in muscle cells, release their hydrolytic enzymes, and the tissue degenerates.

## Gene action in metamorphosis

As we have seen (Chapter 17), changes in ecdysone titer at molting and metamorphosis induce chromosome puffing patterns in larval tissue, visible signs of genes in action. The specific puff sequences induced prior to metamorphosis correlate with the physiological changes underlying puparium formation. No other treatment elicits the same puffing sequence, suggesting that these gene activation patterns are causally related to metamorphosis. Since a puff is a site of specific RNA synthesis, the hormone may activate genes coding for all proteins necessary for the larval-adult tranformation.

Puffing follows a chronological pattern that correlates with metamorphic events. In *Chironomous tentans*, the first cytological changes in puffing pattern of salivary glands at a larval-larval or larval-pupal molt are the appearance of a specific puff on chromosome I at site 18C (Figure 19-16).

Immediately following this puff, a new puff appears at 2B on chromosome IV, and 14A on chromosome I. Later, and only in larval-pupal molts, puffs appear at four additional sites that have never been active in development, while the first set of puffs regresses. This sequence of puffing is found only in the metamorphic molt and suggests that a specific set of genes involved in puparium formation is activated in salivary glands.

The molting hormone ecdysone is responsible for this puffing sequence. Ulrich Clever injected ecdysone into the fourth instar of *Chironomous* and noticed an immediate induction of a puff at site I-18C followed by a second puff at site IV-2B. These puffs would appear at low titers of ecdysone ($10^{-6}$ to $10^{-7}$ M). Each puff seems to respond to its own threshold concentration of ecdysone and this is gradually attained as ecdysone is secreted into the hemolymph of the prepupa. The size of the puff also depends on hormone dose.

Ecdysone induction of puffs at I-18C and IV-2B in *Chironomous* is inhibited by actinomycin D. As the inhibition is released and RNA synthesis recommences, these puffs reappear in normal sequence followed later by the four late puffs. If puromycin is injected 1 hour before ecdysone, then the first two puffs appear on time but later puffs fail to develop. Clever concluded that the activity of the first puffs is essential for activation of puffs at the later sites. Presumably, the RNA product of the first puffs must be translated to induce late puffing, a type of cascading induction (see Chapter 17). Such conclusions are premature, however, in view of the difficulties in interpreting most inhibitor experiments.

Similar difficulties arise when one tries to explain how ecdysone initiates puffing. Some claim the hormone acts within the nucleus, perhaps on the DNA template by derepressing gene sites. Others believe the hormone acts on the cell surface and nuclear membranes and increases permeability to cations such as sodium and potassium. Kroeger claimed that normal puff patterns could be induced by replacing sodium by potassium in an explant

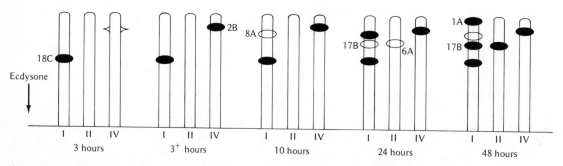

**Figure 19-16**    *Puffing program preceding metamorphosis in* Chironomous. *Three salivary gland chromosomes shown (I, II, and IV). Three hours after injection of ecdysone, a puff appears in chromosome I at 18C in salivary-gland chromosomes. Subsequent puffs appear in sequence shown. One puff, 8A on chromosome I, regresses at 48 hours. (Based on U. Clever, Am. Zool. 6:33–41, 1966.)*

medium. This, too, is in question since ion-induced puffs do not follow the normal sequence produced by ecdysone. Besides, even if ions induce puffing it by no means implies that ecdysone acts on cell surfaces, regulating ion flow. Others have failed to repeat Kroeger's observations, claiming the effects were really nonspecific osmotic changes. This confusion demonstrates that the site of ecdysone action in puffing is still unresolved.

If puffing is evidence of gene activation in metamorphosis, we should be able to correlate changes in physiological function of the salivary gland with puffing patterns. But this, too, has been difficult to do.

The salivary gland secretion, a gluelike mucoprotein, is stored in the lumen until puparium formation, until it is secreted to anchor the puparium to a substrate. The mucoprotein, manufactured in the Golgi, accumulates in cytoplasmic vesicles before secretion. Many salivary gland enzymes also exhibit developmental changes, increasing in amount before pupation. Presumably these are involved in pupation or may even contribute to salivary gland histolysis which takes place at metamorphosis.

Many puffs appearing in the prepupa are also found in larval-larval molts, which means that specificity of puffing patterns is not well-established. It is also uncertain whether secretions of the salivary gland are concerned with the gland's own functional activity. Some have proposed that secretory products are exported for use by differentiating imaginal disc tissues, since salivary glands do histolyze at metamorphosis. What were initially thought to be visible demonstrations of genes in action during metamorphosis have turned out to be a complicated sequence of physiological and genetic changes that are still not clear.

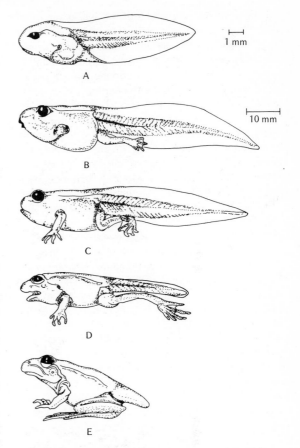

**Figure 19-17** *Stage of anuran metamorphosis (Rana pipiens). (A) Premetamorphic tadpole. (B) Prometamorphic tadpole (growth of hindlimbs). (C) Onset of metamorphic climax (eruption of forelimbs, degeneration of tail). (D, E) Later climax stages showing gradual appearance of frog characteristics. (From E. Witschi,* Development of Vertebrates. *Philadelphia: Saunders, 1956.)*

## MORPHOGENETIC CHANGES DURING AMPHIBIAN METAMORPHOSIS

Many amphibians live a dual existence; they spend the early part of their lives as aquatic larvae and their adult life on land. The transition from aquatic to terrestrial living calls for major physiological adaptations. Basic patterns of feeding and locomotion must be modified. The animal experiences a morphological upheaval, changing from a swimming animal with motile tail and gills to a walking tetrapod living

in air and respiring through lungs (Figure 19-17). Almost every organ system participates in this metamorphic transformation as many larval cells shed one phenotype and take on a new one.

Like insects, amphibian metamorphosis is under neuroendocrine control, involving neurosecretory cells in the brain (the hypothalamus) and two endocrine glands, the pituitary and the thyroid. The trigger to metamorphosis may be an environmental signal impinging on the larval brain through the nervous system, or there may be an endogenous "clock" in the hypothalamus that initiates the process. The exact mechanisms are still obscure, but we assume that neural signals are received by or generated in the hypothalamus, the center for visceral control. Neurosecretory cells in the hypothalamus are stimulated to synthesize a hypothalamic hormone that resembles the mammalian thyrotropin-releasing factor (TRF) which enters the pituitary portal circulation. The anterior pituitary, the target tissue for this brain hormone, is activated and releases thyrotropic-stimulating hormone (TSH), which stimulates the dormant thyroid gland to secrete two thyroid hormones, L-thyroxin ($T_4$ or 3,5,3',5'-tetraiodothyronine) and 3,3'5-triiodothyronine or $T_3$ (see Figure 19-20). The hormones enter the circulation and gradually increase in titer.

The resemblance to insect metamorphosis even extends to the presence in amphibians of a "juvenile hormone." Prolactin, a mammalian hormone that controls milk production in lactating females, is also found in amphibians, where it has a growth-promoting role, among others. It inhibits or delays thyroxin-induced metamorphosis, acting very much like juvenile hormone in sustaining larval characteristics and promoting their growth.

The sequence of events during anuran metamorphosis is divided into three periods (Figure 19-18). The first, a *premetamorphic* period, is marked by tadpole growth with little morphogenetic change

except for initiation of hindlimb growth. During the second *prometamorphic* phase, the body grows at a reduced rate as hindlimbs grow. The ratio of hindlimb length to body length increases rapidly. The third and most dramatic phase, *metamorphic climax*, is marked by cessation of growth accompanied by rapid changes in limbs, tail, head, and internal organs.

Feedback interactions between thyroid, pituitary, and hypothalamus seem to regulate the temporal sequence of events. In a model proposed by Etkin (Figure 19-19), the premetamorphic phase is marked by high titers of prolactin which favor growth, and very low but detectable levels of thyroid hormone. Tissues are sensitive to thyroid hormone since they respond precociously if a tadpole is treated with thyroxin; cells must have thyroxin receptors at this time. The pituitary gland is also able to secrete TSH if properly stimulated during this period. This means that the onset of metamorphosis is limited by low levels of thyroid hormone. The pituitary has not received the hypothalamic signal to secrete thyroid-stimulating hormone. The hypothalamus is immature during the premetamorphic phase but seems to become active just before the prometamorphic phase. Activation by the hypothalamus also depends on development of the portal circulation connecting hypothalamus and pituitary

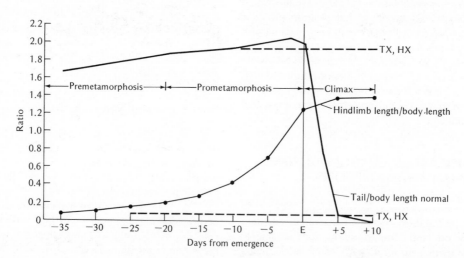

**Figure 19-18** *Pattern of metamorphosis in* Rana pipiens. *Normal animals raised at 23°C, shown by solid line. Comparable data for thyroidectomized (TX) and hypophysectomized animals (HX) shown by broken lines. E = emergence of first forelimbs. (From W. Etkin,* Metamorphosis, *ed. W. Etkin and L. I. Gilbert. New York: Appleton-Century-Crofts, 1968, pp. 313–348.)*

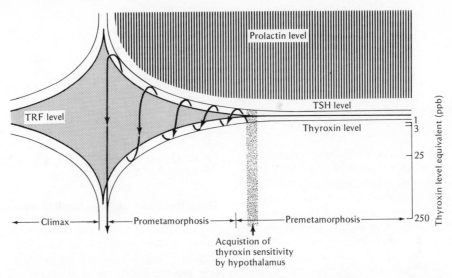

**Figure 19-19**  *Etkin model illustrating the interaction of endocrine factors in determining the time and pattern of anuran metamorphosis. See text for details. (From W. Etkin and A. G. Gona, J. Exp. Zool. 165:249–258, 1967.)*

gland. Gradually, the hypothalamus becomes sensitive to low levels of thyroxin in the circulation and probably begins to secrete thyroid-releasing factor (TRF) which stimulates the pituitary gland to secrete TSH. The thyroid is activated, more thyroid hormone is secreted, and the prometamorphic phase is initiated. At the same time, a prolactin inhibitor is secreted by the hypothalamus, prolactin concentration decreases, and growth declines as metamorphic changes occur.

At metamorphic climax, there is a tenfold increase in thyroid hormone concentration and this initiates the rapid changes that follow. The tadpole tail is quickly resorbed as are the horny teeth and gills. The eyes shift from a lateral position to almost a dorsal-lateral position on the head. Larval pigment in the retina is replaced by adult visual pigment, rhodopsin, better adapted to a terrestrial mode of vision. Meanwhile, other tissues undergo constructive changes. Internal larval tissues such as liver and kidney acquire adult physiological functions. Whole new sets of enzymes appear in the liver to convert ammonia, the principal excretory product of the tadpole, into urea, the adult product. Adult hemoglobins appear in the blood to accommodate to terrestrial respiration. Each tissue responds differently to the same hormonal signal; its response pattern is programmed into its genome.

### The thyroid hormone

The significance of the thyroid in metamorphosis was discovered in 1912 when Gudernatsch showed that tadpoles fed mammalian thyroid exhibited precocious metamorphosis. Proof of the role of the animal's own thyroid in metamorphosis has come from experiments in which the presumptive thyroid gland is removed from the developing embryo; such thyroidectomized animals fail to undergo metamorphosis and grow indefinitely as tadpoles. Exposure of such giant tadpoles to thyroid hormone induces metamorphosis. Furthermore, a tadpole's thyroid gland undergoes developmental changes that correlate with metamorphic events. The thyroxin colloid begins to accumulate in the follicles prior to metamorphosis, and the follicular epithelium thickens at the time of metamorphic climax.

The active compound of the thyroid gland is a hormone, thyroglobulin, a complex of small iodine-containing molecules, $T_4$ and $T_3$, covalently linked to a protein (Figure 19-20). Originally, the iodine was thought to be the active component of these molecules because it could induce metamorphosis when given to thyroidectomized tadpoles. This turned out to be invalid when it was shown that iodine may be replaced by any molecules that give the proper structural configuration to two tyrosine

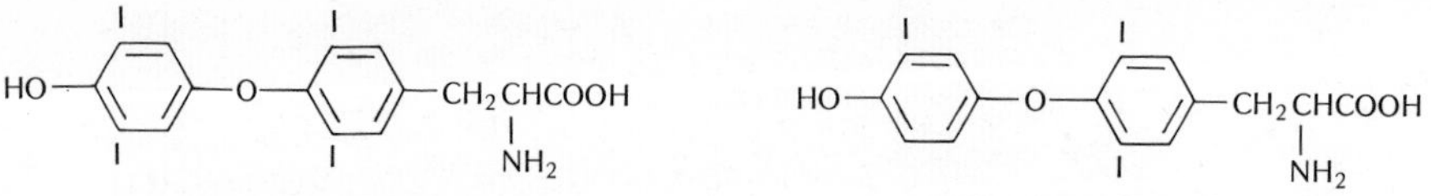

HO—⬡—O—⬡—CH₂CHCOOH
                       NH₂

$$HO{-}⬡{-}O{-}⬡{-}CH_2CHCOOH$$
(with $NH_2$)

(3, 5, 3', 5' -tetraiodothyronine,
thyroxine = $T_4$)

(3, 5, 3' -triiodothyronine = $T_3$)

**Figure 19-20** *Thyroid hormones $T_3$ and $T_4$.*

molecules. Iodine reacts readily with tyrosine residues in many proteins to produce small amounts of thyroxinlike hormone. Many compounds have been synthesized (some without iodine) with far greater activity than the natural product, probably because they penetrate cells more rapidly or stabilize the proper structural configuration to react with cell receptors.

Other compounds which interfere with specific steps of thyroid function inhibit metamorphosis. For instance, thiourea prevents metamorphosis when given to tadpoles. It does not, however, inhibit metamorphosis if given simultaneously with thyroxin to thyroidectomized tadpoles, which means it acts directly on the thyroid gland without antagonizing hormone action at the target site. It is not an anti-metamorphosis agent; rather, it affects thyroid production of hormone.

## Tissue specificity and biochemical response

The responses of tissues and organs are highly local, since implants of thyroxin pellets elicit a response only from adjacent tissues. An implant in the tail only induces resorption in the pellet site, while an implant in the leg promotes bone development. Specificity of response is underscored by the behavior of skeletal muscle. Leg muscles are stimulated to grow and differentiate in a metamorphosing tadpole while skeletal muscles in the tail degenerate. Tail and skeletal muscle, though histotypically identical, exhibit opposite responses to the same

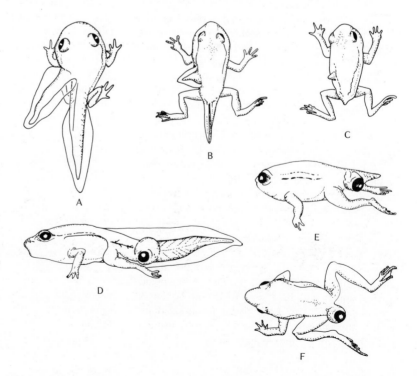

**Figure 19-21** *Organ specificity of metamorphic responses in tadpoles. (A–C) Tail tip grafted to the flank undergoes degeneration along with host tail. (D–F) Eye cup transplanted to the tail remains unaffected by regressing tail tissue. (From E. Geigy, Rev. Suisse Zool. 48:483, 1941; and from J. L. Schwind, J. Exp. Zool. 66:1, 1933.)*

hormonal stimulus. Each tissue has a unique response; an eye implanted in a tail of a metamorphosing tadpole does not regress, although all surrounding tail tissue degenerates. The eye simply shifts into the trunk where it remains, unaffected by thyroid hormone (Figure 19-21).

Some tissues require much greater doses to respond than others. For example, forelimbs require a stronger dose of thyroid hormone in order to break through than do hindlimbs, and an even greater dose is necessary for tail resorption. Each tissue type is responsive to a certain threshold dose for reactivity; once this threshold is reached, the tissue responds. The temporal sequence of normal metamorphosis is probably determined by different thyroxin thresholds in various tissues. In the premetamorphic phase, very low levels of thyroid hormone are detected in the circulation. As the thyroid gland matures, in the prometamorphic period, the level of thyroid hormone increases in the circulation and tissues with low threshold values, such as hindlimbs, respond earlier to the rising level of thyroid hormone than those, such as tail, with much higher thresholds. When high doses of thyroid hormone are given to young tadpoles, all processes start simultaneously. The sequence of events is upset, the patterning is disturbed, and some destructive processes such as tail resorption precede constructive processes; the tail becomes reduced before the legs are able to take over locomotion, and the animal dies.

Three different response patterns are exhibited by larval tissues treated with thyroxin: (1) transformation from larval to adult metabolic pathways; (2) growth and differentiation; and (3) regression and death. All cases, including the last, may depend on the synthesis of new proteins. This means that thyroxin hormone regulates information flow from gene to protein.

## Metabolic transformations: regulation of protein synthesis in metamorphosing tadpole liver

The tadpole liver experiences important metabolic changes as the animal leaves an aquatic life and takes up a terrestrial existence. Most aquatic animals eliminate nitrogenous waste as ammonia because ammonia diffuses easily into large volumes of water. Terrestrial animals, on the other hand, must convert ammonia to urea before it can be excreted. The production of urea depends upon enzymes of the arginine-ornithine cycle and these are low or absent

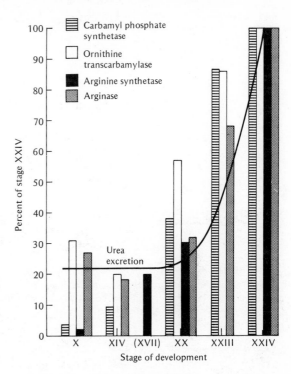

**Figure 19-22** *Activities of urea cycle enzymes and urea excretion during metamorphosis of the liver in* Rana catesbiana. *Assay for arginine synthetase reflects activity of rate-limiting component. All enzymes are coordinately synthesized during this period. (From G. W. Brown and P. P. Cohen,* The Chemical Basis of Development, *ed. W. D. McElroy and B. Glass. Baltimore, Md.: Johns Hopkins University Press, 1958, pp. 495–513.)*

in young tadpole liver but are actively synthesized during metamorphosis (Figure 19-22). While urea cycle enzymes are synthesized, other liver enzymes either do not respond or decrease. Within the same hepatic cells, each enzyme system (and the encoding genes) responds to a common hormone signal.

Thyroxin stimulation of liver protein synthesis is preceded by a long latent period (almost 100 hours or about 5 days), a time when RNA synthesis is stimulated in nuclei and cytoplasm (Figure 19-23). The initial response to the hormone is nuclear RNA synthesis, suggesting hormone action at the level of transcription. Increased RNA synthesis results from enhanced activity of RNA polymerases. Though all species of RNA are equally affected by thyroxin, the bulk of the newly synthesized RNA found in the cytoplasm is ribosomal RNA. Message RNA is also

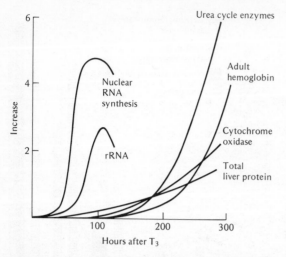

**Figure 19-23**    *Relative increase in several liver cellular activities in* Rana catesbiana *tadpoles after induction of metamorphosis with T$_3$. The initial responses involve activation of nuclear and ribosomal RNA synthesis, possibly as a result of increased RNA polymerase activity. After a long lag, liver protein synthesis begins. Downward trend of curves for RNA synthesis reflects dilution of nucleotide pool as these are released from regressing tail tissues. Rates of RNA synthesis continue high. (Based on data in J. R. Tata, Biochem. J. 105:783, 1967.)*

produced, but these molecules are probably only a small proportion of the total newly synthesized RNA.

The hormone also acts at the translational level; it stimulates polysome accumulation and changes the distribution of ribosomes in the cytoplasm. New ribosomes seem to bind more tenaciously than old ribosomes to membranes of the endoplasmic reticulum before protein synthesis begins.

The hormone may activate both transcription and translation. Inasmuch as low levels of some of the urea cycle enzymes are found in tadpole liver prior to metamorphosis, we can assume the genes are already transcribing. These low enzyme activities may result from low rates of translation, and only in the presence of thyroxin is protein synthesis stimulated because messages are now mobilized into polysomes that attach to the endoplasmic reticulum.

**Growth and differentiation**

A most important change during metamorphosis is limb formation. In contrast to urodeles, where limbs

develop early during embryogenesis (the typical vertebrate pattern), limb morphogenesis in anuran embryos is postponed until metamorphosis. The pattern of limb development is similar to that of the chick, already discussed in Chapter 12.

In addition to the limb, other tissues undergo growth and differentiation in response to thyroxin. New populations of erythrocytes appear containing adult rather than larval hemoglobin.

Five different hemoglobins can be obtained from tadpole erythrocytes, with properties suited to extracting dissolved oxygen from an aqueous milieu. Adult hemoglobins have different electrophoretic properties and are more efficient in extracting and transporting atmospheric oxygen. The differences in these molecules reside, as we have seen, in the combinations of polypeptide chains that make up the tetramer. Stem cells are continuously generating erythrocytes in the tadpole; in low levels of thyroxin, they express larval hemoglobin genes. As thyroxin levels increase, stem cell proliferation is stimulated and new populations of erythroblasts are diverted into pathways of adult hemoglobin synthesis. Meanwhile, old tadpole erythrocytes persist for a time until they die. During metamorphosis, a mixture of adult and tadpole hemoglobins can be extracted from the mixed population of erythrocytes. This model places thyroxin action at the transcriptive level, "selecting" a specific genetic program by derepressing adult hemoglobin genes.

**Destructive changes in metamorphosis: tail regression**

Regression of tail, gut, and gills accounts for the principal destructive changes in amphibian metamorphosis, and these, too, depend upon thyroxin stimulation of protein synthesis. Histolysis of tail tissues is brought about by the action of a variety of hydrolytic enzymes (cathepsin and collagenase), synthesized to 200 times their original levels before the tail is resorbed. We have here a paradoxical phenomenon of a dying tissue actively synthesizing the enzymes necessary for its own destruction.

Tail resorption is studied in an in vitro assay system in which amputated larval tails are cultured in an artificial medium. Untreated tails, though viable, do not change in length over a period of 8 to 12 days. When exposed to a low concentration of T$_3$, the tails regress at a rate corresponding to that observed in vivo. Accompanying this destruction is an increased activity of hydrolytic enzymes such as cathepsin and phosphatase.

The effect of the hormone can be abolished by addition of actinomycin D to inhibit RNA synthesis and puromycin to inhibit protein synthesis. These experiments indicate that regression depends on de novo synthesis of hydrolytic enzymes and does not result from hormone activation of lysosomes as was originally thought. Evidently metamorphic cell death as well as proliferation and differentiation require a genetically determined synthesis of specific hydrolytic enzymes.

### Acquisition of metamorphic competence

We have already seen that larval tissues show different responses to thyroxin. Some, such as brain, exhibit no apparent response except in a few large motor neurons. Others proliferate, synthesize new proteins, differentiate, or die. The ability to respond to thyroxin, or to any hormone for that matter, is not always present in a cell. A cell acquires a competence to react to hormone during the course of its development. We find that the amphibian embryo and larva are not always responsive to thyroxin; they develop metamorphic competence as they mature.

Studies on the action of estrogens on uterine tissues (see Chapter 6) have led to a concept of a target tissue where cells possess hormone receptor sites on their surfaces or in cytoplasm which recognize and "capture" circulating hormone. The receptor-hormone complex is conveyed to the nucleus where it regulates genome expression. Does the acquisition of metamorphic competence by amphibian larval cells require synthesis of specific thyroxin-receptor proteins?

To determine when embryonic cells become competent to respond to thyroid hormone, J. Tata exposed different developmental stages to radioactive thyroxin and followed its binding in the tissues. The embryo is insensitive to the hormone before hatching but soon shows an increase in binding capacity that increases slowly throughout larval life. Presumably embryonic tissues acquire thyroxin binding sites toward the end of embryogenesis. Increased binding also precedes thyroxin stimulation of RNA and DNA synthesis. In other words, there is an almost simultaneous acquisition of metamorphic response to thyroxin with enhanced binding capacity. If it turns out that thyroxin binding underlies the specificity of cell response, then cells may display differences in binding capacity. Moreover, the acquisition of binding capacity of a cell may explain the so-called threshold phenomenon mentioned earlier.

We see that hormonal regulation of metamorphosis is an ontogenetic phenomenon; the regulatory mechanisms must themselves undergo maturation. In order for cells to become competent, they acquire specific receptor sites, perhaps for more than one hormone. The feedback system of neurohumoral controls also matures as endocrine glands and neurosecretory cells differentiate, until the system can respond to environmental signals and trigger the metamorphic program. The temporal sequences we observe in both insects and amphibians reflect these progressive changes.

## SUMMARY

In many organisms, embryogenesis produces a free-living larval stage before the fully mature adult body plan is developed. In species with small eggs, a larval stage is interposed as a feeding adaptation to support continued development of adult structures. As soon as adult tissues mature, the larva undergoes dramatic metamorphic changes as larval tissues are shed or transformed into an adult.

Once a larval stage is introduced, it is subject to natural selection and acquires adaptations for dispersal, site selection, and parasitism. Larvae possess anatomical and physiological adaptations for feeding, movement, and attachment which successfully exploit new habitats and ecological niches, thereby avoiding competition with adults for limited resources. The larva evolves separately from the adult and may, through natural selection and precocious maturity of reproductive function (neotony), give rise to new groups of organisms.

Metamorphosis is controlled by neurosecretory and hormonal feedback systems. Environmental periodicities (diurnal, seasonal) are transduced by neurosecretory cells in the brain into hormonal signals that induce changes in growth and differentiation. Tissues exhibit differential responses to hormones: some do not respond at all; others proliferate; some synthesize new proteins; a few may die. Entirely new enzyme systems adapt the adult organism to a new life, and larval tissues take on adult functions. The nature of a tissue response is built into its genome and probably relates to the presence of mature hormone receptors on cell surfaces or within the cytoplasm. These recognize the hormone and bring it into the nuclear compartment where gene systems are turned on and off. The transformations of larva into adult require tissue-specific programs of gene transcription and translation that account for morphogenetic reprogramming as one life style is replaced by another.

**REFERENCES**    Ashburner, M. 1970. Function and structure of polytene chromosomes during insect development. *Adv. Insect Physiol.* 7:1–95.

———. 1973. Sequential gene activation by ecdysone in polytene chromosomes of *Drosophila melanogaster:* Dependence on ecdysone concentration. *Develop. Biol.* 35:47–61.

Berendes, H. D., and Beermann, W. 1969. Biochemical activity of interphase chromosomes (polytene chromosomes). *Handbook of molecular cytology,* ed. A. Lima-de-Faria. Amsterdam: North-Holland, pp. 500–519.

Berrill, N. J. 1947. Metamorphosis in ascidians. *J. Morphol.* 81:249–267.

———. 1961. *Growth, development and pattern.* San Francisco: Freeman.

Clever, U. 1966. Gene activity patterns and cellular differentiation. *Am. Zool.* 6:33–41.

———. 1972. The control of cellular growth and death in the development of an insect. *Molecular genetic mechanisms in development and aging,* ed. M. Rockstein and G. T. Baker, III. New York: Academic Press, pp. 33–69.

Cloney, R. A. 1963. The significance of the caudal epidermis in ascidian metamorphosis. *Biol. Bull.* 124:241–253.

Cohen, P. P. 1970. Biochemical differentiation during amphibian metamorphosis. *Science* 168: 533–543.

DeBeer, G. R. 1958. *Embryos and ancestors.* 3d ed. Oxford: Clarendon Press.

Doane, W. W. 1973. Role of hormones in insect development. *Developmental systems: Insects,* ed. S. J. Counce and C. H. Waddington. New York: Academic Press, pp. 291–497.

Etkin, W., and Gilbert, L. I. 1968. *Metamorphosis, a problem in developmental biology.* New York: Appleton-Century-Crofts.

Etkin, W., and Gona, A. G. 1967. Antagonism between prolactin and thyroid hormone in amphibian development. *J. Exp. Zool.* 165:249–258.

Frieden, E., and Just, J. J. 1970. Hormonal responses in amphibian metamorphosis. *Biochemical action of hormones,* ed. G. Litwack. Vol. I. New York: Academic Press, pp. 1–52.

Gilbert, L. I. 1974. Endocrine action during insect growth. *Recent Prog. Hormone Res.* 30:347–384.

Mileikovsky, S. A. 1971. Types of larval development in marine bottom invertebrates; their distribution and ecological significance: A re-evaluation. *Mar. Biol.* 10:193–213.

Schwind, J. L. 1933. Tissue specificity at the time of metamorphosis in frog larvae. *J. Exp. Zool.* 66:1–14.

Tata, J. R. 1970. Simultaneous acquisition of metamorphic response and hormone binding in *Xenopus* larvae. *Nature* 227:686–689.

———. 1971. Hormonal regulation of metamorphosis. *Soc. Exp. Biol. Symp.* 25:163–181.

Thomson, J. A. 1975. Major patterns of gene activity during development in holometabolous insects. *Adv. Insect Physiol.* 11:321–398.

Thorson, G. 1950. Reproductive and larval ecology of marine bottom invertebrates. *Biol. Rev.* 25:1–45.

Weber, R. 1967. Biochemistry of amphibian metamorphosis. *The biochemistry of animal development,* ed. R. Weber. New York: Academic Press, pp. 227–302.

Wigglesworth, V. B. 1970. *Insect hormones.* San Francisco: Freeman.

Williams, C. M. 1969. Nervous and hormonal communication in insect development. *Communication in development,* ed. A. Lang. New York: Academic Press, pp. 133–150.

Williams, C. M., and Kafatos, F. C. 1971. Theoretical aspects of the action of juvenile hormone. *Mitt. Schweiz. Entomol. Ges.* 44:151–162.

# DEVELOPMENTAL DEFECTS

Developmental processes are not free of error; mistakes are inevitable in the complex regulatory circuitry of embryogenesis. Genetic information is altered by mutation; cell metabolism suffers random perturbations, and many environmental agents interfere with cell replication and differentiation. Some errors are corrected by cell repair mechanisms, but most errors are cumulative. Lesions produced in early embryonic stages affect differentiation of many cell clones, and congenital abnormalities arise.

Not all developmental defects are expressed before or at birth—many are cryptic and emerge later in life as malignant neoplasms. Carrying genetic lesions in growth regulation, such autonomously proliferating cells eventually invade normal tissues and kill the organism.

Development usually stops when organisms reach reproductive maturity and aging begins. Throughout life organisms suffer a continuous barrage of environmental insults. Homeostatic mechanisms become less efficient, which means aging organisms are less able to cope with stress situations. Aging is a complex interaction of defective macromolecules, declining function of extracellular compartments, and cell death.

In this section we will see how developmental defects lead to congenital abnormalities, how neoplastic cell phenotypes arise, and finally what happens when growth and repair processes become less efficient in the course of aging.

# 20

# Congenital Abnormalities

Every aspect of development is prone to a defect. The sequential nature of development is based on networks of interacting systems; what happens at one stage is vital to all subsequent stages. Defects initiated at an early stage may be expressed at later stages because intermediate steps are abnormal. The magnitude of a defect will depend on the stage of development at which it originates and the developmental process that is the specific target. Lesions arising at early stages usually have more drastic effects since many more cells are affected.

Normal development operates as a precise machine and even slight perturbations in the environment can cause loss of gametes and embryos. We have already mentioned the wastage of oocytes and eggs before maturation (see Chapter 10). Of the thousands of fish oocytes produced, only a small percentage mature and are fertilized. Wastage after fertilization is also the rule, even in humans. It has been estimated that at least 20 percent of fertilized human zygotes are spontaneously lost during early pregnancy. It is difficult to determine what propor-

tions are due to intrinsic congenital abnormalities or to specific environmental agents. Some estimate that approximately 25 percent of spontaneous human abortions in early pregnancy are due to congenital abnormalities. Since 20 percent of all human zygotes abort, this means an incidence of 5 percent of congenital loss before birth. If we add this figure to the incidence that is detected at birth (Table 20-1), we arrive at a total of 10 percent congenital abnormalities in humans, a figure that is surprisingly high.

## EVALUATION OF CONGENITAL ABNORMALITIES

Restrospective studies of mothers of defective children attempt to unravel the causes after the fact of abnormal births. These data, however, are unsatisfactory because human memory is unreliable. A prospective study is a preferred approach; data are collected from pregnant women with respect to diet, drug intake, exposure to radiation, and so on,

659

TABLE 20-1 Estimates of the Incidence of Congenital Defects in Humans

| | | |
|---|---|---|
| Early fetal deaths | | |
| Total incidence of abortions | 20% | |
| Estimated congenital defects | | |
| in aborted embryos | 25% | |
| Total incidence | | 5% |
| Detected at birth | | |
| Minor defects | 1% | |
| Major defects | 0.5% | |
| Lethal defects | 0.5% | |
| Total incidence | | 2% |
| Detected in follow-up studies | | |
| (birth to 6 years) | | 3% |
| Total incidence (all conceptions) | | 10% |

Source: Based on L. Saxen and J. Rapola, *Congenital Defects.* New York: Holt, Rinehart and Winston, 1969.

and any abnormal children produced are correlated with the data obtained during pregnancy. Ideally, information about the causes of congenital abnormalities in humans should come from such studies, but they are difficult to carry out. We are left with the alternative of studying congenital abnormalities in animals.

Only animal test systems can be used to evaluate the effects of thousands of new teratogenic agents introduced into the environment each year. A teratogenic agent is one that causes developmental abnormalities in the embryo or fetus. In testing these agents, we are usually limited to mice, rats, or dogs, but problems arise in extrapolating dose response from animals to humans. For example, cortisone has a wide range of teratogenic effects, from no effect in rats to high incidence in rabbits. In most cases, doses greatly exceeding human therapeutic doses have to be introduced to induce teratogenic effects in animals. And in some cases such innocuous drugs such as caffeine and aspirin have teratogenic effects at very low doses. Moreover, within the same species, genetic strain differences also affect teratogenicity. In Table 20-2 five different mouse strains are compared as to the teratogenic effect of cortisone in producing cleft palate. The incidence varies from 12 percent in the least susceptible CBA strain to 100 percent in the most susceptible A strain. We can see why it is difficult for government agencies to evaluate teratogenic effects of drugs on humans and make decisions about their general use.

A classic demonstration of the problem in extrap-

olating animal results to man is our experience with two related drugs, thalidomide and meclizine.

Thalidomide is a mild sedative that was prescribed for use in pregnancy in many European countries in the late 1950s. In 1961 the German scientist Lenz reported a possible connection between this drug and an increased frequency of a human congenital abnormality of the limbs known as *amelia* (no limbs) and the closely related *phocomelia* (a very reduced appendage with a flipper-like hand or foot) (Figure 20-1). The incidence of this anomaly correlated strongly with the increasing use of thalidomide among pregnant women in England and Japan, whereas in countries like the United States, where thalidomide was not used, no increase was observed.

The drug had been evaluated by the usual toxicity tests before it was prescribed for use in humans. The fact that the human adult could tolerate high doses of the drug did not vitiate its effects on the embryo. Later, after use of the drug was prohibited, it was tested on various standard animals for teratogenicity and some surprising results were obtained. The drug produced no appreciable leg anomalies in the rat embryo. Large quantities of the drug (450 to 1000 mg/kg body weight) caused fetal death but no syndrome comparable to the human amelia. The effects were similar in mice. Only the rabbit fetus showed effects similar to those obtained in humans, but the rabbit is not the usual test animal in such assay procedures. The drug affects some primate embryos (baboons), but high doses cause only sterility and abortion in rhesus monkeys, the common laboratory primate.

In this instance, the human fetus is far more sensitive than the rat or mouse, which means that

TABLE 20-2 Mouse Strain Sensitivity to Cortisone-induced Cleft Palate

| Strain | Percent Incidence |
|---|---|
| CBA | 12 |
| C57BL | 19 |
| C3H | 68 |
| DBA | 92 |
| A | 100 |

Source: From H. Kalter, *Teratology, Principles and Techniques,* ed. J. G. Wilson and J. Warkeny. Copyright © 1965 by The University of Chicago Press.

**Figure 20-1** *Malformed newborn baby with phocomelia in all limbs resulting from exposure to thalidomide during fetal life.*

even had vast numbers of tests been performed to test the dangers of thalidomide, the drug would have been licensed because it would not have produced any obvious anomalies. Fortunately, an alert scientist picked up the information published in Germany early enough to prevent use of the drug in the United States.

The story of meclizine is just as unusual. This antihistamine has been used routinely to combat nausea and vomiting in pregnancy. On the basis of certain restrospective studies, its use in pregnancy was cautioned because of a suspected relationship to certain fetal malformations. Several studies were carried out (restrospective and prospective), all indicating the drug had no teratogenic effect. The percentage of abnormal fetuses recovered from women using the drug was no different from all controls. The drug is now considered safe for human use. Yet the

same drug introduced into pregnant rats and mice (at low doses of 50 to 125 mg/kg) produced a very high proportion of malformations, including cleft palate and skeletal abnormalities. In this case, had this drug been run through the standard tests, it would have been identified immediately as unsuitable for human use.

Although animal test systems are not completely satisfactory, we have no recourse but to continue their use. These studies do furnish useful information about the developmental processes that are most susceptible to certain agents, the stage sensitivities of the embryos, and their possible mechanisms of action.

## DEVELOPMENTAL PROCESSES AND MUTATIONS

Developmental lesions occur at all levels of organization. Any lesion in information flow, from the gene to protein synthesis, can lead to developmental defects. Since all vital processes depend on the genome and its regulation, a mutation, a failure to transcribe certain genes at the proper time, or a lesion in the ribosome could be, and often is, lethal. All cell activities underlying morphogenesis, such as motility, adhesiveness, and shape changes, when disturbed, cause severe anomalies.

### Pituitary dwarfism in mice

Mutations acting in the final phases of differentiation of a single cell type, though highly localized in one tissue, may produce extensive damage to other organ systems. For example, a genetic defect in the differentiation of specific cells in the anterior pituitary in mice causes a severe dwarf syndrome (Figure 20-2). Homozygotes carrying the recessive gene *dw/dw* fail to grow properly. Dwarf animals remain small, bone development is inhibited, and sexual development is retarded because no eosinophiles, or growth-hormone secreting cells, develop in the dwarf pituitary. According to one interpretation, the cell lineage pattern of the eosinophile is disturbed in the mutant and small cells instead of large intermediate growth-hormone secreting cells arise. Though the primary effect of the gene is on pituitary differentiation, its dominant position in the endocrine system means that deficient pituitary function will secon-

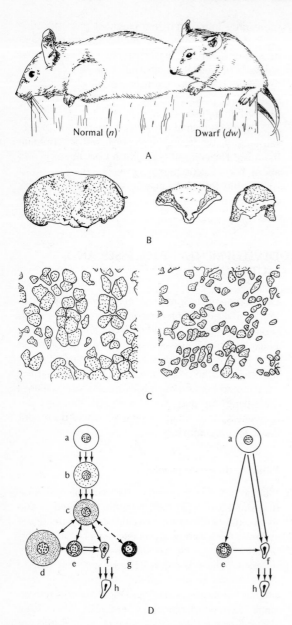

**Figure 20-2** *Comparison of the effect of the gene for pituitary dwarfism (dw) with that for normal development (n) in the mouse. (A) Litter mates at 2½ months. (B) Size and shape of pituitary glands. (C) Size of cells in anterior lobes. (D) Hypothetical cell lineage to explain syndrome. a = typical chromophobe; b = hypoacidophil; c = typical acidophil; d = large acidophil; e = hyperacidophil; f = small pycnotic acidophil; g = basophil; h = small pycnotic chromophobe. (From E. Hadorn,* Developmental Genetics and Lethal Factors. *London: Methuen, 1961.)*

darily affect other endocrine glands. Thyroid, adrenal, and gonads are all affected in this syndrome which can be ameliorated only by frequent implantation of normal pituitary glands into homozygous animals. All other effects of the mutant disappear if the primary cause is treated.

Since all developmental processes are interdependent, we cannot easily distinguish primary from secondary effects of gene mutations. When lesions first occur, which cells are involved, and which processes are the targets, are questions that must be answered for each heritable defect.

### Some human birth defects: mutations

In humans, many birth defects are attributable to recessive or dominant mutations of single genes. Of the thousand genetic disorders in man, more than 50 percent result from dominant gene mutations.

Dominant mutations are the easiest to detect in a human pedigree: they usually afflict several members in a family as illustrated in the pedigree for brachydactyly (short fingers) (Figure 20-3). At least one individual per generation displays the disorder in this pedigree. In the case of dominant genes, a trait will not appear in offspring unless one of the parents expresses that trait. Such pedigrees are typical of mild dominant disorders; more severe abnormalities are generally lethal in the fetus or early in life, before reproductive maturity is attained.

Some dominant genes are not expressed until late in life, well after reproductive maturity. An example is Huntington's chorea, an invariably fatal disease manifested by mental deterioration accompanied by involuntary muscular movements. Although the disease can occur at any age, on the average it appears in both males and females at about forty years. There is no way of predicting whether an individual has the disease until it is manifested. Since most individuals carrying the gene are heterozygous, the chance that a child will be affected if one parent has the disease is 50:50.

The biochemical basis is not understood for most dominant gene disorders because normal enzymes are present in heterozygotes. Mutations in these cases may involve regulatory genes rather than structural genes. On the other hand, recessive genes are more accessible to biochemical or metabolic studies because certain enzymes are lacking or defective in homozygous recessive individuals. The mutations responsible for albinism, sickle cell

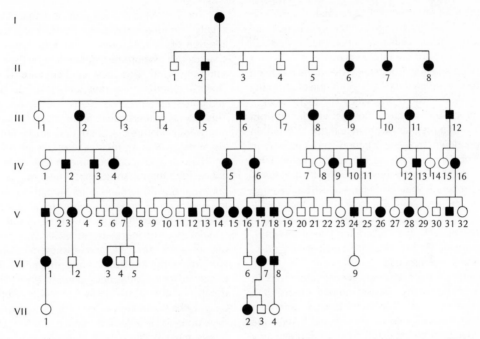

**Figure 20-3**  *Pedigree of a family through seven generations showing inheritance of a dominant mutation for brachydactyly. Note that the abnormal phenotype appears at least once in each generation: the syndrome persists.*

anemia, and phenylketonuria, for example, are concerned with specific proteins—tyrosinase, hemoglobin, and phenylalanine hydroxylase, respectively
—which are either absent or abnormal. In phenylketonuria, only homozygous individuals express the
syndrome while heterozygotes are normal. Absence
of the hydroxylase enzyme in homozygotes means
phenylalanine cannot be metabolized to tyrosine
and is diverted and oxidized to phenylpyruvic and
other phenylketo acids instead that accumulate in
blood and urine. The disease is not evident at birth,
but within the first 6 months the infant shows signs
of impaired mental development and becomes
severely retarded. Apparently the disturbance in
tyrosine metabolism also upsets the metabolism of
neurotransmitters in brain tissue. Since we know the
metabolic basis of the disorder, therapy is possible.
If a pedigree analysis indicates that a couple may
produce offspring with the disease, the infant's blood
is tested after birth for several days to determine the
levels of phenyl derivatives. If high levels are detected, then a diet completely free of phenylalanine
will usually prevent the disorder and a normal child
will develop.

Unfortunately, most recessive disorders in
humans cannot be as readily treated. But, in recent
years, new technologies have been developed for
early detection of these inborn errors of metabolism.
For example, Tay-Sachs disease is caused by a recessive mutation of a gene coding for the enzyme
hexoseamidase A, a lysosomal hydrolase that is
necessary for the hydrolysis of gangliosides in neural
tissue. Absence of the enzyme results in abnormal
accumulations of gangliosides which leads to generalized impairment of mental and muscle development and is fatal. Children appear normal at birth,
but the disease appears after the first 6 months of life.
Such children fail to show normal mental and muscular development. Cells of affected children do not
exhibit hexoseamidase A activity, and tissues from
their heterozygous parents also show reduced levels
of the enzyme. By using cultured cells it is now possible to determine whether parents are heterozygous
for the mutation and calculate the chances that a
homozygous fetus will be conceived. The assay is
carried out on blood sera and families may be
screened if the disease is suspected from previous
history. If only one parent is heterozygous, we can

predict that all offspring will be normal. If, however, both parents are heterozygous, then the chances are one out of four for each conception that a child will be homozygous for the mutant gene.

In vitro screening of genetic disorders has been enhanced by *amniocentesis,* a technique of sampling amniotic fluid early in pregnancy (12 to 14 weeks). In this procedure, amniotic fluid is withdrawn from the uterus of a pregnant woman and the cells within are spun down by centrifugation and cultured in vitro. Since these cells are derived from the fetus (amnion or skin cells), they carry the fetal genome and its properties may be determined from enzyme assays of cultured cells. The cells may also be used for karyotype analyses to detect chromosome anomalies as well as to determine sex of the fetus. We can now diagnose at least 60 metabolic disorders in utero with this procedure.

Amniocentesis has been successfully used in diagnosing Tay-Sachs disease, among others. If a family history shows Tay-Sachs disease and both parents are heterozygous (as determined by enzyme assay), then intrauterine diagnosis of each pregnancy is a recommended procedure. If a fetus is found to be homozygous (it lacks the enzyme), then the chances are high the child will develop the disease and it can be aborted. The success rate has been high in this instance. In almost all cases, if the tests show the fetus to be normal, then the child is born normal. If there is a positive indication of a genetic abnormality, the fetus is usually abnormal when aborted.

## Chromosomal disorders

Chromosomal abnormalities are responsible for approximately 0.5 percent of birth defects in the general population, but this is only a small part of the total zygotic loss due to chromosomal abnormalities. Most zygotes possessing chromosomal abnormalities fail to cleave or abort early in development. Studies of spontaneous human abortions have shown that approximately 25 percent of these are due to chromosomal abnormalities.

In general, abnormalities are manifested as a change in chromosome number resulting from loss or the addition of one or more chromosomes. Some abnormalities arise because of structural changes such as translocations or inversions, or the loss of chromosome fragments. Both autosomes and sex chromosomes are involved.

The loss of any autosome is lethal, and such zygotes are eliminated early. On the other hand, the addition of an autosome is viable but does lead to characteristic syndromes, depending on the chromosome involved. The best studied case is that of Down's syndrome, or mongoloidism, in which an additional chromosome 21 is found, *trisomy 21.* This trisomy is responsible for mental retardation and mongoloid features, a fairly common congenital abnormality with an incidence of one in 700 births.

Loss of a sex chromosome in females is usually not lethal, inasmuch as only one of the two X chromosomes in females is functional. Nevertheless the loss of one X does lead to disorders of sexual development, known as Turner's syndrome. These females develop normally in every way but show few secondary sexual characteristics, develop a male type of body structure, and sometimes show a thickened or weblike neck (Figure 20-4). They are usually of normal mentality but sterile. Apparently both X chromosomes are essential to normal sexual development of a human female. Both X chromosomes must be functional during gonad development in the fetus when meiosis begins.

The addition of a sex chromosome is also a common disorder in humans. Males with an additional X chromosome develop Klinefelter's syndrome, a reduction in male secondary sexual development and sterility. Additional X chromosomes in females (XXX, XXXX, and so on) do not result in "superfemale" development, as one might expect. Because of the Lyon effect, all but one X chromosome are made heterochromatic and nonfunctional. The extra X chromosomes however, do affect sexual development (sterility?) in part and sometimes mental development.

These chromosomal aberrations arise during gametogenesis as a result of *nondisjunction* of homologous sex chromosomes during meiosis. Normally, at the first meiotic metaphase, homologous chromosomes segregate; but if nondisjunction occurs, they do not, and the pair may migrate to one daughter cell, leaving the other daughter without any chromosome of the pair. Two types of gametes are produced in the female, XX and O. If the latter is fertilized by an X-bearing sperm, an XO zygote results that develops Turner's syndrome. Fertilization of an XX egg by a Y-bearing sperm produces an XXY zygote and Klinefelter's syndrome; fertilization by an X sperm produces a tri-X (XXX) female. Male

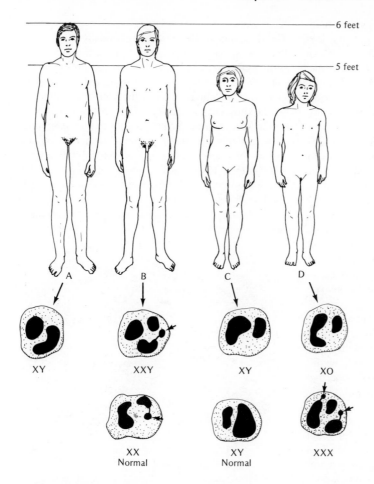

**Figure 20-4** *Chromosomal abnormalities and sex determination in humans. (A) A 24-year-old male eunuch who had been castrated at the age of five. (B) A 30-year-old man with Klinefelter's syndrome. The incidence of this condition is 2:1000 live births. (C) A 17-year-old genetic male with testicular feminization syndrome. Appears as a normal female because he is recessive for a gene that makes all cells unresponsive to testosterone. Female rather than male external genitalia develop as a result. (D) An 18-year-old woman with Turner's syndrome. Only about 1 in 14 such fetuses survive to term, when the incidence is about 1:3000 live births. Intranuclear sex chromosome inclusions of Barr bodies in polymorphonuclear leucocytes are shown.*

nondisjunction may lead to two kinds of sperm, XY and O. An XY sperm fertilizing an X egg leads to the XXY condition of Klinefelter's syndrome, and an O-bearing sperm leads to Turner's syndrome.

Any intrinsic or exogenous factor that can induce chromosome breakage or loss will result in chromosome abnormalities in gametes. Radiation, viruses, and chemicals all cause chromosomal damage, and some human disorders can be attributed to exposure of gametes to these agents at some time during gametogenesis, when sensitivity seems to be greatest.

## STAGE SENSITIVITIES

Waddington coined the term "epigenetic crises" to describe those developmental events that are suscep-

tible to even slight perturbations in the environment. Even momentary distortion of the normal sequence of events at these periods may lead to irreparable embryonic changes. Sensitive periods are detected by treating embryos with a variety of agents. Those stages exhibiting the greatest effects at the lowest concentrations (or doses) are the most sensitive. Amphibian gastrulation, for example, is most sensitive to such diverse agents as anaerobiosis, nitrogen mustard, antimetabolites, and actinomycin D. Cleaving embryos develop normally in their presence but development stops at the beginning of gastrulation. These same agents seem to be most damaging during critical periods of histogenesis or organogenesis in mammalian embryos.

Gametes are particularly prone to damage while developing in the gonad. Overripeness of eggs is a principal source of developmental defects and usually arises as a result of delayed ovulation in mammals. In Chapter 10, we pointed out that the mature egg is an unstable cell, destined to die unless fertilized. If fertilization is delayed, irreversible changes gradually take place within egg organization that affect embryonic processes later on. In many instances, these abnormalities are accompanied by gross chromosomal defects.

Overripeness of eggs has been implicated as a factor in the genesis of chromosomal aberrations in human zygotes. It does increase the frequency of nondisjunction during meiosis of mammalian gametes. The fact that Down's syndrome has a high incidence in infants of older mothers has led some to suggest that these arise from overripe eggs since older women experience less frequent coitus. What is probably more relevant is the fact that oocytes in older women are older, having been in the ovary since birth. Overripeness may be due to delayed ovulation instead.

The sensitivity of cleaving mammalian eggs to teratogenic agents correlates with the frequency of chromosomal aberrations produced. Sensitivity of rat eggs to X-rays, for example, varies greatly within a 24-hour period after fertilization. Most abnormal embryos are produced after irradiation at polar body formation or in early pronuclear stages. Irradiated embryos die early and exhibit a high frequency of sex chromosome loss. DNA synthesis is the target since it is most active in pronuclear stages when sensitivity is highest. After DNA synthesis is completed, the embryos are much less sensitive to X-rays. First cleavage is unaffected, but sensitivity returns when DNA synthesis for second cleavage resumes. In addition to loss of sex chromosomes, autosomes are also lost, and we assume that this loss is the primary cause of embryonic death since sex chromosome loss is not usually lethal.

Stage sensitivity depends on the developmental processes predominating and their importance to organogenesis at that time. Cell division, essential to the development of most organs, is disturbed by many teratogenic agents. The level of cell division in a tissue will affect its response to a chemical teratogen. For instance, treatment of newt larvae with triethylene melamine (TEM), an anticancer agent, induces a high frequency of eye defects, from complete absence of eyes to rudimentary eyes (Figure 20-5). The stage at which treatment begins is critical; larvae treated early are eyeless, while those treated

later possess varying degrees of rudimentary eyes. The drug specifically destroys mitotically active cells because at early stages most cells of the retina are proliferating and TEM destroys the entire retina. At later stages, mitotic activity is restricted to marginal edges of the retina and only these cells are affected by the drug with the formation of reduced eyes.

It seems to be a general rule that proliferating cells are most prone to teratogenic action by all agents since DNA synthesis is readily inhibited. Certain teratogenic viruses also attack proliferating cells, causing embryonic death and fetal abnormalities by inhibiting organ growth.

The rubella virus (measles) in humans is teratogenic when women are infected early in pregnancy.

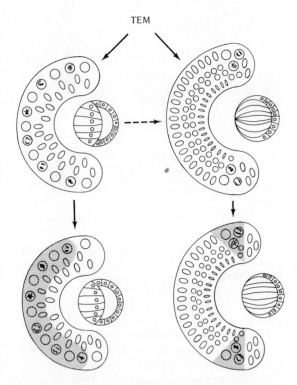

**Figure 20-5** *The effect of triethylene melamine (TEM) on development of the newt eye. In the younger eye on the left the drug inhibits mitosis and causes cell destruction in a wide area at the outer rim of the neural retina. In older eyes the effect is restricted to the margin of the iris where cell proliferation continues. (Based on observations by G. Töndury in Medikamentöse Pathogenese fetale Missbildungen, T. Koller and H. Erb, eds. Basel: Karger, 1964.)*

The fourth to twelfth weeks of pregnancy are most critical and abnormalities of the heart, lens, and inner ear result. In addition, there is also a high frequency of abortion and fetal death in women infected during the first trimester. Infection during the seventh week is serious since a complex syndrome involving all organs results because of overlapping sensitivities.

Rubella virus passes through the placenta into fetal circulation and infects fetal tissues, where it inhibits cell division and causes a reduction in growth. Most organ systems are mildly affected but some, like the heart, show serious anomalies. The auricular chambers of the human heart are normally divided by septa that emerge at the end of organogenesis. During fetal life the auricles are in communication through an opening, the foramen ovale, that shunts blood from the right to the left side of the heart. Late in development this opening is sealed off by growth of a small septum. In most instances of the rubella heart syndrome, the septum fails to grow out because the virus inhibits cell proliferation. This slight inhibition of growth by a virus causes a severe circulatory anomaly since an opening between right and left hearts mixes oxygenated and nonoxygenated blood.

Morphogenetic movements of cell sheets dominate the early stages of organ formation. The most important occur during gastrulation, a period of high teratogenic sensitivity. Even slight changes in salt content of the medium will upset amphibian gastrulation. A hypertonic medium induces exogastrulation and endoderm migrates outwards rather than inwards beneath the ectoderm. No neural tissues are induced because chordamesoderm does not come into contact with presumptive neural ectoderm.

Heart formation depends upon the migration and fusion of two endocardial tubes (see Chapter 12). Treatment of the chick embryo with EDTA, a calcium chelating agent, reduces cell adhesiveness, alters contact relationships, and prevents spreading and fusion of heart primordia to produce *cardia bifida,* a double heart anomaly. Changes in contact relationships between cells alter organ histogenesis when the primary morphogenetic fields are established.

Often the experimentally induced abnormalities mimic genetic syndromes and are called *phenocopies.* For example, insulin injected into chick embryos at 31 hours of incubation produces a high incidence of "rumplessness," an abnormal condition also produced by a mutant gene $rp^2$ when homozygous (Figure 20-6). The tail vertebrae and oil glands

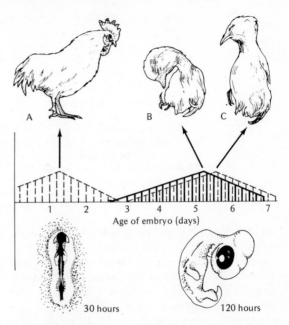

**Figure 20-6**  *Sensitive stages (vertical lines) for insulin-induced phenocopies in the chicken. (A) Insulin treatment during the first period (up to the third day) results in rumplessness, the phenocopy of a genetic mutant syndrome. (B, C) Second period (third to seventh day) when short upper beak (B) and micromelia (C) are produced instead. Periods of maximum sensitivity indicated by arrows and appearance at that time is shown below. (From E. Hadorn,* Developmental Genetics and Lethal Factors. *London: Methuen, 1961.)*

are missing, and other features of feather development are also affected. Insulin probably interferes with carbohydrate metabolism in the target tissues at a critical period in development and mimics the action of certain genes, which suggests the genes themselves may also act by controlling carbohydrate metabolism.

In late stages of organogenesis, tissues undergo differentiation as specific proteins are synthesized. This, as we have seen, depends on the expression of sets of genes. Any agent that interferes with the normal sequence of transcription to translation will perturb gene expression and upset differentiation. Actinomycin D, a powerful teratogenic agent, inhibits RNA transcription but embryos exhibit differential stage sensitivity to this agent, depending on the importance of transcription at the time. (See Chapters 13 and 17). We have seen that early cleavage is in-

TABLE 20-3 The Effect of Sulfanilamide and 6-Aminonicotinamide on Chick Embryos When Injected at 96 Hours of Incubation Singly or in Combination

|  | Experiment 1 | Experiment 2 | Experiment 3 |
|---|---|---|---|
| Sulfanilamide $\mu$g/egg | 300 | 300 |  |
| 6-Aminonicotinamide $\mu$g/egg |  | 1.5 | 1.5 |
| Number treated | 106 | 108 | 105 |
| Survivors of day 13 | 101 | 97 | 99 |
| Normal percentage | 99.0 | 19.6 | 100 |
| Micromelia percentage | 0 | 80.4 | 0 |
| Beak malformations percentage | 0 | 55.7 | 0 |
| Other malformations | 1.0 | 0 | 0 |

Source: After W. Landauer and E. M. Clark, *Nature* 203:527, 1964.

sensitive to this inhibitor because maternal messages stored in the egg are sufficient to carry the embryo through cleavage. Later, during morphogenesis and cytodifferentiation, when new messenger RNAs are required, actinomycin sensitivity increases. In the differentiation of most tissues (pancreas, lens, red blood cells, and so on) the initial stages are most sensitive to the drug. In general, the messages coding for cell-specific proteins are synthesized early during cytodifferentiation, hours or even days before they are translated. Such long-lived messages may be translated over and over again, even in the presence of the inhibitor. This means that stage sensitivity to this drug is difficult to pinpoint because of a delayed-action phenomenon.

Sometimes, in situations where two drugs act together to produce synergistic effects, the time of action is obscure. Each drug has no effect when acting alone, but when the same agents at identical doses are made to act together, a distinct teratogenic effect is evident. Synergisms between two drugs, a virus and a drug, a drug and a hormone, and so on, have been observed in animals and humans. The phenomenon poses serious problems for the identification of human teratogenic agents because testing procedures must be devised to evaluate all possible cases of synergistic interaction. Even such nonteratogenic drugs as aspirin or caffeine may be teratogenic when acting synergistically with other agents. This points up the need for caution during pregnancy.

A case of drug synergism was studied by Landauer in chick embryos. Embryos treated with low doses of sulfanilamide exhibit no effects (Table 20-3), nor do they react to low-dosage injections of 6-aminonicotinamide, a metabolic antagonist of a respiratory coenzyme (NAD). Both drugs are teratogenic at much higher doses. Simultaneous injection of both drugs at the low concentrations results in a high frequency of limb abnormalities (micromelia) and an anomaly known as parrot beak. This synergistic interaction is not well understood, although both drugs are probably acting as antagonists of the same respiratory coenzymes. The reason for their specific action on limbs and beak at the stages involved is also not clear, inasmuch as this coenzyme is found in all cells.

Similar potentiating effects between two agents are also possible in humans. We do not know, for example, whether such apparently innocuous drugs as a mild sedative, commonly ingested by many pregnant women, may in combination with a viral infection, or certain food additives, act synergistically and induce teratogenic responses.

## SUMMARY

The developing organism offers numerous metabolic and morphogenetic targets for teratogenic action. The nature of the response is dependent upon the specific metabolic systems affected, the cell populations involved, and the stage at which the agent acts. Some agents have multiple effects while others interfere with such basic processes as cell division. Sometimes the agent acts immediately, but often, its action is delayed until late in embryogenesis or even much later after birth. Follow-up studies reveal that many congenital abnormalities are not detected at birth but appear in infancy and childhood. This is particularly true of behavorial defects resulting from abnormalities in nervous-system development, a system that continues its development well after birth. We will see in the following chapters that some defects are manifested late in adult life as rapidly growing neoplasms. These arise from abnormal cells that have persisted in an inactive state for years until conditions (hormonal and immunological) favor their sudden emergence as a proliferating cell phenotype, unresponsive to any growth-regulating mechanisms.

**REFERENCES**

Carr, D. H. 1969. Chromosome abnormalities in clinical medicine. *Prog. Med. Genet.* 6:1–61.

———. 1971. Chromosomes and abortion. *Adv. Human Genet.* 2:201.

Cattanach, B. M. 1973. Genetic disorders of sex determination in mice and other mammals. *Birth defects,* ed. A. G. Motulsky and F. J. G. Ebling. New York: Elsevier, pp. 129–141.

Dagg, C. P. 1966. Teratogenesis. *Biology of the laboratory mouse,* ed. E. L. Green. New York: McGraw-Hill, pp. 309–328.

German, J. 1968. Mongolism, delayed fertilization and human sexual behavior. *Nature* 217:516–518.

Hadorn, E. 1961. *Developmental genetics and lethal factors.* London: Methuen.

Hendricks, A. G., Axelrod, L. R., and Clayborn, L. D. 1966. Thalidomide syndrome in baboons. *Nature* 210:958–959.

Ingalls, T. H., Plotkin, S. A., Meyer, H. M., and Parkman, P. D. 1967. Rubella: Epidemiology, virology and immunology. *Am. J. Med. Sci.* 253:349.

Landauer, W. 1948. Hereditary abnormalities and their chemically induced phenocopies. *Growth Symp.* 12:171–200.

Landauer, W., and Clark, E. M. 1964. On the teratogenic nature of sulfanilamide and 3-acetyl-pyridine in chick development. *J. Exp. Zool.* 156:313–322.

Littlefield, J. W., Milunsky, A., and Atkins, L. 1973. An overview of prenatal genetic diagnoses. *Birth defects,* ed. A. G. Motulsky and F. J. G. Ebling. New York: Elsevier, pp. 221–225.

Milunsky, A. 1973. *The prenatal diagnosis of hereditary disorders.* Springfield, Ill.: Thomas.

Saxen, L. 1973. Population surveillance for birth defects. *Birth defects,* A. G. Motulsky and F. J. G. Ebling. New York: Elsevier, pp. 177–186.

Saxen, L., and Rapola, J. 1969. *Congenital defects.* New York: Holt, Rinehart and Winston.

Wilson, J. G., and Warkany. J., eds. 1965. *Teratology, principles and techniques.* Chicago: University of Chicago Press.

Witschi, E. 1952. Overripeness of the egg as a cause of twinning and teratogenesis: A review. *Cancer Res.* 12:763–786.

Zwilling, E. 1955. Teratogenesis. *Analysis of development,* ed. B. H. Willier, P. A. Weiss, and V. Hamburger. Philadelphia: Saunders, pp. 699–719.

Zwilling, E. 1959. Micromelia as a direct effect of insulin. Evidence from in vitro and in vivo experiments. *J. Morphol.* 104:159–179.

# 21

# Neoplasia: An Abnormal Cell Phenotype

Certain cellular lesions transform cells into phenotypes that proliferate in an uncontrolled manner. These cells develop into *neoplasias* (new growths) which may invade and destroy healthy tissue and kill the organism. Transformations of normal to neoplastic cells take place at all stages in the life cycle. Such cells may persist for long periods without detection but can develop into slow-growing benign tumors or vigorously growing malignant tumors after they are stimulated to proliferate. Such potentiating or permissive conditions may take many years to develop, which explains, in part, why most human tumors appear late in life.

Once tumor growth is initiated, its development resembles that of an independent organism, even to the extent of requiring a specific host background. Certain genetic, hormonal, and immunological states of the host are essential to neoplastic development.

Vertebrates seem to be the most common hosts for neoplastic growth; their tumors resemble man's and are relatively easy to detect. Tumors are common in plants but most invertebrates, including insects, are surprisingly tumor-free. Low tumor incidence among invertebrates suggests that tumor development is a slow process with latent periods of years, sometimes decades. Only vertebrates live long enough to favor full expression of the neoplastic cell phenotype.

## NEOPLASTIC CELL PHENOTYPE

Virtually all organs and tissues are sites of tumor formation. Tumors arising in epithelia are usually called *carcinomas;* those appearing in mesodermal derivatives such as muscle or bone are designated as *sarcomas* (for example, a bone tumor is an osteosarcoma). Tumors of blood-forming organs such as spleen and bone marrow are called *lymphomas* or *leukemias.* Most tumors are spontaneous; they develop without any identifiable cause, but experimental tumors can be induced by chemical agents such as hydrocarbons, X-irradiation, or viruses.

Irrespective of the tissue of origin or the inducing

670

agent, all neoplastic cells possess a common set of phenotypic traits that accounts for their abnormal behavior:

1. Uncontrolled cell division — autonomous growth
2. Heritability and transplantability
3. Altered cell surface
4. Invasiveness and metastasis
5. Dedifferentiation — loss of specialized functions

A discussion of each follows.

## Uncontrolled cell division: autonomous growth

Uncontrolled cell division is the primary phenotypic trait of a neoplastic cell. Tumor cells proliferate autonomously, irrespective of every host signal that arrests growth; they seem to be "marching to their own tune." Proliferation need not be rapid. Cleaving blastomeres with short generation times divide more rapidly than most tumor cells. Even the most malignant tumors possess an average 18- to 24-hour cell cycle but they continue to grow while their normal counterparts have stopped. Slow-growing tumors are usually benign, while the more rapidly growing are malignant. Autonomous, unrestrained proliferation identifies a neoplastic cell; certain properties, such as transplantability, are secondarily derived from its high growth potential.

Autonomous growth is not typical of all tumors. Many are *conditional* tumors; that is, they grow in an unrestrained fashion only if certain host conditions prevail, such as hormonal background. At all other times they do not grow. For instance, certain tumors of mice grow cyclicly in females in response to each pregnancy, when circulating progesterone levels are highest (Figure 21-1). After pregnancy, hormone titers diminish and the tumor regresses, but growth resumes in a subsequent pregnancy. Conditional tumors are still responsive to systemic controls, but for some, hormone dependence is short-lived. After several regression cycles, a tumor may escape its hormone dependence, become autonomous, and grow vigorously irrespective of hormonal background. It appears as if such tumors progressively evolve toward an autonomous state.

Autonomous proliferation implies that lesions in the mechanisms that regulate cell division keep cells in a growth-duplication cycle. Division control may be exercised at many different steps in the process, from DNA polymerase to protein synthesis and cytokinesis (see Chapter 5). A lesion at any one or several of these steps may upset the timing and orderliness of cell division.

The most likely place for mistakes is DNA replication. Presumably, at some point during the $G_1$ phase a commitment to replicate is made, and once DNA is replicated, a cell usually divides. Most normal somatic cells are either in a prolonged $G_1$ phase or have left the mitotic cycle to enter a $G_0$ phase of specialized function. For most tumor cells, however, the factors that normally keep a cell in $G_1$ or shift it into a $G_0$ condition do not operate, and the cell pro-

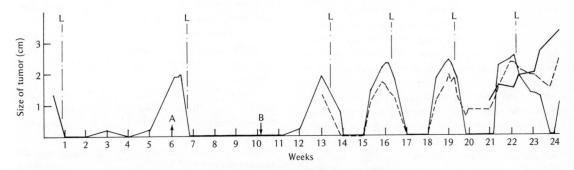

**Figure 21-1** *Tumor progression in mice. Tumor growth is stimulated in pregnant females. L marks the period when lactation has begun at the end of pregnancy. The size of tumors appears in order. The first tumor (solid line) developed in the left vulva, the second appeared in the thirteenth week in the left groin (dashed line), and the third appeared in the twenty-first week in the right vulva. A = male removed; B = male replaced. Note that at the twenty-fourth week two tumors continue to grow after the end of pregnancy; they have become autonomous. (From L. Foulds,* British J. Cancer *3:349, 1949.)*

gresses naturally into the S phase of DNA replication after each mitosis. In fact, the $G_1$ phase of some tumor cells is usually shorter than the $G_1$ of mitotically active cells in comparable normal tissues. The total length of the cell cycle, however, is similar in normal and neoplastic cells; some normal mouse cells may even have a much shorter cell cycle than the most malignant tumors. The intrinsic factors initiating DNA replication in $G_1$ are most affected in neoplastic cells. One or several biochemical events in the prereplicative phase of the cycle may suffer a lesion leading to uncontrolled proliferation. These include synthesis of specific RNA molecules, histones, nonhistone proteins, or enzymes involved in DNA replication.

A lesion may alter a cell's ability to respond to systemic factors regulating cell division activity during normal growth. Since hormone sensitivity depends on specific receptors, loss of these molecules or defects in their synthesis could also obliterate cell division control. One of the first changes in a neoplastic transformation seems to be at the cell surface, suggesting that an abnormal phenotype may arise when a cell loses surface receptors sensitive to growth inhibitory signals.

## Heritability and transplantability

Regulation of cell proliferation relies on a complex network with many switch points. An irreversible change in any one may commit a cell to autonomous proliferation. We can see why each tumor behaves as a "unique organism"; each may acquire a different lesion in the division control network and transmit this stable change to all its progeny.

Many tumors are transplantable; that is, when a small tumor fragment is implanted into a tumor-free host, it grows rapidly into a full-blown tumor and kills the host. Before the host dies, another fragment of the newly developed tumor may be reinoculated into a new host, and this process may be continued indefinitely with the tumor retaining its characteristic morphology, growth rate, and biochemical properties in each host. Most solid tumors are maintained in this manner, but some mouse tumors have been converted into ascites tumors or suspensions of tumor cells growing in the peritoneal fluids of a host, just like cells in vitro. Within 10 days after inoculation, the entire abdominal region is swollen with an ascites tumor. Though tumors retain their original properties through many transfer generations, changes do occur, as a result of mutation or chromosomal aberration.

Tumor cells can be maintained in vitro as permanent cell lines, exhibiting characteristic growth rates and enzymatic constitutions. If reinoculated into host animals they again grow as a tumor. Such tumor lines may be stored in a frozen state indefinitely and thawed at any time to resume growth, either in vitro or in vivo, without loss of any diagnostic properties.

Transplantability is usually host-specific; a tumor must be genetically and immunologically compatible with the host. Otherwise the host's immune system rejects the tumor as foreign. Tumors that arise in strain A mice cannot be transplanted to C3H mice; they are maintained only in strain A mice. It is rare for a mouse tumor (spontaneous or induced) to be transplanted easily to several host strains as soon as it appears. Host specificity may be lost after many transplant generations in one strain, and a tumor will grow as well in other strains. Presumably such tumors have lost histocompatibility antigens that induce graft rejection by other strains. Loss of host specificity is typical of the more malignant neoplasms. Phenotypic stability of transplantable tumors suggests that all tumor cell properties are due to heritable genetic changes, perhaps to mutations in somatic cells.

## Somatic mutation hypothesis

It was Boveri in 1902 who proposed that neoplastic transformation results from mutations in somatic cells. He observed abnormal chromosomes in malignant tissues and concluded that such aberrations were somehow responsible for the neoplastic state. A modern version of this somatic mutation hypothesis says that tumor cells, arising from normal cells, are genetically different because of one or more nuclear gene mutations. This hypothesis excludes extranuclear genetic systems as sites of heritable neoplastic change.

Three lines of evidence in support of the somatic mutation hypothesis will be briefly examined: (1) correlation between malignancy and a high frequency of chromosomal abnormalities in tumor cells, (2) correlation between mutation and carcinogenesis, that is, many chemical mutagens are also carcinogens, (3) the absence in many tumors of cell-specific proteins, suggesting a loss in genetic information.

*Chromosome abnormalities.* Many malignant tumors do have abnormal chromosome complements. Spontaneous and experimentally induced tumors usually have chromosome numbers different from their tissues of origin. They are often hyperdiploid with extra copies of many chromosomes. There is, however, no common pattern of chromosome number that correlates with tumor type; chromosome abnormalities seem to arise at random and are not tumor-specific. Even within the same tumor, chromosome numbers may change during tumor development without any obvious effect on growth rate or malignancy (Figure 21-2). Although an unbalanced karyotype could account for abnormal growth of these cells, the fact that many spontaneously arising tumors in man and animals display normal karyotypes weakens the argument; that is, gross chromosomal damage or aneuploidy is not necessarily causally related to neoplastic transformations. Abnormal karyotypes may arise secondarily from mitotic abnormalities that occur frequently in rapidly proliferating cell populations. Even normal cells, when cultured in vitro and stimulated to high rates of proliferation, show similar karyotypic changes without becoming malignant. Several "nor-

mal" permanent cell lines are hyperdiploid or polyploid, yet they do not develop as tumors when inoculated into animals. On the other hand, many permanent cell lines derived from normal tissues do become malignant after long-term culture in vitro, and this transformation often correlates with the appearance of abnormal karyotypes (Figure 21-3).

In only one known human tumor, *chronic myelocytic leukemia*, does malignancy correlate with a specific abnormal chromosome. One chromosome, the Philadelphia chromosome (because it was discovered in Philadelphia), lacks approximately one-half of the long arm of chromosome 21 and is believed to be a specific morphological marker for these leukemic cells. This chromosome may contain gene sets specifically concerned with the regulation of growth and if some are lost, abnormal proliferation results. The correlation is not always evident, since many leukemic patients possess an intact Philadelphia chromosome; that is, their karyotype is normal.

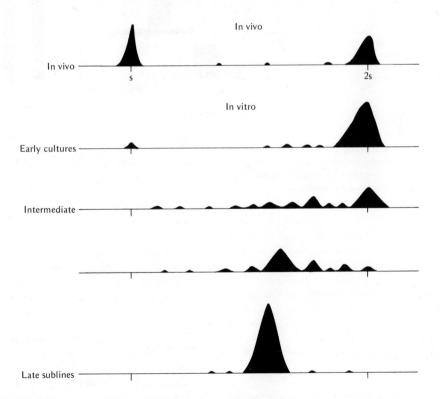

**Figure 21-2** *Chromosomal changes in a population of Novikoff rat hepatoma cells grown in culture. In vivo ascites cells show near-diploid (s) and near-tetraploid (2s) components. Sublines in culture show a predominance of 2s component at first followed by a heteroploid shift to a new equilibrium in late sublines. (Redrawn from T. C. Hsu and O. Clott, J. Nat. Cancer Instit. 22:313–339, 1959.)*

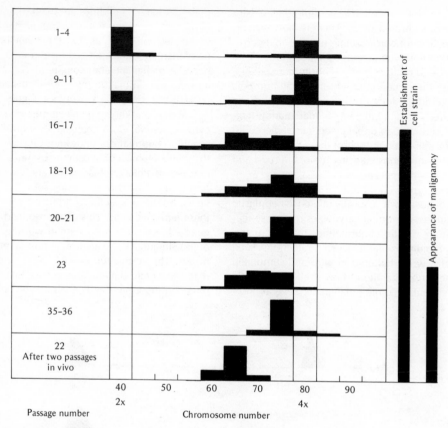

**Figure 21-3**   *Malignant transformation in vitro preceded by karyotypic changes. A population of normal embryonic mouse skin cells in primary culture (passages 1–4) show two modalities, at diploid (2x) and tetraploid (4x) with a few heteroploid cells. Continued passage results in a shift toward the heteroploid, particularly when the cells become a permanent cell line. Only six to seven passages later, the line becomes neoplastic without any change in its karyotype, but after two passages in vivo it assumes a new stable modality. (From A. Levan,* Achste Jaarboek van Kankeronderzoek en Kankerbestrijding in Nederland. *Lund: Carl Blons Boktryckeri, 1958, pp. 110–126.)*

And, in some leukemic patients, many cells derived from bone marrow differentiate and proliferate normally though they possess the Philadelphia chromosome.

These ambiguous results suggest that it is not the unbalanced chromosome condition per se that causes neoplasia; it is the possession of particular altered gene sets that is important. Only if these mutated genes are present does the cell exhibit a malignant phenotype.

*Carcinogens as mutagens.* Many mutagenic agents such as X-irradiation induce chromosomal abnormalities and are also carcinogenic, a result that argues

strongly for a mutation theory of cancer. Here, too, the correlation between mutagenicity and carcinogenicity does not hold up for all mutagens. Many chemicals are mutagenic without being carcinogenic, and vice versa. Since some agents possess both properties, the argument has some validity, but the evidence is ambiguous, even with agents such as radiation. For example, 2 percent of normal hamster cells treated with 300 $\mu$g of X-irradiation in vitro are transformed into tumor cells, and as many as 20 percent may be transformed in vitro with carcinogenic hydrocarbons. Both these transformations are far higher than one would expect from random mutations.

Although a single agent may cause mutations in one organism and induce a neoplastic transformation in another, it cannot be concluded that the tumor is caused by a somatic mutation. One reason for the ambiguity is that most animal systems used for mutagenesis studies are unrelated to those used for studies of carcinogenesis. Moreover, many carcinogenic agents that are nonmutagenic may be metabolized by some organisms and transformed into effective mutagens. To resolve this problem a sensitive prokaryote test system for mutagenesis was devised by Ames at the National Cancer Institute. *Salmonella typhimurium* histidine mutants, which require histidine to grow, exhibit a very low frequency of back mutation; that is, few revertants to histidine independence are found. The frequency of revertants is increased by agents which cause back mutation. Using this system, Ames could test a series of known carcinogens for their ability to increase the frequency of back mutation to histidine autonomy. Many compounds tested were also mutagenic. Moreover, the degree of carcinogenicity correlated with the frequency of induced mutation; the most effective carcinogens produced the highest back-mutation rate.

Most carcinogens acting as mutagens behave like intercalating agents, flat, planar molecules that slip into the spaces between stacked nucleotides in the DNA helix and cause frameshift mutations by upsetting the reading of nucleotide sequences. The fact that carcinogenicity and mutagenicity of the compounds are both enhanced by the addition of side groups that facilitate DNA binding is further argument in support of somatic mutation as a cause of neoplasia.

Some carcinogenic compounds have no apparent mutagenic effect in the *Salmonella* system. What about these? Ames and his group reasoned that such compounds may be transformed into mutagenic agents by detoxifying enzymes in host tissues, particularly the liver. To test this possibility, they prepared an incubation mixture containing a known carcinogen, a homogenate of liver tissue, and the *Salmonella* test organism. If their hypothesis was correct the enzyme systems in liver homogenate might transform the carcinogen into a more effective mutagen. Fifteen different carcinogens were tested in this manner and all became mutagenic only after incubation with the homogenate.

These experiments suggest that neoplastic transformation involves mutations at the DNA template, a change in genetic information that may interfere with regulation of DNA replication and cell division. The transformation probably requires more than one gene mutation. Some have proposed that the long lag for most human tumor development reflects the fact that several mutations must accumulate before a neoplasm acquires autonomous growth behavior.

*Gene deletion and protein loss.* As we have already seen, many chemical carcinogens bind directly with DNA in strong covalent linkages and seem to produce mutations or deletions of genes coding for growth regulation. As a result, tumors lack the specialized organelles or enzyme systems characteristic of their normal progenitors. Even antigenicity of tumor cells is abnormal; some histocompatibility antigens may be lost while new ones are acquired. Such cells are recognized as foreign by host immune systems and may be destroyed. Stable losses of enzymatic and antigenic proteins point to gene mutations, gene deletions, or some other change of genetic information during neoplastic transformation.

If genetic information is lost in tumor cells, then it should be possible to reintroduce the information and "cure" a tumor by fusing a normal cell with a tumor cell by somatic cell hybridization (see Chapter 3). Hybridizing mouse tumor cells with normal mouse fibroblasts should produce a normal hybrid since a normal cell would furnish regulatory factors absent in the tumor. In some experiments, the tumor phenotype is retained; after inoculation into hosts a tumor develops from hybrid cells. Though hybrids exhibit antigenic specificity of the normal cell surface, the principal tumor property of uncontrolled growth is not cured. Such experiments are difficult to interpret, however, because the hybrid cell lines could throw off variants or may even contain contaminating tumor cells. A contaminating malignant cell would be selected in the host and proliferate, thereby giving the impression that a tumor cell phenotype was retained.

Other investigators obtained contradictory results (Figure 21-4). Fusion of a wide range of malignant ascites tumor cells with mouse L cells resulted in nontumorigenic hybrids. When the hybrid cells were injected into hosts, no rapidly growing malignant tumors were obtained. The highly malignant phenotype was suppressed in subsequent generations. The L cell genome contributed something that suppressed tumor cell malignancy; that is, growth control information, lost by the tumor cell, was supplied by the L cell genome.

Passage of hybrid cells through host animals resulted in malignant tumors only when certain chromosomes were lost. If a hybrid cell line retained all or most parental chromosomes, then no malignant tumor appeared. The malignant phenotype acts as a recessive in the presence of normal genes.

These experiments may be interpreted in several ways: the neoplastic transformation could involve gene deletions or recessive mutations that are overcome in the presence of the dominant gene of the normal cell. On the other hand, epigenetic factors controlling gene expression in the normal cell may also control abnormally functioning regulatory genes in tumor cells. In other words, the experiment does not distinguish a purely genetic (somatic mutation) from an epigenetic mechanism for neoplastic transformation.

Most evidence supports the original Boveri hypothesis for somatic mutation origin of neoplasia. Evidently mutation in host genetic material affects regulation of DNA synthesis and leads to neoplasia. A wide range of chemical and physical agents can induce these changes but, in each case, the regulatory genes involved may be those coding for components of the cell surface. We have seen that primary regulation of DNA synthesis and cell division may be mediated by cell surface events (Chapter 5). Could the cell surface be the site of the primary neoplastic lesion?

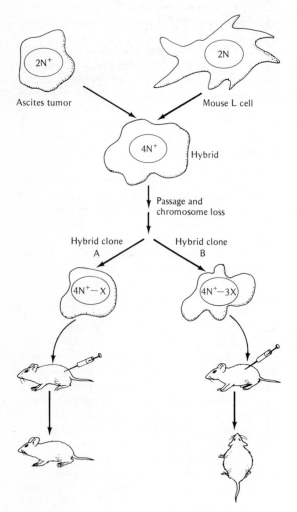

**Figure 21-4** *Somatic cell fusion of normal and ascites tumor cell. Chromosome numbers of cell lines are shown in nuclei. After several passages, hybrid cell lines lose chromosomes and form clones A and B with different chromosome numbers. Clone A, which retains most parental chromosomes, is nonmalignant when injected into a host animal. The B clone, however, which has lost many more parental chromosomes, is malignant. Loss of certain genes may lead to neoplasia. $2N^+$ = hyperdiploid; $2N$ = diploid; $4N^+$ = hypertetraploid; $4N^{\pm}X$ = hypertetraploid minus one chromosome; $4N^{\pm}3X$ = hypertetraploid minus three chromosomes.*

## AN ALTERED CELL SURFACE

Most tumors, whether they are spontaneous or induced, behave as if they have altered cell surfaces. Whereas normal cells exhibit contact inhibition of proliferation and motility when grown in vitro, tumor cells are not contact-inhibited but grow as dense colonies, many cell layers thick, and migrate indiscriminately over their neighbors. Compared to their normal progenitors, tumor cells possess an increased net surface charge on their outer membrane, migrate at different rates in an electrophoretic field, and exhibit different histogenic patterns when cultured in vitro together with normal cells. Normal cells organize into compact parallel arrays while tumor cells tend to spread, at random, as diffuse masses and penetrate normal cells. Enzymatic removal of membrane components such as sialic acid decreases tumor cell surface-charge density; they organize into compact masses and segregate from adjacent normal cells.

Neoplastic cells are also less adhesive; they are more readily detached from a substratum than normal cells and are more motile, which may explain their greater invasive qualities. Antigenicity of the tumor cell surface is also different. Tumor-specific antigens appear in viral-transformed cell membranes, whereas spontaneously arising tumors and those induced chemically possess surface antigens absent in corresponding normal tissues. Little is known about the nature of these surface antigens except that antibodies prepared against them often prevent tumor cell growth. This does not prove that surface antigens play an important part in transformation; it does suggest that neoplastic cell membranes may contain constituents not found in their normal progenitors. Sometimes embryonic or fetal antigens, normally found in embryonic cell surfaces but absent in adult cells, usually reappear on the surface of tumor cells.

Outer surfaces of tumor cells do exhibit a different structure from their normal counterparts; they possess more microvilli, are more "fluid," and seem to be more active. Surface changes could result from (1) the loss or addition of surface constituents or (2) reorganization of existing surface components by conformational changes in membrane proteins or by displacement of protein within the membrane. Moreover, if a surface change is the primary lesion in tumorigenesis, it must somehow relate to mechanisms that regulate DNA synthesis and cell division.

We have already discussed membrane-mediated regulation of DNA synthesis and cell division (see Chapter 5). Though the nature of the membrane-signaling circuit is unknown, we noted that neoplastic cell surfaces, in contrast to normal cells, are sensitive to agglutinating plant lectins such as concanavalin A (Con A). Normal mitotic cells are also agglutinable by lectins, which suggests that proliferating cells undergo a surface change that exposes Con A binding sites. The agglutinin combines with glycoprotein membrane receptors containing sugar moieties. Note that normal cells infected with nontransforming virus also become sensitive to Con A agglutination; a membrane component is altered by virus infection to bring about the change, suggesting that exposure of Con A binding sites and their greater accessibility are not necessary conditions for neoplastic transformation.

Neoplastic cells can be made to behave like normal cells in vitro by treatment with a modified agglutinating agent. Concanavalin A agglutinates cells because it has four reactive sites, each able to combine with a cell surface receptor to form a bridge. Agglutinin molecules with only one reactive site (univalent) are made by enzymatic (trypsin) fragmenta-

tion. Univalent molecules combine with receptors on neoplastic cells but do not agglutinate because no bridges are established. When neoplastic cells are treated with the univalent reagent, they behave as normal cells, exhibiting contact inhibition of growth and organizing into typical monolayers. Apparently, the univalent agglutinin blocks those surface sites which generate positive signals for mitosis and the cells stop dividing on contact.

Mitotic activity correlates with intracellular cyclic AMP levels, inasmuch as its concentration is lowest during mitosis and highest at most other phases of the cell cycle. Though neoplastic cells show some fluctuations, these are at much lower levels of cyclic AMP; in general, tumor cells contain less cyclic AMP. Both contact inhibition of growth and motility are restored when neoplastic cells are treated with dibutyrl derivatives of cyclic AMP. The pattern of cell growth resembles that of normal cells. The transformation by cyclic AMP is inhibited by the addition of colchicine, which suggests that microtubules may be necessary for the changes in cell shape and perhaps for the decline in proliferation.

We do not yet understand how changes at the cell surface control DNA replication in normal and neoplastic cells. The effect of cyclic AMP implicates a surface enzyme, adenylcyclase, in growth regulation by virtue of the diverse regulatory activity of its product. It may regulate gene transcription by its action on histone phosphorylation or it may regulate the assembly of microfilaments and microtubules associated with the cell surface in some manner.

To examine the latter possibility, fluorescent antibodies specific for actin have been applied to normal and neoplastic cells. Normal cells show a complex network of actin bundles, but only a diffuse fluorescence is detected in the cytoplasm of tumor cells, or in normal cells transformed by viruses. Tumor cells that revert to normal rapidly acquire the actin fiber system which shows that the actin is present at all times. Actin extracted from neoplastic cells is structurally similar to normal actin and is present in equivalent amounts. What does change is the relative proportion of soluble to polymerized actin with the former predominating in tumor cells.

Disruption of actin fibers is matched by equivalent changes in microtubules. Here, too, fluorescent antibodies against purified tubulin show a disorganized microtubular network in tumor cells.

These observations suggest that neoplastic transformation is accompanied by a change in the subsurface system of organelles, a surface modulation system that may regulate the movement of membrane constituents (see Chapter 8). The question remains: does this surface change affect those genetic factors that control cell division and growth?

## Invasiveness and metastasis

A change in neoplastic cell surfaces is also expressed as reduced adhesiveness. About 20 years ago, Coman suggested that reduced adhesiveness could explain invasive and metastatic properties of the neoplastic cell. Tumor cells readily penetrate spaces between normal tissue cells, possibly because their surfaces are more motile. Actively moving cells at the surface of a tumor penetrate tissue spaces, attach to normal cells, and pull other tumor cells along by pseudopodial activity. They tend to separate from a primary tumor site, move into neighboring tissues, and destroy them by disrupting tissue organization.

Invasiveness is not an exclusive property of tumor cells. Normal cells, such as lymphocytes, readily penetrate capillary walls and invade tissue spaces. Trophoblast cells of the developing mammalian embryo successfully invade the uterine endometrium. Though initially invasive, these embryonic cells eventually differentiate into a noninvasive syncytium when the placenta develops. Nevertheless, failure of trophoblasts to differentiate may result in formation of a chorioepithelioma, a highly malignant tumor of the trophoblast cell.

Invasiveness can be followed in vitro in mixtures of normal and neoplastic cells. Fibroblasts, grown on a sponge matrix, may be inoculated with a small population of malignant cells. The tumor cells migrate into the matrix and invade the fibroblast tissue, the degree of invasiveness varying with the amount of fibroblastic growth and the orientation of connective tissue cells. Invasion is minimal when fibroblasts lie parallel to the surface of the tumor implant and maximal when they are at right angles to the margin of the implant.

Malignant cells may enter the bloodstream by invading the tumor capillary bed or by directly penetrating the lymphatics. They then become metastatic, that is, freely floating cells, transported to other organ sites by the bloodstream where they establish secondary foci of growth. Metastasizing cells may localize in a variety of vital organs. They retain enough adhesive properties to attach to the endothelial wall of a capillary vessel, and once having done so, they reinvade the surrounding tissues and grow.

We do not know why some tumors are invasive and others are not. Growth rate may be a factor since an increasing bulk of a rapidly growing tumor could, by mechanical pressure, penetrate contiguous tissues. To metastasize, however, tumor cells must invade capillary walls and enter the bloodstream. Since macrophages and lymphocytes do this all the time, any motile cell with certain adhesive properties can easily penetrate endothelial walls. Apparently, the neoplastic transformation confers this property on many tumor cells.

Since metastasis is a more frequent occurrence in humans than in animals, it is less amenable to experimental analysis. When neoplastic cells are detected in primary veins leading from tumors, they are not necessarily metastasizing since many cells die while still in the circulation, and many fail to attach and penetrate other tissues. Most tumor cells detected in the circulation do not establish secondary tumor foci.

Passage of tumor cells through capillary walls is facilitated by local inflammation, since leucocytes easily pass through walls of inflamed venules. Such regions, colonized by leucocytes, might serve as "ports of entry" for malignant cells because conditions at the inflamed site (enzymatic?) are optimal for the entrapment of circulating neoplastic cells.

Changes in membrane permeability, adhesiveness, and contact inhibition may modify growth control. Altered surface receptors, or second messenger systems, may desensitize a cell's responsiveness to exogenous signals and stimulate unrestrained growth.

## DEDIFFERENTIATION: SIMPLIFICATION OF METABOLIC FUNCTION

Many tumor cells often resemble embryonic cells; they have dedifferentiated, losing most specialized functions. According to one hypothesis, neoplastic transformation results from breakdown of normal mechanisms of cytodifferentiation. If differentiation and cell division are, in fact, mutually exclusive activities, then failure to differentiate might very well leave the alternative possibility of uncontrolled proliferation. Such cells may acquire a stable neoplastic phenotype.

Many tumors do become *anaplastic* (unspecialized) and lose all resemblances to their tissue of origin. In general, however, tumors with a common origin exhibit varied morphological patterns. For instance, mammary tumors of mice are diverse, from solid adenocarcinomas without any glandular elements to adenocarcinomas with large macroglands filled with secretory materials (Figure 21-5).

Anaplastic simplification is also found at the biochemical level; tumors lose enzyme systems characteristic of the tissue of origin. Only normal tissues exhibit tissue-specific enzyme profiles; that is, some enzymes are found in only one tissue. For example, kidney has very high alkaline phosphatase

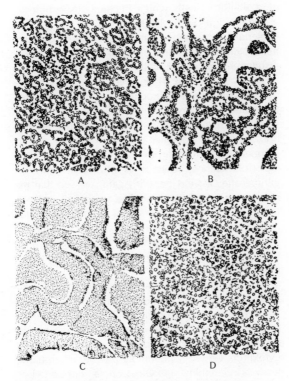

A          B

C          D

**Figure 21-5** *Diversity of morphological structure of mouse mammary tumors. (A) Adenocarcinoma with fine uniform acinar structure. (B) Adenocarcinoma with cyst formation and papillary ingrowth. Small glandular structures are seen in more cellular areas. (C) Adenocarcinoma with macroglandular structure. Necrotic material seen between broad bands of epithelial tissue. (D) Adenocarcinoma with solid cellular formation with no apparent glandular structure. (From T. B. Dunn,* Physiopathology of Cancer, *F. Homberger, ed., 2d ed. New York: Hoeber-Harper, 1959.)*

and liver has high catalase and arginase activity. Tumors, however, lose these distinctive biochemical features and acquire a uniform enzymatic pattern. Enzymatic heterogeneity characteristic of normal tissues is erased, as if all tumors have converged toward a common biochemical pattern. Though tumors lose enzymes, most differences between tumor and normal cells are quantitative rather than qualitative. Hepatomas have lower levels of catalase and arginase, but slightly more acid phosphatase activity than normal liver. Some endocrine tumors lose the ability to synthesize hormones, while others show reduced rates of secretion. There is no common biochemical pattern for all tumors; what is significant, however, is that tumors tend to resemble one another more than they do normal cells.

Many tumors retain specialized cell functions; they secrete hormones, synthesize pigment, and lay down keratin. A chorioepithelioma secretes high amounts of gonadotrophic hormones which can be detected in the urine. Melanoma, one of the·most malignant tumors extant, synthesizes large amounts of pigment. Even sensitivity to hormonal signals is retained by tumors; for example, spontaneous mammary tumors of mice secrete casein when stimulated, just like normal mammary epithelium. Many other aspects of functional regulation do persist in tumor cells. Though neoplastic cells exhibit phenotypic variation, they do so within the repertoire of normal cells from which they are derived.

This means that dedifferentiation per se is not the primary lesion in tumorigenesis nor is it, as we have seen, a general feature of all neoplastic cells. When neoplastic cells dedifferentiate, they do not revert to an embryonic state because embryonic cells respond to morphogenetic cues while tumor cells do not. The biochemical simplification that characterizes many tumors results from intense proliferative activity which, as we have seen, also leads to mitotic abnormalities and chromosomal aberrations. Such unbalanced genomes may produce dedifferentiated cell phenotypes. Even normal cells dedifferentiate when explanted in primary culture because they, too, proliferate actively. Within a short time, normal specialized cells lose enzymatic properties, a sign that biochemical machinery is changing (see Chapter 17).

At one time it was thought that embryonic cells that fail to differentiate, or so-called embryonic rest cells, are the progenitors of all tumors. This hypothesis was based on the fact that many normal cell cultures became neoplastic after long-term growth in vitro. It was assumed that neoplasia resulted from preexisting embryonic rest cells preferentially selected for growth in vitro. Biochemical simplicity and high proliferation potential of tumor cells was attributed to lesions in differentiation of embryonic cells. Most evidence does not support this hypothesis. The kinetics of biochemical simplification are inconsistent with selection of preexisting undifferentiated cells since each biochemical property declines independently in explanted cells and at different rates (see Figure 17-43). If selection was involved, then all properties should decline with the same kinetics during the period of culture. It is, therefore, unlikely that embryonic rest cells are progenitors of tumors.

## EPIGENETIC ORIGIN OF A TUMOR

Although embryonic rest cells are ruled out, defects in epigenetic mechanisms of differentiation might lead to the appearance of neoplastic cell phenotypes. According to this hypothesis, tumors arise just like all other cell phenotypes, by stabilization of the expression of genes that program a proliferative state. In this case, factors that normally repress division genes in functionally specialized cells do not operate and cells remain indefinitely in the cell cycle. This means that no mutation need occur, nor is any genetic information lost; only the fine control system of gene regulation is disturbed. If this is so, then under certain conditions tumors may be induced to revert to normal and differentiate, without benefit of mutation or any other genetic change.

There are some plant tumors that seem to arise as a defect in an epigenetic control mechanism rather than by somatic mutation. These tumors exhibit all neoplastic properties of autonomous proliferation, heritability, transplantability, and dedifferentiation but will differentiate into normal tissues under permissive conditions. This fact alone implies that somatic mutation is not responsible for abnormal growth behavior in these cells.

The crown gall tumor of tobacco plants is probably the best example. Crown gall is initiated on a host plant by a dual stimulus, a tissue wound followed, secondarily, by infection with a crown gall bacterium, *Agrobacterium tumefaciens*. Plants normally respond to wounds by proliferation of a disorganized tissue, known as a *callus*. The wounding reaction stimulates callus proliferation, but the infecting bacteria transform these cells into a crown gall tumor (Figure 21-6). In contrast to normal callus, crown gall tissue grows vigorously in vitro without any supplemental hormones; a basic medium of salts and glucose suffices. It is transplantable to other host plants, does not differentiate into roots or shoots, and exhibits all properties of an autonomously growing tumor.

Transformation involves two steps: a conditioning event (wounding) followed by an induction (bacterial infection). Wounding induces a population of proliferating cells which become susceptible to transformation by the bacterium. If the wound and bacterial inoculation are made simultaneously and the bacteria allowed to act only for 24 hours (a period insufficient to permit cellular transformation), no tumor is produced. If, however, wounding is induced 2 days prior to inoculation, large tumors develop after 24 hours. Cellular proliferation and DNA synthesis are necessary for maximum tumor induction. Only new daughter cells seem to acquire the competence to respond to a tumor-inducing principal. Competence, however, is a transient state, declining after a short period. Again, we see that reprogramming a genome requires an episode of DNA replication, a pattern resembling normal differentiation. In this case cells are reprogrammed for autonomous proliferation instead of differentiation. Once transformed, the new phenotype is stable and transmissible to all progeny.

Immediately after transformation, crown gall cells become biochemically autonomous. Most normal plant cells are unable to grow in vitro on a simple medium of inorganic salts and sucrose but do so after the addition of such hormones as auxin and cytokinin. Crown gall tumor, on the other hand, grows indefinitely in vitro on a hormone-free medium because it synthesizes high levels of auxin and cytokinin and secretes them into the medium. The amounts secreted are enough to support the growth of transplanted pieces of normal callus tissue or pith parenchyma. An ability to synthesize these two hormones is repressed in normal differentiated cells, which suggests that neoplastic transformation is accompanied by derepression of biosynthetic pathways for both auxin and cytokinin.

Autonomy includes activation and stabilization

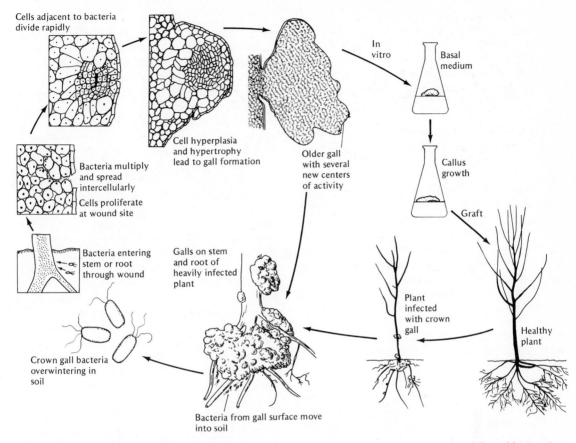

Cells adjacent to bacteria divide rapidly

Cell hyperplasia and hypertrophy lead to gall formation

Older gall with several new centers of activity

In vitro

Basal medium

Callus growth

Bacteria multiply and spread intercellularly

Cells proliferate at wound site

Graft

Bacteria entering stem or root through wound

Galls on stem and root of heavily infected plant

Plant infected with crown gall

Healthy plant

Crown gall bacteria overwintering in soil

Bacteria from gall surface move into soil

**Figure 21-6**  *Natural cycle of crown gall disease caused by* Agrobacterium tumafaciens. *Wounding followed by invasion of the bacterium initiates hyperplasia and formation of the gall. Fragments of crown gall tissue will grow on a simple basal medium without hormones in vitro and can be maintained indefinitely as a callus.*

of other biosynthetic pathways necessary for growth. In fact, the degree to which specific biosynthetic reactions are activated determines the nature of the neoplastic change. If the tumor-inducing principal (bacterium) acts for 34 to 36 hours, only slowly growing tumors develop. A 50-hour exposure results in tumors growing at a moderately fast rate, whereas tumors initiated after 72 to 96 hours grow most rapidly and are the most autonomously developing tumors. Each tumor retains its growth rate indefinitely in culture, which suggests stepwise progression of a normal cell to a fully autonomous tumor cell.

Variable growth rate of tumor clones is matched by biochemical differences. Slowly growing tumors are the most fastidious in vitro, requiring addition to a basal culture medium (salt and glucose) of auxin, glutamine, myoinositol, asparagine, and guanylic and cytidylic acids before fully autonomous growth rates are achieved. Moderately fast-growing tumors require fewer supplements; they need auxin, glutamine, and myoinositol but can dispense with the latter three. The fully autonomous tumor, as we have seen, synthesizes optimal amounts of all growth factors, and can grow profusely on a minimal medium without any supplements. Inasmuch as normal cells require all growth factors plus cytokinin to grow vigorously in vitro, the transformation from normal to the slowly growing tumor involves derepression of cytokinin synthesis. All three tumor types acquire this ability, which is the first step in the progression to autonomy.

In these tumors, heritable changes leading to autonomous growth depend upon derepression of at least seven gene systems coding for specific biosynthetic capabilities usually repressed in normal tissues. In situ growth of normal pith parenchyma is stimulated by hormones and growth factors supplied by distant tissues such as apical bud. Crown gall tumors, on the other hand, are independent of exogenous hormones because they synthesize their own in large amounts. Though the biosynthetic machinery for growth is activated, the "control valves" do not function.

As in animal tumors, crown gall tumor cells possess an altered plasma membrane with increased permeability to all inorganic ions, which suggests that here too the primary lesion may be at the surface. All but one biosynthetic pathway activated in the tumor are stimulated by inorganic ions. Even normal cells grow as a tumor if potassium concentration in the medium is increased. Six of seven essential metabolites synthesized by crown gall tumor cells are also synthesized by normal cells in high potassium. After serial propagation in basal medium with high potassium, a normal callus loses its dependence on potassium and becomes autonomous. Hence transformation does not require a bacterial tumor-inducing principal or any wound hormones. By a process of habituation in culture, normal cells are transformed into tumor cells simply by changing ionic balance!

Genetic information for these metabolic systems, though present in normal cells, is repressed unless the cells are exposed to high concentrations of potassium ions. We assume the tumor-inducing principal modifies the membranes of newly formed daughter cells at the wound site, changes permeability, and permanently activates several growth-factor synthesizing systems. According to this model, regulation of ion traffic is a primary site of growth control.

## Instability of the crown gall tumor phenotype

If the neoplastic phenotype is based on epigenetic changes, then an abnormal phenotype should be reversible to a normal, nonautonomous state under certain conditions; its genetic information is intact and unchanged. A high rate of reversion of neoplastic to normally growing cells would not be compatible with the somatic mutation hypothesis; back mutation is rare in most systems.

Though crown gall tumors are disorganized and do not form roots, shoots, or leaves, some show organizing potentialities. For example, a partially organized tumor, a teratoma, may be induced by treating cut surfaces of a plant stem with a nonvirulent bacterial strain (Figure 21-7). Two types of tumor are produced at the wound sites, depending on the position of the cut surface in relation to an apical bud. An unorganized autonomous tumor usually develops at the upper wound surface in stems with intact apical buds, while teratomas develop at the lower wound surface in stems lacking buds. A teratoma consists of a chaotic assembly of roots, leaves, and shoots.

Isolated fragments of each tumor maintain their respective patterns in the early stages of growth in vitro on simple media. Within a short time, however, the unorganized tumor forms an abnormal collection of organs and resembles the teratoma. The unorganized tumor exhibits two different patterns of growth depending upon its transplantation site in intact hosts. If a fragment of the tumor is placed at the cut surface of an apical tip, it exhibits teratomatous growth, but if grafted into an internode region in a stem with an intact apical bud, then growth is completely disorganized. The apical bud exercises hormonal control over the behavior of the tumor fragment; morphogenetic expression by this tumor depends on its position in the plant axis.

The originally disorganized tumor can be made to revert to normal by repeated grafting of fragments to cut surfaces from which apical buds have been removed (Figure 21-8). After the first graft, abnormal leaves and shoots grow profusely. Fragments of shoots regrafted to a cut tip of another healthy host grow more slowly but are abnormal since they still grow as a tumor in basal medium without any nutritional supplements. At the third graft, reversal to normal is complete; shoots develop into a normal stem, flower, and set fertile seed. Forcing abnormal tumor shoots into organized growth on healthy plants results in a complete reversal to a normal phenotype.

The appearance of a normal phenotype could result from normal cells contaminating the original tumor. These would be preferentially selected under certain conditions of growth. If, however, the tumor is cloned in vitro, no cases of normal callus growth are obtained. Moreover, all cells of the original teratoma are neoplastic, since each grows into a teratoma when isolated as a clone.

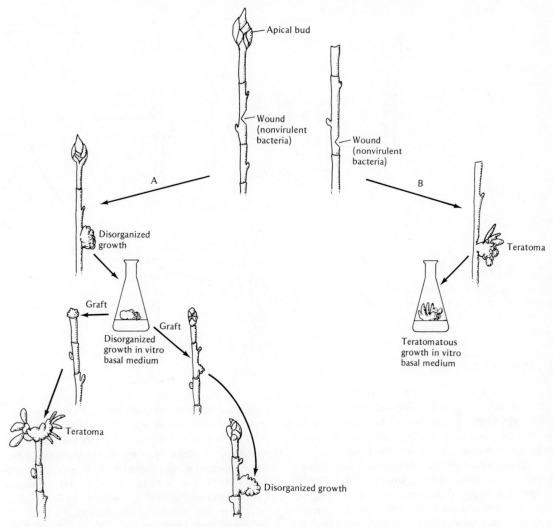

**Figure 21-7** *Induction of teratomatous crown galls with nonvirulent strain of Agrobacterium. In the A pathway, infection of a wound site of a twig with apical bud leads to a disorganized growth. This grows as a disorganized tumor in vitro, but if grafted to an apical bud site, it develops as a teratoma. An internodal graft continues to grow as a disorganized callus. In the B pathway, a teratomatous growth is obtained with disorganized shoots and roots if the apical bud is removed.*

This neoplastic transformation does not involve loss, rearrangement, or mutation of genetic material. Instead, an aspect of gene regulation is modified after transformation. A new genetic state is stabilized and passed on to all future progeny by replication, just as the determined or differentiated state of normal cells is perpetuated.

Summarizing, normal plant cells transformed into a crown gall tumor are stimulated to autonomous proliferation by derepression of several biosynthetic pathways, possibly after a primary change in membrane permeability. The information flow system is affected without modification of the genetic material. The growth behavior of such cells is subject to selection pressures within a host hormonal background and may be channeled to disorganized

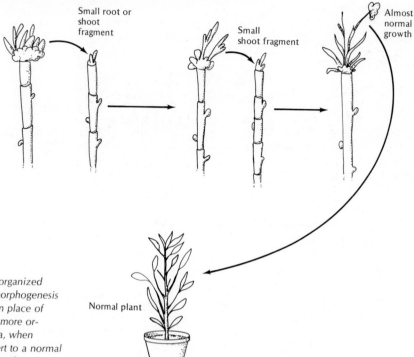

**Figure 21-8**  *Reversion of disorganized crown gall tumor to normal morphogenesis by repeated grafting to twigs in place of apical buds. Fragments of the more organized tissues in the teratoma, when grafted several times, can revert to a normal plant which flowers and sets seeds.*

or teratomatous growth. The relative ease with which disorganized tumors can be made to revert to normal growth patterns indicates that the initial neoplastic change does not involve mutation.

It is uncertain whether this mode of tumor origin applies to all animal tumors. Instances of malignant tumors reverting to more benign conditions with increased differentiation have been reported. A malignant human neuroblastoma grown in vitro, for example, forms populations of undifferentiated neuroblasts, but after 5 months, more mature neural elements such as ganglion cells appear. Epigenetic mechanisms may be involved in animal tumor development but most carcinogenic agents such as chemicals, radiation, and viruses seem to induce changes in genetic information. Of these agents, the mechanism of viral action is best understood.

## VIRAL TRANSFORMATION

Indirect support for the somatic mutation hypothesis comes from viral-induced neoplastic transformations. Incorporation of viral genes into a host cell genome

disrupts regulatory mechanisms and stimulates uncontrolled DNA synthesis and proliferation. In principle, this differs from the somatic mutation hypothesis because the virus is an infectious agent. But the intimate association of viral genomes with host genomes, as lambda phage integrates into the *E. coli* genome, resembles a somatic mutation; that is, host genetic information is changed.

RNA and DNA viruses are known etiologic agents in avian sarcomas and mouse leukemia. These viruses are transmitted between individuals in a population by the usual routes of viral infection (horizontal transmission) and tumors develop in sensitive animals. Other mammalian tumors are viral-mediated but to date no human tumor has been proven to be of viral origin.

### RNA oncogenic viruses: the Rous sarcoma virus

Peyton Rous, in 1911, showed that a virus extracted from chicken sarcomas would induce sarcomas when reinoculated into tumor-free animals. This virus (Rous sarcoma virus) contains a single strand of RNA surrounded by a protein-lipid coat and is one of several RNA viruses associated with tumors in various animals.

Embryonic chick fibroblasts in vitro undergo a neoplastic transformation if infected with Rous virus and exhibit unrestrained growth. Normal fibroblasts show a typical growth curve in culture with growth ceasing after cells are contact-inhibited. Within hours after addition of Rous virus, however, infected cells display different growth kinetics. After a short lag, cells grow exponentially without entering a stationary phase. The transformed cells are no longer contact-inhibited and pile up instead as separate foci. These cells do not produce any detectable virus but develop into a typical sarcoma when inoculated into host animals.

How does the infecting virus alter cellular control mechanisms so that the cell (1) supports continued production of the viral genome and (2) is itself transformed into a stable neoplastic phenotype transmitted to all subsequent progeny? A model for such a mechanism comes from the behavior of lysogenic phage viruses infecting *E. coli*. The DNA of these viruses is integrated into and replicates with the host cell genome without destroying the host. The inactive prophage persists unless switched to the lytic pathway. But Rous virus is an RNA virus. How can it become integrated into host cell DNA, inasmuch as it is normally synthesized and assembled in the cytoplast of an infected cell? Temin and Baltimore, studying the replicative behavior of RNA viruses, have shown that shortly after Rous sarcoma virus enters a cell, DNA synthesis is detected. This DNA is not synthesized on host DNA templates. Instead, it is synthesized on a viral RNA template by a DNA polymerizing enzyme coded by the viral genome, called *RNA transcriptase*. Contrary to the accepted dogma, we have here synthesis of complementary DNA strands on an RNA template. If the template is destroyed by ribonuclease, no DNA synthesis is detected. According to the hypothesis, this newly synthesized DNA strand replicates and is inserted into the host cell chromosome as a provirus (Figure 21-9). Now the chromosome carries all viral genes, some of which induce neoplastic transformation. Information for neoplasia is brought into the host cell genome by an infecting virus; the new information disturbs mechanisms regulating host DNA replication.

Transcriptaselike enzymes have also been found in uninfected rat and embryonic chick cells. According to Temin, these endogenous, RNA-depen-

dent DNA polymerases may be involved in normal somatic cell differentiation. The transcriptases are used to create new DNA sequences or *protoviruses* during the life of the cell. New RNA templates may arise, either as a result of mutation or as a consequence of metabolic changes. Some of these may be copied into DNA and inserted into the genome, thereby creating a modest amount of genetic heterogeneity in a cell population. Selection may favor certain new gene combinations which persist and perhaps become amplified into many tandem copies.

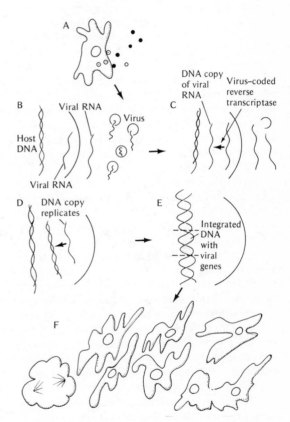

**Figure 21-9** *Model of Rous sarcoma virus transformation of chick fibroblast in vitro. (A) Infection by the virus. (B) Virus in cytoplasm releases RNA chromosome which enters host cell nucleus. (C) Reverse transcriptase enzyme uses viral RNA as a template and synthesizes DNA copies. (D) These replicate into double strands. (E) The strands integrate into host DNA, possibly utilizing viral recombination enzymes. (F) Inclusion of viral genes in host genome alters host regulatory machinery controlling DNA replication, and the cell exhibits uncontrolled proliferation.*

The insertion of protoviruses into specific sites in the genome could have corresponding effects on the cell phenotype. For example, insertion into a region coding for membrane constituents may alter cell surface specificity, or receptor properties. Insertion near genes controlling cell replication may modify cell division. This genetic system is more sensitive to exogenous factors such as carcinogens, and mutant protoviruses incorporated into the genome may induce neoplastic phenotypes by disrupting cell division control elements.

Support for this hypothesis comes from studies showing the presence of Rous viral gene sequences in somatic cells of natural populations of birds. DNA-DNA hybridization analyses show that normal bird cells carry many nucleotide sequences complementary to portions of the viral genome. Up to 50 percent of viral sequences may be found in normal somatic cell genomes, which suggests that they are components of the species genome and undergo evolutionary changes themselves, possibly according to the model proposed by Temin. Transmission of such sequences would occur normally through the germ line.

It is possible that transformation by the Rous virus depends on insertion of only one viral gene. Temperature-sensitive mutants of the virus have been obtained which, after infection, transform chick fibroblasts at the permissive temperature (37°C) but have no transforming effect at the restrictive temperature (41°C). The virus DNA replicates along with the cell at the higher temperature, but some genes are altered so that their product cannot induce transformation. The expression of these mutants can be reversed within a few hours by shifting the temperature. Recombination and complementation studies have shown that roughly 10 percent of the viral genome, about the size of one gene, is enough to transform normal cells to neoplasia.

The transforming sequences of the viral genome seem to be restricted to the normal cells of chickens, the principal host of the Rous sarcoma. Although species such as quail and duck carry viral sequences, they lack those that correspond to the transforming genes, suggesting that the viral genes are, in fact, in process of evolution in natural populations.

Transformation by Rous virus is accompanied by changes in the surface of the chick fibroblast. Although several new proteins appear in the membranes of transformed cells, the most interesting observation is that the organization of the surface modulating system, actin and tubulin, is disrupted. Cells infected with the temperature-sensitive mutant virus at the permissive temperature and showing the neoplastic morphology lack the actin or tubulin fiber structures when stained with fluorescent antibodies. Shortly after these cells are shifted to the restrictive temperature, the cells acquire the typical actin and tubulin bundles and a normal phenotype develops. This process is reversible but can be blocked by factors that inhibit protein synthesis. It seems the product of the transforming gene must be produced before actin and tubulin can polymerize into microfilaments and microtubules. Here we see that a gene product affects the surface modulation system and may secondarily affect the genes controlling DNA replication.

## DNA oncogenic viruses: polyoma

Cell changes associated with neoplastic transformation are better understood in the case of DNA oncogenic viruses. Polyoma virus DNA is circular, double-stranded, and enclosed in a protein coat. The virus infects various rodent tissues, particularly mice and hamsters, and induces multiple tumors. Cells infected by the virus exhibit two phenotypic responses, a lytic phenotype as virus proliferates and destroys the cell, or a host cell transformation as inactive virus integrates into the host cell genome and induces neoplastic changes (Figure 21-10). Again, note the resemblance to lambda phage infections of E. coli.

The lytic cycle of infection follows the usual patterns of most animal DNA viruses: early appearance of viral messenger RNA to code for viral DNA replicating enzymes, then viral DNA synthesis, followed by viral protein synthesis, assembly of infectious virus, and cell lysis. Most of this process takes place in the host cell nucleus.

In the transforming cycle in vitro, less than 5 percent of infected mouse or hamster fibroblast cells transform, each forming a focus of piled up, non-contact-inhibited cells after 10 cell passages in vitro. After additional passages in vitro, these transformed cells develop into typical polyoma tumors when inoculated into host animals.

After infection, polyoma virus enters the host cytoplasm, loses its protein coat, and the now-naked DNA enters the nucleus. In the nucleus, part (or all) of the viral genome is integrated into the host

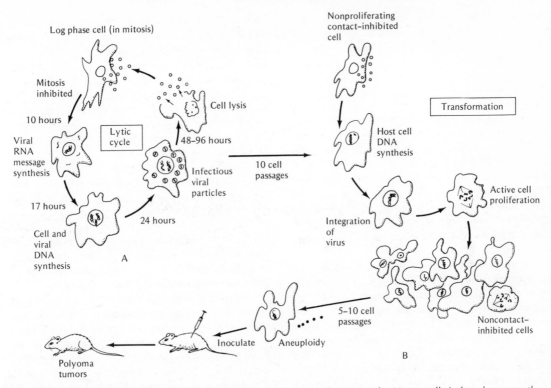

**Figure 21-10**   *Lytic and transforming cycle of polyoma virus in vitro. (A) Lytic cycle. Mouse cells in log phase growth are inoculated with polyoma virus. Mitosis is inhibited, and within 10 hours viral DNA is in cell nuclei and viral message RNAs are detected, coding for DNA-replicating enzymes. By 17 hours, both cell and viral DNA synthesis begin, and viral protein synthesis and capsid assembly are detected by 24 hours. Later, cell lysis begins, and a new infectious cycle is initiated. (B) Transformation. In such mass cultures, lytic cycles persist for about 10 passages with cells becoming confluent. During this period infectious virus disappears and some infected cells (about 5 percent) are transformed. Contact-inhibited cells are stimulated to synthesize DNA after viral infection. Shortly thereafter, viral DNA integrates into host DNA. These cells proliferate and exhibit a surface change that prevents contact inhibition. After additional passages cells become aneuploid and can induce tumors when inoculated into host animals.*

genome. Integration of viral DNA is accompanied by a burst of host DNA synthesis without a concurrent synthesis of viral DNA. Accordingly, host DNA synthesis is first activated in nonproliferating, contact-inhibited cells. This is a critical step in transformation since no neoplastic phenotype develops if host DNA synthesis is inhibited. Insertion of the viral DNA into the host genome releases cells from contact inhibition of growth. Cells pile up in multiple foci instead of growing in monolayers. Meanwhile, viral antigenic proteins become detectable in the host cell surface. The synchrony of these events suggests that changes in the cell surface may be the first step in neoplastic transformation by viruses.

We know that viral DNA is integrated into host cell DNA because (1) viral antigens are found on the surface of transformed cells and their progeny, and (2) DNA-RNA hybridization experiments indicate that virus-free, transformed cells contain RNA molecules that must have been coded by the viral DNA.

One event accompanying transformation is chromosome breakage, leading to abnormal karyotypes, but as we indicated earlier, this in itself is not necessarily responsible for transformation. Chromosomes break because insertion of viral DNA probably depends on enzymes that split host DNA strands and rejoin them to viral DNA.

The transformed state probably depends on the

persistence of viral genetic information in the neoplastic cell. This could mean that the neoplastic cell phenotype is encoded in the viral genome. On the other hand, a viral genome may code for surface proteins that alter host DNA synthesis or repress other host mechanisms of mitotic regulation. Finally, the viral genome may, itself, contribute no information to maintenance of the neoplastic process; the act of infection and integration into the host genome could, by breaking and damaging DNA strands, secondarily disturb host cell mechanisms of growth regulation. In this instance, retention of viral DNA in the host genome is only an accident of the initial viral infection and is unnecessary to maintenance of the transformed state. The phenotype is sustained by the same mechanism that stabilizes normal differentiated phenotypes.

If viral genetic information is necessary for transformation, how much information is actually required and what role does it play in (1) initiating transformation and (2) maintaining stability of the neoplastic phenotype? These questions have been answered, at least in part, by using temperature-sensitive viral mutants for Rous sarcoma and polyoma. These mutated viruses, as we have seen, transform cells at low, permissive temperatures but are inactive at higher, restrictive temperatures. For both polyoma and Rous sarcoma, some viral genetic information must be expressed continuously for the neoplastic phenotype to persist. The evidence suggests that transformation might involve at least two types of viral information, that required for initiation and that needed for maintenance of the neoplastic phenotype.

Other experiments indicate that not all viral information is required for transformation. For example, even defective viral particles produced by ultraviolet radiation or chemicals yield deletion mutants, which do not inhibit transforming ability of the virus.

Some believe that all neoplasia is viral-induced. This "viral oncogene" hypothesis proposes that somatic and germ cell genomes of all vertebrates contain repressed genomes of specific type C RNA viruses that are transmitted through the germ line. We have already seen that Rous sarcoma viral genes are part of normal somatic cell genomes in birds and are transmitted through the germ line. According to the hypothesis viral genes (oncogenes) that code for neoplastic phenotypes are repressed by host factors. Carcinogens such as hydrocarbons or radiation may induce cancer by activating oncogenic information within latent tumor viruses. Only a few viral genes affecting control of cell division need be expressed. Host repressor systems are destroyed, and the cell becomes neoplastic. In a sense this is an epigenetic rather than a genetic model for tumor cell origin because it assumes that viral genes are regulated by the same mechanisms controlling gene expression in differentiated cells.

Most evidence in support of the hypothesis comes from studies of rodent cells in culture. Both normal and neoplastic cells can yield infectious type C RNA viruses which may be seen by electron microscopy in host surface membrane buds or blebs during viral formation. Type C virus genomes are also detected in cells by DNA-RNA hybridization techniques, and in some cases it appears as if the virus is endogenous to the host cell genome.

To prove that endogenous viruses are causal agents in neoplastic transformation is difficult. First, failure to detect a virus is not proof that the tumor is *not* caused by a virus. Nor is presence of a virus proof of viral etiology. Viruses may be secondary invaders; that is, tumor cells might be more susceptible to viruses than are normal cells and readily incorporate most virus if present. Definitive proof requires that normal cells can be transformed into neoplastic cells by the addition of the virus. Recovery of virus from these cells to transform other normal cells must also be shown.

Though some human tumors do contain virus, it is not possible to isolate a virus and transform normal human tissues after inoculation. Certain herpes viruses associated with mononucleosis, for example, do resemble viruses seen in some human tumors, such as Burkitt's lymphoma, a tumor common in young people in parts of West Africa. This tumor contains herpeslike particles which will induce tumors in some animals. Nevertheless, we cannot conclude that the virus is responsible for the human tumor, because we cannot carry out the critical experiment of injecting a normal human with the isolated virus to produce a tumor.

### Environmental carcinogenesis

Viruses are only one of many agents known to cause neoplastic transformations. Some believe that most human tumors are caused by chronic exposure to chemicals and physical agents such as sunlight and radiation which induce mutations in actively growing tissues. The long lag in human tumor development is attributed to a slow accumulation of mutations in growth control mechanisms.

Hydrocarbons in cigarette smoke are implicated as causal agents in lung cancer, and exposure to such environmental agents as coal dust, asbestos, and polyvinyl derivatives, among many others, seems to correlate with a high incidence of specific cancers. Most of these tumors require 15 to 20 years for development, as if the initial exposure to the agent creates cell populations more susceptible to additional genetic lesions and neoplastic transformation.

The rapid rise in lung cancer in humans correlates strongly with the consumption of cigarettes. Nutritional customs and dietary patterns may also induce certain cancers. The frequency of stomach cancer in Japan is generally much higher than in the United States or other Western countries with high meat consumption per capita, but the incidence of stomach cancer among Japanese Americans is about the same as in the U. S. population as a whole. This means that dietary rather than racial factors may be responsible for cancer etiology in this case.

The mode of action of such agents is not understood, although it is generally assumed that the agents are acting as mutagens. Mitotically active cells are most sensitive. Ultraviolet rays in sunlight may induce mutational changes in the basal germinative layers of skin epidermis. The liver, the site for detoxification of a large number of chemical compounds, may also suffer mutational insults as the cells are continuously exposed to these agents. Many skin tumors can be induced in animals by continuous chemical irritation. Even such neutral substances as glass or plastic can induce tumors after implantation.

The problem of cancer control would be much simpler if we could identify the specific chemical agents that are responsible. The incidence of many cancers could be lowered if exposure to these agents was reduced.

The problem of tumor therapy and control is formidable. If cancer is not a single disease but a multifaceted problem of normal development, it is unlikely that it can be eliminated as easily as infectious diseases such as polio or diphtheria. Each tumor has its own origin and developmental program and must be treated accordingly. Classes of tumors, such as viral tumors, may ultimately be controlled by vaccines which enhance the immunological mechanisms of the host organism. Other tumors may be controlled by chemical agents, surgery, or X-irradiation. It is unlikely, however, that a single cancer cure will ever be found.

## SUMMARY

Neoplasia is an inevitable consequence of living systems. It is most prevalent in organisms with extended lives because tumors themselves undergo long-term development. Phenotypic properties of uncontrolled proliferation, transplantability, and dedifferentiation result after several lesions in growth control mechanisms have accumulated. This takes many years since lesions arise as a result of random somatic mutation. In some cases, they may develop as stable changes in the regulation of information flow patterns during development.

Many tumors exhibit abnormal karyotypes, but this is not a necessary condition for tumorogenesis. Often, after long-term proliferation, chromosomal abnormalities arise and create unbalanced genomes that cannot support specialized cell functions. A stable proliferative state is established instead.

Of the many carcinogenic agents, viruses seem to be important in animal tumors where integration of the viral genes into host genomes induces transformation. This is accompanied by changes in surface properties; tumor cells are less adhesive, show no contact inhibition of motility or cell division, and acquire surface-specific antigens. Such surface changes may be transduced into stable alterations in growth regulation.

Surface changes also account for metastatic and invasive properties of tumor cells. Detachment from a primary site and dissemination by the bloodstream to other tissues where new foci of growth are established is a property of the most malignant tumors.

Many environmental agents cause cancer and may be responsible for the high incidence of certain human tumors such as lung cancer. The greater awareness of environmental effects on human and animal populations has alerted us to the possibility that many more agents than those already identified are responsible for other human cancers. Prevention rather than therapy may be the best approach to the cancer problem.

**REFERENCES**    Ames, B. N., Gurney, E. G., Miller, J. A., and Bartsch, H. 1972. Carcinogens as frameshift mutagens: Metabolites and derivatives of 2-acetylaminofluorene and other aromatic amine
carcinogens. *Proc. Nat. Acad. Sci. U.S.* 69:3128.

Ames, B. N., Sims, P., and Grover, P. L. 1972. Epoxides of carcinogenic polycyclic hydrocarbons
are frameshift mutagens. *Science* 176:47.

Baltimore, D. 1975. Tumor viruses. 1974. *Cold Spring Harbor Symp. Quant. Biol.* 39:1187–1200.

Braun, A. C. 1969. *The cancer problem. A critical analysis and modern synthesis.* New York:
Columbia University Press.

Burger, M. M. 1974. Role of the cell surface in growth and transformation. *Macromolecules regulating growth and development,* ed. E. Hay, T. J. King, and J. Papaconstantinou. New York:
Academic Press, pp. 3–24.

Burger, M. M., and Noonen, K. D. 1970. Restoration of normal growth by covering of agglutinin
sites on tumor cell surfaces. *Nature* 228:512–515.

Dulbecco, R. 1969. Cell transformation by viruses. *Science* 166:962–968.

Edelman, G. M., and Yahara, I. 1976. Temperature-sensitive changes in surface modulating
assemblies of fibroblasts transformed by Rous sarcoma virus. *Proc. Nat. Acad. Sci. U.S.*
73:2047–2051.

Fahmy, O. G., and Fahmy, M. J. 1970. Gene elimination in carcinogenesis: Reinterpretation of the
somatic mutation theory. *Cancer Res.* 30:195–205.

Foulds, L. 1969. *Neoplastic development.* Vol. 1. New York: Academic Press.

German, J., ed. 1974. *Chromosomes and cancer.* New York: Wiley.

Graham, F. L. 1977. Biological activity of tumor virus DNA. *Adv. Cancer Res.* 25:1–52.

Harris, M. 1964. *Cell culture and somatic variation.* New York: Holt, Rinehart and Winston.

Hauschka, T. S., and Levan, A. 1958. Cytologic and functional characterization of single cell
clones isolated from Krebs-2 and Ehrlich's ascites tumors. *J. Nat. Cancer Inst.* 21:77–135.

Huebner, R. J., and Todaro, G. J. 1969. Oncogenes of RNA tumor viruses as determinants of
cancer. *Proc. Nat. Acad. Sci. U.S.* 64:1087–1094.

Macpherson, I. 1970. The characteristics of animal cells transformed in vitro. *Adv. Cancer Res.*
13:169–215.

Markert, C. L. 1968. Neoplasia: A disease of cell differentiation. *Cancer Res.* 28:1908–1914.

Meins, F., Jr. 1974. Mechanism underlying tumor transformation and tumor reversal in crown-gall,
a neoplastic disease of higher plants. *Developmental aspects of carcinogenesis and immunity,*
ed. T. J. King. New York: Academic Press, pp. 123–139.

Ohno, S. 1971. Genetic implications of karyological instabilities of malignant somatic cells.
*Physiol. Rev.* 51:496–526.

Ozer, H. L., and Jha, K. K. 1977. Malignancy and transformation: Expression in somatic cell
hybrids and variants. *Adv. Cancer Res.* 25:53–93.

Roblin, R., Chou, I. N., and Black, P. H. 1975. Proteolytic enzymes, cell surface changes and viral
transformation. *Adv. Cancer Res.* 22:203–260.

Sheppard, J. R. 1971. Restoration of inhibited growth to transformed cells by dibutyryl adenosine
3',5'-cyclic monophosphate. *Proc. Nat. Acad. Sci. U.S.* 68:1316–1320.

Stehelen, D., Varmes, H. E., Bishop, J. M., and Vogt, P. K. 1976. DNA related to the transforming
gene(s) of avian sarcoma viruses is present in normal avian DNA. *Nature* 260:171–173.

Temin, H. 1971. The protovirus hypothesis. Speculation on the significance of RNA-directed DNA
synthesis for normal development and for carcinogenesis. *J. Nat. Cancer Inst.* 46:III–V.

Weber, K., Lazarides, E., Goldman, R. D., Vogel, A., and Pollack, R. 1975. Localization and
distribution of actin fibers in normal, transformed and revertant cells. *Cold Spring Harbor
Symp. Quant. Biol.* 39:363–369.

Wiener, F., Klem, G., and Harris, H. 1973. The analysis of malignancy by cell fusion. IV. Hybrids
between tumor cells and a non-malignant L cell derivative. *J. Cell Science* 12:253–261.

# 22

# Aging: A Decline in Growth Potential?

Aging is, in a sense, a continuation of development. It is difficult to determine when development stops and aging begins. We define aging or senescence arbitrarily as a process that begins after an organism has become reproductively mature, though many aging features may appear before that time. If we define senescence as a time-dependent accumulation of biochemical and physiological lesions, then aging can begin very early, even during embryogenesis, inasmuch as mistakes in morphogenesis do occur. Aging and neoplasia are comparable in one sense; both result, in part, from naturally occurring lesions in developmental processes.

## AGING IN POPULATIONS

Aging and death are genetically programmed in the species; they have selective value. Death eliminates the old, nonreproductive individuals, permitting expression of new gene combinations arising in younger generations. As a result, average life span has evolved through selection and is characteristic of a species (Table 22-1).

If aging is an intrinsic process experienced by all members of a population and if death is its end point, then a description of death patterns in a population should tell us something about the kinetics of aging. Two assumptions must be made: (1) death in a population is due entirely to aging factors (thereby eliminating accidents, infant mortality, and so on, as factors); and (2) the population is homogeneous and aging occurs at the same rate in all individuals. A plot of the percentage of survivors at each age in such an idealized population gives the rectangular curve shown in Figure 22-1. It shows that aging begins only after a certain age is reached (reproductive maturity), at which point survivorship declines rapidly. The mean life span of such a population is defined by the 50 percent survival point. We find, however, that natural populations do not display this ideal survivorship curve since death results from

**TABLE 22-1   Maximum Ages of Some Representative Species**

| Phylum | Species | Maximum Age Attained, years |
|---|---|---|
| Porifera | *Suberites carnosus* | 15 |
| Coelenterata | *Cereus pedunculatus* | 85–90 |
| Flatworm | *Taeniorrhynchus saginatus* | 35+ |
| Annelida | *Sabella pavonia* | 10+ |
| Arthropoda | | |
|   Arachnida | Tarantula | 11–20 |
|   Crustacea | *Homarus americanus* | 50 |
|   Insecta | Termites | 25–60 |
| Echinodermata | *Marthasterias glacialis* | 7+ |
| Mollusca | *Venus mercenaria* | 40+ |
| Vertebrates | | |
|   Fish | *Silurus glanis* | 60+ |
|   Amphibia | *Megalobratrachus* | 52+ |
|   Reptiles | *Testudo sumeirei* | 152+ |
|   Birds | *Bubo bubo* | 68+ |
|   Mammals | *Homo sapiens* | 118+ |

factors having nothing to do with aging. Besides, no population is homogeneous with respect to the rate of aging. Consequently, for most animal and human populations the survivorship curves differ significantly from the ideal curve (Figure 22-2).

Each species does have a characteristic survivorship curve. Humans live to about 100 years, elephants to 200, mice to 5 years, and some adult insects may live only a few days. Irrespective of environmental conditions, nutrition, diet, or health standards, longevity seems to be as much a phenotypic trait of the human species as is body shape and hair color. Improvements in diet, sanitation, and control of infectious diseases simply changes the height of the curve and may change its form somewhat, particularly during infancy, but the end point, between 90 and 100 years, seems to be inevitable.

The curves tell us that as we eliminate deaths due to accident, infant mortality, and infectious diseases, we begin to approach the shape of the theoretical curve. For example, the curve for New Zealand almost approximates the ideal curve for aging, suggesting there may indeed be an intrinsic aging process; that is, as we improve public health, medical care, and nutrition we approach a population that dies primarily as a consequence of aging.

There are diseases which seem to be age-dependent. A plot of mortality curves for a variety of human diseases as a function of age is shown in Figure 22-3. Death from all causes appears as a straight line for the most part. Note that few diseases follow the above pattern. Accidents, for example, are most common in the young and account for the peculiar rise seen in early years in the death from all causes curve. The majority of deaths in this U.S. population are caused by cardiovascular diseases,

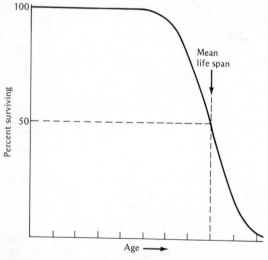

**Figure 22-1** *The percentage of survivors as a function of time in an aging population, assuming that aging is primarily responsible for death.*

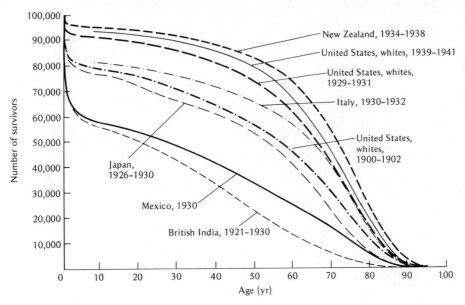

**Figure 22-2**  *Survivorship curves for various human populations. (From* Ageing: The Biology of Senescence *by Alex Comfort. Copyright © 1956, 1964 by Alex Comfort. Reprinted by permission of Holt, Rinehart and Winston.)*

strokes, and neoplasms, the so-called diseases of old age. Of these diseases, arteriosclerosis seems to parallel the curve for total deaths, which suggests that this disease might reflect changes associated with the aging process since it occurs in all members of the population at varying rates. Most respiratory diseases become age-dependent in later years and may reflect an inability of the aged in coping with stress situations, no matter how trivial. Whereas younger people can overcome respiratory infections, the aged cannot recover as readily or as effectively. One feature of the aging process might be a declining ability to restore homeostatic balance in stress situations.

The efficiency of most organ systems (for example, respiratory, cardiovascular, and excretory) gradually does decrease with advancing years (Figure 22-4). With age, there is decreased adaptability to metabolic and environmental insults. Aged animals are less able to maintain body temperature when exposed to heat or cold stress or to restore altered blood glucose levels and pH to normal levels.

Since organ function depends upon a normal vascular supply, the declining function seen in older animals could result from intrinsic vascular changes such as diminishing blood flow in arteriosclerosis.

It becomes almost impossible to separate reduced homeostatic efficiency from nonspecific vascular changes that have widespread effects on overall organ function.

Aging is qualitatively different from other biological processes in that no single biochemical or cellular property is directly implicated. For example, we may define reproduction in terms of DNA replication or growth in terms of cell proliferation or protein synthesis, but aging is a more eclectic phenomenon with no apparent single underlying cellular cause. Sometimes aging is considered the reverse of growth; that is, as rates of protein synthesis and cell proliferation decline, cellular repair mechanisms become less efficient. In this sense, aging is no more than growth decline. Some claim that aging is a distinct, measurable phenomenon like growth, a unique accumulation of aging substances (for example, pigments) in tissues that reduce function, or a decline in membrane function because lipids are irreversibly oxidized. Another view states that aging results from structural changes in macromolecules such as collagen. Since collagen is found in virtually every organ, this theory holds that "aging" collagen molecules diminish organ function.

Any one of these hypotheses has some validity;

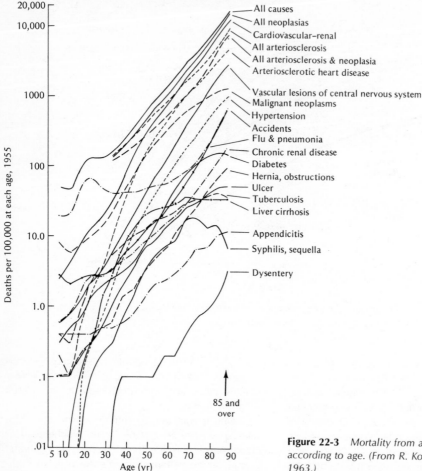

**Figure 22-3**   *Mortality from a variety of causes plotted according to age. (From R. Kohn,* J. Chronic Dis. *16:5–21, 1963.)*

there are no simple paradigms of aging. Nevertheless, a review of a few hypotheses will show how aging is part of development.

First, is aging a cellular phenomenon? Do animals age because their cells age and die?

## THE CELLULAR BASIS OF AGING

The most prevalent view of aging is that it is a cellular process, resulting from somatic cell mutations, or triggered by "aging" genes that program cell death. The basic assumption of this theory of aging is that an organism ages as its cells die, or become nonfunctional. It implies that cells are continuously dying in most tissues, and it is the reduction in cell number

that accounts for decline in organ function and failure of homeostatic mechanisms.

## Somatic mutations and the aging process

The somatic mutation hypothesis proposes that stable genetic lesions impair cell function and eventually lead to cell death. These lesions may be point mutations in DNA or gross chromosomal aberrations. Since most cells in a mature adult are postmitotic, never again to engage in mitosis, these mutations and aberrations, though present, would not be detected. It suggests that aging is directly correlated with the intensity of mutagenic insults. Two approaches have been taken to test this hypothesis: (1) correlate the rate of aging (longevity)

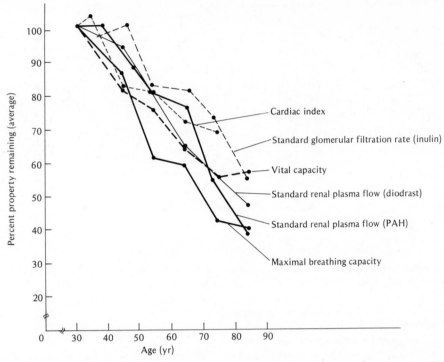

**Figure 22-4** *Physiological function declines with age in human populations. Level at 30 years is assumed to be 100 percent. (Modified from B. L. Strehler,* Aging: Some Social and Biological Aspects, *ed. N. V. Shock, no. 65, pp. 273–303. Copyright © 1960 by the American Association for the Advancement of Science.)*

with exposure to mutagenic agents such as X-rays, and (2) look for evidence of chromosomal aberrations in tissues of aging animals.

Mammals surviving acute effects of ionizing radiation do exhibit decreased longevity and precocious aging. Overall resistance decreases, and many diseases show a higher incidence. On the other hand, chronic irradiation — that is, low doses over a long period (the type of radiation to which most organisms are exposed) — has no aging effect. Differences between acute and chronic irradiation on aging may correlate with effects on chromosome integrity. Much more gross chromosomal damage is noted after acute radiation. If the theory holds, aged animals should show a greater evidence of chromosomal aberrations. Indeed, the number of chromosomal abnormalities in mouse liver increases with age and with X-ray dose. Furthermore, a short-lived strain of mouse was shown to have a greater number of spontaneous chromosomal aberrations in its liver than a long-lived strain. Though both observations

seem to support the somatic mutation hypothesis, closer inspection reveals certain inconsistencies. First, young irradiated animals, though displaying 100 percent chromosomal abnormalities in liver cells, still live longer than older nonirradiated mice which show only 30 percent aberrations. In this instance, there is no correlation between aging and chromosomal damage. Second, the amount of irradiation producing chromosomal aberrations that would normally accumulate in 1 year would not appreciably shorten life span. And, finally, no correlations were observed between aging changes and chromosomal aberrations produced by other mutagenic agents such as nitrogen mustard. It seems that aging is not directly coupled to an accumulation of gross chromosomal aberrations.

This says nothing about somatic point mutations. We find that the frequency of point mutations is higher after acute than it is after chronic X-irradiation, which is consistent with the aging effects of radiation. But X-ray sensitivity differs considerably from one

**Figure 22-5**    *The effect of ploidy on survival of control and irradiated* Habrobracon. *X = diploid males; closed and open circles = haploid males. Dose = 50000 r. (From A. M. Clark and M. A. Rubin,* Rad. Res. *15:244–253, 1961.)*

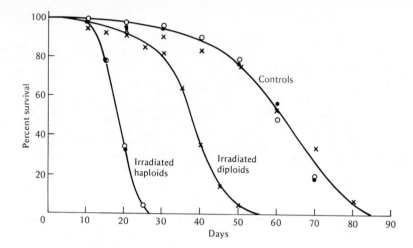

species to another and does not correlate with aging patterns. For example, insects are notoriously radiation-resistant, yet they have short lives compared to the more sensitive mammals. X-irradiation actually increases longevity in some species. Though point mutations may account for radiation-induced aging, they need not be causally related to natural aging.

If point mutations were responsible for aging, we would expect ploidy levels to correlate with aging; that is, haploid animals should age more rapidly than diploids since every recessive point mutation should be expressed immediately in haploids. Yet the natural life span of haploid and diploid *Habrobracon* wasps is not consistent with this expectation. Though all diploid females live longer than haploid males, they also live longer than diploid males. The fact of shorter lives in diploid males seems to correlate with differences in diet, physiological state, and behavior and is unrelated to ploidy. Nevertheless, ploidy does affect radiation-induced aging since haploid males are more sensitive to X-rays than are diploid males (Figure 22-5). The mechanism of radiation-induced aging in this species, however, differs from normal aging patterns. Besides, aging experiments on short-lived insects are difficult to interpret, because they do not live long enough for aging to occur.

Since natural decline of animal populations correlates with nutrition, sexual behavior, and overall physiology, aging has a multiplicity of causes, with mutation only one of many acting at all levels of organization, from the molecular to the organismic.

Organisms age for different reasons; there is no one aging process as there is probably no single cause of neoplastic transformation.

## Somatic cell aging

The cellular theory of aging implies that an organism dies because its somatic cells have a finite life span. Cell loss results in declining physiological function, particularly in organs composed entirely of post-mitotic cells, with no opportunity for repair or renewal. The rate of cell loss is assumed to increase with age. At one time, however, somatic cells were thought to be immortal, a view based largely on the experiments of Alexis Carrell, who maintained a culture of chick heart cells for 34 years. Later it was realized that Carrell was periodically feeding unfiltered chick embryo extract to the cultures, which meant they were continuously replenished with fresh, intact cells. These contaminating cells could seed a declining population and generate new foci of active growth.

Somatic cells from many animals, including man, have a finite life span in vitro. Normal vertebrate tissues explanted as primary cultures follow a predictable pattern of growth. Growth is slow at first as cells adapt to the medium (phase I), and this is followed by exponential growth when cells proliferate actively over a long period (phase II) (Figure 22-6). For most tissues, this latter phase of growth persists for $50 \pm 10$ generations, even when grown as clones. During this finite period of growth, which may last for 1 year, the cell karyotype is diploid and appears normal. Growth rate is constant and high but begins to decline in phase III as cells stop growing and die. In spite of frequent changes of fresh culture medium and removal of all waste products, the cells are unable to grow.

At any time during their growth, cells may be frozen and stored indefinitely. For example, human fetal lung tissue may be stored frozen after it has completed 20 doublings. When thawed the cells resume growth for the number of generations still remaining (30 more doublings), which suggests that total number of cell doublings rather than chronological age is the measure of in vitro aging. This program of a finite number of replications is genetically fixed within somatic cells of the species. We find that embryonic or fetal lung tissues exhibit a greater number of doublings in vitro than do adult tissues, but mature adult tissues are not significantly different from aged tissues.

Cells of different species show characteristic patterns of in vitro growth, often correlating with the average longevity of the species (Figure 22-7). Syrian hamster, chicken, and mouse cells show declining growth patterns very early, whereas those of rat and rabbit usually transform into neoplasms when cultured. Each species is almost unique in its growth behavior.

The reasons for this finite life in vitro are still unexplained. Hayflick and Morehead speculate that an intrinsic process of cellular senescence is responsible, that is, reduction in the rate of synthesis of a factor required for cell survival. Alternatively, dilution of a "rejuvenating" factor synthesized at a rate slower than the rate of cell division could also lead to growth decline.

Contrary to human cells, mouse diploid cells remain in phase II for only 20 passages, or about a month of culturing. But the culture does not die. Within a short time new cell growth is detected and

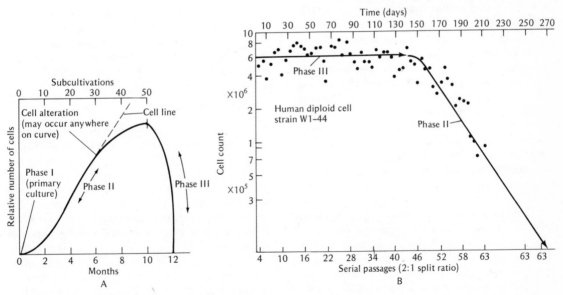

**Figure 22-6**   *In vitro growth pattern of vertebrate cells explanted directly from the animal. (A) Phase I, or primary culture, a period of slow growth, terminates with the formation of first confluent cell sheet. Phase II is a period of proliferation requiring frequent subculture. Cells in phases II and III are cell strains and die out after a finite period of time in phase III. A spontaneous alteration could occur at any time to produce a permanent "cell line" with infinite proliferative potentiality (dashed line). No human diploid fibroblast strain has made this change. Upper abcissa indicates number of cell passages (doublings) expected of a human embryonic diploid strain. (B) An example of an aging human strain of diploid embryonic lung tissue. Cell count is the maximum number of cells achieved at each passage. The plateau in phase II indicates a constant doubling time; all cells grow with the same vigor until phase III, when doubling time increases exponentially. (From L. Hayflick and P. S. Moorhead, Cytogenetics of Cell Culture, ed. R. J. C. Harris, Symp. Int. Soc. Cell Biol., vol. 3. New York: Academic Press, p. 155, 1964.)*

**Figure 22-7** *Growth of normal diploid cells from several species showing behavior in vitro. Cells from short-lived animals have limited life spans in vitro unless they transform to neoplasia, as many do when cultured in bovine serum albumin (BSA). Human cells do not transform but exhibit an extended life in BSA. (From I. MacPherson, Adv. Cancer Res. 13:169–215, 1970.)*

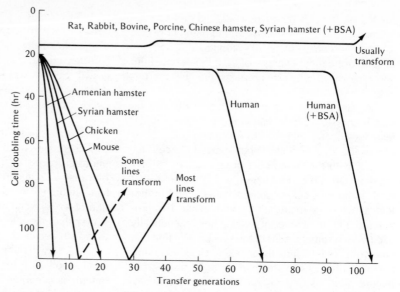

a population appears that grows at an increasing rate, often exceeding the growth rate of the primary cell line. These transformed cells with abnormal karyotypes become established as a permanent cell line and behave as if they are immortal (see Figure 21-3). They have infinite growth potential and often are neoplastic, a spontaneous transformation occurring without addition of carcinogens or viruses. Immortality in vitro seems to be correlated with an unbalanced genome and abnormal growth regulation. It is uncertain whether the abnormal karyotype is causally related to indeterminate growth behavior, inasmuch as karyotypic aberrations are detected within a short time *after* cells exhibit the increase in growth rate.

Cell aging may be an artifact of in vitro conditions. For example, the addition of vitamin E to diploid human embryonic lung cells seems to lengthen their life. Instead of 50 doublings, they complete 100 doublings and continue to grow at the same rate. The cells are clearly diploid at the end of the experiment and do not experience any neoplastic change. The excess vitamin E, approximately 10 times the amount usually found in sera, may have reduced autooxidation of lipids, thereby diminishing the rate of membrane degeneration that occurs in aging cells.

Most cells in an organism are in a nondividing state, and it is doubtful whether senescence of proliferating cells in vitro is comparable to the aging of cells in vivo. More experimental studies of differenti-

ated cells in culture are needed to find out what sustains the differentiated state over a long-term period. Given such systems, we may be able to resolve age-dependent properties and define the factors leading to cellular degeneration and death.

## Programmed cell death

We have already seen that cell death is a normal feature of embryogenesis and metamorphosis (Chapters 12 and 19). Conceivably, aging may occur because all postmitotic cells have a death clock, which is set when cells are in the last division and which runs out during an organism's life. For example, some nerve cells are constantly dying in the vertebrate body, the rate of cell loss varying with the region of the nervous system (Figure 22-8). The most rapid loss is between birth and maturity, presumably when young cells are becoming functional and the nervous system completes its development. The nature of neuronal death is not clear, but it does not behave like programmed cell death during embryogenesis and metamorphosis. Neurons die at random and at declining rates in aging animals.

Most human postmitotic cells survive for many years, until death of the organism. The in vitro experiments suggest that cells have a limited life span set by a constellation of death genes. These may be expressed after a defined period of time in postmitotic cells, but the clock may be accelerated by forcing rapid proliferation in vitro.

Cell death is an autolytic process; various hydrolytic enzymes are released into the cytoplasm to destroy cellular constituents. Presumably *lysosomes*, cytoplasmic vesicles carrying cathepsins, nucleases, lipases, and acid phosphatases, are triggered to release these enzymes into the cytoplasm and thereby initiate autolysis. A correlation between activity of these hydrolytic enzymes and breakdown of a tissue or organ is not always found. For example, no lysosomal enzyme activity is noted before necrosis occurs in the chick limb bud. Instead, DNA replication and mitotic activity decline before degeneration; no other morbid changes are detected before degenerating cells are phagocytosed.

The actual cause of cell death in embryogenesis may be different for each tissue. The important point is that cell death is programmed into the developmental process; it is region- and stage-specific, adapted to morphogenetic events. Of course, the key question is the applicability of these phenomena to the general problem of adult aging. No doubt cells continue to die in adult tissues at prescribed rates; but is this cell loss responsible for organismic aging?

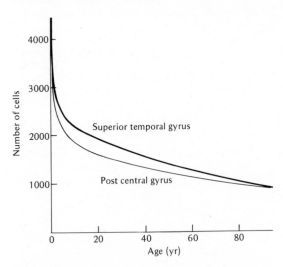

**Figure 22-8** *Cell death in the human cerebral cortex. Cell counts in two areas which show the greatest changes with age. Note that maximum cell loss occurs shortly after birth. (From R. Kohn,* Principles of Mammalian Aging. *Englewood Cliffs, N.J.: Prentice-Hall. Copyright © 1971; based on data in H. Brody,* J. Comp. Neurol. *102: 511, 1955.)*

Those who claim this cannot be so point out that aging is a whole body phenomenon, not limited to any single organ. For example, the human can survive without symptoms with less than 40 percent of his liver, part of one kidney, one lung, and fractions of the stomach and intestine. If aging was a function of equivalent amounts of cell loss, we would expect to find far more cell death and degeneration in most organs than actually exists. Moreover, many cell populations (renewal systems) maintain their youthful vigor even in aged adults; they are capable of continued replication until death of the organism, at rates comparable to those in younger animals. The skin and intestinal germinal epithelium, and of course the stem cell system in bone marrow and spleen, are renewal systems that function efficiently in mature adult animals.

Another theory of aging is based on the accumulation of aging pigment in cells. Do cells die after they accumulate certain metabolic wastes? We know that static postmitotic populations are most prone to aging changes, and these cells do contain aging pigments. Lipofuchsin or chromolipid pigments accumulate linearly with age; they are more prevalent in older tissues than in young tissues. Aging pigments amassed in muscle and nerve are probably the most striking subcellular modification in aging animals. Several forms of these aging pigments are derived from the oxidation of unsaturated lipids. The origin of these pigments is not understood, though it has been suggested that membranes of the endoplasmic reticulum are the primary sources of unsaturated lipids. Accordingly, lipid oxidation and the accumulation of aging pigments may reflect lesions in membrane synthesis and turnover. A breakdown of membrane function would obviously alter membrane permeability and could lead to physiological changes accompanying aging. We have mentioned that the addition of vitamin E to diploid human cells seems to prolong their life, possibly by preventing lipid oxidation in cell membranes.

Aging may reflect imperfections in mechanisms that repair membrane defects induced by free radicals or peroxide accumulation. Damaged membranes may accumulate as aging pigments with no adaptive value to the cell and interfere with normal metabolism. The result could be cell deterioration.

Cells may die because certain cell functions decline with age. Mitochondrial energy production could become less efficient, or lipid metabolism could suffer lesions that result in defective membranes. But most comparisons of enzyme systems involved in these processes show no significant difference between adult and aging animals. True, rapidly growing tissues in embryonic and young animals generally show differences between adult and aged tissues, but these are expected.

Some theories of aging stress defects in the protein synthesis machinery as a primary cause of aging. Cell function declines because abnormal proteins accumulate. For example, aging in human fibroblasts seems to correlate with the appearance of abnormal proteins. The enzyme glucose-6-phosphate dehydrogenase becomes more heat labile in aging fibroblasts; it is more readily inactivated by temperature in aged cells than in young cells (Figure 22-9). Greater heat lability suggests a change in protein structure which may result from mistakes made during transcription, or possibly during replication of DNA. Since heat-labile proteins are found in young and aging cells, a slow buildup of mistakes in protein synthesis may move a cell into an unstable condition where the frequency of mistakes accelerates. This would represent the onset of aging, an irreversible state leading to cell death.

A model of senescence based on errors in protein synthesis is shown in Figure 22-10. The focus is on DNA polymerase and its replication of DNA. If

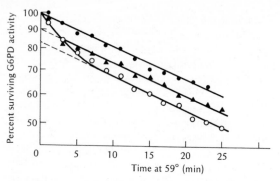

**Figure 22-9**   *Heat inactivation of glucose-6-phosphate dehydrogenase in young, intermediate, and senescent human diploid fibroblast cells. Solid circles = cells sampled at passage 22; triangles = passage 48; open circles = passage 61. (Reprinted from R. Holliday and G. M. Tarrant, Nature 238:26–30, 1972.)*

mutations result in a faulty polymerase, then its efficiency in copying DNA templates may decline and lead to errors in nucleotide sequences that will be translated into abnormal proteins. The dashed lines indicate that these events are uncommon in most normal cells which have long life spans, but these events may increase until they become stabilized (solid arrows) and occur with greater frequency in aged cells. Aging may depend upon a mutation rate

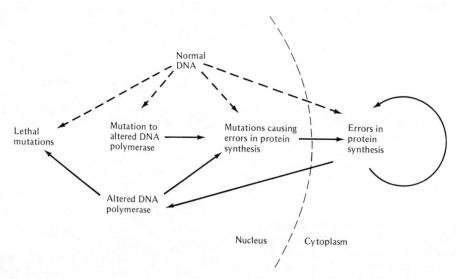

**Figure 22-10**   *Model of senescence based on the accumulation of errors in protein synthesis. See text for details. (Reprinted from R. Holliday and G. M. Tarrant, Nature 238:26–30, 1972.)*

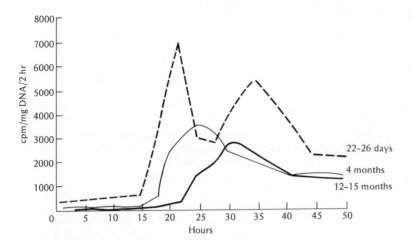

**Figure 22-11**   *Rate of DNA synthesis in rat livers of different ages following partial hepatectomy. (From N. L. R. Bucher et al., Cancer Res. 24:509, 1964.)*

largely due to copy errors made by a modified DNA polymerase.

The efficiency of cellular repair mechanisms is affected by age, but not as drastically as one would imagine. Experiments on liver regeneration in rats indicate that there is no significant difference in the amount of liver regenerated, at least between mature adult and aged animals. Here again, the younger animal displays a greater regenerative capacity, usually restoring 100 percent of the liver after partial hepatectomy, but adult and aged animals restore only 80 percent of the liver removed.

On the other hand, the rate of liver regeneration is affected by age (Figure 22-11). In young animals (22 to 26 days), the lag prior to the onset of DNA replication is only 14 to 15 hours, but this increases to about 17 hours in adults and extends to about 22 hours in aged animals. Surprisingly, it appears that after surgery, the rates of DNA synthesis, as compared with those in normal liver, increase 125-fold in aged animals, only 50-fold in adults, and 20-fold in weanling mice. Although this observation implies that proliferative rates in aged animals are greater, the increased rate could also be attributed to the larger precursor pool in younger animals which dilutes the injected thymidine used to measure these rates.

Because of such ambiguities, it is difficult to assign specific cellular properties to the aging process. Many believe that it is impossible to explain aging entirely on the basis of cellular aging. They stress the multifaceted character of aging, the fact that cells,

tissues, vascular system, and even extracellular matrices are involved, interacting in a network of positive and negative feedback systems that can be understood only at the organismic level.

## AGING AT THE ORGANISMIC LEVEL: EXTRACELLULAR MATRIX AND CONNECTIVE TISSUE

Several theories of aging emphasize the systemic nature of the process; many changes occur simultaneously in different organ systems that could not be entirely due to cell loss. The interdependence of tissues is such that a decline in activity of one system may lead to secondary defects in all dependent systems. Such seems to be the case with connective tissue and extracellular matrices, components of virtually every organ system. The theory states that aging in components of the extracellular matrix can lead to abnormal tissue function and organismic decline.

Many generalized changes in connective-tissue elements correlate with some aspects of mammalian aging. According to this hypothesis, organisms age because essential macromolecules undergo aging changes. One reason for singling out extracellular fibers as a principal factor in organism aging is that they are ubiquitous in all organs and tissues. Collagen is found in basement membranes, and in every tissue as a filler substance, constituting about one-third of total body protein. Its fibers are distributed

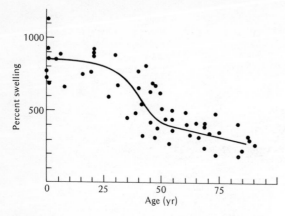

**Figure 22-12**   *Osmotic swelling of human collagen as a function of age. (From R. Kohn and L. Rollerson,* Gerontol. *13:241, 1958.)*

around cells, within and around blood vessel walls, and are, of course, a major component of bone, cartilage, tendon, and skin. It is part of the matrix in which many cells are embedded, the principal microenvironment of all cell surfaces. Any major age-related changes occurring in collagen and the ground substance may modify function of adjacent cells and tissues.

Collagen is a complicated molecule made up of linear tropocollagen precursors aligned to give the characteristic periodicity of the mature collagen fibril (see Chapter 7). The molecule is composed of three polypeptide chains wound about each other in the form of a helix. As these molecules age within the fibril, there is an increase in cross linkages between polypeptide chains of tropocollagen. More stable bonds are established and the molecule becomes crystalline in nature. Aging collagen becomes more insoluble as cross linkages accumulate. For instance, a high proportion of newly synthesized collagen (in young animals) is extractable with neutral solvents, whereas mature collagen from older organisms is less soluble. As long as an animal grows its store of young, soluble collagen is replenished, but after growth ceases, a greater proportion of collagen becomes insoluble. Moreover, mature collagen does not turn over appreciably; it is a stable tissue component, subject to aging changes that are usually measured by the swelling capacity of extracted collagen. In humans, collagen retains a high swelling capacity up to

30 years of age, when it decreases rapidly until 50 years of age and thereafter declines at a slower rate (Figure 22-12). Such changes are consistent with the view that cross linkages within collagen molecules increase with age. Significantly, the changes begin at age 25, after growth has ceased and after new collagen synthesis slows down.

How does collagen age? One possibility is thermal denaturation, since collagen is exposed to sustained body temperatures for many years. Isolated collagen maintained at elevated temperature (56°C) for varying periods produces most molecular age changes. But this alone cannot explain the excessive amount of cross-linking observed in vivo.

The increasing insolubility of collagen may make tissues more rigid, with accompanying deleterious effects, particularly on vascular wall function which must sustain elasticity to facilitate blood flow. Aged collagen in extracellular matrices may inhibit diffusion from blood vessels to surrounding tissue and thereby interfere with cell function. Myocardial tissue would suffer serious changes in contractile activity if much rigid collagen accumulated within tissue spaces. Such changes, in one form or another, have been detected in aging animals and are consistent with the view that collagen aging is directly related to debilitating age-related effects. Decreasing responsiveness to hormones, hardening of blood vessels in arteriosclerosis, and the failure of detoxifying mechanisms in the liver could be attributed to the accumulation of insoluble, cross-linked collagen in the respective tissues.

One difficulty with this concept is that many mammals age within a few years (2 years for the rat). It is unlikely that its collagen ages in 2 years as human collagen does in 70. This may be explained by relating age changes in collagen with the amount of stress imposed on collagen in the tissues. Smaller mammals with high metabolic rates (a high rate of living) impose greater stresses on their collagen, and indeed such animals age rapidly. Consequently rat collagen ages more rapidly than human collagen. There is some evidence that collagen from old rats has properties similar to those of mature collagen from aged humans.

We should not forget the ground substance in connective tissue, since it is the microenvironment of all cells. Cells are not bathed in a tissue fluid pool but are surrounded by matrix with varying degrees of fluid hydration. The life of a cell depends upon the rate of exchange of materials between it and the surrounding extracellular matrix. Collagen and elastic fibers are immersed within this ground substance, which includes plasma proteins, salts, glycoproteins, and water.

What age-related changes occur in these extracellular matrices? One thing that does change with age is the proportion of fibrous elements to ground substance. Collagen increases more rapidly in most tissues than does the matrix. Accompanying these changes are alterations in water compartmentalization, that is, the water-rich phase decreases as the colloid-rich phase increases in proportion. Moreover, certain matrix constituents such as chondroitin sulfate decrease in total mass as well as distribution in the tissues, and matrices become less soluble. This diminishes water-holding capacity and the rate of transfer of fluids through the ground substance. Matrix becomes less permeable with old age and its transport characteristics with respect to oxygen, nutrients, hormones, and antigens may change significantly, at least enough to upset cell function. If the rate of flow of such vital substances as oxygen is seriously inhibited by changes in the extracellular compartment, it is easy to see how irreversible changes could result in oxygen-sensitive tissues such as brain. Senescent changes in physiological and psychological activities are similar to those produced by oxygen lack. In fact, one of the etiological agents proposed for aging is tissue hypoxia. Debilitating changes could stem from a chronic reduction in oxygen availability to tissues.

It seems clear that progressive molecular changes in such ubiquitous macromolecules as collagen could have far-reaching effects on the function of diverse organ systems. Other systems can have equally generalized effects if they too suffer aging lesions. The endocrine or immune systems exercising systemic effects on organ function and stress responses also undergo aging changes, many arising from any of the cellular mechanisms described above. Many aspects of aging resemble autoimmune diseases, such as rheumatoid arthritis, in which antibodies are made against an organism's own tissues. Defects in the immune response system, its recognition mechanisms, or its production of antibodies could lead to such abnormalities. Theories of aging based on autoimmune phenomena have been proposed and seem to apply in some experimental situations. But most aging phenomena do not have an immunological basis.

## SUMMARY

The existence of at least twenty different "theories of aging" is in itself testimony to the fact that aging is not a simple problem. Aging is defined in terms of multiple changes in many systems, all reciprocally interacting and mutually dependent. Because an organism is a complex of interacting subsystems at all levels, what happens to a macromolecule in one system inevitably affects others. We cannot distinguish primary from secondary events because of the tangle of feedback interactions. All aging theories have some validity; each of the factors identified (somatic mutation, rate of living, aging pigments, collagen crosslinking, and so on) does in fact contribute to aging and impinge on other systems that accelerate other facets of aging. The problem, therefore, is a systems problem, of the type now familiar in communications engineering. An array of interlocking systems, beginning at the molecular level and communicating through all levels of organization, must be evaluated together. Such complex systems are best handled by computers. Lesions at one level (for example, a somatic mutation) can become amplified by damaging regulatory tissues in the immune or endocrine systems. These, in turn, could provoke chronic changes in intercellular ground substance and affect its fluid holding capacity. Diffusion of oxygen through intercellular spaces would be impaired and the mild hypoxia induced might precipitate incipient lesions in these tissues. We can easily see how an organism constructed on a "systems within systems" principle is subject to this type of erosion as minor defects and mistakes accumulate in seemingly unrelated sites.

Our insight into the problem of aging is at least an order of magnitude less than our understanding of neoplasia, which means we are still at a primitive state. We are dealing, in both cases, with multifactorial processes with common endpoints and cannot assume that identity of the endpoint implies either common origins or identical action mechanisms. But given our underlying faith in the unity of biological systems, we continue to seek unifying causes to explain these two "negative" aspects of the developmental process.

**REFERENCES**

Cameron, I. L. 1972. Cell proliferation and renewal in aging mice. *J. Gerontol.* 27:162–172.

Carrel, A., and Ebeling, A. H. 1921. Age and multiplication of fibroblasts. *J. Exp. Med.* 34:599–623.

Clark, A. M., and Rockstein, M. 1964. Aging in insects. *The physiology of insects,* ed. M. Rockstein. Vol. 1. New York: Academic Press, pp. 227–283.

Clark, A. M., and Rubin, M. A. 1961. The modification by X-irradiation of the life span of haploid and diploids of the wasp *Habrobracon sp. Radiation Res.* 15:244–253.

Comfort, A. 1956. *Ageing: The biology of senescence.* New York: Rinehart.

Cooper, E. H., Bedford, A. J., and Kenny, T. E. 1975. Cell death in normal and malignant tissues. *Adv. Cancer Res.* 21:59–120.

Cristofalo, W. J. 1972. Animal cell cultures as a model system for the study of aging. *Adv. Gerontol. Res.* 4:45–79.

Curtis, H. J. 1963. Biological mechanisms underlying the aging process. *Science* 141:686–694.

Hahn, H. P. von. 1971. Failures of regulation mechanisms as causes of cellular aging. *Adv. Gerontol. Res.* 3:1–38.

Hayflick, L., and Moorhead, P. S. 1961. The serial cultivation of human diploid cell line. *Exp. Cell. Res.* 25:585–621.

Holliday, R., and Tarrant, G. M. 1972. Altered enzymes in aging human fibroblasts. *Nature* 238:26–30.

Kohn, R. 1971. *Principles of mammalian aging.* Englewood Cliffs, N.J.: Prentice-Hall.

Lockshin, R. A. 1969. Programmed cell death: Activation of lysis by a mechanism involving the synthesis of protein. *J. Insect. Physiol.* 15:1505–1516.

Orgel, L. E. 1963. The maintenance of the accuracy of protein synthesis and its relevance to aging. *Proc. Nat. Acad. Sci. U.S.* 49:517–521.

Sinex, F. M. 1968. The role of collagen in aging. *Treatise on collagen,* ed. B. S. Gould. New York: Academic Press, pp. 409–448.

Sobel, H. 1967. Aging of ground substance in connective tissue. *Adv. Gerontol. Res.* 2:205–283.

Strehler, B. L. 1962. *Time, cells and aging.* New York: Academic Press.

————. 1964. On the histochemistry and ultrastructure of age pigments. *Adv. Gerontol. Res.* 1:343–384.

# INDEX